EMBEDDED SYSTEMS DESIGN AND VERIFICATION

INDUSTRIAL INFORMATION TECHNOLOGY SERIES

EMBEDDED SYSTEMS HANDBOOK
SECOND EDITION

EMBEDDED SYSTEMS DESIGN AND VERIFICATION

Edited by

Richard Zurawski

ISA Corporation
San Francisco, California, U.S.A.

CRC Press
Taylor & Francis Group
Boca Raton   London   New York

CRC Press is an imprint of the
Taylor & Francis Group, an **informa** business

CRC Press
Taylor & Francis Group
6000 Broken Sound Parkway NW, Suite 300
Boca Raton, FL 33487-2742

Library of Congress Cataloging-in-Publication Data

Embedded systems handbook : embedded systems design and verification / edited by Richard
Zurawski. -- 2nd ed.
    p. cm. -- (Industrial information technology series ; 6)
Includes bibliographical references and index.
ISBN-13: 978-1-4398-0755-2 (v. 1)
ISBN-10: 1-4398-0755-8 (v. 1)
ISBN-13: 978-1-4398-0761-3 (v. 2)
ISBN-10: 1-4398-0761-2 (v. 2)
    1. Embedded computer systems--Handbooks, manuals, etc. I. Zurawski, Richard. II. Title. III. Series.

TK7895.E42E64 2009
004.16--dc22
                                      2008049535

Dedication

To Celine, as always.

Contents

Part I   System-Level Design and Verification

Part II   Embedded Processors and System-on-Chip Design

Part III   Embedded System Security and Web Services

Preface

Introduction

Application domains have had a considerable impact on the evolution of embedded systems in terms of required methodologies and supporting tools, and resulting technologies. Multimedia and network applications, the most frequently reported implementation case studies at scientific conferences on embedded systems, have had a profound influence on the evolution of embedded systems with the trend now toward multiprocessor systems-on-chip (MPSoCs), which combine the advantages of parallel processing with the high integration levels of systems-on-chip (SoCs). Many SoCs today incorporate tens of interconnected processors; as projected in the 2006 edition of the *International Technology Roadmap for Semiconductors*, the number of processor cores on a chip will reach over 800 by 2020. The design of MPSoCs invariably involves integration of heterogeneous hardware and software IP components, an activity which still lacks a clear theoretical underpinning, and is a focus of many academic and industry projects.

Embedded systems have also been used in automotive electronics, industrial automated systems, building automation and control (BAC), train automation, avionics, and other fields. For instance, trends have emerged for the SoCs to be used in the area of industrial automation to implement complex field-area intelligent devices that integrate the intelligent sensor/actuator functionality by providing on-chip signal conversion, data and signal processing, and communication functions. Similar trends can also be seen in the automotive electronic systems. On the factory floor, microcontrollers are nowadays embedded in field devices such as sensors and actuators. Modern vehicles employ as many as hundreds of microcontrollers. These areas, however, do not receive, for various reasons, as much attention at scientific meetings as the SoC design as it meets demands for computing power posed by digital signal processing (DSP), and network and multimedia processors, for instance.

Most of the mentioned application areas require real-time mode of operation. So do some multimedia devices and gadgets, for clear audio and smooth video. What, then, is the major difference between multimedia and automotive embedded applications, for instance? Braking and steering systems in a vehicle, if implemented as Brake-by-Wire and Steer-by-Wire systems, or a control loop of a high-pressure valve in offshore exploration, are examples of safety-critical systems that require a high level of dependability. These systems must observe hard real-time constraints imposed by the system dynamics, that is, the end-to-end response times must be bounded for safety-critical systems. A violation of this requirement may lead to considerable degradation in the performance of the control system, and other possibly catastrophic consequences. On the other hand, missing audio or video data may result in the user's dissatisfaction with the performance of the system.

Furthermore, in most embedded applications, the nodes tend to be on some sort of a network. There is a clear trend nowadays toward networking embedded nodes. This introduces an additional constraint on the design of this kind of embedded systems: systems comprising a collection of embedded nodes communicating over a network and requiring, in most cases, a high level of dependability. This extra constraint has to do with ensuring that the distributed application tasks execute in a deterministic way (need for application tasks schedulability analysis involving distributed nodes and the communication network), in addition to other requirements such as system availability, reliability, and safety. In general, the design of this kind of networked embedded systems (NES) is a challenge in itself due to the distributed nature of processing elements, sharing common communication medium, and, frequently, safety-critical requirements.

The type of protocol used to interconnect embedded nodes has a decisive impact on whether the system can operate in a deterministic way. For instance, protocols based on random medium access control (MAC) such as carrier sense multiple access (CSMA) are not suitable for this type of operation. On the other hand, time-triggered protocols based on time division multiple access (TDMA) MAC access are particularly well suited for the safety-critical solutions, as they provide deterministic access to the medium. In this category, TTP/C and FlexRay protocols (FlexRay supports a combination of both time-triggered and event-triggered transmissions) are the most notable representatives. Both TTP/C and FlexRay provide additional built-in dependability mechanisms and services which make them particularly suitable for safety-critical systems, such as replicated channels and redundant transmission mechanisms, bus guardians, fault-tolerant clock synchronization, and membership service.

The absence of NES from the academic curriculum is a troubling reality for the industry. The focus is mostly on a single-node design. Specialized networks are seldom mentioned, and if at all, then controller area network (CAN) and FlexRay in the context of embedded automotive systems—a trendy area for examples—but in a superficial way. Specialized communication networks are seldom included in the curriculum of ECE programs. Whatever the reason for this, some engineering graduates involved in the development of embedded systems in diverse application areas will learn the trade the hard way. A similar situation exists with conferences where applications outside multimedia and networking are seldom used as implementation case studies. A notable exception is the IEEE International Symposium on Industrial Embedded Systems that emphasizes research and implementation reports in diverse application areas.

To redress this situation, the second edition of the *Embedded System Handbook* pays considerable attention to the diverse application areas of embedded systems that have in the past few years witnessed an upsurge in research and development, implementation of new technologies, and deployment of actual solutions and technologies. These areas include automotive electronics, industrial automated systems, and BAC. The common denominator for these application areas is their distributed nature and use of specialized communication networks as a fabric for interconnecting embedded nodes.

In automotive electronic systems [1], the electronic control units are networked by means of one of the automotive communication protocols for controlling one of the vehicle functions, for instance, electronic engine control, antilocking brake system, active suspension, and telematics. There are a number of reasons for the automotive industry's interest in adopting field-area networks and mechatronic solutions, known by their generic name as X-by-Wire, aiming to replace mechanical or hydraulic systems by electrical/electronic systems. The main factors seem to be economic in nature, improved reliability of components, and increased functionality to be achieved with a combination of embedded hardware and software. Steer-by-Wire, Brake-by-Wire, or Throttle-by-Wire systems are examples of X-by-Wire systems. The dependability of X-by-Wire systems is one of the main requirements and constraints on the adoption of these kinds of systems. But, it seems that certain safety-critical systems such as Steer-by-Wire and Brake-by-Wire will be complemented with traditional mechanical/hydraulic backups for reasons of safety. Another equally important requirement for X-by-Wire systems is to observe hard real-time constraints imposed by the system dynamics; the end-to-end response times must be bounded for safety-critical systems. A violation of this requirement may lead to degradation in the performance of the control system, and other consequences as a result. Not all automotive electronic systems are safety critical, or require hard real-time response; system(s) to control seats, door locks, internal lights, etc., are some examples. With the automotive industry increasingly keen on adopting mechatronic solutions, it was felt that exploring in detail the design of in-vehicle electronic embedded systems would be of interest to the readers.

In industrial automation, specialized networks [2] connect field devices such as sensors and actuators (with embedded controllers) with field controllers, programmable logic controllers, as well as man–machine interfaces. Ethernet, the backbone technology of office networks, is increasingly being adopted for communication in factories and plants at the field level. The random and native

CSMA/CD arbitration mechanism is being replaced by other solutions allowing for deterministic behavior required in real-time communication to support soft and hard real-time deadlines, time synchronization of activities required to control drives, and for exchange of small data records characteristic of monitoring and control actions. A variety of solutions have been proposed to achieve this goal [3]. The use of wireless links with field devices, such as sensors and actuators, allows for flexible installation and maintenance and mobile operation required in case of mobile robots, and alleviates problems associated with cabling [4]. The area of industrial automation is one of the fastest-growing application areas for embedded systems with thousands of microcontrollers and other electronic components embedded in field devices on the factory floor. This is also one of the most challenging deployment areas for embedded systems due to unique requirements imposed by the industrial environment which considerably differ from those one may be familiar with from multimedia or networking. This application area has received considerable attention in the second edition.

Another fast-growing application area for embedded systems is building automation [5]. Building automation systems aim at the control of the internal environment, as well as the immediate external environment of a building or building complex. At present, the focus of research and technology development is on buildings that are used for commercial purposes such as offices, exhibition centers, and shopping complexes. Some of the main services offered by the building automation systems typically include climate control to include heating, ventilation, and air conditioning; visual comfort to cover artificial lighting; control of daylight; safety services such as fire alarm and emergency sound system; security protection; control of utilities such as power, gas, and water supply; and internal transportation systems such as lifts and escalators.

This books aims at presenting a snapshot of the state-of-the-art embedded systems with an emphasis on their networking and applications. It consists of 48 contributions written by leading experts from industry and academia directly involved in the creation and evolution of the ideas and technologies discussed here. Many of the contributions are from the industry and industrial research establishments at the forefront of developments in embedded systems. The presented material is in the form of tutorials, research surveys, and technology overviews. The contributions are divided into parts for cohesive and comprehensive presentation. The reports on recent technology developments, deployments, and trends frequently cover material released to the profession for the very first time.

Organization

Embedded systems is a vast field encompassing various disciplines. Not every topic, however important, can be covered in a book of a reasonable volume and without superficial treatment. The topics need to be chosen carefully: material for research and reports on novel industrial developments and technologies need to be balanced out; a balance also needs to be struck in treating so-called "core" topics and new trends, and other aspects. The "time-to-market" is another important factor in making these decisions, along with the availability of qualified authors to cover the topics.

This book is divided into two volumes: "Embedded Systems Design and Verification" (Volume I) and "Networked Embedded Systems" (Volume II). Volume I provides a broad introduction to embedded systems design and verification. It covers both fundamental and advanced topics, as well as novel results and approaches, fairly comprehensively. Volume II focuses on NES and selected application areas. It covers the automotive field, industrial automation, and building automation. In addition, it covers wireless sensor networks (WSNs), although from an application-independent viewpoint. The aim of this volume was to introduce actual NES implementations in fast-evolving areas which, for various reasons, have not received proper coverage in other publications. Different application areas, in addition to unique functional requirements, impose specific restrictions on performance, safety, and quality-of-service (QoS) requirements, thus necessitating adoption of different solutions which in turn give rise to a plethora of communication protocols and systems. For this reason, the discussion of the internode communication aspects has been deferred to this part of the book where the communication aspects are discussed in the context of specific applications of NES.

One of the main objectives of any handbook is to give a well-structured and cohesive description of fundamentals of the area under treatment. It is hoped that Volume I has achieved this objective. Every effort was made to ensure each contribution in this volume contains an introductory material to assist beginners with the navigation through more advanced issues. This volume does not strive to replicate, or replace, university level material. Rather, it tries to address more advanced issues, and recent research and technology developments.

The specifics of the design automation of integrated circuits have been deliberately omitted in this volume to keep it at a reasonable size in view of the publication of another handbook that covers these aspects comprehensively, namely, *The Electronic Design Automation for Integrated Circuits Handbook*, CRC Press, Boca Raton, Florida, 2005, Editors: Lou Scheffer, Luciano Lavagno, and Grant Martin.

The material covered in the second edition of the Embedded Systems Handbook will be of interest to a wide spectrum of professionals and researchers from industry and academia, as well as graduate students from the fields of electrical and computer engineering, computer science and software engineering, and mechatronics engineering.

This edition can be used as a reference (or prescribed text) for university (post) graduate courses. It provides the "core" material on embedded systems. Part II, Volume II, is suitable for a course on WSNs while Parts III and IV, Volume II, can be used for a course on NES with a focus on automotive embedded systems or industrial embedded systems, respectively; this may be complemented with selected material from Volume I.

In the following, the important points of each chapter are presented to assist the reader in identifying material of interest, and to view the topics in a broader context. Where appropriate, a brief explanation of the topic under treatment is provided, particularly for chapters describing novel trends, and for novices in mind.

Volume I. Embedded Systems Design and Verification

Volume I is divided into three parts for quick subject matter identification. Part I, System-Level Design and Verification, provides a broad introduction to embedded systems design and verification covered in 11 chapters: "Real-time in networked embedded systems," "Design of embedded systems," "Models of computation for distributed embedded systems," "Embedded software modeling and design," "Languages for design and verification," "Synchronous hypothesis and polychronous languages," "Processor-centric architecture description languages," "Network-ready, open source operating systems for embedded real-time applications," "Determining bounds on execution times," "Performance analysis of distributed embedded systems," and "Power-aware embedded computing." Part II, Embedded Processors and System-on-Chip Design, gives a comprehensive overview of embedded processors, and various aspects of SoC, FPGA, and design issues. The material is covered in six chapters: "Processors for embedded systems," "System-on-chip design," "SoC communication architectures: From interconnection buses to packet-switched NoCs," "Networks-on-chip: An interconnect fabric for multiprocessor systems-on-chip," "Hardware/software interfaces design for SoC," and "FPGA synthesis and physical design." Part III, Embedded Systems Security and Web Services, gives an overview of "Design issues in secure embedded systems" and "Web services for embedded devices."

Part I. System-Level Design and Verification

An authoritative introduction to real-time systems is provided in the chapter "Real-time in networked embedded systems." This chapter covers extensively the areas of design and analysis with some examples of analysis and tools; operating systems (an in-depth discussion of real-time embedded operating systems is presented in the chapter "Network-ready, open source operating systems for embedded real-time applications"); scheduling; communications to include descriptions of the ISO/OSI reference model, MAC protocols, networks, and topologies; component-based design; as well as testing and debugging. This is essential reading for anyone interested in the area of real-time systems.

A comprehensive introduction to a design methodology for embedded systems is presented in the chapter "Design of embedded systems." This chapter gives an overview of the design issues and stages. It then presents, in some detail, the functional design; function/architecture and hardware/software codesign; and hardware/software co-verification and hardware simulation. Subsequently, it discusses selected software and hardware implementation issues. While discussing different stages of design and approaches, it also introduces and evaluates supporting tools. This chapter is essential reading for novices for it provides a framework for the discussion of the design issues covered in detail in the subsequent chapters in this part.

Models of computation (MoCs) are essentially abstract representations of computing systems and facilitate the design and validation stages in the system development. An excellent introduction to the topic of MoCs, particularly for embedded systems, is presented in the chapter "Models of computation for distributed embedded systems." This chapter introduces the origins of MoCs, and their evolution from models of sequential and parallel computation to attempts to model heterogeneous architectures. In the process it discusses, in relative detail, selected nonfunctional properties such as power consumption, component interaction in heterogeneous systems, and time. Subsequently, it reviews different MoCs to include continuous time models, discrete time models, synchronous models, untimed models, data flow process networks, Rendezvous-based models, and heterogeneous MoCs. This chapter also presents a new framework that accommodates MoCs with different timing abstractions, and shows how different time abstractions can serve different purposes and needs. The framework is subsequently used to study coexistence of different computational models, specifically the interfaces between two different MoCs and the refinement of one MoC into another.

Models and tools for embedded software are covered in the chapter "Embedded software modeling and design." This chapter outlines challenges in the development of embedded software, and is followed by an introduction to formal models and languages, and to schedulability analysis. Commercial modeling languages, Unified Modeling Language and Specification and Description Language (SDL), are introduced in quite some detail together with the recent extensions to these two standards. This chapter concludes with an overview of the research work in the area of embedded software design, and methods and tools, such as Ptolemy and Metropolis.

An authoritative introduction to a broad range of design and verification languages used in embedded systems is presented in the chapter "Languages for design and verification." This chapter surveys some of the most representative and widely used languages divided into four main categories: languages for hardware design, for hardware verification, for software, and domain-specific languages. It covers (1) hardware design languages: Verilog, VHDL, and SystemC; (2) hardware verification languages: OpenVera, the e language, Sugar/PSL, and SystemVerilog; (3) software languages: assembly languages for complex instruction set computers, reduced instruction set computers (RISCs), DSPs, and very-long instruction word processors; and for small (4- and 8-bit) microcontrollers, the C and C^{++} Languages, Java, and real-time operating systems; and (4) domain-specific languages: Kahn process networks, synchronous dataflow, Esterel, and SDL. Each group of languages is characterized for their specific application domains, and illustrated with ample code examples.

An in-depth introduction to synchronous languages is presented in the chapter "The synchronous hypothesis and polychronous languages." Before introducing the synchronous languages, this chapter discusses the concept of synchronous hypothesis, the basic notion, mathematical models, and implementation issues. Subsequently, it gives an overview of the structural languages used for modeling and programming synchronous applications, namely, imperative languages Esterel and

SyncCharts that provide constructs to deal with control-dominated programs, and declarative languages Lustre and Signal that are particularly suited for applications based on intensive data computation and dataflow organization. The future trends section discusses loosely synchronized systems, as well as modeling and analysis of polychronous systems and multiclock/polychronous languages.

The chapter "Processor-centric architecture description languages" (ADL) covers state-of-the-art specification languages, tools, and methodologies for processor development used in industry and academia. The discussion of the languages is centered around a classification based on four categories (based on the nature of the information), namely, structural, behavioral, mixed, and partial. Some specific ADLs are overviewed including Machine-Independent Microprogramming Language (MIMOLA); nML; Instruction Set Description Language (ISDL); Machine Description (MDES) and High-Level Machine Description (HMDES); EXPRESSION; and LISA. A substantial part of this chapter focuses on Tensilica Instruction Extension (TIE) ADL and provides a comprehensive introduction to the language illustrating its use with a case study involving design of an audio DSP called the HiFi2 Audio Engine.

An overview of the architectural choices for real-time and networking support adopted by many contemporary operating systems (within the framework of the IEEE 1003.1-2004 international standard) is presented in the chapter "Network-ready, open source operating systems for embedded real-time applications." This chapter gives an overview of several widespread architectural choices for real-time support at the operating system level, and describes the real-time application interface (RTAI) approach in particular. It then summarizes the real-time and networking support specified by the IEEE 1003.1-2004 international standard. Finally, it describes the internal structure of a commonly used open source network protocol stack to show how it can be extended to handle other protocols besides the TCP/IP suite it was originally designed for. The discussion centers on the CAN protocol.

Many embedded systems, particularly hard real-time systems, impose strict restrictions on the execution time of tasks, which are required to complete within certain time bounds. For this class of systems, schedulability analyses require the upper bounds for the execution times of all tasks to be known to verify statically whether the system meets its timing requirements. The chapter "Determining bounds on execution times" presents architecture of the *aiT* timing-analysis tool and an approach to timing analysis implemented in the tool. In the process, it discusses cache-behavior prediction, pipeline analysis, path analysis using integer linear programming, and other issues. The use of this approach is put in the context of upper bounds determination. In addition, this chapter gives a brief overview of other approaches to timing analysis. The validation of nonfunctional requirements of selected implementation aspects such as deadlines, throughputs, buffer space, and power consumption comes under performance analysis.

The chapter "Performance analysis of distributed embedded systems" discusses issues behind performance analysis, and its role in the design process. It also surveys a few selected approaches to performance analysis for distributed embedded systems such as simulation-based methods, holistic scheduling analysis, and compositional methods. Subsequently, this chapter introduces the modular performance analysis approach and accompanying performance networks, as stated by authors, influenced by the worst-case analysis of communication networks. The presented approach allows to obtain upper and lower bounds on quantities such as end-to-end delay and buffer space; it also covers all possible corner cases independent of their probability.

Embedded nodes, or devices, are frequently battery powered. The growing power dissipation, with the increase in density of integrated circuits and clock frequency, has a direct impact on the cost of packaging and cooling, as well as reliability and lifetime. These and other factors make the design for low power consumption a high priority for embedded systems. The chapter "Power-aware embedded computing" presents a survey of design techniques and methodologies aimed at reducing both static and dynamic power dissipation. This chapter discusses energy and power modeling to include instruction-level and function-level power models, microarchitectural power models, memory and bus models, and battery models. Subsequently, it discusses system/application-level optimizations that explore different task implementations exhibiting different power/energy versus QoS characteristics. Energy-efficient processing subsystems: voltage and frequency scaling, dynamic resource scaling, and processor core selection are addressed next in this chapter. Finally, this chapter discusses energy-efficient memory subsystems: cache hierarchy tuning; novel horizontal and vertical cache partitioning schemes; dynamic scaling of memory elements; software-controlled memories; scratch-pad memories; improving access patterns to on-chip memory; special-purpose memory subsystems for media streaming; and code compression and interconnect optimizations.

Part II. Embedded Processors and System-on-Chip Design

An extensive overview of microprocessors in the context of embedded systems is given in the chapter "Processors for embedded systems." This chapter presents a brief history of embedded microprocessors and covers issues such as software-driven evolution, performance of microprocessors, reduced instruction set computing (RISC) machines, processor cores, and the embedded SoC. After discussing symmetric multiprocessing (SMP) and asymmetric multiprocessing (AMP), this chapter covers some of the most widely used embedded processor architectures followed by a comprehensive presentation of the software development tools for embedded processors. Finally, it overviews benchmarking processors for embedded systems where the use of standard benchmarks and instruction set simulators to evaluate processor cores are discussed. This is particularly relevant to the design of embedded SoC devices where the processor cores may not yet be available in hardware, or be based on user-specified processor configuration and extension.

A comprehensive introduction to the SoC concept, in general, and design issues is provided in the chapter "System-on-chip design." This chapter discusses basics of SoC; IP cores, and virtual components; introduces the concept of architectural platforms and surveys selected industry offerings; provides a comprehensive overview of the SoC design process; and discusses configurable and extensible processors, as well as IP integration quality and certification methods and standards.

On-chip communication architectures are presented in the chapter "SoC communication architectures: From interconnection buses to packet-switched NoCs." This chapter provides an in-depth description and analysis of the three most relevant, from industrial and research viewpoints, architectures to include ARM developed Advanced Micro-Controller Bus Architecture (AMBA) and new interconnect schemes AMBA 3 Advanced eXtensible Interface (AXI), Advanced High-performance Bus (AHB) interface, AMBA 3 APB interface, and AMBA 3 ATB interface; Sonics SMART interconnects (SonicsLX, SonicsMX, and S3220); IBM developed CoreConnect Processor Local Bus (PLB), On-Chip Peripheral Bus (OPB), and Device Control Register (DCR) Bus; and STMicroelectronics developed STBus. In addition, it surveys other architectures such as WishBone, Peripheral Interconnect Bus (PI-Bus), Avalon, and CoreFrame. This chapter also offers some analysis of selected communication architectures. It concludes with a brief discussion of the packet-switched interconnection networks, or Network-on-Chip (NoC), introducing XPipes (a SystemC library of parameterizable, synthesizable NoC components), and giving an overview of the research trends.

Basic principles and guidelines for the NoC design are introduced in the chapter "Networks-on-chip: An interconnect fabric for multiprocessor systems-on-chip." This chapter discusses the rationale for the design paradigm shift of SoC communication architectures from shared busses to NoCs, and briefly surveys related work. Subsequently, it presents details of NoC building blocks to include switch, network interface, and switch-to-switch links. The design principles and the trade-offs are discussed in the context of different implementation variants, supported by the case studies from real-life NoC prototypes. This chapter concludes with a brief overview of NoC design challenges.

The chapter "Hardware/software interfaces design for SoC" presents a component-based design automation approach for MPSoC platforms. It briefly surveys basic concepts of MPSoC design and discusses some related approaches, namely, system-level, platform-based, and component-based.

It provides a comprehensive overview of hardware/software IP integration issues such as bus-based and core-based approaches, integrating software IP, communication synthesis, and IP derivation. The focal point of this chapter is a new component-based design methodology and design environment for the integration of heterogeneous hardware and software IP components. The presented methodology, which adopts automatic communication synthesis approach and uses a high-level API, generates both hardware and software wrappers, as well as a dedicated Operating System for programmable components. The IP integration capabilities of the approach and accompanying software tools are illustrated by redesigning a part of a VDSL modem.

Programmable logic devices, complex programmable logic devices (CPLDs), and field-programmable gate arrays (FPGAs) have evolved from implementing small glue-logic designs to large complete systems that are now the majority of design starts: FPGAs for the higher density design and CPLDs for smaller designs and designs that require nonvolatility targeting. The chapter "FPGA synthesis and physical design" gives an introduction to the architecture of field-programmable date arrays and an overview of the FPGA CAD flow. It then surveys current algorithms for FPGA synthesis, placement, and routing, as well as commercial tools.

Part III. Embedded Systems Security and Web Services

There is a growing trend for networking of embedded systems. Representative examples of such systems can be found in automotive, train, and industrial automation domains. Many of these systems need to be connected to other networks such as LAN, WAN, and the Internet. For instance, there is a growing demand for remote access to process data at the factory floor. This, however, exposes systems to potential security attacks, which may compromise the integrity of the system and cause damage. The limited resources of embedded systems pose considerable challenges for the implementation of effective security policies which, in general, are resource demanding. An excellent introduction to the security issues in embedded systems is presented in the chapter "Design issues in secure embedded systems." This chapter outlines security requirements in computing systems, classifies abilities of attackers, and discusses security implementation levels. Security constraints in embedded systems design discussed include energy considerations, processing power limitations, flexibility and availability requirements, and cost of implementation. Subsequently, this chapter presents the main issues in the design of secure embedded systems. It also covers, in detail, attacks and countermeasures of cryptographic algorithm implementations in embedded systems.

The chapter "Web services for embedded devices" introduces the devices profile for Web services (DPWS). DPWS provides a service-oriented approach for hardware components by enabling Web service capabilities on resource-constraint devices. DPWS addresses announcement and discovery of devices and their services, eventing as a publish/subscribe mechanism, and secure connectivity between devices. This chapter gives a brief introduction to device-centric service-oriented architectures (SOAs), followed by a comprehensive description of DPWS. It also covers software development toolkits and platforms such as the Web services for devices (WS4D), service-oriented architecture for devices (SOA4D), UPnP and DPWS base driver for OSGI, as well as DPWS in Microsoft Vista. The use of DPWS is illustrated by the example of a business-to-business (B2B) maintenance scenario to repair a faulty industrial robot.

Volume II. Networked Embedded Systems

Volume II focuses on selected application areas of NES. It covers automotive field, industrial automation, and building automation. In addition, this volume also covers WSNs, although from an application-independent viewpoint. The aim of this volume was to introduce actual NES implementations in fast-evolving areas that, for various reasons, have not received proper coverage in other publications. Different application areas, in addition to unique functional requirements, impose specific restrictions on performance, safety, and QoS requirements, thus necessitating adoption of different solutions that in turn give rise to a plethora of communication protocols and systems. For this reason, the discussion of the internode communication aspects has been deferred to this volume where the communication aspects are discussed in the context of specific application domains of NES.

Part I. Networked Embedded Systems: An Introduction

A general overview of NES is presented in the chapter "Networked embedded systems: An overview." It gives an introduction to the concept of NES, their design, internode communication, and other development issues. This chapter also discusses various application areas for NES such as automotive, industrial automation, and building automation.

The topic of middleware for distributed NES is addressed in the chapter "Middleware design and implementation for networked embedded systems." This chapter discusses the role of middleware in NES, and the challenges in design and implementation such as remote communication, location independence, reusing existing infrastructure, providing real-time assurances, providing a robust DOC middleware, reducing middleware footprint, and supporting simulation environments. The focal point of this chapter is the section describing the design and implementation of nORB (a small footprint real-time object request broker tailored to a specific embedded sensor/actuator applications), and the rationale behind the adopted approach.

Part II. Wireless Sensor Networks

The distributed WSN is a relatively new and exciting proposition for collecting sensory data in a variety of environments. The design of this kind of networks poses a particular challenge due to limited computational power and memory size, bandwidth restrictions, power consumption restriction if battery powered (typically the case), communication requirements, and unattended mode of operation in case of inaccessible and/or hostile environments. This part provides a fairly comprehensive discussion of the design issues related to, in particular, self-organizing ad-hoc WSNs. It introduces fundamental concepts behind sensor networks; discusses architectures; time synchronization; energy-efficient distributed localization, routing, and MAC; distributed signal processing; security; testing, and validation; and surveys selected software development approaches, solutions, and tools for large-scale WSNs.

A comprehensive overview of the area of WSNs is provided in the chapter "Introduction to wireless sensor networks." This chapter introduces fundamental concepts, selected application areas, design challenges, and other relevant issues. It also lists companies involved in the development of sensor networks, as well as sensor networks-related research projects.

The chapter "Architectures for wireless sensor networks" provides an excellent introduction to the various aspects of the architecture of WSNs. It starts with a description of a sensor node architecture and its elements: sensor platform, processing unit, communication interface, and power source. It then presents two WSN architectures developed around the layered protocol stack approach, and EYES European project approach. In this context, it introduces a new flexible architecture design approach with environmental dynamics in mind, and aimed at offering maximum flexibility while still adhering to the basic design concept of sensor networks. This chapter concludes with a comprehensive discussion of the distributed data extraction techniques, providing a summary of distributed data extraction techniques for WSNs for the actual projects.

The time synchronization issues in sensor networks are discussed in the chapter "Overview of time synchronization issues in sensor networks." This chapter introduces basics of time synchronization for sensor networks. It also describes design challenges and requirements in developing time synchronization protocols such as the need to be robust and energy aware, the ability to operate correctly in the absence of time servers (server-less), and the need to be lightweight and offer a tunable service. This chapter also overviews factors influencing time synchronization such as temperature, phase noise, frequency noise, asymmetric delays, and clock glitches. Subsequently, different time synchronization protocols are discussed, namely, the network time protocol (NTP), timing-sync protocol for sensor networks (TPSN), H-sensor broadcast synchronization (HBS), time synchronization for high latency (TSHL), reference-broadcast synchronization (RBS), adaptive clock synchronization, time-diffusion synchronization protocol (TDP), rate-based diffusion algorithm, and adaptive-rate synchronization protocol (ARSP).

The localization issues in WSNs are discussed in the chapter "Resource-aware localization in sensor networks." This chapter explains the need to know localization of nodes in a network, introduces distance estimation approaches, and covers positioning and navigation systems as well as localization algorithms. Subsequently, localization algorithms are discussed and evaluated, and are grouped in the following categories: classical methods, proximity based, optimization methods, iterative methods, and pattern matching.

The chapter "Power-efficient routing in wireless sensor networks" provides a comprehensive survey and critical evaluation of energy-efficient routing protocols used in WSNs. This chapter begins by highlighting differences between routing in distributed sensor networks and WSNs. The overview of energy-saving routing protocols for WSNs centers on optimization-based routing protocols, data-centric routing protocols, cluster-based routing protocols, location-based routing protocols, and QoS-enabled routing protocols. In addition, the topology control protocols are discussed.

The chapter "Energy-efficient MAC protocols for wireless sensor networks" provides an overview of energy-efficient MAC protocols for WSNs. This chapter begins with a discussion of selected design issues of the MAC protocols for energy-efficient WSNs. It then gives a comprehensive overview of a number of MAC protocols, including solutions for mobility support and multichannel WSNs. Finally, it outlines current trends and open issues.

Due to their limited resources, sensor nodes frequently provide incomplete information on the objects of their observation. Thus, the complete information has to be reconstructed from data obtained from many nodes frequently providing redundant data. The distributed data fusion is one of the major challenges in sensor networks. The chapter "Distributed signal processing in sensor networks" introduces a novel mathematical model for distributed information fusion which focuses on solving a benchmark signal processing problem (spectrum estimation) using sensor networks.

The chapter "Sensor network security" offers a comprehensive overview of the security issues and solutions. This chapter presents an introduction to selected security challenges in WSNs, such as avoiding and coping with sensor node compromise, maintaining availability of sensor network services, and ensuring confidentiality and integrity of data. Implications of the denial-of-service (DoS) attack, as well as attacks on routing, are then discussed, along with measures and approaches that have been proposed so far against these attacks. Subsequently, it discusses in detail the SNEP and μTESLA protocols for confidentiality and integrity of data, the LEAP protocol, as well as probabilistic key management and its many variants for key management. This chapter concludes with a discussion of secure data aggregation.

The chapter "Wireless sensor networks testing and validation" covers validation and testing methodologies, as well as tools needed to provide support that are essential to arrive at a functionally correct, robust, and long-lasting system at the time of deployment. It explains issues involved in testing of WSNs followed by validation including test platforms and software testing methodologies. An integrated test and instrumentation architecture that augments WSN test beds by incorporating the environment and giving exact and detailed insight into the reaction to changing parameters and resource usage is then introduced.

The chapter "Developing and testing of software for wireless sensor networks" presents basic concepts related to software development of WSNs, as well as selected software solutions. The solutions include TinyOS, a component-based operating system, and related software packages such as MATÉ, a byte-code interpreter; TinyDB, a query processing system for extracting information from a network of TinyOS sensor nodes; SensorWare, a software framework for WSNs that provides querying, dissemination, and fusion of sensor data, as well as coordination of actuators; Middleware Linking Applications and Networks (MiLAN), a middleware concept that aims to exploit information redundancy provided by sensor nodes; EnviroTrack, a TinyOS-based application that provides a convenient way to program sensor network applications that track activities in their physical environment; SeNeTs, a middleware architecture for WSNs designed to support the pre-deployment phase; Contiki, a lightweight and flexible operating system for 8-bit computers and integrated microcontrollers. This chapter also discusses software solutions for simulation, emulation, and test of large-scale sensor networks: TinyOS SIMulator (TOSSIM), a simulator based on the TinyOS framework; EmStar, a software environment for developing and deploying applications for sensor networks consisting of 32-bit embedded Microserver platforms; SeNeTs, a test and validation environment; and Java-based J-Sim.

Part III. Automotive Networked Embedded Systems

The automotive industry is aggressively adopting mechatronic solutions to replace, or duplicate, existing mechanical/hydraulic systems. The embedded electronic systems together with dedicated communication networks and protocols play a pivotal role in this transition. This part contains seven chapters that offer a comprehensive overview of the area presenting topics such as networks and protocols, operating systems and other middleware, scheduling, safety and fault tolerance, and actual development tools used by the automotive industry.

This part begins with the chapter "Trends in automotive communication systems" that introduces the area of in-vehicle embedded systems and, in particular, the requirements imposed on the communication systems. Then, a comprehensive review of the most widely used, as well as emerging, automotive networks is presented to include priority busses (CAN and J1850), time-triggered networks (TTP/C, TTP/A, TTCAN), low cost automotive networks (LIN and TTP/A), and multimedia networks (MOST and IDB 1394). This is followed by an overview of the industry initiatives related to middleware technologies, with a focus on OSEK/VDX and AUTOSAR.

The chapter "Time-triggered communication" presents an overview of time-triggered communication, solutions, and technologies put in the context of automotive applications. It introduces dependability concepts and fundamental services provided by time-triggered communication protocols, such as clock synchronization, periodic exchange of messages carrying state information, fault isolation mechanisms, and diagnostic services. Subsequently, the chapter overviews four important representatives of time-triggered communication protocols: TTP/C, TTP/A, TTCAN, and TT Ethernet.

A comprehensive introduction to CANs is presented in the chapter "Controller area network." This chapter overviews some of the main features of the CAN protocol, with a focus on advantages and drawbacks affecting application domains, particularly NESs. CANopen, especially suited to NESs, is subsequently covered to include CANopen device profile for generic I/O modules.

The newly emerging standard and technology for automotive safety-critical communication is presented in the chapter "FlexRay communication technology." This chapter overviews aspects such as media access, clock synchronization, startup, coding and physical layer, bus guardian, protocol services, and system configuration.

The Local Interconnect Network (LIN) communication standard, enabling fast and cost-efficient implementation of low-cost multiplex systems for local interconnect networks in vehicles, is presented in the chapter "LIN standard." This chapter introduces the LIN's physical layer and the LIN protocol. It then focuses on the design process and workflow, and covers aspects such as requirement capture (signal definitions and timing requirements), network configuration and design, and network verification, put in the context of Mentor Graphics LIN tool-chain.

The chapter "Standardized basic system software for automotive applications" presents an overview of the automotive software infrastructure standardization efforts and initiatives. This chapter begins with an overview of the automotive hardware architecture. Subsequently, it focuses on the software modules specified by OSEK/VDX and HIS working groups, followed by ISO and AUTOSAR initiatives. Some background and technical information are provided on the Japanese JasPar, the counterpart to AUTOSAR.

The Volcano concept and technology for the design and implementation of in-vehicle networks using the standardized CAN and LIN communication protocols are presented in the chapter "Volcano technology—Enabling correctness by design." This chapter provides an insight in the design and development process of an automotive communication network

Part IV. Networked Embedded Systems in Industrial Automation

Field-Area Networks in Industrial Automation

The advances in design of embedded systems, tools availability, and falling fabrication costs of semiconductor devices and systems allowed for infusion of intelligence into field devices such as sensors and actuators. The controllers used with these devices provide on-chip signal conversion, data and signal processing, and communication functions. The increased functionality, processing, and communication capabilities of controllers have been largely instrumental in the emergence of a widespread trend for networking of field devices around specialized networks, frequently referred to as field-area networks. One of the main reasons for the emergence of field-area networks in the first place was an evolutionary need to replace point-to-point wiring connections by a single bus, thus paving the road to the emergence of distributed systems and, subsequently, NES with the infusion of intelligence into the field devices.

The part begins with a comprehensive introduction to specialized field-area networks presented in the chapter "Fieldbus systems—Embedded networks for automation." This chapter presents evolution of the fieldbus systems; overviews communication fundamentals and introduces the ISO/OSI layered model; covers fieldbus characteristics in comparison with the OSI model; discusses interconnections in the heterogeneous network environment; and introduces industrial Ethernet. Selected fieldbus systems, categorized by the application domain, are summarized at the end. This chapter is a compulsory reading for novices to understand the concepts behind fieldbuses.

The chapter "Real-time Ethernet for automation applications" provides a comprehensive introduction to the standardization process and actual implementation of real-time Ethernet. Standardization process and initiatives, real-time Ethernet requirements, and practical realizations are covered first. The practical realizations discussed include top of TCP/IP, top of Ethernet, and modified Ethernet solutions. Then, this chapter gives an overview of specific solutions in each of those categories.

The issues involved in the configuration (setting up a fieldbus system in the first place) and management (diagnosis and monitoring, and adding new devices to the network) of fieldbus systems are presented in the chapter "Configuration and management of networked embedded devices." This chapter starts by outlining requirements on configuration and management. It then discusses the approach based on the profile concept, as well as several mechanisms following an electronic datasheet approach, namely, the Electronic Device Description Language (EDDL), the Field Device

Tool/Device Type Manager (FDT/DTM), the Transducer Electronic Datasheets (TEDS), and the Smart Transducer Descriptions (STD) of the Interface File System (IFS). It also examines several application development approaches and their influence on the system configuration.

The chapter "Networked control systems for manufacturing: Parameterization, differentiation, evaluation and application" covers extensively the application of networked control systems in manufacturing with an emphasis on control, diagnostics, and safety. It explores the parameterization of networks with respect to balancing QoS capabilities; introduces common network protocol approaches and differentiates them with respect to functional characteristics; presents a method for networked control system evaluation that includes theoretical, experimental, and analytical components; and explores network applications in manufacturing with a focus on control, diagnostics, and safety in general, and at different levels of the factory control hierarchy. Future trends emphasize migration trend toward wireless networking technology.

Wireless Network Technologies in Industrial Automation

Although the use of wireline-based field-area networks is dominant, wireless technology offers a range of incentives in a number of application areas. In industrial automation, for instance, wireless device (sensor/actuator) networks can provide support for mobile operation required for mobile robots, monitoring and control of equipment in hazardous and difficult to access environments, etc. The use of wireless technologies in industrial automation is covered in five chapters that cover the use of wireless local and wireless personal area network technologies on the factory floor, hybrid wired/wireless networks in industrial real-time applications, a wireless sensor/actuator (WISA) network developed by ABB and deployed in a manufacturing environment, and WSNs for automation.

The issues involving the use of wireless technologies and mobile communication in the industrial environment (factory floor) are discussed in the chapter "Wireless LAN technology for the factory floor: Challenges and approaches." This is comprehensive material dealing with topics such as error characteristics of wireless links and lower layer wireless protocols for industrial applications. It also briefly discusses hybrid systems extending selected fieldbus technologies (such as PROFIBUS and CAN) with wireless technologies.

The chapter "Wireless local and wireless personal area network communication in industrial environments" presents a comprehensive overview of the commercial-off-the-shelf wireless technologies to include IEEE 802.15.1/Bluetooth, IEEE 802.15.4/ZigBee, and IEEE 802.11 variants. The suitability of these technologies for industrial deployment is evaluated to include aspects such as application scenarios and environments, coexistence of wireless technologies, and implementation of wireless fieldbus services.

Hybrid configurations of communication networks resulting from wireless extensions of conventional, wired, industrial networks and their evaluation are presented in the chapter "Hybrid wired/wireless real-time industrial networks." The focus is on four popular solutions, namely, Profibus DP and DeviceNet, and two real-time Ethernet networks: Profinet IO and EtherNet/IP; and the IEEE 802.11 family of WLAN standards and IEEE 802.15.4 WSNs as wireless extensions. They are some of the most promising technologies for use in industrial automation and control applications, and a lot of devices are already available off-the-shelf at relatively low cost.

The chapter "Wireless sensor networks for automation" gives a comprehensive introduction to WSNs technology in embedded applications on the factory floor and other industrial automated systems. This chapter gives an overview of WSNs in industrial applications; development challenges; communication standards including ZeegBee, WirelessHART, and ISA100; low-power design; packaging of sensors and ICs; software/hardware modularity in design, and power supplies. This is essential reading for anyone interested in wireless sensor technology in factory and industrial automated applications.

A comprehensive case study of a factory-floor deployed WSN is presented in the chapter "Design and implementation of a truly wireless real-time sensor/actuator interface for discrete manufacturing automation." The system, known as WISA has been implemented by ABB in a manufacturing cell to network proximity switches. The sensor/actuators communication hardware is based on a standard Bluetooth 2.4 GHz radio transceiver and low-power electronics that handle the wireless communication link. The sensors communicate with a wireless base station via antennas mounted in the cell. For the base station, a specialized RF front end was developed to provide collision-free air access by allocating a fixed TDMA time slot to each sensor/actuator. Frequency hopping (FH) was employed to counter both frequency-selective fading and interference effects, and operates in combination with automatic retransmission requests (ARQ). The parameters of this TDMA/FH scheme were chosen to satisfy the requirements of up to 120 sensor/actuators per base station. Each wireless node has a response or cycle time of 2 ms, to make full use of the available radio band of 80 MHz width. The FH sequences are cell-specific and were chosen to have low cross-correlations to permit parallel operation of many cells on the same factory floor with low self-interference. The base station can handle up to 120 WISAs and is connected to the control system via a (wireline) field bus. To increase capacity, a number of base stations can operate in the same area. WISA provides wireless power supply to the sensors, based on magnetic coupling.

Part V. Networked Embedded Systems in Building Automation and Control

Another fast-growing application area for NES is BAC. BAC systems aim at the control of the internal environment, as well as the immediate external environment of a building or building complex. At present, the focus of research and technology development is on buildings that are used for commercial purposes such as offices, exhibition centers, and shopping complexes. However, the interest in (family type) home automation is on the rise.

A general overview of the building control and automation area and the supporting communication infrastructure is presented in the chapter "Data communications for distributed building automation." This chapter provides an extensive description of building service domains and the concepts of BAC, and introduces building automation hierarchy together with the communication infrastructure. The discussion of control networks for building automation covers aspects such as selected QoS requirements and related mechanisms, horizontal and vertical communication, network architecture, and internetworking. As with industrial fieldbus systems, there are a number of bodies involved in the standardization of technologies for building automation. This chapter overviews some of the standardization activities, standards, as well as networking and integration technologies. Open systems BACnet, LonWorks, and EIB/KNX, wireless IEEE 802.15.4 and ZigBee, and Web Services are introduced at the end of this chapter, together with a brief introduction to home automation.

References

1. N. Navet, Y. Song, F. Simonot-Lion, and C. Wilwert, Trends in automotive communication systems, *Special Issue: Industrial Communication Systems*, R. Zurawski, Ed., *Proceedings of the IEEE*, 93(6), June 2005, 1204–1223.
2. J.-P. Thomesse, Fieldbus technology in industrial automation, *Special Issue: Industrial Communication Systems*, R. Zurawski, Ed., *Proceedings of the IEEE*, 93(6), June 2005, 1073–1101.
3. M. Felser, Real-time Ethernet—Industry perspective, *Special Issue: Industrial Communication Systems*, R. Zurawski, Ed., *Proceedings of the IEEE*, 93(6), June 2005, 1118–1129.

4. A. Willig, K. Matheus, and A. Wolisz, Wireless technology in industrial networks, *Special Issue: Industrial Communication Systems*, R. Zurawski, Ed., *Proceedings of the IEEE*, 93(6), June 2005, 1130–1151.
5. W. Kastner, G. Neugschwandtner, S. Soucek, and H. M. Newman, Communication systems for building automation and control, *Special Issue: Industrial Communication Systems*, R. Zurawski, Ed., *Proceedings of the IEEE*, 93(6), June 2005, 1178–1203.

Locating Topics

To assist readers with locating material, a complete table of contents is presented at the front of the book. Each chapter begins with its own table of contents. Two indexes are provided at the end of the book. The index of authors contributing to the book together with the titles of the contributions, and a detailed subject index.

Richard Zurawski

Acknowledgments

I would like to thank all authors for the effort to prepare the contributions and tremendous cooperation. I would like to express gratitude to my publisher Nora Konopka and other CRC Press staff involved in the book production. My love goes to my wife who tolerated the countless hours I spent on preparing this book.

Richard Zurawski

Editor

Richard Zurawski is with ISA Group, San Francisco, California, involved in providing solutions to 1000 Fortune companies. He has over 30 years of academic and industrial experience, including a regular professorial appointment at the Institute of Industrial Sciences, University of Tokyo, and full-time R&D advisor with Kawasaki Electric, Tokyo. He has provided consulting services to Kawasaki Electric, Ricoh, and Toshiba Corporations, Japan. He has participated in a number of Japanese Intelligent Manufacturing Systems programs.

Dr. Zurawski's involvement in R&D and consulting projects and activities in the past few years included network-based solutions for factory floor control, network-based demand side management, Java technology, SEMI implementations, wireless applications, IC design and verification, EDA, and embedded systems integration.

Dr. Zurawski is the series editor for The Industrial Information Technology (book) Series, CRC Press/Taylor & Francis; and the editor in chief of the *IEEE Transactions on Industrial Informatics*. He has served as editor at large for *IEEE Transactions on Industrial Informatics* (2005–2006); and as an associate editor for *IEEE Transactions on Industrial Electronics* (1994–2005); *Real-Time Systems*; *The International Journal of Time-Critical Computing Systems*, Kluwer Academic Publishers (1997–2003); and *The International Journal of Intelligent Control and Systems*, World Scientific Publishing Company (1996–1997).

Dr. Zurawski was a guest editor of three special issues in *IEEE Transactions on Industrial Electronics* on factory automation and factory communication systems. He was also a guest editor of the special issue on industrial communication systems in the *Proceedings of the IEEE*. He was invited by *IEEE Spectrum* to contribute an article on Java technology to the "Technology 1999: Analysis and Forecast" special issue.

Dr. Zurawski served as a vice president of the Industrial Electronics Society (IES) (1994–1997), as a chairman of the IES Factory Automation Council (1994–1997), and is currently the chairman of the IES Technical Committee on Factory Automation. He was also on a steering committee of the *ASME/IEEE Journal of Microelectromechanical Systems*. In 1996, he received the Anthony J. Hornfeck Service Award from the IEEE IES.

Dr. Zurawski has served as a general co-chair for 13 IEEE conferences and workshops, as a technical program co-chair for 4 IEEE conferences, as a track (co-)chair for 12 IEEE conferences, and as a member of program committees for over 40 IEEE, IFAC, and other conferences and workshops. He has established two major technical events: IEEE Workshop on Factory Communication Systems and IEEE International Conference on Emerging Technologies and Factory Automation.

Dr. Zurawski was the editor of five major handbooks: *The Industrial Information Technology Handbook*, CRC Press, Boca Raton, Florida, 2004; *The Industrial Communication Technology Handbook*, CRC Press, Boca Raton, Florida, 2005; *The Embedded Systems Handbook*, CRC Press/Taylor & Francis, Boca Raton, Florida, 2005; *Integration Technologies for Industrial Automated Systems*, CRC Press/Taylor & Francis, Boca Raton, Florida, 2006; and *Networked Embedded Systems Handbook*, CRC Press/Taylor & Francis, Boca Raton, Florida, 2008.

Dr. Zurawski received his MEng in electronics from the University of Mining and Metallurgy, Krakow and PhD in computer science from LaTrobe University, Melbourne, Australia.

Contributors

Nupur Andrews
Tensilica Inc.
Santa Clara, California

José L. Ayala
Department of Computer
 Architecture
Complutense University of Madrid
Madrid, Spain

Luca Benini
Department of Electrical
 Engineering and Computer
 Science
University of Bologna
Bologna, Italy

Davide Bertozzi
Engineering Department
University of Ferrara
Ferrara, Italy

Vaughn Betz
Altera Corporation
Toronto, Ontario, Canada

Hendrik Bohn
Institute of Applied
 Microelectronics and
 Computer Science
University of Rostock
Rostock, Germany

Wander O. Cesário
System-Level Synthesis Group
Techniques of Informatics and
 Microelectronics for Integrated
 Systems Architecture (TIMA)
 Laboratory
Grenoble, France

Ivan Cibrario Bertolotti
Institute of Electronics and
 Information Engineering
 and Telecommunications
National Research Council
Turin, Italy

Stephen A. Edwards
Department of Computer Science
Columbia University
New York, New York

**Anastasios G.
Fragopoulos**
Department of Electrical and
 Computer Engineering
University of Patras
Patras, Greece

Francisco Gilabert
Department of Computer
 Systems and Computation
Polytechnic University of
 Valencia
Valencia, Spain

Frank Golatowski
Center for Life Science
 Automation
Rostock, Germany

Wolfgang Haid
Department of Information
 Technology and Electrical
 Engineering
Swiss Federal Institute of
 Technology
Zurich, Switzerland

Hans Hansson
School of Innovation, Design
 and Engineering
Mälardalen University
Västerås, Sweden

Mike Hutton
Altera Corporation
San Jose, California

Margarida F. Jacome
Department of Electrical
 and Computer Engineering
University of Texas at Austin
Austin, Texas

Axel Jantsch
Department for Microelectronics
 and Information Technology
Royal Institute of Technology
Stockholm, Sweden

A. A. Jerraya
Electronics and Information
 Technology Laboratory
Atomic Energy Commission,
 Minatec
Grenoble, France

Luciano Lavagno
Department of Electronics
Polytechnic University of Turin
Turin, Italy

Steve Leibson
Tensilica Inc.
Santa Clara, California

Marisa Lopez-Vallejo
Department of Electronic
 Engineering
ETSI Telecomunicacion
Ciudad Universitaria
Madrid, Spain

Grant Martin
Tensilica Inc.
Santa Clara, California

Giovanni De Micheli
Institute of Electrical
 Engineering
Ecole Polytechnique Fédérale de
 Lausanne
Lausanne, Switzerland

Marco Di Natale
Sant'Anna School of Advanced
  Studies
Pisa, Italy

Thomas Nolte
School of Innovation, Design
  and Engineering
Mälardalen University
Västerås, Sweden

Claudio Passerone
Department of Electronics
Polytechnic University of Turin
Turin, Italy

Katalin Popovici
System-Level Synthesis Group
Techniques of Informatics and
  Microelectronics for Integrated
  Systems Architecture (TIMA)
  Laboratory
Grenoble, France

Dumitru Potop-Butucaru
National Institute for
  Research in Computer Science
  and Control (INRIA)
Rocquencourt, France

Anand Ramachandran
Department of Electrical
  and Computer Engineering
University of Texas at Austin
Austin, Texas

Himanshu Sanghavi
Tensilica Inc.
Santa Clara, California

Dimitrios N. Serpanos
Department of Electrical and
  Computer Engineering
University of Patras
Patras, Greece

Robert de Simone
National Institute for
  Research in Computer Science
  and Control (INRIA)
Sophia Antipolis, France

Mikael Sjödin
School of Innovation, Design
  and Engineering
Mälardalen University
Västerås, Sweden

Daniel Sundmark
School of Innovation, Design
  and Engineering
Mälardalen University
Västerås, Sweden

Jean-Pierre Talpin
National Institute for
  Research in Computer Science
  and Control (INRIA)
Rennes, France

Lothar Thiele
Department of Information
  Technology and Electrical
  Engineering
Swiss Federal Institute of
  Technology
Zurich, Switzerland

Artemios G. Voyiatzis
Department of Electrical and
  Computer Engineering
University of Patras
Patras, Greece

Flávio R. Wagner
Institute of Informatics
Federal University of
  Rio Grande do Sul
Porto Alegre, Brazil

Ernesto Wandeler
Computer Engineering and
  Networks Laboratory
Department of Information
  Technology and Electrical
  Engineering
Swiss Federal Institute of
  Technology
Zurich, Switzerland

Reinhard Wilhelm
Department of Computer Science
University of Saarland
Saarbrücken, Germany

and

AbsInt
Saarbrücken, Germany

International Advisory Board

I

System-Level Design and Verification

1

Real-Time in Networked Embedded Systems

Hans Hansson
Mälardalen University

Thomas Nolte
Mälardalen University

Mikael Sjödin
Mälardalen University

Daniel Sundmark
Mälardalen University

In this chapter, we provide an introduction to issues, techniques, and trends in networked embedded real-time systems (RTSs). We specifically discuss design of RTSs, real-time operating systems (RTOSs), real-time scheduling, real-time communication, and real-time analysis, as well as testing and debugging of RTSs. For each of these areas, state-of-the-art tools and standards are presented.

1.1 Introduction

Consider the air bag in the steering wheel of your car. It should, after the detection of a crash (and only then), inflate just in time to softly catch your head to prevent it from hitting the steering wheel; not too early—since this would make the air bag deflate before it can catch you; nor too late—since the exploding air bag then could injure you by blowing up in your face and/or catch you too late to prevent your head from banging into the steering wheel.

The computer-controlled air bag system is an example of a real-time system (RTS). But RTSs come in many different flavors, including vehicles, telecommunication systems, industrial automation systems, household appliances, etc.

There is no commonly agreed upon definition of what a RTS is, but the following characterization is (almost) universally accepted:

- RTSs are computer systems that physically interact with the real world.
- RTSs have requirements on the timing of these interactions.

Typically, the real-world interactions are via sensors and actuators rather than the keyboard and screen of your standard PC.

Real-time requirements typically express that an interaction should occur within a specified time bound. It should be noted that this is quite different from requiring the interaction to be as fast as possible.

Essentially all RTSs are embedded in products, and the vast majority of embedded computer systems are RTSs. RTSs are the dominating application of computer technology, as more than 99% of the manufactured processors are used in embedded systems.

Returning to the air bag system, we note that in addition to being an RTS it is a safety-critical system, i.e., a system which due to severe risks of damage has strict quality of service (QoS) requirements, including requirements on the functional behavior, robustness, reliability, and timeliness.

A typical strict timing property could be that a certain response to an interaction must always occur within some prescribed time, e.g., the charge in the air bag must detonate between 10 and 20 ms from the detection of a crash; violating this must be avoided at any cost, since it would lead to something unacceptable, i.e., you having to spend a couple of months in hospital. A system that is designed to meet strict timing requirements is often referred to as a hard RTS. In contrast, systems for which occasional timing failures are acceptable—possibly because this will not lead to something terrible—are termed soft RTS.

An illustrative comparison between hard and soft RTSs that highlights the difference between the extremes is shown in Table 1.1. A typical hard RTS could in this context be an engine control system, which must operate with microsecond-precision, and which will severely damage the engine if timing requirements fail by more than a few milliseconds. A typical soft RTS could be a banking system, for which timing is important, but where there are no strict deadlines and some variations in timing are acceptable.

Unfortunately, it is impossible to build real systems that satisfy hard real-time requirements, since due to the imperfection of hardware (and designers) any system may break. The best that can be achieved is a system that with very high probability provides the intended behavior during a finite interval of time.

TABLE 1.1   Typical Characteristics of Hard and Soft RTSs

Characteristic	Hard Real-Time	Soft Real-Time
Timing requirements	Hard	Soft
Pacing	Environment	Computer
Peak-load performance	Predictable	Degraded
Error detection	System	User
Safety	Critical	Noncritical
Redundancy	Active	Standby
Time granularity	Millisecond	Second
Data files	Small	Large
Data integrity	Short term	Long term

Source: Kopetz, H., Introduction in real-time systems: Introduction and overview, *Part XVIII of Lecture Notes from ESSES 2003—European summer school on Embedded systems*, Västerås, Sweden, September 2003.

However, on the conceptual level hard real-time makes sense, since it implies a certain amount of rigor in the way the system is designed, e.g., it implies an obligation to prove that the strict timing requirements are met, at least under some simplifying, but realistic, assumptions.

Since the early 1980s a substantial research effort has provided a sound theoretical foundation (e.g., [75,139]) and many practically useful results for the design of hard RTSs. Most notably, hard RTS scheduling has evolved into a mature discipline, using abstract, but realistic, models of tasks executing on a CPU, a multiprocessor, or distributed computer systems, together with associated methods for timing analysis. Such schedulability analysis, e.g., the well-known rate monotonic (RM) analysis [11,73,94], has also found significant use in some industrial segments.

However, hard real-time scheduling is not the cure for all RTSs. Its main weakness is that it is based on analysis of the worst possible scenario. For safety-critical systems this is of course a must, but for other systems, where general customer satisfaction is the main criterion, it may be too costly to design the system for a worst-case scenario that may not occur during the system's lifetime.

If we look at the other end of the spectrum, we find a best-effort approach, which is still the dominating approach in industry. The essence of this approach is to implement the system using some best practice, and then to use measurements, testing, and tuning making sure that the system is of sufficient quality. Such a system will hopefully satisfy some soft real-time requirement, the weakness being that we do not know which. On the other hand, compared to the hard real-time approach, the system can be better optimized for the available resources. Another difference is that hard RTS methods are essentially applicable to static configurations only, whereas it is less problematic to handle dynamic task creation, etc. in best-effort systems.

Having identified the weaknesses of the hard real-time and best-effort approaches, major efforts are now put into more flexible techniques for soft RTSs. These techniques provide analyzability (like hard real-time), together with flexibility and resource efficiency (like best-effort). The bases for the flexible techniques are often quantified QoS characteristics. These are typically related to nonfunctional aspects, such as timeliness, robustness, dependability, and performance. To provide a specified QoS, some sort of resource management is needed. Such a QoS management is handled by the application, by the operating system, by some middleware, or by a mix of the above. The QoS management is often a flexible online mechanism that dynamically adapts the resource allocation to balance between conflicting QoS requirements.

1.2 Design of Real-Time Systems

The main issue in designing RTSs is timeliness, i.e., the system performs its operations at proper instances in time. Not considering timeliness at the design phase will make it virtually impossible to analyze and predict the timely behavior of the RTS. This section presents some important architectural issues for embedded RTSs, together with some supporting commercial tools.

1.2.1 Reference Architecture

A generic system architecture for an RTS is depicted in Figure 1.1. This architecture is a model of any computer-based system interacting with an external environment via sensors and actuators.

Since our focus is on the RTS we will look more into different organizations of that part of the generic architecture in Figure 1.1. The simplest RTS is a single processor, but nowadays also multicore architectures are becoming more common. Moreover, in many cases, the RTS is a distributed computer system consisting of a set of processors interconnected by a communications network. There could be several reasons for making an RTS distributed, including

- Physical distribution of the application
- Computational requirements which may not be conveniently provided by a single CPU

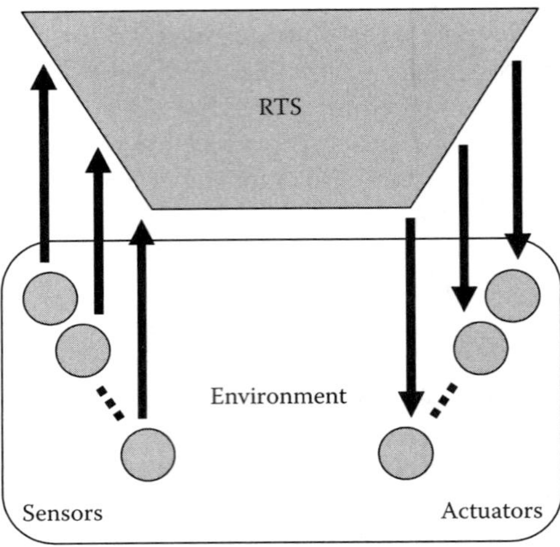

FIGURE 1.1   Generic RTS architecture.

- Need for redundancy to meet availability, reliability, or other safety requirements
- To reduce the cabling in the system

Figure 1.2 shows an example of a distributed RTS. In a modern car, like the one depicted in the figure, there are some 20–100 computer nodes (which in the automotive industry are called electronic control units [ECUs]) interconnected with one or more communication networks. The initial motivation for this type of electronic architecture in cars was the need to reduce the amount of cabling. However, the electronic architecture has also led to other significant improvements, including substantial pollution reduction and new safety mechanisms, such as computer-controlled electronic stabilization programs (ESPs). The current development is toward making the most safety-critical vehicle functions, like braking and steering, completely computer controlled. This is done by replacing the mechanical connections (e.g., between steering wheel and front wheels and between brake pedal and brakes), with computers and computer networks. Meeting the stringent safety requirements for such functions will require careful introduction of redundancy mechanisms in hardware and communication, as well as software, i.e., a safety-critical system architecture is needed (an example of such an architecture is time-triggered architecture (TTA) [82]).

1.2.2   Models of Interaction

In the previous section, we presented the physical organization of an RTS, but for an application programmer this is not the most important aspect of the system architecture. Actually, from an application programmer's perspective the system architecture is more given by the execution paradigm (execution strategy) and the interaction model used in the system. In this section, we describe what an interaction model is and how it affects the real-time properties of a system, and in Section 1.2.3, we discuss the execution strategies used in RTSs.

A model of interaction describes the rules by which components interact with each other (in this section we use the term component to denote any type of software unit, such as a task or a module). The interaction model can govern both control flow and data flow between system components. One of the most important design decisions, for all types of systems, is choosing interaction models to use (sadly, however, this decision is often implicit and hidden in the system's architectural description).

When designing RTSs, attention should be paid to the timing properties of the interaction models chosen. Some models have a more predictable and robust behavior with respect to timing than other

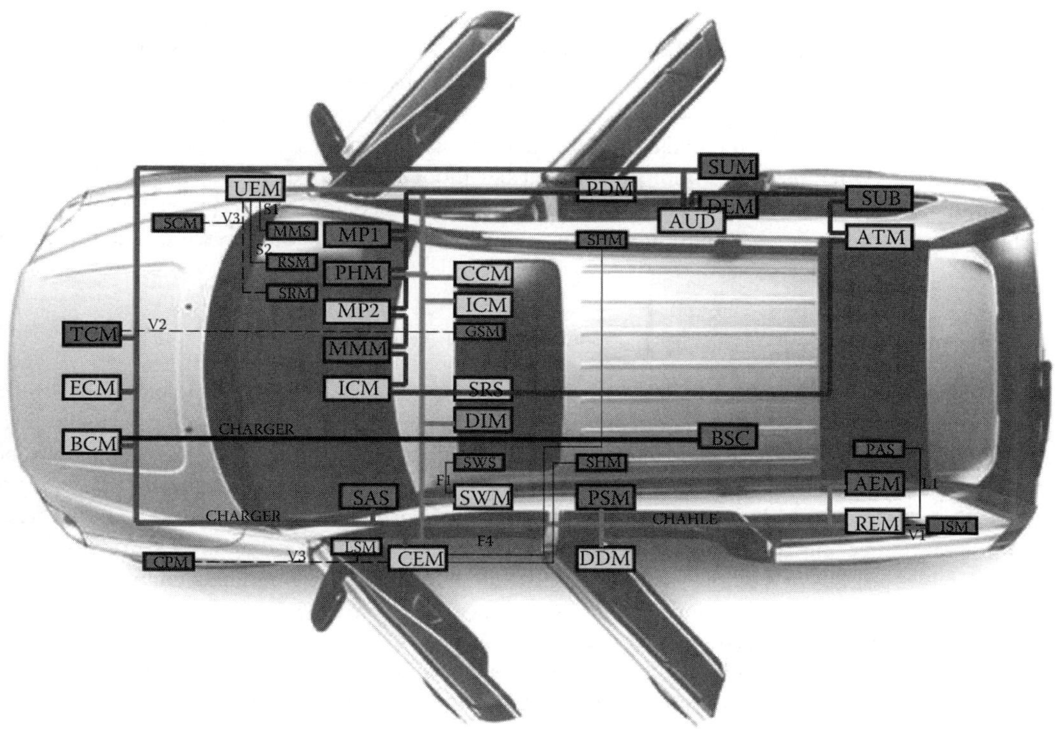

FIGURE 1.2 Network infrastructure of Volvo XC90.

models. Examples of some of the more predictable models that are also commonly used in RTSs design are given as follows.

1.2.2.1 Pipes-and-Filters

In this model, both data and control flow are specified using input and output ports of components. A component becomes eligible for execution when data arrives on its input ports and when the component finishes execution, and then it produces output on its output ports.

This model fits for many types of control programs very well, and control laws are easily mapped to this interaction model. Hence, it has gained widespread use in the real-time community. The real-time properties of this model are also quite nice. Since both data and control flow are unidirectional through a series of components, the order of execution and end-to-end timing delay usually become highly predictable. The model also provides a high degree of decoupling in time; that is, components can often execute without having to worry about timing delays caused by other components. Hence, it is usually straightforward to specify the compound timing behavior of set of components.

1.2.2.2 Publisher–Subscriber

The publisher–subscriber model is similar to the pipes-and-filters model but it usually decouples data and control flow. That is, a subscriber can usually choose different forms for triggering its execution. If the subscriber chooses to be triggered on each new published value, the publisher–subscriber model takes on the form of the pipes-and-filters model. However, in addition, a subscriber could choose to ignore the timing of the published values and decide to use the latest published value. Also, for the publisher–subscriber model, the publisher is not necessarily aware of the identity, or even the existence, of its subscribers. This provides a higher degree of decoupling of components.

Similar to the pipes-and-filters model, the publisher–subscriber model provides good timing properties. However, a prerequisite for analysis of systems using this model is that subscriber-components make explicit which values they will subscribe to (this is not mandated by the model itself). However, when using the publisher–subscriber model for embedded systems, it is the norm that subscription information is available (this information is used, for instance, to decide the values that are to be published over a communications network and to decide the receiving nodes of those values).

1.2.2.3 Blackboard

The blackboard model allows variables to be published in a globally available blackboard area. Thus, it resembles the use of global variables. The model allows any component to read or write values to variables in the blackboard. Hence, the software engineering qualities of the blackboard model is questionable. Nevertheless, this model is commonly used, and in some situations it provides a pragmatic solution to problems that are difficult to address with more stringent interaction models.

Software engineering aspects aside, the blackboard model does not introduce any extra elements of unpredictable timing. On the other hand, the flexibility of the model does not help engineers achieve predictable systems. Since the model does not address the control flow, components can execute relatively undisturbed and decoupled from other components.

On the other end of the spectrum of interaction models there are models that increase the (timing) unpredictability of the system. These models should, if possible, be avoided when designing RTSs. The two most notable, and commonly used, are

Client–Server. In the client–server model, a client asynchronously invokes a service of a server. The service invocation passes the control flow (plus any input data) to the server, and control stays at the server until it has completed the service. When the server is done, the control flow (and any return data) is returned to the client who in turn resumes execution.

The client–server model has inherently unpredictable timing. Since services are invoked asynchronously, it is very difficult to a priori assess the load on the server for a certain service invocation. Thus, it is difficult to estimate the delay of the service invocation and, in turn, it is difficult to estimate the response time of the client. This matter is furthermore complicated by the fact that most components often behave both as clients and servers (a server often uses other servers to implement its own services); leading to very complex and unanalyzable control flow paths.

Message Boxes. A component can have a set of message boxes, and components communicate by posting messages in each others message boxes. Messages are typically handled in first-in first-out (FIFO) order, or in priority order (where the sender specifies a priority). Message passing does not change the flow of control for the sender. A component that tries to receive a message from an empty message box, however, blocks on that message box until a message arrives (often the receiver can specify a timeout to prevent indefinite blocking).

From a sender's point of view, the message box model has similar problems as the client–server model. The data sent by the sender (and the action that the sender expects the receiver to perform) may be delayed in an unpredictable way when the receiver is highly loaded. Also, the asynchronous nature of the message passing makes it difficult to foresee the load of a receiver at any particular moment.

Furthermore, from the receiver's point of view, the reading of message boxes is unpredictable in the sense that the receiver may or may not block on the message box. Also, since message boxes often are of limited size, a highly loaded receiver risk to lose some message. Which messages are lost is another source of unpredictability.

1.2.3 Execution Strategies

There are two main execution paradigms for RTSs: time-triggered and event-triggered. When using timed-triggered execution, activities occur at predefined instances of time, e.g., a specific sensor value is read exactly every 10 ms and exactly 2 ms later the corresponding actuator receives an updated control parameter. In an event-triggered execution, on the other hand, actions are triggered by event occurrences, e.g., when the level of toxic fluid in a tank reaches a certain level an alarm will go off. It should be noted that the same functionality typically can be implemented in both paradigms, e.g., a time-triggered implementation of the above alarm would be to periodically read the level-measuring sensor and set off the alarm when the read level exceeds the maximum allowed. If alarms are rare, the time-triggered version will have much higher computational overhead than the event-triggered one. On the other hand, the periodic sensor readings will facilitate detection of a malfunctioning sensor.

Time-triggered executions are used in many safety-critical systems with high dependability requirements (such as avionic control systems), whereas the majority of other systems are event triggered. Dependability can be guaranteed also in the event-triggered paradigm, but due to the observability provided by the exact timing of time-triggered executions, most experts argue for using time-triggered in ultra-dependable systems. The main argument against time-triggered is its lack of flexibility and the requirement of preruntime schedule generation (which is a nontrivial and possibly time-consuming task).

Time-triggered systems are mostly implemented by simple proprietary table-driven dispatchers [186] (see Section 1.4.2 for a discussion on table-driven execution), but complete commercial systems including design tools are also available [81,176]. For the event-triggered paradigm, a large number of commercial tools and operating systems are available (examples are given in Section 1.3.3). There are also examples of systems integrating the two execution paradigms, thereby aiming at getting the better of two worlds: time-triggered dependability and event-triggered flexibility. One example is the Basement system [53] and its associated real-time kernel Rubus [8].

Since computations in time-triggered systems are statically allocated in both space (to a specific processor) and time, some sort of configuration tool is often used. This tool assumes that the computations are packaged into schedulable units (corresponding to tasks or threads in an event-triggered system). Typically, for example, in Basement, computations are control-flow based, in the sense that they are defined by sequences of schedulable units, each unit performing a computation based on its inputs and producing outputs to the next unit in sequence. The system is configured by defining the sequences and their timing requirements. The configuration tool will then automatically (if possible*) generate a schedule, which guarantees that all timing requirements are met.

Event-triggered systems typically have a richer and more complex application programmer interfaces (APIs), defined by the user operating system and middleware, which will be elaborated on in Section 1.3.

1.2.4 Tools for Design of Real-Time Systems

In industry the term "real-time system" is highly overloaded and can mean anything from interactive systems to superfast systems, or embedded systems. Consequently, it is not easy to judge what tools are suitable for developing RTSs (as we define real-time in this chapter).

For instance, unified modeling language (UML) [116] is commonly used for software design. However, UML's focus is mainly on client–server solutions, and many tools are inapt for RTSs design. As a consequence, UML-based tools that extend UML with constructs suitable for real-time programs

---

* This scheduling problem is theoretically intractable, so the configuration tool will have to rely on some heuristics which works well in practice, but which does not guarantee to find a solution in all cases when there is a solution.

have emerged. The two most known such products are IMB's Rational Rose Technical Developer [133] and Telelogic's Rhapsody [164]. These tools provide UML-support with the extension of real-time profiles. While giving real-time engineers access to suitable abstractions and computational models, these tools do not provide means to describe timing properties or requirements in a formal way; thus they do not allow automatic verification of timing requirements.

For resource-constrained hard RTSs, design tools are provided by Arcticus System [8], ETAS [41], TTTech [176], and Vector [180]. These tools are instrumental during both system design and implementation and also provide some timing analysis techniques which allow timing verification of the system (or parts of the system). However, these tools are based on proprietary formats and processes and have as such reached a limited customer base (mainly within the automotive industry).

When it comes to functional design of RTSs, model-based development has recently become popular. Using high-level tools, functions are described in a modeling language. From these models code is generated. This code is typically not timing aware and is thus subject to later integration in an RTS. Thus, these approaches support in generating the functional content of an RTS, but they cannot generate the whole system. A popular example is control engineers modeling control-function in Simulink [104] or Stateflow [105] and automatically generate code for the function with TargetLink [35].

1.3 Real-Time Operating Systems

An RTOS provides services for resource access and resource sharing, very much similar to a general-purpose operating system. An RTOS, however, provides additional services suited for real-time development and also supports the development process for embedded-systems development. Using a general-purpose operating system when developing RTSs has several drawbacks:

- High resource utilization, e.g., large RAM and ROM footprints and high internal CPU-demand
- Difficult to access hardware and devices in a timely manner, e.g., no application-level control over interrupts
- Lack of services to allow timing sensitive interactions between different processes

1.3.1 Typical Properties of RTOSs

The state of practice in RTOSs is reflected in [43]. Not all operating systems are RTOSs. An RTOS is typically multithreaded and preemptible; there has to be a notion of thread priority, predictable thread synchronization has to be supported, priority inheritance should be supported, and the OS behavior should be known [9]. This means that the interrupt latency, worst-case execution time (WCET) of system calls, and maximum time during which interrupts are masked must be known. A commercial RTOS is usually marketed as a runtime component of an embedded development platform.

As a general rule of thumb one can say that RTOSs are

Suitable for resource-constrained environments. RTSs typically operate in such environments. Most RTOSs can be configured preruntime (e.g., at compile time) to only include a subset of the total functionality. Thus, the application developer can choose to leave out unused portions of the RTOS in order to save resources.

RTOSs typically store much of their configuration in ROM. This is done for mainly two purposes: (1) minimize use of expensive RAM memory and (2) minimize the risk that critical data is overwritten by an erroneous application.

Giving the application programmer easy access to hardware features. These include interrupts and devices. Most often the RTOSs give the application programmer means to install interrupt service routines during compile time and/or runtime. This means that the RTOS leaves all interrupt handing to the application programmer, allowing fast, efficient, and predictable handling of interrupts.

In general-purpose operating systems, memory-mapped devices are usually protected from direct access using the Memory Management Unit of the CPU, hence forcing all device accesses to go through the operating system. RTOSs typically do not protect such devices and allow the application to directly manipulate the devices. This gives faster and more efficient access to the devices. (However, this efficiency comes at the price of an increased risk of erroneous use of the device.)

Providing services that allow implementation of timing sensitive code. An RTOS typically has many mechanisms to control the relative timing between different processes in the system. Most notably an RTOS has a real-time process scheduler whose function is to make sure that the processes execute in the way the application programmer intended them to. We will elaborate more on the issues of scheduling in Section 1.4.

An RTOS also provides mechanisms to control the processes relative performance when accessing shared resources. This can, for instance, be done by priority queues, instead of plain FIFO-queues as is used in general-purpose operating systems. Typically an RTOS supports one or more real-time resource-locking protocols, such as priority inheritance or priority ceiling (Section 1.3.2 discusses resource-locking protocols further).

Tailored to fit the embedded systems development process. RTSs are usually constructed in a host environment that is different from the target environment, so-called cross platform development. Also, it is typical that the whole memory image, including both RTOS and one or more applications, is created at the host platform and downloaded to the target platform. Hence, most RTOSs are delivered as source code modules or precompiled libraries that are statically linked with the applications at compile time.

1.3.2 Mechanisms for Real-Time

One of the most important functions of an RTOS is to arbitrate access to shared resources in such a way that the timing behavior of the system becomes predictable. The two most obvious resources that the RTOS manages to access are

- CPU—That is, the RTOS should allow processes to execute in a predictable manner
- Shared memory areas—That is, the RTOS should resolve contention to shared memory in a way that gives predictable timing

The CPU access is arbitrated with a real-time scheduling policy. Section 1.4 describes real-time scheduling policies in more depth. Examples of scheduling policies that can be used in RTSs are priority scheduling, deadline scheduling, and rate scheduling. Some of these policies directly use timing attributes (like deadline) of the tasks to perform scheduling decisions, whereas other policies use scheduling parameters (like priority, rate, and bandwidth) which indirectly affect the timing of the tasks.

A special form of scheduling, which is also very useful for RTSs, is table-driven (static) scheduling. Table-driven scheduling is described further in Section 1.4.2. To summarize, in table-driven scheduling all arbitration decisions have been made off-line and the RTOS scheduler just follows a simple table. This gives very good timing predictability, albeit at the expense of system flexibility.

The most important aspect of a real-time scheduling policy is that it should provide means to á priori analyze the timing behavior of the system, hence giving a predictable timing behavior of the system. Scheduling in general-purpose operating systems normally emphasizes properties like fairness, throughput, and guaranteed progress; these properties may all be adequate in their own respect; however, they are usually in conflict with the requirement that an RTOS should provide timing predictability.

Shared resources (such as memory areas, semaphores, and mutexes) are also arbitrated by the RTOS. When a task locks a shared resource it will block all other tasks that subsequently try to lock the resource. In order to achieve predictable blocking times special real-time resource-locking protocols have been proposed (Refs. [21,24] provide more details about the protocols).

1.3.2.1 Priority Inheritance Protocol

Priority Inheritance Protocol (PIP) makes a low-priority task inherit the priority of any higher-priority task that becomes blocked on a resource locked by the lower-priority task.

This is a simple and straightforward method to lower the blocking time. However, it is computationally intractable to calculate the worst-case blocking (which may be infinite since the protocol does not prevent deadlocks). Hence, for hard RTSs or when timing performance needs to be calculated a priori, the PIP is not adequate.

1.3.2.2 Priority Ceiling Inheritance Protocol

Priority Ceiling Inheritance Protocol (PCP) associates to each resource a ceiling value that is equal to the highest priority of any task that may lock the resource. By cleaver use of the ceiling values of each resource, the RTOS scheduler will manipulate task priorities to avoid the problems of PIP.

PCP guarantees freedom from deadlocks, and the worst-case blocking is relatively easy to calculate. However, the Computational complexity of keeping track of ceiling values and task priorities gives the PCP high run-time overhead.

1.3.2.3 Immediate Ceiling PIP

Immediate inheritance PIP (IIP) also associates to each resource a ceiling value which is equal to the highest priority of any task that may lock the resource. However, different from the PCP, in IIP a task is immediately assigned the ceiling priority of the resource it is locking.

IIP has the same real-time properties as the PCP (including the same worst-case blocking time*). However, IIP is significantly easier to implement. It is, in fact, because single-node systems are easier to implement than any other resource-locking protocol (including non-real-time protocols). In IIP no actual locks need to be implemented; it is enough for the RTOS to adjust the priority of the task that locks or releases a resource. IIP has other operational benefits; notably it paves the way for letting multiple tasks use the same stack area. Operating systems based on IIP can be used to build systems with footprints that are extremely small [10,96].

1.3.3 Commercial RTOSs

There are an abundance of commercial RTOSs. Most of them provide adequate mechanisms to enable development of RTSs. Some examples are Tornado/VxWorks [185], LYNX [100], OSE [160], QNX [128], RT-Linux [138], and ThreadX [98]. However, the major problem with these tools is the rich set of primitives provided. These systems provide both primitives that are suitable for RTSs and primitives that are unfit for RTSs (or that should be used with great care). For instance, they usually provide multiple resource-locking protocols; some of which are suitable and some of which are not suitable for real-time.

This richness becomes a problem when these tools are used by inexperienced engineers and/or when projects are large and project management does not provide clear design guidelines/rules. In these situations, it is very easy to use primitives that will contribute to timing unpredictability of the

---

* However, the average blocking time will be higher in IIP than in PCP.

developed system. Rather, an RTOS should help the engineers and project managers by providing mechanisms that help designing predictable systems. However, there is an obvious conflict between the desire/need of RTOS manufacturers to provide rich interfaces and stringency needed by designers of RTSs.

There is a smaller set of RTOSs that has been designed to resolve these problems, and at the same time also allows extreme lightweight implementations of predictable RTSs. The driving idea is to provide a small set of primitives that guides the engineers toward good design of their system. Typical examples are the research RTOS Asterix [10] and the commercial RTOS SSX5 [96]. These systems provide a simplified task model, in which tasks cannot suspend themselves (e.g., no `sleep()` primitive) and tasks are restarted from their entry point on each invocation. The only resource-locking protocol that is supported is IIP, and the scheduling policy is fixed-priority scheduling. These limitations make it possible to build an RTOS that is able to run, e.g., 10 tasks using less that 200 bytes of RAM, and at the same time to give predictable timing behavior [7]. Other commercial systems that follow a similar principle of reducing the degrees of freedom and hence promote stringent design of predictable RTSs include Arcticus Systems' Rubus OS [8].

Many of the commercial RTOSs provide standard APIs. The most important RTOS-standards are RT-POSIX [62], OSEK [50], and APEX [4]. We will here only deal with POSIX since it is the most widely adopted RTOS standard, and those interested in automotive and avionic systems should take a closer look at OSEK and APEX, respectively.

The POSIX standard is based on Unix, and its goal is portability of applications at the source code level. The basic POSIX services include task and thread management, file system management, input and output, and event notification via signals. The POSIX real-time interface defines services facilitating concurrent programming and providing predictable timing behavior. Concurrent programming is supported by synchronization and communication mechanisms that allow predictability. Predictable timing behavior is supported by preemptive fixed-priority scheduling, time management with high resolution, and virtual memory management. Several restricted subsets of the standard intended for different types of systems, as well as specific language bindings, for example, Ada [65], have been defined.

1.4 Real-Time Scheduling

A real-time scheduler schedules real-time tasks sharing a resource (e.g., a CPU or a network link). The goal of the real-time scheduler is to ensure that the timing constraints of these tasks are satisfied. The scheduler decides, based on the task timing constraints, which task to execute or to use the resource at any given time.

Traditionally, real-time schedulers are divided into off-line and online schedulers. Off-line schedulers make all scheduling decisions before the system is executed. At runtime a simple dispatcher is used to activate tasks according to the schedule generated before run-time. Online schedulers, on the other hand, make scheduling decisions based on the system's timing constraints during runtime.

As there are many different schedulers developed in the research community [23,143], only the basic concepts of different types of schedulers are presented in this chapter. We have divided real-time schedulers into three categories: time-driven schedulers, priority-driven schedulers, and share-driven schedulers. This classification of real-time schedulers is depicted in Figure 1.3.

1.4.1 Introduction to Scheduling

In order to reason about an RTS, a number of RTS models that more or less accurately capture the temporal behavior of the system have been developed.

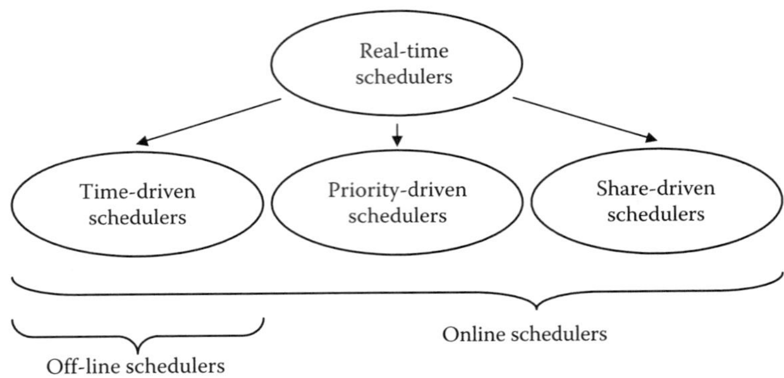

FIGURE 1.3   Real-time schedulers.

A typical RTS can be modeled as a set of real-time programs, each of which in turn consists of a set of tasks. These tasks typically control a system in an environment of sensors, control functions, and actuators, all with limited resources in terms of computation and communication capabilities. Resources such as memory and computation time are limited, imposing strict requirements on the tasks in the system (the system's task set). The execution of a task is triggered by events generated by time (time events), other tasks (task events), or input sensors (input events). The execution delivers data to output actuators (output events) or to other tasks. Tasks have different properties and requirements in terms of time, e.g., WCETs, periods, and deadlines. Several tasks can be executing on the same processor, i.e., sharing the same CPU. An important issue is to determine whether all tasks can execute as planned during peak-load. By enforcing some task model and calculating the total task utilization in the system (e.g., the total utilization of the CPU by all tasks in the system's task set), or the response time of all tasks in the worst-case scenarios (at peak-load), it can be determined if they will fulfill the requirement of completing their executions within their deadlines. Examples of this are given in Section 1.6.

As tasks are executing on a CPU, when there are several tasks to choose from (ready for execution), it must be decided which task to execute. Tasks can have different priorities in order to, for example, let a more important task execute before a less important task. Moreover, an RTS can be preemptive or nonpreemptive. In a preemptive system, tasks can preempt each other, allowing for the task with the highest priority to execute as soon as possible. However, in a nonpreemptive system a task that has been allowed to start will always execute until its completion, thus deferring execution of any higher priority tasks. The difference between preemptive and nonpreemptive execution in a priority scheduled system is shown in Figure 1.4. Here, two tasks, task A and task B, are executing on a CPU.

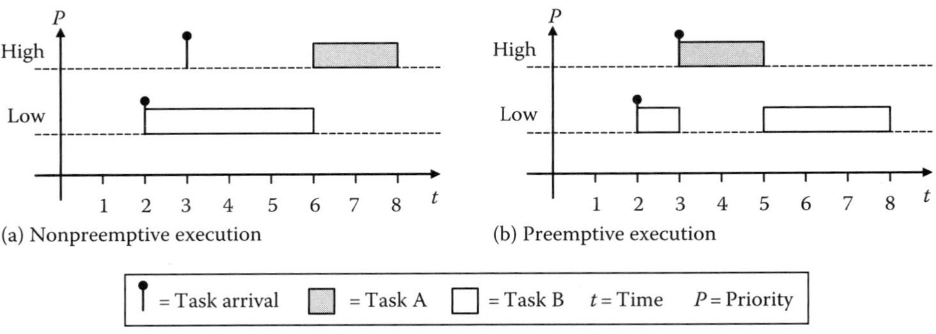

FIGURE 1.4   Difference between (a) nonpreemptive and (b) preemptive systems.

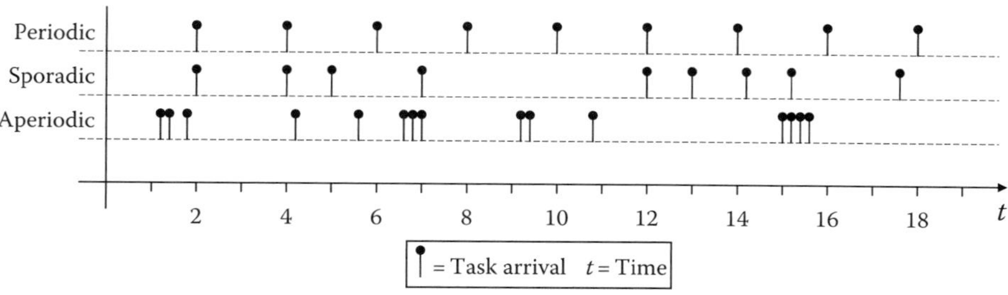

FIGURE 1.5   Periodic, sporadic, and aperiodic task arrival.

Task A has higher priority than task B. Task A is arriving at time 3 and task B is arriving at time 2. Scenarios for both nonpreemptive execution (Figure 1.4a) and preemptive execution (Figure 1.4b) are shown. In Figure 1.4a, the high-priority task arrives at time 3 but is blocked by the lower-priority task and cannot start its execution until time 6 when the low-priority task has finished its execution. In Figure 1.4b, the high-priority task executes directly on its arrival at time 3, preempting the low-priority task.

Before a task can start to execute it has to be triggered. Once a task is triggered it will be ready for execution in the system. Tasks are either event- or time-triggered, triggered by events that are either periodic, sporadic, or aperiodic in their nature. Due to this behavior, tasks are modeled as either periodic, sporadic, or aperiodic. The instant when a task is triggered and ready for execution in the system is called the task arrival. The time between two arrivals of the same task (between two task arrivals) is called the task interarrival time. Periodic tasks are ready for execution periodically with a fixed interarrival time (called period). Aperiodic tasks have no specific interarrival time and may be triggered at any time, usually triggered by interrupts. Sporadic tasks, although having no period, have a known minimum interarrival time. The difference between periodic, sporadic, and aperiodic task arrivals is illustrated in Figure 1.5. In Figure 1.5, the periodic task has a period equal to 2, i.e., interarrival time is 2; the sporadic task has a minimum interarrival time of 1; and the aperiodic task has no known interarrival time.

The choice between implementing a particular part of the RTS using periodic, sporadic, or aperiodic tasks is typically based on the characteristics of the function. For functions dealing with measurements of the state of the controlled process (e.g., its temperature), a periodic task is typically used to sample the state. For the handling of events (e.g., an alarm) a sporadic task can be used if the event is known to have a minimum interarrival time. The minimum interarrival time can be constrained by physical laws, or it can be enforced by some mechanical mechanism. If the minimum time between two consecutive events is unknown, an aperiodic task is required for the handling of the event. While it can be impossible to guarantee a performance of an individual aperiodic task, the system can be constructed such that aperiodic tasks will not interfere with the sporadic and periodic tasks executing on the same resource.

Moreover, real-time tasks can be classified as tasks with hard or soft real-time requirements. The real-time requirements of an application spans a spectrum, as depicted in Figure 1.6 showing some example applications having non, soft, and hard real-time requirements [154]. Hard real-time tasks have high demands on their ability to meet their deadlines, and violation of these requirements may have severe consequences. If the violation may be catastrophic, the task is classified as being safety-critical. However, many tasks have real-time requirements although violation of these is not so severe, and in some cases a number of deadline violations can be tolerated. Examples of RTSs including such tasks are robust control systems and systems that contain audio/video streaming. Here, the real-time constraints must be met in order for the video and/or sound to appear good to the end user, and a violation of the temporal requirements will be perceived as a decrease in quality.

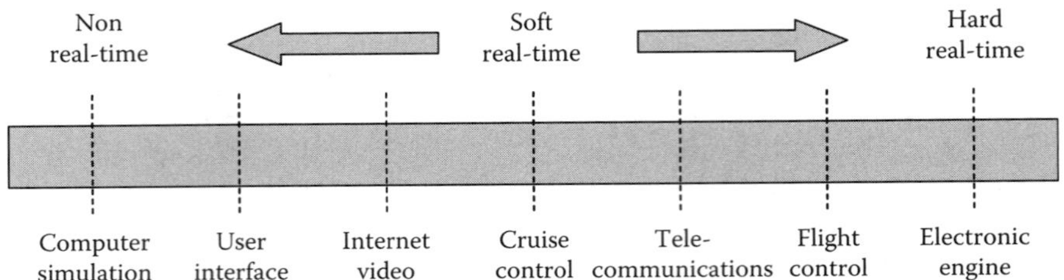

FIGURE 1.6   The real-time spectrum.

A central problem when dealing with RTS models is to determine how long a real-time task will execute, in the worst case. The task is usually assumed to have a WCET. The WCET is part of the RTS models used when calculating worst-case response times of individual tasks in a system, or to determine if a system is schedulable using a utilization-based test. Determining the WCET is a research problem of its own, which has not yet been satisfactorily solved. However, there exist several emerging techniques for the estimation of the WCET [39,127].

1.4.2   Time-Driven Scheduling

Time-driven schedulers [77] work in the following way: The scheduler creates a schedule (sometimes called the table). Usually the schedule is created before the system is started (off-line), but it can also be done during runtime (online). At runtime, a dispatcher follows the schedule and makes sure that tasks are only executing in their predetermined time slots.

By creating a schedule off-line, complex timing constraints, such as irregular task arrival patterns and precedence constraints, can be handled in a predictable manner that would be difficult to do online during runtime (tasks with precedence constraints require a special order of task executions, e.g., task A must execute before task B). The schedule that is created off-line is the schedule that will be used at runtime. Therefore, the online behavior of time-driven schedulers is very predictable. Because of this predictability, time-driven schedulers are the more commonly used schedulers in applications that have very high safety-critical demands, e.g., in avionics. However, since the schedule is created off-line, the flexibility is very limited, in the sense that as soon as the system changes (due to adding of functionality or change of hardware), a new schedule has to be created and given to the dispatcher. To create a new schedule can be nontrivial and sometimes very time consuming motivating the usage of priority-driven schedulers described in Section 1.4.3.

1.4.3   Priority-Driven Scheduling

Scheduling policies that make their scheduling decisions during runtime are classified as online schedulers. These schedulers make their scheduling decisions online based on the system's timing constraints, such as task priority. Schedulers that base their scheduling decisions on task priorities are called priority-driven schedulers.

Using priority-driven schedulers the flexibility is increased (compared to time-driven schedulers), since the schedule is created online based on the currently active tasks' properties. Hence, priority-driven schedulers can cope with changes in workload as well as adding and removing of tasks and functions, as long as the schedulability of the complete task set is not violated. However, the exact behavior of priority-driven schedulers is harder to predict. Therefore, these schedulers are not used often in the most safety-critical applications.

Priority-driven scheduling policies can be divided into fixed-priority schedulers (FPSs) and dynamic-priority schedulers (DPSs). The difference between these scheduling policies is whether the priorities of the real-time tasks are fixed or whether they can change during execution (i.e., dynamic priorities).

1.4.3.1 Fixed-Priority Schedulers

When using FPS, once priorities are assigned to tasks they are not changed. Then, during execution, the task with the highest priority among all tasks that are available for execution is scheduled for execution. Priorities can be assigned in many ways, and depending on the system requirements some priority assignments are better than others. For instance, using a simple task model with strictly periodic noninterfering tasks with deadlines equal to the period of the task, an RM priority assignment has been shown by Liu and Layland [95] to be optimal in terms of schedulability. In RM, the priority is assigned based on the period of the task. The shorter the period, the higher will be the priority assigned to the task.

1.4.3.2 Dynamic-Priority Schedulers

The most well-known DPS is the earliest deadline first (EDF) scheduling policy [95]. Using EDF, the task with the nearest (earliest) deadline among all tasks ready for execution gets the highest priority. Therefore, the priority is not fixed; it changes dynamically over time. For simple task models, it has been shown that EDF is an optimal scheduler in terms of schedulability [32]. Also, the EDF allows for higher schedulability compared with FPSs. Schedulability is in the simple scenario guaranteed as long as the total load in the scheduled system is $\leq 100\%$, whereas a FPS in these simple cases has a schedulability bound of about 69%. For a good comparison between RM and EDF interested readers are referred to [22].

1.4.4 Share-Driven Scheduling

Another way of scheduling a resource is to allocate a share [156] of the resource to a user or task. This is useful, for example, when dealing with aperiodic tasks when their behavior is not completely known. Using share-driven scheduling it is possible to allocate a fraction of the resource to these aperiodic tasks, preventing them from interfering with other tasks that might be scheduled using time-driven or priority-driven scheduling techniques.

In order for the priority-driven schedulers to cope with aperiodic tasks, different service methods have been presented. The objective of these service methods is to give a good average response time for aperiodic requests, while preserving the timing constraints of periodic and sporadic tasks. These services can be implemented as share-driven scheduling policies, either based on general processor sharing (GPS) [118,119] algorithms, or using special server-based schedulers, e.g., [1,2,89,90,130–132, 144,148–150,157]. In the scheduling literature, many types of servers are described, implementing server-based schedulers. In general, each server is characterized partly by its unique mechanism for assigning deadlines and partly by a set of parameters used to configure the server. Examples of such parameters are priority, bandwidth, period, and capacity.

1.5 Real-Time Communications

Real-time communication aims at providing timely and deterministic communication of data between devices in a distributed system. In many cases, there are requirements on providing guarantees of the real-time properties of these transmissions. The communication is carried out over a communications network relying on either a wired or a wireless medium.

No	ISO/OSI layers	TCP/IP layers
7	Application layer	Application
6	Presentation layer	
5	Session layer	
4	Transport layer	Transport
3	Network layer	Internet
2	Data link layer	Network interface
1	Physical layer	Hardware

FIGURE 1.7   ISO/OSI reference model.

1.5.1   ISO/OSI Reference Model

The objective of the ISO/OSI reference model [31,193] is to manage the complexity of communication protocols. The model contains seven layers, depicted in Figure 1.7, together with one of its most-common implementation, the TCP/IP protocol.

The lowest three layers are network dependent, where the physical layer is responsible for the transmission of raw data on the medium used. The data link layer is responsible for the transmission of data frames and to recognize and correct errors related to this. The network layer is responsible for the setup and maintenance of network wide connections. The upper three layers are application oriented, and the intermediate layer (the transport layer) isolates the upper three and the lower three layers from each other, i.e., all layers above the transport layer can transmit messages independent of the underlying network infrastructure.

In this chapter, the lower layers of the ISO/OSI reference model are of great importance, where for real-time communications, the medium access control (MAC) protocol determines the degree of predictability of the network technology. Usually, the MAC protocol is considered a sublayer of the physical layer or the data link layer. In Section 1.5.2, a number of relevant (for real-time communications) MAC protocols are described.

1.5.2   MAC Protocols

A node with networking capabilities has a local communications adapter that mediates access to the medium used for message transmissions. Tasks that send messages send their messages to the local communications adapter. Then, the communications adapter takes care of the actual message transmission. Also, the communications adapter receives messages from the medium, delivering them to the corresponding receiving tasks (via the ISO/OSI protocol stack). When data is to be sent from the communications adapter to the wired or wireless medium, the message transmission is controlled by the MAC protocols.

Common MAC protocols used in real-time communication networks can be classified into random access protocols, fixed-assignment protocols, and demand-assignment protocols. Examples of these MAC protocols are given below:

1. Random access protocols such as

    a. CSMA/CD (carrier sense multiple access / collision detection)

    b. CSMA/CR (carrier sense multiple access / collision resolution)

    c. CSMA/CA (carrier sense multiple access / collision avoidance)

   2. Fixed-assignment protocols such as

      a. TDMA (time division multiple access)

      b. FTDMA (flexible TDMA)

   3. Demand-assignment protocols such as

      a. Distributed solutions relying on tokens

      b. Centralized solutions by the usage of masters

These MAC protocols are all used both for real-time and non-real-time communications, and each of them have different timing characteristics. In the following sections, all of these MAC protocols (together with the random access related MAC protocols ALOHA and CSMA) are presented.

1.5.2.1 ALOHA

The classical random access MAC protocol is the ALOHA protocol [3]. Using ALOHA, messages arriving at the communications adapter are immediately transmitted on the medium, without prior checking the status of the medium, i.e., if it is idle or busy. Once the sender has completed its transmission of a message, it starts a timer and waits for the receiver to send an acknowledgment message, confirming the correct reception of the transmitted message at the receiver end. If the acknowledgment is received at the transmitter before the timer is expired, the timer is stopped and the message is considered successfully transmitted. If the timer expires, the transmitter selects a random backoff time and waits for this time until the message is retransmitted.

ALOHA is a primitive random access MAC protocol with primary strength in its simplicity. However, due to the simplicity it is not very efficient and predictable, hence not suitable for real-time communications.

1.5.2.2 CSMA

Improving the above-mentioned approach of ALOHA is to check the status of the medium before transmitting [74], i.e., check if the medium is idle or busy before starting transmitting (this process is called carrier sensing). CSMA protocols do this and allow for ongoing message transmissions to be completed without disturbance of other message transmissions. If the medium is busy, CSMA protocols wait for some time (the backoff time) before a transmission is tried again (different approaches exist, e.g., nonpersistent CSMA, p-persistent CSMA, and 1-persistent CSMA). CSMA relies (as ALOHA) on the receiver to transmit an acknowledgment message to confirm the correct reception of the message.

However, the number of collisions is still high when using CSMA (although lower compared with ALOHA). Using pure CSMA the performance of ongoing transmissions is improved but still it is a delicate task to initiate a transmission when several communication adapters want to start transmitting at the same time. If several transmissions are started in parallel, all transmitted messages are corrupted which is not detected until the lacking reception of a corresponding acknowledge message. Hence, time and bandwidth are lost.

1.5.2.3 CSMA/CD

In CSMA/CD networks, collisions between messages on the medium are detected by simultaneously writing the message and reading the transmitted signal on the medium. Thus, it is possible to verify if the transmitted signal is the same as the signal currently being transmitted. If they are not the same, one or more parallel transmissions are going on. Once a collision is detected, the transmitting stations stop their transmissions and wait for some time (generated by the backoff algorithm) before retransmitting the message in order to reduce the risk of the same messages colliding again. However, due to the possibility of successive collisions, the temporal behavior of CSMA/CD networks can be somewhat hard to predict. CSMA/CD is used, e.g., for Ethernet (see Section 1.5.3).

1.5.2.4   CSMA/CR

CSMA/CR does not go into a backoff mode (as the above-mentioned approaches) once there is a collision detected. Instead, CSMA/CR resolves collisions by determining one of the message transmitters involved in the collision that is allowed to go on with an uninterrupted transmission of its message. The other messages involved in the collision are retransmitted at another time, e.g., directly after the transmission of the first message. The same scenario using the CSMA/CD MAC protocol would cause all messages involved in the collision to be retransmitted.

Due to the collision resolution feature of CSMA/CR, it has the possibility to become more predictable in its temporal behavior compared to CSMA/CD. An example of a network technology that implements CSMA/CR is CAN [66].

1.5.2.5   CSMA/CA

In some cases it is not possible to detect collisions although it might still be desirable to try to avoid them. For example, using a wireless medium often makes it impossible to simultaneously read and write (send and receive) to the medium, as (at the communications adapter) the signal sent is so much stronger than (and therefore overwrites) the signal received. CSMA/CA protocols can avoid collisions by the usage of some handshake protocol in order to guarantee a free medium before the initiation of a message transmission. CSMA/CA is used by, e.g., ZigBee [63,192].

1.5.2.6   TDMA

TDMA is a fixed-assignment MAC protocol where time is used to achieve temporal partitioning of the medium. Messages are sent at predetermined instances in time, called message slots. Often, a schedule of slots is created off-line (before the system is running), and this schedule is then followed and repeated online, but schedules can also be created online.

Due to the time-slotted nature of TDMA networks, their temporal behavior is very predictable and deterministic. TDMA networks are therefore very suitable for safety-critical systems with hard real-time guarantees. A drawback of TDMA networks is that they are somewhat inflexible, as a message cannot be sent at an arbitrary time. A message can only be sent in one of the message's predefined slots, which affects the responsiveness of the message transmissions. Also, if a message is shorter than its allocated slot, bandwidth is wasted since the unused portion of the slot cannot be used by another message. For example, suppose a message requires only half of its slot, then 50% of the bandwidth in that slot is wasted, compared with a CSMA/CR network that is available for any message as soon as the transmission of the previous message is finished. One example of a TDMA real-time network is TTP/C [79,173], where off-line created schedules allow for the usage of TTP/C in safety-critical applications. One example of an online scheduled TDMA network is the GSM network.

1.5.2.7   FTDMA

Another fixed-assignment MAC protocol is the FTDMA. As regular TDMA networks, FTDMA networks avoid collisions by dividing time into slots. However, FTDMA networks are using a mini-slotting concept in order to make more efficient use of the bandwidth, compared to a TDMA network. FTDMA is similar to TDMA with a difference in runtime slot size. In a FTDMA schedule, the size of a slot is not fixed, but will vary depending on whether the slot is used or not. In case all slots are used in a FTDMA schedule, the FTDMA operates the same way as the TDMA. However, if a slot is not used within a small time offset Δ after its initiation, the schedule will progress to its next slot. Hence, unused slots will be shorter compared to a TDMA network where all slots have fixed size. However, used slots have the same size in both FTDMA and TDMA networks. Variants of mini slotting can be found in, e.g., Byteflight [14] and FlexRay [46].

1.5.2.8 Tokens

An alternative way of eliminating collisions on the network is to achieve mutual exclusion by the usage of token-based demand assignment MAC protocols. Token-based MAC protocols provide a fully distributed solution allowing for exclusive usage of the communications network to one transmitter (communications adapter) at a time.

In token networks, only the owner of the (unique within the network) token is allowed to transmit messages on the network. Once the token holder is done transmitting messages, or has used its allotted time, the token is passed to another node. Examples of token protocols are the timed token protocol (TTP) [101] or the IEEE 802.5 Token Ring Protocol [64]. Also, tokens are used by PROFIBUS [61,69,125].

1.5.2.9 Master/Slave

Another example of demand assignment MAC protocols is the centralized solutions relying on a specialized node called the master node. The other nodes in the system are called slave nodes. In master/slave networks, elimination of message collisions is achieved by letting the master node control the traffic on the network, deciding which messages are allowed to be sent and when. This approach is used in, e.g., LIN [91,92], TTP/A [76,79,174], and PROFIBUS.

1.5.3 Networks

Communication network technologies presented in this chapter are either wired networks or wireless networks. The medium can be either wired, transmitting electrical or optical signals in cables or optical fibres, or wireless, transmitting radio signals or optical signals.

1.5.3.1 Wired Networks

Wired networks, which are the more common type of networks in embedded systems, are represented by two categories of networks: fieldbus networks and Ethernet networks.

Fieldbus networks. Fieldbuses are a family of factory communication networks that have evolved during the 1980s and 1990s as a response to the demand to reduce cabling costs in factory automation systems [140]. By moving from a situation in which every controller has its own cables connecting its sensors to the controller (parallel interface), to a system with a set of controllers sharing a single network (serial interface), costs could be cut and flexibility could be increased. Pushing for this evolution of technology was due to the fact that the number of cables in the system increased as the number of sensors and actuators grew, together with controllers moving from being specialized with their own microchip, to sharing a microprocessor (CPU) with other controllers.

Fieldbuses were soon ready to handle the most demanding applications on the factory floor. Several fieldbus technologies, usually very specialized, were developed by different companies to meet the demands of their applications. Different fieldbuses are used in different application domains. A comprehensive overview of the history and evolution of fieldbuses is given in [140].

Ethernet networks. In parallel with the development of various (specialized) fieldbus technologies providing real-time communications for avionics, trains, industrial and process automation, and building and home automation, Ethernet established itself as the de facto standard for non-real-time communications. Of all networking solutions for automation networks and office networks, fieldbuses were initially the choice for the former. At the same time, Ethernet evolved as the standard for office automation, and due to its popularity, prices on Ethernet-based networking solutions dropped. A lower price on Ethernet controllers made it interesting to develop additions and modifications to Ethernet for real-time communications, allowing Ethernet to compete with established real-time networks.

Ethernet is not very suitable for real-time communications due to its handling of message collisions. Automation networks typically require timing guarantees for individual messages. Several proposals to minimize or eliminate the occurrence of collisions on Ethernet have been proposed over the years. The strongest candidate today is the usage of a switched-based infrastructure, where the switches separate collision domains to create a collision-free network providing real-time message transmissions over Ethernet [57,58,67,68,145].

Other proposals providing real-time predictability using Ethernet include, making Ethernet more predictable using TDMA [80], off-line scheduling [187], or token algorithms [124,181]. Note that a dedicated network is usually required when using tokens, where all nodes sharing the network must obey the token protocol (e.g., the TTP [101] or the IEEE 802.5 Token Ring Protocol [64]). A different approach for predictability is to modify the collision resolution algorithm [85,110,129].

Other predictable approaches are, e.g., the usage of a master/slave concept as flexible time-triggered (FTT)-Ethernet [122] (part of the FTT framework [6]), or the usage of the virtual-time CSMA (VTCSMA) [38,111,190] protocol, where packets are delayed in a predictable way in order to eliminate the occurrence of collisions. Moreover, window protocols [191] are using a global window (synchronized time interval) that also removes collisions. The window protocol is more dynamic and somewhat more efficient in its behavior compared to the VTCSMA approach.

Without modifications to the hardware or networking topology (infrastructure), the usage of traffic smoothing [25,83,84,97] can eliminate bursts of traffic, which have severe impact on the timely delivery of message packets on the Ethernet. By keeping the network load below a given threshold, a probabilistic guarantee of message delivery can be provided.

For more information on real-time Ethernet interested readers are referred to [121].

1.5.3.2 Wireless Networks

The wireless medium is often unpredictable compared to a wired medium in terms of the temporal behavior of message transmissions. Therefore, the temporal guarantees that can be provided by a wireless network are usually not as reliable as the guarantees provided by a wired link. The reason for the lack of reliable timing guarantees is that the interference on the medium cannot be predicted (and analytically taken into consideration) as accurately for a wireless medium as for a wired medium, especially interference from sources other than the communications network itself. Due to this unpredictability, no commercially available wireless communication networks provide hard real-time guarantees.

1.5.4 Network Topologies

There are different ways to connect the nodes in a wired distributed system. Looking at the applications targeted in this chapter, three different network topologies are the more commonly used: bus, ring, and star topology. Using a bus topology, all nodes in the distributed system are connected directly to the network. In a ring topology, each node in the distributed system is connected to exactly two other nodes in a specific way forming a ring of connected nodes. Finally, in a star topology, all nodes in the distributed system are connected to a specific central node, forming a star. These three network topologies are depicted in Figure 1.8. Note that combinations of network topologies can exist, for example, a ring or a star might be connected to a bus together with other nodes.

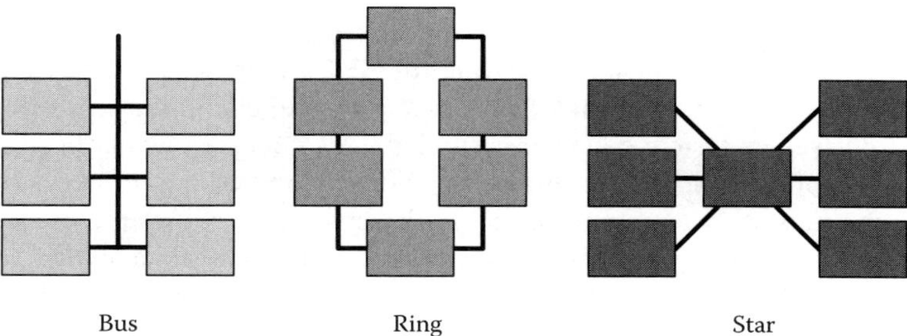

FIGURE 1.8 Network topologies: bus, ring, and star.

1.6 Analysis of Real-Time Systems

The most important property to analyze in an RTS is its temporal behavior, i.e., the timeliness of the system. The analysis should provide strong evidence that the system performs as intended at the correct time. This section gives an overview of the basic properties that are analyzed of an RTS. The section concludes with a presentation of trends and tools in the area of RTS analysis.

1.6.1 Timing Properties

Timing analysis is a complex problem. Not only are the used techniques sometimes complicated, but the problem itself is elusive as well; for instance, what is the meaning of the term "program execution time"? Is it the average time to execute the program, or the worst possible time, or does it mean some form of "normal" execution time? Under what conditions does a statement regarding program execution times apply? Is the program delayed by interrupts or higher priority tasks? Does the time include waiting for shared resources? etc.

To straighten out some of these questions and to be able to study some existing techniques for timing analysis, we structure timing analysis into four major types. Each type has its own purpose, benefits, and limitations. The types are listed as follows.

Execution time. This refers to the execution time of a singe task (or program, or function, or any other unit of single threaded sequential code). The result of an execution-time analysis is the time (i.e., the number of clock cycles) the task takes to execute, when executing undisturbed on a single CPU, i.e., the result should not account for interrupts, preemption, background DMA transfers, DRAM refresh delays, or any other types of interfering background activities.

At a first glance, leaving out all types of interference from the execution-time analysis would give us unrealistic results. However, the purpose of the execution-time analysis is not to deliver estimates on "real-world" timing when executing the task. Instead, the role of execution-time analysis is to find out how much computing resources is needed to execute the task. (Hence, background activities that are not related to the task should not be accounted for.)

There are some different types of execution times that can be of interest:

- Worst-case execution time (WCET)—This is the worst possible execution time a task could exhibit or equivalently the maximum amount of computing resources required to execute the task. The WCET should include any possible atypical task execution such as exception handling or cleanup after abnormal task termination.
- Best-case execution time (BCET)—During some types of real-time analysis, not only the WCET is used, but as we will describe later, having knowledge about the BCET of tasks is useful as well.

- Average execution time (AET)—The AET can be useful in calculating throughput figures for a system. However, for most RTS analysis the AET is of less importance, simply since a reasonable approximation of the average case is easy to obtain during testing (where, typically, the average system behavior is studied). Also, only knowing the average, and not knowing any other statistical parameters such as standard deviation or distribution function, makes statistical analysis difficult. For analysis purposes a more pessimistic metric such as the 95%-quartile would be more useful. However, analytical techniques using statistical metrics of execution time are scarce and not very well developed.

Response time. The response time of a task is the time it takes from the invocation of the task to the completion of the task. In other words, it is the time from when the task first is placed in the operating-system's ready queue to the time when it is removed from the running state and placed in the idle or sleeping state.

Typically, for analysis purposes it is assumed that a task does not voluntarily suspend itself during its execution. That is, the task may not call primitives such as `sleep()` or `delay()`. When a program voluntarily suspends itself, that program should be broken down into two (or more) tasks during the analysis. However, involuntarily suspension, such as blocking on shared resources, is allowed. That is, primitives such as `get_semaphore()` and `lock_database_tuple()` are allowed.

The response time is typically a system level property, in that it includes interference from other, unrelated, tasks and parts of the system. The response time also includes delays caused by contention on shared resources. Hence, the response time is only meaningful when considering a complete system, or in distributed systems, a complete node.

End-to-end delay. The previously described "execution time" and "response time" are useful concepts since they are relatively easy to understand and have well-defined scopes. However, when trying to establish the temporal correctness of a system, knowing the WCET and/or the response times of tasks is often not enough. Typically, the correctness criterion is stated using end-to-end latency timing requirements, for instance, an upper bound on the delay between the input of a signal and the output of a response.

In a given implementation, there may be a chain of events taking place between the input of a signal and the output of a response. For instance, one task may be in charge of reading the input and another task, of generating the response, and the two tasks may have to exchange messages on a communications link before the response can be generated. The end-to-end timing denotes timing of externally visible events.

Jitter. The term "jitter" is used as a metric for variability in time. For instance, the jitter in execution time of a task is the difference between the task's BCET and WCET. Similarly, the response-time jitter of a task is the difference between its best-case response time and its worst-case response time. Often, control algorithms have requirements that the jitter of the output should be limited. Hence, the jitter is sometimes a metric equally important as the end-to-end delay.

Also input to the system can have jitter. For instance, an interrupt which is expected to be periodic may have a jitter (due to some imperfection in the process generating the interrupt). In this case the jitter-value is used as a bound on the maximum deviation from the ideal period of the interrupt. Figure 1.9 illustrates the relation between the period and the jitter for this example.

Note that jitter should not accumulate over time. For our example, even though two successive interrupts could arrive sooner than one period, in the long run, the average interrupt interarrival time will be that of the period.

In the above list of types of time, we only mentioned time to execute programs. However, in many RTSs, other timing properties may also exist. Most typical are delays on a communications network, but also other resources such as hard disk drives may be causing delays and hence need to be analyzed.

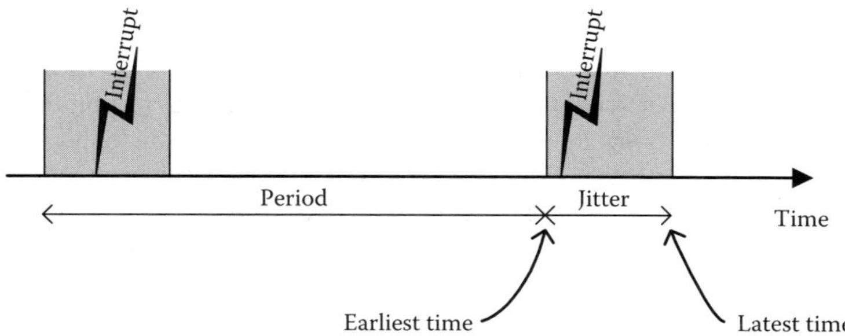

FIGURE 1.9 Jitter used as a bound on variability in periodicity.

The above introduced times can all be mapped to different types of resources, for instance, the WCET of a task corresponds the maximum size of a message to be transmitted, and the response time of message is defined analogous to the response time of a task.

1.6.2 Methods for Timing Analysis

When analyzing hard RTSs, it is essential that the estimates obtained during timing analysis are safe. An estimate is considered safe if it is guaranteed that it is not an underestimation of the actual worst-case time. It is also important that the estimate is tight, meaning that the estimated time is close to the actual worst-case time.

For the previously defined types of timings (Section 1.6.1) there are different methods available:

Execution-time estimation. For real-time tasks, the WCET is the most important execution time measure to obtain. Sadly, however, it is also often the most difficult measure to obtain.

Methods to obtain the WCET of a task can be divided into two categories: (1) static analysis and (2) dynamic analysis. Dynamic analysis is essentially equivalent to testing (i.e., executing the task on the target hardware) and has all the drawbacks/problems that testing exhibits (such as being tedious and error prone). One major problem with dynamic analysis is that it does not produce safe results. In fact, the result can never exceed the true WCET and it is very difficult to be sure that the estimated WCET is really the true WCET.

Static analysis, on the other hand, can give guaranteed safe results. Static analysis is performed by analyzing the code (source or object code is used) and basically counting the number of clock cycles that the task may use to execute (in the worst possible case). Static analysis uses models of the hardware to predict execution times for each instruction. Hence, for modern hardware it may be very difficult to produce static analyzers that give good results. One source of pessimism in the analysis (i.e., overestimation) is hardware caches; whenever an instruction or data-item cannot be guaranteed to reside in the cache, a static analyzer must assume a cache miss. And since modeling the exact state of caches (sometimes of multiple levels), branch predictors, etc. is very difficult and time consuming, only few tools that give adequate results for advanced architectures exist. Also, to perform a program flow and data analysis that exactly calculates, e.g., the number of times a loop iterates or the input parameters for procedures is difficult.

Methods for good hardware and software modeling do exist in the research community; however, combining these methods into good-quality tools has proven tedious.

Schedulability analysis. The goal of schedulability analysis is to determine whether or not a system is schedulable. A system is deemed schedulable if it is guaranteed that all task deadlines will always be met. For statically scheduled (table-driven) systems, calculations of response times are trivially given from the static schedule. However, for dynamically scheduled systems (such as fixed priority or deadline scheduling) more advanced techniques have to be used.

There are two main classes of schedulability analysis techniques: (1) response-time analysis and (2) utilization analysis. As the name suggest, a response-time analysis calculates a (safe) estimate of the worst-case response time of a task. This estimate can then be compared to the deadline of the task and if it does not exceed the deadline then the task is schedulable. Utilization analysis, in contrast, does not directly derive the response times for tasks; rather, they give a boolean result for each task telling whether or not the task is schedulable. This result is based on the fraction of utilization of the CPU for a relevant subset of the tasks, hence the term utilization analysis.

Both types of analysis are based on similar types of task models. However, typically the task models used for analysis are not the task models provided by commercial RTOSs. This problem can be resolved by mapping one or more OS-tasks to one or more analysis task. However, this mapping has to be performed manually and requires an understanding of the limitations of the analysis task model and the analysis technique used.

End-to-end delay estimation. The typical way to obtain an end-to-end delay estimate is to calculate the response time for each task/message in the end-to-end chain and to summarize these response times. When using a utilization-based analysis technique (in which no response time is calculated) one has to resort to using the task/message deadlines as safe upper bounds on the response times.

However, when analyzing distributed RTSs, it may not be possible to calculate all response times in one pass. The reason for this is that delays on one node will lead to jitter on another node, and that this jitter may in turn affect the response times on that node. Since jitter can propagate in several steps between nodes, in both directions, there may not exist a right order to analyze the nodes. (If A sends a message to B, and B sends a message to A, which node should one analyze first?) Solutions to this type of problems are called holistic schedulability analysis methods (since they consider the whole system). The standard method for holistic response-time analysis is to repeatedly calculate response time for each node (and update jitter values in the nodes affected by the node just analyzed) until response times do not change (i.e., a fix-point is reached).

Jitter estimation. To calculate the jitter one needs to perform not only a worst-case analysis (of, for instance, response-time or end-to-end delay) but also a best-case analysis.

However, even though best-case analysis techniques often are conceptually similar to worst-case analysis techniques, there has been little attention paid to best-case analysis. One reason for not spending too much time on best-case analysis is that it is quite easy to make a conservative estimate of the best case: the best-case time is never less than zero (0). Hence, in many tools it is simply assumed that the BCET (for instance) is zero, whereas great efforts can be spent on analyzing the WCET.

However, it is important to have tight estimates of the jitter and to keep the jitter as low as possible. It has been shown that the number of execution paths in a multitasking RTS can dramatically increase if jitter increases [166]. Unless the number of possible execution paths is kept as low as possible it becomes very difficult to achieve good coverage during testing.

1.6.3   Example of Analysis

Here we give simple examples of schedulability analysis. We show a very simple example of how a set of tasks running on a single CPU can be analyzed, and we also give an example of how the response times for a set of messages sent on a CAN-bus can be calculated.

Analysis of tasks. This example is based on some 30-year-old task models and is intended to give the reader a feeling for how these types of analysis work. Today's methods allow for far richer and more realistic task models, with the resulting increase of complexity of the equations used (hence they are not suitable for use in our simple example).

TABLE 1.2 Example Task Set for Analysis

Task	T	C	D	Prio
X	30	10	30	High
Y	40	10	40	Medium
Z	52	10	52	Low

TABLE 1.3 Result of RM Test

Task	T	C	D	Prio	U
X	30	10	30	High	0.33
Y	40	10	40	Medium	0.25
Z	52	10	52	Low	0.23
				Total:	0.81
				Bound:	0.78

In the first example, we will analyze a small task set described in Table 1.2, where T, C, and D denote the tasks' period, WCET, and deadline, respectively. In this example $T = D$ for all tasks and priorities have been assigned in RM order, i.e., the highest rate gives the highest priority.

For the task set in Table 1.2 original analysis techniques of Liu and Layland [94] and Joseph and Pandya [70] are applicable, and we can perform both utilization-based and response-time-based schedulability analyses.

We start with the utilization-based analysis; for this task model, Liu and Layland's result is that a task set of n tasks is schedulable if its total utilization, U_{tot}, is bounded by the following equation:

$$U_{\text{tot}} \leq n\left(2^{1/n} - 1\right)$$

Table 1.3 shows the utilization calculations performed for the schedulability analysis. For our example, task set $n = 3$ and the bound is approximately 0.78. However, the utilization ($U_{\text{tot}} = \sum_{i=1}^{n} \frac{C_i}{T_i}$) for our task set is 0.81, which exceeds the bound. Hence, the task set fails the RM test and cannot be deemed schedulable.

Joseph and Pandya's response-time analysis allows us to calculate worst-case response-time, R_i, for each task, i, in our example (Table 1.2). This is done using the following formula

$$R_i = C_i + \sum_{j \in hp(i)} \left\lceil \frac{R_i}{T_j} \right\rceil C_j \tag{1.1}$$

where $hp(i)$ denotes the set of tasks with priority higher than i.

The observant reader may have noticed that Equation 1.1 is not in the closed form, in that R_i is not isolated on the left-hand side of the equality. As a matter of fact, R_i cannot be isolated on the left-hand side of the equality; instead Equation 1.1 has to be solved using fix-point iteration. This is done with the recursive formula in Equation 1.1, starting with $R_i^0 = 0$ and terminating when a fix-point has been reached (i.e., when $R_i^{m+1} = R_i^m$).

$$R_i^{m+1} = C_i + \sum_{j \in hp(i)} \left\lceil \frac{R_i^m}{T_j} \right\rceil C_j \tag{1.2}$$

For our example task set, Table 1.4 shows the results of calculating Equation 1.1. From the table, we can conclude that no deadlines will be missed and that the system is schedulable.

Remarks. As we could see for our example task set in Table 1.2, the utilization-based test could not deem the task set as schedulable whereas the response-time-based test could. This situation is symptomatic for the relation between utilization-based and response-time-based schedulability tests. That is, the response-time-based tests find more task sets schedulable than the utilization-based tests.

TABLE 1.4 Result of Response-Time
Analysis for Tasks

Task	T	C	D	Prio	R	$R \leq D$
X	30	10	30	High	10	Yes
Y	40	10	40	Medium	20	Yes
Z	52	10	52	Low	52	Yes

TABLE 1.5 Example CAN-Message Set

Message	T	S	D	Id
X	350	8	300	00010
Y	500	6	400	00100
Z	1000	5	800	00110

However, also as shown by the example, the response-time-based test needs to perform more calculations than the utilization-based test does. For this simple example the extra computational complexity of the response-time test is insignificant. However, when using modern task-models (that are capable of modeling realistic systems), the computational complexity of response-time-based tests is significant. Unfortunately, for these advanced models, utilization-based tests are not always available.

Analysis of messages. In our second example, we show how to calculate the worst-case response times for a set of periodic messages sent over the CAN-bus. We use a response-time analysis technique similar to the one we used when we analyzed the task set in Table 1.2. In this example, our message set is given in Table 1.5, where T, S, D, and Id denote the messages' period, data size (in bytes), deadline, and CAN-identifier, respectively. (The time-unit used in this example is "bit-time", i.e., the time it takes to send 1 bit. For a 1 Mbit CAN this means that 1 time-unit is 10^{-6} s.)

Before we attack the problem of calculating response times we extend Table 1.5 with two columns. First, we need the priority of each message; in CAN this is given by the identifier, the lower the numerical value the higher the priority. Second, we need to know the worst-case transmission time of each message. The transmission time is given partly by the message data-size but we also need to add time for the frame-header and for any *stuff bits*.* The formula to calculate the transmission time, C_i, for a message i is

$$C_i = S_i + 47 + \left\lfloor \frac{34 + 8S_i + 1}{5} \right\rfloor \tag{1.3}$$

In Table 1.6, the two columns Prio and C show the priority assignment and the transmission times for our example message set.

Now we have all the data needed to perform the response-time analysis. However, since CAN is a nonpreemptive resource the structure of the equation is slightly different from Equation 1.1 which we used for analysis of tasks. The response-time equation for CAN is given in Equation 1.4.

TABLE 1.6 Result of Response-Time Analysis for CAN

Message	T	S	D	Id	Prio	C	w	R	$R \leq D$
X	350	8	300	00010	High	130	130	260	Yes
Y	500	6	400	00100	Medium	111	260	371	Yes
Z	1000	5	800	00110	Low	102	612	714	Yes

---

* CAN adds stuff bits, if necessary, to avoid the two reserved bit patterns 000000 and 111111. These stuff bits are never seen by the CAN-user but have to be accounted for in the timing analysis.

$$R_i = w_i + C_i$$

$$w_i = 130 + \sum_{\forall j \in hp(i)} \left\lceil \frac{w_i + 1}{T_j} \right\rceil C_j \qquad (1.4)$$

In Equation 1.4, $hp(i)$ denotes the set of messages with higher priority than message i. Note that (similar to Equation 1.1) w_i is not isolated on the left-hand side of the equation, and its value has to be calculated using fix-point iteration (compare to Equation 1.2).

Applying Equation 1.4 we can now calculate the worst-case response time for our example messages. In Table 1.6, the two columns, w and R, show the results of the calculations, and the final column shows the schedulablilty verdict for each message.

As we can see from Table 1.6, our example message set is schedulable, meaning that the messages will always be transmitted before their deadlines. Note that this analysis was made assuming that there would not be any retransmissions of broken messages. CAN normally automatically retransmits any message that is broken due to interference on the bus. To account for such automatic retransmissions an error model needs to be adopted and the response-time equation adjusted accordingly, as shown by Tindell et al. [169].

A note of caution: with respect to existing literature on scheduling, analysis of CAN messages is due. It has recently been shown by Bril et al. that early analysis techniques for CAN, e.g., [169] above, give potentially unsafe results [18]. However, using the equations in this chapter gives safe (but not exact) results.

1.6.4   Trends and Tools

As pointed our earlier, and also illustrated by our example in Table 1.2, there is a mismatch between the analytical task models and the task models provided by commonly used RTOSs. One of the basic problems is that there is no one-to-one mapping between analysis tasks and RTOS tasks. In fact, for many systems there is a N-to-N mapping between the task types. For instance, an interrupt handler may have to be modeled as several different analysis tasks (one analysis task for each type of interrupt it handles), and one OS task may have to be modeled as several analysis tasks (for instance, one analysis task per call to `sleep()` primitives).

Also, current schedulability analysis techniques cannot adequately model other types of task-synchronization than locking/blocking on shared resources. Abstractions such as message queues are difficult to include in the schedulability analysis.* Furthermore, tools to estimate the WCET are also scarce. Currently only a handful of tools that give safe WCET estimates are commercially available; see Wilhelm et al. for a survey [184].

These problems have led to low penetration of schedulability analysis in industrial software-development processes. However, in isolated domains, such real-time networks, some commercial tools, that are based on real-time analysis do exist. For instance, Volcano [26,183] provides tools for the CAN bus that allow system designers to specify signals on an abstract level (giving signal attributes such as size, period, and deadline) and automatically derive a mapping of signals to CAN-messages where all deadlines are guaranteed to be met.

On the software side tool-suites provided by, for instance, Arcticus Systems [8], ETAS [42], and TTTech [176] can provide system development environments with timing analysis as an integrated part of the tool suite. However, these tools require that the software development processes are under

---

* Techniques to handle more advanced models include timed logic and model checking. However, the computational and conceptual complexity of these techniques has limited their industrial impact. There are, however, examples of existing tools for this type of verification, e.g., The Times Tool [168].

complete control of the respective tool. More general-purpose analysis tools include Rapid RMA provided by Tri-Pacific [170] and SymTA/S by Symta Vision [159].

The widespread use of UML [116] in software design has led to some specialized UML products for real-time engineering [133,164]. However, these products, as of today, do not support timing analysis of the designed systems. Within UML, the profile, "modeling and analysis of real-time and embedded (MARTE) systems" [102] allows specification of both timing properties and requirement in a standardized way. It is expected that this will lead to products that can analyze UML models conforming to the MARTE profile for timing properties. However, building complete timing models from UML models is nontrivial, and the models are highly dependent on code-generation strategies. Hence, no commercial tool for analyzing MARTE models exists yet.

1.7   Component-Based Design of RTS

Component-based design (CBD) is a current trend in software engineering [30,54,161]. In a CBD, a software component is used to encapsulate some functionality. That functionality is only accessed though the interface of the component. A system is composed by assembling a set of components and connecting their interfaces. In the desktop area, component technologies like COM [16,107], .NET [29,108], and Java Beans [112,158] have gained widespread use. These technologies give substantial benefits, in terms of reduced development time and software complexity, when designing complex and/or distributed systems.

Though originally not developed for RTSs, the CBD has a strong potential for such systems as well. By extending components with introspective interfaces to retrieve information about extra-functional properties of the component, means for handling and reasoning about the essential properties and attributes such as memory consumption, execution times, task periods, etc. can be integrated in component-based frameworks. For RTSs, timing properties are of course of particular interest.

Unlike the functional interfaces of components, the introspective interfaces can be available off-line, i.e., during the component assembly phase. This way, the timing attributes of the system components can be obtained at design time and tools to analyze the timing behavior of the system could be used. If the introspective interfaces are also available online they could be used in, for instance, admission control algorithms. An admission control could query new components for their timing behavior and resource consumption before deciding to accept new component to the system.

Unfortunately, many industry standard software techniques are based on the client–server or the message-box models of interaction, which we deemed, in Section 1.2.2, unfit for RTSs. This is true especially for the most commonly used component models. For instance, the Corba Component Model (CCM) [117], Microsoft's COM [107] and .NET [108] models, and Java Beans [158] all have the client–server model as their core model. Also, none of these component technologies allow the specification of extra-functional properties through introspective interfaces. Hence, from the real-time perspective, biggest advantage of CBD is void for these technologies.

However, there are numerous research projects addressing the CBD for real-time and embedded systems (e.g., [5,19,20,36,44,51,56,113,126,136,146,152,155]) and also some industrial initiatives (e.g., [12,52,99,178,179]). These projects are addressing the issues left behind by most desktop component technologies, such as timing predictability (using suitable computational models), support for off-line analysis of component assemblies, and better support for resource-constrained systems. Often, these projects strive to remove the considerable runtime flexibility provided by existing technologies, since the flexibility contributes to unpredictability (and is also adding to the runtime complexity and prevents the CBD for resource-constrained systems).

As stated before, the main challenge of designing RTSs is the need to consider issues that do not typically apply to general-purpose computing systems. These issues include

- Constraints on extra-functional properties, such as timing, QoS, and dependability
- The need to statically predict (and verify) these extra-functional properties
- Scarce resources, including processing power, memory, and communication bandwidth

In the remainder of this chapter, we discuss how these issues can be addressed in the context of CBD. In doing so, we also highlight the challenges in designing a CBD process and component technology for development of RTSs.

1.7.1 Timing Properties and CBD

In general, for systems where timing is crucial there will necessarily be at least some global timing requirements that have to be met. If the system is built from components, this will imply the need for timing parameters/properties of the components and some proof that the global timing requirements are met.

In Section 1.6, we introduced the following four types of timing properties:

- Execution time
- Response time
- End-to-end delay
- Jitter

So, how are these related to the use of a CBD methodology?

1.7.1.1 Execution Time

For a component used in a real-time context, an execution time measure will have to be derived. This is, as discussed in Section 1.6, not an easy or satisfactorily solvable problem. Furthermore, since execution time is inherently dependent on the target hardware and reuse is the primary motivation for CBD, it is highly desirable if the execution time for several targets are available. (Alternatively, the execution time for new hardware platforms is automatically derivable.)

The nature of the applied component model may also make execution-time estimation more or less complex. Consider, for instance, a client–server oriented component model, with a server-component that provides services of different types, as illustrated in Figure 1.10a. What does "execution time" mean for such a component? Clearly, a single execution time is not appropriate;

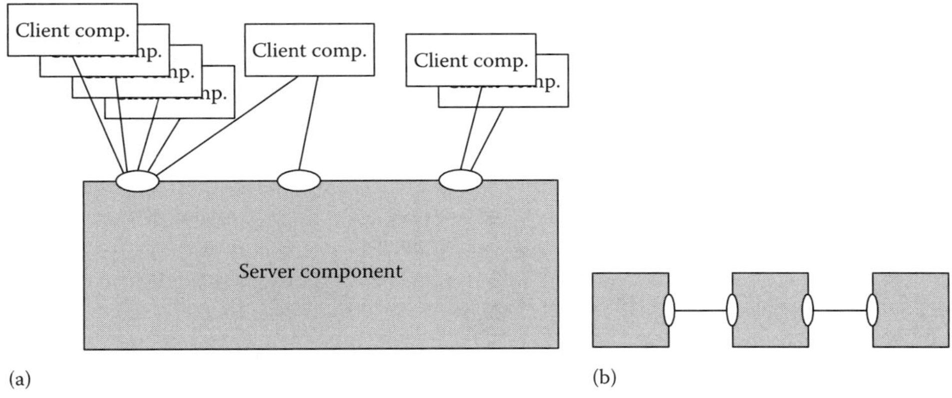

FIGURE 1.10 (a) A complex server component, providing multiple services to multiple users and (b) a simple chain of components implementing a single thread of control.

rather, the analysis will require a set of execution times related to servicing different requests. On the other hand, for a simple port-based object component model [155] in which components are connected in sequence to form periodically executing transactions (illustrated in Figure 1.10b), it could be possible to use a single execution-time measure, corresponding to the execution time required for reading the values at the input ports, performing the computation and writing values to the output ports.

1.7.1.2 Response Time

Response times denote the time from invocation to completion of tasks, and response-time analysis is the activity to statically derive response-time estimates.

The first question to ask from a CBD perspective is: What is the relation between a "task" and a "component?"

This is obviously highly related to the component model used. As illustrated in Figure 1.11a, there could be a 1-to-1 mapping between components and tasks, but in general, several components could be implemented in one task (Figure 1.11b) or one component could be implemented by several tasks (Figure 1.11c), and hence there is a many-to-many relation between components and tasks. In principle, there could even be more irregular correspondence between components and tasks, as illustrated in Figure 1.11d. Furthermore, in a distributed system there could be a many-to-many relation between components and processing nodes, making the situation even more complicated.

Once we have sorted out the relation between tasks and components, we can calculate the response times of tasks, given that we have an appropriate analysis method for the used execution paradigm, and that relevant execution-time measures are available. However, how to relate these response times

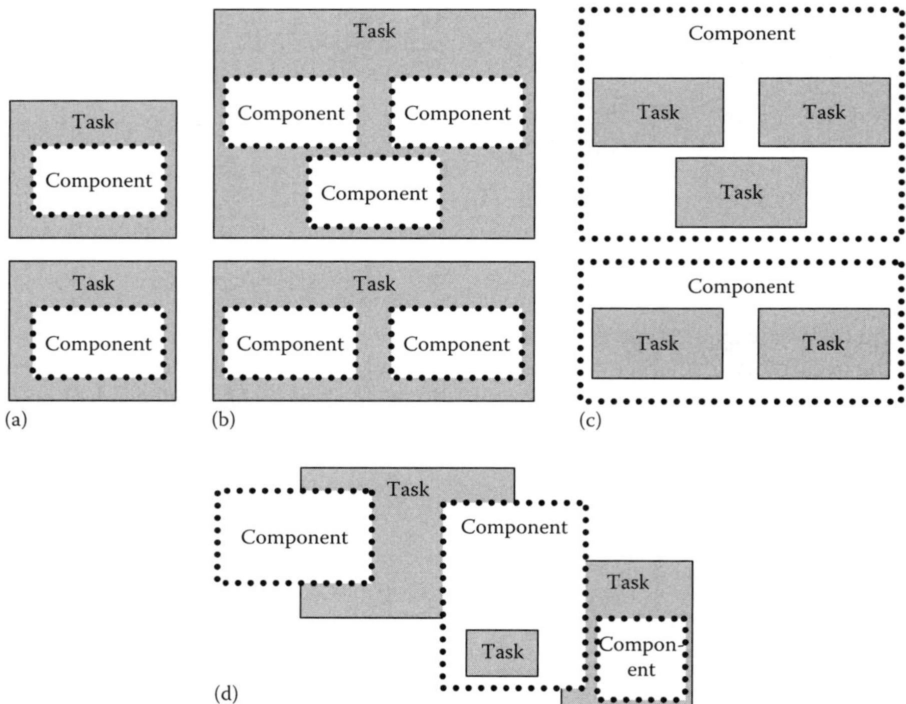

FIGURE 1.11 Tasks and components: (a) 1-to-1 correspondence, (b) 1-to-many correspondence, (c) many-to-1 correspondence, (b and c) many-to-many correspondence, and (d) irregular correspondence.

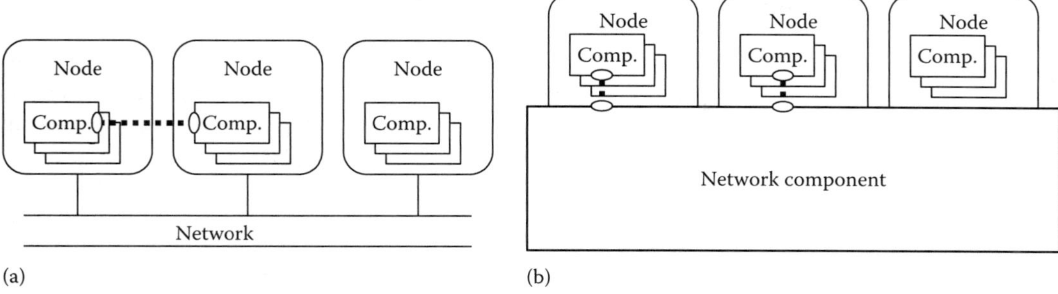

FIGURE 1.12 Components and communication delays: (a) communication delays can be part of the intercomponent communication properties and (b) communication delays can be timing properties of components.

to components and the application-level timing requirements may not be straightforward, but this is an issue for the subsequent end-to-end analysis.

Another issue with respect to response times is how to handle communication delays in distributed systems. In essence there are two ways to model the communication, as depicted in Figure 1.12. In Figure 1.12a, the network is abstracted away and the intercomponent communication is handled by the framework. In this case, response-time analysis is made more complicated since it must account for different delays in intercomponent communication, depending on the physical location of components. In Figure 1.12b, on the other hand, the network is modeled as a component itself, and network delays can be modeled as delays in any other component (and intercomponent communication can be considered instantaneous).

However, the choice of how to model network delays also has an impact on the software engineering aspects of the component model. In Figure 1.12a, the communication is completely hidden from the components (and the software engineers), hence giving optimizing tools many degrees of freedom with respect to component allocation, signal mapping, and scheduling parameter selection. Whereas, in Figure 1.12b, the communication is explicitly visible to the components (and the software engineers), hence putting a larger burden on the software engineers to manually optimize the system.

1.7.1.3 End-to-End Delay

End-to-end delays are application-level timing requirements relating the occurrence in time of one event to the occurrence of another event. As pointed out above, how to relate such requirements to the lower-level timing properties of components discussed above is highly dependent on both the component model and the timing analysis model.

When designing an RTS using CBD the component structure gives excellent information about the points of interaction between the RTS and its environment. Since the end-to-end delay is about timing estimates and timing requirements on such interactions, CBD gives a natural way of stating timing requirements in terms of signals received or generated. (In traditional RTS development, the reception and generation of signals are embedded into the code of tasks and are not externally visible, hence making it difficult to relate response times of tasks to end-to-end requirements.)

1.7.1.4 Jitter

Jitter is an important timing parameter that is related to execution time, and that will affect response times and end-to-end delays. There may also be specific jitter requirements. Jitter has the same relation to CBD as end-to-end delay.

1.7.1.5   Summary of Timing and CBD

As described above, there is no single solution for how to apply the CBD to an RTS. In some cases, timing analysis is made more complicated when using CBD, e.g., when using client–server-oriented component models, whereas in other cases, CBD actually helps timing analysis, e.g., identifying interfaces/events associated with end-to-end requirements is facilitated when using CBD.

Further, the characteristics of the component model have a great impact on the analyzability of CBDed RTSs. For instance, interaction patterns like client–server do not map well to established analysis methods and make analysis difficult, whereas pipes-and-filter-based patterns (such as the port-based objects component model [155]) map very well to existing analysis methods and allow for tight analysis of timing behavior. Also, the execution semantics of the component model has impact on the analyzability. The execution semantics gives restrictions on how to map components to tasks, e.g., in the Corba component model [117] each component is assumed to have its own thread of execution, making it difficult to map multiple components to a single thread. On the other hand, the simple execution semantics of pipes-and-filter-based models allows for automatic mapping of multiple components to a single task, simplifying timing analysis and making better use of system resources.

1.7.2   Real-Time Operating Systems

There are two important aspects regarding CBD and RTOSs: (1) the RTOS itself may be component based, and (2) the RTOS may support or provide a framework for CBD.

Component-based RTOSs. Most RTOSs allow for off-line configuration where the engineer can choose to include or exclude large parts of functionality. For instance, which communications protocols to include is typically configurable. However, this type of configurability is not the same as the RTOS being component based (even though the unit of configuration is often referred to as components in marketing material). For an RTOS to be component based it is required that the components conform to a component model, which is typically not the case in most configurable RTOSs.

There has been some research on component-based RTOSs, for instance, the research RTOS VEST [151,152]. In VEST, schedulers, queue managers, and memory management are built up out of components. Furthermore, special emphasis has been put on predictability and analyzability. However, VEST is currently still in the research stage and has not been released to the public. Publicly available is, however, the eCos RTOS [37,103] which provides a component-based configuration tool. Using eCos components the RTOS can be configured by the user, and third-party extension can be provided.

RTOSs that support CBD. Looking at component models in general and those intended for embedded systems in particular, we observe that they are all supported by some runtime executive or simple RTOS. Many component technologies provide frameworks that are independent of the underlying RTOS, and hence RTOS can be used to support CBD using such an RTOS-independent framework. Examples include Corba's ORB [115] and the framework for PECOS [109,120].

Other component technologies have a tighter coupling between the RTOS and the component framework, in that the RTOS explicitly supports the component model by providing the framework (or part of the framework). Such technologies include

- Koala [178] is a component model and architectural description language from Philips, providing high-level APIs to the computing and audio/video hardware. The computing layer provides a simple proprietary real-time kernel with priority-driven preemptive scheduling. Special techniques for thread-sharing are used to limit the number of concurrent threads.
- Chimera RTOS provides an execution framework for the port-based object component model [155]. Chimera is intended for development of sensor-based control systems,

specifically reconfigurable robotics applications. Chimera has multiprocessor support and handles both static and dynamic scheduling, the latter being EDF-based.

- Rubus is an RTOS supporting a component model in which behaviors are defined by sequences of port-based objects. The Rubus kernel supports predictable execution of statically scheduled periodic tasks (termed red tasks in Rubus) and dynamically fixed-priority preemptive scheduled tasks (termed Blue). In addition, support for handling interrupts is provided. In Rubus, support is provided for transforming sets of components into sequential chains of executable code. Each such chain is implemented as a single task. Support is also provided for analysis of response times and end-to-end deadlines, based on execution-time measures that have to be provided, i.e., execution-time analysis is not provided by the framework.

- Time-triggered operating system (TTOS) is an adapted and extended version of the MARS OS [80] and marketed by TTTech under the trade name TTP OS [175]. Task scheduling in TTOS is based on an off-line generated scheduling table and relies on the global time base provided by the TTP/C communication system. All synchronizations are handled by the off-line scheduling. TTOS, and in general the entire TTA, is (just as IEC61131-3) well suited for the synchronous execution paradigm.

  In a synchronous execution the system is considered sequential computing in each step (or cycle) a global output based on a global input. The effect of each step is defined by a set of transformation rules. Scheduling is done statically by compiling a set of rules into a sequential program implementing these rules and executing them in some statically defined order. A uniform timing bound for the execution of global steps is assumed. In this context, a component is a design-level entity.

  TTA defines a protocol for extending the synchronous language paradigm to distributed platforms, allowing distributed components to interoperate, as long as they conform to imposed timing requirements.

1.7.3 Real-Time Scheduling

Ideally, from a CBD perspective, the response time of a component should be independent of the environment in which it is executing (since this would facilitate reuse of the component). However, this is in most cases highly unrealistic for the following reasons:

1. Execution time of the task will be different in different target environments
2. Response time is additionally dependent on the other tasks competing for the same resources (CPU, etc.) and the scheduling method used to resolve the resource contention

Rather than aiming for the nonachievable ideal, a realistic ambition could be to have a component model and a framework which allow for analysis of response times based on abstract models of components and their compositions. Time-triggered systems go one step toward the ideal solution, in the sense that components can be timely isolated from each other. While not having a major impact on the component model, time-triggered systems simplify the implementation of the component framework since all synchronizations between components are resolved off-line. Also, from a safety perspective, the time-triggered paradigm gives benefits in that it reduces the number of possible execution scenarios (due to the static order of execution of components and the lack of preemtion).

  Also, in time-triggered component models it is possible to use the structure given by the component composition to synthesize scheduling parameters. For instance, in Rubus [8] and TTA [82] this is done by generating the static schedule using the components as schedulable entities.

In theory, a similar approach could also be used for dynamically scheduled systems, using a scheduler/task configuration-tool to automatically derive mappings of components to tasks and scheduling parameters (such as priorities or deadlines) for the tasks. However, this approach is still in the research stage.

1.8   Testing and Debugging of RTSs

Testing is the process of dynamically investigating the behavior of software in order to reveal inconsistencies with the specification or the requirements (i.e., failures). Debugging, on the other hand, refers to the process of revealing and removing the causes of these failures (i.e., the bugs). There are extensive studies suggesting that up to 80% of the life cycle cost for software is spent on testing and debugging (see, e.g., NIST [177]). Despite the importance of testing and debugging, there are few results concerning testing and debugging of RTSs.

Considering RTS analyses, testing and debugging are more pragmatic approaches for determining quality. While being more generally applicable, they do not provide the same level of quality assurance as safe analysis results do. In fact, a fundament of software testing is the impracticability of exhaustive testing, i.e., testing of the entire behavioral space of the software is generally impossible. Hence, testing can prove only the presence of bugs, not their absence [33].

In Section 1.8.1, we investigate relevant issues concerning testing and debugging of RTSs. This section is concluded with a brief discussion of state-of-practice for industrial systems in this area.

1.8.1   Issues in Testing and Debugging of RTSs

Testing and debugging of RTSs are difficult, time-consuming, and challenging tasks. The main reason for this is that RTSs are timing critical, often embedded, and that they interact with the real world. An additional problem for most RTSs is that the system consists of several concurrently executing threads. This concurrency will per se lead to a problematic nondeterminism. Reasons for this nondeterminism include race conditions caused by, e.g., slight variations in execution time of tasks. In turn, this leads to variations of the preemtion points (when tasks preempt each other), causing unpredictability in terms of the number of preemption scenarios that are possible. Consequently, it is hard to predict which scenario will actually be executed in a specific situation. All in all, these issues pose problems regarding the observability, controllability, and reproducibility required by testing and debugging.

Observability. In testing, observability (i.e., the ability to observe the effects of stimuli provided to the system under test) is crucial. This is even more true for the traditional debugging process, where the inspection of system variables and states is the very fundamental for pinpointing the cause of a failure. However, due to the embedded nature of the vast majority of RTSs, the observability is inherently low.

The act of observation can be achieved either by using nonintrusive hardware monitoring devices (hardware-based monitoring), or by instrumenting statements inserted in the system code (software-based monitoring).

Nonintrusive hardware recorders use in-circuit emulators (ICEs) with dual-port RAM. An ICE replaces the ordinary CPU; it is plugged onto the CPU target socket and works together with the rest of the system. The difference between an ICE and an ordinary CPU is the amount of auxiliary output available. If the ICE (like those from, e.g., Lauterbach [87]) has RTOS awareness, this type of history recorder needs no instrumentation of the target system. The only input needed is the location of the data to monitor. Some hardware monitors are attached to specialized microcontroller debug and trace ports, such as JTAG [153] and BDM [59] ports. Certain microcontrollers also provide trace ports

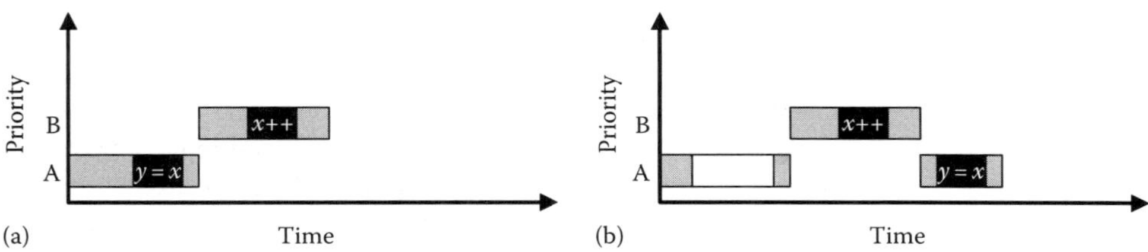

FIGURE 1.13   (a) Execution without software probe and (b) execution with software probe, give different results with respect to the value of variable y.

that allow each machine code instruction to be recorded. Hardware-based recorders are potentially nonintrusive, since they do not steal any CPU-cycles or target memory. However, due to price and space limitations they cannot usually be delivered with the product. These types of recorders are consequently best suited for predeployment laboratory testing and debugging.

Software probes, on the other hand, are intrusive with respect to the resources of the system being instrumented. Software instrumentation steals processor time and memory from the application, which may be a problem in resource-constrained systems. Furthermore, in a multitasking system, the very act of observing may alter the behavior of the observed system (see the example in Figure 1.13). Such effects on system behavior are known as probe effects and their presence in concurrent software was first described by Gait [47].

Setting debugging breakpoints in concurrent systems may introduce probe effects, since they may stop one task from executing while allowing all others to continue their execution, thereby invalidating system execution orderings. The same goes for instrumentation for facilitating measurement of coverage of different test criteria during testing. If the system probing code is altered or removed in between the testing and the software deployment, this may manifest in the form of probe effects. Consequently, the test cases that were passed during testing may very well lead to failures in the deployed system, and tests that failed may very well not cause any problem at all in the deployed system. Hence, some level of probing code is often left resident in the deployed code.

Controllability. Similar to the issue of observability, due to the embedded nature of RTSs, the number of resources for controlling these systems is generally very low, making the high interactivity needed for the testing and debugging processes troublesome to achieve. For example, during testing or debugging of embedded RTSs, it might be required to provide the system with interrupts with a very high temporal precision, in order to enforce desired behaviors. This is in contrast to testing and debugging in desktop system environments, where command line inputs or test scripts might suffice.

The hardware solutions discussed above (e.g., ICEs, JTAG, and BDM) may aid testers and debuggers in this task. Alternative solutions include simulator-based embedded testing and debugging, where the problem of lacking peripheral resources is solved by using highly interactive software target simulators or hardware target emulators instead of actual target machines during debugging sessions. Running software simulators can be either RTOS-level simulators (e.g., the VxWorks simulator in the WindRiver Workbench [185]) or hardware-level simulators (e.g., those provided by gdb [49] or IAR EW [60]).

Reproducibility. For debugging, being able to reliably reproduce executions is very important. In a typical debugging scenario, the execution of a test case has pointed out the existence of a bug in the system under test, and a debugging process is initiated in order to find the location of the bug. However, if the original erroneous execution cannot be reproduced, investigation of the sequence of events that led to the failure is not possible.

Typically, a reproduction of a particular execution requires an identical starting state, identical initial and intermediate input, and identical provision of asynchronous events (e.g., interrupts). As RTSs are incorporated in different temporal and environmental contexts, system behavior reproducibility is low. Interactions with external contexts and multitasking will reduce the probability of traditional execution reproducibility to a negligible level. Events occurring in the temporal external context will actively or passively have an impact on the RTS execution. For example, entering a breakpoint during runtime in an RTS will stop the execution for an unspecified time. The problem with this is that, while the RTS is halted, the controlled external process will continue to evolve (e.g., a car will not momentarily stop by terminating the execution of the controlling software).

1.8.2 RTS Testing

Testing of RTSs typically focuses on testing for timeliness and testing for detecting interleaving failures (even though recent contributions have taken alternative approaches to handle the growing complexity of RTSs, e.g., statistical verification [15]). In the following sections we list the main areas of RTS testing.

1.8.2.1 Testing for RTS Timeliness

Testing for timeliness could be seen as a complement to execution-time analysis and response-time analysis. One motivation here is that execution-time analyses make certain assumptions regarding the RTS and its tasks hold in order for the analysis results to be correct. For timeliness testing, on the other hand, the aim is to detect, enforce, and test the worst-case response times of the actual system tasks.

For testing of temporal correctness, approaches include mutation-based testing using genetic algorithms [114], generation of test cases from Time Reachability Trees, i.e., symbolical representations of time behaviors derived through timed Petri nets [17], and nonintrusive monitoring techniques that record runtime information, which is then used to analyze if temporal constraints are violated [171].

Related to this, Cheung et al. [27] describe how to perform timeliness testing of multimedia software systems with soft real-time requirements.

1.8.2.2 Testing for Interleaving Errors

In multitasking RTSs, as in any concurrent system, the execution and preemption order of tasks may cause alterations in the system behavior. The aim of testing for interleaving errors is to cover, and specifically to enforce, the task interleaving sequences that give rise to system failures. When testing for interleaving errors, one typically differs between event-triggered and time-triggered systems, since the number of potential interleaving sequences in the former poses a major problem when testing for interleaving errors. Consequently, the testability (i.e., the probability for failures to be observed during testing when errors are present [165]) is lower in event-triggered RTSs than in time-triggered RTSs [142].

Thane et al. [123,166] propose a method for deterministic testing of time-triggered multitasking RTSs with sporadic interrupts. The key element here is to identify the different execution orderings (serializations of the concurrent system) and to treat each of these orderings as a sequential program. The main weakness of this approach is the potentially exponential blowup of the number of execution orderings.

On the event-triggered RTSs side, there has been some notable work on improving the testability of these systems. Regehr [134] has proposed the use of random testing and shown how to perform such testing without suffering from the unwanted effects of aberrant interrupts (i.e., interrupts that are only feasible in a test laboratory environment). Furthermore, Lindström et al. [93] suggest some selective restrictions on the runtime environment in order to increase the testability while maintaining the flexibility of event-triggered RTSs.

1.8.2.3 Model-Based RTS Testing

Using modeling for RTS testing is established within the testing community [45,55,71,86,147,163,182]. Most often, the system specification is modeled using some formal modeling tool or method supporting temporal formalisms. The model is then analyzed with respect to different test criteria, and through searches of the possible state space, test suites (i.e., sets of test cases aimed at fulfilling a specific test criterion) are generated. Model-based RTS testing can be used both in order to test nonfunctional aspects (e.g., timeliness) and functional aspects.

One inherent drawback of functional model-based testing is that the quality and thoroughness of the generated test suite are highly dependent on the quality and thoroughness of the system specification model. If the models are manually generated, the skills of the person modeling the system will have a serious impact on the quality assuring abilities of the test suite generated by the model.

1.8.2.4 Testing of Distributed RTSs

When testing distributed RTSs, not only problems of intertask dependencies and interference make life hard for testers, but the interaction between the different nodes in the distributed RTS also has to be taken into account.

In this area, Schütz [141] proposed a testing strategy for distributed RTSs. The strategy is tailored for the time-triggered MARS system [80]. Furthermore, Khoumsi [72] proposes a centralized architecture, for testing distributed RTSs, that ensures controllability and optimizes observability. Also, Thane's approach for testing of interleaving errors is extended to distributed RTSs [166].

1.8.3 RTS Debugging

The possibility of traditional debugging of RTSs is highly impaired by the lack of reproducibility and observability in these systems. Not surprisingly, this is also reflected in the research contributions available in this area. In order to provide reproducibility (and, in some sense, also observability) for RTS debugging, the prominent approach is based on record/replay [13,28,34,88,106,162,167,172,188].

Record/replay debugging. The basic idea of record/replay is to, during an original reference execution, observe and record events causing nondeterminism during this execution. Next, these events are used in order to enforce the same execution behavior in a controlled environment (often in a debugger), thereby achieving reproducibility. What is common for all execution replay methods is the possibility of cyclic debugging of otherwise nondeterministic and nonreproducible systems during the replay execution.

Debugging by means of execution replay was pioneered in 1987 by LeBlanc and Mellor-Crummey [88], who proposed a method that focuses on logging the sequence of accesses to shared objects in concurrent executions. Record/replay is one of the primary debugging methods for general-purpose concurrent systems [28,88,106,162,188], and it has also been adopted by the real-time community. Here, the recording of nondeterministic events and data can be performed by nonintrusive dedicated hardware [13,172], or by intrusive software probes [34,167], where the recording code could be left resident in the deployed system. The latter comes at a cost in memory space and execution time, but gives the additional benefit that it becomes possible to debug the deployed system as well in case of a failure [137].

Alternative approaches. More recently, non-replay-based alternatives for debugging of RTSs have been proposed. Examples of such approaches include a method for stochastic analysis and visualization of execution-time measurements in order to find the causes of timing errors [40], and a method for automatic state reaching (i.e., given a specific (reachable) state in a program, the method automatically provides the input required for reaching that state) in a debugger for reactive systems [48].

1.8.4 Industrial Practice

Testing and debugging of multitasking RTSs are time-consuming activities. At best, companies use hardware emulators, e.g., [87], to get some level of observability without interfering with the observed system. In situations where hardware and software, or different software components that depend on each other, are developed concurrently, it is common to use hardware/software in-the-loop simulation in order to be able to conduct preliminary tests even though the system is not complete. Here, the missing hardware or software is emulated by mathematical representations. This might also include modeling of the environment that is to be controlled by the RTS. While this provides some level of verification, it is impossible to accurately model the exact behavior of the simulated missing hardware or software.

More often, testing and debugging of RTSs are ad hoc activities, using intrusive instrumentations of the code either to observe test results or to try to track down intricate timing errors. However, some tools using the above record/replay method are now emerging on the market, e.g., [189] for industrial systems and [135] for games and game consoles.

1.9 Summary

This chapter has presented the most important issues, methods, and trends in the area of embedded RTSs. A wide range of topics has been covered, from the initial design of embedded RTSs to analysis and testing. Important issues discussed and presented are design tools, operating systems, and major underlying mechanisms such as architectures, models of interactions, real-time mechanisms, executions strategies, and scheduling. Moreover, communications, analysis, and testing techniques are presented.

Over the years, the academics have put an effort in increasing the various techniques used to compose and design complex embedded RTSs. Standards and industry are following a slower pace, while also adopting and developing area-specific techniques. Today, we can see diverse techniques used in different application domains, such as automotive, aero, and trains. In the area of communications, an effort is made in the academic, and also in some parts of industry, toward using Ethernet. This is a step toward a common technique for several application domains.

Different real-time demands have led to domain-specific operating systems, architectures, and models of interactions. As many of these have several commonalities, there is a potential for standardization across several domains. However, as this takes time, we will most certainly stay with application-specific techniques for a while, and for specific domains, with extreme demands on safety or low cost, specialized solutions will most likely be used in the future as well. Therefore, knowledge of the techniques used in and suitable for the various domains will remain important.

References

1. L. Abeni and G. Buttazzo. Integrating multimedia applications in hard real-time systems. In *Proceedings of the 19th IEEE International Real-Time Systems Symposium (RTSS'98)*, pp. 4–13, Madrid, Spain, December 1998. IEEE Computer Society.
2. L. Abeni and G. Buttazzo. Resource reservations in dynamic real-time systems. *Real-Time Systems*, 27(2):123–167, July 2004.
3. N. Abramson. Development of the ALOHANET. *IEEE Transactions on Information Theory*, 31(2): 119–123, March 1985.
4. Airlines Electronic Engineering Committee (AEEC). ARINC 653: Avionics Application Software Standard Interface (Draft 15), June 1996.
5. M. Åkerholm, J. Carlson, J. Fredriksson, H. Hansson, J. Håkansson, A. Möller, P. Pettersson, and M. Tivoli. The save approach to component-based development of vehicular systems. *Journal of Systems and Software*, 80(5):655–667, May 2007.

6. L. Almeida, P. Pedreiras, and J.A. Fonseca. The FTT-CAN protocol: Why and how. *IEEE Transaction on Industrial Electronics*, 49(6):1189–1201, December 2002.

7. Northern Real-Time Applications. Total time predictability, 1998. Whitepaper on SSX5.

8. Arcticus Systems, homepage. The Rubus Operating System. http://www.arcticus.se

9. Roadmap—Adaptive Real-Time Systems for Quality of Service Management. ARTIST—Project IST-2001-34820, May 2003. http://www.artist-embedded.org/Roadmaps/

10. The Asterix Real-Time Kernel. http://www.mrtc.mdh.se/projects/asterix/

11. N.C. Audsley, A. Burns, R.I. Davis, K. Tindell, and A.J. Wellings. Fixed priority preemptive scheduling: An historical perspective. *Real-Time Systems*, 8(2/3):129–154, 1995.

12. Autosar project. http://www.autosar.org/

13. V.P. Banda and R.A. Volz. Architectural support for debugging and monitoring real-time software. In *Proceedings of Euromicro Workshop on Real Time*, pp. 200–210, Como, Italy, June 1989. IEEE.

14. J. Berwanger, M. Peller, and R. Griessbach. Byteflight—A new high-performance data bus system for safety-related applications. *BMW AG*, February 2000.

15. P. Binns, M. Elgersma, S. Ganguli, V. Ha, and T. Samad. Statistical verification of two non-linear real-time UAV controllers. In *Proceedings of IEEE Real-Time and Embedded Technology and Applications Symposium (RTAS'04)*, pp. 341–350, Toronto, Canada, May 2004.

16. D. Box. *Essential COM*. Addison-Wesley, Reading, MA, 1998. ISBN: 0-201-63446-5.

17. V. Braberman, M. Felder, and M. Marre. Testing timing behavior of real-time software. In *Proceedings of the International Software Quality Week*, pp. 143–155, San Francisco, CA, May 1997.

18. R.J. Bril, J.J. Lukkien, R.I. Davis, and A. Burns. Message response time analysis for ideal Controller Area Network (CAN) refuted. In J.-D. Decotignie, editor, *Proceedings of the 5th International Workshop on Real-Time Networks (RTN'06) in Conjunction with the 18th Euromicro International Conference on Real-Time Systems (ECRTS'06)*, Dresden, Germany, July 2006.

19. E. Bruneton, T. Coupaye, M. Leclercq, V. Quema, and J.B. Stefani. An open component model and its support in Java. In *Proceedings of the International Symposium on Component-Based Software Engineering (CBSE2004)*, Edinburgh, Scotland, 2004.

20. T. Bures, J. Carlson, S. Sentilles, and A. Vulgarakis. Towards component modelling of embedded systems in the vehicular domain. Technical Report ISSN 1404-3041, ISRN MDH-MRTC-226/2008-1-SE, Mälardalen University, April 2008.

21. A. Burns and A. Wellings. *Real-Time Systems and Programming Languages*, 2nd edn. Addison-Wesley, Reading, MA, 1996. ISBN 0-201-40365-X.

22. G.C. Buttazzo. Rate Monotonic vs. EDF: Judgment day. *Real-Time Systems*, 29(1):5–26, January 2005.

23. G.C. Buttazzo, editor. *Hard Real-Time Computing Systems*, 2nd edn. Springer-Verlag New York, 2005.

24. G.C. Buttazzo. *Hard Real-Time Computing Systems*. Kluwer Academic, Boston, MA, 1997. ISBN 0-7923-9994-3.

25. A. Carpenzano, R. Caponetto, L. Lo Bello, and O. Mirabella. Fuzzy traffic smoothing: An approach for real-time communication over Ethernet networks. In *Proceedings of the 4th IEEE International Workshop on Factory Communication Systems (WFCS'02)*, pp. 241–248, Västerås, Sweden, August 2002. IEEE Industrial Electronics Society.

26. L. Casparsson, A. Rajnak, K. Tindell, and P. Malmberg. Volcano—A revolution in on-board communications. *Volvo Technology Report*, 1:9–19, 1998.

27. S.-C. Cheung, S.T. Chanson, and Z. Xu. Toward generic timing tests for distributed multimedia software systems. In *ISSRE '01: Proceedings of the 12th International Symposium on Software Reliability Engineering (ISSRE'01)*, pp. 210, Washington, D.C., 2001. IEEE Computer Society.

28. J.D. Choi, B. Alpern, T. Ngo, M. Sridharan, and J. Vlissides. A pertrubation-free replay platform for cross-optimized multithreaded applications. In *Proceedings of the 15th International Parallel and Distributed Processing Symposium*, San Francisco, CA, April 2001. IEEE Computer Society.

29. J. Conard, P. Dengler, B. Francis, J. Glynn, B. Harvey, B. Hollis, R. Ramachandran, J. Schenken, S. Short, and C. Ullman. *Introducing .NET*. Wrox Press Ltd. Birmingham, U.K., 2000. ISBN: 1-861004-89-3.

30. I. Crnkovic and M. Larsson. *Building Reliable Component-Based Software Systems*. Artech House Publisher, Norwood, MA, 2002. ISBN 1-58053-327-2.

31. J.D. Day and H. Zimmermann. The OSI reference model. *Proceedings of the IEEE*, 71(12):1334–1340, December 1983.

32. M.L. Dertouzos. Control robotics: The procedural control of physical processes. In *Proceeding of International Federation for Information Processing (IFIP) Congress*, pp. 807–813, Stockholm, Sweden, August 1974.

33. E.W. Dijkstra. Notes on Structured Programming. In *Structured Programming*. Academic Press, London, U.K., 1972.

34. P. Dodd and C.V. Ravishankar. Monitoring and debugging distributed real-time programs. *Software – Practice and Experience*, 22(10):863–877, October 1992.

35. dSpace. TargetLink. http://www.dspace.de/ww/en/inc/home/products/sw/pcgs/targetli.cfm

36. EAST, Embedded Electronic Architecture Project. http://www.east-eea.net/

37. eCos Home Page. http://sources.redhat.com/ecos

38. M. El-Derini and M. El-Sakka. A CSMA protocol under a priority time constraint for real-time communication systems. In *Proceedings of 2nd IEEE Workshop on Future Trends of Distributed Computing Systems (FTDCS'90)*, pp. 128–134, Cairo, Egypt, September 1990. IEEE Computer Society.

39. J. Engblom. *Processor Pipelines and Static Worst-Case Execution Time Analysis*. PhD thesis, Uppsala University, Department of Information Technology, Uppsala, Sweden, April 2002.

40. J. Entrialgo, J. Garcia, J.L. Diaz, and D.F. Garcia. Stochastic metrics for debugging the timing behaviour of real-time systems. In *RTAS '07: Proceedings of the 13th IEEE Real Time and Embedded Technology and Applications Symposium*, pp. 183–192, Washington, D.C., 2007. IEEE Computer Society.

41. ETAS. ASCET. http://www.etas.com/en/products/ascet_software_products.php

42. ETAS. RTA-OSEK. http://www.etas.com/en/products/rta_software_products.php

43. Comp.realtime FAQ. http://www.faqs.org/faqs/realtime-computing/faq/

44. J.P. Fassino, J.B. Stefani, J.L. Lawall, and G. Muller. Think: A software framework for component-based operating system kernels. In *Proceedings of the General Track: 2002 USENIX Annual Technical Conference Table of Contents*, pp. 73–86, Monterey, CA, June 2002.

45. M.A. Fecko, M.Ü. Uyar, A.Y. Duale, and P.D. Amer. A technique to generate feasible tests for communications systems with multiple timers. *IEEE/ACM Transactions on Networking*, 11(5):796–809, 2003.

46. FlexRay Consortium. FlexRay communications system—protocol specification, Version 2.0, June 2004.

47. J. Gait. A probe effect in concurrent programs. *Software—Practice and Experience*, 16(3): 225–233, March 1986.

48. F. Gaucher, E. Jahier, B. Jeannet, and F. Maraninchi. Automatic state reaching for debugging reactive programs. In *AADEBUG'2003—Fifth International Workshop on Automated Debugging*, Ghent, September, 2003.

49. GNU Debugger Home Page, 2008. http://www.gnu.org

50. OSEK Group. OSEK/VDX Operating System Specification 2.2.1. http://www.osek-vdx.org/

51. H. Hansson, H. Lawson, O. Bridal, C. Norström, S. Larsson, H. Lönn, and M. Strömberg. Basement: An architecture and methodology for distributed automotive real-time systems. *IEEE Transactions on Computers*, 46(9):1016–1027, September 1997.

52. K. Hänninen, J. Mäki-Turja, M. Nolin, M. Lindberg, J. Lundbäck, and K.-L. Lundbäck. The rubus component model for resource constrained real-time systems. In *3rd IEEE International Symposium on Industrial Embedded Systems*, Montpellier, France, June 2008.

53. H. Hansson, H. Lawson, and M. Strömberg. BASEMENT a distributed real-time architecture for vehicle applications. *Real-Time Systems*, 3(11):223–244, November 1996.

54. G.T. Heineman and W.T. Councill. *Component-Based Software Engineering, Putting the Pieces Together*. Addison–Wesley Professional, Reading, MA, 2001. ISBN: 0-201-70485-4.

55. A. Hessel and P. Pettersson. Cover—A real-time test case generation tool. In *19th IFIP International Conference on Testing of Communicating Systems and 7th International Workshop on Formal Approaches to Testing of Software 2007*, pp. 31–34, Tallin, Estonia, June 2007.

56. S. Hissam, J. Ivers, D. Plakosh, and K.C. Wallnau. Pin component technology (V1. 0) and its C interface, Technical note CMU/SEI-2005-TN-001, April 2006.

57. H. Hoang and M. Jonsson. Switched real-time Ethernet in industrial applications: Asymmetric deadline partitioning scheme. In *Proceedings of the 2nd International Workshop on Real-Time LANs in the Internet Age (RTLIA'03) in Conjunction with the 15th Euromicro International Conference on Real-Time Systems (ECRTS'03)*, pp. 55–58, Porto, Portugal, June 2003. Polytechnic Institute of Porto, ISBN 972-8688-12-1.

58. H. Hoang, M. Jonsson, U. Hagstrom, and A. Kallerdahl. Switched real-time Ethernet with earliest deadline first scheduling: Protocols and traffic handling. In *Proceedings of the IEEE International Parallel and Distributed Processing Symposium (IPDPS'02)*, pp. 94–99, Fort Lauderdale, FL, April 2002. IEEE Computer Society.

59. S. Howard. A background debugging mode driver package for modular microcontroller. Semiconductor Application Note AN1230/D, Motorola Inc., 1996.

60. IAR Systems Home Page, 2008. http://www.iar.com

61. IEC 61158/61784. Digital data communications for measurement and control: Fieldbus for use in industrial control systems, 2003, IEC, http://www.iec.ch

62. IEEE. Standard for Information Technology—Standardized Application Environment Profile—POSIX Realtime Application Support (AEP), 1998. IEEE Standard P1003.13-1998.

63. IEEE 802.15. Working Group for Wireless Personal Area Networks (WPANs), http://www.ieee802.org/15/

64. Cisco Systems Inc. Token ring/IEEE 802.5. In *Internetworking Technologies Handbook*, 3rd edn. Cisco Press, 2000, pp. 9-1–9-6. ISBN 1-58705-001-3.

65. ISO. Ada95 Reference Manual, 1995. ISO/IEC 8652:1995(E).

66. ISO 11898. Road vehicles—Interchange of digital information—controller area network (CAN) for high-speed communication. International Standards Organisation (ISO), ISO Standard-11898, November 1993.

67. J. Jasperneite and P. Neumann. Switched Ethernet for factory communication. In *Proceedings of the 8th IEEE International Conference on Emerging Technologies and Factory Automation (ETFA'01)*, pp. 205–212 (vol. 1), Antibes–Juan les Pins, France, October 2001. IEEE Industrial Electronics Society.

68. J. Jasperneite, P. Neumann, M. Theis, and K. Watson. Deterministic real-time communication with switched Ethernet. In *Proceedings of the 4th IEEE International Workshop on Factory Communication Systems (WFCS'02)*, pp. 11–18, Västerås, Sweden, August 2002. IEEE Industrial Electronics Society.

69. U. Jecht, W. Stripf, and P. Wenzel. PROFIBUS: Open solutions for the world of automation. In R. Zurawski, editor, *The Industrial Communication Technology Handbook*, pp. 10-1–10-23. CRC Press, Taylor & Francis Group, Boca Raton, FL, 2005.

70. M. Joseph and P. Pandya. Finding response times in a real-time system. *The Computer Journal*, 29(5):390–395, 1986.

71. A.A. Julius, G.E. Fainekos, M. Anand, I. Lee, and G. Pappas. Robust test generation and coverage for hybrid systems. *Hybrid Systems: Computation and Control*, Proceedings of 10th International Conference HSCC 2007, Lecture Notes in Computer Science, 4416, pp. 329–342, Pisa, Italy, April 2007.

72. A. Khoumsi. Testing distributed real-time systems : An efficient method which ensures controllability and optimizes observability. In *Proceedings of the 8th International Conference on Real-Time Computing Systems and Applications (RTCSA 2002)*, pp. 99–107, Tokyo, Japan, March 2002.

73. M.H. Klein, T. Ralya, B. Pollak, R. Obenza, and M.G. Harbour. *A Practitioners Handbook for Rate-Monotonic Analysis*. Kluwer, Dordrecht, the Netherlands, 1998.

74. L. Kleinrock and F.A. Tobagi. Packet switching in radio channels. Part I. Carrier sense multiple access models and their throughput-delay characteristic. *IEEE Transactions on Communications*, 23(12):1400–1416, December 1975.

75. Kluwer. *Real-Time Systems (Journal)*. http://www.wkap.nl/kapis/CGI-BIN/WORLD/journalhome.htm?0922-6443

76. H. Kopetz. TTP/A—A time-triggered protocol for body electronics using standard UARTs. In *SAE World Congress*, pp. 1–9, Detroit, MI, 1995. SAE.

77. H. Kopetz. The time-triggered model of computation. In *Proceedings of the 19th IEEE International Real-Time Systems Symposium (RTSS'98)*, pp. 168–177, Madrid, Spain, December 1998. IEEE Computer Society.

78. H. Kopetz. Introduction in real-time systems: Introduction and overview. *Part XVIII of Lectures Notes from ESSES 2003—European Summer School on Embedded Systems*, Västerås, Sweden, September 2003.

79. H. Kopetz and G. Bauer. The time-triggered architecture. *Proceedings of the IEEE*, 91(1):112–126, January 2003.

80. H. Kopetz, A. Damm, C. Koza, M. Mulazzani, W. Schwabl, C. Senft, and R. Zainlinger. Distributed fault-tolerant real-time systems: The MARS approach. *IEEE Micro*, 9(1):25–40, February 1989.

81. H. Kopetz and G. Grünsteidl. TTP—A protocol for fault-tolerant real-time systems. *IEEE Computer*, 27(1):14–23, January 1994.

82. H. Kopetz and G. Bauer. The time-triggered architecture. *Proceedings of the IEEE, Special Issue on Modeling and Design of Embedded Software*, 91(1):112–126, January 2003.

83. S.-K. Kweon, K.G. Shin, and G. Workman. Achieving real-time communication over Ethernet with adaptive traffic smoothing. In *Proceedings of the 6th IEEE Real-Time Technology and Applications Symposium (RTAS'00)*, pp. 90–100, Washington, D.C., May–June 2000. IEEE Computer Society.

84. S.-K. Kweon, K.G. Shin, and Z. Zheng. Statistical real-time communication over Ethernet for manufacturing automation systems. In *Proceedings of the 5th IEEE Real-Time Technology and Applications Symposium (RTAS'99)*, pp. 192–202, Vancouver, BC, Canada, June 1999. IEEE Computer Society.

85. G. Lann and N. Riviere. Real-time communications over broadcast networks: The CSMA/DCR and the DOD-CSMA/CD protocols. Technical report, Rapport de Recherche RR-1863, INRIA, Le Chesnay Cedex, France, 1993.

86. K.G. Larsen, M. Mikucionis, B. Nielsen, and A. Skou. Testing real-time embedded software using UPPAAL-TRON: An industrial case study. In *EMSOFT '05: Proceedings of the 5th ACM International Conference on Embedded Software*, pp. 299–306, New York, 2005. ACM Press.

87. Lauterbach. Lauterbach. http://www.laterbach.com

88. T.J. LeBlanc and J.M. Mellor-Crummey. Debugging parallel programs with instant replay. *IEEE Transactions on Computers*, 36(4):471–482, April 1987.

89. J.P. Lehoczky and S. Ramos-Thuel. An optimal algorithm for scheduling soft-aperiodic tasks fixed-priority preemptive systems. In *Proceedings of the 13th IEEE International Real-Time Systems Symposium (RTSS'92)*, pp. 110–123, Phoenix, AZ, December 1992.

90. J.P. Lehoczky, L. Sha, and J.K. Strosnider. Enhanced aperiodic responsiveness in hard real-time environments. In *Proceedings of 8th IEEE International Real-Time Systems Symposium (RTSS'87)*, pp. 261–270, San Jose, CA, December 1997. IEEE Computer Society.

91. LIN Consortium. LIN Protocol Specification, Revision 1.3, December 2002. http://www.lin-subbus.org/

92. LIN Consortium. LIN Protocol Specification, Revision 2.0, September 2003. http://www.lin-subbus.org/

93. B. Lindström, J. Mellin, and S. Andler. Testability of dynamic real-time systems. In *Proceedings of the 8th International Conference on Real-Time Computing Systems and Applications (RTCSA 2002)*, pp. 201–210, Tokyo, Japan, March 2002.

94. C. Liu and J. Layland. Scheduling algorithms for multiprogramming in a hard-real-time environment. *Journal of the ACM*, 20(1):46–61, 1973.

95. C.L. Liu and J.W. Layland. Scheduling algorithms for multiprogramming in a hard real-time environment. *Journal of the ACM*, 20(1):40–61, January 1973.

96. LiveDevices. Realogy Real-Time Architect, SSX5 Operating System, 1999. http://www.livedevices.com/realtime.shtml

97. L. Lo Bello, G.A. Kaczynski, and O. Mirabella. Improving the real-time behavior of Ethernet networks using traffic smoothing. *IEEE Transactions on Industrial Informatics*, 1(3):151–161, August 2005.

98. Express Logic. Threadx. http://www.expresslogic.com

99. K.-L. Lundbäck, J. Lundbäck, and M. Lindberg. Development of dependable real-time applications. Arcticus Systems, December 2004. http://www.arcticus.se

100. Lynuxworks. http://www.lynuxworks.com

101. N. Malcolm and W. Zhao. The timed token protocol for real-time communication. *IEEE Computer*, 27(1):35–41, January 1994.

102. The official OMG MARTE Web site—Modeling and analysis of real-time and embedded systems. http://www.omgmarte.org/

103. A. Massa. *Embedded Software Development with eCos*. Prentice Hall, Upper Saddle River, NJ 2002. ISBN: 0130354732.

104. Mathworks. Mathlab/Simulink. http://www.mathworks.com/products/simulink/

105. Mathworks. Mathlab/Stateflow. http://www.mathworks.com/products/stateflow/

106. J. Mellor-Crummey and T. LeBlanc. A software instruction counter. In *Proceedings of the 3rd International Conference on Architectural Support for Programming Languages and Operating Systems*, pp. 78–86. ACM, Boston, MA, April 1989.

107. Microsoft. Microsoft COM Technologies. http://www.microsoft.com/com/

108. Microsoft. .NET Home Page. http://www.microsoft.com/net/

109. P.O. Müller, C.M. Stich, and C. Zeidler. Component based embedded systems. *Building Reliable Component-Based Software Systems*, pp. 303–323. Artech House, Norwood, MA, 2002. ISBN 1-58053-327-2.

110. M. Molle. A new binary logarithmic arbitration method for Ethernet. Technical report, TR CSRI-298, CRI, University of Toronto, Canada, 1994.

111. M. Molle and L. Kleinrock. Virtual time CSMA: Why two clocks are better than one. *IEEE Transactions on Communications*, 33(9):919–933, September 1985.

112. R. Monson-Haefel. *Enterprise JavaBeans*, 3rd edn. O'Reilly & Assiciates, Inc., Sebastopol, CA, 2001. ISBN: 0-596-00226-2.

113. O. Nierstrass, G. Arevalo, S. Ducasse, R. Wuyts, A. Black, P. Müller, C. Zeidler, T. Genssler, and R. van den Born. A component model for field devices. In *Proceedings of the 1st International IFIP/ACM Working Conference on Component Deployment*, pp. 200–209, Berlin, Germany, June 2002.

114. R. Nilsson, J. Offutt, and J. Mellin. Test case generation for mutation-based testing of timeliness. In *Proceedings of the 2nd International Workshop on Model-based Testing*, pp. 102–121, Vienna, Austria, March 2006.

115. OMG. CORBA Home Page. http://www.omg.org/corba/

116. OMG. Unified Modeling Language (UML). http://www.omg.org/spec/UML/

117. OMG. CORBA Component Model 3.0, June 2002. http://www.omg.org/technology/documents/formal/components.htm

118. A.K. Parekh and R.G. Gallager. A generalized processor sharing approach to flow control in integrated services networks: The single-node case. *IEEE/ACM Transactions on Networking*, 1(3):344–357, June 1993.

119. A.K. Parekh and R.G. Gallager. A generalized processor sharing approach to flow control in integrated services networks: The multiple-node case. *IEEE/ACM Transactions on Networking*, 2(2):137–150, April 1994.

120. PECOS Project Web Site. http://www.pecos-project.org

121. P. Pedreiras, L. Almeida, and J.A. Fonseca. The quest for real-time behavior in Ethernet. In R. Zurawski, editor, *The Industrial Information Technology Handbook*, pp. 48-1–48-14. CRC Press, Boca Raton, FL, 2005.

122. P. Pedreiras, L. Almeida, and P. Gai. The FTT-Ethernet protocol: Merging flexibility, timeliness and efficiency. In *Proceedings of the 14th Euromicro Conference on Real-Time Systems (ECRTS'02)*, pp. 134–142, Vienna, Austria, June 2002. IEEE Computer Society.

123. A. Pettersson and H. Thane. Testing of multi-tasking real-time systems with critical sections. In *Proceedings of 9th International Conference on Real-Time and Embedded Computing Systems amd Applications*, Tainan City, Taiwan, R.O.C, February 18–20, 2003.

124. D.W. Pritty, J.R. Malone, D.N. Smeed, S.K. Banerjee, and N.L. Lawrie. A real-time upgrade for Ethernet based factory networking. In *Proceedings of the 21st IEEE International Conference on Industrial Electronics, Control, and Instrumentation (IECON'95)*, pp. 1631–1637, Orlando, FL, November 1995. IEEE Press.

125. PROFIBUS International. PROFInet—Architecture description and specification. No. 2.202, 2003.

126. PROGRESS Project. http://www.mrtc.mdh.se/progress/

127. P. Puschner and A. Burns. A review of worst-case execution-time analysis. *Real-Time Systems*, 18(2/3):115–128, May 2000.

128. QNX Software Systems. QNX realtime OS. http://www.qnx.com

129. K.K. Ramakrishnan and H. Yang. The Ethernet capture effect: Analysis and solution. In *Proceedings of 19th IEEE Local Computer Networks Conference (LCNC'94)*, pp. 228–240, Minneapolis, MN, October 1994. IEEE Press.

130. S. Ramos-Thuel and J.P. Lehoczky. A correction note to: On-line scheduling of hard deadline aperiodic tasks in fixed priority systems. *Handout at the 14th IEEE International Real-Time Systems Symposium (RTSS'93)*, Raleigh Durham, NC, December 1993.

131. S. Ramos-Thuel and J.P. Lehoczky. On-line scheduling of hard deadline aperiodic tasks in fixed-priority systems. In *Proceedings of 14th IEEE International Real-Time Systems Symposium (RTSS'93)*, pp. 160–171, Raleigh Durham, NC, December 1993. IEEE Computer Society.

132. S. Ramos-Thuel and J.P. Lehoczky. Algorithms for scheduling hard aperiodic tasks in fixed priority systems using slack stealing. In *Proceedings of 15th IEEE International Real-Time Systems Symposium (RTSS'94)*, pp. 22–33, San Juan, Puerto Rico, December 1994. IEEE Computer Society.

133. Rational. Rational Rose Technical Developer. http://www.ibm.com/software/awdtools/developer/technical/

134. J. Regehr. Random testing of interrupt-driven software. In *EMSOFT '05: Proceedings of the 5th ACM International Conference on Embedded Software*, pp. 290–298, New York, 2005. ACM.

135. Replay Solutions Home Page, 2008, www.replaysolutions.com

136. Robocop project. www.extra.research.philips.com/euprojects/robocop/

137. M. Ronsse, K. De Bosschere, M. Christiaens, J. Chassin de Kergommeaux, and D. Kranzlmüller. Record/replay for nondeterministic program executions. *Communications of the ACM*, 46(9):62–67, September 2003.

138. List of real-time Linux variants. http://www.realtimelinuxfoundation.org/variants/variants.html

139. IEEE Computer Society, Technical Committee on Real-Time Systems Home Page. http://www.cs.bu.edu/pub/ieee-rts/

140. T. Sauter. Fieldbus systems: History and evolution. In R. Zurawski, editor, *The Industrial Communication Technology Handbook*, pp. 7-1–7-39. CRC Press, Taylor & Francis Group, Boca Raton, FL, 2005.

141. W. Schütz. Fundamental issues in testing distributed real-time systems. *Real-Time Systems*, 7:129–157, 1994. Kluwer.

142. W. Schutz. *The Testability of Distributed Real-Time Systems*. Kluwer Academic, Norwell, MA, 1993.

143. L. Sha, T. Abdelzaher, K.-E. Årzén, A. Cervin, T.P. Baker, A. Burns, G. Buttazzo, M. Caccamo, J.P. Lehoczky, and A.K. Mok. Real time scheduling theory: A historical perspective. *Real-Time Systems*, 28(2/3):101–155, November/December 2004.

144. L. Sha, J.P. Lehoczky, and R. Rajkumar. Solutions for some practical problems in prioritized preemptive scheduling. In *Proceedings of 7th IEEE International Real-Time Systems Symposium (RTSS'86)*, pp. 181–191, New Orleans, LA, December 1986. IEEE Computer Society.

145. T. Skeie, S. Johannessen, and O. Holmeide. Switched Ethernet in automation networking. In R. Zurawski, editor, *The Industrial Communication Technology Handbook*, pp. 21-1–21-15. CRC Press, Taylor & Francis Group, Boca Raton, FL, 2005.

146. Space4u project. www.extra.research.philips.com/euprojects/space4u/

147. J. Springintveld, F. Vaandrager, and P.R. D'Argenio. Testing timed automata. *Theoretical Computer Science*, 254(1–2):225–257, 2001.

148. B. Sprunt, L. Sha, and J.P. Lehoczky. Aperiodic task scheduling for hard real-time systems. *Real-Time Systems*, 1(1):27–60, June 1989.

149. M. Spuri and G.C. Buttazzo. Efficient aperiodic service under earliest deadline scheduling. In *Proceedings of the 15th IEEE International Real-Time Systems Symposium (RTSS'94)*, pp. 2–11, San Juan, Puerto Rico, December 1994. IEEE Computer Society.

150. M. Spuri and G.C. Buttazzo. Scheduling aperiodic tasks in dynamic priority systems. *Real-Time Systems*, 10(2):179–210, March 1996.

151. J. Stankovic, P. Nagaraddi, Z. Yu, Z. He, and B. Ellis. Exploiting prescriptive aspects: A design time capability. In *Proceedings of the 4th ACM International Conference on Embedded Software (EMSOFT)*, pp. 165–174, Pisa, Italy, September 2004. ACM.

152. John A. Stankovic. VEST—A toolset for constructing and analyzing component based embedded systems. Lecture Notes in Computer Science, 2211:390–402, 2001.

153. IEEE Std. IEEE standard test access port and boundary-scan architecture. Technical report 1532-2001, IEEE, 2001.

154. D.B. Stewart. Introduction to real-time. Embedded Systems Design, 2001, http://www. embedded.com/

155. D.B. Stewart, R.A. Volpe, and P.K. Khosla. Design of dynamically reconfigurable real-time software using port-based objects. *IEEE Transactions on Software Engineering*, 23(12):759–776, 1997.

156. I. Stoica, H. Abdel-Wahab, K. Jeffay, S.K. Baruah, J.E. Gehrke, and C.G. Plaxton. A proportional share resource allocation algorithm for real-time, time-shared systems. In *Proceedings of 17th IEEE International Real-Time Systems Symposium (RTSS'96)*, pp. 288–299, Washington, D.C., December 1996. IEEE Computer Society.

157. J.K. Strosnider, J.P. Lehoczky, and L. Sha. The deferrable server algorithm for enhanced aperiodic responsiveness in hard real-time environments. *IEEE Transactions on Computers*, 44(1):73–91, January 1995.

158. SUN Microsystems. Introducing Java Beans. http://developer.java.sun.com/developer/online Training/Beans/Bea nsl/index.html

159. Symta Vision. http://www.symtavision.com/

160. Enea OSE Systems. Ose. http://www.ose.com

161. C. Szyperski. *Component Software—Beyond Object-Oriented Programming*, 2nd edn. Pearson Education Limited, Essex, England, 2002. ISBN 0-201-74572-0.

162. K.C. Tai, R. Carver, and E. Obaid. Debugging concurrent ADA programs by deterministic execution. *IEEE Transactions on Software Engineering*, 17(1):280–287, January 1991.

163. L. Tan, O. Sokolsky, and I. Lee. Specification-based testing with linear temporal logic. In *Proceedings of the 2004 IEEE International Conference on Information Reuse and Integration*, pp. 483–498, Las Vegas, NV, November 2004.

164. Telelogic. Rhapsody. http://modeling.telelogic.com/products/rhapsody/

165. H. Thane. Monitoring, testing and debugging of distributed real-time systems. Doctoral thesis, Royal Institute of Technology, Stockholm, Sweden, May 2000. Mechatronic Laboratory, Department of Machine Design.

166. H. Thane and H. Hansson. Towards systematic testing of distributed real-time systems. In *Proceedings of the 20th IEEE Real-Time Systems Symposium (RTSS)*, pp. 360–369, Phoenix, AZ, December 1999.

167. H. Thane and H. Hansson. Using deterministic replay for debugging of distributed real-time systems. In *Proceedings of the 12th Euromicro Conference on Real-Time Systems*, pp. 265–272, Stockholm, Sweden, June 2000. IEEE Computer Society.

168. The Times Tool. http://www.docs.uu.se/docs/rtmv/times

169. K. Tindell, H. Hansson, and A. Wellings. Analysing real-time communications: Controller area network (CAN). In *Proceedings of the 15th IEEE Real-Time Systems Symposium (RTSS)*, pp. 259–263, San Juan, Puerto Rico, December 1994. IEEE Computer Society Press.

170. Tri-Pacific. http://www.tripac.com/

171. J.J.P. Tsai, K.-Y. Fang, and Y.-D. Bi. On real-time software testing and debugging. In *Proceedings of 14th Annual International Computer Software and Application Conference*, pp. 512–518, Chicago, IL, November 1990.

172. J.J.P. Tsai, Y. Bi, and R. Smith. A noninterference monitoring and replay mechanism for real-time systems. *IEEE Transaction on Software Engineering*, 16: 897–916, 1990.

173. TTA-Group. Specification of the TTP/C protocol, 2003. http://www.ttagroup.org

174. TTA-Group. Specification of the TTP/A protocol, 2005. http://www.ttagroup.org

175. TTTech. Operating system for fault-tolerance and real-time. http://www.ttagroup.org/technology/doc/TTTech-TTP-OS-Flyer.pdf

176. Time Triggered Technologies. http://www.tttech.com

177. U.S. Department of Commerce. The Economic Impacts of Inadequate Infrastructure for Software Testing. NIST Report, May 2002.

178. R. van Ommering. The Koala component model. *Building Reliable Component-Based Software Systems*, pp. 223–236. Artech House Publishers, Norwood, MA, July 2002. ISBN 1-58053-327-2.

179. R. van Ommering, F. van der Linden, K. Kramer, and J. Magee. The Koala component model for consumer electronics software. *IEEE Computer*, 33(3):78–85, March 2000.

180. Vector. DaVinci Tool Suite. http://www.vector-worldwide.com/vi_davinci_en.html

181. C. Venkatramani and T. Chiueh. Supporting real-time traffic on Ethernet. In *Proceedings of 15th IEEE Real-Time Systems Symposium (RTSS'94)*, pp. 282–286, San Juan, Puerto Rico, December 1994. IEEE Computer Society.

182. E.R. Vieira and A. Cavalli. Towards an automated test generation with delayed transitions for timed systems. In *RTCSA '07: Proceedings of the 13th IEEE International Conference on Embedded and Real-Time Computing Systems and Applications*, pp. 226–231, Washington, D.C., 2007. IEEE Computer Society.

183. Volcano automotive group. http://www.volcanoautomotive.com

184. R. Wilhelm, J. Engblom, A. Ermedahl, N. Holsti, S. Thesing, D. Whalley, G. Bernat, C. Ferdinand, R. Heckmann, T. Mitra, F. Mueller, I. Puaut, P. Puschner, J. Staschulat, and P. Stenström. The worst-case execution time problem—Overview of methods and survey of tools. *ACM Transactions on Programming Languages and Systems*, 7(3): 1–53, April 2008.

185. Wind River Systems Inc. VxWorks Programmer's Guide. http://www.windriver.com/

186. J. Xu and D.L. Parnas. Scheduling processes with release times, deadlines, precedence, and exclusion relations. *IEEE Transactions on Software Engineering*, 16(3):360–369, March 1990.

187. R. Yavatkar, P. Pai, and R.A. Finkel. A reservation-based CSMA protocol for integrated manufacturing networks. Technical report, CS-216-92, Department of Computer Science, University of Kentucky, Lexington, KY, 1992.

188. F. Zambonelli and R. Netzer. An efficient logging algorithm for incremental replay of message-passing applications. In *Proceedings of the 13th International and 10th Symposium on Parallel and Distributed Processing*, pp. 392–398, San Juan, Puerto Rico, April 1999. IEEE.

189. ZealCore. ZealCore Embedded Solutions AB. http://www.zealcore.com

190. W. Zhao and K. Ramamritham. A virtual time CSMA/CD protocol for hard real-time communication. In *Proceedings of 7th IEEE International Real-Time Systems Symposium (RTSS'86)*, pp. 120–127, New Orleans, LA, December 1986. IEEE Computer Society.

191. W. Zhao, J.A. Stankovic, and K. Ramamritham. A window protocol for transmission of time-constrained messages. *IEEE Transactions on Computers*, 39(9):1186–1203, September 1990.

192. ZigBee Alliance. ZigBee Specification, version 1.0, December 14, 2000.

193. H. Zimmermann. OSI reference model: The ISO model of architecture for open system interconnection. *IEEE Transactions on Communications*, 28(4):425–432, April 1980.

2

Design of Embedded Systems

Luciano Lavagno
Polytechnic University of Turin

Claudio Passerone
Polytechnic University of Turin

2.1 The Embedded System Revolution

The world of electronics has witnessed a dramatic growth in its applications in the last few decades. From telecommunications to entertainment, from automotive to banking, almost every aspect of our everyday life employs some kind of electronic components. In most cases, these components are computer-based systems, which are not, however, used or perceived as a computer. For instance, they often do not have a keyboard or a display to interact with the user, and they do not run standard operating systems and applications. Sometimes, these systems constitute a self-contained product themselves (e.g., a mobile phone), but they are frequently embedded inside another system, for which they provide better functionalities and performance (e.g., the engine control unit of a motor vehicle). We call these computer-based systems *embedded systems*.

The huge success of embedded electronics has several causes. The main one in our opinion is that embedded systems bring the advantages of Moore's Law into everyday life, that is, an exponential increase in performance and functionality at an ever-decreasing cost. This is possible because of the capabilities of integrated circuit technology and manufacturing, which allow one to build more and more complex devices, and because of the development of new design methodologies, which allow one to efficiently and cleverly use those devices. Traditional steel-based mechanical development, on the other hand, has reached a plateau near the middle of the twentieth century, and thus it is not a significant source of innovation any longer, unless coupled to electronic manufacturing technologies (MEMS) or embedded systems, as argued above.

There are many examples of embedded systems in the real world. For instance, a modern car contains tens of electronic components (control units, sensors, and actuators) that perform very different tasks. The first embedded systems that appeared in a car were related to the control of mechanical aspects, such as the control of the engine, the antilock brake system, and the control of suspension and transmission. However, nowadays, cars also have a number of components which are not directly related to mechanical aspects but are mostly related to the use of the car as a vehicle for moving around, or the communication needs of the passengers: navigation systems, digital audio and video players, and phones are just a few examples. Moreover, many of these embedded systems are connected together using a network, because they need to share information regarding the state of the car.

Other examples come from the communication industry: a cellular phone is an embedded system whose environment is the mobile network. These are very sophisticated computers whose main task is to send and receive voice but are also currently used as personal digital assistants, for games, to send and receive images and multimedia messages, and to wirelessly browse the Internet. They have been so successful and pervasive that in just a decade they became essential in our life. Other kinds of embedded systems have significantly changed our life as well: for instance, ATM and point-of-sale machines modified the way we do payments, and multimedia digital players have changed how we listen to music and watch videos.

We are just at the beginning of a revolution that will have an impact on every other industrial sector. Special-purpose embedded systems will proliferate and will be found in almost any object that we use. They will be optimized for the application and show a natural user interface. They will be flexible, in order to adapt to a changing environment. Most of them will also be wireless, in order to follow us wherever we go and keep us constantly connected with the information we need and the people we care for. Even the role of computers will have to be reconsidered, as many of the applications for which they are used today will be performed by specially designed embedded systems.

What are the consequences of this revolution in the industry? Modern car manufacturers today need to acquire a significant amount of skills in hardware and software designs, in addition to the mechanical skills that they already have in house, or they should outsource the requirements they have to an external supplier. In either case, a broad variety of skills needs to be mastered, from the design of software architectures for implementing the functionality to being able to model the performance, because real-time aspects are extremely important in embedded systems, especially those related to safety critical applications. Embedded system designers must also be able to architect and analyze the performance of networks, as well as validate the functionality that has been implemented in a particular architecture and the communication protocols that are used.

A similar revolution has happened or is about to happen to other industrial and socioeconomical areas as well, such as entertainment, tourism, education, agriculture, government, and so on. It is therefore clear that new, more efficient, and easy to use embedded electronics design methodologies need to be developed, in order to enable the industry to make use of the available technology.

2.2 Design of Embedded Systems

Embedded system are informally defined as a collection of programmable parts surrounded by application-specific integrated circuits (ASICs) and other standard components (application-specific standard parts, ASSPs) that interact continuously with an environment through sensors and actuators. The collection can be physically a set of chips on a board, or a set of modules on an integrated circuit. Software is used for features and flexibility, while dedicated hardware is used for increased performance and reduced power consumption. An example of an architecture of an embedded system is shown in Figure 2.1. The main programmable components are microprocessors and digital signal processors (DSPs), that implement the software partition of the system. One can view reconfigurable

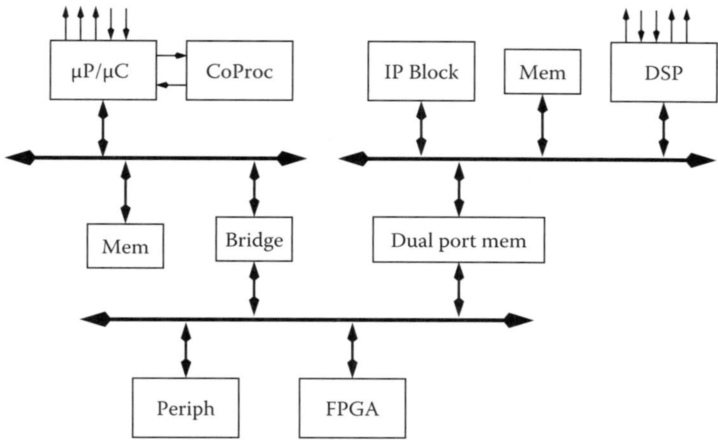

FIGURE 2.1 Reactive real-time embedded system architecture.

components, especially if they can be reconfigured at runtime, as programmable components in this respect. They exhibit area, cost, performance, and power characteristics that are intermediate between dedicated hardware and processors. Custom and programmable hardware components, on the other hand, implement application-specific blocks and peripherals. All components are connected through standard and/or dedicated buses and networks, and data is stored on a set of memories. Often several smaller subsystems are networked together to control, e.g., an entire car, or to constitute a cellular or wireless network.

We can identify a set of typical characteristics that are commonly found in embedded systems. For instance, they are usually not very flexible and are designed to perform always the same task: if you buy an engine control embedded system, you cannot use it to control the brakes of your car, or to play games. A PC, on the other hand, is much more flexible because it can perform several very different tasks. An embedded system is often part of a larger controlled system. Moreover, cost, reliability, and safety are often more important criteria than performance, because the customer may not even be aware of the presence of the embedded system, and so he looks at other characteristics, such as the cost, the ease of use, and the lifetime of a product.

Another common characteristic of many embedded systems is that they need to be designed in an extremely short time to meet their time-to-market. Only a few months should elapse from the conception of a consumer product to the first working prototypes. If these deadlines are not met, the result is a concurrent increase in design costs and decrease of the profits, because fewer items will be sold. So delays in the design cycle may make a huge difference between a successful product and an unsuccessful one.

In the current state of the art, embedded systems are designed with an ad hoc approach that is heavily based on earlier experience with similar products and on manual design. Often the design process requires several iterations to obtain convergence, because the system is not specified in a rigorous and unambiguous fashion, and the level of abstraction, details, and design style in various parts are likely to be different. But as the complexity of embedded systems scales up, this approach is showing its limits, especially regarding design and verification time.

New methodologies are being developed to cope with the increased complexity and enhance designers' productivity. In the past, a sequence of two steps has always been used to reach this goal: *abstraction* and *clustering*. Abstraction means describing an object (i.e., a logic gate made of MOS transistors) using a model where some of the low-level details are ignored (i.e., the Boolean expression representing that logic gate). Clustering means connecting a set of models at the same level of abstraction, to get a new object, which usually shows new properties that are not part of the isolated

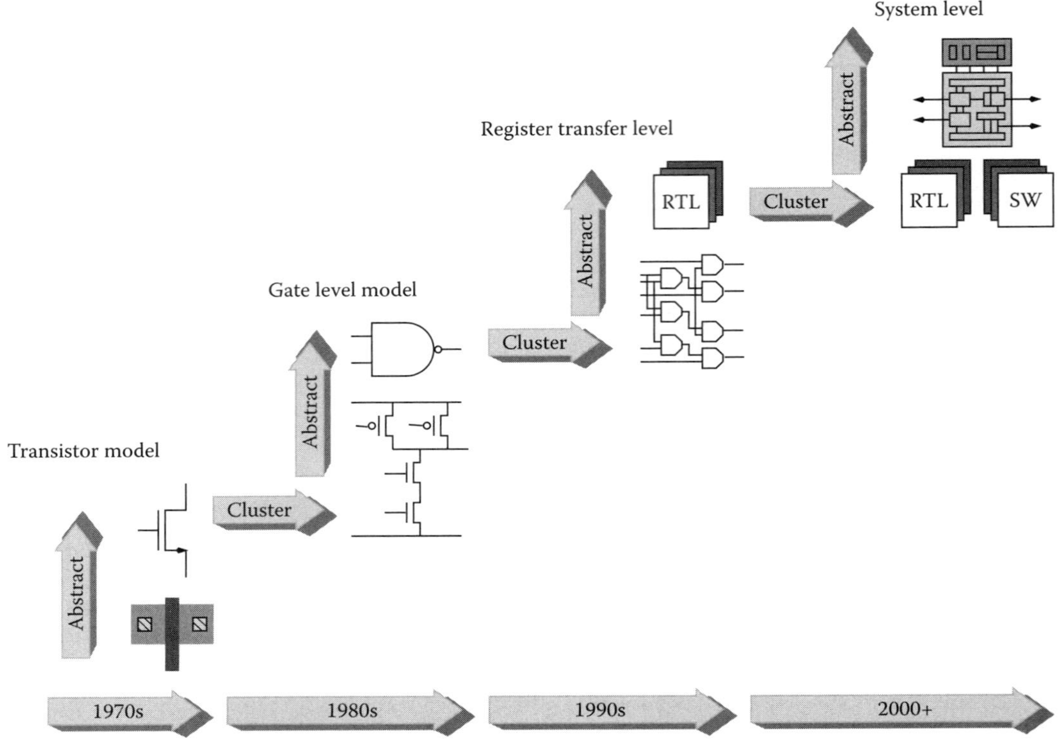

FIGURE 2.2 Abstraction and clustering levels in hardware design.

models that constitute it. By successively applying these two steps, digital electronic design went from drawing layouts to transistor schematics to logic gate netlists to register transfer level descriptions, as shown in Figure 2.2.

The notion of *platform* is key to the efficient use of abstraction and clustering. A platform is a single abstract model that hides the details of a set of different possible implementations as clusters of lower level components. The platform, e.g., a family of microprocessors, peripherals, and bus protocols, allows developers of designs at the higher level (generically called "applications" in the following) to operate without detailed knowledge of the implementation (e.g., the pipelining of the processor or the internal implementation of the serial port, UART). At the same time, it allows platform implementors to share design and fabrication costs among a broad range of potential users, broader than if each design was a one-of-a-kind type.

Today we are witnessing the appearance of a new higher level of abstraction as a response to the growing complexity of integrated circuits. Objects can be functional descriptions of complex behaviors or architectural specifications of complete hardware platforms. They make use of formal high level models that can be used to perform an early and fast validation of the final system implementation, although with reduced details with respect to a lower level description.

The relationship between an application and elements of a platform is called a *mapping*. This exists, e.g., between logic gates and geometric patterns of a layout, as well as between register transfer level statements and gates. At the system level, the mapping is between functional objects with their communication links and platform elements with their communication paths. Mapping at the system level means associating a functional behavior (e.g., an FFT or a filter) to an architectural element that can implement that behavior (e.g., a CPU or DSP or piece of dedicated hardware). It can also associate a communication link (e.g., an abstract FIFO) to some communication services available in the architecture (e.g., a driver, a bus, and some interfaces). The mapping step may also need to specify parameters for these associations (e.g., the priority of a software task or the size of a FIFO), in order to completely describe it. The object that we obtain after mapping shows properties that were not directly exposed in the separate descriptions, such as the performance of the selected system implementation. Performance is not just timing, but any other quantity that can be defined to characterize an embedded system, either physical (area, power consumption, etc.) or logical (quality of service [QOS], fault tolerance, etc.).

Since the system-level mapping operates on heterogeneous objects, it also allows one to nicely separate different and orthogonal aspects such as

1. Computation and communication. This separation is important because refinement of computation is generally done by hand, or by compilation and scheduling, while communication makes use of patterns.
2. Application and platform implementation. This is also called functionality and architecture (e.g., in [8]) because they are often defined and designed independently by different groups or companies.
3. Behavior and performance. This should be kept separate because performance information can either represent nonfunctional requirements (e.g., maximum response time of an embedded controller) or the result of an implementation choice (e.g., the worst-case execution time [WCET] of a task). Nonfunctional constraint verification can be performed traditionally by simulation and prototyping or with static formal checks, such as schedulability analysis.

All these separations result in better reuse, because they decouple independent aspects that would otherwise tie, e.g., a given functional specification to low-level implementation details, by modeling it as assembler or Verilog code. This in turn allows one to reduce design time, by increasing the productivity and decreasing the time needed to verify the system.

A schematic representation of a methodology that can be derived from these abstraction and clustering steps is shown in Figure 2.3. At the functional level, a behavior for the system to be implemented is specified, designed, and analyzed either through simulation or by proving that certain

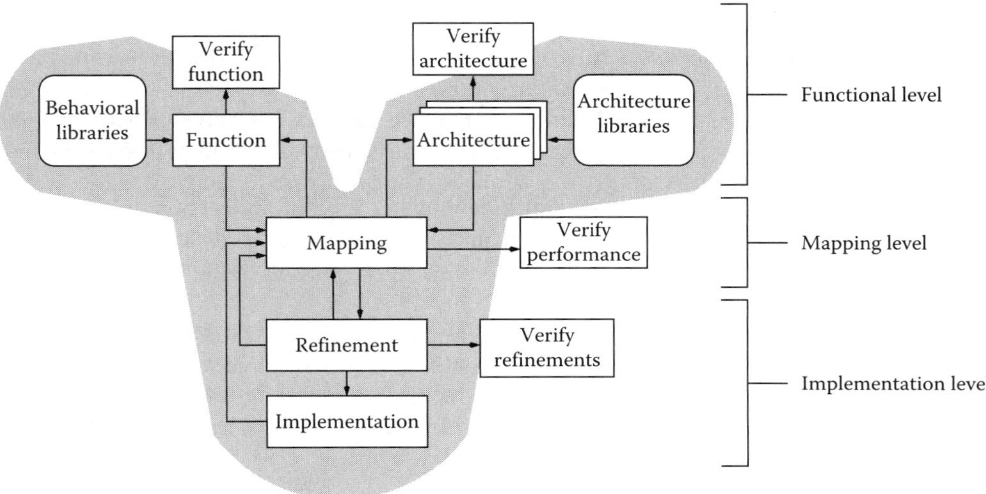

FIGURE 2.3 Design methodology for embedded system.

properties are satisfied (the algorithm always terminates, the computation performed satisfies a set of specifications, the complexity of the algorithm is polynomial, etc.). In parallel, a set of architectures is composed from a clustering of platform elements and is selected as candidates for the implementation of the behavior. These components may come from an existing library or may be specifications of components that will be designed later.

Now functional operations are assigned to the various architecture components, and patterns provided by the architecture are selected for the defined communications. At this level we are now able to verify the performance of the selected implementation, with much richer details than at the pure functional level. Different mappings to the same architecture, or mapping to different architectures, allow one to explore the design space to find the best solutions to important design challenges. These kinds of analysis let the designer identify and correct possible problems early in the design cycle, thus reducing drastically the time to explore the design space and weed out potentially catastrophic mistakes and bugs. At this stage it is also very important to define the organization of the data storage units for the system. Various kinds of memories (e.g., ROM, SRAM, DRAM, Flash, etc.) have different performance and data persistency characteristics and must be used judiciously to balance cost and performance. Mapping data structures to different memories and even changing the organization and layout of arrays can have a dramatic impact on the satisfaction of a given latency in the execution of an algorithm. In particular, a System-On-Chip designer can afford to do a very fine tuning of the number and sizes of embedded memories (especially SRAM, but now also Flash) to be connected to processors and dedicated hardware [15].

Finally, at the implementation level, the reverse transformation of abstraction and clustering occurs, i.e., a lower level specification of the embedded system is generated. This is obtained through a series of manual or automatic refinements and modifications that successively add more details, while checking their compliance with the higher level requirements. This step does not need to generate directly a manufacturable final implementation, but rather produces a new description that in turn constitutes the input for another (recursive) application of the same overall methodology at a lower level of abstraction (e.g., synthesis, placement and routing for hardware, compilation and linking for software). Moreover, the results obtained by these refinements can be back-annotated to the higher level, to perform a better and more accurate verification.

2.3 Functional Design

As discussed in the previous section, system-level design of embedded electronics requires two distinct phases. In the first phase, functional and nonfunctional constraints are the key aspects. In the second phase, the available architectural platforms are taken into account, and detailed implementation can proceed after a mapping phase that defines the architectural component on which every functional model is implemented. This second phase requires a careful analysis of the trade-offs between algorithmic complexity, functional flexibility, and implementation costs.

In this section we describe some of the tools that are used for requirements capture, focusing especially on those that permit executable specification. Such tools generally belong to two broad classes.

The first class is represented, for example, by Simulink® [67], MATRIXx [49], Ascet-SD [5], SPW [19], SCADE [70], and SystemStudio [73]. It includes block-level editors and libraries using which the designer composes *data-dominated* digital signal processing and embedded control systems. The libraries include simple blocks, such as multiplication, addition, and multiplexing, as well as more complex ones, such as FIR filters, FFTs, and so on.

The second class is represented by tools such as Tau [76], StateMate [68], Esterel Studio [70], StateFlow [67]. It is oriented to *control-dominated* embedded systems. In this case, the emphasis is placed on the decisions that must be taken by the embedded system in response to environment and user inputs, rather than on numerical computations. The notation is generally some form of Har'el's statecharts [35].

The Unified Modeling Language (UML), as standardized by the Object Management Group [77], is in a class by itself, since first of all it focused historically more on general-purpose software (e.g., enterprise and commercial software) than on embedded real-time software. Only recently some embedded aspects such as performance and time have been incorporated in UML 2.0 and SysML [44,77] and emphasis has been placed on model-based software generation. However, tool support for UML 2.0 is still limited (Tau [76], Real Time Studio [69], and Rose RealTime [62] provide some), and UML-based hardware design is still in its infancy. Furthermore, the UML is a collection of notations, some of which (especially statecharts) are supported by several of the tools listed above in the control-dominated class.

Simulink and its related tools and toolboxes, both from Mathworks and from third parties such as dSPACE [23], are the workhorse of modern "model-based embedded system design." In model-based design, a functional executable model is used for algorithm development. This is made easier in the case of Simulink by its tight integration with MATLAB®, the standard tool in DSP algorithm development. The same functional model, with added annotations such as bit widths and execution priorities, is then used for algorithmic refinements such as floating-point to fixed-point conversion and real-time task generation.

Then automated software generators such as Real-Time Workshop, Embedded Coder [67], and TargetLink [23] are used to generate task code and sometimes to customize a real-time operating system (RTOS) on which the tasks will run. Ascet-SD, for example, automatically generates a customization of the OSEK automotive RTOS [57] for the tasks that are generated from a functional model. In all these cases, a task is typically generated from a set of blocks that is executed at the same rate or triggered by the same event in the functional model.

Task formation algorithms can use either direct user input (e.g., the execution rate of each block in discrete time portions of a Simulink or Ascet-SD design) or static scheduling algorithms for dataflow models (e.g., based on relative block-to-block rate specifications in SPW or SystemStudio [10,45]).

Simulink is also tightly integrated with StateFlow, a design tool for control-dominated applications, in order to ease the integration of decision making and computation code. It also allows one to smoothly generate both hardware and software from the very same specification. This capability, as well as the integration with some sort of statechart-based Finite State machine editor, is available in most tools from the first class above. The difference in market share can be attributed to the availability of Simulink "toolboxes" for numerous embedded system design tasks (from fixed-point optimization to FPGA-based implementation) and their widespread adoption in undergraduate university courses, which makes them well known to most of today's engineers.

The second class of tools either plays an ancillary role in the design of embedded control systems (e.g., as StateFlow and EsterelStudio) or is devoted to inherently control-dominated application areas, such as telecommunication protocols. In the latter market the clear dominator today is Tau, which also has code generation capabilities for both application code and customization of real-time kernels on which the FSM-generated code will run. The use of Tau for embedded code generation (model-based design) significantly predates that of Simulink-based code generators, mostly due to the highly complex nature of telecom protocols and the less demanding memory and computing power constraints that switches and other networking equipment have.

Tau has links to the requirements capture tool Doors [76], which allows one to trace dependencies between multiple requirements written in English, and connect them to aspects of the embedded system design files, which implement these requirements. The state of the art of such requirement tracing, however, is far from satisfactory, since there is no formal means in Doors to automatically check for violations. Similar capabilities are provided by Reqtify [63].

Techniques for automated functional constraint validation, starting from formal languages, are described in several books, e.g., [43,52]. Deadline, latency, and throughput constraints are one special kind of nonfunctional requirements which have received extensive treatment in the real-time scheduling community. They are also covered in several books, e.g., [12,31,34].

While model-based functional verification is quite attractive, due to its high abstraction level, it ignores cost and performance implications of algorithmic decisions. These are taken into account by the tools described in the next section.

2.4   Function/Architecture and Hardware/Software Codesign

In this section we describe some of the tools that are available to help embedded system designers optimally architect the implementation of the system, and choose the best solution for each functional component. After these decisions have been made, detailed design can proceed using the languages, tools, and methods described in the following chapters in this book.

This step of the design process, whose general structure has been outlined in Section 2.2 by using the platform-based design paradigm, has received various names in the past. Early work [26,32] called it hardware/software codesign (or cosynthesis), because one of the key decisions at this level is what functionality has to be implemented in software vs. dedicated hardware, and how the two partitions of the design interact together with minimum cost and maximum performance.

Later on, people came to realize that hardware/software was too coarse a granularity, and that more implementation choices had to be taken into account. For example, one could trade off single vs. multiple processors, general-purpose CPUs vs. specialized DSPs and Application-Specific Instruction-set Processors (ASIPs), dedicated ASIC vs. ASSP (e.g., an MPEG coprocessor or an Ethernet Medium Access Controller) and standard cells vs. FPGA. Thus the term function/architecture codesign was coined [8] to refer to the more complex partitioning problem of a given functionality onto a heterogeneous architecture such as the one in Figure 2.1.

The term electronic system-level (ESL) design also had some popularity in the industry [13,19], to indicate "the level of design above register transfer, at which software and hardware interact." Other terms, such as timed functional model, have also been used [37].

The key problems that are tackled by tools acting as a bridge between the system-level application and the architectural platform are

1. How to model the performance impact of making mapping decisions from a virtually "implementation-independent" functional specification to an architectural model.
2. How to efficiently drive downstream code generation, synthesis, and validation tools to avoid redoing the modeling effort from scratch at the RTL, C, or assembly code levels, respectively. The notion of automated implementation generation from a high-level functional model is called "model-based design" in the software world.

In both cases, the notion of what is an implementation-independent functional specification, which can be retargeted indifferently to hardware and software implementations, must be carefully evaluated and considered. Taken in its most literal terms, this idea has often been taunted as a myth. However, current practice shows that it is already a reality, at least for some application domains (automotive electronics and telecommunication protocols). It is intuitively very appealing, since it can be considered as a high-level application of the platform-based design principle, by using a formal "system-level platform." Such a platform, embodied in one of the several models of computation that are used in embedded system design, is a perfect candidate to maximize design reuse and to optimally exploit different implementation options.

In particular, several of the tools which have been mentioned in the previous section (e.g., Simulink, TargetLink, StateFlow, SPW, System Studio, Tau, Ascet-SD, StateMate, and Esterel Studio) have code generation capabilities that are considered good enough for "implementation" and not just for rapid prototyping and simulation acceleration. Moreover, several of them (e.g., Simulink, StateFlow, SPW, System Studio, StateMate, and Esterel Studio) can generate indifferently C for software implementation and synthesizable VHDL or Verilog for hardware implementation. Unfortunately, these code generation capabilities, in the data-dominated case, often require the laborious creation of implementation models for each block on each target platform (e.g., software in C or assembler for a given DSP, synthesizable VHDL or macroblock netlist for ASIC or FPGA, etc.). However, since these careful implementations are instances of the system-level platform mentioned above, their development cost can be shared among a multitude of designs performed using the tool.

Most block diagram or statechart-based code generators work in a syntax-directed fashion. A piece of C or synthesizable VHDL code is generated for each block and connection or for each hierarchical state and transition. Thus the designer has tight control over the complexity of the generated software or hardware. While this is a convenient means to bring manual optimization capabilities within the model-based design flow, it has a potentially significant disadvantage in terms of cost and performance (like disabling optimizations in the case of a C compiler). On the other hand, more recent tools like EsterelStudio and SystemStudio take a more radical approach to code generation based on aggressive optimizations [9]. These optimizations, based on logic synthesis techniques in the case of software implementation, destroy the original model structure and thus make debugging and maintenance much harder. However, they can result in an order of magnitude improvement in terms of cost (memory size) and performance (execution speed) with respect to their syntax-directed counterparts [24].

Assuming that good automated code generation, or manual design, is available for each block in the functional model of the application, we are now faced with the function-architecture codesign problem. This essentially means tuning the functional decomposition, as well as the algorithms employed by the overall functional model and each block within it, to the available architecture, and vice versa.

Several design environments, for example,

- POLIS [8], COSYMA [26], Vulcan [32], COSMOS [39], and Roses [16] in the academic world, as well as
- Real Time Studio [69], Foresight [75], and CARDtools [14] in the commercial world,

help the designer in this task by somehow using the notion of independence between functional specification on one side and hardware/software partitioning or architecture mapping choices on the other.

The step of performance evaluation is performed in an abstract, approximate manner by the tools listed above. Some of them use estimators to evaluate the cost and performance of mapping a functional block to an architectural block. Others (e.g., POLIS) rely on cycle-approximate simulation to perform the same task in a manner which better reflects real-life effects, such as burstiness of resource occupation and so on. Techniques for deriving both abstract static performance models (e.g., the WCET of a software task) and performance simulation models are discussed below.

In all cases, the cost of both computation and communication must be taken into account. This is because the best implementation, especially in the case of multimedia systems that manipulate large amounts of image and sound data, is often one that reduces the amount of transferred data between multiple memory locations, rather than one that finds the absolute best trade-off between software flexibility and hardware efficiency. In this area, the Atomium project at IMEC [6,15] has focused on finding the best memory architecture and schedule of memory transfers for data-dominated applications on mixed hardware/software platforms. By exploiting array access models based on polyhedra, they identify the best reorganization of inner loops of DSP kernels and the best embedded memory architecture. The goal is to reduce memory traffic due to register spills, and maximize the overall performance by accessing several memories in parallel (many DSPs offer this opportunity even in the embedded software domain). A very interesting aspect of Atomium, which distinguishes it from most other optimization tools for embedded systems, is the ability to return "a set of Pareto-optimal" solutions (i.e., solutions which are not strictly better than one another in at least one aspect of the cost function), rather than a single solution. This allows the designer to pick the best point based on the various aspects of cost and performance (e.g., silicon area versus power and performance), rather than forcing him or her to "abstract" optimality into a single number.

Performance analysis can be based on simulation, as mentioned above, or rely on automatically constructed models that reflect the WCET of pieces of software (e.g., RTOS tasks) running on an embedded processor. Such models, which must be both provably conservative and reasonably accurate, can be constructed by using an execution model called "abstract interpretation" [18]. This technique traverses the software code, while building a symbolic model, often in the form of linear inequalities [3,46], which represents the requests that the software makes to the underlying hardware (e.g., code fetches, data loads and stores, and code execution). A solution to these inequalities then represents the total "cost" of one execution of the given task. It can be combined then with processor, bus, cache, and main memory models that in turn compute the cost of each of these requests in terms of time (clock cycles) or energy. This finally results in a complete model for the cost of mapping that task to those architectural resources.

Another technique for software performance analysis, which does not require detailed models of the hardware, uses an approximate compilation step from the functional model to an executable model (rather than a set of inequalities as above) annotated with the same set of fetch, load, store, and execute requests. Then simulation is used, in a more traditional setting, to analyze the cost of implementing that functionality on a given processor, bus, cache, and memory configuration. Simulation is more effective than WCET analysis in handling multiprocessor implementations, in which bus conflicts and cache pollution can be difficult, if not utterly impossible, to predict statically in a manner which is not too conservative. However, its success in identifying the true worst case depends on the designers ability to provide the appropriate simulation scenarios. Coverage enhancement techniques from the hardware verification world [7,38] can be extended to help in this case.

Similar abstract models can be constructed in the case of implementation as dedicated hardware, by using high-level synthesis techniques. Such techniques may sometimes not yet be good enough to generate production-quality RTL code but can always be considered as a reasonable estimator of area, timing, and energy costs for both ASIC and FPGA implementations [21,71,78].

SystemC [37] and SpecC [28,29], on the other hand, are more traditional modeling and simulation languages, for which the design flow is based on successive refinement aided by synthesis, rather than codesign or mapping. Finally, OPNET [56] and NS [55] are simulators with a rich modeling library specialized for wireline and wireless networking applications. They help the designer with the more abstract task of generic performance analysis, without the notion of function/architecture separation and codesign.

Communication performance analysis, on the other hand, is generally not done using approximate compilation or WCET analysis techniques like those outlined above. Communication is generally implemented not by synthesis but by "refinement" using patterns and "recipes," such as interrupt-based, DMA-based, and so on. Thus several design environments and languages at the function/architecture level, such as POLIS, COSMOS, Roses, SystemC and SpecC, as well as N2C [19], provide mechanisms to replace abstract communication, e.g., FIFO-based or discrete event-based, with detailed protocol stacks using buses, interrupt controllers, memories, drivers, and so on. These refinements can then be estimated by either using a library-based approach (they are generally part of a library of implementation choices anyway) or sometimes using the approaches described above for computation. Their cost and performance can thus be combined in an overall system-level performance analysis.

However, approximate performance analysis is often not good enough, and a more detailed simulation step is required. This can be achieved by using tools such as Seamless [65], CoMET [17],

MaxSim [50], and N2C [19]. They work at a lower abstraction level, by cosimulating software running on Instruction Set Simulators (ISSs) and hardware running in a Verilog or VHDL simulator. While the simulation is often slower than with more abstract models, and dramatically slower than with static estimators, the precision can now be at the cycle level. Thus it permits close investigation of detailed communication aspects, such as interrupt handling and cache behavior. These approaches are further discussed in the next section.

The key advantage of using the mapping-based approach over the traditional design-evaluate-redesign one is the speed with which design space exploration can be performed. This is done by setting up experiments that change either mapping choices or parameters of the architecture (e.g., cache size, processor speed, or bus bandwidth). Key decisions, such as the number of processors and the organization of the bus hierarchy, can thus be based on quantitative application-dependent data, rather than on past experience.

If mapping can then be used to drive synthesis, in addition to simulation and formal verification, advantages in terms of time-to-market and reduction of design effort are even more significant. Model-based code generation, as we mentioned in the previous section, is reasonably mature, especially for embedded software in application areas, such as avionics, automotive electronics, and telecommunications. In these areas, considerations other than absolute minimum memory footprint and execution time, e.g., safety, sheer complexity, and time-to-market, dominate the design criteria.

At the very least, if some form of automated model-based synthesis is available, it can be used to rapidly generate FPGA- and processor-based prototypes of the embedded system. This significantly speeds up verification, with respect to workstation-based simulation. It permits even some hardware-in-the-loop validation for cases (e.g., the notion of "driveability" of a car) in which no formalization or simulation is possible, but a real physical experiment is required.

2.5 Hardware/Software Coverification and Hardware Simulation

Traditionally the term "hardware/software codesign" has been identified with the ability to execute a simulation of hardware and software at the same time. We prefer to use the term "hardware/software coverification" for this task, and leave codesign for the synthesis- and mapping-oriented approaches outlined in the previous section. In the form of simultaneously running an ISS and a Hardware Description Language (HDL) simulator, while keeping the timing of the two synchronized, the area is not new [64]. In recent years, however, we have seen a number of approaches to speeding up the task, in order to tackle platforms with several processors, and the need to boot an operating system in order to cover a platform with a processor and its peripherals.

Recent techniques have been devoted to the three main ways in which cosimulation speed can be increased:

1. Accelerate the hardware simulator. Coverification generally works at the "clock cycle accurate" level, meaning that both the hardware simulator and the ISS view time as a sequence of discrete clock cycles, ignoring finer aspects of timing (sometimes clock phases are considered, e.g., for DSP systems, in which different memory banks are accessed in different phases of the same cycle). This allows one to speed up simulation with respect to traditional event-driven logic simulation, and yet retain enough precision to identify, e.g., bottlenecks such as interrupt service latency or bus arbitration overhead.

   Native-code hardware simulation (e.g., NCSim [13]) and emulation (e.g., Quick-Turn [13] and Mentor Emulation [65]) can be used to further speed up hardware simulation, at the expense of longer compilation times and much higher costs, respectively.

2. Accelerate the instruction set simulator. Compiled-code simulation has been a popular topic in this area as well [79]. The technique compiles a piece of assembler or C code for a target processor into object code that can be run on a host workstation. This code generally also contains annotations counting clock cycles by modeling the processor pipeline. The speedup that can be achieved with this technique over a traditional ISS, which fetches, decodes, and executes each target instruction individually, is significant (at least one order of magnitude). Unfortunately this technique is not suitable for self-modifying code, such as that of a RTOS. This means that it is difficult to adapt to modern embedded software, which almost invariably runs under RTOS control, rather than on the bare CPU. However, hybrid techniques involving partial compilation on the fly are reportedly used by companies selling fast ISSs [17,50].

3. Accelerate the interface between the two simulators. This is the area where the earliest work has been performed. For example, Seamless [65] uses sophisticate filters to avoid sending requests for memory accesses over the CPU bus. This allows the bus to be used only for peripheral access, while memory data are provided to the processor directly by a "memory server," which is a simulation filter sitting in between the ISS and the HDL simulator. The filter reduces stimulation of the HDL simulator, and thus can result in speedups of one or more orders of magnitude, when most of the bus traffic consists of filtered memory accesses. Of course, also precision of analysis drops, since it becomes harder to identify an overload in the processor bus due to a combination of memory and peripheral accesses, since no simulator component sees both.

In the HDL domain, as mentioned above, progress in the levels of performance has been achieved essentially by raising the level of abstraction. A "cycle-based" simulator, i.e., one that ignores the timing information within a clock cycle, can be dramatically faster than one that requires the use of a timing queue to manage time-tagged events. This is mainly due to two reasons. The first one is that now most of the simulation can be executed always, at every simulation clock cycle. This means that it is much more parallelizable, while event-driven simulators do not fit well over a parallel machine due to the presence of the centralized timing queue. Of course, there is a penalty if most of the hardware is generally idle, since it has to be evaluated anyway, but clock gating techniques developed for low power consumption can obviously be applied here. The second one is that the overhead of managing the time queue, which often accounts for 50%–90% of the event-driven simulation time, can now be completely eliminated.

Modern HDLs are either totally cycle-based (e.g., SystemC 1.0 [37]) or have a "synthesizable subset" which is fully synchronous and thus fully compilable to cycle-based simulation. The same synthesizable subset, by the way, is also supported by hardware emulation techniques for obvious reasons.

Another interesting area of cosimulation in embedded system design is analog–digital cosimulation. This is because such systems quite often include analog components (amplifiers, filters, A/D and D/A converters, demodulators, oscillators, PLLs, etc.), and models of the environment quite often involve only continuous variables (distance, time, voltage, etc.). Simulink includes a component for simulating continuous-time models, employing a variety of numerical integration methods, which can be freely mixed with discrete-time sampled-data subsystems. This is very useful when modeling and simulating, e.g., a control algorithm for automotive electronics, in which the engine dynamics are modeled with differential equations, while the controller is described as a set of blocks implementing a sampled-time subsystem.

Simulink is still mostly used to drive software design, despite good toolkits implementing it in reconfigurable hardware [11,30]. Simulators in the hardware design domain, on the other hand, generally use HDLs as their input languages. Analog extensions of both VHDL [36] and Verilog [59] are available. In both cases, one can represent quantities that satisfy Kirchhoff's laws (i.e., are conserved over cycles or nodes). Thus one can easily build netlists of analog components interfacing with the digital portion, modeled using traditional Boolean or multivalued signals. The simulation environment will then take care of synchronizing the event-driven portion and the continuous time portion. A key problem here is to avoid causality errors, when an event that happens later in "host workstation" time (because the simulator takes care of it later) has an effect on events that preceded it in "simulated time." In this case, one of the simulators has to "roll back" in time, undoing any potential changes in the state of the simulation, and restart with the new information that something has happened in the past (generally the analog simulator does it, since it is easier to reverse time in that case).

Also in this case, as we have seen for hardware/software cosimulation, execution is much slower than in the pure event-driven or cycle-based case, due to the need to take small simulation steps in the analog part. There is only one case in which the performance of the interface between the two domains or of the continuous time simulator is not problematic. It is when the continuous time part is much slower in reality than the digital part. A classical example is automotive electronics, in which mechanical time constants are larger by several orders of magnitude than the clock period of a modern integrated circuit. Thus the performance of continuous time electronics and mechanical cosimulation may not be the bottleneck, except in the case of extremely complex environment models with huge systems of differential equations (e.g., accurate combustion engine models). In that case, hardware emulation of the differential equation solver is the only option (see e.g., [23]).

2.6 Software Implementation

The next two sections provide an overview of traditional design flows for embedded hardware and software. They are meant to be used as a general introduction to the topics described in the rest of the book, and also as a source of references to standard design practice.

The software components of an embedded system are generally implemented using the traditional design-code-test-debug cycle, which is often represented using a V-shaped diagram to illustrate the fact that every implementation level of a complex software system must have a corresponding verification level (Figure 2.4). The parts of the V-cycle which relate to system design and partitioning

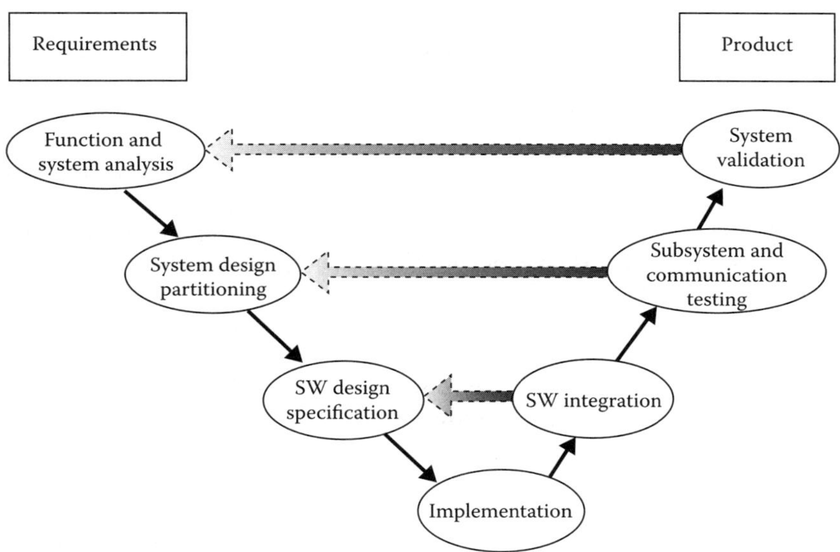

FIGURE 2.4 V-cycle for software implementation.

was described in the previous sections. Here we outline the tools that are available to the embedded software developer.

2.6.1   Compilation, Debugging, and Memory Model

Compilation of mathematical formulas into binary machine-executable code followed almost immediately the invention of electronic computers. The first Fortran compiler dates back to 1954, and subroutines were introduced in 1958, resulting in the creation of the Fortran II language. Languages have since evolved a little; more structured programming methodologies have been developed; and compilers have improved quite a bit, but the basic method has remained the same. In particular the C language, originally designed by Ritchie [40] between 1969 and 1972, and used extensively for programming the UNIX operating system, is now dominant in the embedded system world, almost replacing the more flexible but much more cumbersome and less portable assembler. Its descendants Java and C++ are now beginning to make some inroads but are still viewed as requiring too much memory and computing power for widespread embedded use. Java, although originally designed for embedded applications [4,72], has a memory model based on garbage collection, that still defies effective embedded real-time implementation [27].

The first compilation step in a high-level language is the conversion of the human-written or machine-generated code into an internal format, called Abstract Syntax Tree [2], which is then translated into a representation that is closer to the final output (generally assembler code) and is suitable for a host of optimizations. This representation can take the form of a Control/Data Flow Graph or a sequence of register transfers. The internal format is then mapped, generally via a graph-matching algorithm, to the set of available machine instructions, and written out to a file. A set of assembler files, in which references to data variables and to subroutine names are still based on symbolic labels, is then converted into an absolute binary file, in which all addresses are explicit. This phase is called assembly and loading. Relocatable code generation techniques, which basically permit code and its data to be placed anywhere in memory, without requiring recompilation, are now also being used in the embedded system domain, thanks to the availability of index registers and relative addressing modes in modern microprocessors.

Debuggers for modern embedded systems are much more vital than for general-purpose programming, due to the more limited accessibility of the embedded CPU (often no file system, limited display and keyboard, etc.). They must be able to show several concurrent threads of control, as they interact with each other and with the underlying hardware. They must also be able to do so by minimally disrupting normal operation of the system, since it often has to work in real time, interacting with its environment. Both hardware and operating system support are essential, and the main RTOS vendors, such as WindRiver, all provide powerful interactive multitask debuggers. Hardware support takes the form of breakpoint and watchpoint registers, which can be set to interrupt the CPU when a given address is used for fetching, loading, or storing without requiring one to change the code (which may be in ROM) or to continuously monitor data accesses, which would dramatically slow down execution.

A key difference between most embedded software and most general-purpose software is the memory model. In the latter case, memory is viewed as an essentially infinite uniform linear array, and the compiler provides a thin layer of abstraction on top of it, by means of arrays, pointers, and records (or `struct`s). The operating system generally provides virtual memory capabilities, in the form of user functions to allocate and deallocate memory, by swapping less frequently used pages of main memory with disk. This provides the illusion of a memory as large as the disk area allocated to paging, but with the same direct addressability characteristics as main memory. In embedded systems, however, memory is an expensive resource, in terms of both size and speed. Cost, power, and physical size constraints generally forbid the use of virtual memory, and performance constraints force the designer to always carefully lay out data in memory and match its characteristics (SRAM, DRAM, Flash, ROM) to the access patterns of data and code. Scratchpads [60], i.e., manually managed areas of small and fast memory (on-chip SRAM) are still dominant in the embedded world. Caches are frowned upon in the real-time application domain, since the time when a computation is performed often matters much more than the accuracy of its result. This is due to the fact that, despite a large body of research devoted to timing analysis of software code in the presence of caches (see e.g., [47,54]), their performance must still be assumed to be worst case, rather than average case as in general-purpose and scientific computing, thus leading to poor performance at a high cost (large and power-hungry tag arrays).

However, compilers that traditionally focused on code optimizations for various underlying architectural features of the processor [48] now offer more and more support for memory-oriented optimizations, in terms of scheduling data transfers, sizing memories of various types, and allocating data to memory, sometimes moving it back and forth between fast and expensive and slow and cheap storage [15,60].*

2.6.2 Real-Time Scheduling

Another key difference with respect to general-purpose software is the real-time characteristics of most embedded software, due to its continual interaction with an environment that seldom can wait. In "hard real-time" applications, results produced after the deadline are totally useless. On the other hand, in "soft real-time" applications a merit function measures QOS, allowing one to evaluate trade-offs between missing various deadlines and/or degrading the precision or resolution with which computations are performed. While the former is often associated with safety-critical (e.g., automotive or avionics) applications and the latter is associated with multimedia and telecommunication applications, algorithm design can make a difference even within the very same domain. Consider, for example, a frame decoding algorithm that generates its result at the end of each execution, and that is scheduled to be executed in real-time every 50th of a second. If the CPU load does not allow it to complete each execution before the deadline, the algorithm will not produce any results, and thus behave as a hard real-time application, without being life-threatening. On the other hand, a smarter algorithm or a smarter scheduler would just reduce the frame size or the frame rate, whenever the CPU load due to other tasks increases and thus produce a result that has lower quality but is still viewable.

A huge amount of research, summarized in excellent books such as [12,31,34], has been devoted to solving the problems introduced by real-time constraints on embedded software. Most of this work models the system (application, environment, and platform) in very abstract terms, as a set of tasks, each with a release time (when the task becomes ready), a deadline (by which the task must complete), and a WCET. In most cases tasks are periodic, i.e., release times and deadlines of multiple instances of the same task are separated by a fixed period. The job of the scheduler is to find an execution order such that each task can complete by its deadline, if it exists. The scheduler, depending on the underlying hardware and software platform (CPU, peripherals, and RTOS), may or may not be able to preempt an executing task in order to execute another one. Generally the scheduler bases its preemption decision, and the choice of which task must be run next, on an integer rank assigned to each task, which is called "priority." Priorities may be assigned statically at compile time or dynamically at runtime. The trade-off is between the usage of precious CPU resources for runtime

---

* While this may seem similar to virtual memory techniques, it is generally done explicitly by a DMA, always keeping cost, power, and performance under tight control.

(also called online) priority assignment, based on an observation of the current execution conditions, and the waste of resources inherent in the compile-time definition of a priority assignment. A scheduling algorithm is also supposed in general to be able to tell conservatively if a set of tasks is schedulable with a given platform and a set of modeling assumptions (e.g., availability of preemption, fixed or stochastic execution time, and so on). Unschedulability may occur, for example, because the CPU is not powerful enough and the WCETs are too long to satisfy some deadline. In this case the remedy could be the choice of a faster clock frequency or a change of CPU or the transfer of some functionality to a hardware coprocessor or the relaxation of some of the constraints (periods, deadlines, etc.).

A key distinction in this domain is between "time-triggered" and "event-triggered" scheduling [41]. The former (also called Time-Division Multiple Access in telecommunications) relies on the fact that the start, preemption (if applicable), and end times of all instances of all tasks are decided a priori, based on worst-case analysis. The resulting system implementation is very predictable, is easy to debug, and allows one to guarantee some service even under fault hypotheses [42]. The latter decides start and preemption times based on the actual time of occurrence of the release events and possibly on the actual execution time (shorter than worst case). It is more efficient than time-triggering in terms of CPU utilization, especially when release and execution times are not known precisely but subject to jitter. It is, however, more difficult to use in practice because it requires some form of conservative schedulability analysis a priori, and the dynamic nature of event arrival makes troubleshooting much harder.

Some models and languages listed above, such as synchronous languages and dataflow networks, lend themselves well to time-triggered implementations. Some form of time-triggered scheduling is being or will most likely be used for both CPUs and communication resources for safety-critical applications. This is already state of the art in avionics ("fly-by-wire," as used, e.g., in the Boeing 777 and in all Airbus models), and it is being seriously considered for automotive applications (X-by-wire, where X can stand for brake, drive, or steer). It is considered, coupled with certified high-level language compilers and standardized code review and testing processes, to be the only mechanism to comply with the rules imposed by various governmental certification agencies. Moving such control functions to embedded hardware and software, thus replacing older mechanical parts, is considered essential in order to both reduce costs and improve safety. Embedded electronic systems can analyze continuously possible wearing and faults in the sensors and the actuators and thus warn drivers or maintenance teams.

The simple task-based model outlined above can also be modified in various ways in order to take the following into account:

- Cost of various housekeeping operations, such as recomputing priorities, context switch between tasks, accessing memory, and so on
- The availability of multiple resources (processors)
- Fact that a task may need more than one resource (e.g., the CPU, a peripheral, and a lock on a given part of memory) and possibly may have different priorities and different preemptability characteristics on each such resource (e.g., CPU access may be preemptable, while disk or serial line access may not)
- Data or control dependencies between tasks

Most of these refinements of the initial model can be taken into account by appropriately modifying the basic parameters of a task set (release time, execution time, priority, and so on). The only exception is the extension to multiple concurrent CPUs, which makes the problem substantially more complex. We refer the interested reader to [12,31,34] for more information about this subject. This formal real-time schedulability analysis is currently replacing manual trial and error and extensive simulation as a means to ensure satisfaction of deadlines or a given QOS requirement.

2.7 Hardware Implementation

The modern hardware implementation process [53,61] in most cases starts from the so-called register transfer level. At this level of abstraction the required functionality and timing of the circuit are modeled with the accuracy of a clock cycle; that is, it is known in which clock cycle each operation, such as addition or data transfer, occurs, but the actual delay of each operation, and hence the stabilization time of data on the inputs of the registers, is not known. At this level the number of registers and their bit widths are also precisely known. The designer usually writes the model using a HDL such as Verilog or VHDL, in which registers are represented using special kinds of "clock-triggered" assignments, and combinational logic operations are represented using the standard arithmetic, relational, and Boolean operators, which are familiar to software programmers using high-level languages.

The target implementation generally is not in terms of individual transistors and wires but uses the Boolean gate abstraction as a convenient hand-off point between logic designer and layout engineer. Such abstraction can take the form of a "standard cell," i.e., an interconnection of transistors realized and well characterized on silicon, which implements a given Boolean function and exhibits a specific propagation delay from inputs to outputs, under given supply, temperature, and load conditions. It can also be a combinational logic block (CLB) in a field-programmable gate array. The former, which is the basis of the modern Application-Specific Integrated Circuit (ASIC) design flow, is much more efficient than the latter;* however, it requires a very significant investment in terms of EDA† tools, mask production costs, and engineer training.

The advantage of ASICs over FPGAs in terms of area, power, and performance efficiency comes from two main factors. The first one is the broader choice of basic gates: an average standard cell library includes about 100–500 gates, with both different logic functions and different drive strengths, while a given FPGA contains only one type of CLB. The second one is the use of static interconnection techniques, i.e., wires and contact vias versus the transistor-based dynamic interconnects of FPGAs.

The much higher nonrecurrent engineering cost of ASICs comes first of all from the need to create at least a set of masks for each design (assuming it is correct the first time, i.e., there is no need to respin), which is over 1M$ for current technologies and is growing very fast, and from the long fabrication times, which can be up to several weeks. Design costs are also higher, again in the million dollar range, both due to the much greater flexibility, requiring skilled personnel and sophisticated implementation tools, and due to the very high cost of design failure, requiring sophisticated verification tools. Thus ASIC designs are the most economically viable solution only for very high volumes. The rising mask costs and manufacturing risks are making the FPGA option viable for larger and larger production counts as technology evolves. A third alternative, structured ASICs, has been proposed recently. It features fixed layout schemes, similar to FPGAs, but also implements interconnect using contact vias and hence reduces the number of masks required to implement a design. A comparison of the alternatives, for a given design complexity and varying production volumes, is shown in Figure 2.5 (the exact points at which each alternative is best are still subject to debate, and they are moving to the right over time).

2.7.1 Logic Synthesis and Equivalence Checking

The semantics of HDLs and of languages such as C or Java are very different from each other. HDLs were born in the 1970s in order to model highly concurrent hardware systems, built using registers

---

* The difference is about one order of magnitude in terms of area, power, and performance for the current fabrication technology, and the ratio is expected to remain constant over future technology generations.

† The term EDA, which stands for electronic design automation, is often used to distinguish this class of tools from the CAD tools used for mechanical and civil engineering design.

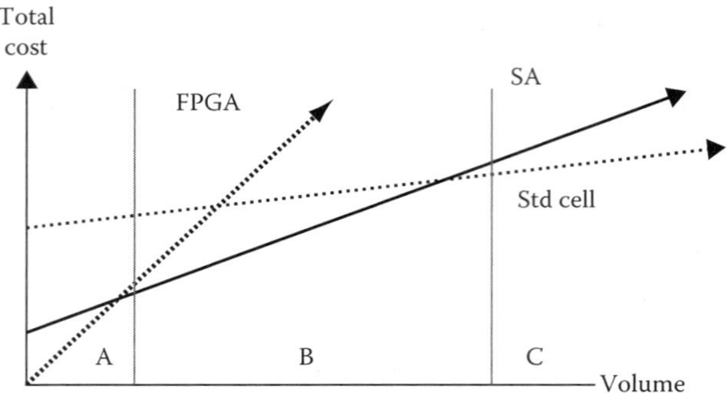

FIGURE 2.5   Comparison between ASIC, FPGA, and structured ASIC production costs.

and Boolean gates. They and the associated simulators which allow one to analyze the behavior of the modeled design in detail are very efficient in handling fine-grained concurrency and synchronization, which is necessary when simulating huge Boolean netlists. However, they often lack constructs found in modern programming languages, such as recursive functions and complex data types (only recently introduced in Verilog), or objects, methods, and interfaces. An HDL model is essentially meant to be simulated under a variety of timing models (generally at the register transfer or gate level, even though cosimulation with analog components or continuous time models is also supported, e.g., in Verilog-AMS and AHDL).

Synthesis from an HDL into an interconnection of registers and gates normally consists of two substeps. The first one, called RTL synthesis and module generation, transforms high-level operators such as adders, multiplexers, and so on into Boolean gates using an appropriate architecture (e.g., ripple carry or carry lookahead). The second one, called logic synthesis, optimizes the combinational logic resulting from the above step, under a variety of cost and performance constraints [22,33].

It is well known that, given a function to be implemented (e.g., 32-bit two's-complement addition), one can use the properties of Boolean algebra in order to find alternative implementations with different characteristics in terms of

1. Area, e.g., estimated as the number of gates or as the number of gate inputs or as the number of literals in the Boolean expression representing each gate function or using a specific value for each gate selected from the standard cell library or even considering an estimate of interconnect area. This sequence of cost functions increases estimation precision but is more and more expensive to compute.
2. Delay, e.g., estimated as the number of levels or more precisely as a combination of levels and fanout of each gate or even more precisely as a table that takes into account gate type, transistor size, input transition slope, output capacitance, and so on.
3. Power, e.g., estimated as transition activity times capacitance, using the well-known equation valid for CMOS transistors.

It is also well known that generally Pareto-optimal solutions to this problem exhibit an area-delay product which is approximately constant for a given function.

Modern EDA tools, such as Design Compiler from Synopsys [73], RTL Compiler from Cadence [13], Leonardo Spectrum from Mentor Graphics [65], Synplify from Synopsis [74], Blast Create from Magma Design Automation [20] and others, perform such task efficiently for designs that today may include a few million gates. Their widespread adoption has enabled designers to tackle huge designs in a matter of months, which would have been unthinkable or extremely inefficient using either manual or purely block-based design techniques. Such logic synthesis systems take into account the required functionality, the target clock cycle, and the set of physical gates which are available for implementation (the standard-cell library or the CLB characteristics, e.g., number of inputs), as well as some estimates of capacitance and resistance of interconnection wires[*] and generate efficient netlists of Boolean gates, which can be passed on to the following design steps.

While synthesis is performed using precise algebraic rules, bugs can creep into any program. Thus, in order to avoid extremely costly respins due to an EDA tool bug, it is essential to verify that the functionality of the synthesized gate netlist is the same as that of the original RTL model. This verification step was traditionally performed using a multilevel HDL simulator, comparing responses to designer-written stimuli in both representations. However, multimillion gate circuits would require too many very slow simulation steps (a large circuit today can be simulated at the speed of a handful of clock cycles per second). Formal verification is thus used to prove, using algorithms that are based on the same laws as synthesis techniques but which have been written by different people and thus hopefully have different bugs that indeed the responses of the two circuit models are identical under all legal input sequences.

This verification, however, solves only half of the problem. One must also check that all combinational logic computations complete within the required clock cycle. This second check can be performed using timing simulators; however, complexity considerations also suggest to use a more static approach. Static timing analysis, based on worst-case longest-path search within combinational logic, is today a workhorse of any logic synthesis and verification framework. It can be based on purely topological information or consider only so-called true paths along which a transition can propagate [51] or even include the effects of cross talk on path delay. Cross talk may alter the delay of a "victim" wire, due to simultaneous transitions of temporally and spatially close "aggressor" wires, as analyzed by tools such as PrimeTime from Synopsys [73] and CeltIc from Cadence [13]. This kind of coupling of timing and geometry makes cross-talk-aware timing analysis very hard and essentially contributes to the breaking of traditional boundaries between synthesis, placement, and routing.

Tools performing these task are available from all major EDA vendors (e.g., Synopsys, Cadence) as well as from a host of startups. Synthesis has become more or less a commodity technology, while formal verification, even in its simplest form of equivalence checking, as well as in other emerging forms, such as property checking, that are described below, is still an emerging technology, for which disruptive innovation occurs mostly in smaller companies.

2.7.2 Placement, Routing, and Extraction

After synthesis (and sometimes during synthesis) gates are placed on silicon, either at fixed locations (the positions of CLBs) for FPGAs and Structured ASICs or with a row-based organization for standard cell ASICs. Placement must avoid overlaps between cells, while at the same time satisfying clock cycle time constraints, avoiding excessively long wires on critical paths.[†]

Placement, especially for multimillion-gate circuits, is an extremely difficult problem, which requires complex constrained combinatorial optimization. Modern algorithms [66] drastically simplify the model, in order to ensure reasonable runtimes. For example, the Quadratic Placement model used in several modern EDA tools minimizes the sum of squares of net lengths. This permits very efficient derivation of the cost function and fast identification of a minimum cost solution. However,

---

[*] Some such tools also include rough placement and routing steps, which will be described below, in order to increase the precision of such interconnect estimates for current technologies.

[†] Power density has recently become a prime concern for placement as well, implying the need to avoid "hot spots" of very active cells, where power dissipation through the silicon substrate would lead to excessive heating.

this quadratic cost only approximately correlates with the true objective, which is the minimization of the clock period, due to parasitic capacitance. True cost first of all depends on the actual interconnect, which is designed only later by the routing step, and second depends on the maximum among a set of sums (one for each register-to-register path), rather than on the sum over all gate-to-gate interconnects. For this reason, modern placers iterate steps solved using fast but approximate algorithms, with more precise analysis phases, often involving actual routing, in order to recompute the actual cost function at each step.

Routing is the next step, which involves generating (or selecting from the available prelaid-out tracks in FPGAs) the metal and via geometries that will interconnect placed cells. It is extremely difficult in modern submicron technologies, not only due to the huge number of geometries involved (10 million gates can easily involve a billion wire segments and contacts), but also due to the complexity of modern interconnect modeling. A wire used to be modeled, in CMOS technology, essentially as a parasitic capacitance. This (or minor variations considering also resistance) is still the model used by several commercial logic synthesis tools. However, nowadays a realistic model of a wire, to be used when estimating the cost of a placement or routing solution, must take the following into account

- Realistic resistance and capacitance, e.g., using the Elmore model [25], considering each wire segment and via separately, due to the very different resistance and capacitance characteristics of different metal layers[*]
- Cross-talk noise due to capacitive coupling[†]

This means that exactly as in placement (and sometimes during placement) one needs to alternate between fast routing using approximate cost functions and detailed analysis steps that refine the value of the cost function. Again, all major EDA vendors offer solutions to the routing problem, which are generally tightly integrated with the placement tool, even though in principle the two perform separate functions. The reason for the tight coupling lies in the above-mentioned need for the placer to accurately estimate the detailed route taken by a given interconnect, rather than just estimating it with the square of the distance between its terminals.

Exactly as in the case of synthesis, a verification step must be performed after placement and routing. This is required in order to verify that

- All design rules are satisfied by the final layout
- All and only the desired interconnects have been realized by placement and routing

This step is done by extracting electrical and logic models from layout masks and comparing these models with the input netlist (already verified for equivalence with the RTL). Note that within each standard cell, design rules are verified independently, since the ASIC designer for reason of Intellectual Property protection generally does not see the actual layout of the standard cells, but only an external "envelope" of active (transistor) and interconnect areas, which is sufficient to perform this kind of verification. The layout of each cell is known and used only at the foundry, when masks are finally produced.

2.7.3   Simulation, Formal Verification, and Test Pattern Generation

The steps mentioned above create a layout implementation from RTL, while checking simultaneously that no errors are introduced either due to programming errors or due to manual modifications, and

---

[*] Layers that are farther away from silicon are best for long-distance wires, due to the smaller substrate and mutual capacitance, as well as due to the smaller sheet resistance [58].

[†] Inductance fortunately is not yet playing a significant role, and many doubt that it ever will for digital integrated circuits.

that performance and power constraints are satisfied. However, they do nothing to ensure either that the original RTL model satisfies the customer-defined requirements, or that the circuit after manufacturing does not have any flaws compromising either its functionality or its performance.

The former problem is tackled by simulation, prototyping, and formal verification. None of these techniques is sufficient to ensure that an ill-defined problem has a solution: customer needs are inherently nonformalizable.[*] However, they help building up confidence that the final product will satisfy the requirements. Simulation and prototyping are both trial-and-error procedures, similar to the compile-debug cycle used for software. Simulation is generally cheaper, since it only requires a general-purpose workstation (nowadays often a PC running Linux), while prototyping is faster (it is based on synthesizing the RTL model into one or several FPGAs). Cost and performance of these options differ by several orders of magnitude. Prototyping on emulation platforms, such as those offered by Quickturn, is thus limited to the most expensive designs, such as microprocessors.[†]

Unfortunately both simulation and prototyping suffer from a basic *capacity* problem. It is true that cost decreases exponentially and performance increases exponentially over technology generations for the simulation and prototyping platforms (CPUs and FPGAs). However, the complexity of the verification problem grows as a "double or even triple exponential" (approximately) with technology. The reason is that the number of potential states of a digital design grows exponentially with the number of memory-holding components (flip-flops and latches), and the complexity of the verification problem for a sequential entity (e.g., a Finite State machine) grows even more than exponentially with its state space. For this reason, the growth in the number of input patterns which are required to prove up to a given level of confidence that a design is correct grows "triply exponentially" with each technology generation, while capacity and performance grow "only" as a single exponential. This is clearly an untenable situation, given that the number of engineers is finite, and the size of the verification teams is already much larger than that of the design teams.

Formal verification, defined as proving semiautomatically that under a set of assumptions a given property holds for a design, is a means of alleviating at least the human aspect of the "verification complexity explosion" problem. Formal verification allows one to state a property, such as "this protocol never deadlocks" or "the value of this register, is never overwritten before being read," using relatively simple mathematical formulas. Then one can automatically check that the property holds over "all possible input sequences." The problem, unfortunately, is inherently extremely complex (the triple exponential mentioned above affects this formulation as well). However, the complexity is now relegated to the automated portion of the flow. Thus manual generation and checking of individual pattern sequences are no longer required. Several EDA companies on the market, such as Cadence, Mentor Graphics, Synopsys, as well as several silicon startups, currently offer such tools. The key barriers to adoption are twofold:

1. The complexity of the task, as mentioned above, is just shifted. While a workstation costs much less than an engineer, exponential growth is never tenable in the long term, regardless of the constant factors. This means that significant human intervention is still required in order to keep within acceptable limits the time required to check each individual property. This involves both breaking properties into simpler subproperties and abstracting away aspects of the system which are not relevant for the property at hand.

---

[*] For example, what is the definition of "a correct phone call?" Does this refer to not dropping the communication or to transferring exactly a certain number of voice samples per second or to setting up quickly a communication path? Since all these desirable characteristics have a cost, what is the maximum price various classes of customers are willing to pay for them, and what is the maximum degree of violation that can be admitted by each class?

[†] Nowadays, even microprocessors are mostly designed using a modified ASIC-like flow, except for memories, register files, and sometimes portions of the ALU, which are still designed by hand down to the polygon level, at least for leading edge CPUs.

Abstraction, however, hides aspects of the real design from the automated prover and thus implies the risk of "false positive" results, i.e., of declaring a system correct even when it is not.

2. Specification of properties is much more difficult than identification of input patterns. A property must encompass a variety of possible scenarios and state explicitly all assumptions made (e.g., there is no deadlock in the bus access protocol only if no master makes requests at every clock cycle). The language in which properties are specified is often a form of mathematical logics and thus is even less familiar than software languages to a typical design engineer.

However, significant progress is being made in this area every year by researchers, and adoption of such automated or semiautomated formal verification techniques in the specification verification domain is growing.

Testing a manufactured circuit to verify that it operates correctly according to the RTL model is a closely related problem. In principle, one would need to prove equivalent behavior under all possible input/output sequences, which is clearly impossible. In practice, test engineers either use a "naturally orthogonal" architecture, such as that of a microprocessor, in order to functionally test small sequences of instructions, or decompose testing into that of combinational and sequential logic. Combinational logic testing is a relatively "easy" task, as compared to the formal verification described above. If one considers only Boolean functionality (i.e., delay is not tested), its complexity (assuming that no polynomial algorithm exists for NP-complete problems) is just a single exponential in the number of combinational circuit inputs.

While a priori there is no reason why testing only Boolean equivalence between the specification and the manufactured circuit should be enough to ensure correct functionality, empirically there is a significant amount of evidence that fully testing for a relatively small class of Boolean manufacturing faults, namely "stuck-at faults," coupled with some functional at-speed testing is sufficient to ensure satisfactory yield for ASICs. The stuck-at fault model assumes that the only problem that can occur during manufacturing is that some gate inputs are fixed at logical 0 or 1. This may have been a physically realistic model in the early days of bipolar-based Transistor–Transistor Logic. However, in CMOS a host of physical defects may short wires together, increase or decrease their resistance and/or capacitance, short a transistor gate to its source or drain, and so on. At the logic level, a combinational function may become sequential (even worse, it may exhibit dynamic behavior, i.e., slowly change output values over time, without changing inputs), or it may become faster or slower. Still, full checking for stuck-at faults is in practice sufficient to ensure that none of these complex physical problems has occurred or will affect the operation of the circuit.

For this reason, today testing is mostly accomplished by first of all reducing sequential testing to combinational testing using special memory elements, the so-called scan flip-flops and latches. Secondly, combinational test pattern generation is performed only at the Boolean level, using the above-mentioned stuck-at model. Test pattern generation is similar to equivalence checking, because it amounts to proving that two copies of the same circuit, one with and one without a given fault, are equivalent. The witness to this nonequivalence is the pattern to be applied to the circuit inputs to identify the fault.

The problem of actually applying the pattern to the physical fragment of combinational logic and then observing its outputs to verify if the fault is present is solved by converting all or most of the registers of the sequential circuit into one (or a handful of) giant shift registers called scan registers, each including several hundred thousand bits. The pattern (and several others used to test several CLBs in parallel) is first loaded serially through the shift register. Then a multiplexer at the input of each flip-flop is switched, transforming the serial loading mode into parallel loading mode, using as register inputs the outputs of each CLB. Finally, serial conversion is performed again, and the outputs of the logic are checked for correctness by the test equipment. Figure 2.6 shows

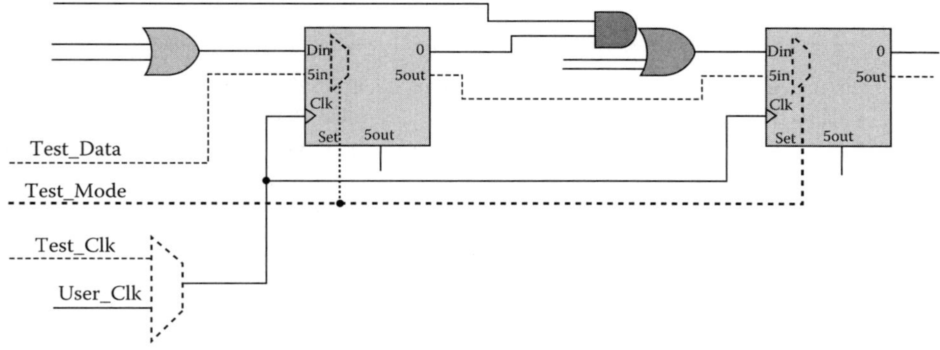

FIGURE 2.6 Two scan flip-flops with combinational logic.

an example of this sort of arrangement, in which the flip-flop clock is also changed from normal operation (in which it can be gated, for example) to test mode. The only drawback of this elegant solution, due to the IBM engineers in the 1970s, is the additional time that the circuit needs to spend on very expensive testing machines, in order to shift in and out patterns through very long flip-flop chains. Test pattern generation for combinational circuits is a very well-established area of research, and again the reader is referred to one of many books in the area for a more extensive description [1].

Note that memories are not tested using this mechanism, both because it would be too expensive to convert each cell into a scan register, and because the stuck-at fault model does not apply to this kind of circuits. Memories are tested using appropriate input/output pattern sequences, which are generated, applied and verified on-chip, using either self-test software running on the embedded processor, or some form of Built-In Self-test (BIST) logic circuitry. Modern RAM generators, which produce directly the layout in a given process, based on the requested number of rows and columns, also often produce directly the BIST circuitry.

2.8 Conclusions

This chapter discussed several aspects of embedded system design, including both methodologies that allow one to perform judicious algorithmic and architectural decisions and tools supporting various steps of these methodologies. One must not forget, however, that often embedded systems are complex compositions of parts that have been implemented by various parties, and thus the task of physical board or chip integration can be as difficult as, and much more expensive than, the initial architectural decisions.

In order to support the integration and system testing tasks one must use formal models throughout the design process and if possible perform early evaluation of the difficulties of integration by virtual integration and rapid prototyping techniques. These allow one to find or completely avoid subtle bugs and inconsistencies earlier in the design cycle and thus reduce overall design time and cost.

Thus the flow and tools that are described in this chapter help not only with the initial design, but also with the final integration. This is because they are based on executable specifications of the whole system (including models of its environment), early virtual integration, and systematic (often automated) refinement toward implementation.

The last part of the chapter summarized the main characteristics of the current hardware and software implementation flows. While complete coverage of this huge topic is beyond our scope, a lightweight introduction can hopefully serve to direct the interested reader who has only a general electrical engineering or computer science background toward the most appropriate source of information.

References

1. M. Abramovici, M. A. Breuer, and A. D. Friedman. *Digital Systems Testing and Testable Design.* Computer Science Press, New York, 1990.
2. A.V. Aho, J.E. Hopcroft, and J.D. Ullman. *The Design and Analysis of Computer Algorithms.* Addison-Wesley, Reading, MA, 1974.
3. AbsInt Worst-Case Execution Time Analyzers. http://www.absint.com.
4. K. Arnold and J. Gosling. *The Java Programming Language.* Addison Wesley, Reading, MA, 1996.
5. ETAS Ascet-SD. http://www.etas.de.
6. IMEC ATOMIUM. http://www.imec.be/design/atomium/.
7. 0-In Design Automation. http://www.0-in.com/.
8. F. Balarin, E. Sentovich, M. Chiodo, P. Giusto, H. Hs ieh, B. Tabbara, A. Jurecska, L. Lavagno, C. Passerone, K. Suzuki, and A. Sangiovanni-Vincentelli. *Hardware-Software Co-design of Embedded Systems – The POLIS approach.* Kluwer Academic Publishers, Dordrecht, the Netherlands, 1997.
9. G. Berry. The foundations of esterel. In *Proof, Language and Interaction: Essays in Honour of Robin Milner.* MIT Press, Cambridge, MA, 2000.
10. J. Buck and R. Vaidyanathan. Heterogeneous modeling and simulation of embedded systems in El Greco. In *Proceedings of the International Conference on Hardware Software Codesign,* May 2000.
11. Altera DSP Builder. http://www.altera.com.
12. G. Buttazzo. *Hard Real-Time Computing Systems: Predictable Scheduling Algorithms and Applications.* Kluwer Academic Publishers, Boston, MA, 1997.
13. CeltIc Cadence Design Systems RTL Compiler and Quickturn. http://www.cadence.com.
14. CARDtools. http://www.cardtools.com.
15. F. Catthoor, S. Wuytack, E. De Greef, F. Balasa, L. Nachtergaele, and A. Vandecapelle. *Custom Memory Management Methodology: Exploration of Memory Organisation for Embedded Multimedia System Design.* Kluwer Academic Publishers, Boston, MA, 1998.
16. W. Cesario, A. Baghdadi, L. Gauthier, D. Lyonnard, G. Nicolescu, Y. Paviot, S. Yoo, A. A. Jerraya, and M. Diaz-Nava. Component-based design approach for multicore socs. In *Proceedings of the Design Automation Conference,* June 2002.
17. VAST Systems CoMET. http://www.vastsystems.com/.
18. P. Cousot and R. Cousot. Abstract interpretation: A unified lattice model for static analysis of programs by construction of approximation of fixpoints. In *Proceedings of the ACM Symposium on Principles of Programming Languages.* ACM, 1977.
19. N2C CoWare SPW and LISATek. http://www.coware.com.
20. Magma Design Automation Blast Create. http://www.magma-da.com.
21. Forte Design Systems Cynthesizer. http://www.forteds.com.
22. S. Devadas, A. Ghosh, and K. Keutzer. *Logic synthesis.* McGraw-Hill, New York, 1994.
23. dSPACE TargetLink and Prototyper. http://www.dspace.de.
24. S.A. Edwards. Compiling Esterel into sequential code. In *International Workshop on Hardware/ Software Codesign.* ACM Press, May 1999.
25. W.C. Elmore. The transient response of damped linear network with particular regard to wideband amplifiers. *Journal of Applied Physics,* 19:55–63, 1948.
26. R. Ernst, J. Henkel, and T. Benner. Hardware-software codesign for micro-controllers. *IEEE Design and Test of Computers,* 10(3):64–75, September 1993.
27. Real-Time for Java Expert Group. The real time specification for Java. https://rtsj.dev.java.net/, 1998.
28. D. Gajski, J. Zhu, and R. Domer. *The SpecC language.* Kluwer Academic Publishers, Boston, MA, 1997.
29. D. Gajski, J. Zhu, R. Domer, A. Gerstlauer, and S. Zhao. *SpecC: Specification Language and Methodology.* Kluwer Academic Publisher, Dordrecht, the Netherlands, 2000.
30. Xilinx System Generator. http://www.xilinx.com.

31. H. Gomaa. *Software Design Methods for Concurrent and Real-Time Systems*. Addison-Wesley, Boston, MA, 1993.

32. R.K. Gupta and G. De Micheli. Hardware-software cosynthesis for digital systems. *IEEE Design and Test of Computers*, 10(3):29–41, September 1993.

33. G.D. Hachtel and F. Somenzi. *Logic Synthesis and Verification Algorithms*. Kluwer Academic Publishers, Norwell, MA, 1996.

34. W.A. Halang and A.D. Stoyenko. *Constructing Predictable Real Time Systems*. Kluwer Academic Publishers, Norwell, MA, 1991.

35. D. Har'el, H. Lachover, A. Naamad, A. Pnueli, M. Politi, R. Sherman, A. Shtull-Trauring, and M.B. Trakhtenbrot. STATEMATE: A working environment for the development of complex reactive systems. *IEEE Transactions on Software Engineering*, 16(4):403–414, April 1990.

36. IEEE. Standard 1076.1, vhdl-ams. http://www.eda.org/vhdl-ams.

37. Open SystemC Initiative. http://www.systemc.org.

38. C. Norris Ip. Simulation coverage enhancement using test stimulus transformation. In *Proceedings of the International Conference on Computer Aided Design*, November 2000.

39. T.B. Ismail, M. Abid, and A.A. Jerraya. COSMOS: A codesign approach for communicating systems. In *International Workshop on Hardware/Software Codesign*. ACM Press, 1994.

40. B. Kernighan and D. Ritchie. *The C Programming Language*. Prentice-Hall, Upper Saddle River, NJ, 1988.

41. H. Kopetz. Should responsive systems be event-triggered or time-triggered? *IEICE Transactions on Information and Systems*, E76-D(11):1325–32, November 1993.

42. H. Kopetz and G. Grunsteidl. TTP – A protocol for fault-tolerant real-time systems. *IEEE Computer*, 27(1):14–23, January 1994.

43. R. P. Kurshan. *Automata-Theoretic Verification of Coordinating Processes*. Princeton University Press, Princeton, NJ, 1994.

44. L. Lavagno, G. Martin, and B. Selic, editors. *UML for Real: Design of Embedded Real-Time Systems*. Kluwer Academic Publishers, New York, 2003.

45. E. A. Lee and D. G. Messerschmitt. Synchronous data flow. *IEEE Proceedings*, 75(9):1235–1245, September 1987.

46. Y.T.S. Li and S. Malik. Performance analysis of embedded software using implicit path enumeration. In *Proceedings of the Design Automation Conference*, June 1995.

47. Y.T.S. Li, S. Malik, and A. Wolfe. Performance estimation of embedded software with instruction cache modeling. In *Proceedings of the International Conference on Computer-Aided Design*, November 1995.

48. P. Marwedel and G. Goossens, editors. *Code Generation for Embedded Processors*. Kluwer Academic Publishers, Norwell, MA, 1995.

49. National Instruments MATRIXx. http://www.ni.com/matrixx/.

50. Axys Design Automation MaxSim and MaxCore. http://www.axysdesign.com/.

51. P. McGeer. *On the Interaction of Functional and Timing Behavior of Combinational Logic Circuits*. PhD thesis, University of California Berkeley, California, November 1989.

52. K. McMillan. *Symbolic Model Checking*. Kluwer Academic Publishers, Dordrecht, the Netherlands, 1993.

53. G. De Micheli. *Synthesis and Optimization of Digital Circuits*. McGraw-Hill, New York, 1994.

54. F. Mueller and D.B̃. Whalley. Fast instruction cache analysis via static cache simulation. In *Proceedings of the 28th Annual Simulation Symposium*, April 1995.

55. Network Simulator NS-2. http://www.isi.edu/nsnam/ns/.

56. OPNET. http://www.opnet.com.

57. OSEK/VDX. http://www.osek-vdx.org/.

58. R.H.J.M. Otten and R.K. Brayton. Planning for performance. In *Proceedings of the Design Automation Conference*, June 1998.

59. OVI. Verilog-a standard. http://www.ovi.org.

60. P. Panda, N. Dutt, and A. Nicolau. Efficient utilization of scratch-pad memory in embedded processor applications. In *Proceedings of Design Automation and Test in Europe (DATE)*, February 1997.

61. Jan M. Rabaey, A. Chandrakasan, and Borivoje Nikolic. *Digital Integrated Circuits* (2nd edition). Prentice-Hall, Upper Saddle River, NJ, 2003.

62. IBM Rational Rose RealTime. http://www.rational.com/products/rosert/.

63. TNI Valiosys Reqtify. http://www.tni-valiosys.com.

64. J. Rowson. Hardware/software co-simulation. In *Proceedings of the Design Automation Conference*, pp. 439–440, 1994.

65. Mentor Graphics Seamless and Emulation. http://www.mentor.com.

66. Naveed A. Sherwani. *Algorithms for VLSI Physical Design Automation* (3rd edition). Kluwer Academic Publishers, Norwell, MA, 1999.

67. The Mathworks Simulink and StateFlow. http://www.mathworks.com.

68. I-Logix Statemate and Rhapsody. http://www.ilogix.com.

69. Artisan Software Real Time Studio. http://www.artisansw.com/.

70. Esterel Technologies Esterel Studio. http://www.esterel-technologies.com.

71. Celoxica DK Design suite. http://www.celoxica.com.

72. Sun Microsystem, Inc. Embedded Java Specification. http://java.sun.com, 1998.

73. Design Compiler Synopsys SystemStudio and PrimeTime. http://www.synopsys.com.

74. Synplicity Synplify. http://www.synplicity.com.

75. Foresight Systems. http://www.foresight-systems.com.

76. Telelogic Tau and Doors. http://www.telelogic.com.

77. The Object Management Group UML. http://www.omg.org/uml/.

78. K. Wakabayashi. Cyber: High level synthesis system from software into ASIC. In R. Camposano and W. Wolf, editors, *High Level VLSI Synthesis*. Kluwer Academic Publisher, Norwell, MA, 1991.

79. V. Zivojnovic and H. Meyr. Compiled HW/SW co-simulation. In *Proceedings of the Design Automation Conference*, 1996.

3

Models of Computation for Distributed Embedded Systems

Axel Jantsch
Royal Institute of Technology

3.1 Introduction

A model of computation is an abstraction of a real computing device. Different computational models serve different objectives and purposes. Thus, they always suppress some properties and details that are irrelevant for the purpose at hand, and they focus on other properties that are essential. Consequently, models of computation have been evolving during the history of computing. In the early decades between 1920 and 1950 the main focus was on the question: "What is computable?" The Turing machine and the lambda calculus are prominent examples of computational models developed to investigate that question.* It turned out that several, very different models of computation such as the Turing machine, the lambda calculus, partial recursive functions, register machines, Markov algorithms, Post systems, etc. [Tay98] are all equivalent in the sense that they all denote the same set of computable mathematical functions. Thus, today the so-called Church–Turing thesis is widely accepted.

---

* The term "model of computation" came in use only much later in the 1970s, but conceptually the computational models of today can certainly be traced back to the models developed in the 1930s.

Church-Turing Thesis: If function f is effectively calculable, then f is Turing-computable. If function f is not Turing-computable, then f is not effectively calculable. [Tay98, p. 379]

It is the basis for our understanding of today what kind of problems can be solved by computers, and what kind of problems principally are beyond a computer's reach. A famous example of what cannot be solved by a computer is the halting problem for Turing machines. A practical consequence is that there cannot be an algorithm that, given a function f and a C++ program P (or a program in any other sufficiently complex programming language) could determine if P computes f. This illustrates the principal difficulty of programming language teachers in correcting exams and of verification engineers in validation programs and circuits.

Later the focus changed to the question: "What can be computed in reasonable time and with reasonable resources?" which spun off the theories of algorithmic complexity based on computational models exposing timing behavior in a particular but abstract way. This resulted in a hierarchy of complexity classes for algorithms according to their "asymptotic complexity." The computation time (or other resources) for an algorithm is expressed as a function of some characteristic figure of the input, e.g., the size of the input. For instance we can state that the function $f(n) = 2n$, for natural numbers n can be computed in $p(n)$ time steps by any computer for some polynomial function, $p(n)$. By contrast, the function $g(n) = n!$ cannot be computed in $p(n)$ time steps on any sequential computer for any polynomial function, $p(n)$, and arbitrary, n. With growing n the time steps required to compute $g(n)$ grows faster than can be expressed by any polynomial function.

This notion of asymptotic complexity allows us to express properties about algorithms in general disregarding details of the algorithms and the computer architecture. This comes at the cost of accuracy. We may only know that there exists some polynomial function, $p(n)$, for every computer, but we do not know $p(n)$ since it may be very different for different computers. To be more accurate one needs to take into account more details of the computer architecture. As a matter of fact, the complexity theories rest on the assumption that one kind of computational model, or machine abstraction, can simulate another one with a bounded and well-defined overhead. This simulation capability has been expressed in the following thesis.

Invariance Thesis: "Reasonable" machines can simulate each other with a polynomially bounded overhead in time and a constant overhead in space. [vEB90]

This thesis establishes an equivalence between different machine models and make results for a particular machine more generally useful. However, some machines are equipped with considerably more resources and cannot be simulated by a conventional Turing machine according to the invariance thesis. Parallel machines have been the subject of a huge research effort and the question, how parallel resources increase the computational power of a machine has led to a refinement of computational models and an accuracy increase for estimating computation time. The fundamental relation between sequential and parallel machines has been captured by the following thesis.

Parallel Computation Thesis: Whatever can be solved in polynomially bounded space on a reasonable sequential machine model can be solved in polynomially bounded time on a reasonable parallel machine, and vice versa. [vEB90]

Parallel computers prompted researchers to refine computational models to include the delay of communication and memory access, which we review briefly in Section 3.1.1.

Embedded systems require a further evolution of computational models due to new design and analysis objectives and constraints. The term "embedded" triggers two important associations. First, an embedded component is squeezed into a bigger system, which implies constraints on size, the form factor, weight, power consumption, cost, etc. Second, it is surrounded by real-world components, which implies timing constraints and interfaces to various communication links, sensors, and actuators. As a consequence, the computational models used and useful in embedded system design are different from those in general-purpose sequential and parallel computing.

The difference comes from the nonfunctional requirements and constraints and from the heterogeneity.

3.1.1 Models of Sequential and Parallel Computation

Arguably, general-purpose sequential computing had for a long time a privileged position, in that it had a single, very simple, and effective model of computation. Based on the van Neumann machine, the random access machine (RAM) model [CR73] is a sufficiently general model to express all important algorithms and reflects the salient nonfunctional characteristics of practical computing engines. Thus, it can be used to analyze performance properties of algorithms in a hardware architecture and implementation independent way. This favorable situation for sequential computing has been eroded over the years as processor architectures and memory hierarchies became ever more complex and deviated from the ideal RAM model.

The parallel computation community has been searching for a similarly simple and effective model in vain [MMT95]. Without a universal model of parallel computation, the foundations for the development of portable and efficient parallel applications and architectures were lacking. Consequently, parallel computing has not gained as wide acceptance as sequential computing and is still confined to niche markets and applications. The parallel random access machine (PRAM) [FW78] is perhaps the most popular model of parallel computation and closest to its sequential counterpart with respect to simplicity. A number of processors execute in a lock-step way, i.e., synchronized after each cycle governed by a global clock, and access global, shared memory simultaneously within one cycle. The PRAM model's main virtue is its simplicity but it captures poorly the costs associated with computing. Although the RAM model has a similar cost model, there is a significant difference. In the RAM model the costs (execution time, program size) are in fact well reflected and grow linearly with the size of the program and the length of the execution path. This correlation is in principle correct for all sequential processors. The PRAM model does not exhibit this simple correlation because in most parallel computers the cost of memory access, communication, and synchronization can be vastly different depending on which memory location is accessed and which processors communicate. Thus, the developer of parallel algorithms does not have sufficient information from the PRAM model alone to develop efficient algorithms. He or she has to consult the specific cost models of the target machine.

Many PRAM variants have been developed to more realistically reflect real cost. Some made the memory access more realistic. The exclusive read–exclusive write (EREW) and the concurrent read–exclusive write (CREW) models [FW78] serialize access to a given memory location by different processors but still maintain the unit cost model for memory access. The local memory PRAM (LPRAM) model [ACS90] introduces a notion of memory hierarchy while the queued read–queued write (QRQW) PRAM [GMR94] models the latency and contention of memory access. A host of other PRAM variants have factored in the cost of synchronization, communication latency, and bandwidth. Other models of parallel computation, many of which are not directly derived from the PRAM machine, focus on memory. There, either the distributed nature of memory is the main concern [Upf84], or various cost factors of the memory hierarchy are captured [ACS90,AACS87,ACFS94]. An introductory survey of models of parallel computation has been written by Maggs et al. [MMT95].

3.1.2 Nonfunctional Properties

A main difference between sequential computation and parallel computation comes from the role of time. In sequential computing, time is solely a performance issue which is moreover captured fairly well by the simple and elegant RAM model. In parallel computing the execution time can be captured only by complex cost functions that depend heavily on various details of the parallel computer. In addition, the execution time can also alter the functional behavior, because the changes in the relative timing of different processors and the communication network can alter the overall functional behavior. To counter this danger, different parts of the parallel program must be synchronized properly.

In embedded systems the situation is even more delicate if real-time deadlines have to be observed. A system that responds slightly too late may be as unacceptable as a system that responds with incorrect data. Even worse, it is entirely context dependent if it is better to respond slightly too late or incorrectly or not at all. For instance when transmitting a video stream, incorrect data arriving on time may be preferable to correct data arriving too late. Moreover, it may be better not to send data that arrives too late to save resources. On the other hand, control signals to drive the engine or the brakes in a car must always arrive, a tiny delay may be preferable to no signal at all. These observations lead to the distinction of different kinds of real-time systems, e.g., hard versus soft real-time systems, depending on the requirements on the timing.

Since most embedded systems interact with real-world objects they are subject to some kind of real-time requirements. Thus, time is an integral part of the functional behavior and cannot be abstracted away completely in many cases. So it should not come as a surprise that models of computation have been developed to allow the modeling of time in an abstract way to meet the application requirements while at the same time avoiding the unnecessary burden of too detailed timing. We will discuss some of these models below. In fact the timing abstractions of different models of computation is a main organizing principle in this chapter.

Designing for low power is a high priority for most, if not all, embedded systems. However, power has been treated in a limited way in computational models because of the difficulty in abstracting the power consumption from the details of architecture and implementation. For VLSI circuits computational models have been developed to derive lower and upper bounds with respect to complexity measures that usually include both circuit area and computation time for a given behavior. AT^2 has been found to be a relevant and interesting complexity measure, where A is the circuit area and T is the computation time either in clock cycles or in physical time. These models have also been used to derive bounds on the energy consumption by usually assuming that the consumed energy is proportional to the state changes of the switching elements. Such analysis shows for instance that AT^2 optimal circuits, i.e., circuits which are optimal up to a constant factor with respect to the AT^2 measure for a given Boolean function, utilize their resources to a high degree, which means that on average a constant fraction of the chip changes state. Intuitively this is obvious since if large parts of a circuit are not active over a long period (do not change state), it can presumably be improved by making it either smaller or faster and thus utilizing the circuit resources to a higher degree on average. Or, to conclude the other way round, an AT^2 optimal circuit is also optimal with respect to energy consumption for computing a given Boolean function. One can spread out the consumed energy over a larger area or a longer time period, but one cannot decrease the asymptotic energy consumption for computing a given function. Note that all these results are asymptotic complexity measures with respect to a particular size metric of the computation, e.g., the length in bit of the input parameter of the function. For a detailed survey of this theory see Lengauer [Len90].

These models have several limitations. They make assumptions about the technology. For instance in different technologies the correlation between state switching and energy consumption is different. In n-channel metal oxide semiconductor (NMOS) technologies the energy consumption is more correlated with the number of switching elements. The same is true for complementary metal oxide semiconductor (CMOS) technologies if leakage power dominates the overall energy consumption. Also, they provide asymptotic complexity measures for very regular and systematic implementation styles and technologies with a number of assumptions and constraints. However, they do not expose relevant properties for complex modern microprocessors, VLIW processors, digital signal processors (DSPs), field programmable gate arrays (FPGAs), or application specific integrated circuits (ASIC) designs in a way useful for system-level design decisions. And we are again back at our original question about what exactly the purpose of a computational model is and how general or how specific it should be.

In principle, there are two alternatives to integrate nonfunctional properties such as power, reliability, and also time in a model of computation.

- First, we can include these properties in the computational model and associate every functional operation with a specific quantity of that property. For example, an add operation takes 2 ns and consumes 60 pW. During simulation or some other analysis we can calculate the overall delay and power consumption.

- Second, we can allocate abstract budgets for all parts of the design. For instance, in synchronous design styles, we divide the time axis in slots or cycles and assign every part of the design to exactly one slot. Later on during implementation, we have to find the physical time duration of each slot, which determines the clock frequency. We can optimize for high clock frequency by identifying the critical path and optimizing that design part aggressively. Alternatively, we can move some of the functionality from the slot with the critical part to a neighboring slot, thus balancing the different slots.
  This budget approach can also be used for managing power consumption, noise, and other properties.

The first approach suffers from inefficient modeling and simulation when all implementation details are included in a model. Also, it cannot be applied to abstract models since these implementation details are not available there. Recall that a main idea of computational models is that they should be abstract and general enough to support analysis of a large variety of architectures. The inclusion of detailed timing and power consumption data would obstruct this objective. Even the approach to start out with an abstract model and later on back-annotate the detailed data from realistic architectural or implementation models does not help, because the abstract model does not allow to draw concrete conclusions and the detailed, back-annotated model is valid only for a specific architecture.

The second approach with abstract budgets is slightly more appealing to us. On the assumption that all implementations will be able to meet the budgeted constraints, we can draw general conclusions about performance or power consumption on an abstract level valid for a large number of different architectures. One drawback is that we do not know exactly for which class of architectures our analysis is valid, since it is hard to predict which implementations will at the end be able to meet the budget constraints. Another complication is that we do not know the exact physical size of these budgets and it may indeed be different for different architectures and implementations. For instance an ASIC implementation of a given architecture may be able to meet a cycle constraint of 1 ns and run at 1 GHz clock frequency, while an FPGA implementation of exactly the same algorithms requires a cycle budget of 5 ns. But still, the abstract budget approach is promising because it divides the overall problem into more manageable pieces. At the higher level we make assumptions about abstract budgets and analyze a system based on these assumptions. Our analysis will then be valid for all architectures and implementations that meet the stated assumptions. At the lower level we have to ensure and verify that these assumptions are indeed met.

3.1.3 Heterogeneity

Another salient feature of many embedded systems is heterogeneity. It comes from various environmental constraints on the interfaces, from heterogeneous applications, and from the need to find different trade-offs between performance, cost, power consumption, and flexibility for different parts of the system. Consequently, we see analog and mixed signal parts, digital signal processing parts, image and video processing parts, control parts, and user interfaces coexist in the same system or even on the same VLSI device. We also see irregular architectures with microprocessors, DSPs, VLIWs, custom hardware coprocessors, memories, and FPGAs connected via a number of different segmented and hierarchical interconnection schemes. It is a formidable task to develop a uniform model of computation that exposes all relevant properties while nicely suppressing irrelevant details.

Heterogeneous models of computation are one way to address heterogeneity at the application, architecture, and implementation level. Different computational models are connected and integrated into a hierarchical, heterogeneous model of computation that represents the entire system. Many different approaches have been taken to either connect two different computational models or provide a general framework to integrate a number of different models. It turns out that issues of communication, synchronization, and time representation pose the most formidable challenges. The reason is that the communication, and in particular the synchronization semantics between different MoC domains, correlates the time representation between the two domains. As we will see below, connecting a timed model of computation with an untimed model leads to the import of a time structure from the timed to the untimed model resulting in a heterogeneous, timed model of computation. Thus the integration cannot stop superficially at the interfaces leaving the interior of the two computational domains unaffected.

Due to the inherent heterogeneity of embedded systems, different models of computation will continue to be used and thus different MoC domains will coexist within the same system. There are two main possible relations; one is due to refinement and the other due to partitioning. One more abstract model of computation can be refined into a more detailed model. In our framework, time is the natural parameter that determines the abstraction level of a model. The untimed MoC is more abstract than the synchronous MoC, which in turn is more abstract than the timed MoC. It is in fact common practice that a signal processing algorithm is first modeled as an untimed data flow algorithm, which is then refined into a synchronous circuit description, which in turn is mapped onto a technology-dependent netlist of fully timed gates.

However, this is not a natural flow for all applications. Control-dominated systems or subsystems require some notion of time already at the system level and sensor and actuator subsystems may require a continuous time (CT) model right from the start. Thus, different subsystems should be modeled with different MoCs.

3.1.4   Component Interaction

A troubling issue in complex, heterogeneous systems is an unexpected behavior of the system due to subtle and complex ways of interaction of different MoCs parts. Eker et al. [EJL$^+$03] call this phenomenon emergent behavior. Some examples illustrate this important point:

Priority inversion: Threads in a real-time operating system may use two different mechanism of resource allocation [EJL$^+$03]. One is based on priority and preemption to schedule the threads. The second is based on monitors. Both are well defined and predictable in isolation. For instance, priority- and preemption-based scheduling means that a higher priority thread cannot be blocked by a lower priority thread. However, if the two threads also use a monitor lock, the lower priority thread may block the high priority thread via the monitor for an indefinite amount of time.

Performance inversion: Assume there are four CPUs on a bus. CPU_1 sends data to CPU_2; CPU_3 sends data to CPU_4 over the bus [Ern03]. We would expect that the overall system performance improves when we replace one CPU with a faster processor, or at least that the system performance does not decrease. However, replacing CPU_1 with a faster CPU_1' may mean that the data is sent from CPU_1' to CPU_2 with a higher frequency, at least for a limited amount of time. This means that the bus is more loaded by this traffic, which may slow down the communication from CPU_3 to CPU_4. If this communication performance has a direct influence on the system performance, we will see a decreased overall system performance.

Over synchronization: Assume that the upper and lower branches in Figure 3.1 have no mutual functional dependence as the data flow arrows indicate. Assume further that process B is blocked when

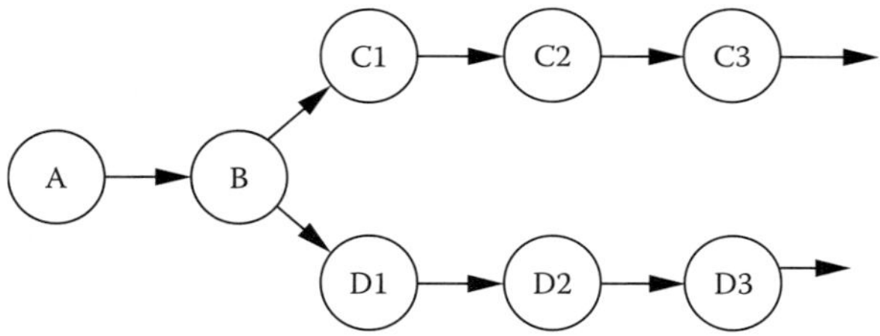

FIGURE 3.1 Over synchronization between functionally independent subsystems.

it tries to send data to C1 or D1, but the receiver is not ready to accept the data. Then, a delay or deadlock in branch D will propagate back through process B to both A and the entire C branch.

These examples are not limited to situations when different MoCs interact. They show that, when separate, seemingly unrelated subsystems interact via a nonobvious mechanism, which is often a shared resource, the effects can be hard to analyze. When the different subsystems are modeled in different MoCs, the problem is even more pronounced and harder to analyze due to different communication semantics, synchronization mechanisms, and time representation.

3.1.5 Time

The treatment of time will serve for us as the most important dimension to distinguish MoCs. We can identify at least four levels of accuracy, which are continuous time, discrete time, clocked time, and causality. In the sequel, we only cover the last three levels.

When time is not modeled explicitly, events are only partially ordered with respect to their causal dependences. In one approach, taken for instance in deterministic data flow networks [Kah74,LP95], the system behavior is independent of delays and timing behavior of computation elements and communication channels. These models are robust with respect to time variations in that any implementation, no matter how slow or fast it is, will exhibit the same behavior as the model. Alternatively, different delays may affect the system's behavior and we obtain an inherently nondeterministic model since time behavior, which is not modeled explicitly, is allowed to influence the observable behavior. This approach has been taken both in the context of data flow models [Bro83,BA81,Kos78,Par83] and process algebras [Mil89,Hoa78]. In this chapter we follow the deterministic approach, which however can be generalized to approximate nondeterministic behavior by means of stochastic processes as shown in Jantsch et al. [JSW01].

To exploit the very regular timing of some applications, the synchronous data flow (SDF) [LM87a] has been developed. Every process consumes and emits a statically fixed number of events in each evaluation cycle. The evaluation cycle is the reference time. The regularity of the application is translated into a restriction of the model which in turn allows efficient analysis and synthesis techniques that are not applicable for more general models. Scheduling buffer size optimization and synthesis has been successfully developed for the SDF.

One facet related to the representation of time is the dichotomy between data flow-dominated and control flow-dominated applications. Data flow-dominated applications tend to have events that occur in very regular intervals. Thus, explicit representation of time is not necessary and in fact often inefficient. In contrast, control-dominated applications deal with events occurring at very irregular time instants. Consequently, explicit representation of time is a necessity because the timing of events cannot be inferred. Difficulties arise in systems which contain both elements. Unfortunately, these kinds of systems become more common since the average system complexity steadily increases. As a consequence, several attempts to integrate data flow and control-dominated modeling concepts have emerged.

In the synchronous piggybacked data flow model [PJH02], control events are transported on data flow streams to represent a global state without breaking the locality principal of data flow models.

The composite signal flow [JB00] distinguishes between control and data flow processes and puts significant effort to maintain the frame-oriented processing, which is so common in data flow and signal processing applications for efficiency reasons. However, conflicts occur when irregular control events must be synchronized with data flow events inside frames. The composite signal flow addresses this problem by allowing an approximation of the synchronization and defines conditions when approximations are safe and do not lead to erroneous behavior.

Time is divided into time slots or clock cycles by various synchronous models. According to the perfect synchrony assumption [Hal00,BB91] neither communication nor computation takes any noticeable time and the time slots or evaluation cycles are completely determined by the arrival of input events. This assumption is useful because designer and tools can concentrate solely on the functionality of the system without mixing this activity with timing considerations. Optimization of performance can be done in a separate step by means of static timing analysis and local retiming techniques. Even though timing does not appear explicitly in synchronous models, the behavior is not independent of time. The model constrains all implementations such that they must be fast enough to process input events properly and to complete an evaluation cycle before the next events arrive. When no events occur in an evaluation cycle, a special token called "absent event" is used to communicate the advance of time. In our framework we use the same technique in Sections 3.3.4 and 3.3.5 for both the synchronous MoC and the fully timed MoC.

Discrete timed models use a discrete set, usually integers or natural numbers, to assign a time stamp to each event. Many discrete event models fall into this category [Sev01,LK00,Cas93] as well as most popular hardware description languages such as VHDL and Verilog. Timing behavior can be modeled most accurately, which makes it the most general model we consider here and makes it applicable to problems such as detailed performance simulation where synchronous and untimed models cannot be used. The price for this is the intimate dependence of functional behavior on timing details and significantly higher computation costs for analysis, simulation, and synthesis problems. Discrete timed models may be nondeterministic, as are mainly used in performance analysis and simulation (see, e.g., [Cas93]), or deterministic, as are more desirable for hardware description languages such as VHDL.

The integration of these different timing models into a single framework is a difficult task. Many attempts have been made on a practical level keeping a concrete design task, mostly simulation, in mind [BJ01,EKJ$^+$98,MVH$^+$98,JO95,LM93]. On a conceptual level Lee and Sangiovanni-Vincentelli [LSV98] have proposed a tagged time model in which every event is assigned a time tag. Depending on the tag domain we obtain different models of computation. If the tag domain is a partially ordered set, it results in an untimed model according to our definition. Discrete, totally ordered sets lead to timed MoCs and continuous sets result in CT MoCs. There are two main differences between the tagged time model and our proposed framework. First, in the tagged time model processes do not know how much time has progressed when no events are received since global time is only communicated via the time stamps of ordinary events. For instance, a process cannot trigger a time-out if it has not received events for a particular amount of time. Our timed model in Section 3.3.5 does not use time tags but absent events to globally order events. Since absent events are communicated between processes whenever no other event occurs, processes are always informed about the advance of global time. We chose this approach because it resembles better the situation in design languages such as VHDL, C, or SDL where processes always can experience time-outs. Second, one of our main motivations was the separation of communication and synchronization issues from the computation part of processes. Hence, we strictly distinguish between process interfaces and process functionality. Only the interfaces determine to which MoC a process belongs while the core functionality is independent of the MoC. This feature is absent from the tagged token model. This separation of concerns has been inspired by the concept of firing cycles in data flow process networks [Lee97].

Our mechanism for consuming and emitting events based on signal partitionings as described in Sections 3.3.2 and 3.3.3.1 is only slightly more general than the firing rules described by Lee [Lee97] but it allows a useful definition of process signatures based on the way processes consume and emit events.

3.1.6 Purpose of a Model of Computation

As mentioned several times, the purpose of a computational model determines how it is designed, what properties it exposes, and what properties it suppresses.

We argue that models of computation for embedded systems should not address principle questions of computability or feasibility but should rather aid the design and validation of concrete systems. How this is accomplished best remains a subject of debate, but for this chapter we assume a model of computation should support the following properties:

Implementation independence: An abstract model should not expose too much details of a possible implementation, e.g., which kind of processor used, how much parallel resources available, what kind of hardware implementation technology used, details of the memory architecture, etc. Since a model of computation is a machine abstraction, it should by definition avoid unnecessary machine details. Practically speaking, the benefits of an abstract model include that analysis and processing are faster and more efficient, that analysis results are relevant for a larger set of implementations, and that the same abstract model can be directed to different architectures and implementations. On the downside, we note diminished analysis accuracy and a lack of knowledge of the target architecture that can be exploited for modeling and design. Hence, the right abstraction level is a fine line that is also changing over time. While many embedded system designer could long safely assume a purely sequential implementation, current and future computational models should avoid such an assumption. Resource sharing and scheduling strategies become more complex, and a model of computation should thus either allow the explicit modeling of such a strategy or restrict the implementations to follow a particular, well-defined strategy.

Composability: Since many parts and components are typically developed independently and integrated into a system, it is important to avoid unexpected interferences. Thus some kind of composability property [JT03] is desirable. One step in this direction is to have a deterministic computational model, such as Kahn process networks, that guarantees a particular behavior independent of the time individual activities and the amount of available resources in general.

This is of course only a first step since, as argued above, time behavior is often an integral part of the functional behavior. Thus, resource sharing strategies that greatly influence timing will still have a major impact on the system behavior even for fully deterministic models. We can reconcile good system composability with shared resources by allocating a minimum but guaranteed amount of resources for each subsystem or task. For instance, two tasks get a fixed share of the communication bandwidth of a bus. This approach allows for ideal composability but has to be based on worst-case behavior. It is very conservative and hence does not utilize resources efficiently.

We can relax this approach by allocating abstract resource budgets as part of the computational model. Then we require from the implementation to provide the requested resources and at the same time to minimize the abstract budgets and thus the required resources. For example consider two tasks that have a particular communication need per abstract time slot, where the communication need may be different for different slots. The implementation has to fulfill the communication requirements of all tasks by providing the necessary bandwidth in each time slot, tuning the length of the individual time slots, or by moving communication from one slot to another. These optimizations will have to consider also global timing and resource constraints. In any case, in the abstract model we can deal with abstract budgets and assume they will be provided by any valid implementation.

Analyzability: A general trade-off exists between the expressiveness of a model and its analyzability. By restricting models in clever ways, one can apply powerful and efficient analysis and synthesis methods. For instance, the SDF model allows all actors only a constant amount of input and output tokens in each activation cycle. While this restricts the expressiveness of the model, it allows to efficiently compute static schedules when they exist. For general data flow graphs this may not be possible because it could be impossible to ensure that the amount of input and output is always constant for all actors, even if they are in a particular case. Since SDF covers a fairly large and important application domain, it has become a very useful model of computation. The key is to understand the important properties (finding static schedules, finding memory bounds, finding maximum delays, etc.) and to devise a model of computation that allows to handle these properties efficiently but does not restrict the modeling power too much.

In the following sections we will discuss a framework to study different models of computation. The idea is to use different types of process constructors to instantiate processes of different MoCs. Thus, one type of process constructors would yield only untimed processes, while another type results in timed processes. The elements for process construction are simple functions and are in principle independent of a particular MoC. However, the independence is not complete since some MoCs put specific constraints on the functions. But still the separation of the process interfaces from the internal process behavior is fairly far reaching. The interfaces determine the time representation, synchronization, and communication, hence the MoC.

In this chapter we will not elaborate all interesting and desirable properties of computational models. Rather we will use the framework to introduce four different MoCs, which only differ in their timing abstraction. Since time plays a very prominent role in embedded systems, we focus on this aspect and show how different time abstractions can serve different purposes and needs. Another defining aspect of embedded systems is heterogeneity, which we address by allowing different MoCs to coexists in a model. The common framework makes this integration semantically clean and simple. We study two particular aspects of this coexistence, namely, the interfaces between two different MoCs and the refinement of one MoC into another.

Other central issues of embedded systems such as power consumption, global analysis, and optimization are not covered, mostly because they are not very well understood in this context and few advanced proposals exist on how to deal with them from a MoC perspective.

3.2 Models of Computation

We systematically review different models of computation by organizing them according to their time abstraction [Jan03,Jan05]. We distinguish between untimed models, synchronous time, discrete time, and continuous time. This is consistent with the tagged-signal model proposed by Lee and Sangiovanni-Vincentelli [LSV98]. There each event has a time tag and different time tag structures result in different MoCs. For example, if the time tags correspond to real numbers we have a CT model; integer time tags result in discrete time models; time tags drawn from a partially ordered set result in an untimed MoC.

MoCs can be organized along other criteria, e.g., along the kinds of elements manipulated in a MoC, which leads Paul and Thomas [PT04] to a grouping of MoCs for hardware artifacts, MoCs for software artifacts, and MoCs for design artifacts. However, an organization along properties that are not inherent properties of MoCs is of limited use because it changes when MoCs are used in different ways.

A drawback of an organization along the time abstraction is that all strictly sequential models such as finite state machines and sequential algorithms all fall into the same class of MoCs, where the representation of time is irrelevant. However, this is of minor concern to us, since we focus on parallel MoCs.

3.2.1 Continuous Time Models

When time is represented by a continuous set, usually the real numbers, we talk of a CT MoC. Prominent examples of CT MoC instances are Simulink [DH98], VHDL–AMS, and Modelica [EMO98]. The behavior is typically expressed as equations over real numbers. Simulators for CT MoCs are based on differential equation solver that compute the behavior of a model including arbitrary internal feedback loops.

Due to the need to solve differential equations, simulations of CT models are very slow. Hence, only small parts of a system are usually modeled with continuous time such as analog and mixed signal components.

To be able to model and analyze a complete system that contains analog components, mixed-signal languages and simulators such as VHDL–AMS have been developed. They allow to model the pure digital parts in a discrete time MoC and the analog parts in a CT MoC. This allows for complete system simulations with acceptable simulation performance. It is also a typical example where heterogeneous models based on multiple MoCs have a clear benefit.

SystemC–AMS [VGE05,VGE03] avoids differential equations by addressing a restricted set of CT models. By assuming that all CT signals at the system's inputs are sampled periodically, the system is modeled as an SDF graph [LM87a] which in fact is a special case of an untimed MoC. The real-time property is maintained only implicitly by associating the evaluation cycle of the system model with a sampling period. Thus, the simulation is conducted with an essentially untimed model and the real time is tracked by counting evaluation cycles. This approach has the benefit of very fast simulation, as fast as with an untimed model, but is restricted to a set of CT systems that can be expressed as a regular computation over periodically sampled and digitized input signals. Consequently, SystemC–AMS can be used for abstract, system-level simulation while detailed, component-level modeling and simulation is best done with VHDL–AMS or full-fledged CT simulators.

3.2.2 Discrete Time Models

Models, where all events are associated with a time instant and the time is represented by a discrete set, such as the integer or natural numbers, are called discrete time models. Sometimes this group of MoCs is denoted as discrete event MoC. Strictly speaking "discrete event" and "discrete time" are independent, orthogonal concepts. In discrete event models the values of events are drawn from a discrete set. All four combinations occur in practice: continuous time/continuous event models, continuous time/discrete event models, discrete time/continuous event models, and discrete time/discrete event models. See for instance Cassandras [Cas93] for a good coverage of discrete event models.

Discrete time models are often used for performance analysis or the simulation of hardware. Both VHDL [APT02], Verilog [TM02], and SystemC [GLMS02] use a discrete time model for their simulation semantics. A simulator for discrete time MoCs is usually implemented with a global event queue that sorts all occurring events. Discrete time models may suffer from nondeterminism and from causality problems.

The causality problem due to zero-delay feedback is illustrated in Figure 3.2a. If the NAND gate is modeled as a zero-delay component, a logic True on the upper input leads to inconsistency at the output. If the second input is True, the output has to be False; if the second input is False, the output has to be True.

Zero-delay components also lead to nondeterminism as illustrated in Figure 3.2b. If block B is a zero-delay component and block A emits events e_1 and e_2 simultaneously, it is not obvious in which sequence component C experiences its input events e_2 and e_3. Which block should be evaluated, after A has been evaluated and events e_1 and e_2 have been generated? If B is evaluated first, block C would see both events e_2 and e_3 simultaneously and would consume both during its next evaluation. On the other hand, if C is evaluated first, it would only see and use event e_2. Only after B has been evaluated,

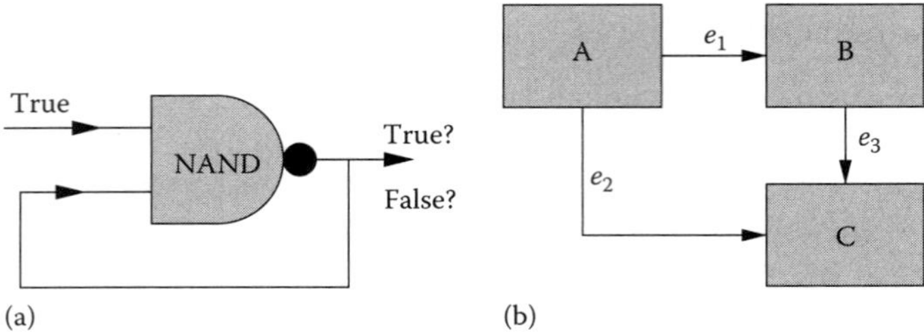

FIGURE 3.2 Nondeterminism due to zero-delay components: (a) a zero-delay feedback loop results in inconsistencies and (b) discrete time models may be nondeterministic.

block C would observe event e_3 and consume it in a second evaluation round. In general, there is no natural evaluation sequence since we do not know what blocks B and C will emit. Hence, the model is nondeterministic depending on the evaluation sequence chosen.

To avoid these undesired features of discrete time models, VHDL, Verilog, and SystemC have been equipped with a δ-delay model that does not allow zero-delay components. Between two real-time instants there are potentially infinitely many δ-time instants as shown in Figure 3.3. Each component takes at least one δ to evaluate. Hence, the output event of a component occurs never at the same time as the input events but at least one δ later. Based on this model, the evaluation sequence of components is determined by the time stamps of their input events, where a time stamp consists of the real time and the δ time part. In Figure 3.2b the component C would first see event e_2 and then, one δ later, it would see event e_3 and evaluate a second time. This will be the case no matter if B or C is evaluated first. The two evaluation orders A, B, C, C and A, C, B, C will lead to the same system behavior. Consequently, the system is deterministic.

In Figure 3.2a the δ delay of the NAND gate will lead to an infinite sequence of True, False, True, False, … output events, each separated by a δ. This is a perfectly consistent and deterministic model but may lead to infinite simulation loops because the next real-time instant $t+1$ is never reached even though the simulation continues to progress along the δ time axis.

The fact that discrete time MoCs offer a general technique to model arbitrary systems has made them very popular and widespread. They have become a universal tool for many engineering disciplines and they are employed to model almost everything from tiny integrated circuits to fleets of trucks and airplanes. Also, they are used for every possible design task such as design specification, functional analysis, verification, performance analysis, documentation, etc.

However, this generality of the time model also implies inefficiencies in cases that do not require the most general solution. In the design of embedded systems there are a number of situations and tasks that allow to exploit specific assumptions that make modeling, simulation, and analysis orders of magnitude faster than what is possible with discrete time models.

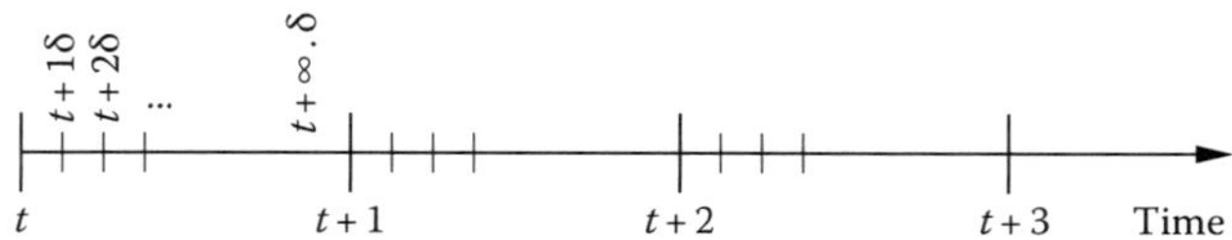

FIGURE 3.3 δ-time model.

3.2.3 Synchronous Models

Synchronous models were inspired by synchronous circuits, an example of which is shown in Figure 3.4. The main idea is that all computation is divided into clock cycles by separating combinational bocks from each other with registers. Registers propagate values from their inputs to their outputs only at either the rising or falling edge of the clock signal. The behavior of a combinational block is defined by the values of its outputs at the clock edge. All transient values that appear during the computational cycle are irrelevant and have no impact on the system behavior. Consequently, the combinational block can be represented as a set of Boolean equations or as a truth table, as indicated in Figure 3.5. By assuming that each combinational block computes its outputs within a clock period, all timing and delay issues are separated from the circuit behavior. Static timing analysis [Sap06] can verify that this assumption is indeed met in a specific implementation and retiming techniques [LS91] can be used to balance delays in neighboring combinational blocks for maximizing the clock frequency. This is a very convenient abstraction for synthesis, verification, and simulation. A simulator

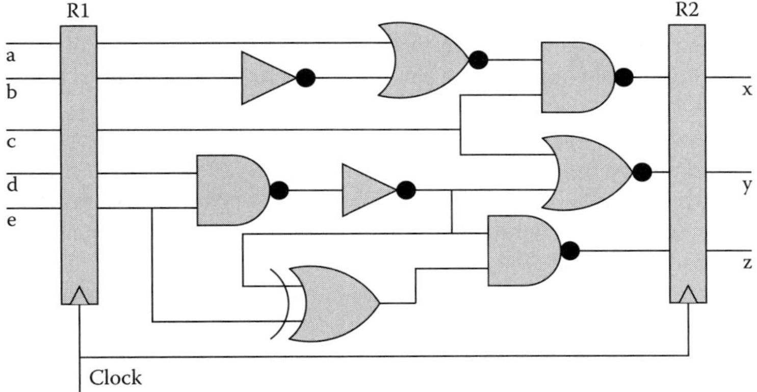

FIGURE 3.4 Blocks of combinational logic are separated by clocked registers in a synchronous circuit design.

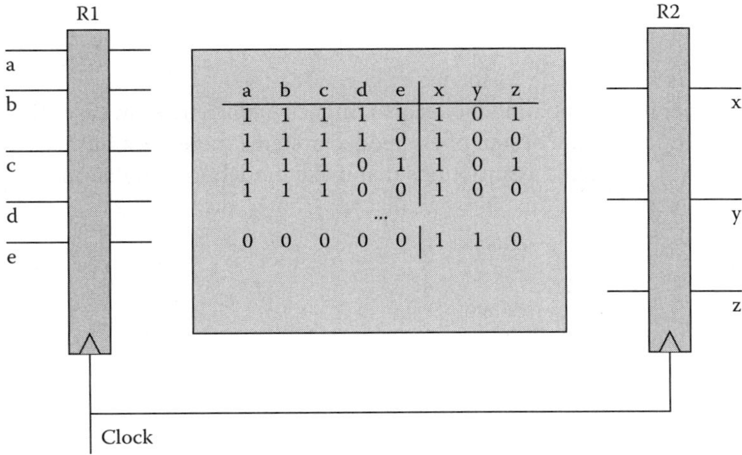

FIGURE 3.5 Combinational logic can be models as truth tables or Boolean functions.

based on the synchronous model can run an order of magnitude faster than a discrete time-based simulator because far fewer internal events are generated, sorted, and processed. In hardware design, almost all synthesis and formal verification tools use the synchronous model as a basis. Even when discrete time-based languages such as VHDL or Verilog are used as input, the tools interpret the circuit descriptions according to the clocked synchronous model and ignore all specific timing information that may be part of the input. This may potentially lead to synthesized hardware that behaves differently from the synthesis input model. Thus, all synthesis tools require that all input descriptions comply to modeling guidelines, which ensures that an interpretation of the model according to the synchronous MoC is reasonable.

While the synchronous MoC is almost universally accepted in hardware design, it is confined to niche domains in software development. Although not mainstream, several successful programming languages have been developed based on the "perfect synchrony assumption," which states that neither communication nor computation takes any noticeable time [Hal00,BB91]. This means that a system reacts immediately and instanteously to inputs from the environment. Hence, the timing is completely determined by the environment. Again, this implies that all timing details of a system are irrelevant to the behavior under the assumption that the system reacts sufficiently fast to inputs from the environment. "Sufficiently fast" means that the system has completed its reaction to a set of input events before the next set of input events appear. Figure 3.6 shows the resulting execution cycle of a synchronous program. Every implementation that is sufficiently fast exhibits the same behavior as the ideal program.

Synchronous languages are used successfully in safety critical, real-time embedded systems such as aerospace, naval, and rail control software [est]. Esterel [PBEB07], the best-known synchronous language, is an imperative language that is suitable for the design of control systems. An Esterel program consists of a set concurrent threads that communicate with each other by means of signals. Esterel offers a large number of statements for handling time-outs, exceptions, and preemptions. Esterel has a thoroughly design formal semantics [Ber98,Tar07] that facilitates formal analysis, verification, and synthesis. Other synchronous languages, such as Signal [lGGlBlM91] and Lustre [HLR92], target data flow and signal processing applications.

Even though the synchrony assumption seems to suggest a tight coupling of the different parts of the system in terms of their timing behavior, several techniques for implementing a synchronous system onto a distributed architecture [CGP99,Sch99,LSSJ07]. Intuitively this is natural because the essence of the synchronous MoC is an intermediate timing abstraction that allows for modeling of time in a precise and formal way without the need to deal with physical time in all its hairy details.

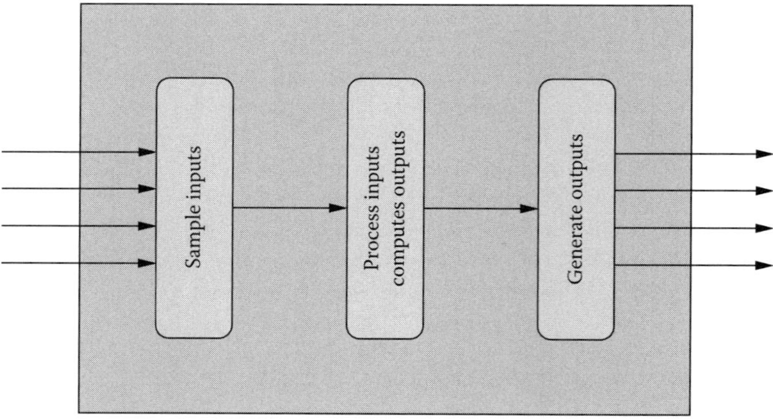

FIGURE 3.6   Execution cycle of a synchronous program.

3.2.4   Untimed Models

In untimed MoCs no timing or delay information is included and the order of events and activities is solely determined by the order of input data arrival and data dependencies. Consequently, untimed models are well suited to focus on ideal algorithmic behavior. They are widely used in signal processing heavy data processing domains to develop and study abstract behavior and algorithms.

3.2.4.1   Data Flow Process Networks

Data flow process networks [LP95] are a special case of Kahn process networks [Kah74]. In a Kahn process, network processes communicate with each other via unbounded FIFO channels. Writing to these channels is "non-blocking," i.e., it always succeed and does not stall the process, while reading from these channels is *blocking*, i.e., a process that reads from an empty channel will stall and can only continue when the channel contains sufficient data items (*tokens*). Processes in a Kahn process network are "monotonic," which means that they only need partial information of the input stream to produce partial information of the output stream. Monotonicity allows parallelism, since a process does not need the whole input signal to start the computation of output events. Processes are not allowed to test an input channel for existence of tokens without consuming them. In a Kahn process network there is a total order of events inside a signal. However, there is no order relation between events in different signals. Thus Kahn process networks are only partially ordered, which classifies them as an "untimed model."

A data flow program is a directed graph consisting of nodes ("actors") that represent computation and arcs that represent ordered sequences of events as illustrated in Figure 3.7a. Data flow networks can be hierarchical since a node can represent a data flow graph.

The execution of a data flow process is a sequence of "firings or evaluations." For each firing tokens are consumed and produced. The number of tokens consumed and produced may vary for each firing and is defined in the "firing rules" of a data flow actor. Data flow process networks have been shown very valuable in digital signal processing applications. When implementing a data flow process network on a single processor, a sequence of firings, also called a "schedule," has to be found. For general data flow models it is undecidable whether such a schedule exists because it depends on the input data.

Synchronous data flow [LM87b,LM87a] puts further restrictions on the data flow model, since it requires that a process consumes and produces a fixed number of tokens for each firing. With this restriction it can be tested efficiently, if a finite static schedule exists. If one exists it can be effectively computed. Figure 3.7b shows an SDF process network. The numbers on the arcs show how many tokens are produced and consumed during each firing. A possible schedule for the given SDF network is {A,B,A,B,A,B,C} and it requires that there are at least three data samples buffered at each input of process A at the start of the schedule. SDF also allows to analyze buffer requirements in the system

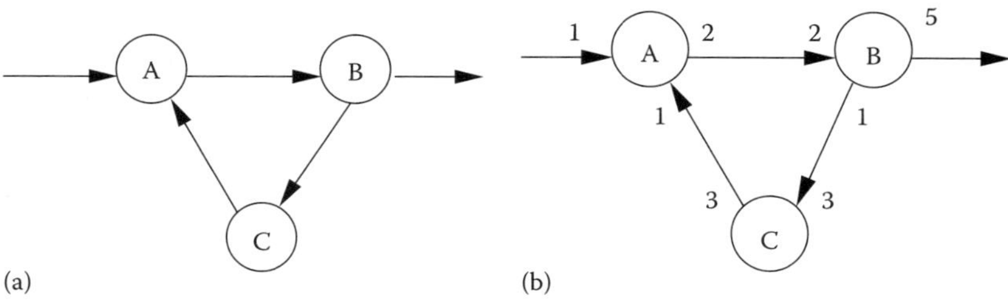

FIGURE 3.7   Data flow networks: (a) a data flow process network and (b) a synchronous data flow process network.

and efficient heuristics for buffer minimization have been developed [BML96]. There exists a variety of different data flow models; for an excellent overview see [LP95].

3.2.4.2 Rendezvous-Based Models

A rendezvous-based model consists of concurrent sequential processes. Processes communicate with each other only at synchronization points. In order to exchange information, processes must have reached this synchronization point; otherwise, they have to wait for each other. In the tagged signal model each sequential process has its own set of tags. Only at synchronization points processes share the same tag. Thus there is a partial order of events in this model. The process algebra community uses rendezvous-based models. The communicating sequential processes (CSP) model of Hoare [Hoa78] and the calculus of communicating systems (CCS) model of Milner [Mil89,Mil80] are prominent examples. The language Ada [BB94] has a communication mechanism based on rendezvous.

3.2.5 Heterogeneous Models of Computation

A lot of effort has been spent to mix different models of computation. This approach has the advantage that a suitable model of computation can be used for each part of the system. On the other hand, as the system model is based on several computational models, the semantics of the interaction of fundamentally different models has to be defined, which is no simple task. This even amplifies the validation problem, because the system model is not based on a single semantics. There is little hope that formal verification techniques can help and thus simulation remains the only means of validation. In addition, once a heterogeneous system model is specified, it is very difficult to optimize systems across different models of computation. In summary, while heterogeneous MoCs provide a very general, flexible, and useful simulation and modeling environment, cross-domain validation and optimization will remain elusive for many years for any heterogeneous modeling approach. In the following an overview of related work on mixed models of computation is given.

In *charts [GLL99] hierarchical finite state machines are embedded within a variety of concurrent models of computations. The idea is to decouple the concurrency model from the hierarchical FSM semantics. An advantage is that modular components, e.g., basic FSMs, can be designed separately and composed into a system with the model of computation that best fits the application domain. It is also possible to express a state in an FSM by a process network of a specific model of computation. *charts has been used to describe hierarchical FSMs that are composed using data flow, discrete event, and synchronous models of computations.

The composite data flow [JB00] integrates data and control flow. Vectors and the conversion from scalar values to vectors and vice versa are integral parts of the model. This allows to capture the timing effects of these conversions without resorting to a synchronous or timed MoC. Timing of processes is represented only to the level to determine if sufficient data are available to start a computation. In this way the effects of control and timing on data flow processing are considered at the highest possible abstraction level because they only appear as data dependency problems. The model has been implemented to combine Matlab and SDL into an integrated system specification environment [BJ01].

The most well-known heterogeneous modeling framework is Ptolemy [Lee03,EJL+03]. It allows to integrate a wide range of different MoCs by defining the interaction rules of different MoC domains. In Ptolemy each MoC domain has a "director" that governs the execution, e.g., scheduling, of all processes in that domain. Different MoC domains are connected via ports. Ports are responsible for the necessary data conversion between domains while the directors are responsible for exchanging timing information and maintaining a consistent, global notion of time.

3.3   MoC Framework

In the remainder of this chapter we discuss a framework that accommodates models of computation with different timing abstractions. It is based on "process constructors," which are a mechanism to instantiate processes. A process constructor takes one or more pure functions as arguments and creates a process. The functions represent the process behavior and have no notion of time or concurrency. They simply take arguments and produce results. The process constructor is responsible for establishing communication with other processes. It defines the time representation, communication, and synchronization semantics. A set of process constructors determines a particular model of computation. This leads to a systematic and clean separation of computation and communication. A function that defines the computation of a process can in principle be used to instantiate processes in different computational models. However, a computational model may put constraints on functions. For instance, the synchronous MoC requires a function to take exactly one event on each input and produce exactly one event for each output. The untimed MoC does not have a similar requirement.

After some preliminary definitions in this section we introduce the untimed processes, give a formal definition of a MoC, and define the untimed MoC (Section 3.3.3), the perfectly synchronous and the clocked synchronous MoC (Section 3.3.4) and the discrete time MoC (Section 3.3.5). Based on this we introduce interfaces between MoCs and present an interface refinement procedure in the next section. Furthermore, we discuss the refinement from an untimed MoC to a synchronous MoC and to a timed MoC.

3.3.1   Processes and Signals

Processes communicate with each other by writing to and reading from signals. Given is a set of values V, which represents the data communicated over the signals. "Events," which are the basic elements of signals, are or contain values. We distinguish between three different kinds of events.

"Untimed events" $\dot{E}$ are just values without further information, $\dot{E} = V$. "Synchronous events" $\bar{E}$ include a pseudo value $\perp$ in addition to the normal values, and hence $\bar{E} = V \cup \{\perp\}$. "Timed events" $\hat{E}$ are identical to synchronous events, $\hat{E} = \bar{E}$. However, since it is often useful to distinguish them, we use different symbols. Intuitively, timed events occur at much finer granularity than synchronous events and they would usually represent physical time units such as a nanosecond. By contrast, synchronous events represent abstract time slots or clock cycles. This model of events and time can only accommodate discrete time models. Continuous time would require a different representation of time and events. We use the symbols $\dot{e}$, $\bar{e}$, and $\hat{e}$ to denote individual untimed, synchronous, and timed events, respectively. We use $E = \dot{E} \cup \bar{E} \cup \hat{E}$ and $e \in E$ to denote any kind of event.

Signals are sequences of events. Sequences are ordered and we use subscripts as in e_i to denote the ith event in a signal. For example, a signal may be written as $\langle e_0, e_1, e_2 \rangle$. In general signals can be finite or infinite sequences of events and S is the set of all signals. We also distinguish between three kinds of signals and $\dot{S}$, $\bar{S}$, and $\hat{S}$ denote the untimed, synchronous, and timed signal sets, respectively, and $\dot{s}$, $\bar{s}$, and $\hat{s}$ designate individual untimed, synchronous, and timed signals, respectively.

$\langle \rangle$ is the empty signal and $\oplus$ concatenates two signals. Concatenation is associative and has the empty signal as its neutral element: $s_1 \oplus (s_2 \oplus s_3) = (s_1 \oplus s_2) \oplus s_3$, $\langle \rangle \oplus s = s \oplus \langle \rangle = s$. To keep the notation simple we often treat individual events as one-event sequences, e.g., we may write $e \oplus s$ to denote $\langle e \rangle \oplus s$.

We use angle brackets, "$\langle$" and "$\rangle$," to denote ordered sets or sequences of events, and also for sequences of signals if we impose an order on a set of signals.

$\#s$ gives the length of signal s. Infinite signals have infinite length and $\#\langle \rangle = 0$.

$[\,]$ is an index operation to extract an event on a particular position from a signal. For example $s[2] = e_2$ if $s = \langle e_1, e_2, e_3 \rangle$.

Processes are defined as functions on signals

$$p : S \to S.$$

"Processes" are functions in the sense that for a given input signal we always get the same output signal, i.e., $s = s' \Rightarrow p(s) = p(s')$. Note that this still allows processes to have an internal state. Thus, a process does not necessarily react identically to the same event applied at different times. But it will produce the same, possibly infinite, output signal when confronted with identical, possibly infinite, input signals provided it starts with the same initial state.

3.3.2 Signal Partitioning

We shall use the partitioning of signals into subsequences to define the portions of a signal that is consumed or emitted by a process in each evaluation cycle.

A "partition" $\pi(v, s)$ of a signal s defines an ordered set of signals, $\langle r_i \rangle$, which, when concatenated together, form "almost" the original signal s. The function $v : \mathbb{N}_0 \to \mathbb{N}_0$ defines the lengths of all elements in the partition. $v(0) = \#r_0$ gives the length of the first element in the partition; $v(1) = \#r_1$ gives the length of the second element, etc.

Example 3.1

Let $s_1 = \langle 1, 2, 3, 4, 5, 6, 7, 8, 9, 10 \rangle$ and $v_1(0) = v_1(1) = 3$, $v_1(2) = 4$. Then we get the partition $\pi(v_1, s_1) = \langle \langle 1, 2, 3 \rangle, \langle 4, 5, 6 \rangle, \langle 7, 8, 9, 10 \rangle \rangle$.
Let $s_2 = \langle 1, 2, 3, \ldots \rangle$ be the infinite signal with ascending integers. Let $v_2(i) = 2$ for all $i \geq 0$. The resulting partition is infinite: $\pi(v_2, s_2) = \langle \langle 1, 2 \rangle, \langle 3, 4 \rangle, \ldots \rangle$.

The function $v(i)$ defines the length of the subsignals r_i. If it is constant for all i we usually omit the argument and write v. Figure 3.8 illustrates a process with an input signal s and an output signal s'. s is partitioned into subsignals of length 3 and s' into subsignals of length 2.

3.3.3 Untimed Models of Computation

3.3.3.1 Process Constructors

Our aim is to relate functions of events to processes, which are functions of signals. Therefore we introduce process constructors that can be considered as higher order functions. They take functions on events as arguments and return processes. We define only a few basic process constructors that can be used to compose more complex processes and process networks. All untimed process constructors and processes operate exclusively on untimed signals.

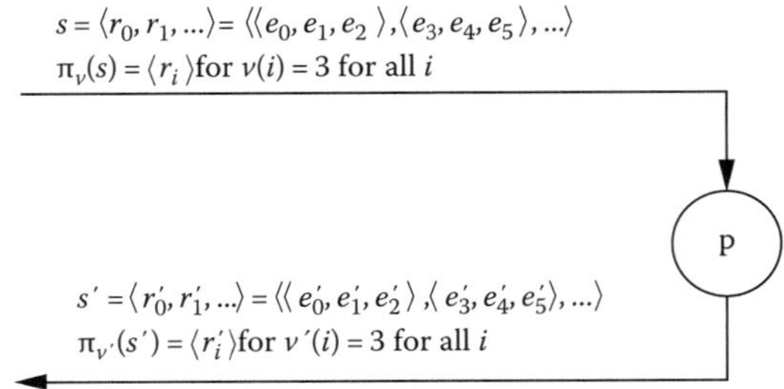

FIGURE 3.8 Input signal of process p is partitioned into an infinite sequence of subsignals, each of which contains three events, while the output signal is partitioned into subsignals of lengths 2.

Processes with arbitrary number of input and output signals are cumbersome to deal with in a formal way. To avoid this inconvenience we mostly deal with processes with one input and one output. To handle arbitrary processes, we introduce "zip" and "unzip" processes which merge two input signals into one and split one input signal into two output signals, respectively. These processes together with appropriate process composition allow us to express arbitrary behavior.

Processes instantiated with the `mealyU` constructor resemble Mealy state machines in that they have a next state function and an output encoding function that depend on both the input and the current state.

DEFINITION 3.1   Let V be an arbitrary set of values, let $g, f : (V \times \dot{S}) \to \dot{S}$ the next state and output encoding functions, let $\gamma : V \to \mathbb{N}$ be a function defining the input partitioning, and let $w_0 \in V$ be an initial state. `mealyU` is a process constructor which, given γ, f, g, and w_0 as arguments, instantiates a process $p : \dot{S} \to \dot{S}$. The function γ determines the number of events consumed by the process in the current evaluation cycle. γ is dependent on the current state. p repeatedly applies g to the current state and the input events to compute the next state. Further it repeatedly applies f to the current state and the input events to compute the output events.

Processes instantiated by `mealyU` are general state machines with one input and one output. To create processes with arbitrary inputs and outputs, we also use the following constructors:

- `zipU` instantiates a process with two inputs and one output. In every evaluation cycle this process takes one event from the left input and one event from the right input and packs them into an event pair that is emitted at the output.

- `unzipU` instantiates a process with one input and two outputs. In every evaluation cycle this process takes one event from the input. It requires it to be an event pair. The first event of this pair is emitted to the left output; the second event of the event pair is emitted to the right output.

For truly general process networks we would in fact need more complex zip processes, but for the purpose of this chapter the simple constructors are sufficient and we refer the reader for details to Jantsch [Jan03].

3.3.3.2 Composition Operators

We consider only three basic composition operators, namely, sequential composition, parallel composition, and feedback.

We give the definitions only for processes with one or two input and output signals, because the generalization to arbitrary numbers of inputs and outputs is straightforward.

DEFINITION 3.2   Let $p_1, p_2 : \dot{S} \to \dot{S}$ be two processes with one input and one output each, and let $s_1, s_2 \in \dot{S}$ be two signals. Their parallel composition, denoted as $p_1 \parallel p_2$, is defined as follows.

$$(p_1 \parallel p_2)(\langle s_1, s_2 \rangle) = \langle p_1(s_1), p_2(s_2) \rangle.$$

Since processes are functions we can easily define sequential composition in terms of functional composition.

DEFINITION 3.3   Let again $p_1, p_2 : \dot{S} \to \dot{S}$ be two processes and let $s \in \dot{S}$ be a signal. The sequential composition, denoted as $p_1 \circ p_2$, is defined as follows.

$$(p_2 \circ p_1)(s) = p_2(p_1(s)).$$

DEFINITION 3.4   Given a process $p : (S \times S) \rightarrow (S \times S)$ with two input signals and two output signals we define the process $\mu p : S \rightarrow S$ by the equation

$$(\mu p)(s_1) = s_2 \text{ where } p(s_1, s_3) = (s_2, s_3).$$

The behavior of the process μp is defined by the least fixed point semantics based on the prefix order of signals.

The μ operator gives feedback loops a well-defined semantics. Moreover, the value of the feedback signal can be constructed by repeatedly simulating the process network starting with the empty signal until the values on all feedback signals stabilize and do not change any more (see Figure 3.9) [Jan03].
   Now we are in a position to define precisely what we mean with a model of computation.

DEFINITION 3.5   A *Model of Computation (MoC)* is a 2-tuple MoC= (C, O), where C is a set of process constructors, each of which, when given constructor specific parameters, instantiates a process. O is a set of process composition operators, each of which, when given processes as arguments, instantiates a new process.

DEFINITION 3.6   The *Untimed Model of Computation (Untimed MoC)* is defined as Untimed MoC=(C, O), where

$$C = \{\texttt{mealyU}, \texttt{zipU}, \texttt{unzipU}\}$$
$$O = \{\|, \circ, \mu\}$$

In other words, a process or a process network belongs to the *Untimed MoC Domain* iff all its processes and process compositions are constructed either by one of the named process constructors or by one of the composition operators. We call such processes *U-MoC processes*.

Because the process interface is separated from the functionality of the process, interesting transformations can be done. For instance, a process can be mechanically transformed into a process that consumes and produces a multiple number of events of the original process. Processes can be easily merged into more complex processes. Moreover, there may be the opportunity to move functionality from one process to another. For more details on this kind of transformations see [Jan03].

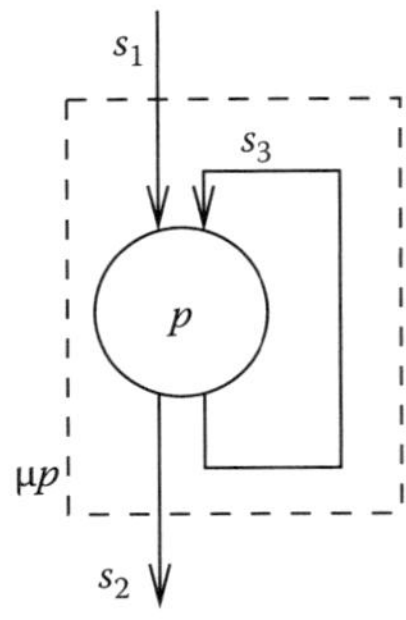

FIGURE 3.9   Feedback composition of a process.

3.3.4 Synchronous Model of Computation

The synchronous languages, i.e., StateCharts [Har87], Esterel [BCG88], Signal [lGGlBlM91], Argos, Lustre [HCRP91], and some others, have been developed on the basis of the perfect synchrony assumption.

Perfect synchrony hypothesis: Neither computation nor communication takes time.

Timing is entirely determined by the arriving of input events because the system processes input samples in zero time and then waits until the next input arrives. If the implementation of the system is fast enough to process all input before the next sample arrives, it will behave exactly as the specification in the synchronous language.

3.3.4.1 Process Constructors

Formally, we develop synchronous processes as a special case of untimed processes. This will later allow us to easily connect different domains.

Synchronous processes have two specific characteristics. First, all synchronous processes consume and produce exactly one event on each input or output in each evaluation cycle, i.e., the signature is always $\langle \{1, \ldots\}, \{1, \ldots\} \rangle$. Second, In addition to the value set V events can carry the special value $\bot$, which denotes the absence of an event; this is the way we defined synchronous events, $\bar{E}$, and signals, $\bar{S}$, in Section 3.3.1. Both the processes and their contained functions must be able to deal with these events.

All synchronous process constructors and processes operate exclusively on synchronous signals.

DEFINITION 3.7   Let V be an arbitrary set of values, $\bar{E} = V \cup \{\bot\}$, let $g, f : (\bar{E} \times \bar{S}) \rightarrow \bar{S}$, and let $w_0 \in V$ be an initial state. $mealyS$ is a process constructor, which, given f, g, and w_0 as arguments, instantiates a process $p : \bar{S} \rightarrow \bar{S}$. p repeatedly applies g on the current state and the input event to compute the next state. Further it repeatedly applies f on the current state and the input event to compute the output event. p consumes exactly one input event in each evaluation cycle and emits exactly one output event.

We only require that g and f are defined for absent input events and that the output signal partitioning is constant 1.

When we merge two signals into one we have to decide how to represent the absence of an event in one input signal in the compound signal. We choose to use the $\bot$ symbol for this purpose as well, which has the consequence that $\bot$ also appears in tuples together with normal values. Thus, it is essentially used for two different purposes. Having clarified this, $zipS$ and $unzipS$ can be defined straightforward. $zipS$- based processes pack two events from the two inputs into an event pair at the output, while $unzipS$ performs the inverse operation.

3.3.4.2 Perfectly Synchronous Model of Computation

Again, we can now make precise what we mean by synchronous model of computation.

DEFINITION 3.8   The *Synchronous Model of Computation (Synchronous MoC)* is defined as Synchronous MoC$=(C, O)$, where

$$C = \{mealyS, zipS, unzipS\}$$
$$O = \{\|, \circ, \mu_S\}$$

In other words, a process or a process network belongs to the *Synchronous MoC Domain* iff all its processes and process compositions are constructed either by one of the named process constructors or by one of the composition operators. We call such processes *S-MoC processes*.

Note that we do not use the same feedback operator for the synchronous MoC. μ_S defines the semantics of the feedback loop based on the Scott order of the values in $\bar{E}$. It is also based on a fixed point semantics but it is resolved "for each event" and not over a complete signal. We have adopted μ_S to be consistent with the zero-delay feedback loop semantics of most synchronous languages. For our purpose here this is not significant and we do not need to go into more details. For precise definitions and a thorough motivation the reader is referred to [Jan03]

Merging of processes and other related transformations are very simple in the synchronous MoC because all processes have essentially identical interfaces. For instance, the merge of two `mealyS`-based processes can be formulated as follows.

$$mealyS(g_1, f_1, v_0) \circ mealyS(g_2, f_2, w_0) = mealyS(g, f, (v_0, w_0))$$

where
$$g((v, w), \bar{e}) = (g_1(v, f_2(w, \bar{e})), g_2(w, \bar{e}))$$
$$f((v, w), \bar{e}) = f_1(v, f_2(w, \bar{e}))$$

3.3.4.3 Clocked Synchronous Model of Computation

It is useful to define a variant of the perfectly synchronous MoC, the clocked synchronous MoC which is based on the following hypothesis.

Clocked Synchronous Hypothesis: There is a global clock signal controlling the start of each computation in the system. Communication takes no time and computation takes one clock cycle.

First, we define a delay process Δ, which delays all inputs by one evaluation cycle.

$$\Delta = mealyS(f, g, \perp)$$

where
$$g(w, \bar{e}) = \bar{e}$$
$$f(w, \bar{e}) = w$$

Based on this delay process we define the constructors for the clocked synchronous model.

DEFINITION 3.9

$$mealyCS(g, f, w_0) = mealyS(g, f, w_0) \circ \Delta$$
$$zipCS()(\bar{s}_1, \bar{s}_2) = zipS()(\Delta(\bar{s}_1), \Delta(\bar{s}_2)) \tag{3.1}$$
$$unzipCS() = unzipS() \circ \Delta$$

Thus, elementary processes are composed of a combinatorial function and a delay function, which essentially represents a latch at the inputs.

DEFINITION 3.10   The *Clocked Synchronous Model of Computation (Clocked Synchronous MoC)* is defined as Clocked Synchronous MoC=(C, O), where

$$C = \{mealyCS, zipCS, unzipCS\}$$
$$O = \{\|, \circ, \mu\}$$

In other words, a process or a process network belongs to the *Clocked Synchronous MoC Domain* iff all its processes and process compositions are constructed either by one of the named process constructors or by one of the composition operators. We call such processes *CS-MoC processes*.

3.3.5 Discrete Timed Models of Computation

Timed processes are a blend of untimed and synchronous processes in that they can consume and produce more than one event per cycle and they also deal with absent events. In addition, they have to comply with the constraint that output events cannot occur before the input events of the same evaluation cycle. This is achieved by enforcing an equal number of input and output events for each evaluation cycle, and by prepending an initial sequence of absent events. Since the signals also represent the progression of time, the prefix of absent events at the outputs corresponds to an initial delay of the process in reaction to the inputs. Moreover, the partitioning of input and output signals corresponds to the duration of each evaluation cycle.

DEFINITION 3.11 $mealyT$ is a process constructor which, given γ, f, g, and w_0 as arguments, instantiates a process $p : \hat{S} \rightarrow \hat{S}$. Again, γ is a function of the current state and determines the number of input events consumed in a particular evaluation cycle. Function g computes the next state and f computes the output events with the constraint that the output events do not occur earlier than the input events on which they depend.

This constraint is necessary because in the timed MoC each event corresponds to a time stamp and we have a globally total order of time, relating all events in all signals to each other. To avoid causality flaws every process has to abide by this constraint.

Similarly $zipT$-based processes consume events from their two inputs and pack them into tuples of events emitted at the output. $unzipT$ performs the inverse operation. Both have to comply with the causality constraint as well.

Again, we can now make precise what we mean by Timed Model of Computation.

DEFINITION 3.12 The *Timed Model of Computation (Timed MoC)* is defined as Timed MoC= (C, O), where

$$C = \{mealyT, zipT, unzipT\}$$
$$O = \{\|, \circ, \mu\}$$

In other words, a process or a process network belongs to the *Timed MoC Domain* iff all its processes and process compositions are constructed either by one of the named process constructors or by one of the composition operators. We call such processes *T-MoC processes.*

Merging and other transformations as well as analysis of time process networks is more complicated than for synchronous or untimed MoCs, because the timing may interfere with the pure functional behavior. However, we can further restrict the functions used in constructing the processes, to more or less separate behavior from timing in the timed MoC. To illustrate this we discuss a few variants of the Mealy process constructor.

$mealyPT$: In $mealyPT(\gamma, f, g, w_0)$ based processes the functions, f and g, are not exposed to absent events and they are only defined on untimed sequences. The interface of the process strips off all absent events of the input signal, hands over the result to f and g, and inserts absent events at the output as appropriate to provide proper timing for the output signal. The function γ, which may depend on the process state as usual, defines how many events are consumed. Essentially, it represents a timer and determines when the input should be checked the next time.

$mealyST$: In $mealyST(\gamma, f, g, w_0)$ based processes γ determines the number of nonabsent events that should be handed over to f and g for processing. Again, f and g never see or produce absent events and the process interface is responsible for providing them with the appropriate input data and for synchronization and timing issues on inputs and outputs. Unlike *mealyPT* processes, functions f and g in *mealyST* processes have no influence on when they are invoked. They only control how many nonabsent events have appeared before their invocation. f and g in *mealyPT* processes on the other hand determine the time instant of their next invocation independent of the number of nonabsent events.

mealyTT: However, a combination of these two process constructors is *mealyTT*, which allows to control the number of nonabsent input events and a maximum time period, after which the process is activated in any case independent of the number of nonabsent input events received. This allows to model processes that wait for input events but can set internal timers to provide time-outs.

These examples illustrate that process constructors and models of computation could be defined that allow to precisely define to which extent communication issues are separated from the purely functional behavior of the processes. Obviously, a stricter separation greatly facilitates verification and synthesis but may restrict expressiveness.

3.3.6   Continuous Time Model of Computation

Since the time domain in the synchronous and the discrete time MoCs are a countable set isomorph to the integers, signals in these MoCs can be conveniently represented as an infinite stream of events with values at discrete times instants. By contrast, signals in the CT MoC are based on a CT domain isomorph to the real numbers. Hence, it is not sufficient to enumerate the signal values at discrete time instants; signals must be defined at all time instants of the CT domain. Therefore, a signal is represented as a function over the time domain. To be more precise, a signal is a sequence of (Function, Interval) pairs to allow for different functions to define a signal at different points in time. For instance,

$$\tilde{s} = \langle (F_0, I_0), (F_1, I_1), \ldots, (F_m, I_m) \rangle$$

is a CT signal that is defined by function F_0 in time interval I_0, by function F_1 in time interval I_1, etc. It is required that the signal described in this way is completely and consistently defined in the entire interval defined collectively by all intervals $I_0, \ldots, I_m$.

A CT process is a function that maps a sequence of (Function, Interval) pairs onto a new sequence of (Function, Interval) pairs. Process constructors take functions as arguments and apply them to the input signals to obtain the output signals.

DEFINITION 3.13   A stateless, combinatorial process constructor *mapCT* takes arguments c and f and instantiates a CT process $p : \tilde{S} \rightarrow \tilde{S}$. c is a real number that determines the period of each process evaluation cycle, and f is the function that transforms the input signal of a given period of length c into an output signal of the same duration.

For instance, if $(\int dt)$ is the integral function, $\tilde{s} = \langle (\sin(t), [0, \infty)) \rangle$ and $p = mapCT(1, (\int dt))$, then $p(\tilde{s}) = -\cos(t)$ for $t \in [0, \infty)$.

Another example is a scaling process. Let $f_a(F)$ be a function that multiplies the result of function F by a, i.e., $(f_a(F))(t) = a\dot{F}(t)$. Then, $p = combCT(1, f_a)$ is a process that scales a signal by a factor of a.

As a final example let us construct a process that adds up two signals. Let $zipCT(\tilde{s}_1, \tilde{s}_2)$ be a process that merges two signals into one compound signal that contains both original signals. Further, let f_+ be a function that adds the values of two functions pointwise, i.e., $(f_+(F_1, F_2))(t) = F_1(t) + F_2(t)$. Then, $p = mapCT(1, f_+)(zipCT(\tilde{s}_1, \tilde{s}_2))$ is a process that adds up the two signals $\tilde{s}_1$ and $\tilde{s}_2$ pointwise.

Statefull process constructors are defined correspondingly.

DEFINITION 3.14 $mealyCT$ is a process constructor which, given γ, f, g, w_0, and G_0 as arguments, instantiates a process $p : \tilde{S} \to \tilde{S}$. γ is a function of the current state and determines the period pf the next evaluation cycle. Function g computes the next state; function f transforms the input signal during the current period into an output signal but depends on the current state value. G_0 is the output signal during the initial interval.

With an analog and proper definition of $zipCT$ and $unzipCT$ process constructors and with appropriate process combinators, we can instantiate any CT process and, hence, we can define a CT MoC.

DEFINITION 3.15 The *Continuous Time Model of Computation (CT MoC)* is defined as CT MoC=(C, O), where

$$C = \{mealyCT, zipT, unzipT\}$$
$$O = \{\|, \circ, \mu\}$$

In other words, a process or a process network belongs to the *CT-MoC Domain* iff all its processes and process compositions are constructed either by one of the named process constructors or by one of the composition operators. We call such processes *CT-MoC processes*.

In the CT MoC signals are never evaluated fully; they are only evaluated at certain points when needed. It becomes necessary to partially evaluate a signal when a designer wants to print certain signal values or display a signal graph. Furthermore, it partial signal evaluation is necessary if a signal is an input to a synchronous, a discrete time or an untimed MoC domain. In all these cases, we refer to partial signal evaluation as sampling. Apparently, the designer or the domain interface process has to specify the sampling points.

3.4 Integration of Models of Computation

3.4.1 MoC Interfaces

Interfaces between different MoCs determine the relation of the time structure in the different domains and they influence the way a domain is triggered to evaluate inputs and produce outputs. If a MoC domain is time triggered, the time signal is made available through the interface. Other domains are triggered when input data is available. Again, the input data appears through the interfaces.

We introduce a few simple interfaces for the MoCs of the previous sections, in order to be able to discuss concrete examples.

DEFINITION 3.16 A $stripS2U$ process constructor takes no arguments and instantiates a process $p : \bar{S} \to \dot{S}$ that takes a synchronous signal as input and generates an untimed signal as output. It reproduces all data from the input in the output in the same order with the exception of the absent event, which is translated into the value 0.

DEFINITION 3.17 A $insertU2S$ process constructor takes no arguments and instantiates a process $p : \dot{S} \to \bar{S}$ that takes an untimed signal as input and generates a synchronous signal as output. It reproduces all data from the input in the output in the same order without any change.

These interface processes between the synchronous and the untimed MoCs are very simple. However, they establish a strict and explicit time relation between two connected domains.

Connecting processes from different MoCs requires also a proper semantic basis, which we provide by defining a hierarchical MoC:

DEFINITION 3.18   A hierarchical model of computation (HMoC) is a 3-tuple HMoC = (M, C, O) where

M is a set of HMoCs or simple MoCs, each capable of instantiating processes or process networks
C is a set of process constructors
O is a set of process composition operators that governs the process composition at the highest hierarchy level but not inside process networks instantiated by any of the HMoCs of M

In the following examples and discussion we will use a specific but rather simple HMoC:

DEFINITION 3.19 $H = (M, C, O)$ with

$$M = \{\text{U-MoC}, \text{S-MoC}\}$$
$$C = \{stripS2U, insertU2S\}$$
$$O = \{\|, \circ, \mu\}$$

Example 3.2

As example consider the equalizer system of Figure 3.10 [Jan03]. The control part consists of two S-MoC processes and the data flow part, modeled as U-MoC processes, filters and analyzes an audio stream. Depending on the analysis results of the Analyzer process, the Distortion control will modify the filter parameters. The Button control also takes user input into account to steer the filter. The purpose of Analyzer and Distortion control is to avoid dangerously strong signals that could jeopardize the loud speakers.

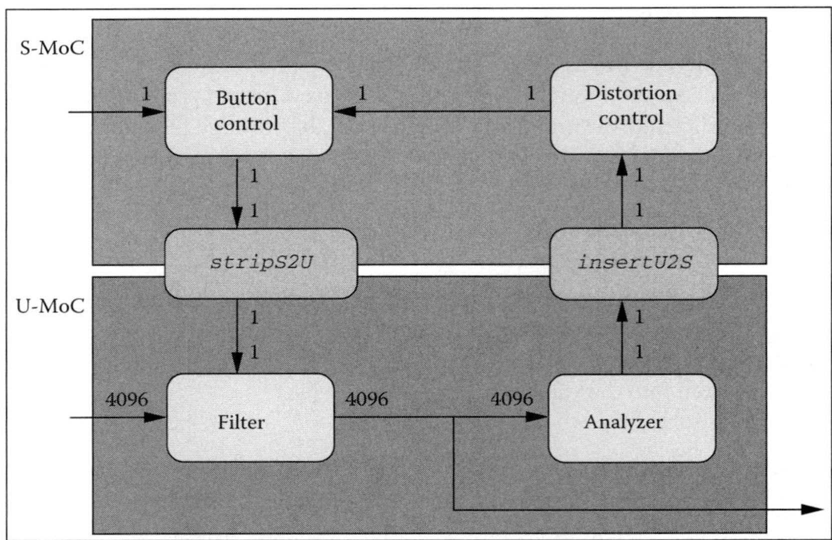

FIGURE 3.10   Digital equalizer consisting of a data flow part and control. The numbers annotating process inputs and outputs denote the number of tokens consumed and produced in each evaluation cycle.

Control and data flow parts are connected via two interface processes. The data flow processes can be developed and verified separately in the untimed MoC domain, but as soon as they are connected to the S-MoC control part, the time structure of the S-MoC domain gets imposed on all the U-MoC processes. With the simple interfaces of Figure 3.10 the Filter process consumes 4096 data tokens from the primary input, 1 token from the *stripS2U* process, and it emits 4096 tokens in every S-MoC time slot. Similarly, the activity of the Analyzer is precisely defined for every S-MoC time slot. Also, the activities of the two control processes are related precisely to the activities of the data flow processes in every time slot. Moreover, the timing of the two primary inputs and the primary output are now related timewise. Their timing must be consistent because the timing of the primary input data determines the timing of the entire system. For example, if the input signal to the Button control process assumes that each time slot has the same time duration, the 4096 data samples of the Filter input in each evaluation cycle must correspond to the same constant time period. It is the responsibility of the domain interfaces to correctly relate the timing of the different domains to each other. It is required that the time relation established by all interfaces is consistent with each other and with the timing of the primary inputs. For instance if the *stripS2U* takes 1 token as input and emits 1 token as output in each evaluation cycle, the *insertU2S* process cannot take 1 token as input and produce 2 tokens as output.

The interfaces in Figure 3.10 are very simple and lead to a strict coupling between the two MoC domains. Could more sophisticated or nondeterministic interfaces avoid this coupling effect? The answer is "no" because even if the input and output tokens of the interfaces vary from evaluation cycle to evaluation cycle in complex or nondeterministic ways, we still have a very precise timing relation in each and every time slot. Since in every evaluation cycle all interface processes must consume and produce a particular number of tokens, this determines the time relation in that particular cycle. Even though this relation may vary from cycle to cycle, it is still well defined for all cycles and hence for the entire execution of the system.

The possibly nondeterministic communication delay between MoC domains, as well as between any other processes, can be modeled, but this should not be confused with establishing a time relation between two MoC domains.

3.4.2 Interface Refinement

In order to show this difference and to illustrate how abstract interfaces can be gradually refined to accommodate channel delay information and detailed protocols, we propose an interface refinement procedure.

Add a time interface: When we connect two different MoC domains, we always have to define the time relation between the two. This is even the case if the two domains are of the same type, e.g., both are S-MoC domains, because the basic time unit may or may not be identical in the two domains.

In our MoC framework the occurrence of events represents time in both the S-MoC and T-MoC domains. Thus, setting the time relation means to determine the number of events in one domain that corresponds to one event in the other domain. For example, in Figure 3.10 the interfaces establish a one-to-one relation while the interface in Figure 3.11 represents a 3/2 relation.

In other frameworks establishing a time relation will take a different form. For instance if languages, like SystemC or VHDL, are used, the time of the different domains has to be related to the common time base of the simulator.

Refine the protocol: When the time relation between the two domains is established, we have to provide a protocol that is able to communicate over the final interface at that point. The two domains may represent different clocking regimes on the same chip or one may end up as software while the other

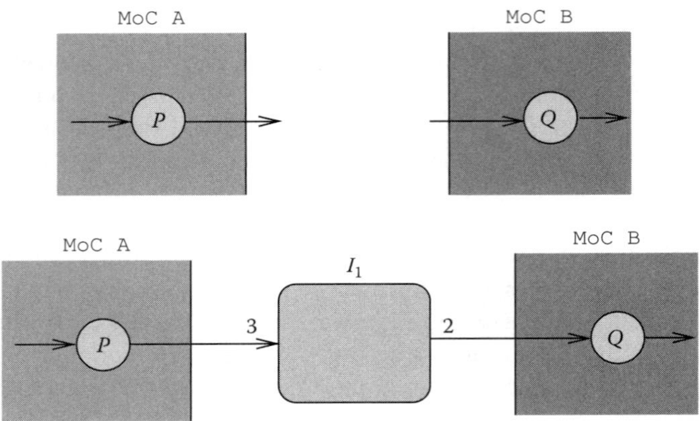

FIGURE 3.11   Determining the time relation between two MoC domains.

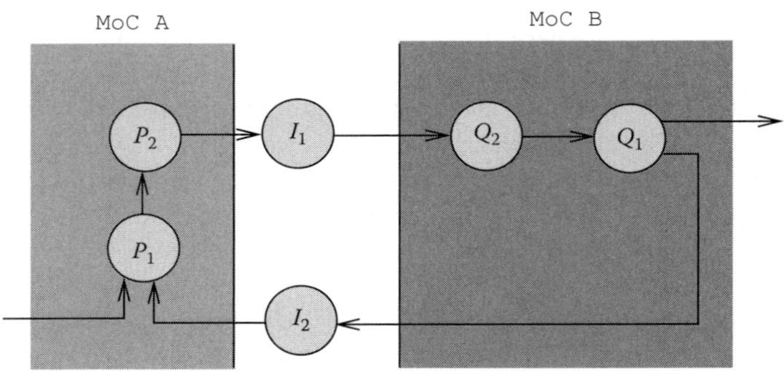

FIGURE 3.12   Simple handshake protocol.

is implemented as hardware or both may be implemented as software on different chips or cores, etc. Depending on the final implementations we have to develop a protocol fulfilling the requirements of the interface such as buffering and error control.

In our example in Figure 3.12 we have selected a simple handshake protocol with limited buffering capability. Note, however, that this assumes that for every three events arriving from MoC A there are only two useful events to be delivered to MoC B . The interface processes I_1 and I_2 and the protocol processes P_1, P_2, Q_1 and Q_2 must be designed carefully to avoid both losing data and deadlock.

Model the channel delay: In order to have a realistic channel behavior, the delay can be modeled deterministically or stochastically. In Figure 3.13 we have added a stochastic delay varying between two and five MoC B cycles. The protocol will require more buffering to accommodate the varying delays. To dimension the buffers correctly we have to identify the average and the worst-case behavior that we should be able to handle.

This refinement procedure proposed here is consistent with and complementary to other techniques proposed, e.g., in the context of SystemC [GLMS02]. We only want to emphasize here on separating the time relation between domains from channel delay and protocol design. Often these issues are not separated clearly making interface design more complicated than necessary. More details about this procedure and the example can be found in [Jan03].

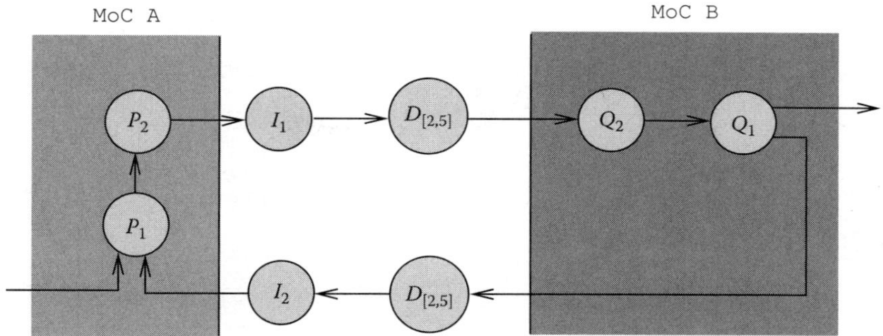

FIGURE 3.13   Channel delay can vary between two and five cycles measured in MoC B cycles.

3.4.3   MoC Refinement

The three introduced models of computation represent three time abstractions and, naturally, design often starts with higher time abstractions and gradually leads to lower abstractions. It is not always appropriate to start with an untimed MoC because when timing properties are an inherent and crucial part of the functionality, a synchronous model is more appropriate to start with. But if we start with an untimed model, we need to map it onto an architecture with concrete timing properties. Frequently, resource sharing makes the consideration of time functionally relevant, because of deadlock problems and complicated interaction patterns. All three phenomenon discussed in Section 3.1.4, priority inversion, performance inversion, and over-synchronization, emerged due to resource sharing.

Example 3.3

We discuss therefore an example for MoC refinement from the untimed through the synchronous to the timed MoC, which is driven by resource sharing. In Figure 3.14 we have two U-MoC process pairs, which are functionally independent from each other. At this level, under the assumption of infinite buffers and unlimited resources, we can analyze and develop the core functionality embodied by the process internal functions, f and g.

In the first refinement step, shown in Figure 3.15, we introduce finite buffers between the processes. $B_{n,2}$ and $B_{M,2}$ represent buffers of size n and m, respectively. Since the untimed MoC assumes implicitly infinite buffers between two communicating processes, there is no point in modeling finite buffers in the U-MoC domain. We just would not see any effect. In the S-MoC domain, however, we can analyze the consequences of finite buffers. The processes need to be refined. Processes P_2 and R_2 have to be able to handle full buffers while processes Q_2 and S_2 have to handle empty buffers. In the U-MoC processes always block on empty input buffers. This behavior can also be modeled in S-MoC processes easily. In addition more complicated behavior such as time-outs can be modeled and analyzed. To find the minimum buffer sizes while avoiding deadlock and ensuring the original

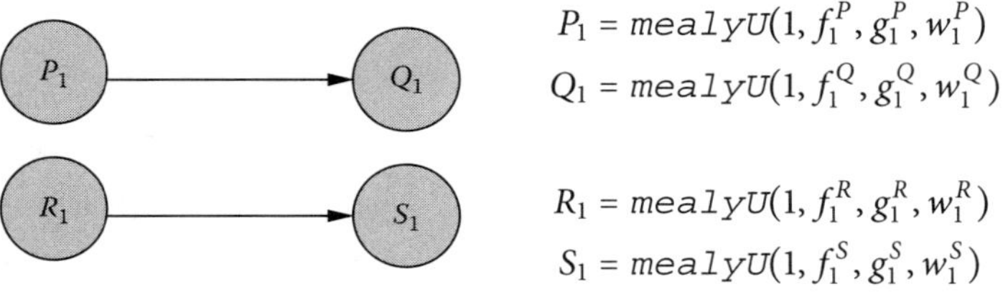

$$P_1 = \mathtt{mealyU}(1, f_1^P, g_1^P, w_1^P)$$
$$Q_1 = \mathtt{mealyU}(1, f_1^Q, g_1^Q, w_1^Q)$$

$$R_1 = \mathtt{mealyU}(1, f_1^R, g_1^R, w_1^R)$$
$$S_1 = \mathtt{mealyU}(1, f_1^S, g_1^S, w_1^S)$$

FIGURE 3.14   Two independent process pairs.

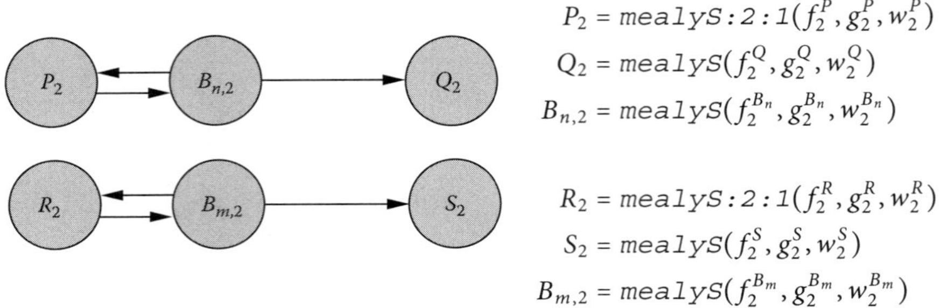

$$P_2 = \mathtt{mealyS:2:1}(f_2^P, g_2^P, w_2^P)$$
$$Q_2 = \mathtt{mealyS}(f_2^Q, g_2^Q, w_2^Q)$$
$$B_{n,2} = \mathtt{mealyS}(f_2^{B_n}, g_2^{B_n}, w_2^{B_n})$$

$$R_2 = \mathtt{mealyS:2:1}(f_2^R, g_2^R, w_2^R)$$
$$S_2 = \mathtt{mealyS}(f_2^S, g_2^S, w_2^S)$$
$$B_{m,2} = \mathtt{mealyS}(f_2^{B_m}, g_2^{B_m}, w_2^{B_m})$$

FIGURE 3.15   Two independent process pairs with explicit buffers.

system behavior is by itself a challenging task. Basten and Hoogerbrugge [BH01] propose a technique to address this. More frequently, the buffer minimization problem is formulated as part of the process scheduling problem [SB00,BML96].

The communication infrastructure is typically shared among many communicating actors. In Figure 3.16 we map the communication links onto one bus, represented as process I_3. It contains an arbiter that resolves conflicts when both processes $B_{n,3}$ and $B_{m,3}$ try to access the bus at the same time. It also implements a bus access protocol that has to be followed by connecting processes. The S-MoC model in Figure 3.16 is cycle true and the effect of bus sharing on system behavior and performance can be analyzed. A model checker can use the soundness and fairness of the arbitration algorithm, and performance requirements on the individual processes can be derived to achieve a desirable system performance.

Sometimes it is a feasible option to synthesis the model of Figure 3.16 directly into a hardware or software implementation provided we can use standard templates for the process interfaces. Alternatively we can refine the model into a fully timed model. However, we still have various option depending on what exactly we would like to model and analyze. For each process we can decide how much of the timing and synchronization details should be explicitly taken care of by the process and how much can be handled implicitly by the process interfaces. For instance in Section 3.3.5 we have introduced constructors `mealyST` and `mealyPT`. The first provides a process interface that strips off all absent events and inserts absent events at the output as needed. The internal functions have only to deal with the functional events but they have no access to timing information. This means that an untimed `mealyU` process can be directly refined into a timed `mealyST` process with exactly the same functions, f and g. Alternatively, the constructor `mealyPT` provides an interface that invokes

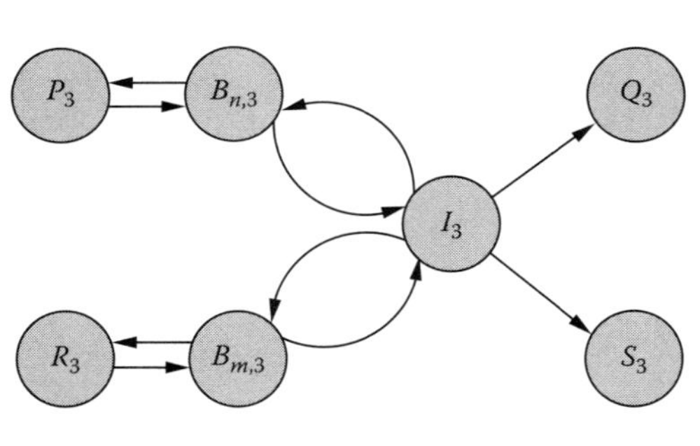

$$P_3 = \mathtt{mealyS}(f_3^P, g_3^P, w_3^P)$$
$$Q_3 = \mathtt{mealyS}(f_3^Q, g_3^Q, w_3^Q)$$
$$B_{n,3} = \mathtt{mealyS:2:1}(f_3^{B_n}, g_3^{B_n}, w_3^{B_n})$$

$$R_3 = \mathtt{mealyS}(f_3^R, g_3^R, w_3^R)$$
$$S_3 = \mathtt{mealyS}(f_3^S, g_3^S, w_3^S)$$
$$B_{m,3} = \mathtt{mealyS:2:1}(f_3^{B_m}, g_3^{B_m}, w_3^{B_m})$$

$$I_3 = \mathtt{mealyS:4:2}(f_3^I, g_3^I, w_3^I)$$

FIGURE 3.16   Two independent process pairs with explicit buffers.

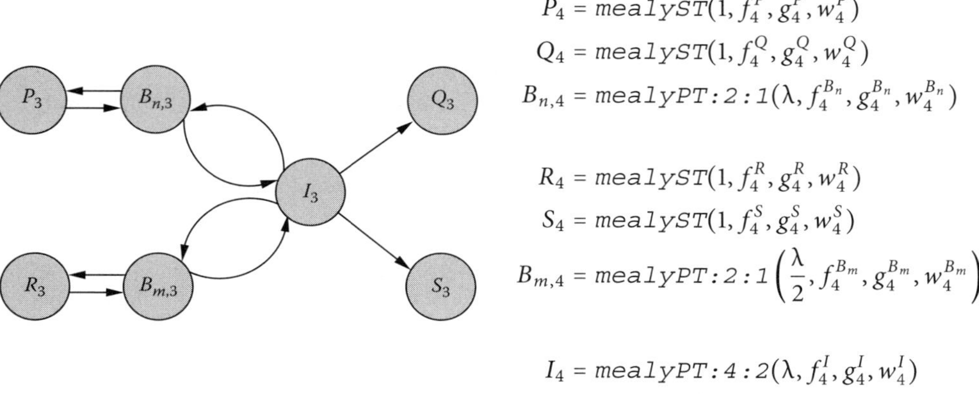

$$P_4 = \mathtt{mealyST}(1, f_4^P, g_4^P, w_4^P)$$

$$Q_4 = \mathtt{mealyST}(1, f_4^Q, g_4^Q, w_4^Q)$$

$$B_{n,4} = \mathtt{mealyPT:2:1}(\lambda, f_4^{B_n}, g_4^{B_n}, w_4^{B_n})$$

$$R_4 = \mathtt{mealyST}(1, f_4^R, g_4^R, w_4^R)$$

$$S_4 = \mathtt{mealyST}(1, f_4^S, g_4^S, w_4^S)$$

$$B_{m,4} = \mathtt{mealyPT:2:1}\left(\frac{\lambda}{2}, f_4^{B_m}, g_4^{B_m}, w_4^{B_m}\right)$$

$$I_4 = \mathtt{mealyPT:4:2}(\lambda, f_4^I, g_4^I, w_4^I)$$

FIGURE 3.17   All processes are refined into the T-MoC but with different synchronization interfaces.

the internal functions at regular time intervals. If this interval corresponds to a synchronous time slot, an S-MoC process can be easily mapped onto a $\mathtt{mealyPT}$ type of process with the only difference, that the functions in a $\mathtt{mealyPT}$ process may receive several nonabsent events in each cycle. But in both cases the processes experience a notion of time based on cycles.

In Figure 3.17 we have chosen to refine processes P, Q, R, and S into $\mathtt{mealyST}$-based processes to keep them similar to the original untimed processes. Thus, the original f and g functions can be used without major modification. The process interfaces are responsible to collect the inputs, present them to the f and g functions, and emit properly synchronized output.

The buffer and the bus processes however have been mapped onto $\mathtt{mealyPT}$ processes. The constants λ and $\lambda/2$ represent the cycle time for the processes. Process $B_{m,4}$ operates with half the cycle time of the other processes, which illustrates that the modeling accuracy can be arbitrarily selected. We can also choose other process constructors and hence interfaces if desirable. For instance, some processes can be mapped onto $\mathtt{mealyT}$ type processes in a further refinement step to expose them to even more timing information.

3.5   Conclusion

We tried to motivate that model of computation for embedded systems should be different from the many computational models developed in the past. The purpose of model of embedded computation should be to support analysis and design of concrete systems. Thus, it needs to deal with salient and critical features of embedded systems in a systematic way. These features include real-time requirements, power consumption, architecture heterogeneity, application heterogeneity, and real-world interaction.

We have proposed a framework to study different MoCs, which allows to appropriately capture some, but unfortunately not all, of these features. In particular, power consumption and other nonfunctional properties are not covered. Time is of central focus in the framework but CT models are not included in spite of their relevance for the sensors and actuators in embedded systems.

Despite the deficiencies of this framework we hope that we were able to argue well for a few important points:

- Different computational models should and will continue to coexist for a variety of technical and nontechnical reasons.

- To use the "right" computational model in a design and for a particular design task can greatly facilitate the design process and the quality of the result. What is the "right" model depends on the purpose and objectives of a design task.
- Time is of central importance and computational models with different timing abstractions should be used during system development.

From a MoC perspective several important issues are open research topics and should be addressed urgently to improve the design process for embedded systems:

- We need to identify efficient ways to capture a few important nonfunctional properties in models of computation. At least power and energy consumption and perhaps signal noise issues should be attended to.
- The effective integration of different MoCs will require (1) the systematic manipulation and refinement of MoC interfaces and interdomain protocols; (2) the cross-domain analysis of functionality, performance, and power consumption; (3) the global optimization and synthesis including migration of tasks and processes across MoC domain boundaries.
- In order to make the benefits and the potential of well-defined MoCs available in the practical design work, we need to project MoCs into design languages such as VHDL, Verilog, SystemC, C++, etc. This should be done by properly subsetting a language and by developing pragmatics to restrict the use of a language. If accompanied by tools to enforce the restrictions and to exploit the properties of the underlying MoC, this will be accepted quickly by designers.

In the future we foresee a continuous and steady further development of MoCs to match future theoretical objectives and practical design purposes. But we also hope that they become better accepted as practically useful devices for supporting the design process just like design languages, tools, and methodologies.

References

[AACS87]   A. Aggarwal, B. Alpern, A.K. Chandra, and M. Snir. A model for hierarchical memory. In *19th Annual ACM Symposium on Theory of Computing*, New York, pp. 305–314, May 1987.

[ACFS94]   B. Alpern, L. Carter, E. Feig, and T. Selker. The uniform memory hierarchy model of computation. *Algorithmica*, 12(2/3):72–109, 1994.

[ACS90]   A. Aggarwal, A.K. Chandra, and M. Snir. Communication complexity of PRAMs. *Theoretical Computer Science*, 71(1):3–28, March 1990.

[APT02]   P.J. Ashenden, G.D. Peterson, and D.A. Teegarden. *Designers Guide to VHDL–AMS*. Morgan Kaufman, San Francisco, CA, September 2002.

[BA81]   J.D. Brock and W.B. Ackerman. Scenarios: A model of non-determinate computation. In J. Diaz and I. Ramos, editors, *Formalism of Programming Concepts*, volume 107 of Lecture Notes in Computer Science, pp. 252–259. Springer-Verlag, Berlin, 1981.

[BB91]   A. Benveniste and G. Berry. The synchronous approach to reactive and real-time systems. *Proceedings of the IEEE*, 79(9):1270–1282, September 1991.

[BB94]   G. Booch and D. Bryan. *Software Engineering with Ada*. The Benjamin/Cummings Publishing Company, Mento Park, CA, 1994.

[BCG88]   G. Berry, P. Couronne, and G. Gonthier. Synchronous programming of reactive systems: An introduction to Esterel. In K. Fuchi and M. Nivat, editors, *Programming of Future Generation Computers*, pp. 35–55. Elsevier, North Holland, 1988.

[Ber98]   G. Berry. The foundations of Esterel. In G. Plotkin, C. Stirling, and M. Tofte, editors, *Proof, Language and Interaction: Essays in Honour of Robin Milner*. MIT Press, Cambridge, MA, 1998.

[BH01]   T. Basten and J. Hoogerbrugge. Efficient execution of process networks. In A. Chalmers, M. Mirmehdi, and H. Muller, editors, *Communicating Process Architectures*. IOS Press, Amsterdam, the Netherlands, 2001.

[BJ01]   P. Bjuréus and A. Jantsch. Modeling of mixed control and dataflow systems in MASCOT. *IEEE Transactions on Very Large Scale Integration (VLSI) Systems*, 9(5):690–704, October 2001.

[BML96]   S.S. Bhattacharyya, P.K. Murthy, and E.A. Lee. *Software Synthesis from Dataflow Graphs*. Kluwer Academic, Norwell, MA, 1996.

[Bro83]   J. D. Brock. *A Formal Model for Non-deterministic Dataflow Computation*. PhD thesis, Massachusets Institute of Technology, Cambridge, MA, 1983.

[Cas93]   C.G. Cassandras. *Discrete Event Systems*. Aksen Associates, Bosten, MA, 1993.

[CGP99]   P. Caspi, A. Girault, and D. Pilaud. Automatic distribution of reactive systems for asynchronous networks of processors. *IEEE Transactions on Software Engineering*, 25(3):416–427, May/June 1999.

[CR73]   S. Cook and R. Reckhow. Time bounded random access machines. *Journal of Computer and System Sciences*, 7:354–375, 1973.

[DH98]   J. Dabney and T.L. Harman. *Mastering SIMULINK 2*. Prentice-Hall, Lebanon, IN, 1998.

[EJL$^+$03]   J. Eker, J.W. Janneck, E.A. Lee, J. Liu, X. Liu, J. Ludvig, S. Neuendorffer, S. Sachs, and Y. Xiong. Taming heterogeneity—the Ptolemy approach. *Proceedings of the IEEE*, 91(1):127–144, January 2003.

[EKJ$^+$98]   P. Ellervee, S. Kumar, A. Jantsch, B. Svantesson, T. Meincke, and A. Hemani. IRSYD: An internal representation for heterogeneous embedded systems. In *Proceedings of the 16th NORCHIP Conference*, Lund, Sweden, 1998.

[EMO98]   H. Elmqvist, S.E. Mattsson, and M. Otter. Modelica—the new object-oriented modeling language. In *Proceedings of the 12th European Simulation Multiconference*, Manchester, UK, June 1998.

[Ern03]   R. Ernst. MPSOC performance modeling and analysis. Presentation at the 3rd International Seminar on Application-Specific Multi-Processor SoC, Chamonix, France, July 2003.

[est]   Esterel Technologies. http://www.esterel-technologies.com/

[FW78]   S. Fortune and J. Wyllie. Parallelism in random access machines. In *Proceedings of the 10th Annual Symposium on Theory of Computing*, San Diego, CA, 1978.

[GLL99]   A. Girault, B. Lee, and E.A. Lee. Hierarchical finite state machines with multiple concurrency models. *Integrating Communication Protocol Selection with Hardware/ Software Codesign*, 18(6):742–760, June 1999.

[GLMS02]   T. Grötker, S. Liao, G. Martin, and S. Swan. *System Design with SystemC*. Kluwer Academic, Boston, MA, 2002.

[GMR94]   P.B. Gibbons, Y. Matias, and V. Ramachandran. The QRQW PRAM: Accounting for contention in parallel algorithms. In *Proceedings of the 5th Annual ACM-SIAM Symposium on Discrete Algorithms*, pp. 638–648, Arlington, VA, January 1994.

[Hal00]   N. Halbwachs. Synchronous programming of reactive systems. In *Proceedings of Computer Aided Verification (CAV)*, Chicago, IL, 2000.

[Har87]   D. Harel. Statecharts: A visual formalism for complex systems. *Science of Computer Programming*, 8:231–274, 1987.

[HCRP91]   N. Halbwachs, P. Caspi, P. Raymond, and D. Pilaud. The synchronous data flow programming language LUSTRE. *Proceedings of the IEEE*, 79(9):1305–1320, September 1991.

[HLR92]   N. Halbwachs, F. Lagnier, and C. Ratel. Programming and verifying real-time systems by means of the synchronous data-flow language LUSTRE. *IEEE Transactions on Software*

Engineering, September 1992. Special issue on the Specification and Analysis of Real-Time Systems.

[Hoa78]   C.A.R. Hoare. Communicating sequential processes. *Communications of the ACM*, 21(8):666–676, August 1978.

[Jan03]   A. Jantsch. *Modeling Embedded Systems and SoCs—Concurrency and Time in Models of Computation*. Systems on Silicon. Morgan Kaufmann, San Francisco, CA, June 2003.

[Jan05]   A. Jantsch. Models of embedded computation. In Richard Zurawski, editor, *Embedded Systems Handbook*. CRC Press, Boca Raton, FL, 2005. Invited contribution.

[JB00]   A. Jantsch and P. Bjuréus. Composite signal flow: A computational model combining events, sampled streams, and vectors. In *Proceedings of the Design and Test Europe Conference (DATE)*, Royal Institute of Technology, Sweden, 2000.

[JO95]   A.A. Jerraya and K. O'Brien. Solar: An intermediate format for system-level modeling and synthesis. In J. Rozenblit and K. Buchenrieder, editors, *Codesign: Computer-Aided Software/Hardware Engineering*, Chapter 7, pp. 145–175. IEEE Press, Piscataway, NJ, 1995.

[JSW01]   A. Jantsch, I. Sander, and W. Wu. The usage of stochastic processes in embedded system specifications. In *Proceedings of the 9th International Symposium on Hardware/Software Codesign*, Copenhagen, Denmark, April 2001.

[JT03]   A. Jantsch and H. Tenhunen. Will networks on chip close the productivity gap? In Axel Jantsch and Hannu Tenhunen, editors, *Networks on Chip*, Chapter 1, pp. 3–18. Kluwer Academic, Hingham, MA, February 2003.

[Kah74]   G. Kahn. The semantics of a simple language for parallel programming. In *Proceedings of the IFIP Congress 74*, Stockholm, Sweden, 1974.

[Kos78]   P. R. Kosinski. A straight forward denotational semantics for nondeterminate data flow programs. In *Proceedings of the 5th ACM Symposium on Pronciples of Programming Languages*, ACM, New York, pp. 214–219, 1978.

[Lee97]   E.A. Lee. A denotational semantics for dataflow with firing. Technical Report UCB/ERL M97/3, Department of Electrical Engineering and Computer Science, University of California, Berkeley, CA, January 1997.

[Lee03]   E. A. Lee. Overview of the ptolemy project. Technical Report UCB/ERL M03/25, University of California, Berkeley, CA, July 2003.

[Len90]   T. Lengauer. VLSI theory. In J. van Leeuwen, editor, *Handbook of Theoretical Computer Science*, volume A: *Algorithms and Complexity*, Chapter 16, pp. 835–868. Elsevier Science, North Holland, 2nd edn, 1990.

[lGGlBlM91] P. le Guernic, T. Gautier, M. le Borgne, and C. le Maire. Programming real-time applications with SIGNAL. *Proceedings of the IEEE*, 79(9):1321–1336, September 1991.

[LK00]   A.M. Law and W.D. Kelton. *Simulation, Modeling and Analsysis*. Industrial Engineering Series. McGraw-Hill, New York, 3rd edn, 2000.

[LM87a]   E.A. Lee and D.G. Messerschmitt. Static scheduling of synchronous data flow programs for digital signal processing. *IEEE Transactions on Computers*, C-36(1):24–35, January 1987.

[LM87b]   E.A. Lee and D.G. Messerschmitt. Synchronous data flow. *Proceedings of the IEEE*, 75(9):1235–1245, September 1987.

[LM93]   E.A. Lee and D.G. Messerschmitt. An overview of the ptolemy project. Report from Department of Electrical Engineering and Computer Science, University of California, Berkeley, CA, Jannuary 1993.

[LP95]   E.A. Lee and T.M. Parks. Dataflow process networks. *Proceedings of the IEEE*, 83:773–801, May 1995.

[LS91]   C.E. Leiserson and J.B. Saxe. Retiming synchronous circuitry. *Algorithmica*, 6(1):5–35, 1991.

[LSSJ07]   Z. Lu, J. Sicking, I. Sander, and A. Jantsch. Using synchronizers for refining synchronous communication onto hardware/software architectures. In *Proceedings of the 18th IEEE/IFIP International Workshop on Rapid System Prototyping*, Porto Alegre, Brazil, May 2007.

[LSV98]   E.A. Lee and A. Sangiovanni-Vincentelli. A framework for comparing models of computation. *IEEE Transactions on Computer-Aided Design of Integrated Circuits and Systems*, 17(12):1217–1229, December 1998.

[Mil80]   R. Milner. *A Calculus of Communicating Systems*, volume 92 of *Lecture Notes of Computer Science*. Springer-Verlag, New York, 1980.

[Mil89]   R. Milner. *Communication and Concurrency*. International Series in Computer Science. Prentice-Hall, 1989.

[MMT95]   B.M. Maggs, L.R. Matheson, and R.E. Tarjan. Models of parallel computation: A survey and synthesis. In *Proceedings of the 28th Hawaii International Conference on System Sciences (HICSS)*, volume 2, pp. 61–70, Hawaii, 1995.

[MVH+98]   P. Le Marrec, C.A. Valderrama, F. Hessel, A.A. Jerraya, M. Attia, and O. Cayrol. Hardware, software and mechanical cosimulation for automotive applications. In *Proceedings of the 9th International Workshop on Rapid System Prototyping*, Lauven, Belgium, pp. 202–206, 1998.

[Par83]   D. Park. The 'fairness' problem and nondeterministic computing networks. In J.W. De Baker and J. van Leeuwen, editors, *Foundations of Computer Science IV, Part 2: Semantics and Logic*, volume 159, pp. 133–161. Mathematical Centre Tracts, Amsterdam, the Netherlands, 1983.

[PBEB07]   D. Potop-Butucaru, S.A. Edwards, and G. Berry. *Compiling Esterel*. Springer, New York, 2007.

[PJH02]   C. Park, J. Jung, and S. Ha. Extended synchronous dataflow for efficient DSP system prototyping. *Design Automation for Embedded Systems*, 6(3):295–322, March 2002.

[PT04]   J.M. Paul and D.E. Thomas. Models of computation for systems-on-chip. In Ahmed Jerraya and Wayne Wolf, editors, *MultiProcessor Systems-on-Chip*, Chapter 15. Morgan Kaufman, San Francisco, CA, 2004.

[Sap06]   S. Sapatnekar. Static timing analysis. In L. Lavagno, G. Martin, and L. Scheffer, editors, *Electronic Design Automation for Integrated Circuits Handbook*, volume 2, Chapter 8. CRC Press, Boca Raton, FL, 2006.

[SB00]   S. Sriram and S.S. Bhattacharyya. *Embedded Multiprocessors: Scheduling and Synchronization*. Marcel Dekker, New York, January 2000.

[Sch99]   P. Scholz. From synchronous specifications to asynchronous distributed implementations. In Franz J. Rammig, editor, *Distributed and Parallel Embedded Systems*, pp. 39–50. Kluwer Academic, Hingham, MA, 1999.

[Sev01]   F.L. Severance. *System Modeling and Simulation*. John Wiley & Sons, New York, 2001.

[Tar07]   O. Tardieu. A deterministic logical semantics for pure Esterel. *ACM Transactions on Programming Languages and Systems*, 29(2), 2007.

[Tay98]   R.G. Taylor. *Models of Computation and Formal Language*. Oxford University Press, New York, 1998.

[TM02]   D.E. Thomas and P.R. Moorby. *The Verilog hardware description language*. Springer, New York, 2002.

[Upf84]   E. Upfal. Efficient schemes for parallel communication. *Journal of the ACM*, 31:507–517, 1984.

[vEB90]   P. van Embde Boas. Machine models and simulation. In J. van Leeuwen, editor, *Handbook of Theoretical Computer Science*, volume A: *Algorithms and Complexity*, Chapter 1, pp. 1–66. Elsevier Science Publishers B.V., North Holland 1990.

[VGE03]   A. Vachoux, C. Grimm, and K. Einwich. SystemC–AMS requirements, design objectives and rationale. In *Proceedings of the Design Automation and Test Europe Conference*, München, März, 2003.

[VGE05]   A. Vachoux, C. Grimm, and K. Einwich. Extending SystemC to support mixed discrete-continuous system modeling and simulation. In *Proceedings of the IEEE International Symposium on Circuits and Systems*, Kobe, Japan, 2005.

4

Embedded Software Modeling and Design

Marco Di Natale

Sant'Anna School of Advanced Studies

4.1 Introduction

The increasing cost necessary for the design and fabrication of application-specific integrated circuits (ASICs) together with the need for the reuse of functionality, adaptability, and flexibility is among the causes for an increasing share of software-implemented functions in embedded projects. Figure 4.1 represents a typical architectural framework for embedded systems, where application software runs on top of a real-time operating system (RTOS) (and possibly a middleware layer), which abstracts from the hardware and provides a common application programmer interface (API) for reuse of functionality (such as the OSEK standard in the automotive domain).

Unfortunately, mechanisms for improving the reuse of software at the level of programming code, such as RTOS- or middleware-level APIs are still not sufficient to achieve the desired levels of productivity, and the error rate of software programs is exceedingly high. Today, model-based design of software bears the promise for a much needed step up in productivity and reuse.

The use of abstract software models may significantly increase the chances that the design and its implementation are correct, when used at the highest possible level in the development process. Correctness can be achieved in many ways. Ideally, it should be mathematically proved by formal reasoning upon the model of the system and its desired properties, provided the model is built on solid mathematical foundations and its properties are expressed by some logic predicate(s). Unfortunately, in many cases, formal model checking is not possible or simply impractical. In this case, the

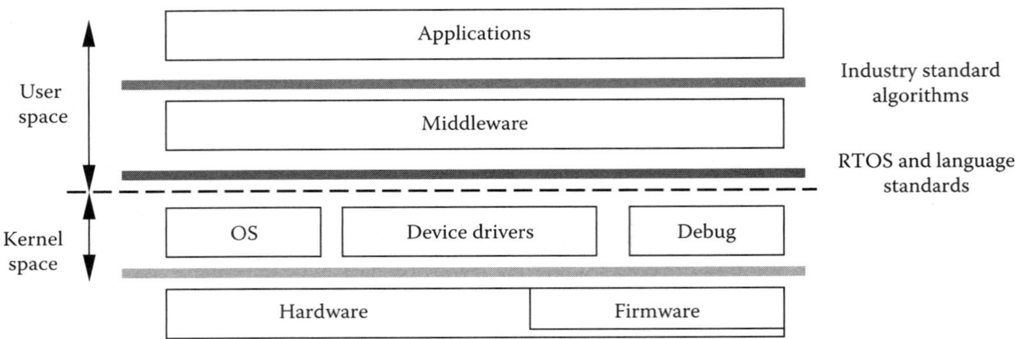

FIGURE 4.1   Common architecture for embedded software.

modeling language should at least provide abstractions for the specification of reusable components, so that software artifacts can be clearly identified with the provided functions or services.

Also, when exhaustive proof of correctness cannot be achieved, a secondary goal of the modeling language should be providing support for simulation and testing. In this view, formal methods can also be used to guide the generation of the test suite and guarantee some degree of coverage. Finally, modeling languages and tools should ensure that the model of the software, after being checked by means of formal proof or by simulation, is correctly implemented in a programming language executed on the target hardware. (This requirement is usually satisfied by automatic code generation tools.)

Industry and research groups have been working now for decades in the software engineering area looking for models, methodologies, and tools to improve the correctness and increase the reusability of software components. Traditionally, software models and formal specifications have had their focus on behavioral properties and have been increasingly successful in the verification of functional correctness. However, embedded software is characterized by concurrency and resource constraints and by nonfunctional properties, such as deadlines or other timing constraints, which ultimately depend upon the computation platform.

This chapter attempts at providing an overview of (visual and textual) languages and tools for embedded software modeling and design. The subject is so wide, rapidly evolving, and is encompassing so many different issues that only a short survey is possible in the limited space allocated to this chapter. As of today, despite all efforts, existing methodologies and languages fall short in achieving most of the desirable goals and yet they are continuously being extended in order to allow for the verification of at least a subset of the properties of interest.

The objective is to provide the reader with an understanding of what are the principles for functional and nonfunctional modeling and verification, what are the languages and tools available on the market, and what can be possibly achieved with respect to practical designs. The description of (some) commercial languages, models, and tools is supplemented with a survey of the main research trends and the results that may open new possibilities in the future. The reader is invited to refer to the cited papers in the bibliography section for wider discussion upon each issue.

The organization of the chapter is the following: the introduction section defines a reference framework for the discussion of the software modeling problem and provides a short review of abstract models for functional and temporal (schedulability) analysis. The second section provides a quick glance at the two main categories of available languages and models: synchronous as opposed to asynchronous models. Then, an introduction to the commercial modeling languages unified modeling language (UML) and specification and description language (SDL) is provided. A discussion of what can be achieved with both, with respect to formal analysis of functional properties, schedulability analysis, simulation, and testing follows. The chapter also discusses the recent extensions of existing methodologies to achieve the desirable goal of component-based design. Finally, a quick glance at the research work in the area of embedded software design, methods, and tools closes the chapter.

4.1.1 Challenges in the Development of Embedded Software

According to a typical development process (represented in Figure 4.2), an embedded system is the result of multiple refinement stages encompassing several levels of abstractions, from user requirements, to system testing, and sign-off. At each stage, the system is described using an adequate formalism, starting from abstract models for the user domain entities at the requirements level. Lower-level models, typically developed in later stages, provide an implementation of the abstract model by means of design entities, representing hardware and software components. The implementation process is a sequence of steps that constrain the generic specification by leveraging the possible options (nondeterminism) available from higher levels.

The designer's problem is making sure that the models of the system developed at the different stages satisfy the properties required from the system and that low level descriptions of the system are correct implementations of higher-level specifications. This task can be considerably easier if the models of the system at the different abstraction levels are homogeneous, that is, if the computational models on which they are based share a common semantics and, possibly, a common notation.

The problem of correct mapping from a high-level specification, employing an abstract model of the system, to a particular software and hardware architecture or platform is one of the key aspects of the design of embedded systems.

The separation of the two main concerns of functional and architectural specification and the mapping of functions to architecture elements are among the founding principles of many design

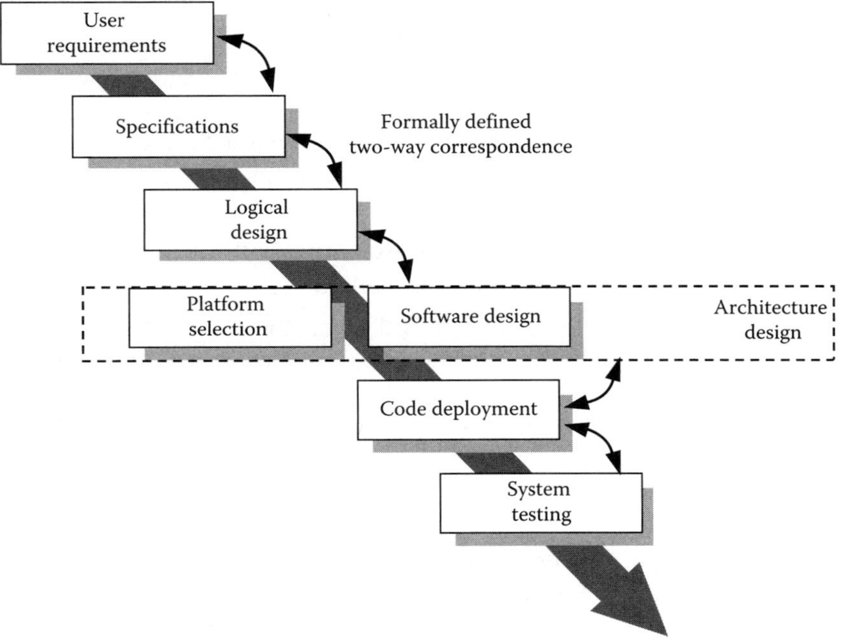

FIGURE 4.2 Typical embedded software development process.

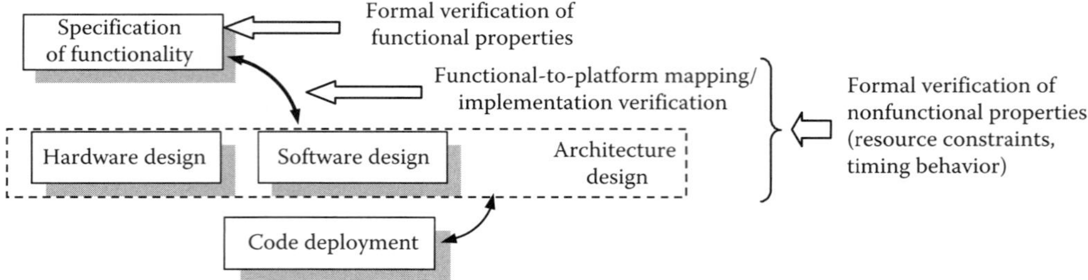

FIGURE 4.3   Mapping formal specifications to an hardware/software platform.

methodologies such as the platform-based design [1] and tools like the Ptolemy and Metropolis frameworks [2,3], as well as of emerging standards and recommendations, such as the UML Profile for schedulability, performance, and time (SPT) from the object management group, (OMG) [4], and industry best practices, such as the V-cycle of software development common in the automotive industry [5]. A keyhole view of the corresponding stages is represented in Figure 4.3.

The main design activities taking place at this stage and the corresponding challenges can be summarized as follows:

- Specification of functionality is concerned with the development of logically correct system functions. If the specification is defined using a formal model, formal verification allows checking that the functional behavior satisfies a given set of properties.
- System software and hardware platform components are defined in the Architecture design level.
- After the definition of logical and physical resources available for the execution of the functional model and the definition of the mapping of functional model elements into the platform (architecture) elements executing them, formal verification of nonfunctional properties, such as timing properties and schedulability analysis, but also reliability analysis, may take place.

Complementing the above two steps, implementation verification is the process of checking that the implementation of the functional model, after mapping onto the architecture model, preserves the semantics (and the properties) of the high-level formal specifications.

4.1.2   Short Introduction to Formal Models and Languages and to Schedulability Analysis

A short review of the most common models of computation (formal languages) proposed by academia or industry, and possibly supported by tools, with the objective of formal or simulation-based verification is fundamental for understanding commercial models and languages and it is also important for understanding today's and future challenges (please refer to [6,7] for more detail).

4.1.2.1   Formal Models

Formal models are mathematical-based languages that specify the semantics of computation and communication (also defined as model of computation or MOC [6]). MOCs may be expressed, for example, by means of a language or automaton formalisms.

System-level MOCs are used to describe the system as a (possibly hierarchical) collection of design entities (blocks, actors, tasks, processes) performing units of computations represented as transitions or actions; characterized by a state and communicating by means of events (tokens) and data values carried by signals. Composition and communication rules, concurrency models, and time representation are among the most important characteristics of an MOC.

Once the system specifications are given according to a formal MOC, formal methods can be used to achieve design-time verification of properties and implementation as in Figure 4.3. In general, properties of interest go under the two general categories of "ordered execution" and "timed execution":

- Ordered execution relates to the verification of event and state ordering. Properties such as safety, liveness, absence of deadlock, fairness, or reachability belong to this category.

- Timed execution relates to event enumeration, such as checking that no more than n events (including time events) occur between any two events in the system. "Timeliness" and some notions of "fairness" are examples.

Verification of desirable system properties may be quite hard or even impossible to achieve by logical reasoning on formal models. Formal models are usually classified according to the decidability of properties. Decidability of properties in timed and untimed models depends on many factors, such as the type of logic (propositional or first-order) for conditions on transitions and states, the real-time semantics, including the definition of the time domain (discrete or dense) and the linear or branching time logic that is used for expressing properties (the interested reader may refer to [7] for a survey on the subject.) In practice [6], decidability should be carefully evaluated. In some cases, even if it is decidable, the problem cannot be practically solved since the required run time may be prohibitive and, in other instances, even if undecidability applies to the general case, it may happen that the problem at hand admits a solution.

Verification of models properties can take many forms. In the deductive approach, the system and the property are represented by statements (clauses) written in some logic (e.g., properties can be expressed in the linear temporal logic (LTL) [8] or in the branching-time computation tree logic [9]) and a theorem proving tool (usually under the direction of a designer or some expert) applies deduction rules until (hopefully) the desired property reduces to a set of axioms or a counterexample is found. In model checking, the system and possibly the desirable properties are expressed by using an automaton or some other kind of executable formalism. The verification tool ensures that neither executable transition nor any system state violates the property. To do so, it can generate all the potential (finite) states of the system (exhaustive analysis). When the property is violated, the tool usually produces the (set of) counterexample(s).

The first model checkers worked by constructing the whole structure of states prior to property checking, but modern tools are able to perform verification as the states are produced. This means that the method not necessarily requires the construction of the whole state graph. On the fly model, checking and the SPIN [10] toolset provide, respectively, an instance and an implementation of this approach.

To give some examples (Figure 4.4), checking a system implementation **I** against a specification of a property **P** in case both are expressed in terms of automata (homogeneous verification) requires the following steps. The implementation automaton $\mathbf{A_I}$ is composed with the complementary automaton $\neg\mathbf{A_P}$ expressing the negation of the desired property. The implementation **I** violates the specification property if the product automaton $\mathbf{A_I}||\neg\mathbf{A_P}$ has some possible run and it is verified if the composition has no runs. "Checking by observers" can be considered as a particular instance of this method, very popular for synchronous models.

In the very common case in which the property specification consists of a logical formula and the implementation of the system is given by an automaton, the verification problem can be solved algorithmically or deductively by transforming it into an instance of the previous cases, for example, by transforming the negation of a specification formula f_S into the corresponding automaton and by using the same techniques as in homogeneous verification.

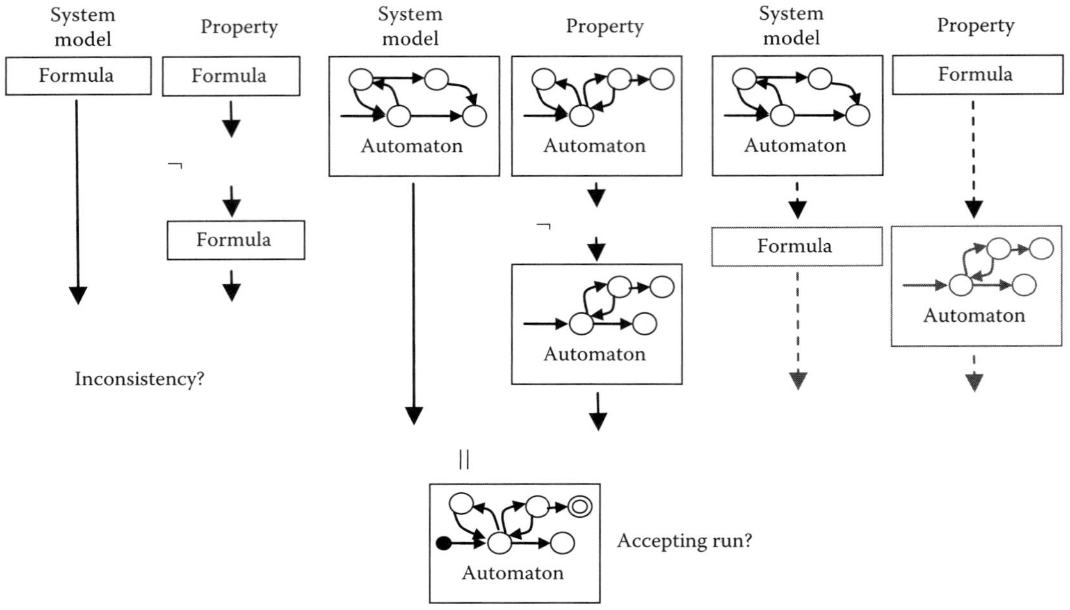

FIGURE 4.4   Checking the system model against some property.

Verification of implementation is a different problem, in which the set of possible behaviors of a model refinement must be compared against the behaviors of the higher-level model. In this case, verification is usually performed by leveraging simulation and bisimulation properties.

Following, a very short survey of formal system models is provided, starting with finite state machines (FSMs), probably the most popular and the basis for many extensions.

In FSM, process behavior is specified by enumerating the (finite) set of possible system states and the transitions among them. Each transition connects two states and it is labeled with the subset of input signals (and possibly the guard condition upon their values) that triggers or enables its execution. Furthermore, each transition can produce output variables. In Mealy FSMs, outputs depend on both state and input variables, while in the Moore model outputs only depend on the process state. Guard conditions can be expressed according to different logics, for example, propositional logic, first-order logic, or even (turing-complete) programming code.

In the synchronous FSM model, transitions occur for all components at the same time on the set of signal values present as input. Signal propagation is assumed to be instantaneous. Transitions and the evaluation of the next state happen for all the system components at the same time. Synchronous languages, such as Esterel and Lustre, are based on this model. In the asynchronous model, two asynchronous FSMs never execute a transition at the same time except when a rendezvous is explicitly specified (a pair of transitions of the communicating FSMs occur simultaneously). The SDL process behavior is an instance of this general model.

Composition of FSMs is obtained by the construction of a product transition system, that is, a single FSM where the set of states is the Cartesian product of the sets of the states of the component machines. The difference between synchronous and asynchronous execution semantics is quite clear when compositional behaviors are compared. Figure 4.5 portrays an example showing the differences in the synchronous and asynchronous composition of two FSMs.

When there is a cyclic dependency among variables in interconnected synchronous FSMs, the Mealy model, where outputs are instantaneously produced based on the input values, may result in a

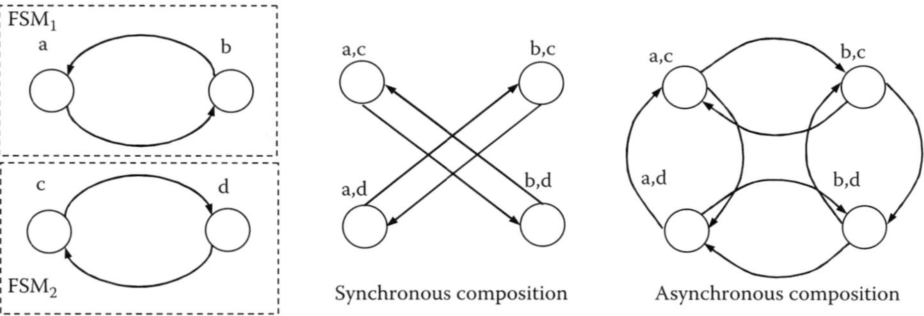

FIGURE 4.5   Composition of synchronous and asynchronous FSMs.

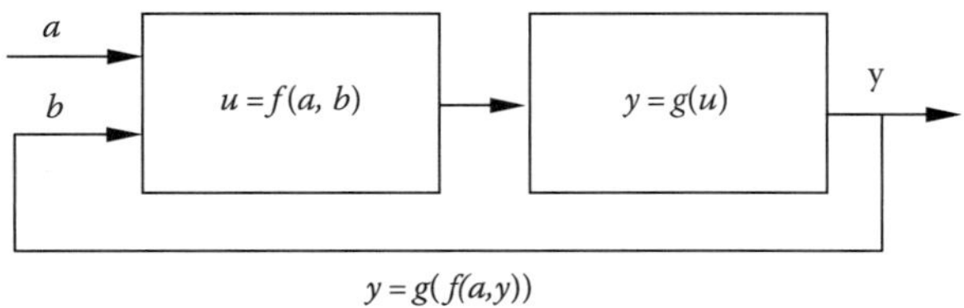

FIGURE 4.6   Fixed-point problem arising from composition and cyclic dependencies.

fixed-point problem and possibly inconsistency (Figure 4.6 shows a simple functional dependency). The existence of a unique fixed-point solution (and its evaluation) is not the only problem resulting from the composition of FSMs.

In large, complex systems, composition may easily result in a huge number of states. This phenomenon is known with the name of "state explosion".

In its statecharts extension [11], Harel proposed three mechanisms to reduce the size of FSM for modeling practical systems: state hierarchy, simultaneous activity, and nondeterminism. In statecharts, a state can possibly represent an enclosed state machine. In this case, the machine is in one of the states enclosed by the superstate (or-states) and concurrency is achieved by enabling two or more state machines to be active simultaneously (and-states, such as `elapsed` and `play` in Figure 4.7).

In petri net (PN) models, the system is represented by a graph of places connected by transitions. Places represent unbounded channels that carry "tokens" and the state of the system is represented at any given time by the number of tokens existing in a given subset of places. Transitions represent the elementary reaction of the system. A transition can be executed (fired) when it has a fixed, prespecified number of tokens in its input places. When fired, it consumes the input tokens and produces a fixed number of tokens on its output places. Since more than one transition may originate from the same place, one transition can execute while blocking another one by removing the tokens from shared input places. Hence, the model allows for nondeterminism and provides a natural representation of concurrency by allowing simultaneous execution of multiple transitions (Figure 4.8, left side).

The FSM and PN models have been originally developed with no reference to time or time constraints, but the capability of expressing and verifying timing requirements is key in many design domains (including embedded systems). Hence, both have been extended in order to allow time-related specifications. Time extensions differ according to the time model that is assumed. Models that represent time with a discrete time base are said to belong to the family of discrete time models, while the others are based on continuous (dense) time. Furthermore, proposals of extensions differ

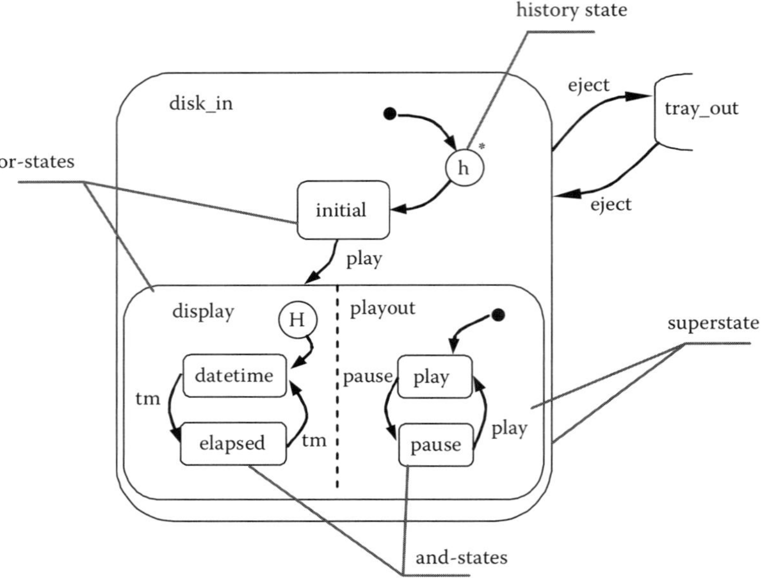

FIGURE 4.7 Example of statechart.

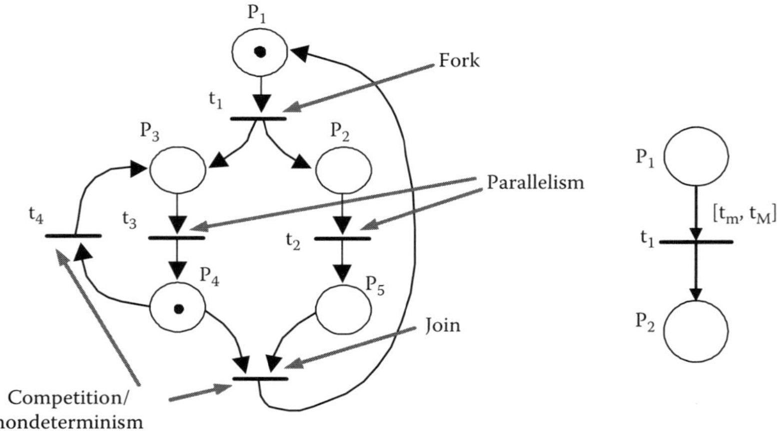

FIGURE 4.8 Sample PN showing examples of concurrent behavior and nondeterminism (left side) and notations for the TPN model (right side).

in how time references should be used in the system, whether a global clock or local clock should be used and how time should be used in guard conditions on transitions or states, inputs and outputs.

When building the set of reachable states, the time value adds another dimension, further contributing to the state explosion problem. In general, discrete time models are easier to analyze if compared with dense time models, but synchronization of signals and transitions results in fixed-point evaluation problems whenever the system model contains cycles without delays.

Please note, discrete-time systems are naturally prone to an implementation based on the time-triggered paradigm, where all actions are bound to happen at multiples of a time reference (usually implemented by means of a response to a timer interrupt or a cyclic RTOS task) and continuous time (CT) (asynchronous systems) conventionally correspond to implementations based on the event-based design paradigm, where system actions can happen at any time instant. This does not imply a correspondence between time-triggered systems and synchronous systems. The latter are characterized by the additional constraints that all system components must perform an action synchronously (at the same time) at each tick in a periodic time base.

Many models have been proposed in the research literature for time-related extensions. Among those, time petri nets (TPNs) [12,13] and timed automata (TA) [14] are probably the best known.

TA (an example in Figure 4.9) operates with a finite set of locations (states) and a finite set of real-valued clocks. All clocks proceed at the same rate and measure the amount of time that passed since they were started (reset). Each transition may reset some of the clocks and each defines a restriction on the value of the symbols as well as on the clock values required for it to happen. A state may be reached only if the values of the clocks satisfy the constraints and the proposition clause defined on the symbols evaluates to true.

Timed PNs [15] and TPNs are extensions of the PN formalism allowing for the expression of time-related constraints. The two differ in the way time advances: in Timed PNs time advances in transitions, thus violating the instantaneous nature of transitions (which makes the model much less prone to formal verification). In the TPN model, time advances while token(s) is are in place (Figure 4.10). Enabling and deadline times can be associated to transitions, the enabling time being the time a transition must be enabled before firing and the deadline being the time instant by which the transition must be taken (Figure 4.8, right side).

The additional notion of stochastic time allows the definition of the (generalized) stochastic PNs [16,17] used for the purpose of performance evaluation.

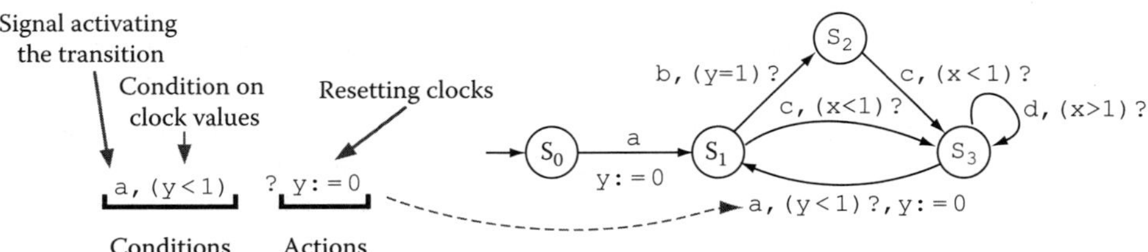

FIGURE 4.9　Example TA.

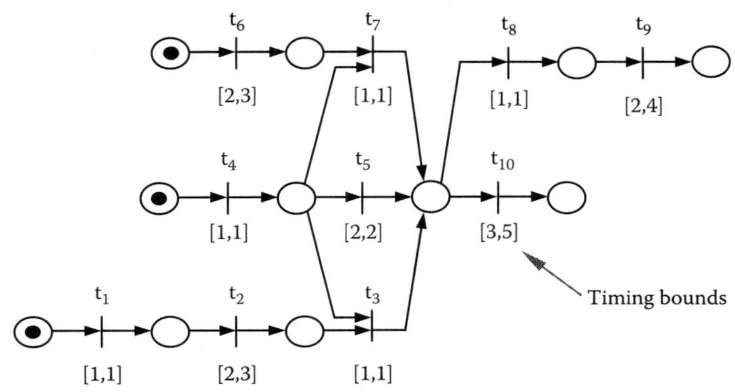

FIGURE 4.10　Sample TPN.

Many further extensions have been proposed for both TAs and TPNs. The task of comparing the two models for expressiveness should take into account all the possible variants and is probably not particularly interesting in itself. For most problems of practical interest, however, both models are essentially equivalent when it comes to expressive power and analysis capability [18].

A few tools based on the TA paradigm have been developed and are very popular. Among those, we cite Kronos [19] and Uppaal [20]. The Uppaal tool allows modeling, simulation, and verification of real-time systems modeled as a collection of nondeterministic processes with finite control structure and real-valued clocks, communicating through channels or shared variables [20,21]. The tool is free for no profit and academic institutions.

TAs and TPNs allow the formal expression of requirements for logical-level resources, timing constraints, and timing assumptions, but timing analysis only deals with abstract specification entities, typically assuming infinite availability of physical resources (such as memory or CPU speed). If the system includes an RTOS, with the associated scheduler, the model needs to account for preemption, resource sharing and the nondeterminism resulting from them. Dealing with these issues requires further evolution of the models.

For example, in TA, we may want to use clock variables for representing the execution time of each action. In this case, however, only the clock associated with the action scheduled on the CPU should advance, with all the others being stopped.

The hybrid automata model [22] combines discrete transition graphs with continuous dynamical systems. The value of system variables may change according to a discrete transition or it may change continuously in system states according to a trajectory defined by a system of differential equations. Hybrid automata have been developed for the purpose of modeling digital systems interacting with (physical) analog environments, but the capability of stopping the evolution of clock variables in states (first derivative equal to 0) makes the formalism suitable for the modeling of systems with preemption.

TPNs and TA can also be extended to cope with the problem of modeling finite computing resources and preemption. In the case of TA, the extension consists in the Stopwatch Automata model, which handles suspension of the computation due to the release of the CPU (because of real-time scheduling), implemented in the HyTech [23] (for linear hybrid automata) tool. Alternatively, the scheduler is modeled with an extension to the TA model, allowing for clock updates by subtraction inside transitions (besides normal clock resetting). This extension, available in the Uppaal tool, avoids the undecidability of the model when clocks associated with the actions not scheduled on the CPU are stopped.

Likewise, TPNs can be extended to the preemptive TPN model [24], as supported by the ORIS tool [25]. A tentative correspondence between the two models is traced in [26]. Unfortunately, in all these cases, the complexity of the verification procedure caused by the state explosion poses severe limitations upon the size of the analyzable systems.

Before moving on to the discussion of formal techniques for the analysis of time-related properties at the architecture level (schedulability), the interested reader is invited to refer to [27] for a survey on formal methods, including references to industrial examples.

4.1.2.2 Schedulability Analysis

If specification of functionality aims at producing a logically correct representation of system behavior, architecture-level design is where physical concurrency and schedulability requirements are expressed. At this level, the units of computation are the processes or threads (the distinction between these two operating system [OS] concepts is not relevant for the purpose of this chapter and in the following, the generic term "task" will be optionally used for both), executing concurrently in response to environment stimuli or prompted by an internal clock. Threads cooperate by exchanging data and synchronization or activation signals and contend for use of the execution resource(s) (the processor) as well as for the other resources in the system. The physical architecture level is also the place where the concurrent entities are mapped onto target hardware. This activity entails the selection of an appropriate scheduling policy (e.g., offered by a RTOS), and possibly support by timing or schedulability analysis tools.

Formal models, exhaustive analysis techniques, and model checking are now evolving toward the representation and verification of time and resource constraints together with the functional behavior. However, applicability of these models is strongly limited by state explosion. In this case, exhaustive analysis and joint verification of functional and nonfunctional behavior can be sacrificed for the lesser goal of analyzing only the worst-case timing behavior of coarse-grain design entities representing concurrently executing threads.

Software models for time and schedulability analysis deal with preemption, physical and logical resource requirements and resource management policies, and are typically limited to a quite simplified view of functional (logical) behavior, mainly limited to synchronization and activation signals.

To give an example if, for sake of simplicity, we limit discussion to single processor systems, the scheduler assigns the execution engine (the CPU) to threads (tasks) and the main objective of real-time scheduling policies is to formally guarantee the timing constraints (deadlines) on the thread response to external events.

In this case, the software architecture can be represented as a set of concurrent tasks (threads). Each task τ_i executes periodically or according to a sporadic pattern and it is typically represented by a simple set of attributes, such as the tuple $(C_i, \theta_i, p_i, D_i)$, representing the worst-case computation time, the period (for periodic threads) or minimum interarrival time (for sporadic threads), the priority and the relative (to the release time r_i) deadline of each thread instance.

Fixed Priority Scheduling and rate monotonic analysis (RMA) [28,29] are by far the most common real-time scheduling and analysis methodologies. RMA provides a very simple procedure for assigning static priorities to a set of independent periodic tasks together with a formula for checking schedulability against deadlines.

The highest priority is assigned to the task having the highest rate and schedulability is guaranteed by checking the worst-case scenario that can possibly happen. If the set of tasks is schedulable in that condition, then it is schedulable under all circumstances. For RMA, the critical condition happens when all tasks are released at the same time instant initiating the largest busy period (CT interval when the processor is busy executing tasks of a given priority level).

By analyzing the busy period (from $t = 0$), it is possible to derive the worst-case completion time W_i for each task τ_i. If the task can be proven to be complete before or at the deadline ($W_i \leq D_i$) then it can be guaranteed. The iterative formula for computing W_i (in case $\theta_i, \leq D_i$) is

$$W_i = C_i + \sum_{\forall j \in he(i)} \frac{W_i}{\theta_j} C_j$$

where $he(i)$ are the indices of those tasks having a priority higher than or equal to p_i.

Rate monotonic (RM) scheduling was developed starting from a very simple model where all tasks are periodic and independent. In reality, tasks require access to shared resources (apart from the processor) that can only be used in an exclusive way, for example, communication buffers shared among asynchronous threads.

In this case, it is possible that one task is blocked because another task holds a lock on the shared resources. When the blocked task has a priority higher than the blocking task, priority inversion occurs and finding the optimal priority assignment becomes an NP-hard problem. Real-time scheduling theory settles at finding resource assignment policies that provide at least a worst-case bound on the blocking time. The priority inheritance (PI) and the (Immediate) priority ceiling (PC) protocols [30] belong to this category.

The essence of the PC protocol (which has been included in the real-time OS OSEK standard issued by the automotive industry) consists in raising the priority of a thread entering a critical section to the highest among the priorities of all threads that may possibly request access to the same critical section. The thread returns to its nominal priority as soon as it leaves the critical section. The PC protocol ensures that each thread can be blocked at most once and bounds the duration of the blocking time to the largest critical section shared between itself or higher priority threads and lower priority threads.

When the blocking time due to priority inversion is bound for each task and its worst-case value is B_i, the evaluation of the worst-case completion time in the schedulability test becomes

$$W_i = C_i + \sum_{\forall j \in he(i)} \frac{W_i}{\theta_j} C_j + B_i$$

4.1.2.3 Mapping the Functional Model into the Architectural Model

The mapping of the actions defined in the functional model onto architectural model entities is the critical design activity where the two views are reconciled. In practice, the actions or transitions defined in the functional part must be executed in the context of one or more system threads. The definition of the architecture model (number and attributes of threads) and the selection of resource management policies, the mapping of the functional model into the corresponding architecture model, and the validation of the mapped model against functional and nonfunctional constraints is probably one of the major challenges in software engineering.

Single thread implementations are quite common and an easy choice that allows for (practical) verification of implementation and schedulability analysis, meaning that there exist CASE tools that can provide both, at least in the context of synchronous reactive MOCs. The entire functional specification is executed in the context of a single thread performing a never ending cycle where it serves events in a noninterruptable fashion according to the run-to-completion paradigm. The thread waits for an event (either external, like an interrupt from an I/O interface, or internal, like a call or signal from one object or FSM to another); fetches the event and the associated parameters and, finally, it executes the corresponding code.

All the actions defined in the functional part need be scheduled (statically or dynamically) for execution inside the thread. The schedule is usually driven by the partial order in the execution of the actions, as defined by the MOC semantics. Commercial implementations of this model range from code produced by the Esterel compiler [31] to single thread implementations by the Embedded Coder toolset from Mathworks and TargetLink from DSpace (of Simulink models) [32,33] or the single thread code generated by rational rose technical developer [34] for the execution of UML models.

The scheduling problem is much simpler than it is in the multithreaded case, since there is no need to account for thread scheduling and preemption and resource sharing usually results in trivial problems.

On the other extreme, one could define one thread for every functional block or every possible action. Each thread can be assigned its own priority, depending on the criticality and on the deadline of the corresponding action. At run time, the OS scheduler properly synchronizes and sequentializes the tasks so that the order of execution respects the functional specification.

Both approaches may be inefficient. The single thread implementation suffers from scheduling problems, due to the need for completing the processing of each system reaction within the base period of the system (the rate at which fastest events are processed/produced by the system). The one-to-one mapping of functions or actions to threads suffers from at least two problems. It may be difficult to provide a function-to-task mapping and a scheduling model that guarantees value- and time-determinism and preserves the semantics of the functional model. In addition, this implementation may introduce excessive scheduler overhead caused by the need for a context switch at each action. Considering that the action specified in a functional block can be very short and that the number of functional blocks is usually quite high (in many applications it is in the order of hundreds), the overhead of the OS could easily prove unbearable.

The designer essentially tries to achieve a compromise between these two extremes, balancing responsiveness with schedulability, flexibility of the implementation, and performance overhead.

4.1.3 Paradigms for Reuse: Component-Based Design

One more dimension can be added to the complexity of the software design problem if the need for maintenance and reuse is considered. To this purpose, component-based and object-oriented (OO) techniques have been developed for constructing and maintaining large and complex systems.

A component is a product of the analysis, design, or implementation phases of the life cycle and represents a prefabricated solution that can be reused to meet subsystem requirement(s). A component is commonly used as a vehicle for the reuse of two basic design properties:

- Functionality: The functional syntax and semantics of the solution the component represents.
- Structure: The structural abstraction the component represents. These can range from "small grain" to architectural features, at the subsystem or system level.

The generic requirement for "reusability" maps into a number of issues. Probably the most relevant property that components should exhibit is "abstraction," meaning the capability of hiding implementation details and describing relevant properties only. Components should also be easily adaptable to meet changing processing requirements and environmental constraints through controlled modification techniques (like "inheritance" and "genericity") and "composition" rules must be used to build higher-level components from existing ones. Hence an ideal component-based modeling language should ensure that properties of components (functional properties, such as liveness, reachability, deadlock avoidance or nonfunctional properties such as timeliness and schedulability) are preserved or at least decidable after composition. Additional (practical) issues include support for implementation, separate compilations, and imports.

Unfortunately, reconciling the standard issues of software components, such as context-independence, understandability, adaptability, and composability, with the possibly conflicting requirements of timeliness, concurrency, and distribution, typical of hard real-time system development, is not an easy task and still an open problem.

OO design of systems has traditionally embodied the (far from perfect) solution to some of these problems. While most (if not all) OO methodologies, including the UML, offer support for inheritance and genericity, adequate abstraction mechanisms and especially composability of properties are still subject of research.

With its latest release, the UML has reconciled the abstract interface abstraction mechanism with the common box-port-wire design paradigm. Lack of an explicit declaration of required interface and absence of a language feature for structured classes were among the main deficiencies of classes and objects, if seen as components. In UML 2.0, ports allow for a formal definition of a required as well as a provided interface. Association of protocol declaration with the ports further improves clarify the semantics of interaction with the component. In addition, the concept of a structured class allows for a much better definition of a component.

Of course, port interfaces and the associated protocol declarations are not sufficient for specifying the semantics of the component. In UML 2.0, the object constraint language (OCL) can also be used to define behavioral specifications in the form of invariants, preconditions, and postconditions, in the style of the contract-based design methodology (implemented in Eiffel [35]).

Recently, automotive companies have promoted the AUTOSAR consortium, to develop a standard for software components in automotive systems. To achieve the technical goals of modularity, scalability, transferability, and reusability of functions, AUTOSAR provides a common software infrastructure based on standardized interfaces for the different layers.

The current version of the AUTOSAR model includes a reference architecture and interface specifications. Also, the AUTOSAR consortium recently acknowledged that the specification was lacking a formal model of components for design-time verification of their properties and for the development of virtual platforms. As a result, the definition of the AUTOSAR metamodel was started.

The AUTOSAR project has been focused on the concepts of location independence, standardization of interfaces, and portability of code. While these goals are undoubtedly of extreme importance, their achievement will not necessarily be a sufficient condition for improving the quality of software systems. As for most other embedded system, car electronics are characterized by functional as well as nonfunctional properties, assumptions, and constraints. The current specification has at least two major shortcomings that prevent achieving the desired goals. The AUTOSAR metamodel, as of now, is affected by the lack of a clear and unambiguous communication and synchronization semantics and the lack of a timing model.

4.2 Synchronous vs. Asynchronous Models

Verification of functional and nonfunctional properties of software demands for a formal semantics and a strong mathematical foundation of the models. Many argue that a fully analyzable model cannot be constructed unless shedding generality and restricting the behavioral model to simple and analyzable semantics.

Among the possible choices, the *SR* model enforces determinism and provides a sound methodology for checking functional and nonfunctional properties at the price of expensive implementation and performance limitations. Moreover, the synchronous model is built on assumptions (computation times neglectable with respect to the environment dynamics and synchronous execution) that not always apply to the controlled environment and to the architecture of the system.

Asynchronous or general models typically allow for (controlled) nondeterminism and more expressiveness, at the price of strong limitations in the extent of the functional and nonfunctional verification that can be performed.

Some modeling languages, such as UML, are deliberately general enough, so that they can be possibly used for specifying a system according to a generic asynchronous or synchronous paradigm provided that a suitable set of extensions (semantics restrictions) are defined.

By the end of this chapter, it should be clear how neither of the two design paradigms (synchronous or asynchronous) is currently capable of providing the complete solution to all the implementation challenges of complex systems. The requirements of the synchronous assumption (on the environment and the execution platform) are difficult to meet and component-based design is very difficult (if not impossible). The asynchronous paradigm, on the other hand, results in implementations, which are very difficult to analyze for logical and time behavior.

4.3 Synchronous Models

In the *SR* model, time advances at discrete instants and the program progresses according to successive atomic reactions (sets of synchronously executed actions), which are performed instantaneously (zero computation time) meaning that the reaction is fast enough with respect to the environment. The resulting discrete-time model is quite natural to many domains, such as control engineering and (hardware) synchronous digital logic design (VHDL).

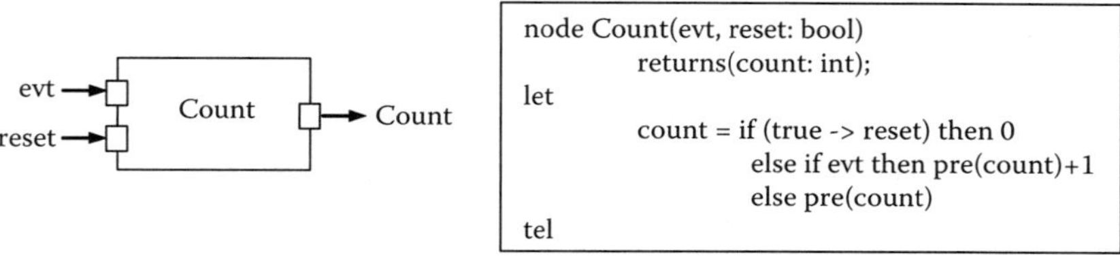

FIGURE 4.11 Example Lustre node and its program.

Composition of system blocks implies product combination of the states and the conjunction of the reactions for each component. In general, this results in a fixed-point problem and the composition of the function blocks is a relation, not a function, as outlined in the previous Section 4.1.

The French synchronous languages Signal, Esterel, and Lustre are probably the best representatives of the synchronous modeling paradigm.

Lustre [36,37] is a declarative language based on the dataflow model where nodes are the main building blocks. In Lustre, each flow or stream of values is represented by a variable, with a distinct value for each tick in the discrete time base. A node is a function of flows: it takes a number of typed input flows and defines a number of output flows by means of a system of equations.

A Lustre node (an example in Figure 4.11) is a pure functional unit except for the `pre` and initialization $(->)$ expressions, which allow referencing the previous element of a given stream or forcing an initial value for a stream. Lustre allows streams at different rates, but in order to avoid nondeterminism it forbids syntactic cyclic definitions.

Esterel [38] is an imperative language, more suited for the description of control. An Esterel program consists of a collection of nested, concurrently running threads. Execution is synchronized to a single, global clock. At the beginning of each reaction, each thread resumes its execution from where it paused (e.g., at a pause statement) in the last reaction, executes imperative code (e.g., assigning the value of expressions to variables and making control decisions), and finally either terminates or pauses waiting for the next reaction.

Esterel threads communicate exclusively through signals representing globally broadcast events. A signal does not persist across reactions and it is present in a reaction if and only if it is emitted by the program or by the environment.

Although tool support for this feature has now been discontinued, Esterel formally allows cyclic dependencies and treats each reaction as a fixed-point equation, but the only legal programs are those that behave functionally in every possible reaction. The solution of this problem is provided by "constructive causality" [39], which amounts at checking if, regardless of the existence of cycles, the output of the program (the binary circuit implementing it) can be formally proven to be causally dependent from the inputs for all possible input assignments.

The language allows for conceptually sequential (operator;) or concurrent (operator ‖) execution of reactions, defined by language expressions handling signal identifiers (as in the example of Figure 4.12). All constructs take 0 time except `await` and `loop ... each ...`, which explicitly produce a program pause. Esterel includes the concept of preemption, embodied by the `loop ... each R` statement in the example of Figure 4.12 or the `abort` *action* when *signal* statement. The reaction contained in the body of the loop is preempted (and restarted) when the signal R is set. In case of an `abort` statement, the reaction is preempted and the statement terminates.

Formal verification was among the original objectives of Esterel. In synchronous languages, verification of properties may be performed with the definition of a special program called "observer" that observes the variables or signals of interest and at each step decides if the property is

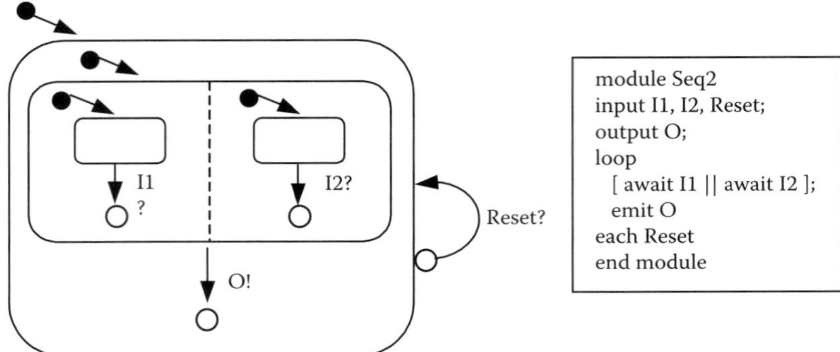

FIGURE 4.12 An example showing features of the Esterel language as an equivalent statechart-like visual formalization.

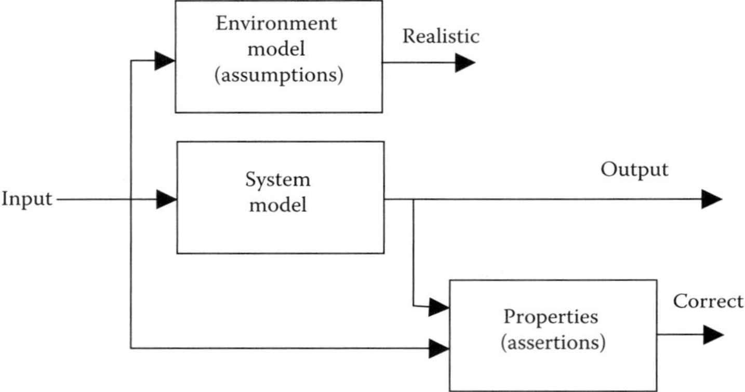

FIGURE 4.13 Verification by observers.

fulfilled (Figure 4.13). A program satisfies the property if and only if the observer never complains during any execution.

The verification tool takes the program implementing the system, an observer of the desired property, and another program modeling the assumptions on the environment. The three programs are combined in a synchronous product, and the tool explores the set of reachable states. If the observer never reaches a state where the system property is not valid before reaching a state where the assumption observer declares violation of the environment assumptions, then the system is correct. The process is described in detail in [40].

Finally, the commercial package Simulink by Mathworks [41] allows modeling and simulation of control systems according to a SR MOC, although its semantics is not formally nor completely defined. Rules for translating a Simulink model into Lustre have been outlined in [42], and in [43], the very important problem of how to map a zero-execution time Simulink semantics into a software implementation of concurrent threads where each computation necessarily requires a finite execution time is discussed.

4.3.1 Architecture Deployment and Timing Analysis

Synchronous models are typically implemented as a single task that executes according to an event server model. Reactions decompose into atomic actions that are partially ordered by the causality analysis of the program. The scheduling is generated at compile time trying to exploit the partial causality order of functions in order to make the best possible use of hardware and shared resources. The main concern is checking that the synchrony assumption holds, that is, ensuring that the longest chain of reactions ensuing from any internal or external event is completed within the step duration. Static scheduling means that critical applications are deployed without the need for any OS (and the corresponding overhead). This reduces system complexity and increases predictability avoiding preemption, dynamic contention over resources, and other nondeterministic OS functions.

4.3.2 Tools and Commercial Implementations

Lustre is implemented by the commercial toolset Scade, which offers an editor that manipulates both graphical and textual descriptions; two code generators, one of which is accepted by certification authorities for qualified software production; a simulator; and an interface to verification tools such as the plug-in from Prover [44].

The early Esterel compilers had been developed by Gerard Berry's group at INRIA/CMA and freely distributed in binary form. The commercial version of Esterel was first marketed in 1998 and it is now available from Esterel Technologies, which later acquired the Scade environment. Scade has been used in many industrial projects, including integrated nuclear protection systems (Schneider Electric), flight control software (Airbus A340–600), and track control systems (CS Transport). Dassault Aviation was one of the earliest supporters of the Esterel project, and has long been one of its users.

Several verification tools use the synchronous observer technique for checking Esterel programs [40]. It is also possible to verify implementation of Esterel programs with tools leveraging explicit state space reduction and bisimulation minimization (FC2Tools) and, finally, tools can also be used to automatically generate test sequences with guaranteed state/transition coverage.

The very popular Simulink tool by Mathworks was developed with the purpose of simulating control algorithms and has been since its inception extended with a set of additional tools and plug-ins, such as, for example, the Stateflow plug-in for the definition of the FSM behavior of a control block, allowing modeling of hybrid systems, and a number of automatic code generation tools, such as the Real-time Workshop and Embedded Coder by Mathworks and TargetLink by DSpace.

4.3.3 Challenges

The main challenges and limitations that the Esterel language must face when applied to complex systems are the following:

- Despite improvements, the space- and time-efficiency of the compilers is still not satisfactory.
- Embedded applications can be deployed on architectures or control environments that do not comply with the SR model.
- Designers are familiar with other dominant methods and notations. Porting the development process to the synchronous paradigm and languages is not easy.

Efficiency limitations are mainly due to the formal compilation process and the need to check for constructive causality. The first three Esterel compilers used automata-based techniques and produced efficient code for small programs, but they did not scale to large-scale systems because of state explosion. Versions 4 and 5 are based on translation into digital logic and generate smaller executables at the price of slow execution. (The program generated by these compilers requires time for evaluating each gate at every clock cycle.) This inefficiency can produce code 100 times slower than that from previous compilers [45].

The version 5 of the compiler allows cyclic dependencies by exploiting Esterel's constructive semantics. Unfortunately, this requires evaluating all the reachable states by symbolic state space traversal [46], which makes it extremely slow.

As for the difficulty in matching the basic paradigm of synchrony with system architectures, the main reasons of concern are

- Bus and communication lines, if not specified according to a synchronous (time triggered) protocol and the interfaces with the analog world of sensors and actuators
- Dynamics of the environment, which can possibly invalidate the instantaneous execution semantics

The former has been discussed at length in a number of papers (such as [47,48]), giving conditions for providing a synchronous implementation in distributed systems.

Finally, in order to integrate synchronous languages with the mainstream commercial methodologies and languages, translation and import tools are required. For example, it is possible from Scade to import discrete time Simulink diagrams and Sildex allows importing Simulink/Stateflow discrete time diagrams. Another example is UML, with the attempt at an integration between Esterel Studio and Rational Rose and the proposal for an Esterel/UML coupling drafted by Dassault [49] and adopted by commercial Esterel tools.

4.4   Asynchronous Models

UML and SDL are languages developed, respectively, in the context of general-purpose computing and in the context of (large) telecommunication systems. UML is the merger of many OO design methodologies aimed at the definition of generic software systems. Its semantics is not completely specified and intentionally retains many variation points in order to adapt to different application domains. These are (among others), the reasons for which, to be practically applicable to the design of embedded systems, further characterization (a specialized profile in UML terminology) is required. In the 2.0 revision of the language, the system is represented by a (transitional) model where active and passive components, communicating by means of connections through port interfaces, cooperate in the implementation of the system behavior. Each reaction to an internal or external event results in the transition of a statechart automaton describing the object behavior.

SDL (standard ISO-ITU) has a more formal background, since it was developed in the context of software for telecommunication systems for the purpose of easing the implementation of verifiable communication protocols. An SDL design consists of blocks cooperating by means of asynchronous signals. The behavior of each block is represented by one or more (conceptually concurrent) processes. Each process, in turn, implements an extended FSM.

Until the development of the UML profile for schedulability, performance, and time (standard), UML did not provide any formal means for specifying time or time-related constraints, nor for specifying resources and resource management policies. The deployment diagrams were the only (inadequate) means for describing the mapping of software onto the hardware platform and tool vendors had tried to fill the gap by proposing nonstandard extensions. The upcoming release of the new modeling and analysis for real-time and embedded (MARTE) system profile attempts at filling some of these gaps.

The situation with SDL is not much different, although SDL offers at least the notion of global and external time. Global time is made available by means of a special expression and can be stored in variables or sent in messages.

Implementation of asynchronous languages, typically (but not necessarily) relies on an OS. The latter is responsible for scheduling, which is necessarily based on static (design time) priorities, if a commercial OS is used. Unfortunately, as it will be clear in the following, real-time schedulability analysis techniques are only applicable to very simple models and are extremely difficult to generalize to most models of practical interest or even to the implementation model assumed by most (if not all) commercial tools.

4.4.1 UML

The UML represents a collection of engineering practices that have proven successful in the modeling of large and complex systems and has emerged as the software industry's dominant OO-modeling language.

Born at Rational in 1994, UML was taken over in 1997 at version 1.1 by the object management group (OMG) revision task force (RTF), which became responsible for its maintenance. The RTF released UML version 1.4 in September 2001 and a major revision, UML 2.0, which also aims to address the embedded or real-time dimension, has been adopted in late 2003, and it is posted on the OMG's Web site as "UML 2.0 Final Adopted Specification" [4].

UML has been designed as a wide-ranging, general-purpose modeling language for specifying, visualizing, constructing, and documenting the artifacts of software systems. It has been successfully applied to a wide range of domains, ranging from health and finance to aerospace and e-commerce, and its domains go even beyond software, given recent initiatives in areas as systems engineering, testing and hardware design. A joint initiative between OMG and INCOSE (International Council on Systems Engineering) is working on a profile for systems engineering and the SysML consortium was established and defined a systems modeling language based on UML (SysML), now a standard OMG profile. At the time of this writing, over 60 UML CASE tools can be listed from the OMG resource page (http://www.omg.org).

After revision 2.0, the UML specification consists of four parts:

- UML 2.0 Infrastructure, defining the foundational language constructs and the language semantics in a more formal way than it was in the past
- UML 2.0 Superstructure, which defines the user level constructs
- OCL 2.0 Object Constraint Language, which is used to describe expressions (constraints) on UML models
- UML 2.0 Diagram Interchange, including the definition of the XML-based XMI format, for model interchange among tools

UML comprises of a metamodel definition and a graphical representation of the formal language, but it intentionally refrains from including any design process. The UML language in its general form is deliberately semiformal and even its state diagrams (a variant of statecharts) retain sufficient semantics variation points in order to ease adaptability and customization.

The designers of UML realized that complex systems cannot be represented by a single design artifact. According to UML, a system model is seen under different views, representing different aspects. Each view corresponds to one or more diagrams, which taken together, represent a unique model. Consistency of this multiview representation is ensured by the UML metamodel definition. The diagram types included in the UML 2.0 specification are represented in Figure 4.14, as organized in the two main categories that relate to "structure" and "behavior."

When domain-specific requirements arise, more specific (more semantically characterized) concepts and notations can be provided as a set of stereotypes and constraints and packaged in the context of a profile.

Structure diagrams show the static structure of the system, that is, specifications that are valid irrespective of time. Behavior diagrams show the dynamic behavior of the system. The main diagrams are

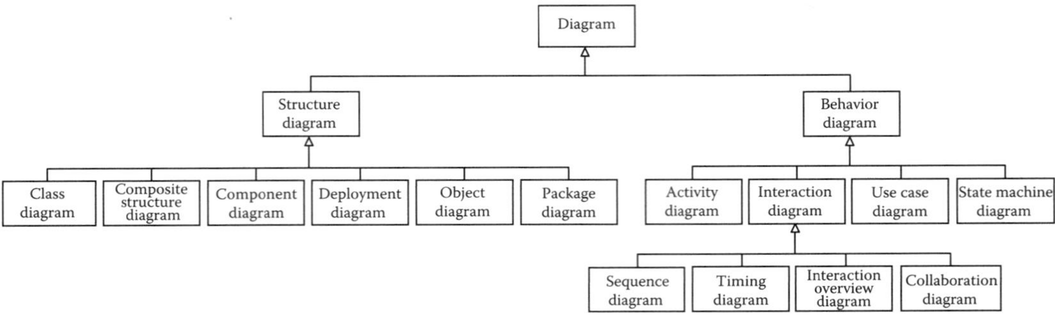

FIGURE 4.14   Taxonomy of UML 2.0 diagrams.

- Use case diagram, a high-level (user requirements-level) description of the interaction of the system with external agents
- Class diagram, representing the static structure of the software system, including the OO description of the entities composing the system, and of their static properties and relationships
- Behavior diagrams (including sequence diagrams and state diagrams as variants of message sequence charts [MSCs] and statecharts), providing a description of the dynamic properties of the entities composing the system, using various notations
- Architecture diagrams (including composite and component diagrams, portraying a description of reusable components), a description of the internal structure of classes and objects and a better characterization of the communication superstructure, including communication paths and interfaces
- Implementation diagrams, containing a description of the physical structure of the software and hardware components

The class diagram is typically the core of a UML specification, as it shows the logical structure of the system. The concept of classifiers (class) is central to the OO design methodology. Classes can be defined as user-defined types consisting of a set of attributes defining the internal state and a set of operations (signature) that can be possibly invoked on the class objects resulting in an internal transition. As units of reuse, classes embody the concepts of encapsulation (or information) hiding and abstraction. The signature of the class abstracts the internal state and behavior, and restricts possible interactions with the environment. Relationships exist among classes and relevant relationships are given special names and notations, such as, aggregation and composition, use and dependency. The generalization (or refinement) relationship allows controlled extension of the model by letting a derived class specification inherit all the characteristics of the parent class (attributes and operations, but also, selectively, relationships) while providing new ones (or redefining the existing).

Objects are instances of the type defined by the corresponding class (or classifier.) As such, they embody all of the classifier attributes, operations, and relationships. Several books [50,51] have been dedicated to the explanation of the full set of concepts in OO design. The interested reader is invited to refer to the literature on the subject for a more detailed discussion.

All diagram elements can be annotated with constraints, expressed in OCL or in any other formalism that the designer sees as appropriate. A typical class diagram showing dependency, aggregation, and generalization associations is shown in Figure 4.15.

UML 2.0 finally acknowledged the need for a more formal characterization of the language semantics and for better support for component specifications. In particular, it became clear that simple

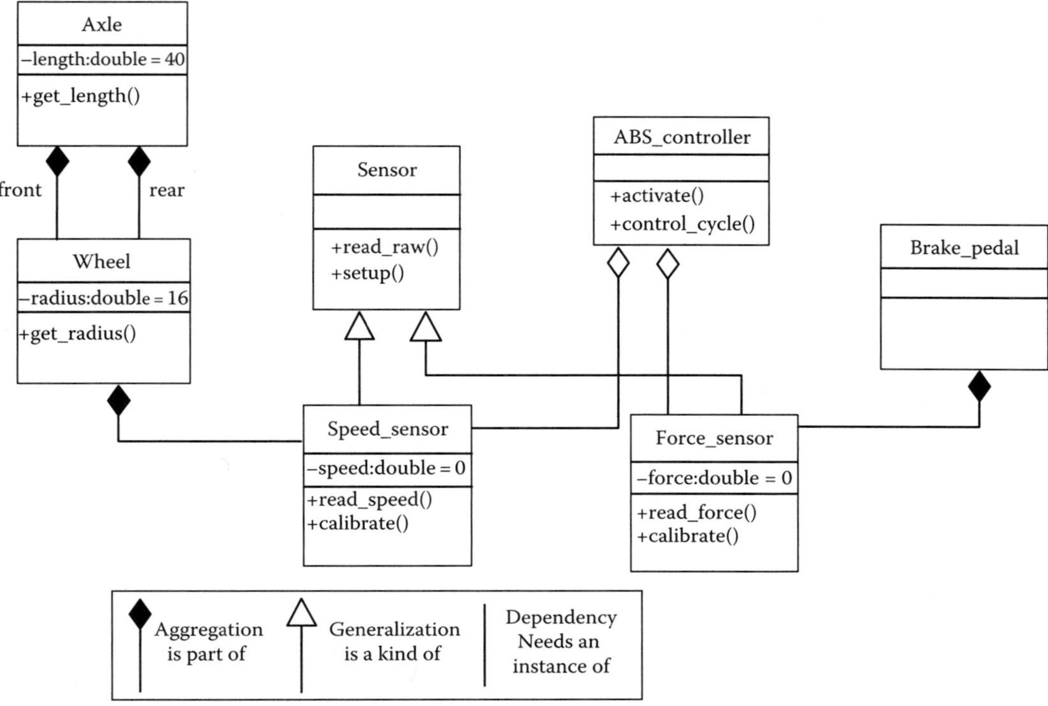

FIGURE 4.15   Sample class diagram.

classes provide a poor match for the definition of a reusable component (as outlined in previous sections).

As a result, necessary concepts, such as the means to clearly identify provided and (especially) required interfaces have been added by means of the port construct. An interface is an abstract class declaring a set of functions with their associated signature. Furthermore, structured classes and objects allow the designer to formally specify the internal communication structure of a component configuration.

UML 2.0 classes, structured classes, and components are now encapsulated units that model active system components and can be decomposed into contained classes communicating by signals exchanged over a set of ports, which model communication terminals (Figure 4.16). A port carries both structural information on the connection between classes or components and protocol information that specify what messages can be exchanged across the connection. A state machine and/or a UML Sequence Diagram may be associated to a protocol to express the allowable message exchanges. Two components can interact if there is a connection between any two ports that they own and that support the same protocol in complementary (or conjugated) roles. The behavior or reaction of a component to an incoming message or signal is typically specified by means of one or more statechart diagrams.

Behavior diagrams comprise statechart diagrams, sequence diagrams, and collaboration diagrams.

Statecharts [11] describe the evolution in time of an object or an interaction between objects by means of a hierarchical state machine. UML statecharts are extensions of Harel's statecharts, with the possibility of defining actions upon entering or exiting a state as well as actions to be executed when a transition is taken. Actions can be simple expressions or calls to methods of the attached object (class) or entire programs. Unfortunately, not only the turing-completeness of actions prevents decidability

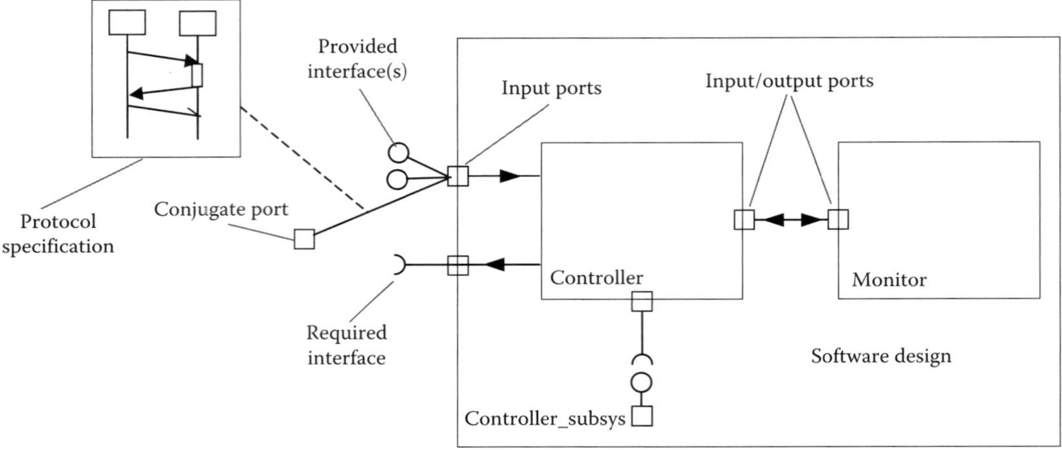

FIGURE 4.16   Ports and Components in UML 2.0.

of properties in the general model, but UML does not even clarify most of the semantics variations left open by the standard Statecharts formalism.

Furthermore, the UML specification explicitly gives actions a run-to-completion execution semantics, which makes them non-preemptable and makes the specification (and analysis) of typical RTOS mechanisms such as interrupts and preemption impossible.

To give an example of UML Statecharts, Figure 4.17 shows a sample diagram where, upon entry of the composite state (the outermost rectangle), the subsystem finds in three concurrent (and-type) states, named `Idle`, `WaitForUpdate`, and `Display_all`, respectively. Upon entry in the `WaitForUpdate` state, the variable `count` is also incremented. In the same portion of the diagram, reception of message `msg1` triggers the exit action setting the variable `flag` and the

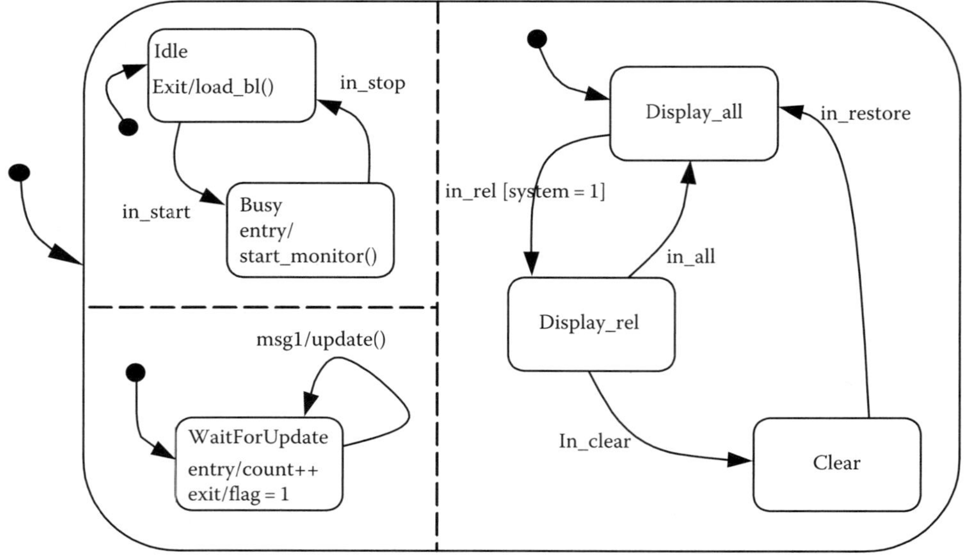

FIGURE 4.17   Example of UML Statechart.

(unconditioned) transition with the associated "call action" `update()`. The `count` variable is finally incremented upon reentry in the state `WaitForUpdate`.

Statechart diagrams provide the description of the state evolution of a single object or class, but are not meant to represent the emergent behavior deriving from the cooperation of more objects, neither are appropriate for the representation of timing constraints. "Sequence diagrams" partly fill this gap. Sequence diagrams show the possible message exchanges among objects, ordered along a time axis. The timepoints corresponding to message related events can be labeled and referred to in constraint annotations. Each sequence diagram focuses on one particular scenario of execution and provides an alternative to temporal logic for expressing timing constraints in a visual form (Figure 4.18).

"Collaboration diagrams" also show message exchanges among objects, but they emphasize structural relationships among objects (i.e., who talks with whom) rather than time sequences of messages.

Collaboration diagrams are also the most appropriate way for representing logical resource sharing among objects. Labeling of messages exchanged across links defines the sequencing of actions in a similar (but less effective) way to what can be specified with sequence diagrams (Figure 4.19).

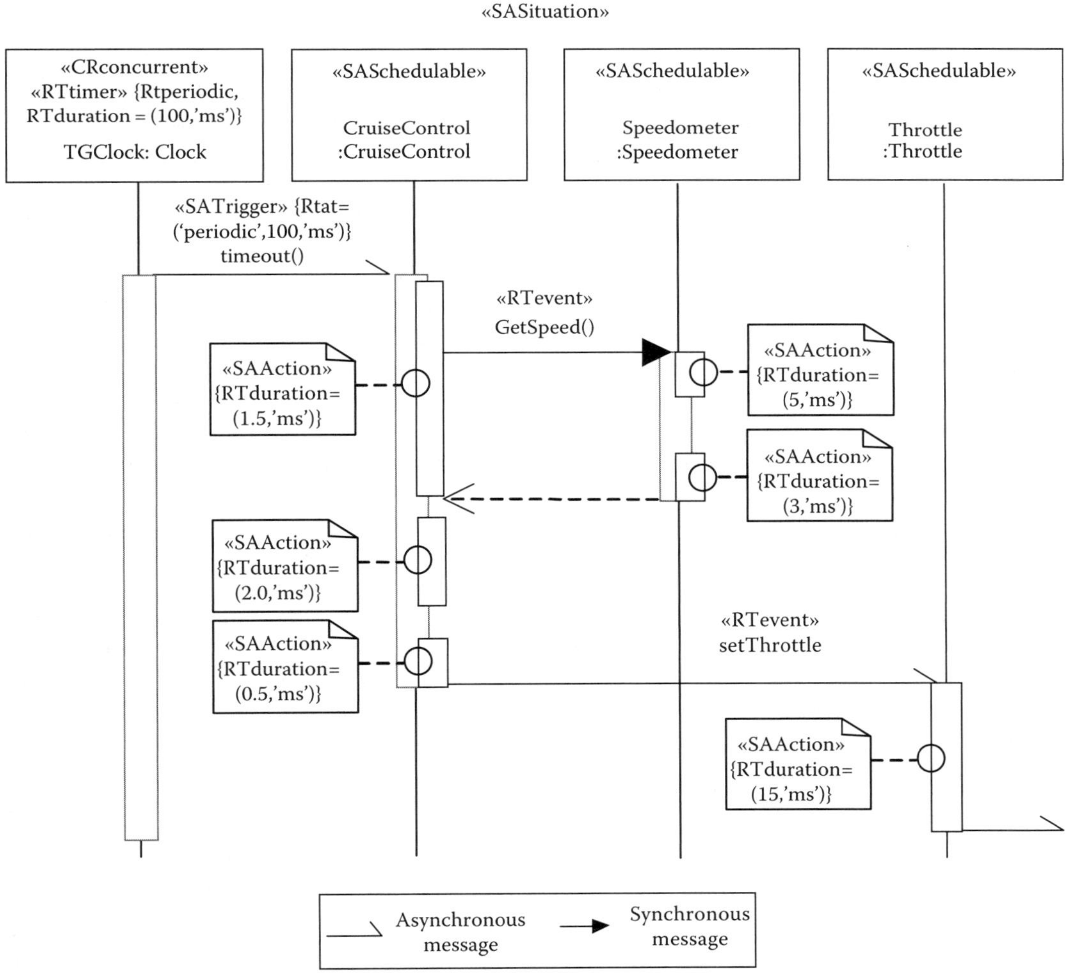

FIGURE 4.18 Sample sequence diagram with annotations showing timing constraints.

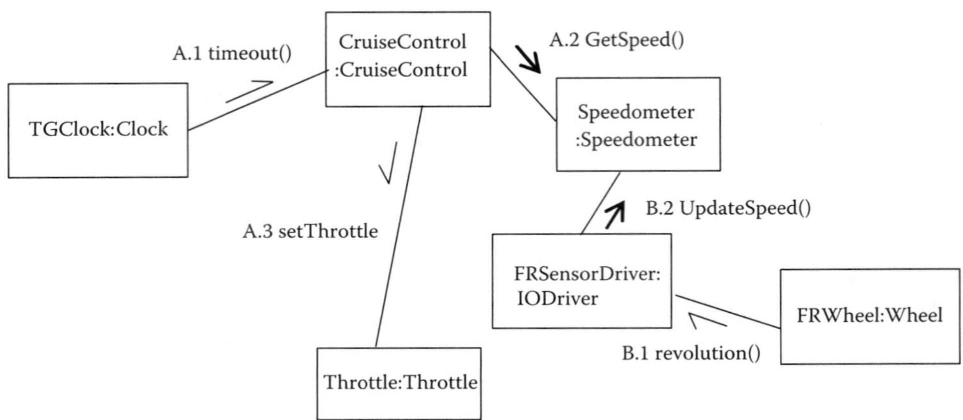

FIGURE 4.19   Defining action sequences in a collaboration diagram.

Despite availability of multiple diagram types (or maybe because of it), the UML metamodel is quite weak when it comes to the specification of dynamic behavior. The UML metamodel concentrates on providing structural consistency among the different diagrams and provides sufficient definition for the static semantics, but the dynamic semantics is never adequately addressed, up to the point that a major revision of the UML action semantics has become necessary. UML is currently headed in a direction where it will eventually become an executable modeling language, which would for example allow early verification of system functionality. Within the OMG, a standardization action has been purposely defined with the goal of providing a new and more precise definition of actions. This activity goes under the name of "action semantics for the UML". Until UML actions are given a more precise semantics, a faithful model, obtained by combining the information provided by the different diagrams is virtually impossible. Of course, this also nullifies the chances for formal verification of functional properties on a standard UML model.

However, simulation or verification of (at least) some behavioral property and (especially) automatic production of code are features that tool vendors cannot ignore if UML is not to be relegated at the role of simply documenting software artifacts. Hence, CASE tools provide an interpretation of the variation points. This means that validation, code generation, and automatic generation of test cases are tool-specific and depend upon the semantics choices of each vendor.

Concerning formal verification of properties, it is important to point out that UML does not provide any clear means for specifying the properties that the system (or components) is expected to satisfy, neither any means for specifying assumptions on the environment. The proposed use of OCL in an explicit contract section to specify assertions and assumptions acting upon the component and its environment (its users) can hopefully fill this gap.

As of today, research groups are working on the definition of a formal semantic restriction of UML behavior (especially by means of the statecharts formalism), in order to allow for formal verification of system properties [52,53]. After the definition of such restrictions, UML models can be translated into the format of existing validation tools for timed MSCs or TA.

Finally, the last type of UML diagrams are implementation diagrams, which can be either component diagrams or deployment diagrams. Component diagrams describe the physical structure of the software in terms of software components (modules) related with each other by dependency and containment relationships. Deployment diagrams describe the hardware architecture in terms of processing or data storage nodes connected by communication associations, and show the placement of software components onto the hardware nodes.

The need to express in UML timeliness-related properties and constraints, and the pattern of hardware and software resource utilization as well as resource allocation policies and scheduling algorithms found a (partial) response only in 2001 with the OMG issuing a standard SPT profile, which is now being replaced by the MARTE profile for real-time and embedded systems. The specification of timing attributes and constraints in UML designs will be discussed in the following subsection 4.4.3. Finally, the OMG has developed also a testing profile for UML 2.0 in 2005, with the objective of deriving and validating test specifications from a formal UML model.

Also, UML profiles for quality of service (QoS) and fault-tolerance characteristics and for systems-on-a-chip have been defined.

4.4.1.1 OCL

The OCL [54] is a formal language used to describe (constraint) expressions on UML models. An OCL expression is typically used to specify invariants or other type of constraint conditions that must hold for the system. OCL expressions refer to the contextual instance, that is, the model element to which the expression applies, such as classifiers, e.g., types, classes, interfaces, associations (acting as types), and datatypes. Also, all attributes, association-ends, methods, and operations without side effects that are defined on these types can be used.

OCL can be used to specify invariants associated with a classifier. In this case, it returns a Boolean type and its evaluation must be true for each instance of the classifier at any moment in time (except when an instance is executing an operation).

Preconditions and postconditions are other types of OCL constraints that can be possibly linked to an operation of a classifier and their purpose is to specify the conditions or contract under which the operation executes (Figure 4.20). If the caller fulfills the precondition before the operation is called, then the called object ensures the postcondition to hold after execution of the operation, but of course, only for the instance that executes the operation.

4.4.2 Specification and Description Language

The SDL is an International Telecommunications Union (ITU-T) standard promoted by the SDL Forum Society for the specification and description of systems [55].

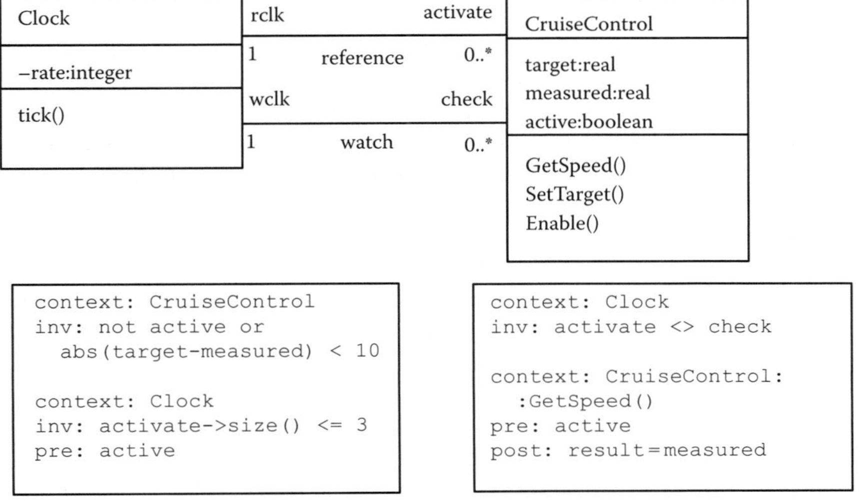

FIGURE 4.20 OCL examples adding constraints or defining behavior of operations in a class diagram.

Since its inception, a formal semantics has been part of the SDL standard (Z.100), including visual and textual constructs for the specification of both the architecture and the behavior of a system. The behavior of (active) SDL objects is described in terms of concurrently operating and asynchronously communicating abstract state machines. SDL provides the formal behavior semantics that UML is currently missing: it is a language based on communicating state machines, enabling tool support for simulation, verification, validation, testing, and code generation.

In SDL, systems are decomposed into a hierarchy of block agents communicating via (unbounded) channels that carry typed signals. Agents may be used for structuring the design and can in turn be decomposed into subagents until leaf blocks are decomposed into process agents. Block and process agents differ since blocks allow internal concurrency (subagents), while process agents only have an interleaving behavior.

The behavior of process agents is specified by means of extended finite and communicating state machines (SDL services) represented by a connected graph consisting of states and transitions. Transitions are triggered by external stimuli (signals, remote procedure calls) or conditions on variables. During a transition, a sequence of actions may be performed, including the use and manipulation of data stored in local variables or asynchronous interaction with other agents or the systems environment via signals that are placed into and consumed from channel queues.

Channels are asynchronous (as opposed to synchronous or rendezvous) FIFO queues (one for each process) and provide a reliable, zero- or finite-delay transmission of communication elements from a sender to a receiver agent.

Signals sent to an agent will be delivered to the input port of the agent. Signals are consumed in the order of their arrival either as a trigger of a transition or by being discarded in case there is no transition defined for the signal in the current state. Actions executed in response to the reception of input signals include expressions involving local variables or calls to procedures.

In summary, an agent definition consists of

- Behavior specification given by the agent extended FSM
- Structure and interaction diagram detailing the hierarchy of system agents with their internal communication infrastructure
- Variables (attributes) under the control of each agent
- Black box or component view of the agent defining the interaction points (ports) with the provided and required interfaces

Figure 4.21 shows an example of SDL notation with five interacting blocks and one leaf block (C2) decomposed into processes. The behavior of the process is specified in terms of its states and reactions as represented in the right diagram, where rectangles with concave points represent messages received, and rectangles with convex points denote messages sent. The SDL notation is a representation of the behavior of the equivalent extended state machine (an example in the right side of Figure 4.22).

Figure 4.22 shows a process behavior and a matching representation by means of an extended FSM (right side). The figure shows the use of a variable (`var1`) as a continuous input signal for the state S3. A continuous input is an SDL Boolean expression in terms of one or more variables. If the expression evaluates to true and no transition on regular signals is available, the corresponding branch is taken. Please note that continuous signals are evaluated after all other signals are checked (i.e., they have lowest priority).

The state with asterisk reached from the continuous input signal transition is a placeholder, representing the starting state of the transition (S3) itself.

Agents combine behavior and structure in a single entity and fit nicely into the concept of active objects in OO languages (active objects are objects that possess a thread of control and may initiate a reaction).

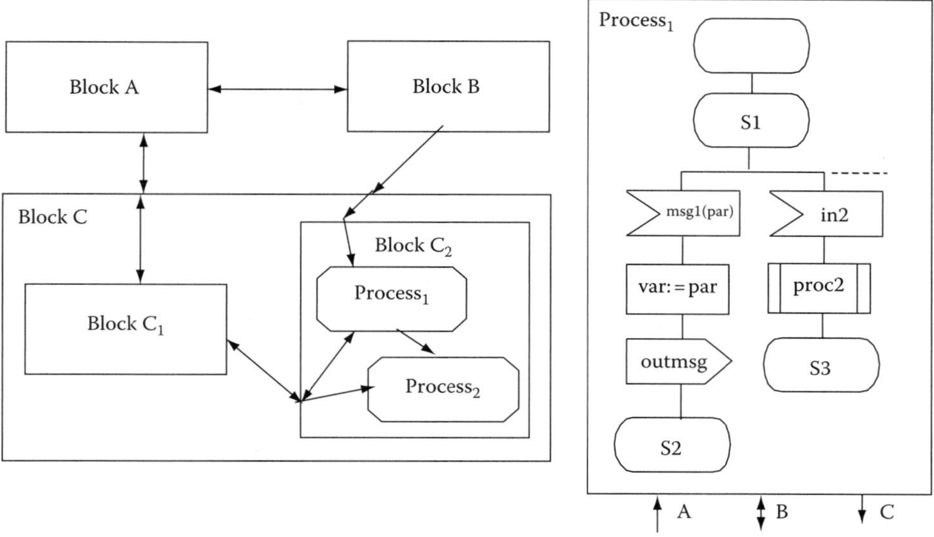

FIGURE 4.21 Interacting SDL blocks, block decomposition, and the description of the behavior of a process.

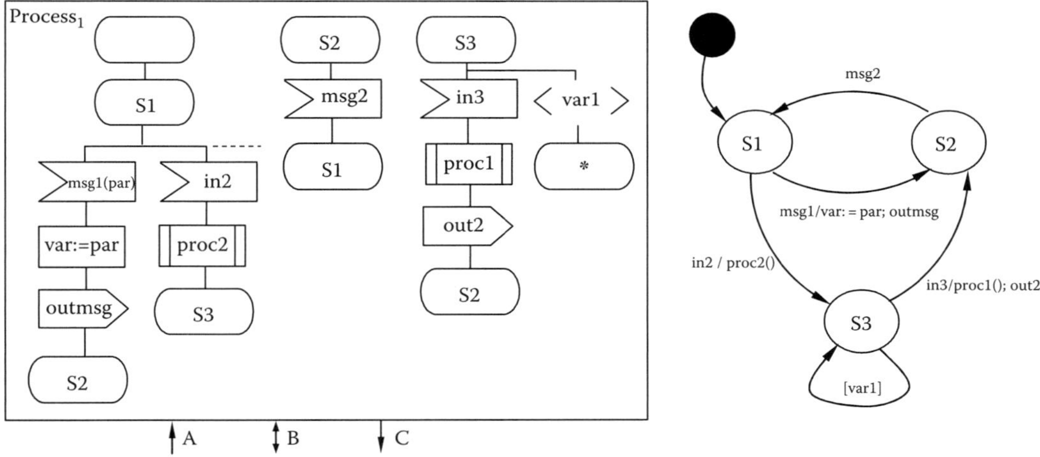

FIGURE 4.22 Process behavior and the corresponding extended FSM.

The latest SDL version, SDL 2000, extends the language by including typical features of OO and component-based modeling techniques, including encapsulation of services by ports (gates) and interfaces, classifiers and associations, and specialization (refinement) of virtual class structure and behavior. In SDL 2000, the user can define classifier (types in SDL) for blocks, processes and services, and use instances of types for block, process, service, procedure, and signal description. Specification of behavior for automatic generation of implementations is further improved by support for exception handling and state machine decomposition.

In SDL 2000, two communicating agents must provide gates with matching interfaces (required or supported) and a connection between gates must exist.

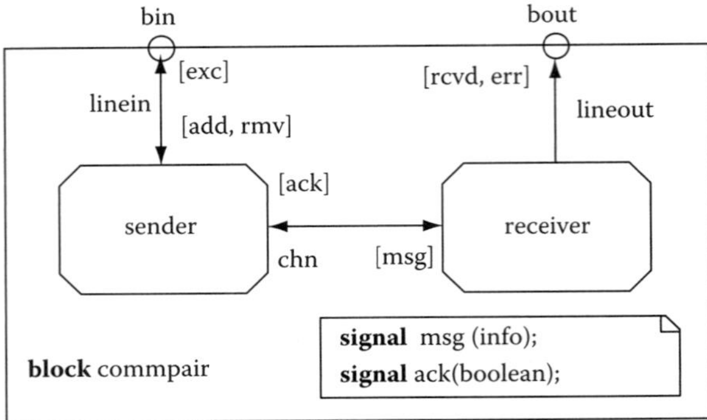

FIGURE 4.23   SDL blocks communicating by ports implementing a signal interface.

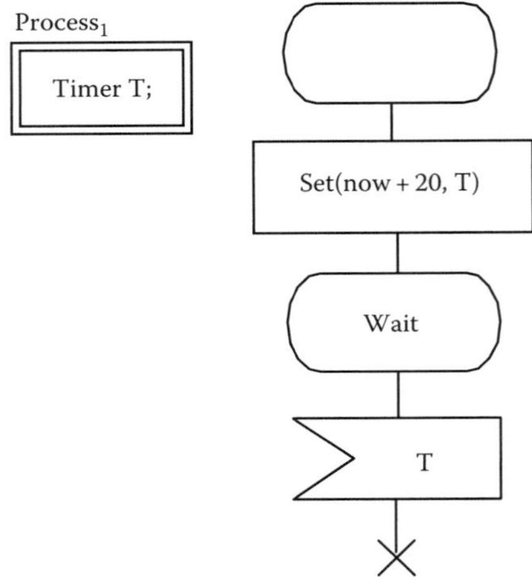

FIGURE 4.24   Example of timer use.

In the example of Figure 4.23, an SDL 2000 block exposing two gates (`bin` of input/output type and the output gate `bout`) is shown. The block defines two internal signals msg and ack exchanged on the channel connecting the internal processes `sender` and `receiver`.

SDL types can be specialized, leading to subtypes by exploiting the inheritance mechanism. Virtual objects or transitions can be redefined, and additional objects or actions can be added to the specification of virtual objects. A redefined transition or object is subject to further specialization (if virtual) unless it is marked as finalized.

SDL offers native mechanisms for representing external (global) time. Time is available by means of the predefined variable now, the `now()` primitive and timer constructs. Process actions can set timers, that is, the specification of a signal at a predefined point in time, wait, and eventually receive a timer expiry signal (as in Figure 4.24). SDL timer timeouts are always received in the form of asynchronous messages and timer values are only meant to be minimal bounds, meaning that any timer signal may remain enqueued for an unbounded amount of time.

In SDL, processes inside a block are meant to be executed concurrently, and no specification for a sequential implementation by means of a scheduler, necessary when truly concurrent hardware is not available, is given. Activities implementing the processes behavior (transitions between states) are executed in a run-to-completion fashion. From an implementation point of view, this raises the same concerns that hold for implementation of UML models. In SDL, however, the exception mechanism is available to model preemption and interrupt handling.

Other language characteristics make verification of time-related properties impossible: the Z.100 SDL semantics says that external time progresses in both states and actions. However, each action may take an unbounded amount of time to execute, and each process can remain in a state for an indeterminate amount of time before taking the first available transition.

Furthermore, for timing constraints specification, SDL does not include the explicit notion of event, therefore it is impossible to define a time tagging of most events of interest such as transmission and reception of signals, although MSCs (similar in nature to UML sequence diagrams, defined in part Z.120 of the standard [56]) are typically used to fill this gap since they allow expression of constraints on time elapsing between events.

Incomplete specification of time events and constraints prevents timed analysis of SDL diagrams, but the situation is not much better for untimed models. Properties of interest are in general undecidable because of infiniteness of data domains, unbounded message buffers, and the semiformal modeling style, where SDL is mixed with code fragments inside conditions and actions (the formal semantics of SDL is aimed at code generation rather than at simulation and verification). Untimed verification of properties is only possible in restricted models and seldom performed by commercial or research tools dealing with SDL code. The SDL description, (often restricted to a subset of the language) is provided as an input to transformation tools, which convert it to other formalisms, such as for example, PNs (analyzed by [25,57]) or TA (for analysis by the Kronos/IF [19,58,59] or Uppaal [20] toolsets).

Research on verification of SDL designs has been done in academic and industrial laboratories and centers. For example, [60] describes the verification of a layer of the GSM protocol (6 processes and 1013 reachable states) performed at Siemens by using BDD-based model checking and [61] reports on several case studies related to telecommunication protocols.

4.4.3 Architecture Deployment, Timing Specification, and Analysis

Both UML and SDL have been developed outside the context of embedded systems design and it is clear from previous sections how they do not cope with the modeling of resource allocation and sharing, nor with the minimum requirements for timing analysis. In fact, almost nothing exists in standard UML and SDL for modeling (or analyzing) nonfunctional aspects, nor scheduling or placement of software components on hardware resources can be specified and analyzed.

Several proposals exist, most notably the OMG profile for schedulability, performance, and time, and the new MARTE profile for UML, and others for SDL, which enhance the standard language defining timed models of systems, including time assumptions on the environment and platform-dependent aspects like resource availability and scheduling. Such model extensions allow formal or simulation-based validation of the timing behavior of the software.

The RT-UML profile is currently implemented by commercial tools, and OMG reported that proof of feasibility has been obtained by interaction with schedulability analysis tools such as TimeWiz by TimeSys and Pert by TriPacific.

4.4.3.1 UML

The OMG Profile for SPT, voted for adoption in November 2001, aims at substituting a number of proposals for time-related extensions that have appeared in recent years. For example, in order to better support the mapping of active objects into concurrent threads, many research and commercial systems introduced additional nonstandard diagrams. One such example is Artisan's Real-time Studio, which features the Concurrency diagram, capturing the multitasking and concurrency aspects of the system, and a System Architecture diagram.

The SPT profile defines a comprehensive conceptual framework that extends the UML metamodel with a much broader scope than any other real-time extension and applies to all diagrams. Currently, it consists mostly of a notation framework or vocabulary, with the purpose of providing the necessary concepts for schedulability and performance analysis of (timed) behavioral diagrams or scenarios. However, the SPT inherits from UML the deficiencies related to its incomplete semantics and, at least as of today, it lacks a sufficiently established practice. A major revision of the profile has been issued by OMG following the request for a profile proposal for real-time embedded systems (MARTE).

The profile is based on extensions (stereotyped model elements, tagged values, and constraints) belonging to three main framework packages, further divided into subpackages (as in Figure 4.25).

Of the three frameworks, the general resource modeling framework contains the fundamental definitions for modeling time, in the RTtimeModeling subpackage, generic resource specification and usage patterns, in the RTresourceModeling profile, and extensions for modeling concurrency in the RTconcurrencyModeling profile. The analysis models package contains specialized concepts for modeling schedulability (RAprofile) and performance (PAprofile) analysis. Finally, the infrastructure models package deals with real-time middleware specification (RealTimeCORBAModel).

In the SPT, the time model provides for both continuous and discrete time models, as well as global and local clock, including drift and offset specifications. The profile allows referencing to time instances (associated to events), of "Time" type and to the time interval between any two instances of time of "duration" type in attributes or constraints specifications inside any UML diagram.

The time package not only contains definitions for a formal model of time, but also stereotyped definitions for the two basic mechanisms of timer and clock. Timers can be periodic (as in SDL);

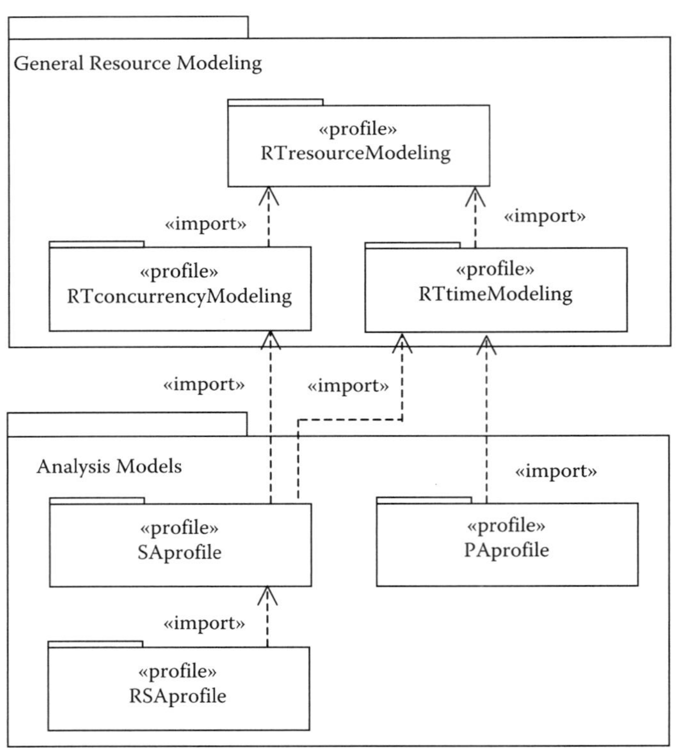

FIGURE 4.25   Framework packages and the subpackages in the OMG SPT.

they can be set or reset, paused or restarted and, when expired, they send timeout signals. Clocks are specialized periodic timers capable of generating Tick events.

Time values are typically associated to events, defined in UML as a "specification of a type of observable occurrence" (change of state). A pairing of an event with the associated time instance (time tag) is defined in the SPT profile as a TimedEvent. MARTE further extended the timing model allowing for logical clocks and defining a value specification language for the expression of timing constraints and attributes.

The resource model package defines schedulability analysis as a QoS matching problem between a client and the resource it uses. The profile provides two interpretations: in the "peer interpretation," the client and the resource coexist at the same level; in the "layered interpretation," the resources are used to implement the client. This is an attempt at capturing the distinctive aspect of schedulability analysis as a match between the timing (or, in general, QoS) requirements of logical entities in the logical layer and the corresponding QoS offers of an implementation or engineering layer.

In the OMG Profile the mapping between the logical entities and the physical architecture supporting their execution is a form of realization layering (synonymous of deployment). The semantics of the mapping provides a further distinction between the "deploys" mapping, indicating that instances of the supplier are located on the client (an example being the relation between a package and the hardware node on which it executes), and the "requires" mapping, which is a specialization indicating that the client provides a minimum deployment environment as required by the supplier, typically in the context of a QoS contract (Figure 4.26).

Although a visual representation of both is possible, it is not clear in which of the existing diagrams it should appear. Hence, the SPT recommendation for expressing the mappings consists in a table of relationships expressing the deploy and requires associations among logical and engineering components.

The QoS match is a quite simplified view of the schedulability problem with very few practical instances. To give one possible example (right side of Figure 4.26), consider the sufficient schedulability condition for RM, based on processor utilization [28]

$$U = \sum_{i=1,n} \frac{C_i}{\theta_i} \leq U_{\text{lub}} = n\left(2^{(1/n)} - 1\right) \quad \text{where } U_i = \frac{C_i}{\theta_i}$$

The QoS demand of each thread consists of a demand for utilization U_i to the OS/scheduling policy and a demand for a target worst-case computation time C_i to the hardware. The QoS offer from the RM scheduling policy is the guaranteed least upper bound on the utilization U_{lub} and the QoS matching rule easily follows.

The schedulability analysis model is based on stereotyped scenarios (scheduling situations in the SPT language). Each scheduling situation is in practice a sequence diagram, or a collaboration

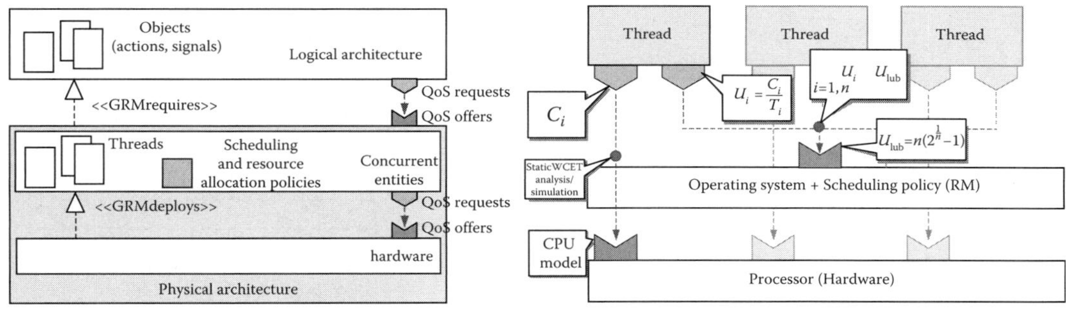

FIGURE 4.26 Schedulability analysis requires creating a correspondence between logical and physical architecture.

diagram (better suited for representing resource sharing), where one or more trigger events result in actions to be scheduled within the deadline associated with the trigger.

RMA is the method of choice for analyzing simple models with a restricted semantics, conforming to the so-called task-centric design paradigm. This requires updating the current definition of UML actions, in order to allow for preemption (which is a necessary prerequisite of RMA). Such a recommendation has been explicitly formulated in the SPT profile and is among the recommended changes in the agenda of the OMG action semantics group.

In this task-centric approach, the behavior of each active object or task consists of a combination of reading input signals, performing computation, and producing output signals (Figure 4.27, left side). Each active object can request the execution of actions of other passive objects in a synchronous or asynchronous fashion.

Each task is logically concurrent; it is activated periodically, and handles a single event. Active objects cooperate only by means of pure asynchronous messages, possibly implemented by means of memory mailboxes, a kind of protected (shared resource) object (Figure 4.28).

In the SPT, the TimedEvent that triggers the execution of the task action is identified with the stereotype "SATrigger", and it is the model element to which the deadline constraint (for the execution of the action) and the schedulability response are associated (Figure 4.28).

Unfortunately, the task-centric model, even if simple and effective in some contexts (see also other design methodologies [62–64] where the analysis of UML models is made possible by restricting the semantics of the analyzable models), only allows analysis of simple models where active objects do not interact with each other. In general UML models, each action is part of multiple end-to-end computations (with the associated timing constraints), and the system consists of a complex network of cooperating active objects, implementing state machines and exchanging asynchronous messages.

In their work [65–67], Saksena, Karvelas, and Wang present an integrated methodology that allows dealing with more general OO models where multiple events can be provided as input to a single thread. According to their model, each thread has an incoming queue of events possibly representing real-time transactions and the associated timing constraints (Figure 4.27).

Consequently, scheduling priority (usually a measure of time urgency or criticalness) is attached to events rather than threads. This design and analysis paradigm is called "event-centric design." Each event has an associated priority (e.g., deadline monotonic priority assignment can be used). Event queues are ordered by priority (i.e., threads process events based on their priority) and threads inherit the priority of the events they are currently processing. This model entails a two-level scheduling: the

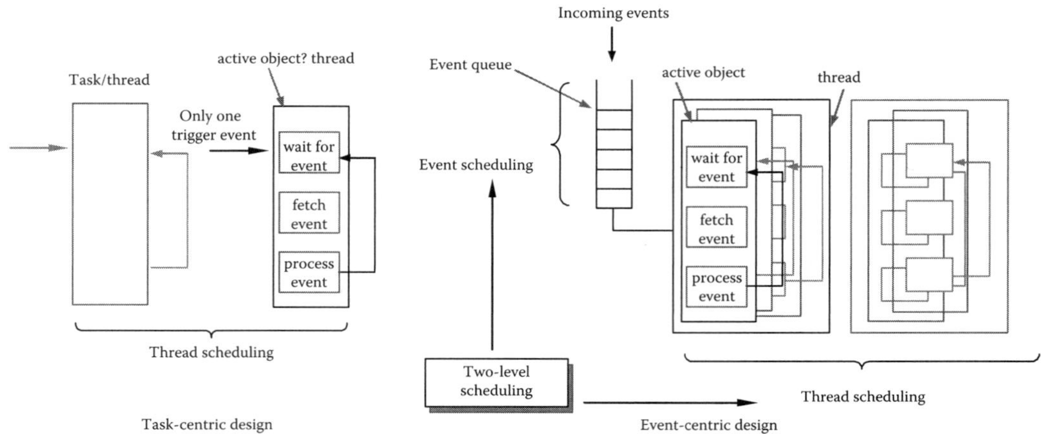

FIGURE 4.27   Dual scheduling problem in the event-centric analysis.

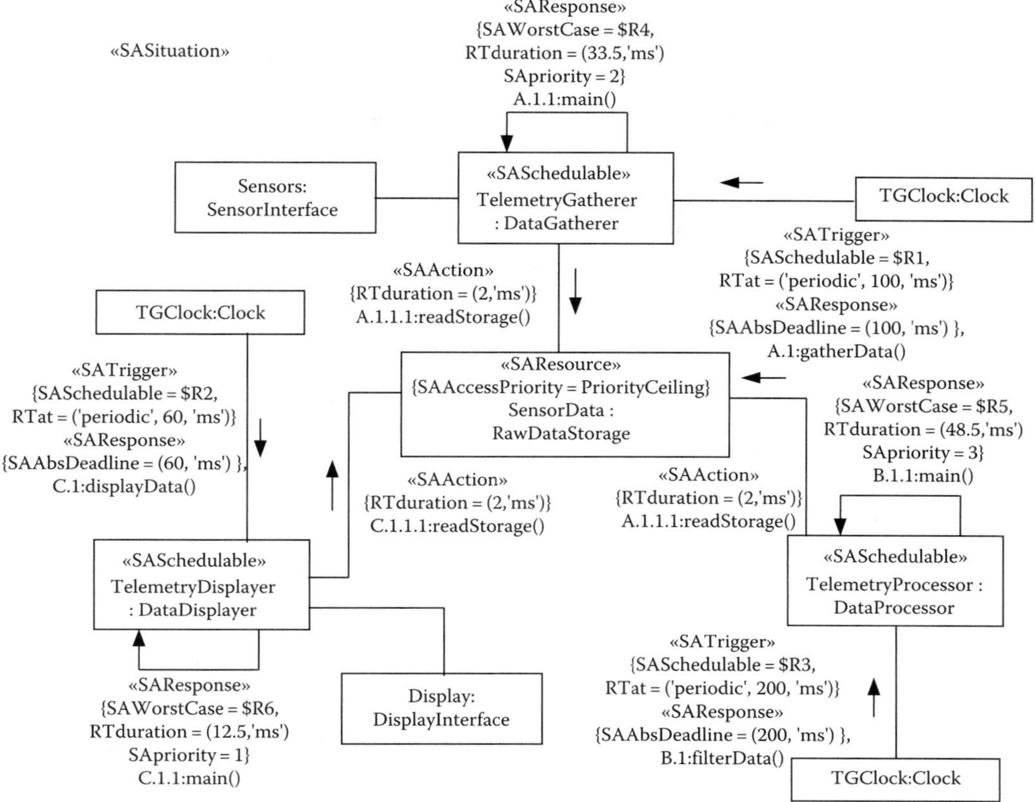

FIGURE 4.28 Sample collaboration instance representing a situation suitable for RMA. (Adapted from the OMG profile document.)

events enqueued as input to a thread need to be scheduled to find their processing order (Figure 4.27). At system level, the threads are scheduled by the underlying RTOS (a preemptive priority-based scheduler is assumed).

Schedulability analysis of the general case is derived from the analysis methodology for generic deadlines (possibly greater than the activation rates of the triggering events) by computing response times of actions relative to the arrival of the external event that triggers the end-to-end reaction (transaction) T containing the action. The analysis is based on the standard concepts of critical instant and busy-period for task instances with generic deadlines (adapted to the transaction model).

In [66], many implementation options for the event-centric model are discussed, namely, single thread implementation, multithread implementation with fixed priority or dynamic priority multi-threaded implementation, and a formula or procedure for schedulability analysis is provided or at least discussed for each of these models. The interested reader should refer to [66] for details.

From a time analysis point of view, it makes sense to define the set of (nested) actions that are called synchronously in response to an event. This set is called "synchronous set."

In a single-thread implementation, the only application thread processes pending events in priority order. Since there is only one thread, there is only one level of scheduling. Actions inherit their priority from the priority of the triggering event. Also, a synchronously triggered action inherits its priority from its caller. In single-threaded implementations, any synchronous set that starts executing is completed without interruption. Hence, the worst-case blocking time of an action starting at

$t = 0$ is bound by the longest synchronous set of any lower priority action that started prior to $t = 0$, and interference can be computed as the sum of the interference terms from other transactions and the interference from actions in the same transaction.

The single thread model can be analyzed for schedulability (details in [66]) and it is also conformant to the non-preemptable semantics of UML actions. Most of the existing CASE tools support a single threaded implementation and some of them (such as Rational Rose RT) support the priority-based enqueuing of activation events (messages).

In multithreaded implementations, each event represents a request for an end-to-end sequence of actions, executed in response to its arrival. Each action is executed in the context of a thread. Conceptually, we can reverse the usual meaning of thread scheduling, considering events (and the end-to-end chain of actions) as the main scheduling item and the threads required for the execution of actions as special mutual exclusion (mutex) resources, because of the impossibility of having a thread preempt itself. This insight allows using threads and thread priorities in a way that facilitates response time analysis.

If threads behave as mutexes, then it makes sense to associate with each thread a ceiling priority as the priority at which the highest priority event is served. As prescribed by PI or PC inheritance Protocol (PCP), threads inherit the priority of the highest priority event in their waiting queue and this allows bounding priority inversion. In this case, the worst-case blocking time is restricted to the processing of a lower priority event. Furthermore, before processing an event, a thread locks the active object within which the event is to be processed (this is necessary when multiple threads may handle events that are forwarded to the same object), and a ceiling priority and a PI rule must be defined for active objects as well as for threads.

For the multithreaded implementation, a schedulability analysis formula is currently available for the aforementioned case in which threads inherit the priority of the events. The schedulability analysis formula generally results in a reduced priority inversion with respect to the single thread implementation, but its usefulness is hindered by the fact that (to the author's knowledge) there is no CASE tool supporting generation of code with the possibility of assigning priorities to events and run-time support for executing threads inheriting the priority of the event they are currently handling.

Assigning static priorities to threads is tempting (static priority scheduling is supported by most RTOSes) and it is a quite common choice for multithreaded implementations (Rational Rose RT and other tools use this method). Unfortunately, when threads are required to handle multiple events, it is not clear what priority should be assigned to them and the computation of the worst-case response time is in general very hard because of the two-level scheduling and because of the difficulty of constructing an appropriate critical instant. The analysis needs to account for multiple priority inversion that arises from all the lower priority events handled by a higher priority thread. Still, however, it may be possible to use static thread priorities in very simple cases when it is easy to estimate the amount of priority inversion.

The discussion of the real-time analysis problem bypasses some of the truly fundamental problems in designing a schedulable software implementation of a functional (UML) model. The three main degrees of freedom in a real-time UML design subject to schedulability analysis are

- Assigning priorities to events
- Defining the optimal number of tasks
- Defining a mapping of methods (or entire objects) to tasks for their execution

and timing analysis (as discussed here) is only the last stage, after the mapping decisions have been taken.

What is truly required is a set of design rules, or even better an automatic synthesis procedure that helps in the definition of the set of threads and especially in the mapping of the logical entities to threads and to the physical platform.

The problem of synthesizing a schedulable implementation is actually a complex combinatorial optimization problem. In [67], the authors propose an automatic procedure for synthesizing the three main degrees of freedom in a real-time UML design. The synthesis procedure uses a heuristic strategy based on a decomposition approach, where priorities to events/actions are assigned in a first stage and mapping is performed in a separate stage.

This problem, probably the most important in the definition of a software implementation of real-time UML models, is still an open problem.

4.4.3.2 SDL

SDL contains primitives for dealing with global time, but nevertheless, when the specification of time behavior and resource usage is an issue, it is impossible to capture most time specifications and constraints.

For example, SDL does not offer the possibility of defining local time; timers cannot be periodic and it is difficult to express clocks, where periodic ticks are instantaneously consumed or lost. The main reason for these deficiencies is that the SDL notion of time was aimed at code generation, rather than at simulation and verification.

As stated in the previous section, SDL timer values only represent minimal bounds. This means that the SDL semantics provides no means to specify constraints on the time passed neither in states nor actions, whereas the verification or simulation of a property often depends on finite lower and upper bounds. Also, the delays in the transmission of signals are unbounded.

The use of MSC and annotations to express the timing constraints in which standard SDL is missing is commonplace in commercial tools and research proposals. In particular, some tools allow the expression of duration constraints on actions or other simple constraints in the form of special comments (with the problem that each tool defines its own syntax and semantics). Other proposals allow the definition of upper and lower time bounds for the delays applied to messages conveyed by a channel, or the definition of the delay probabilities.

Concerning the expressivity problem, a general real-time framework for SDL is presented in [68, 69] (the first one is aimed at hardware software codesign). The extension of SDL with more powerful timers and clocks as well as local clocks has been proposed by many (see e.g. [70]). Other proposals for expression of timing requirements can be found in [71,72].

Another problem of SDL is that it does not provide deployment and resource usage information, neither the notion of preemptable nor non-preemptable actions that is necessary for schedulability analysis. One proposal [73] consists in defining a list of resources with an attribute defining their preemptability. Block, Process, and Task definitions need to be complemented with the specification of the resource on which they are executed. Scheduling policies are defined by keywords, such as RMA or earliest deadline first (EDF), or by priority rules attached to resources or agents. Priorities may be static or dynamic and in the latter case, they may be specified based on preconditions, or by means of an observer which explicitly updates priorities depending on the observed states and events.

Some tools, for example, time enhanced versions of ObjectGeode [74] as presented in [75] and Tau [76,77] and the tool and methods based on Queuing SDL [78], propose to attach explicit timing information with SDL constructs such as tasks, and to provide some minimal deployment information. The ObjectGeode simulator, for example, allows associating an execution time (interval) with each action. Schedulability of SDL models is also discussed in [79].

Deficiencies in the specification of time also clearly affect the simulation of SDL models with time constraints. Any rigorous attempt to construct the simulation graph of an SDL system (the starting point for simulation and verification) must account for all possible combinations of execution times, timer expirations, and consumptions. Since no control over time progress is possible, many undesirable executions might be obtained during this exhaustive simulation.

In practice, existing simulation and verification tools make simplifying assumptions on execution and idle times. The usual convention is that statements or entire transactions are atomic; actions take zero time to execute, and that the system executes immediately whatever it can execute (time passes only where explicitly required by waiting conditions). This option is justified by the fact that it generates the highest degree of determinism, thus reducing the state space by an important factor (making the system practically analyzable).

In [80], an interesting integration of SDL model simulation with resource scheduling is proposed: annotations related to timing constraints and resource usage are translated into scheduler and resource manager models that can be used to simulate the time behavior of the model with respect to schedulability constraints on a standard SDL tool. Performance evaluation of SDL models is the subject of other research work [75,81,82].

The verification of time-related properties of SDL models can often be done on a finite control abstraction, that is a system with finite data domains, bounded message buffers, and bounded agent creation. A number of interesting verification problems are decidable on such a finite control abstraction under the condition that in the TA, obtained by translation, clocks can only be reset to zero, stopped and restarted, and that the only allowed tests are comparisons of clock values or differences of clock values with constants, which, in practice, allows transformation of the model into the standard semantics of TA.

Some SDL tools provide the facility to define observers as a means for defining constraints and properties, but SDL does not provide any standard notation for them.

4.4.4   Tools and Commercial Implementations

There are a large number of specialized CASE tools for UML modeling of embedded systems. Among those, Rhapsody of ILogix (now IBM) [83], Real-time Studio of Artisan [84], TAU Generation 2 from Telelogic (now IBB) [76], and Rational RoseRT (now Rational Rose Technical Developer [34], which is in essence a profile implementing ROOM in the UML framework) of IBM (Rational) are probably the most common. In contrast with general-purpose UML tools, all these try to answer the designer's need for automated support for model simulation, verification, and testing. In order to do so, they all provide an interpretation of the dynamic semantics of UML. Common features include interactive simulation of models and automatic generation of code. Nondeterminism is forbidden or eliminated by means of semantics restrictions. Some timing features and design of architecture level models are usually provided, although in nonstandard form if compared to the specifications of the SPT profile. Furthermore, support for OCL and user-defined profiles are often quite weak.

Formal validation of untimed and timed models is typically not provided, since commercial vendors focus on providing support for simulation, automatic code generation, and (partly) automated testing. Third party tools interoperating by means of the standard XMI format provide schedulability analysis and research work on restricted UML semantics models demonstrates that formal verification techniques can be applied to UML behavioral models. This is usually done by transformation of UML models into a formal MOC and subsequent formal verification by existing tools (for both untimed and timed systems). A number of tools, including those previously cited, allow the definition of UML models including real-time constraints and, later, possible verification of schedulability properties.

SDL is supported by powerful development environments integrating advanced facilities (like simulation, model checking, test generation, and auto-coding). Tools like ObjectGeode or the Telelogic Tau2 series support many phases of software development, ranging from some restricted form of analysis to implementation (code generation) and on-target deployment. Properties are expressed by MSC or by means of a tool language allowing the definition of observers. No schedulability analysis is currently available from commercial vendors.

The most recent versions of some of these tools attempt to combine various modeling approaches in order to unite the best of both worlds. One example is SDL–UML and the Telelogic TAU series of products [76], where the TTCN language (set to become one of the UML test profile recommended notations) is used to formally describe the test cases and test sequences [85].

4.5 Research on Models for Embedded Software

The models and tools described in the previous sections are representatives of a larger number of methods, models, languages, and tools (at different levels of maturity) that are currently being developed to face the challenges posed by embedded software design.

This section provides an insight on other (recent) proposals for solving advanced problems related to the modeling, simulation, and/or verification of functional and nonfunctional properties, including support for compositionality and possibly for integration of heterogeneous models, where heterogeneity applies to both the execution model and the semantics of the component communications or, in general, interactions.

In particular, the request for heterogeneity arises from the observation that there is no clear winner among the different models and tools and that different parts of complex-embedded systems may be better suited to different modeling and analysis paradigms (such as dataflow models for data-handling blocks and FSMs for control blocks). The first objective, in this case, is to reconcile by unification the synchronous and asynchronous execution paradigms.

The need to handle timing and resource constraints (hence preemption) together with the modeling of system functionality is another major requirement for modern methodologies.

Performance analysis by simulation of timed (control) systems with scheduling and resource constraints is, for example, the main goal of the TrueTime Toolset from the LTH Technical University in Lund [86].

The TrueTime toolset is one example of modeling paradigm where scheduling and resource management policies can be handled as separate modules to be plugged in together with the blocks expressing the functional behavior. TrueTime consists of a set of Simulink blocks that simulate real-time kernel policies as found in commercial RTOSs. The system model is obtained by connecting kernel blocks, network blocks, and ordinary Simulink blocks representing the functional behavior of the control application. The toolbox is capable of simulating the system behavior of the real-time system with interrupts, preemption, and scheduling of resources.

The Times tool [87] is another research framework (built on top of Uppaal and available at http://www.timestool.org) attempting an integration of the TA formalism with methods and algorithms for task and resource scheduling. In Times, a timed automaton can be used to define an arbitrary complex activation model for a set of deadline-constrained tasks, to be scheduled by fixed or dynamic (EDF [28]) priorities. Model checking techniques are used to verify the schedulability of the task set, with the additional advantage (if compared to standard worst-case analysis) or providing the possible runs that fail the schedulability test.

Other design methodologies bring further distinction among (at least) three different design dimensions representing

- The (functional) behavior of the system and the timing constraints imposed by the environment and/or resulting from design choices
- The communication network and the semantics of communication defining interactions upon each link
- The platform onto which the system is mapped and the timing properties of the system blocks resulting from the binding

The Ptolemy and the Metropolis environments are probably the best-known examples of design methodologies founded on the previous principles.

Ptolemy (http://www.ptolemy.eecs.berkeley.edu/) is a simulation and rapid prototyping framework for heterogeneous systems developed at the Center for Hybrid and Embedded Software Systems (CHESS) in the Department of Electrical Engineering and Computer Sciences of the University of California at Berkeley [2]. Ptolemy (now at version II) targets complex heterogeneous systems encompassing different operations, such as signal processing, feedback control, sequential decision making, and possibly even user interfaces.

Prior to UML 2.0, Ptolemy II had already introduced the concept of actor-oriented design, by complementing traditional object-orientation with concurrency and abstractions of communication between components.

In Ptolemy II, the model of the application is built as a hierarchy of interconnected actors (most of which are domain polymorphic). Actors are units of encapsulation (components) that can be composed hierarchically producing the design tree of the system. Actors have an interface abstracting their internals and providing bidirectional access to functions. The interface includes ports that represent points of communication, and parameters which are used to configure the actor operations.

Communication channels pass data from one port to another according to some communication semantics. The abstract syntax of actor-oriented design can be represented concretely in several ways, one example being the graphical representation provided by the Ptolemy II Vergil GUI (Figure 4.29).

Ptolemy is not built on a single, uniform, MOC, but it rather provides a finite library of "directors" implementing different models of computations. A director, when placed inside a Ptolemy II actor, defines its abstract execution semantics (MOC) and the execution semantics of the interactions among its component actors (its domain).

A component that has a well-defined behavior under different models of computation is called a domain polymorphic component; this means that its behavior is polymorphic with respect to the domain or MOC that is specified at each node of the design tree.

The available MOC in Ptolemy II includes

- Communicating sequential processes with synchronous rendezvous
- CT (where behavior is specified by a set of algebraic or differential equations)
- Discrete events (consisting of a value and a time stamp)

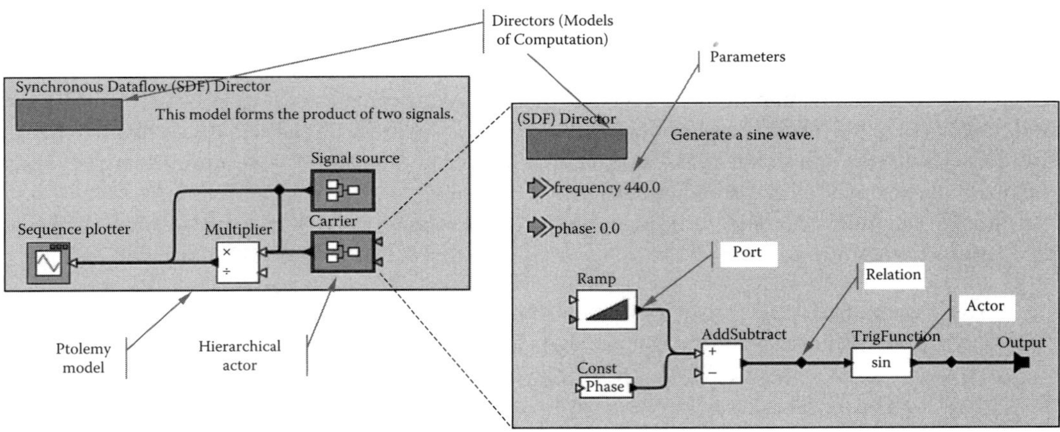

FIGURE 4.29   Ptolemy actor and its black box (higher-level) representation.

- Finite State Machines (FSMs)
- (Kahn) Process networks (PN)
- Synchronous dataflows (SDF)
- Synchronous/Reactive (SR) models and time-triggered synchronous execution (the Giotto framework [88])

The CT and SR domain have fixed-point semantics, meaning that in each iteration, the domain may repeatedly fire the components (execute the available active transitions inside them) until a fixed point is found.

Of course, key to the implementation of polymorphism is proving that an aggregation of components under the control of a domain defines in turn a polymorphic component. This has been proven possible for a large number of possible combinations of MOCs [89].

In Ptolemy II, the implementation language is Java and an (experimental) module for automatic Java code generation from a design is now available at http://ptolemy.eecs.berkeley.edu/. The source code of the Ptolemy framework is available for free from the same Web site. Currently, the software has hundreds of active users at various sites worldwide in industry and academia.

The Metropolis environment embodies the Platform-based design methodology [1] for design representation, analysis, and synthesis under development at the University of California at Berkeley, under the sponsorship of the MARCO Gigascale System Research Center.

In the Metropolis toolset [90], system design is seen as the result of a progressive refinement of high level specifications into lower level models, possibly "etherogeneous" in nature. The environment deals with all phases of the design from conception to final implementation.

Metropolis is designed as a very general infrastructure based on a metamodel with precise semantics that is general enough to support existing computation models and accommodate new ones. The metamodel supports not only functionality capture and analysis through simulation and formal verification, but also architecture description and the mapping of functionality to architectural elements (Figure 4.30).

The metropolis metamodel (MMM) language provides basic building blocks that are used to describe communication and computation semantics in a uniform way. These components represent computation, communication, and coordination or synchronization:

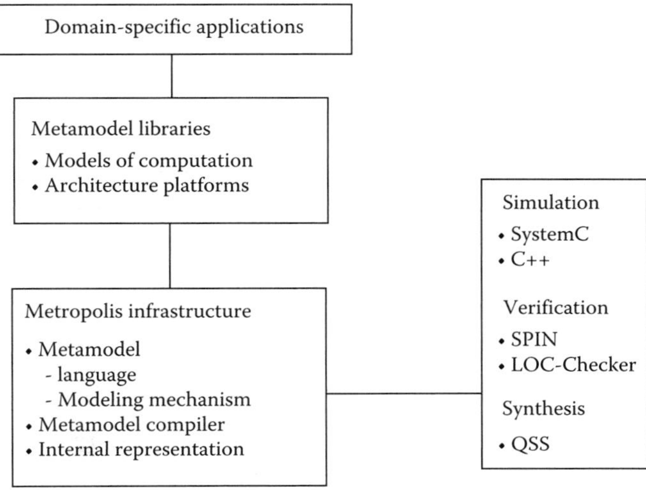

FIGURE 4.30 Metropolis framework.

- Processes for describing computation
- Media for describing communication between processes
- Quantity managers for enforcing a scheduling policy of processes
- Netlists for describing interconnection of objects and for instantiating and interconnecting quantity managers and state-media

A process is an active object (it possesses its own thread of control) that executes a sequence of actions (instructions, subinstructions, function calls, and awaits [91]) and generates a sequence of events. Each process in a system evolves by executing one action after the other. At each step (which is formally described in [91] in terms of a global execution index), each process in the system executes one action and generates the corresponding event. This synchronous execution semantics is relaxed to asynchronicity by means of a special event called NOP that can be freely interleaved in the execution of a process (providing nondeterminism).

Each "process" defines its "ports" as the possible interaction points. Ports are typed objects associated with an interface, which declares the services that can be called by the process or called from external processes (bidirectional communication) in a way similar to what is done in UML 2.0.

Processes cannot connect directly to other processes but the interconnection has to go through a "Medium," which has to define (i.e., implement) the services declared by the interface associated with the ports. The separation of the three orthogonal concepts of Process, Port, and Medium provides maximum flexibility and reusability of behaviors, that is, the meaning of the communication can be changed and refined without changing the computation description that resides in the processes (Figure 4.31).

A Metropolis program can be seen as a variant of a system-encompassing timed automaton where the transition between states is labeled by event vectors (one event per process). At each step, there is a set of event vectors that could be executed to make a transition from the current state to the next state. Unless a suitably defined quantity manager restricts the set of possible executions by means of scheduling constraints, the choice among all possible transitions is performed nondeterministically.

Quantity managers define the scheduling of actions by assigning tags to events. A tag is an abstract quantity from a partially order set (such as, Time). Multiple requests for action execution can be issued to a quantity manager that has to resolve them and schedule the processes in order to satisfy the ordering relation on the set of tags. The tagged-signal model (Lee and Sangiovanni-Vincentelli [92]) is the formalism that defines the unified semantic framework of signals and processes that stands at the foundation of the MMM.

By defining different communication primitives and different ways of resolving concurrency, the user can, in effect, specify different MOCs. For instance, in a synchronous MOC all events in an event vector have the same tag.

The mapping of a functional model into a platform is performed by enforcing a synchronization of function and architecture. Each action in the function side is correlated with an action on the architecture side using synchronization constraints.

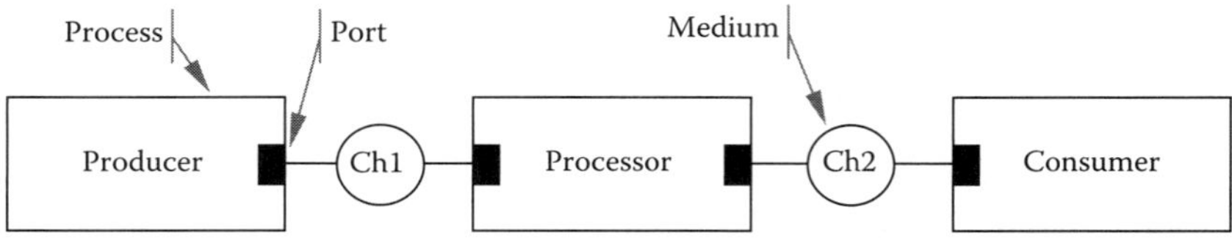

FIGURE 4.31   Processes, Ports, and Media in Metropolis.

The precise semantics of Metropolis allows for system simulation, synthesis, and formal analysis. The metamodel includes constraints that represent in abstract form requirements not yet implemented or assumed to be satisfied by the rest of the system and its environment.

Metropolis uses a logic language to capture nonfunctional constraints (time or energy constraints, for example). Constraints are declarative formulas of two kinds: LTL and LOC (logic of constraints).

Although two verification tools are currently provided to check that constraints are satisfied at each level of abstraction, the choice of the analysis and synthesis method or tool design depends on the application domain and the design phase. Metropolis clearly cannot possibly provide algorithms and tools for all possible design configurations and phases, but it provides mechanisms to compactly store and communicate all design information, including a parser that reads metamodel designs and a standard API that lets developers browse and modify design information so that they can plug in the required algorithms for a given application domain. This mechanism has been exploited to integrate the spin software verification tool [10].

Finally, as a last example, separation of behavior, communication or interaction and execution models are among the founding principles of the component-based design and verification procedure outlined in [93,94]. The methodology is based on the assumption that the system model is a transition system with dynamic priorities restricting nondeterminism in the execution of system actions. Priorities define a strict partial order among actions ($a_1 \prec a_2$). A priority rule is a set of pairs $\{(C^j, \prec^j)\}_{j \in J}$ such that $\prec^j$ is a priority order and C^j is a state constraint specifying when the rule applies.

In [95], a component is defined as the superposition of three models defining its behavior, its interactions, and the execution model. The transition-based formalism is used to define the behavior, and interaction specification consists of a set of connectors, defining a maximal set of interacting actions and specification of actions that must occur simultaneously in the context of an interaction. The set of actions that are allowed to occur in the system consists of the complete actions (triangles in Figure 4.32). Incomplete actions (circles in Figure 4.32) can only occur in the context of an interaction defining a higher-level complete action (as in the definition of $IC[K_1]^+$ in Figure 4.32, which defines the complete action $a_5|a_9$, meaning that incomplete action a_9 can only occur synchronously with a_5). This definition of interaction rules allows for a general model of asynchronous components, which can be possibly synchronized upon a subset of their actions if and when needed. Finally, the execution model consists of a set of dynamic priority rules.

Composing a set of components means applying a set of rules at each of the three levels. The rules defined in [95] define an associative and commutative composition operator. Further, the composition model allows for composability (properties of components are preserved after composition) and compositionality (properties of the compound can be inferred from properties of the components) with respect to deadlock freedom (liveness).

In [95], the authors propose a methodology for analysis of timed systems (and a framework for composition rules for component-based real-time models) based on the framework represented in Figure 4.33. According to the proposed design methodology, the system architecture consists of a number of layers, each capturing a different set of properties.

At the lowest level, threads (processes) are modeled according to the TA formalism, capturing the functional behavior, and the timing properties and constraints.

At the preemption level, system resources and preemptability are accounted for by adding one or more preemption transitions (right-hand side of Figure 4.34), one for each preemptable resource and mutual exclusion rules are explicitly formulated as a set of constraints acting upon the transitions of the processes.

Finally, the resource management policies and the scheduling policies are represented as additional constraints $K_{\text{pol}} = K_{\text{adm}} \wedge K_{\text{res}}$ where K_{adm} are the admission control constraints and K_{res} are the constraints specifying how resource conflicts are to be resolved.

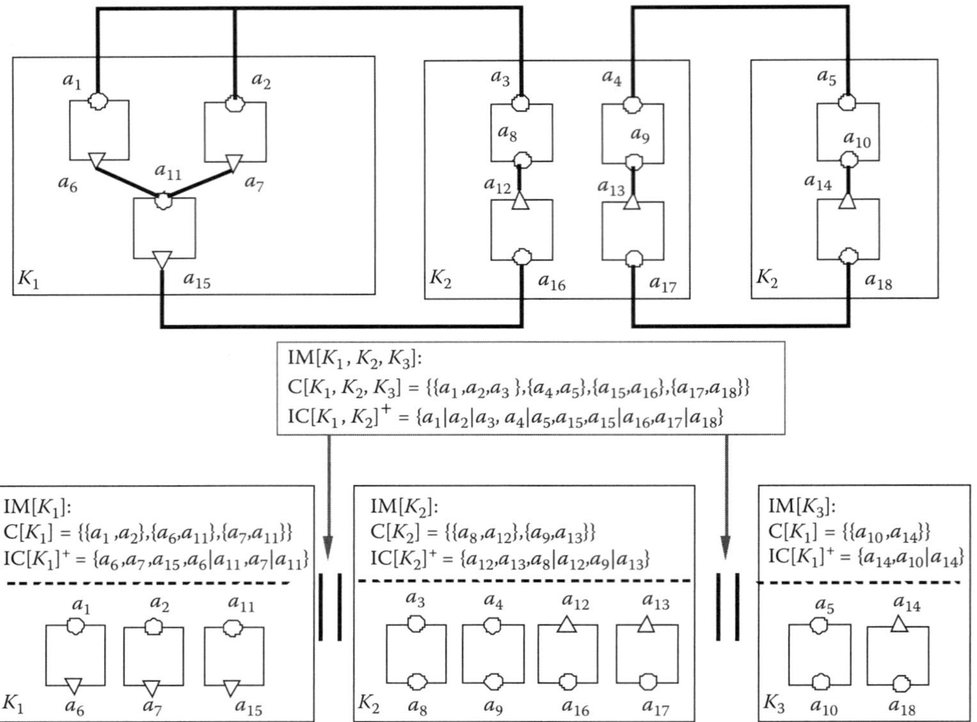

FIGURE 4.32 Interaction model among system components. (Adapted from Gossler G. and Sifakis J., *Proceedings of the FMCO Conference*, Leiden, the Netherlands, November 5–8, 2002.)

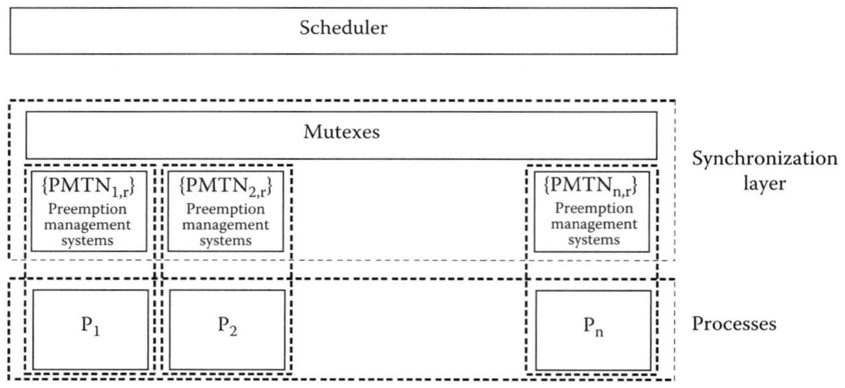

FIGURE 4.33 Architecture of the system model. (Adapted from Altisen K., et al., *J. Real-Time Syst.*, 23, 55, 2002.)

Once scheduling policies and schedulability requirements are in place, getting a correctly scheduled system amounts at finding a nonempty control invariant K such that $K \Rightarrow K_{\text{sched}} \wedge K_{\text{pol}}$.

Finally, the last example of research framework for embedded software modeling is the generic modeling environment or GME, developed at Vanderbilt University [96] is a configurable toolkit offering a meta–metamodeling facility for creating domain-specific models and program synthesis environments. The configuration of a domain-specific metamodel can be achieved by defining the syntactic, semantic, and presentation information regarding the domain. This implies defining the

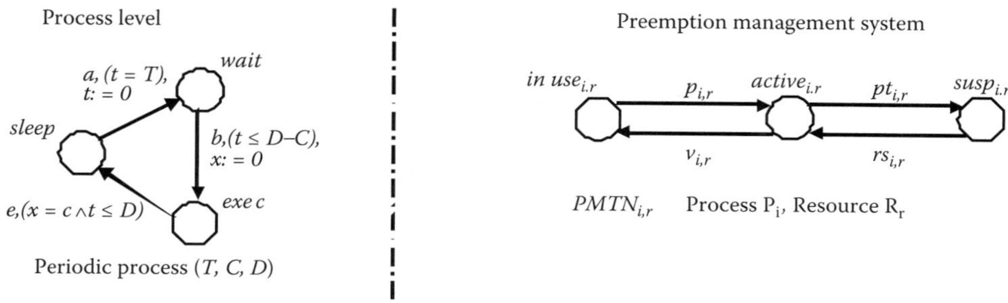

FIGURE 4.34   Process and preemption modeling. (Adapted from Altisen K., et al., *J. Real-Time Syst.*, 23, 55, 2002.)

main domain concepts, the possible relationships, and the constraints restricting the possible system configurations as well as the visibility rules of object properties.

The vocabulary of the domain-specific languages implemented by different GME configurations is based on a set of generic concepts built into GME itself. These concepts include hierarchy, multiple aspects, sets, references, and constraints. Models, atoms, references, connections, and sets are first-class objects.

Models are compound objects that can have parts and inner structure. Each part in a container is characterized by a role. The modeling instance determines what parts are allowed and in which roles. Models can be organized in a hierarchy, starting with the root module. Aspects provide visibility control. Relationships can be (directed or undirected) connections, further characterized by attributes. The model specification can define several kinds of connections, which objects can participate in a connection and further explicit constraints. Connections only appear between two objects in the same model entity. References (to model-external objects) help establish connections to external objects as well.

4.6   Conclusion

This chapter discusses the use of software models for the design and verification of embedded software systems. It attempts at a classification and a survey of existing formal models of computation, following the classical divide between synchronous and asynchronous models and between models for functionality as opposed to models for software architecture specification. Problems like formal verification of system properties, both timed and untimed, and schedulability analysis are discussed. The chapter also provides an overview of the commercially relevant modeling languages UML and SDL and discusses recent extensions to both these standards. The discussion of each topic is supplemented with an indication of the available tools that implement the available methodologies and analysis algorithms. Finally, the chapter contains a short survey of recent research results and a discussion of open issues and future trends.

References

1. Sangiovanni-Vincentelli A., Defining platform-based design, *EEDesign of EETimes*, February 2002, http://www.eedesign.com/showArticle.jhtml?articleID=16504380.
2. Lee E. A., Overview of the Ptolemy project, *Technical Memorandum UCB/ERL M03/25*, July 2, 2003, University of California, Berkeley, CA.

3. Balarin F., Hsieh H., Lavagno L., Passerone C., Sangiovanni-Vincentelli A., and Watanabe Y., Metropolis: An integrated environment for electronic system design, *IEEE Computer*, 36(4):45–52, 2003.

4. *UML Profile for Schedulability, Performance and Time Specification.* OMG Adopted Specification, July 1, 2002, http://www.omg.org

5. Beck T., Current trends in the design of automotive electronic systems, *Proceedings of the Design Automation and Test in Europe Conference*, Munich, Germany, pp. 38–44, 2001.

6. Edwards S., Lavagno L., Lee E. A., and Sangiovanni-Vincentelli A., Design of embedded systems: Formal models, validation and synthesis, *Proceedings of the IEEE*, 366–390, March 1997.

7. Alur R. and Henzinger T. A., Logics and models of real time: A survey. In *Real-Time: Theory in Practice, REX Workshop*, LNCS 600, pp. 74–106, 1991.

8. Pnueli A., The temporal logic of programs. In *Proceedings of the 18th Annual Symposium on the Foundations of Computer Science*, pp. 46–57. IEEE, Providence, RI, November 1977.

9. Emerson E. A., Temporal and modal logics. In J. van Leeuwen, Ed., *Handbook of Theoretical Computer Science*, volume B, pp. 995–1072. Elsevier, 1990.

10. Holzmann G. J., *Design and Validation of Computer Protocols.* Prentice-Hall, Englewood Cliffs, NJ, 1991.

11. Harel D., Statecharts: A visual approach to complex systems, *Science of Computer Programming*, 8:231–275, 1987.

12. Merlin P. M. and Farber D. J. Recoverability of communication protocols, *IEEE Transactions of Communications*, 24(9):36–103, September 1976.

13. Sathaye A. S. and Krogh B. H. Synthesis of real-time supervisors for controlled time Petri nets, *Proceedings of the 32nd IEEE Conference on Decision and Control*, vol. 1, San Antonio, pp. 235–238, 1993.

14. Alur R. and Dill D. L., A theory of timed automata, *TCS*, 126:183–235, 1994.

15. Ramchandani C., Analysis of Asynchronous Concurrent Systems by Timed Petri Nets. Cambridge, MA: MIT, Department of Electrical Engineering, PhD Thesis, 1974.

16. Molloy M. K., Performance analysis using stochastic Petri nets, *IEEE Transactions on Computers*, 31(9), 913–917, 1982.

17. Ajmone Marsan M., Conte G., and Balbo G., A class of generalized stochastic Petri nets for the performance evaluation of multiprocessor systems, *ACM Transactions on Computer Systems*, 2(2), 93–122, 1984.

18. Haar S., Kaiser L., Simonot-Lion F., and Toussaint J., On equivalence between timed state machines and time petri nets, *Rapport de recherche de l'INRIA—Lorraine*, November 2000.

19. Yovine S., Kronos: A verification tool for real-time systems, *Springer International Journal of Software Tools for Technology Transfer*, 1(1–2): 123–133, 1997.

20. Larsen K. G., Pettersson P., and Yi W. Uppaal in a nutshell, *Springer International Journal of Software Tools for Technology Transfer*, 1(1–2): 134–152, 1997.

21. Yi W., Pettersson P., and Daniels M., Automatic verification of real-time communicating systems by constraint solving. In *Proceedings of the 7th International Conference on Formal Description Techniques*, Berne, Switzerland, October 4–7, 1994.

22. Henzinger T. A., The theory of hybrid automata, *Proceedings of the 11th Annual Symposium on Logic in Computer Science (LICS)*, pp. 278–292. IEEE Computer Society Press, New Brunswick, NJ, July 27–30, 1996.

23. Henzinger T. A., Ho P.-H., and Wong-Toi H., HyTech: A model checker for hybrid systems, *Software Tools for Technology Transfer* 1:110–122, 1997.

24. Vicario E., Static analysis and dynamic steering of time-dependent systems using time petri nets, *IEEE Transactions on Software Engineering*, 27(8), 728–748, July 2001.

25. The ORIS tool, http://www.dsi.unifi.it/~vicario/Research/ORIS/oris.html

26. Lime D. and Roux O. H., A translation based method for the timed analysis of scheduling extended time petri nets, *The 25th IEEE International Real-Time Systems Symposium*, December 5–8, 2004 Lisbon, Portugal.

27. Clarke E. M. and Wing J. M., Formal methods: State of the art and future directions, Technical Report CMU-CS-96-178, Carnegie Mellon University (CMU), September 1996.

28. Liu C. and Layland J., Scheduling algorithm for multiprogramming in a hard real-time environment, *Journal of the ACM*, 20(1):46–61, January 1973.

29. Klein M. H., Ralya T., Pollak B., Obenza R., González Harbour M., *A Practitioner's Handbook for Real-Time Analysis: Guide to Rate Monotonic Analysis for Real-Time Systems.* Kluwer Academic Publishers, Hingham, MA, 1993.

30. Rajkumar R., *Synchronization in Multiple Processor Systems, Synchronization in Real-Time Systems: A Priority Inheritance Approach*, Kluwer Academic Publishers, Hingham, MA, 1991.

31. Benveniste A., Caspi P., Edwards S. A., Halbwachs N., Le Guernic P., and de Simone R., The synchronous languages 12 years later, *Proceedings of the IEEE*, 91(1), 64–83, January 2003.

32. Real-Time Workshop Embedded Coder 4.2, http://www.mathworks.com/products/rtwembedded/.

33. dSPACE Produkte: Production Code Generation Software, www.dspace.de/ww/de/pub/products/sw/targetli.htm

34. Rational Rose Technical Developer, http://www-306.ibm.com/software/awdtools/developer/technical/.

35. Meyer B., An overview of Eiffel. In *The Handbook of Programming Languages*, vol. 1, *Object-Oriented Languages*, Peter H. Salus, Ed., Macmillan Technical Publishing, Indianapolis, IN, 1998.

36. Caspi P., Pilaud D., Halbwachs N., and Plaice J. A., LUSTRE: A declarative language for programming synchronous systems. In *ACM Symposium on Principles Programming Languages (POPL)*, Munich, Germany, 1987, pp. 178–188.

37. Halbwachs N., Caspi P., Raymond P., and Pilaud D., The synchronous data flow programming language LUSTRE, *Proceedings of the IEEE*, 79, 1305–1320, September 1991.

38. Boussinot F. and de Simone R., The Esterel language, *Proceedings of the IEEE*, 79, 1293–1304, September 1991.

39. Berry G., The constructive semantics of pure Esterel, *5th Algebraic Methodology and Software Technology Conference*, Munich, Germany, July 1–5, 1996, pp. 225–232.

40. Westhead M. and Nadjm-Tehrani S., Verification of embedded systems using synchronous observers. In LNCS 1135, *Formal Techniques in Real-Time and Fault-Tolerant Systems.* Heidelberg, Germany: Springer-Verlag, 1996.

41. The Mathworks Simulink and StateFlow. http://www.mathworks.com

42. Scaife N., Sofronis C., Caspi P., Tripakis S., and Maraninchi F. Defining and translating a "safe" subset of Simulink/Stateflow into Lustre. In *Proceedings of 2004 Conference on Embedded Software, EMSOFT'04*, Pisa, Italy, September 2004. Springer.

43. Scaife N. and Caspi P., Integrating model-based design and preemptive scheduling in mixed time- and event-triggered systems. In *16th Euromicro Conference on Real-Time Systems (ECRTS'04)*, pp. 119–126, Catania, Italy, June–July 2004.

44. Prover Technology, http://www.prover.com/.

45. Edwards S. A., An Esterel compiler for large control-dominated systems, *IEEE Transactions on Computer-Aided Design of Integrated Circuits and Systems*, 21(2), 169–183, February 2002.

46. Shiple T. R., Berry G., and Touati H. Constructive analysis of cyclic circuits, *European Design and Test Conference*, Paris, France, March 11–14, 1996.

47. Benveniste A., Caspi P., Le Guernic P., Marchand H., Talpin J.-P., and Tripakis S. A protocol for loosely time-triggered architectures. In *Proceedings of 2002 Conference on Embedded Software, EMSOFT'02*, J. Sifakis and A. Sangiovanni-Vincentelli, Eds., LNCS 2491, pp. 252–265, Springer Verlag, Grenoble, France, October 7–9.

48. Benveniste A., Caillaud B., Carloni L., Caspi P., and Sangiovanni-Vincentelli A. Heterogeneous reactive systems modeling: Capturing causality and the correctness of loosely time-triggered architectures (LTTA), *Proceedings of 2004 Conference on Embedded Software, EMSOFT'04*, G. Buttazzo and S. Edwards, Eds., Pisa, Italy, September 27–29, 2004.

49. Biannic Y. L., Nassor E., Ledinot E., and Dissoubray S. UML object specification for real-time software, *RTS Show 2000*, Paris, France.

50. Selic B., Gullekson G., and Ward P. T., *Real-Time Object-Oriented Modeling*. John Wiley & Sons, New York, 1994.

51. Douglass B. P. *Doing Hard Time: Developing Real-Time Systems with Objects, Frameworks, and Patterns.* Addison-Wesley, Reading, MA, 1999.

52. Latella D., Majzik I., and Massink M. Automatic verification of a behavioural subset of UML statechart diagrams using the SPIN modelchecker, *Formal Aspects of Computing* 11(6):637–664, 1999.

53. del Mar Gallardo M., Merino P., and Pimentel E. Debugging UML designs with model checking, *Journal of Object Technology*, 1(2):101–117, July–August 2002.

54. UML 2.0 OCL Final adopted specification, http://www.omg.org/cgi-bin/doc?ptc/2003–10–14.

55. ITU-T. Recommendation Z.100. Specification and Description Language (SDL). Z-100, International Telecommunication Union Standard. Section, 2000.

56. ITU-T. Recommendation Z.120. Message Sequence Charts. Z-120, International Telecommunication Union Standard. Section, 2000.

57. The PEP tool (Programming Environment based on Petri Nets), Documentation and user guide, http://parsys.informatik.uni-oldenburg.de/~pep/Paper/PEP1.8_doc.ps.gz.

58. Bozga M., Ghirvu L., Graf S., and Mounier L., IF: A validation environment for timed asynchronous systems. In *Computer Aided Verification, CAV*, LNCS 1855, 2000.

59. Bozga M., Graf S., and Mounier L., IF-2.0: A validation environment for component-based real-time systems. In *Comp. Aided Verification, CAV*, LNCS 2404, pp. 343–348, 2002.

60. Franz Regensburger and Aenne Barnard. Formal verification of SDL systems at the Siemens mobile phone department. In *Tools and Algorithms for the Construction and Analysis of Systems. 4th International Conference, TACAS'98*, LNCS 1384, pp. 439–455. Springer Verlag, 1998.

61. Bozga M., Graf S., and Mounier L., Automated validation of distributed software using the IF environment. In *Workshop on Software Model Checking*, Electronic Notes in Theoretical Computer Science, 55(3). Elsevier, 2001.

62. Gomaa H. *Software Design Methods for Concurrent and Real-Time Systems*. Addison-Wesley, Reading, MA, 1993.

63. Burns A. and Wellings A. J. HRT-HOOD: A design method for hard real-time, *Journal of Real-Time Systems*, 6(1):73–114, 1994.

64. Awad M., Kuusela J., and Ziegler J. *Object-Oriented Technology for Real-Time Systems: A Practical Approach Using OMT and Fusion*. Prentice Hall, NJ, 1996.

65. Saksena M., Freedman P., and Rodziewicz P. Guidelines for automated implementation of executable object oriented models for real-time embedded control systems, *Proceedings, IEEE Real-Time Systems Symposium* 1997, pp. 240–251, San Francisco, CA, December 3–5, 1997.

66. Saksena M. and Karvelas P. Designing for schedulability: Integrating schedulability analysis with object-oriented design, *Proceedings of the Euromicro Conference on Real-Time Systems*, Stockholm, Sweden, June 2000.

67. Saksena M., Karvelas P., and Wang Y. Automatic synthesis of multi-tasking implementations from real-time object-oriented models. *Proceeding of the IEEE International Symposium on Object-Oriented Real-Time Distributed Computing*, Newport Beach, CA, March 2000.

68. Slomka F., Dörfel M., Münzenberger R., and Hofmann R. Hardware/Software codesign and rapid-prototyping of embedded systems, *IEEE Design and Test of Computers, Special Issue: Design Tools for Embedded Systems*, 17(2), 28–38, April–June 2000.

69. Bozga M., Graf S., Mounier L., Ober I., Roux J.-L., and Vincent D. Timed extensions for SDL, *Proceedings of the SDL Forum 2001*, LNCS 2078, Copenhagen, Denmark, June 27–29, 2001.

70. Münzenberger R., Slomka F., Dörfel M., and Hofmann R. A general approach for the specification of real-time systems with SDL. In R. Reed and J. Reed, Eds., *Proceedings of the 10th International SDL Forum*, Springer LNCS 2078, Copenhagen, Denmark, June 27–29, 2001.

71. Algayres B., Lejeune Y., and Hugonnet F., GOAL: Observing SDL behaviors with Geode, *Proceedings of SDL Forum* 95, Amsterdam, the Netherlands.

72. Dörfel M., Dulz W., Hofmann R., and Münzenberger R. SDL and non-functional requirement, Internal Report IMMD7 05/01, University of Erlangen, August 20, 2001.

73. Mitschele-Thiel S. and Muller-Clostermann B. Performance engineering of sdl/msc systems, *Computer Networks*, 31(17):1801–1815, June 1999.

74. Telelogic ObjectGeode, http://www.telelogic.com/products/additional/objectgeode/index.cfm

75. Roux J. L., SDL Performance analysis with ObjectGeode, *Workshop on Performance and Time in SDL*, 1998.

76. Telelogic TAU Generation2, http://www.telelogic.com/products/tau/tg2.cfm

77. Bjorkander M. Real-Time Systems in UML (and SDL), *Embedded System Engineering* October/November 2000.

78. Diefenbruch M., Heck E., Hintelmann J., and Müller-Clostermann B. Performance evaluation of SDL systems adjunct by queuing models, *Proceedings of the SDL-Forum '95*, Amsterdam, the Netherlands, 1995.

79. Alvarez J. M., Diaz M., Llopis L. M., Pimentel E., and Troya J. M. Deriving hard-real time embedded systems implementations directly from SDL specifications. In *International Symposium on Hardware/Software Codesign CODES*, Copenhagen, Denmark, 2001.

80. Bucci G., Fedeli A., and Vicario E., Specification and simulation of real time concurrent systems using standard SDL tools, *11th SDL Forum*, Stuttgart, July 2003.

81. Spitz S., Slomka F., and Dörfel M. SDL*—An annotated specification language for engineering multimedia communication systems, *Workshop on High Speed Networks*, Stuttgart, October 1999.

82. Malek M. PerfSDL: Interface to protocol performance analysis by means of simulation, *Proceedings of the SDL Forum* 99, Montreal, Canada, June 21–25, 1999.

83. I-Logix Rhapsody, http://www.ilogix.com/rhapsody/rhapsody.cfm

84. Artisan Real-Time Studio, http://www.artisansw.com/products/professional_overview.asp

85. Telelogic TAU TTCN Suite, http://www.telelogic.com/products/tau/ttcn/index.cfm

86. Henriksson D., Cervin A. and Årzén K.-E. TrueTime: Simulation of control loops under shared computer resources. In *Proceedings of the 15th IFAC World Congress on Automatic Control*, Barcelona, Spain, July 2002.

87. Amnell T. et al. Times—A tool for modelling and implementation of embedded systems. In *Proceedings of 8th International Conference, TACAS* 2002, Grenoble, France, April 8–12, 2002.

88. Henzinger T. A. Giotto: A time-triggered language for embedded programming. In *Proceedings on the 1st International Workshop on Embedded Software (EMSOFT'2001)*, Tahoe City, CA, October 2001, LNCS 2211, pp. 166–184. Springer Verlag, 2001.

89. Lee E. A. and Xiong Y. System-level types for component-based design. In *Proceedings on the 1st International Workshop on Embedded Software (EMSOFT'2001)*, Tahoe City, CA, October 2001, LNCS 2211, pp. 237–253. Springer Verlag, 2001.

90. Balarin F., Lavagno L., Passerone C., and Watanabe Y. Processes, interfaces and platforms. Embedded software modeling in metropolis, *Proceedings of the EMSOFT Conference 2002*, Grenoble, France, pp. 407–416.

91. Balarin F., Lavagno L., Passerone C., Sangiovanni Vincentelli A., Sgroi M., and Watanabe Y. *Modeling and Design of Heterogeneous Systems*, LNCS 2549, pp. 228, 273, Springer Verlag 2002.

92. Lee E. A. and Sangiovanni-Vincentelli A. A framework for comparing models of computation. In *IEEE Transactions on CAD*, 17(12):1217–1229, December 1998.

93. Gossler G. and Sifakis J. Composition for component-based modeling, *Proceedings of FMCO'02*, Leiden, the Netherlands, LNCS 2852, pp. 443–466, November 2002.

94. Altisen K., Goessler G., and Sifakis J. Scheduler modeling based on the controller synthesis paradigm, *Journal of Real-Time Systems* 23, 55–84, 2002. [Special issue on Control Approaches to Real-Time Computing.]

95. Gossler G. and Sifakis J., Composition for component-based modeling, *Proceedings of the FMCO Conference*, Leiden, the Netherlands, November 5–8, 2002.
96. Ledeczi A., Maroti M., Bakay A., Karsai G., Garrett J., Thomason IV C., Nordstrom G., Sprinkle J., and Volgyesi P. The generic modeling environment, *Workshop on Intelligent Signal Processing*, Budapest, Hungary, May 17, 2001.

5

Languages for Design and Verification

Stephen A. Edwards
Columbia University

5.1 Introduction

An embedded system is a computer masquerading as something else that must perform a small set of tasks cheaply and efficiently. A typical system might have communication, signal processing, and user interface tasks to perform.

Because the tasks must solve diverse problems, a language general-purpose enough to solve them all would be difficult to write, analyze, and compile. Instead, a variety of languages has evolved, each best suited to a particular problem domain. The most obvious divide is between languages for software and hardware, but there are others. For example, a signal-processing language might be superior to assembly for a data-dominated problem, but not for a control-dominated one.

While these languages can be distinguished by their place in the Chomsky hierarchy (e.g., hardware languages tend to be regular (finite-state) and those for software are usually Turing complete), the more practical differences tend to be in the sort of algorithms they can describe most elegantly. Hardware languages tend to excel at describing highly parallel algorithms consisting of fine-grained operators and data movement; software languages are better for describing algorithms that consist of a complex sequence of steps. This reflects the "physics" of the targeted computational elements (e.g., wires and transistors vs. stored-program computers), but the influence also goes the other way: a design language has profound effect on a designer's style of thinking. Thinking beyond the domain of a single language is a key motivation for studying many of them.

This chapter describes popular hardware, software, dataflow, and hybrid languages, each of which excels at certain problems in embedded systems.

5.2   Hardware Design Languages

Hardware description languages (HDLs) are now the preferred way to enter the design of an integrated circuit, having supplanting graphical schematic capture programs in the early 1990s. A typical chip design in 2008 starts with some back-of-the-envelope calculations that lead to a rough architectural plan. This architecture is refined and tested for functional correctness using a model of it implemented in C or C++, perhaps using SystemC libraries. Once this high-level model is satisfactory, designers recode it in a register-transfer-level (RTL) dialect of VHDL or Verilog—the two industry-dominant HDLs. The RTL model is usually simulated to compare it to the higher-level model, and then is fed to a logic synthesis system such as Synopsys' Design Compiler, which translates the RTL into an efficient gate-level netlist. Finally, this netlist is given to a place-and-route system that generates the list of polygons that will become wires and transistors on the chip.

None of these steps, of course, is as simple as it might sound. Translating a C model of a system into an RTL requires adding many details, ranging from protocols to cycle-level scheduling. Despite many years of research, this step remains stubbornly manual in most flows. Synthesizing a netlist from an RTL dialect of an HDL has been automated, but it is the result of many years of university and industrial research, as are all the automated steps after it.

Compared to the software languages discussed later in this chapter, concurrency and the notion of control are fundamental differences between hardware and software languages. In hardware, every part of the "program" is always running, but in software, exactly one part of the program is running at any one time. Software languages naturally focus on sequential algorithms, while hardware languages enable concurrent function evaluation, speculation, and concurrency.

Ironically, efficient simulation in software is a main focus of these hardware languages, so their discrete-event semantics are a compromise between what would be ideal for hardware and what simulates efficiently.

Verilog [36,66] and VHDL [4,21,35,60] are the most popular languages for hardware description, but SystemC [32], essentially a modeling library built atop the C++ programming language, is gaining ground as a higher-level hardware modeling language. Each model systems with discrete-event semantics that ignore idle portions of the design for efficient simulation. Each describe systems with structural hierarchy: a system consists of blocks that contain instances of primitives, other blocks, or sequential processes. Connections are listed explicitly.

Verilog provides more primitives geared specifically toward hardware simulation. VHDL's primitives are assignments such as $a = b + c$ or procedural code. Verilog adds transistor and logic gate primitives and allows new ones to be defined with truth tables. SystemC's primitives are more software-like: vectors, arithmetic operators, and other familiar C++ constructs.

All three languages allow concurrent processes to be described procedurally. Such processes sleep until awakened by an event that causes them to run, read, and write variables and suspend. Processes may wait for a period of time (e.g., `#10` in Verilog, `wait for 10ns` in VHDL), a value change (`@(a or b)`, `wait on a, b`), or an event (`@(posedge clk)`, `wait on clk until clk='1'`). SystemC has analogous constructs.

VHDL communication is more disciplined and flexible. Verilog communicates through "wires," which behave like their namesake; and "regs," which are shared memory locations that can cause race conditions. VHDL's signals behave like wires but the resolution function (applied when a wire has multiple drivers) may be user-defined. VHDL's variables are local to a single process unless declared shared. SystemC provides communication channels more like VHDL's and also has facilities for building more complex abstractions.

Verilog's type system models hardware with four-valued bit vectors and arrays for modeling memory. VHDL does not include four-valued vectors, but its type system allows them to be added. Furthermore, composite types such as C *structs* can be defined. SystemC, since it is built on C++, allows the use of C++'s more elaborate, object-oriented type system.

Overall, Verilog is a leaner language more directly geared toward simulating digital integrated circuits. VHDL is a much larger, more verbose language capable of handing a wider class of simulation and modeling tasks. SystemC is even more flexible as a modeling platform but has far less mature synthesis support and can be even more verbose.

5.2.1 History

Many credit Reed [61] with the first HDL. His formalism, simply a list of Boolean functions that define the inputs to a block of flip-flops driven by a single clock (i.e., a synchronous digital system), captures the essence of an HDL: a semiformal way of modeling systems at a higher level of abstraction. Reed's formalism does not mention the wires and vacuum tubes that would actually implement his systems, yet it makes clear how these components should be assembled.

In the decades since Reed, both the number and the need for HDLs have increased. In 1973, Omohundro [54] could list nine languages and dozens more have been proposed since.

The main focus of HDLs has shifted as the cost of digital hardware has dropped. In the 1950s and 1960s, the cost of digital hardware remained high and was used primarily for general-purpose computers. Chu's CDL [19] is representative of the languages of this era: it uses a programming-language-like syntax; has a heavy bias toward processor design; and includes the notions of arithmetic, registers and register transfer, conditionals, concurrency, and even microprograms. Bell and Newell's influential ISP (described in their 1971 book [7]) was also biased toward processor design.

The 1970s saw the rise of many more design languages [15,18]. One of the more successful was ISP'. Developed by Charles Rose and his student Paul Drongowski at Case Western Reserve in 1975–1976, ISP' was based on Bell and Newell's ISP and was used in a design environment for multiprocessor systems called N.mPc [57]. Commercialized in 1980, it enjoyed some success, but starting in 1985 the Verilog simulator and the accompanying language began to dominate the market.

The 1980s brought Verilog and VHDL, which remain the dominant HDLs to this day. Initially successful because of its superior gate-level simulation speed and its ability to model both circuits and testbenches for them, Verilog started life in 1984 as a proprietary language in a commercial product, while VHDL, the VHSIC (very high-speed integrated circuit) HDL, was designed at the behest of the U.S. Department of Defense as a unifying representation for electronic design [23].

While the 1980s was the decade of the widespread commercial use of HDLs for simulation, the 1990s brought them an additional role as input languages for logic synthesis. While the idea of automatically synthesizing logic from an HDL dates back to the 1960s, it was only the development of multilevel logic synthesis in the 1980s [16] that made them practical for specifying hardware, much as compilers for software require optimization to produce competitive results. Synopsys was one of the first to release a commercially successful logic synthesis system that could generate efficient hardware from RTL Verilog specifications and by the end of the 1990s, virtually every large integrated circuit was designed this way.

Verifying that an RTL model of a design is functionally correct is the main challenge in chip design. Simulation continues to be the dominant way of raising confidence in the correctness of RTL models but has many drawbacks. One of the more serious is the need for simulation to be driven by appropriate test cases. These need to exercise the design, preferably the difficult cases that expose bugs, and be both comprehensive and relatively short since simulation takes time.

Knowing when simulation has exposed a bug and estimating how complete a set of test cases actually are two other major issues in a simulation-based functional verification methodology. Clearly articulated in features recently added to SystemVerilog, it is now common to automatically generate simulation test cases using biased random variables (e.g., in which "reset" occurs very little), check that these cases thoroughly exercise the design (e.g., by checking whether certain values or transitions have been overlooked), and check whether invariants have been violated during the simulation process (e.g., making sure that each request is followed by an acknowledgment). HDLs are expanding to accommodate such methodologies.

5.2.2 Verilog and SystemVerilog

The Verilog HDL [1,36,37] was designed and implemented by Phil Moorby at Gateway Design Automation in 1983–1984 (see Moorby's history of the language [15]). The Verilog product was very successful, buoyed largely by the speed of its "XL" gate-level simulation algorithm. Cadence bought Gateway in 1989 and largely because of the pressure from the competing, open VHDL, made the language public in 1990. Open Verilog International was formed shortly thereafter to maintain and promote the standard, IEEE adopted it in 1995, and ANSI in 1996.

The first Verilog simulator was event-driven and very efficient for gate-level circuits, the fashion of the time, but the opening of the Verilog language in the early 1990s paved the way for other companies to develop more efficiently compiled simulators, which traded up-front compilation time for simulation speed.

Like tree rings, the syntax and semantics of the Verilog language embody a history of simulation technologies and design methodologies. At its conception, gate- and switch-level simulations were in fashion, and Verilog contains extensive support for these modeling styles that is now little used. (Moorby had worked with others on this problem before designing Verilog [28].)

Like many HDLs, Verilog supports hierarchy, but was originally designed assuming modules would have at most tens of connections. Hundreds or thousands of connections are now common, and Verilog-2001 [37] added a more succinct connection syntax to address this problem.

Procedural or behavioral modeling, once intended mainly for specifying testbenches, was pressed into service first for RTL specifications and later for so-called behavioral specifications. Again, Verilog-2001 added some facilities to enable this (e.g., `always @*` to model combinational logic procedurally) and SystemVerilog has added additional support (e.g., `always_comb`, `always_ff`).

The syntax and semantics of Verilog are a compromise between modeling clarity and simulation efficiency. A "reg" in Verilog, the variable storage class for behavioral modeling, is exactly a shared variable. This not only means that it simulates very efficiently (e.g., writing to a reg is just an assignment to memory), but also means that it can be misused (e.g., when written to by two concurrently-running processes) and misinterpreted (e.g., its name suggests a memory element such as a flip-flop, but it often represents purely combinational logic).

Thomas and Moorby [66] has long been the standard text on the language. The language reference manual [36], since it was adopted from the original Verilog simulator user manual, is also very readable. Other references include Palnitkar [55] for an overall description of the language, and Mittra [52] and Sutherland [65] for the programming language interface. Smith [63] compares Verilog and VHDL. French et al. [29] discuss how to accelerate compiled Verilog simulation.

5.2.2.1 Coding in Verilog

A Verilog description is a list of modules. Each module has a name; an interface consisting of a list of named ports, each with a type, such as a 32-bit vector, and a direction; a list of local nets and regs; and a body that can contain instances of primitive gates such as ANDs and ORs, instances of other modules (allowing hierarchical structural modeling), continuous assignment statements, which can be used to model combinational datapaths, and concurrent processes written in an imperative style.

Figure 5.1 illustrates the various modeling styles supported in Verilog. The two-input multiplexer circuit in Figure 5.1a can be represented in Verilog using primitive gates (Figure 5.1b), a continuous assignment (Figure 5.1c), a concurrent process (Figure 5.1d), and a user-defined primitive (a truth table, Figure 5.1e). All of these models roughly exhibit the same behavior (minor differences occur when some inputs are undefined) and can be mixed freely within a design.

One of Verilog's strengths is its ability to represent testbenches within the model being tested as well. Figure 5.1f illustrates a testbench for this simple mux, which applies a sequence of inputs over time and prints a report of the observed behavior.

Communication within and among Verilog processes takes place through two distinct types of variables: *nets* and *regs*. Nets model wires and must be driven either by gates or by continuous assignments. Regs are exactly shared memory locations and can be used to model memory elements.

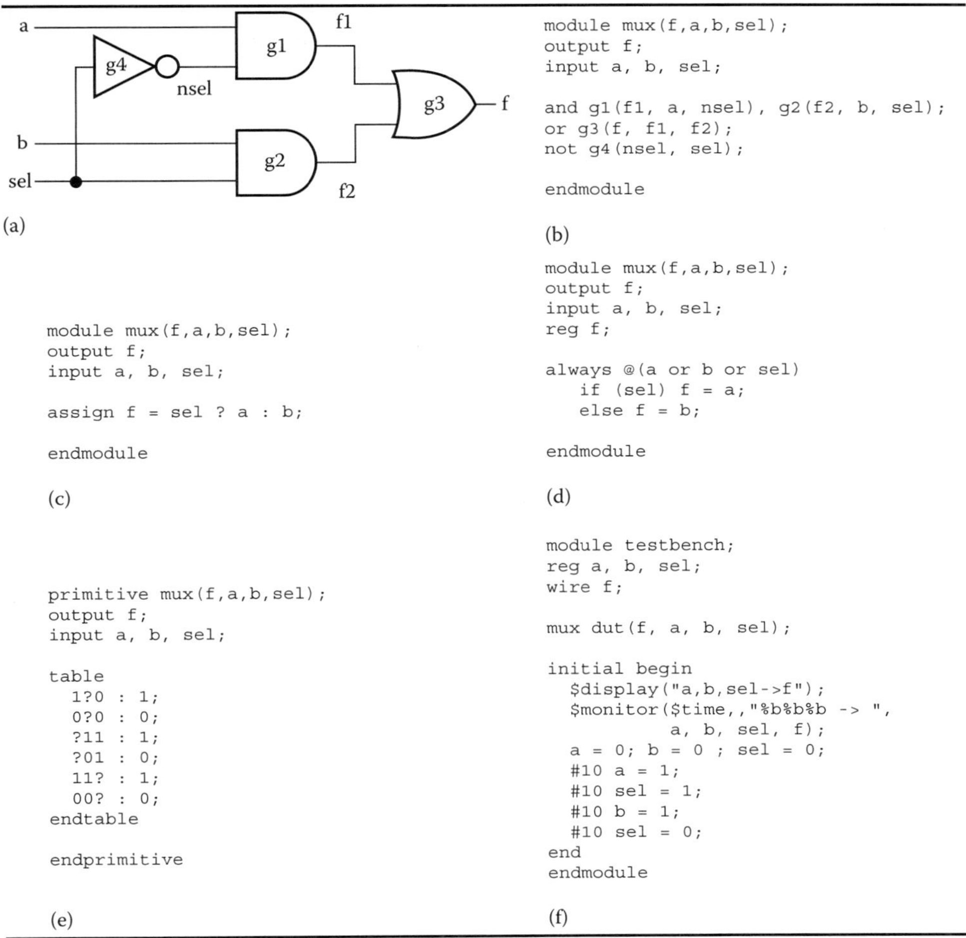

(a)

```
module mux(f,a,b,sel);
output f;
input a, b, sel;

and g1(f1, a, nsel), g2(f2, b, sel);
or g3(f, f1, f2);
not g4(nsel, sel);

endmodule
```

(b)

```
module mux(f,a,b,sel);
output f;
input a, b, sel;

assign f = sel ? a : b;

endmodule
```

(c)

```
module mux(f,a,b,sel);
output f;
input a, b, sel;
reg f;

always @(a or b or sel)
   if (sel) f = a;
   else f = b;

endmodule
```

(d)

```
primitive mux(f,a,b,sel);
output f;
input a, b, sel;

table
  1?0 : 1;
  0?0 : 0;
  ?11 : 1;
  ?01 : 0;
  11? : 1;
  00? : 0;
endtable

endprimitive
```

(e)

```
module testbench;
reg a, b, sel;
wire f;

mux dut(f, a, b, sel);

initial begin
   $display("a,b,sel->f");
   $monitor($time,,"%b%b%b -> ",
            a, b, sel, f);
   a = 0; b = 0 ; sel = 0;
   #10 a = 1;
   #10 sel = 1;
   #10 b = 1;
   #10 sel = 0;
end
endmodule
```

(f)

FIGURE 5.1 Verilog examples. (a) A multiplexer circuit. (b) The multiplexer as a Verilog structural model. (c) The multiplexer using continuous assignment. (d) The multiplexer in imperative code. (e) A user-defined primitive for the multiplexer. (f) A testbench for the multiplexer.

Regs can be assigned only by imperative assignment statements that appear in *initial* and *always* blocks. Both nets and regs can be single bits or bit vectors, and regs can also be arrays of bit vectors to model memories. Verilog also has limited support for integers and floating-point numbers. Figure 5.2 shows a variety of declarations.

The distinction between regs and nets in Verilog is pragmatic: nets have complicated semantics (e.g., they can be assigned a capacitance to model charge storage and they can be connected to multiple tristate drivers to model buses); regs behave exactly like memory locations and are therefore easier to simulate quickly. Unfortunately, the semantics of regs make it easy to inadvertently introduce nondeterminism (e.g., when two processes simultaneously attempt to write to the same reg, the result is a race whose outcome is undefined). This will be discussed in more detail in the next section.

Figure 5.3 illustrates the syntax for defining and instantiating models. Each module has a name and a list of named ports, each of which has a direction and a width. Instantiating such a module involves giving the instance a name and listing the signals or expressions to which it is connected. Connections can be made positionally or by port name, the latter being preferred for modules with many (perhaps ten or more) connections.

Continuous assignments are a simple way to model both Boolean and arithmetic datapaths. A continuous assignment uses Verilog's extensive expression syntax to define a function to be computed and its semantics are such that the value of the expression on the right of a continuous expression is always copied to the net on the left (regs are not allowed on the left of a continuous

```
wire a;                  // Simple wire
tri [15:0] dbus;         // 16-bit tristate bus
tri #(5,4,8) b;          // Wire with delay
reg [-1:4] vec;          // Six-bit register
trireg (small) q;        // Wire stores charge
integer imem[0:1023];    // Array of 1024 integers
reg [31:0] dcache[0:63]; // A 32-bit memory
```

FIGURE 5.2　Various Verilog *net* and *reg* definitions.

```
module mymod(out1, out2, in1, in2);
output out1;        // Outputs first by convention
output [3:0] out2; // four-bit vector
input in1;
input [2:0] in2;

// Module body: instances, continuous assignments, initial and always blocks

endmodule

module usemymod;
reg a;
reg [2:0] b;
wire c, e, g;
wire [3:0] d, f, h;

mymod m1(c, d, a, b);                           // simple instance
mymod m2(e, f, c, d[2:0]),                      // instance with part-select input
      m3(.in1(e), .in2(f[2:0]), .out1(g), .out2(h));  // connect-by-name

endmodule
```

FIGURE 5.3　Verilog structure: An example of a module definition and another module containing three instances of it.

assignment). Practically, Verilog simulators implement this by recomputing the expression on the right whenever there is a change in any variable it references. Figure 5.4 illustrates some continuous assignments.

Behavioral modeling in Verilog uses imperative code enclosed in *initial* and *always* blocks that write to reg variables to maintain state. Each block effectively introduces a concurrent process that is awakened by an event and runs until it hits a delay or a wait statement. The example in Figure 5.5 illustrates basic behavioral modeling.

```verilog
module add8(sum, a, b, carryin);
output [8:0] sum;
input [7:0] a, b;
input carryin;

assign sum = a + b + carryin;    // unsigned arithmetic
endmodule

module datapath(addr_2_0, icu_hit, psr_bm8, hit);
output [2:0]    addr_2_0
output          icu_hit
input           psr_bm8, hit;
wire    [31:0] addr_qw_align;
wire    [3:0]  addr_qw_align_int;
wire    [31:0] addr_d1;
wire           powerdown, pwdn_d1;

assign addr_qw_align =
   { addr_d1[31:4], addr_qw_align_int[3:0] };    // part select + vector concat
assign addr_offset =
  psr_bm8 ? addr_2_0[1:0] : 2'b00;    // if-then-else operator
assign icu_hit = hit & !powerdown & !pwdn_d1;  // Boolean operators

endmodule
```

FIGURE 5.4   Verilog modules illustrating continuous assignment. The first is a simple 8-bit full adder producing a 9-bit result. The second is an excerpt from a processor datapath.

```verilog
module behavioral;
reg [1:0] a, b;

initial begin
  a = 'b1;
  b = 'b0;
end

always begin
  #50 a = ~a;  // Toggle a every 50 time units
end

always begin
  #100 b = ~b;  // Toggle b every 100 time units
end

endmodule
```

FIGURE 5.5   Simple Verilog behavioral model. The code in the *initial* block runs once at the beginning of simulation to initialize the two registers. The code in the two *always* blocks runs periodically: once every 50 and 100 time units, respectively.

```verilog
module FSM(o, a, b, reset);
output o;
reg o;  // declared reg: o is assigned procedurally
input a, b, reset;
reg [1:0] state; // only "state" holds state
reg [1:0] nextstate;

always @(a or b or state)     // Combinational block: sensitive to all inputs;
outputs always assigned case (state)
    2'b00: begin
      o = a & b;
      nextState = a ? 2'b00 : 2'b01;
    end
    2'b01: begin
      o = 0; nextState = 2'b10;
    end
    default: begin
      o = 0; nextState = 2'b00;
    end
  endcase

always @(posedge clk or reset)  // Sequential block: sensitive to clock edge
   if (reset)
     state <= 2'b00;
   else
     state <= nextState;

endmodule
```

FIGURE 5.6   Verilog behavioral model for a state machine illustrating the common practice of separating combinational and sequential blocks.

Figure 5.6 shows a more complicated behavioral model: a state machine written in the common style where the combinational and sequential parts of a state machine are written as two separate processes. The first process is combinational. The `@(a or b or state)` directive triggers its execution when signal *a*, *b*, or *state* changes. The code consists of a multiway choice—a case statement—and performs procedural assignments to the *o* and *nextState* registers. This illustrates one of the odder aspects of Verilog modeling: both *o* and *nextState* are declared reg, yet neither corresponds to the output of a latch or flip-flop. Instead, this reflects the Verilog requirement that procedural assignment can only be performed on regs. A logic synthesis system will recognize that both of these regs are always assigned when any input changes and thus correspond to combinational logic.

The second process models a pair of flip-flops that holds the state between cycles. The `@(posedge clk or reset)` directive makes the process sensitive to the rising edge of the clock or a change in the reset signal. At the positive edge of the clock, the process captures the value of the `nextState` variable and copies it to `state`.

The example in Figure 5.6 illustrates the two types of behavioral assignments. The assignments used in the first process are so-called blocking assignments, written =, and take effect immediately. Nonblocking assignments are written <= and have somewhat subtle semantics. Instead of taking effect immediately, the right-hand-sides of nonblocking assignments are evaluated when they are executed, but the assignment itself does not take place until the end of the current time instant. Such behavior effectively isolates the effect of nonblocking assignments to the next clock cycle, much like the output of a flip-flop is only visible after a clock edge. In general, nonblocking assignments are preferred when writing to state-holding elements for exactly this reason. See Figure 5.7 and the next section for a more extensive discussion of blocking vs. nonblocking assignments.

```
module bad_sr(o, i, clk);          module good_sr(o, i, clk);
output o;                          output o;
input i, clk;                      input i, clk;
reg a, b, o;                       reg a, b, o;

always @(posedge clk) a = i;       always @(posedge clk) a <= i;
always @(posedge clk) b = a;       always @(posedge clk) b <= a;
always @(posedge clk) o = b;       always @(posedge clk) o <= b;

endmodule                          endmodule

(a)                                (b)
```

FIGURE 5.7   Verilog examples illustrating the difference between blocking and nonblocking assignments. (a) An erroneous implementation of a three-stage shift register that may or may not work depending on the order in which the simulator chooses to execute the three *always* blocks. (b) A correct implementation using nonblocking assignments, which make the variables take on their new values after all three blocks are done for the instant.

5.2.2.2   Verilog Shortcomings

Compared to VHDL, Verilog does a poor job at protecting users from themselves. Verilog's `regs` are exactly shared variables and the language permits all the standard pitfalls associated with them, such as races and nondeterministic behavior. Most users avoid such behavior by following rules (e.g., by restricting assignments to a shared variable to a single concurrent process), but Verilog does not flag dangerous usage. Tellingly, a number of EDA companies exist solely to provide lint-like tools for Verilog that report such poor coding practices. Gordon [30] provides a more detailed discussion of the semantic challenges of Verilog.

Nonblocking assignments are one way to ameliorate most problems with nondeterminism caused by shared variables, but they, too, can lead to unexpected behavior. To illustrate the hazards of shared variables, consider a three-stage shift register. The implementation in Figure 5.7a appears to be correct but in fact may not behave as expected because the language says the simulator is free to execute the three *always* blocks in any order when they are triggered. If the processes execute top-to-bottom, the module behaves like a single flip-flop, but if they execute bottom-to-top, the module behaves like a shift register. Even worse, the simulator may behave one way and synthesized logic another.

Figure 5.7b shows a correct implementation of the shift register that uses nonblocking assignments to avoid this problem. The semantics of these assignments are such that the value on the right-hand side of the assignment is captured when the statement runs, but the change to the reg on the left is only done at the "end" of each time instant, i.e., after all three blocks have finished executing. As a result, the order in which the three assignments are executed does not affect the result; this code behaves like a three-stage shift register.

However, the behavior of nonblocking assignments can also be surprising. In most languages, the effect of an assignment can be felt by the instruction immediately following it, but the delayed-assignment semantics of a nonblocking assignment goes against this. Consider the erroneous decimal counter in Figure 5.8a. In most languages, the counter would count from 0 to 9, but in Verilog it counts to 10 before being reset because the test of o by the *if* statement gets the value of o from the previous clock cycle, and not the results of the `o <= o + 1` statement. A corrected version is shown in Figure 5.8b, which uses a local variable `count` to maintain the count, blocking assignments to touch it, and finally a nonblocking assignment to o to send the count outside the module.

Coupled with the rules for register inference, the circuit implied by the counter in Figure 5.8b is actually just fine. While both o and `count` are marked as `reg`, only the `count` variable will actually become a state-holding element.

```
module bad_counter(o, clk);          module good_counter(o, clk);
output o;                            output o;
input clk;                           input clk;
reg [3:0] o;                         reg [3:0] o;
                                     reg [3:0] count;

always @(posedge clk) begin
  o <= o + 1;                        always @(posedge clk) begin
  if (o == 10)                         count = count + 1;
    o <= 0;                            if (count == 10)
end                                      count = 0;
                                       o <= count;
endmodule                            end

(a)                                  endmodule

                                     (b)
```

FIGURE 5.8 Verilog examples illustrating a pitfall of nonblocking assignments. (a) An erroneous implementation of a counter, which counts to 10, and not to 9. (b) A correct implementation using a combination of blocking and nonblocking assignments.

5.2.3 VHDL

Although VHDL and Verilog have grown to be nearly interchangeable, they could not have had more different histories. Unlike Verilog, VHDL was deliberately designed to be a standard HDL after a lengthy, deliberate process. As Dewey explains [23], VHDL was created at the behest of the U. S. Department of Defense in response to the desire to incorporate integrated circuits (specifically very high-speed integrated circuits, and hence the name of the program from which VHDL evolved, VHSIC) in military hardware. Starting with a summer study at Woods Hole, Massachusetts in 1981, requirements for and scope of the language were first established, and then after a bidding process, a contract to develop the language was awarded in 1983 to three companies: Intermetrics, which was the prime contractor for Ada, the software programming language developed for the U.S. military in the early 1980s; Texas Instruments; and IBM. Dewey and de Geus describe this history in more detail [24].

The VHDL was created in 1983 and 1984, essentially concurrent with Verilog, and first released publicly in 1985. Interest in an IEEE standard HDL was high at the time, and VHDL was eventually adapted and adopted as IEEE standard 1076 in 1987 [39] and revised in 1993 [35]. Verilog, meanwhile, remained proprietary until 1990. The standardization and growing popularity of VHDL was certainly instrumental in Cadence's decision to make Verilog public.

The original objectives of the VHDL [23] were to provide a means of documenting hardware (i.e., as an alternative to imprecise English descriptions) and of verifying it through simulation. As such, a VHDL simulator was developed along with the language.

VHDL has spawned a plethora of books discussing its proper usage. Basic texts include Lipsett, Schaefer, Ussery [49] (one of the earliest), Dewey [25], Bhasker [10], Perry [60], and Ashenden [4,5]. More advanced is Cohen [21], which suggests preferred idioms in VHDL, and Harr and Stancluescu [34], which discusses using VHDL for a variety of modeling tasks, and not just RTL.

The VHDL is vast, complicated and has a verbose syntax derived from Ada. It does not have many built-in features such as Verilog's four-valued bit vectors, gate, and transistor-level models. Instead, it has very flexible type and package systems that allow such things to be described in libraries. While its popularity as a means of formal documentation is questionable, it has succeeded as a modeling language for hardware simulation and, like Verilog, more recently as a specification language for RTL logic synthesis.

Although there are many features absent in Verilog that are unique to VHDL (and vice versa), in practice most designers use a nearly identical subset because this is what the synthesis tools accept. Thus although the syntax of the two languages differs greatly, and the semantics appear different, their usage has converged.

5.2.3.1 Coding in VHDL

Unlike Verilog, VHDL draws a strong distinction between the interface to a hierarchical object and its implementation. VHDL interfaces are called *entities* and their implementations are called *architectures*. Figure 5.9 shows code for the same two-input multiplexer as in the Verilog example of Figure 5.1. Figure 5.9a is the entity declaration for the multiplexer, which defines its input and output ports. Figure 5.9b is a structural description of the multiplexer: it defines internal signals, the interface to the Inverter, AndGate, and OrGate components and instantiates four of these entities. The name of the architecture, "`structural`" is arbitrary; it is used to distinguish among different architectures. The Verilog equivalent in Figure 5.1b did not define the internal signals, relying instead on Verilog's rule of automatically treating undeclared signals as single-bit wires.

Figure 5.9c illustrates a dataflow model for the multiplexer with each logic gate made explicit. VHDL does have built-in logical operators.

Finally, Figure 5.9d shows a behavioral implementation of the mux. It defines a concurrently running process sensitive to the three mux inputs (a, b, and sel) and uses an if-then-else statement (VHDL provides most of the usual control-flow statements) to select between copying the a and the b signal to the output f.

A VHDL behavioral process runs until it reaches a `wait` statement or the equivalent. This suspends the process until a particular event occurs, which may be an event on a signal, a condition on a signal, a timeout, or any combination of these. By itself, `wait` terminates a process. At the other extreme, `wait on Clk until Clk = '1' for 5ns;` waits for the clock to rise or for 5 ns, whichever comes first.

Combinational processes, which always run in response to a change on any of their inputs, are common enough to warrant a shorthand. Thus, `process(a, b, sel)` effectively executes a `wait on a, b, sel` statement at the end.

One of the design philosophies behind the VHDL was to maximize its flexibility by making most things user-definable. As a result, unlike Verilog, it has only the most rudimentary built-in types (e.g., Boolean variables, but nothing to model four-valued logic) but has a much more powerful type and package system that allows such types to be defined. For example, the standard logic library for VHDL includes types for representing wire states and standard functions such as AND and OR that operate on these types.

This approach has advantages and disadvantages. While it allows more things to be added to the language later, it also makes for a few infelicities. For example, the predicate in the *if* statement in Figure 5.9d must be written "`sel = '1'`" instead of "`sel = 1`" or even "`sel`" because the argument must be of type Boolean, and not std_logic. While not a serious issue, it is yet another thing that contributes to VHDL's verbosity.

Enumerated literals may be single characters or identifiers. Identifiers are useful for FSM states and single characters are useful for Boolean wire values. Typical declarations

```
type Bit is ('0', '1');
type FourV is ('0', '1', 'X', 'Z');
type State is (Reset, Running, Halted);
```

Objects in VHDL such as types, variables, and signals have attributes such as size, base, and range. Such information can be useful for, say, iterating over all elements in an array. For example, if `type Index is range 31 downto 0`, then `Index'LOW` is 0. Access to information about

```vhdl
entity mux is
  port (
    a, b, sel : in  std_logic;
    f         : out std_logic);
end mux;
```

(a)

```vhdl
architecture structural of mux is
  signal nsel, f1, f2 : std_logic;   -- local signals

  component Inverter
    port (a : in std_logic; y : out std_logic);
  end component;

  component AndGate
    port (a1, a2 : in std_logic; y : out std_logic);
  end component;

  component OrGate
    port (a1, a2 : in std_logic; y : out std_logic);
  end component;

begin
  G1 : AndGate port map (a, nsel, f1);   -- positional
  G2 : AndGate port map (
      a1 => b, a2 => sel, y => f2);      -- by name
  G3 : OrGate port map (a1 => f1, a2 => f2, y => f);
  G4 : Inverter port map (a => sel, y => nsel);
end structural;
```

(b)

```vhdl
architecture dataflow of mux is          architecture procedural of mux is
                                         begin
  signal nsel, f1, f2 : std_logic;
                                           process (a, b, sel)     -- sensitivity list
begin                                      begin
                                             if sel = '1' then
  nsel <= not sel;                            f <= a;
  f1 <= a and nsel;                          else
  f2 <= b and sel;                            f <= b;
  f <= f1 or f2;                             end if;
                                           end process;
end dataflow;
                                         end procedural;
```

(c)   (d)

FIGURE 5.9   VHDL examples. Compare with Figure 5.1. (a) The entity declaration for the multiplexer, which defines its interface. (b) A structural description of the multiplexer from Figure 5.1a. (c) A dataflow description with one equation per gate. (d) An imperative behavioral description.

signals can be used for collecting simulation statistics. For example, if Count is a signal, then Count ′ EVENT is true when there is an event on the signal.

Figure 5.10 shows a VHDL implementation of the classic traffic-light controller from Mead and Conway [51]. This is written in a synthesizeable dialect, using sythe common practice of separating the output and next-state logic from the state-holding element. The first process is sensitive only to the clock signal. The *if* statement in the first process contains checks for an event on the clock and the clock being high, i.e., the rising edge of the clock. The second process is sensitive only to the inputs

```vhdl
library ieee;
use ieee.std_logic_1164.all;

entity tlc is
  port (clk, reset                   : in  std_logic;
        cars, short, long            : in  std_logic;
        highway_yellow, highway_red  : out std_logic;
        farm_yellow, farm_red        : out std_logic;
        start_timer                  : out std_logic);
end tlc;

architecture imp of tlc is
type states is (HG, HY, FY, FG);
signal state, next_state : states;
begin
  process (clk)      -- Sequential process
  begin
    if rising_edge(clk) then
      state <= next_state;
    end if;
  end process;
process (state, reset, cars, short, long)
begin
  if reset = '1' then start_timer <= '1'; next_state <= HG;
  else
    case state is
      when HG =>
        highway_yellow <= '0'; highway_red <= '0';
        farm_yellow <= '0'; farm_red <= '1';
        if cars = '1' and long = '1' then start_timer <= '1'; next_state <= HY;
        else start_timer <= '0'; next_state <= HG;
        end if;
      when HY =>
        highway_yellow <= '1'; highway_red <= '0';
        farm_yellow <= '0'; farm_red <= '1';
        if short = '1' then start_timer <= '1'; next_state <= FG;
        else start_timer <= '0'; next_state  <= HY;
        end if;
      when FG =>
        highway_yellow <= '0'; highway_red <= '1';
        farm_yellow <= '0'; farm_red <= '0';
        if cars = '0' or long = '1' then start_timer <= '1'; next_state <= FY;
        else start_timer <= '0'; next_state <= FG;
        end if;
      when FY =>
        highway_yellow <= '0'; highway_red <= '1';
        farm_yellow <= '1'; farm_red <= '0';
        if short = '1' then start_timer <= '1'; next_state <= HG;
        else start_timer <= '0'; next_state  <= FY;
        end if;
    end case;
  end if;
end process;

end imp;
```

FIGURE 5.10   Traffic-light controller from Mead and Conway [51] implemented in VHDL, illustrating the common practice of separating combinational and state-holding processes.

and present state of the machine, not the clock, and thus models combinational logic. It illustrates the multiway conditional *case* statement, constants, and bit vectors. It employs types and operators from the ieee.std_logic_1164 library, the IEEE standard library [40] for modeling logic that can represent unknown values ("X") as well as 0s and 1s.

5.2.3.2   VHDL Shortcomings

One shortcoming of VHDL is its verbosity: the use of begin/end pairs instead of braces, the need to always separate entities and their architectures, the need to spell out things like ports, its lengthy names for standard logic types (e.g., `std_logic_vector`), and its requirement of enclosing Boolean values and vectors in quotes. Many of these are artifacts of its roots in the Ada language, another verbose language commissioned by the U.S. Department of Defense, but others are due to questionable design decisions. Consider the separation of entity/architecture pairs. While separating these concepts is a boon to abstraction and simplifies the construction of simulations of the same system in different configurations (e.g., to run a simulation using a gate-level architecture in place of a behavioral one for more precise timing estimation), in practice most designers only ever write a single architecture for a given entity and such pairs are usually written together.

The flexibility of VHDL also has advantages and disadvantages. Its type system is much more flexible than Verilog's, providing things such as aggregate types and overloaded functions and operators, but this flexibility also comes with a need for standardization and tends to increase the verbosity of the language as well. Some of the need for standardization was recognized early, resulting in libraries such as the widelysupported IEEE 1164 library for multivalued logic modeling. However, a standard for signed and unsigned arithmetic on logic vectors was slower in coming (it was eventually standardized in 1997 [41]), prompting both Synopsys and Mentor to introduce similar but incompatible versions of a similar library.

Many of the problems stem from a desire to make the language too general. Aspects of the type system suffer from this as well. While the ability to define new enumerated types for multivalued logic modeling is powerful, it seems a little odd to require virtually every VHDL program (its main use has long been specification for RTL synthesis) to include one or more standard libraries. This also leads to the need to be constantly comparing signals to the literal `'1'` instead of just using a signal's value directly and requires a user to carefully manage the types of subexpressions.

5.2.4   SystemC

SystemC is a relative latecomer to the HDL wars. Developed at Synopsys in the late 1990s, primarily by Stan Liao, SystemC was originally called Scenic [47] and was intended to replace Verilog and VHDL as the input language for logic synthesis. See Arnout [3] for some of the arguments for SystemC. SystemC is not a true language; it is a C++ library along with a set of coding rules, but this is exactly its strength. It evolved from the common practice of first writing a high-level simulation model in C or C++, refining it, and finally recoding it in RTL Verilog or VHDL. SystemC was intended to smooth the refinement process by removing the need for a separate HDL.

SystemC can be thought of as a dialect of C++ for modeling digital hardware. Like Verilog and VHDL, it supports hierarchical models whose blocks consist of input/output ports, internal signals, concurrently running imperative processes, and instances of other blocks. The SystemC libraries make two main contributions: an inexpensive mechanism for running many processes concurrently (originally based on a lightweight thread package; see Liao et al. [47]), and an extensive set of types for modeling hardware systems, including bit vectors and fixed-point numbers. A SystemC model consists of a series of class definitions, each of which defines a block. Methods defined for such a class become concurrently running processes, and the constructor for each class starts these processes running by passing them to the simulation kernel. Simulating a SystemC model starts by calling the constructors for all blocks in the design and then invoking the scheduler, which executes the concurrent processes as needed.

The computational model behind earlier versions of SystemC was cycle-based instead of the event-driven model of Verilog and VHDL. This meant that the simulation was driven by a collection of potentially asynchronous but periodic clocks. Later versions (SystemC 2.0 and higher) adopted an event-driven model much like VHDL's.

A number of SystemC books have now appeared. Grötker et al. [32] provide a nice introduction to SystemC 2.0. Bhasker [11] is also an introduction. The volume edited by Muller et al. [53] surveys more advanced SystemC modeling techniques.

5.2.4.1 Coding in SystemC

Figure 5.11 shows a small SystemC model for a 0–99 counter driving a pair of seven-segment displays. It defines two modules (the `decoder` and `counter` *structs*) and an `sc_main` function that defines some internal signals, instantiates two decoders and a counter, and runs the simulation while printing out what it does.

The two modules in Figure 5.11 illustrate two of the three types of processes possible in SystemC. The `decoder` module is the simpler one: it models a purely combinational process by defining a method (called, arbitrarily, "`compute`") that will be invoked by the scheduler every time the `number` input changes, as indicated by the `sensitive << number;` statement beneath the definition of `compute` as an `SC_METHOD`.

The second module `counter` is an `SC_CTHREAD` process: a method (here called "`tick`") that is invoked in response to a clock edge (here, the positive edge of the `clk` input, as defined by the `SC_CTHREAD(tick, clk.pos());` statement) and can suspend itself with the `wait()` statement. Specifically, the scheduler resumes the method when the clock edge occurs, and the method runs until it encounters a `wait()` statement, at which point its state is saved and control passes back to the scheduler.

This example illustrates only a very small fraction of the SystemC type libraries. It uses unsigned integers (`sc_uint`), bit vectors (`sc_bv`), and a clock (`sc_clock`). The nonclock types are wrapped in `sc_signals`, which behave like VHDL signals. In particular, when an `SC_CTHREAD` method assigns a value to a signal, the effect of this assignment is felt only after all the processes triggered by the same clock edge have been run. Thus, such assignments behave like blocking assignments in Verilog to ensure that the nondeterministic order in which such processes are invoked (the scheduler is allowed to invoke them in any order) does not affect the ultimate outcome of simulating the system.

For speed, SystemC also supports transaction-level modeling, in which bus transactions are modeled as function calls rather than on a per-cycle basis as would be done in, e.g., Verilog. For example, a burst-mode bus transfer would be modeled with a function that marks the bus as in use, advances simulation time according to the number of bytes to be transferred, actually copies the data in the simulator, and marks the bus as unused. Nowhere in the simulation would the actual sequence of signals and bits transferred over the bus appear.

5.2.4.2 SystemC Shortcomings

Like many languages, the most common use of SystemC has diverged from its designers' original intentions—an input for hardware synthesis in the case of SystemC. A number of synthesis tools for the language do exist, but SystemC is now used primarily (and successfully) for system modeling. This does mean, however, that it does not solve the "separate language for synthesis" problem.

A big disadvantage of SystemC is that C++ was never intended for modeling digital hardware and as a result is even more lax about enforcing electrical rules than Verilog. Its syntax can be awkward and relies on some very tricky macro preprocessor definitions. On detailed models, the simulation speed

```cpp
#include "systemc.h"
#include <stdio.h>

struct decoder : sc_module {
  sc_in<sc_uint<4> > number;
  sc_out<sc_bv<7> > segments;

  void compute() {
    static sc_bv<7> codes[10] =
        { 0x7e, 0x30, 0x6d, 0x79, 0x33, 0x5b, 0x5f, 0x70, 0x7f, 0x7b };
    if (number.read() < 10) segments = codes[number.read()];
  }

  SC_CTOR(decoder) {
    SC_METHOD(compute);
    sensitive << number;
  }
};

struct counter : sc_module {
  sc_out<sc_uint<4> > ones, tens;
  sc_in_clk clk;

  void tick() {
    int one = 0, ten = 0;
    for (;;) {
      if (++one == 10) {
        one = 0;
        if (++ten == 10) ten = 0;
      }
      ones = one; tens = ten;
      wait();
    }
  }

  SC_CTOR(counter) { SC_CTHREAD(tick, clk.pos()); }
};

int sc_main(int argc, char *argv[])
{
  sc_signal<sc_uint<4> > ones, tens;
  sc_signal<sc_bv<7> > ones_segments, tens_segments;
  sc_clock clk;

  decoder decoder1("decoder1"); decoder1(ones, ones_segments);
  decoder decoder2("decoder2"); decoder2(tens, tens_segments);

  counter counter1("counter1"); counter1(tens, ones, clk);

  for (int i = 0 ; i < 12 ; i++) {
    sc_start(clk, 1);
    printf("%d %d %x %x\n", (int)tens.read(), (int)ones.read(),
                            (int)(sc_uint<7>)tens_segments.read(),
                            (int)(sc_uint<7>)ones_segments.read());
  }
}
```

FIGURE 5.11   SystemC model for a two-digit decimal counter driving two seven-segment displays.

of a good compiled-code Verilog or VHDL simulator may be better, although SystemC is much faster for higher-level models. For such systems, which consist of complex processes, SystemC is superior since the simulation becomes nearly a normal C++ program. However, the context-switching cost in SystemC is higher than that of a good Verilog or VHDL simulator when running a more detailed model, so a system with many small processes may not simulate as quickly.

Another issue is the ease with which a SystemC model can inadvertently be made nondeterministic. Although carefully following the discipline of communicating only among processes through signals can ensure deterministic behavior, any slight deviation from this will cause problems. For example, library functions that use global variables may cause nondeterminism if called from different processes. Accidentally holding state in an SC_METHOD process (easily done if class variables are assigned) can also cause problems since such method are invoked in an undefined order.

Many argue that nondeterministic behavior in a language can be desirable for modeling nondeterministic systems, which certainly exist and need to be modeled. However, the sort of nondeterminism in a language such as SystemC or Verilog creeps in unexpectedly and is difficult to use for modeling. For the simulation of a nondeterministic model to be interesting, there needs to be some way of seeing the different possible behaviors, yet a nondeterministic artifact such as an SC_METHOD process that holds state provides no mechanism for ensuring that it is not, in fact, predictable. As a result, a designer has a hard time answering whether a model of a nondeterministic system can exhibit undesired behavior, even through a careful selection of test cases.

5.3 Hardware Verification Languages

The time required to validate a hardware design greatly outstrips the time required to design or fabricate it. Although many novel techniques have been proposed to address the validation problem, simulation remains the preferred method. So-called formal verification techniques—essentially efficient exhaustive simulation—have been gaining ground, but they suffer from capacity problems.

Simulation applies a stimulus to a model to predict the behavior of the fabricated system. Naturally, there is a trade-off between highly detailed models that can predict many attributes, say, logical values, timing, and power consumption; and simplified models that can only predict logical behavior but execute much faster.

Because the size of the typical design has grown exponentially over time, functional simulation, which only predicts the logical behavior of a synchronous circuit at clock-cycle boundaries, has become the preferred form of simulation because of its superior speed. Furthermore, designers have shied away from more timing-sensitive circuitry such as gated clocks and transparent latches because they require more detailed simulation models and are therefore more costly to simulate.

Simulation-based validation raises three important questions: what the stimulus should be, whether it exposes any errors, and whether the stimulus is long and varied enough to have exposed "all" design errors. Historically, these three questions have been answered manually, i.e., by having a test engineer write test cases, check the results of simulation, and make some informed guesses about how comprehensive the test suite actually is.

A manual approach has many shortcomings. Writing testcases is tedious and the number necessary for "complete" verification grows faster than the size of the system description. Manually checking the output of simulation is similarly tedious and subtle errors can easily be overlooked. Finally, it is difficult to judge quantitatively how much of a design has really been tested.

More automated methodologies, and ultimately languages, have evolved to address some of these challenges, although the verification problem remains one of the most difficult. Biased random test case generation has become standard practice, although it has only supplemented, not completely supplanted, manual test case generation. Designer-inserted assertions, long standard practice for software development, have also become standard for hardware, although the sort of assertions needed in hardware are more complicated than the typical "the argument must be nonzero" sort of checking that works well in software. Finally, automated "coverage" checking, which attempts to quantify how much of a design's behavior has been exercised in simulation, has also become standard.

All of these techniques are improvements, and not a panacea. While biased random test case generation can quickly generate many interesting tests, it provides no guarantee of completeness; bugs may go unnoticed. Because they must often check temporal properties (e.g., "acknowledge arrives within three cycles of every request"), assertions in hardware systems can be more difficult to write than those in software systems (which most often check data structure consistency) and again there is no way to know when enough assertions have been added; it is possible that the assertions themselves have flaws (e.g., they let bugs by). Finally, test cases that achieve "complete" coverage can also let bugs by because the criteria for coverage is necessarily weak. Coverage typically checks what states particular variables have been in, but it cannot consider all combinations because they grow exponentially with design size. As a result, certain important combinations may not be checked even though coverage checks report "complete coverage."

While the utility of biased random test generation and coverage metrics is mostly limited to simulation, assertion specification techniques are useful for, and have been heavily influenced by, formal verification. Pure formal techniques consider all possible behaviors by definition and therefore do not require explicit test cases (implicitly, they consider all possible test cases) and also do not need to consider coverage. But knowing what behavior is unwanted is crucial for formal techniques, whose purpose is to either expose unwanted behavior or formally prove it cannot occur.

Recently, a sort of renaissance has occurred in verification languages. Temporal logics, specifically Linear Temporal Logic (LTL) and Computation Tree Logic (CTL), form the mathematical basis for most assertion checking, but their mathematical syntax is awkward. Instead, a number of more traditional computer languages, which combine a more human-readable syntax for the basic logic with a lot of "syntactic sugar" for more naturally expressing common properties, have been proposed for expressing properties in these logics. Languages from Intel (ForSpec) and IBM (Sugar) have emerged as the most complete.

Meanwhile, some EDA companies have produced languages designed for writing testbenches and checking simulation coverage. Vera, originally designed by Systems Science and since acquired by Synopsys, and e, designed and sold by Verisity, have been the two most commercially successful. Bergeron [8] discusses how to use the two languages.

All four of these languages have recently undergone extensive crossbreeding. Vera has been made public, rechristened OpenVera, had Intel's ForSpec assertions grafted onto it, and added almost in its entirety to SystemVerilog. Sugar, meanwhile, has been adopted by the Accelera standards committee, rechristened the Property Specification Language (PSL), and also added in part to SystemVerilog. Verisity's e has changed the least, only recently having been made public.

There are obvious advantages in having a single industry-standard language for assertions, so it seems likely that most of these languages will eventually disapper. SystemVerilog, in part because it has taken a melting-pot approach, appears to be in the lead, but things can always change.

5.3.1　OpenVera

OpenVera began life around 1995 as Vera, a proprietary language implemented by Systems Science, Inc. mainly for creating testbenches for Verilog simulations. As such, its syntax was heavily influenced by Verilog. VHDL support was added later. Synopsys bought the company in 1998, released the language to the public in 2001, and rechristened it OpenVera.

OpenVera is a concurrent, imperative language designed for writing testbenches. It executes in concert with a Verilog or VHDL simulator and can both provide stimulus to the simulator and observe the results. In addition to the usual high-level imperative language constructs, such as conditionals, loops, functions, strings, and associative arrays, it provides extensive facilities for generating biased random patterns (designed to be applied to the hardware design under test) and monitoring what values particular variables take on during the simulation (for checking coverage).

In the process of making OpenVera public, Synopsys added the assertion specification capabilities of Intel's ForSpec language, making it easy to check whether certain behaviors ever appear during the simulation.

As further crossbreeding, much of Vera was incorporated into SystemVerilog 3.1 around 2003. In particular, its styles of generating biased random variables and checking for behavior coverage have been adopted more-or-less verbatim.

The following is a quick overview of OpenVera 1.3. References for the language include the OpenVera language reference manual, available from the OpenVera web site, and Haque et al. [33].

OpenVera has three main facilities that separate it from more traditional programming languages: biased random variable generation subject to constraints, monitoring facilities for reporting coverage of state variables, and the ability to specify temporal assertions.

Figure 5.12 shows a simple OpenVera program that demonstrates the language's ability to generate biased random variables. First, a Java/C++-like class is defined containing the 16-bit field `addr` and the 32-bit field `data`. These are marked `rand`, meaning their values will be set by a call to the `randomize()` method.

Following the definition of the fields, a constraint (named arbitrarily `word_align`) is defined for objects of this class. Such constraints restrict the possible values the `randomize()` method may assign to various fields. This particular constraint simply restricts the two lowest bits of `addr` to be zero, i.e., aligned on a four-byte boundary.

The "demo" program creates a new `Bus` object and then, for 50 times, invokes `randomize()` to generate a new set of "random" values (in fact, they are taken from a pseudorandom sequence guaranteed to be the same each time the program runs) for the two fields in the `bus` object that conform to the given constraint.

Overly constrained or inconsistent constraints may lead `randomize()` to fail. It signals this with a non-OK return value, here assigned to the local integer variable `result`. For this simple example, randomization will always succeed. The constraint solver guarantees that it will only fail if there is no consistent solution to the supplied constraints.

This example only illustrates a fraction of OpenVera's facilities for biased random variable generation. It can also add constraints on the fly, impose set membership constraints, follow user-defined

```
class Bus {
  rand bit[15:0] addr;
  rand bit[31:0] data;

  constraint world_align { addr[1:0] == '2b0; }
}

program demo {
  Bus bus = new();
  repeat (50) {
    integer result = bus.randomize();
    if (result == OK) printf("addr = %16h data = %32h\n", bus.addr, bus.data);
    else
      printf("randomization failed\n");
  }
}
```

FIGURE 5.12 Simple Vera program illustrating its ability to generate biased random variables and its mix of imperative and object-oriented styles. From the OpenVera 1.3 LRM.

distributions, impose conditional constraints, impose constraints between variables, selectively randomize and disable randomization of user-specified variables, and generate random sequences from a language specified by a grammar.

OpenVera supports so-called functional coverage checking, which can monitor state variables and state transitions (unlike software coverage, which usually monitors which statements and branches have been executed). It uses a bin model: each bin represents a particular state or transition and when a matching activity occurs, the counter for that bin is incremented. The number of bins that remain empty after simulation therefore gives a rough idea of what behavior is yet to be exercised.

The `coverage_group` construct defines a type of monitor. Each specifies a set of variables to monitor and an event that triggers a check of the variables, typically the positive edge of a clock. Like a class, these constructs must be explicitly instantiated and there may be multiple copies of each, especially useful when a coverage group has parameters. Figure 5.13 illustrates a simple coverage group.

Like the earlier example, Figure 5.13 shows only the most basic coverage functionality. OpenVera can also monitor cross coverage, i.e., the combinations of values taken by two or more variables; selectively disable coverage checking, e.g., during system reset; allow the user to explicitly specify bins and the values mapped to them; and monitor transition coverage, i.e., combinations of values taken by the same variable in successive cycles.

OpenVera's assertions provide a way to check temporal properties such as "an acknowledge signal must occur within three clock cycles after any request." The syntax for assertions is a little unusual.

Figure 5.14 shows the assertion language. It defines a checker ("mychecker") that takes a single integer parameter, p, an enable signal, a clock, and an 8-bit bus and looks for the sequence p, 4, 6, 9, 3 on the bus. The `clock` construct defines a collection of events (patterns) synchronized to the positive edge of the clock signal. Events `e0` through `e3` are simple properties: they look for patterns on the result signal. The `myseq` pattern is the interesting one: it looks for the appearance of the four events separated by a single clock cycle (a one-cycle delay is written #1). The `assert` directive means to check for the `myseq` event and report an error otherwise.

Finally, `bind` indicates where to instantiate the checker, gives it the name `myinst`, passes the parameter 4, and connects the checker to the enable, clock, and result signals.

The example in Figure 5.14 barely scratches the surface. Sequences can also contain consecutive and nonconsecutive repetition, explicit delays, simultaneous and disjoint sequences, and sequence containment.

```
bit clk;
bit [15:0] addr;
bit [7:0] data;

coverage_group MyChecker {
  sample_event = @(posedge clk);
  sample addr, data;
}

program demo_coverage {
  MyChecker checker = new();
  ...
}
```

FIGURE 5.13   Simple Vera program containing a single coverage group that monitors which values appear on the addr and data state variables. From the OpenVera 1.3 LRM.

```
/* Checks for the sequence p, 6, 9, 3 */
unit mychecker
  #(parameter integer p = 0)
  (logic en, logic clk, logic [7:0] result);

  clock posedge (clk) {
    event e0: (result == p);
    event e1: (result == 6);
    event e2: (result == 9);
    event e3: (result == 3);
    event myseq: if (en) then (e0 #1 e1 #1 e2 #1 e3);
  }

  assert myassert: check(myseq, "Missed a step.");
endunit

/* Watch for the sequence 4, 6, 9, 3 on outp */
bind instances cnt_top.dut: mychecker myinst #(4) (enable, clk, outp);
```

FIGURE 5.14 Sample of OpenVera's assertion language. It defines a checker that watches for the sequence p, 6, 9, 3 and creates an instance of it that checks for the sequence 4, 6, 9, 3 on the outp bus. From the OpenVera 1.3 LRM.

5.3.2 The e Language

The e language was developed by Verisity as part of its Specman product as a tool for efficiently writing testbenches. Like Vera, it is an imperative object-oriented language with concurrency, the ability to generate constrained random values, mechanisms for checking functional (variable value) coverage, and a way to check temporal properties (assertions). Books on e include Palnitkar [56] and Iman and Joshi [38].

The syntax of e is unusual. First, all code must be enclosed in < ' and ' > symbols; otherwise, it is considered a comment. Eschewing the C syntax, e declarations are written "name : type." The syntax for fields in compound types (e.g., structs) includes particles such as % and !, which indicate when a field is to be driven on the device-under-test and not randomly computed, respectively.

Figure 5.15 shows a fragment of an e program that defines an abstract test strategy for a simple microprocessor, specifically how to generate instructions for it. It illustrates the type system of the language as well as the utility of constraints. It defines two enumerated scalar types, `opcode` and `reg`, and defines the width of each. The `struct instr` defines a new compound type (`instr`) that represents a single instructions. First is the `op` field, which is one of the opcodes defined earlier. The `op1` and the `op2` fields are marked with %, indicating that they should be considered by the `pack` built-in procedure, which marshals data to send to the simulation.

The `kind` field is also an enumerated scalar but is used here as a type tag. It is not marked with %, which means its value will not be included when the structure is packed and sent to the simulation. The two when directives define two subtypes, i.e., "`reg instr`" and "`imm instr`." Such subtypes are similar to derived classes in object-oriented programming languages. Here, the value of the `kind` field, which can be either `imm` or `reg`, determines the subtype.

The two `keep` directives impose constraints between the `kind` field and the opcode, ensuring, e.g., that `ADD` and `SUB` instructions are of the `reg` type. Although these constraints are simple, e is able to impose much more complicated constraints on the values of fields in a struct.

The final `when` directive further constrains the `JMP` and `CALL` instructions, i.e., by restricting what values the `op2` field may take for these instructions.

The `extend sys` directive adds a field named `instrs` to the `sys` built-in structure, which is the basic environment. The leading ! makes the system create an empty list of instructions, which will be filled in later.

```
Instruction encoding for a very simple processor
<'
type opcode: [ADD, SUB, ADDI, JMP, CALL] (bits: 4);
type reg: [REG0, REG1, REG2, REG3] (bits: 2);

struct instr {
  %op  : opcode;       // Four-bit opcode
  %op1 : reg;          // Two-bit operand

  kind : [imm, reg]; // Second operand type
  when reg instr { %op2 : reg; }  // reg operand
  when imm instr { %op2 : byte; } // imm operand

  // Constrain certain instructions to reg or imm
  keep op in [ ADD, SUB ] => kind == reg;
  keep op in [ ADDI, JMP, CALL ] => kind == imm;

  // Constrain JMP, CALL second operand
  when imm instr { keep opcode in [JMP, CALL] => op2 < 16; }
};

extend sys {
  // Add a non-generated field called instrs
  !instrs : list of instr;
};
'>
```

FIGURE 5.15   e code defining instruction encoding for a simple 8-bit microprocessor. After an example in the Specman tutorial.

```
<'
extend instr {
  keep opcode in [ADD, ADDI];
  keep op1 == REG0;
  when reg instr { keep op2 == REG1; }
  when imm instr { keep op2 == 0x3; }
};

extend sys {
  keep instrs.size() == 10;
};
'>
```

FIGURE 5.16   e code that uses the instruction encoding of Figure 5.15 to randomly generate 10 instructions. After an example in the Specman tutorial.

Figure 5.16 illustrates how the definition of Figure 5.15 can be used to generate tests that exercise the ADD and ADDI instructions. It first adds constraints to the instr class (the template for instructions defined in Figure 5.15) that restrict the opcodes to either ADD or ADDI and then imposes a constraint on the top level (sys) that makes it generate exactly 10 instructions. Running the source code of Figures 5.15 and 5.16 together makes the system generate a sequence of 10 pseudorandom instructions.

5.3.3   PSL

The PSL evolved from the proprietary Sugar language developed at IBM. Its focus is narrower than either OpenVera or e, since its goal is purely to specify temporal properties to be checked in hardware designs, but it is more disciplined and has more formal semantics.

Beer et al. [6] provide a nice introduction to an earlier version of the language, which they explain evolved over many years. It has been used within IBM in the RuleBase formal verification system since 1994 and was also pressed into service as a checker generator for simulators in 1997. Accelera, an EDA standards group, adopted it as their formal property language in 2002. Cohen [22] provides a tutorial.

The PSL is based on CTL [20], a powerful but rather cryptic temporal logic for specifying properties of finite-state systems. It is able to specify both safety properties ("this bad thing will never happen") as well as liveness properties ("this good thing will eventually happen"). Liveness properties can only be checked formally because it makes a statement about all the possible behaviors of a system; safety properties can also be tested in simulation. LTL, a subset of CTL, expresses only safety properties and can therefore be turned into checking automata meant to be run in concert with a simulation to look for unwanted behavior. The PSL carefully defines which subset of its properties is purely LTL and is therefore a candidate for use in simulation-based checking.

The PSL is divided into four layers. The lowest, Boolean, consists of instantaneous Boolean expressions on signals in the design under test. The syntax of this layer follows that of the HDL to which the PSL is being applied and can be Verilog, SystemVerilog, VHDL, or GDL. For example, `a[0:3] & b[0:3]` and `a(0 to 3) and b(0 to 3)` represent the bit-wise AND of the four most significant bits of vectors a and b in the Verilog and VHDL flavors, respectively.

The second layer, temporal, is where the PSL gets interesting. It allows a designer to state properties that hold across multiple clock cycles. The `always` operator, which states that a Boolean expression holds in every clock cycle, is one of the most basic. For example, `always !(ena & enb)` states that the signals `ena` and `enb` will never be true simultaneously in any clock cycle.

More interesting are operators that specify delays. The `next` operator is the simplest. The property `always (req -> next ack)` states that in every cycle that the `req` signal is true, the `ack` signal is true in the next cycle. The `->` symbol denotes implication, i.e., if the expression to the left is true, that on the right must also be true. The `next` operator can also take an argument, e.g., `always req -> next[2] ack` means that `ack` must be true two cycles after each cycle in which `req` is true.

The PSL provides an extended form of regular expressions convenient for specifying complex behaviors. Although it would be possible to write

```
always (req -> next (ack-> next !cancel))
```

to indicate that `ack` must be true after any cycle in which `req` is true and `cancel` must be false in the cycle after that, it is easier to write `always {req ; ack ; !cancel}`. This illustrates a basic principle of the PSL: most operators are actually just "syntactic sugar"; the set of fundamental operators is quite small.

The PSL draws a clear distinction between "weak" operators, which can be checked in simulation (i.e., safety properties) and "strong" operators, which express liveness properties and can only be checked formally. Strong operators are marked with a trailing exclamation point (`!`), and some operators come in both strong and weak varieties.

The `eventually!` operator illustrates the meaning of strong operators. The property `always (req -> eventually! ack)` states that after `req` is asserted, `ack` will always be asserted eventually. This is not something that can be checked in simulation: if a particular simulation saw `req` but did not see `ack`, it would be incorrect to report that this property failed because running that particular simulation longer might have produced `ack`.

Another subtlety is that it is possible to express properties in which time moves backward through a property. A simple example is `always ((a && next[3](b)) -> c`, which states that when a is true and b is true three clock cycles later, c is true in the first cycle, i.e., when a was true. While it is possible to check this in simulation (for each cycle in which a is true, remember whether c is true and look three clock cycles later for b), it is more difficult to build automata that check such properties in simulation.

The third layer of the PSL, the verification layer, instructs a verification tool what tests to perform on a particular design. It amounts to a binding between properties defined with expressions from the Boolean and temporal layer and modules in the design under test. The following simple example

```
vunit ex1a(top_block.i1.i2) {
  A1: assert never (ena && enb);
}
```

declares a "verification unit" called `ex1a`, binds it to the instance named `top_block.i1.i2` in the design under test, and declares (the assertion named `A1`) that the signals `ena` and `enb` in that instance are never true simultaneously.

In addition to `assert`, verification units may also include `assume` directives, which allow the tool to assume a property; `assume_guarantee`, which both assumes and tests a property; `restrict`, which constrains the tool to only consider those behaviors that have a property; `cover`, which asks the tool to check whether a certain property was ever observed; and `fairness`, which instructs the tool to only consider paths in which the given property occurs infinitely often, e.g., only when the system does not wait indefinitely.

The fourth (modeling) layer of the PSL allows Verilog, SystemVerilog, VHDL, or GDL code to be included in a PSL specification. The intention here is to include additional details about the system under test in the PSL source file.

5.3.4 SystemVerilog

Many aspects of the Vera, Sugar, and ForSpec verification languages have been merged into Verilog along with the higher-level programming constructs of Superlog [27] (which were taken nearly verbatim from C and C++) to produce SystemVerilog [1]. As a result, Verilog has become the English of the HDL world: voraciously assimilating parts of other languages and making them its own.

The C- and C++-like features added to SystemVerilog read like the list of features in those languages. SystemVerilog adds enumerated types, record types (*structs*), *typedef*s, type casting, a variety of operators such as `+=`, operator overloading, control-flow statements such as `break` and `continue`, as well as object-oriented programming constructs such as classes, inheritance, and dynamic object creation and deletion. At the very highest level, it also adds strings, associative arrays, concurrent process control (e.g., fork/join), semaphores, and mailboxes, giving it features found in concurrent software languages such as Java.

Perhaps the most interesting features added to SystemVerilog are those directly related to verification. Specifically, SystemVerilog includes constrained, biased random variable generation, user-defined functional coverage checking, and temporal assertions, much like those in Vera, e, PSL, and ForSpec.

Figure 5.17 illustrates some of the random test-generation constructs in SystemVerilog, which were largely taken from Vera. Compare this with Figure 5.12.

Figure 5.18 illustrates some of the coverage constructs in SystemVerilog. In general, one defines "covergroups," which are collections of bins that sample values on a given event, typically a clock edge. Each covergroup defines the sorts of values it will be observing (e.g., values of a single variable, combinations of multiple variables, and sequences of values on a single variable) and rules that define the "bins" each of whose values will be placed in. In the end, the simulator reports which bins were empty, indicating that none of the matching behavior was observed. Again, much of this machinery was taken from Vera (cf. Figure 5.13).

```
class Bus;
  rand bit[15:0] addr;
  rand bit[31:0] data;

  constraint world_align { addr[1:0] = 2'b0; }
endclass

initial begin
  Bus bus = new;

  repeat (50) begin
    if (bus.randomize() == 1)
        $display("addr = %16h data = %h\n", bus.addr, bus.data);
    else
        $display("overconstrained\n");
  end
end

typedef enum { low, mid, high } AddrType;

class MyBus extends Bus;
  rand AddrType atype; // Random variable

  // Additional address constraint
  constraint addr_range {
    (atype == low ) -> addr inside { [0:15] };
    (atype == mid ) -> addr inside { [16:127] };
    (atype == high) -> addr inside { [128:255] };
  }
endclass

task exercise_bus;
  int res;

  // Restrict to low addresses
  res = bus.randomize() with { atype == low; };
  // Restrict to address range
  res = bus.randomize() with { 10 <= addr && addr <= 20 };
  // Restrict data to powers of two
  res = bus.randomize() with { data & (data - 1) == 0 };


  bus.word_align.constraint_mode(0);   // Disable word alignment
  res = bus.randomize with { addr[0] || addr[1] };
  bus.word_align.constraint_mode(1);   // Re-enable word alignment
endtask
```

FIGURE 5.17 Constrained random variable constructs in SystemVerilog. The example starts with a simple definition of a Bus class that constraints the two least-significant bits of the address to be zero and then invokes the randomize method to randomly generate address/data pairs and print the result. Next is a refined version of the Bus class that adds a field taken from an enumerated type that further constrains the address depending on its value. The example ends with a task that illustrates various ways to control the constraints. After examples in the SystemVerilog LRM [1].

Figure 5.19 shows some of SystemVerilog's assertion constructs. In addition to signaling an error when an "instantaneous" condition does not hold (e.g., a set of variables are taking on mutually-incompatible values), SystemVerilog has the ability to describe temporal sequences such as "ack must rise between one and five cycles after req rises" and check whether they appear during simulation. Much of the syntax comes from PSL/Sugar.

```
enum { red, green, blue } color;

bit [3:0] adr, offset;

covergroup g2 @(posedge clk);
  Hue:     coverpoint color;
  Offset: coverpoint offset;
  // Consider (color, adr) pairs, e.g., (red, 3'b000), (red, 3'b001), (blue, 3'b111)
  AxC:     cross color, adr;
  // Consider (color, adr, offset) triplets.  Creates 3 * 16 * 16 = 768 bins
  all:     cross color, adr, Offset;
endgroup

g2 g2_inst = new; // Create a watcher
bit [9:0] a;         // Takes values 0--1023

covergroup cg @(posedge clk);

  coverpoint a {
    bins a = { [0:63], 65 };        // place values 0--63 and 65 in bin a
    // create 65 bins, one for 127, 128, ..., 191
    bins b[] = { [127:150], [148:191] };
    bins c[] = { 200, 201, 202 };   // create three bins: 200, 201, and 202
    bins d = { [1000:$] };          // place values 1000--1023 in bin d
    bins others[] = default;
    // place all other values (e.g., 64, 66, ..., 126, 192, ...) in their own bin
    bins others[] = default;
  }

endgroup

bit [3:0] a;

covergroup cg @(posedge clk);
  coverpoint a {
    // Place 4->5->6, 7->11, 8->11, 9->11, 10->11, 7->12, 8->12, 9->12,
    // and 10->12 into sa.
    bins sa = (4 => 5 => 6), ([7:9],10 => 11,12);
    // Create bins for 4->5->6, 7->10, 8->10, and 9->10
    bins sb[] = (4 => 5 => 6), ([7:9] => 10);
    // Look for the sequence 3->3->3->3
    bins sc = 3 [* 4];
    // Look for the sequences 5->5, 5->5->5, or 5->5->5->5
    bins sd = 5 [* 2:4];

    // Look for sequence of the form 6->...->6->...->,
    // where ... represents any sequence excluding 6
    bins se = 6 [->3];
  }
endgroup
```

FIGURE 5.18 SystemVerilog coverage constructs. The example begins with a definition of a "covergroup" that considers the values taken by the color and offset variables as well as combinations. Next is a covergroup illustrating the variety of ways "bins" may be defined to classify values for coverage. The final covergroup illustrates SystemVerilog's ability to look for and classify sequences of values, not just simple values. After examples in the SystemVerilog LRM [1].

```
// Make sure req1 or req2 is true if we are in the REQ state
always @(posedge clk)
  if (state == REQ) assert (req1 || req2);

// Same, but report the error ourselves
always @(posedge clk)
  if (state == REQ) assert (req1 || req2)
  else $error("In REQ; req1 || req2 failed (\%0t)", $time);

property req_ack;
  @(posedge clk) // Sample req, ack at rising edge
    // After req is true, ack must rise between 1 and 3 cycles later
    req ##[1:3] $rose(ack);
endproperty

// Assert that this property holds, i.e., create a checker
as_req_ack: assert property (req_ack);

// The own_bus signal goes high in 1 to 5 cycles,
// then the breq signal goes low one cycle later.
sequence own_then_release_breq;
  ##[1:5] own_bus ##1 !breq
endsequence

property legal_breq_handshake;
  @(posedge clk)                           // On every clock,
  disable iff (reset)                      // unless reset is true,
  $rose(breq) |-> own_then_release_breq;   // once breq has risen, own_bus
                                           // should rise; breq should fall.
endproperty

assert property (legal_breq_handshake);
```

FIGURE 5.19 SystemVerilog assertions. The first two *always* blocks check simple safety properties, i.e., that `req1` and `req2` are never true at the positive edge of the clock. The next property checks a temporal property: that `ack` must rise between one and three cycles after each time `req` is true. The final example shows a more complex property: when reset is not true, a rising `breq` signal must be followed by `own_bus` rising between one and five cycles later and `breq` falling.

5.4 Software Languages

Software languages describe sequences of instructions for a processor to execute. As such, most consist of sequences of imperative instructions that communicate through memory: an array of numbers that hold their values until changed.

Each machine instruction typically does little more than, say, adding two numbers, so high-level languages aim to specify many instructions concisely and intuitively. Arithmetic expressions are typical: coding an expression such as $ax^2 + bx + c$ in machine code is straightforward, tedious, and best done by a compiler. The C language provides such expressions, control-flow constructs such as loops and conditionals, and recursive functions. The C++ language adds classes as a way to build new data types, templates for polymorphic code, exceptions for error handling, and a standard library of common data structures. Java is a still higher-level language that provides automatic garbage collection, threads, and monitors to synchronize them (Table 5.1).

TABLE 5.1   Software Language Features Compared

	C	C++	Java
Expressions	●	●	●
Control-flow	●	●	●
Recursive functions	●	●	●
Exceptions	○	●	●
Classes and inheritance		●	●
Templates		●	
Namespaces		●	●
Multiple inheritance		●	○
Threads and locks			●
Garbage collection		○	●

Note: ●, full support;   ○, partial support.

```
                                          mov   %i0, %o1
                                          b     .LL3
                                          mov   %i1, %i0
                                          mov   %i0, %o1
                                          b     .LL3
                                          mov   %i1, %i0
      jmp   L2                           .LL5:
  L1:                                     mov   %o0, %i0
      movl  %ebx, %eax                   .LL3:
      movl  %ecx, %ebx                    mov   %o1, %o0
  L2:                                     call  .rem, 0
      xorl  %edx, %edx                    mov   %i0, %o1
      divl  %ebx                          cmp   %o0, 0
      movl  %edx, %ecx                    bne   .LL5
      testl %ecx, %ecx                    mov   %i0, %o1
      jne   L1

         (a)                                   (b)
```

FIGURE 5.20   Euclid's algorithm (a) i386 assembly (CISC) and (b) SPARC assembly (RISC). SPARC has more registers and must call a routine to compute the remainder (the i386 has division instruction). The complex addressing modes of the i386 are not shown in this example.

5.4.1 Assembly Languages

An assembly language program (Figure 5.20) is a list of processor instructions written in a symbolic, human-readable form. Each instruction consists of an operation such as addition along with some operands. For example, add r5, r2, r4 might add the contents of registers r2 and r4 and write the result to r5. Such arithmetic instructions are executed in order, but branch instructions can perform conditionals and loops by changing the processor's program counter—the address of the instruction being executed.

A processor's assembly language is defined by its opcodes, addressing modes, registers, and memories. The opcode distinguishes, say, addition from conditional branch, and an addressing mode defines how and where data is gathered and stored (e.g., from a register or from a particular memory location). Registers can be thought of as small, fast, easy-to-access pieces of memory.

There are roughly four categories of modern assembly languages (Table 5.2). The oldest are those for the so-called complex instruction set computers (CISC). These are characterized by a rich set of instructions and addressing modes. For example, a single instruction in Intel's x86 family, a typical CISC processor, can add the contents of a register to a memory location whose address is the sum of two other registers and a constant offset. Such instruction sets are usually convenient for human

TABLE 5.2   Typical Modern Processor Architectures

CISC	RISC	DSP	Microcontroller
x86	SPARC	TMS320	8051
68000	MIPS	DSP56000	PIC
	ARM	ASDSP-21xx	AVR

programmers, who are usually good at using a heterogeneous collection of constructs, and the code itself is usually quite compact. Figure 5.20a illustrates a small program in x86 assembly.

By contrast, reduced instruction set computers (RISC) tend to have fewer instructions and much simpler addressing modes. The philosophy is that while you generally need more RISC instructions to accomplish something, it is easier for a processor to execute them because it does not need to deal with the complex cases and easier for a compiler to produce them because they are simpler and more uniform. Patterson and Ditzel [59] were among the first to argue strongly for such a style. Figure 5.20b illustrates a small program for a RISC, in SPARC assembly.

The third category of assembly languages arises from more specialized processor architectures such as digital signal processors (DSPs) and very long instruction word processors. The operations in these instruction sets are simple like those in RISC processors (e.g., add two registers), but they tend to be very irregular (only certain registers may be used with certain operations) and support a much higher degree of instruction-level parallelism. For example, Motorola's DSP56001 can, in a single instruction, multiply two registers, add the result to the third, load two registers from memory, and increase two circular buffer pointers. However, the instruction severely limits which registers (and even which memory) it may use. Figure 5.21a shows a filter implemented in 56001 assembly.

The fourth category includes instruction sets on small (4- and 8-bit) microcontrollers. In some sense, these combine the worst of all worlds: there are few instructions and each cannot do much, much like a RISC processor, and there are also significant restrictions on which registers can be used, much like a CISC processor. The main advantage of such instruction sets is that they can

```
move #samples, r0                          START:
move #coeffs, r4                               MOV    SP, #030H
move #n-1, m0                                  ACALL  INITIALIZE
move m0, m4                                     ORL    P1,#0FFH
movep y:input, x:(r0)                           SETB   P3.5
clr   a                                     LOOP:
      x:(r0)+, x0 y:(r4)+, y0                   CLR    P3.4
rep   #n-1                                      SETB   P3.3
mac   x0,y0,a                                   SETB   P3.4
      x:(r0)+, x0 y:(r4)+, y0               WAIT:
macr  x0,y0,a   (r0)-                           JB     P3.5, WAIT
movep a, y:output

                                               CLR    P3.3
                                               MOV    A,P1
                                               ACALL  SEND
                                               SETB   P3.3
                                               AJMP   LOOP

      (a)                                          (b)
```

FIGURE 5.21   (a) Finite impulse response filter in DSP56001 assembly. The *mac* instruction (multiply and accumulate) does most of the work, multiplying registers X0 and Y0, adding the result to accumulator A, fetching the next sample and coefficient from memory, and updating circular buffer pointers R0 and R4. The *rep* instruction repeats the *mac* instruction in a zero-overhead loop. (b) Writing to a parallel port in 8051 microcontroller assembly. This code takes advantage of the 8051's ability to operate on single bits.

be implemented very cheaply. Figure 5.21b shows a routine that writes to a parallel port in 8051 assembly.

5.4.2 The C Language

C is the most popular language for embedded system programming. C compilers exist for virtually every general-purpose processor, from the lowliest 4-bit microcontroller to the most powerful 64-bit processor for compute servers.

C was originally designed by Dennis Ritchie [62] as an implementation language for the Unix operating system being developed at Bell Labs on a 24K DEC PDP-11. Because the language was designed for systems programming, it provides direct access to the processor through such constructs as untyped pointers and bit-manipulation operators, which are appreciated today by embedded systems programmers. Unfortunately, the language also has many awkward aspects, such as the need to define everything before it is used, that are holdovers from the cramped execution environment in which it was first implemented.

A C program (Figure 5.22) consists of functions built from arithmetic expressions structured with loops and conditionals. Instructions in a C program run sequentially, but control-flow constructs such as loops of conditionals can affect the order in which instructions execute. When control reaches a function call in an expression, control is passed to the called function, which runs until it produces a result, and control returns to continue evaluating the expression that called the function.

C derives its types from those a processor manipulates directly: signed and unsigned integers ranging from bytes to words, floating-point numbers, and pointers. These can be further aggregated into arrays and structures—groups of named fields.

C programs use three types of memory. Space for global data is allocated when the program is compiled, the stack stores automatic variables allocated and released when their function is called and returns, and the heap supplies arbitrarily sized regions of memory that can be deallocated in any order.

The C language is an ISO standard; the book by Kernighan and Ritchie [44] remains an excellent tutorial.

C succeeds because it can be compiled into efficient code and because it allows the programmer almost arbitrarily low-level access to the processor as desired. As a result, virtually every kind of code can be written in C (exceptions include those that manipulate specific processor registers) and can be

```c
#include <stdio.h>

int main(int argc, char *argv[])
{
  char *c;
  while (++argv, --argc > 0) {
    c = argv[0] + strlen(argv[0]);
    while (--c >= argv[0])
      putchar(*c);
    putchar('\n');
  }
  return 0;
}
```

FIGURE 5.22 C program that prints each of its arguments backwards. The outermost while loop iterates through the arguments (count in argc, array of strings in argv), while the inner loop starts a pointer at the end of the current argument and walks it backwards, printing each character along the way. The ++ and -- prefixes increment the variable they are attached to before returning its value.

expected to be fairly efficient. C's simple execution model also makes it easy to estimate the efficiency of a piece of code and improve it if necessary.

While C compilers for workstation-class machines usually comply closely to ANSI/ISO standard C, C compilers for microcontrollers are often much less standard. For example, they often omit support for floating-point arithmetic and certain library functions. Many also provide language extensions that, while often very convenient for the hardware for which they were designed, can complicate porting the code to a different environment.

5.4.3   C++

C++ (Figure 5.23) [64] extends C with structuring mechanisms for big programs: user-defined data types, a way to reuse code with different types, namespaces to group objects and avoid accidental name collisions when program pieces are assembled, and exceptions to handle errors. The C++ standard library includes a collection of efficient polymorphic data types such as arrays, trees, and strings for which the compiler generates custom implementations.

A C++ class defines a new data type by specifying its representation and the operations that may access and modify it. Classes may be defined by inheritance, which extends and modifies existing classes. For example, a rectangle class might add length and width fields and an area method to a shape class.

A template is a function or class that can work with multiple types. The compiler generates custom code for each different use of the template. For example, the same *min* template could be used for both integers and floating-point numbers.

C++ also provides exceptions, a mechanism intended for error recovery. Normally, each method or function can only return directly to its immediate caller. Throwing an exception, however, allows control to return to an arbitrary caller, usually an error-handling mechanism in a distant caller, such as *main*. Exceptions can be used, for example, to gracefully recover from out-of-memory conditions regardless of where they occur without the tedium of having to check the return code of every function.

```
class Cplx {
  double re, im;
public:
  Cplx(double v)  :  re(v), im(0) {}
  Cplx(double r, double i)  :  re(r), im(i) {}
  double abs() const {
    return sqrt(re*re + im*im);
  }
  void operator+= (const Cplx& a) {
    re += a.re; im += a.im;
  }
};

int main() {
  Cplx a(5), b(3,4);
  b += a;
  cout << b.abs() << '\n';
  return 0;
}
```

FIGURE 5.23   C++ fragment illustrating a partial complex number type and how it can be used (the C++ library has a complete version). This class defines how to create a new complex number from either a scalar or by specifying the real and imaginary components, how to compute the absolute value of a complex number, and how to add a complex number to an existing one.

Memory consumption is a disadvantage of C++'s exception mechanism. While most C++ compilers do not generate slower code when exceptions are enabled, they do generate larger executables by including tables that record the location of the nearest exception handler. For this reason, many compilers, such as GNU's gcc, have a flag that turns off the exception mechanism.

C++ is being used more and more within embedded systems, but it is sometimes a less suitable choice than C for a number of reasons. First, C++ is a much more complicated language that demands a much larger compiler, so C++ has been ported to fewer architectures than C. Second, certain language features such as dynamic dispatch (virtual function calls) and exceptions can be too costly to implement in small embedded systems. It is a more difficult language to learn and use properly, so there may be fewer qualified C++ programmers. Also, it is often more difficult to estimate the cost of a certain construct in C++ because the object-oriented programming style encourages many more function calls than the procedural style of C, and the cost of these is harder to estimate.

5.4.4 Java

Sun's Java language [2,31,48] resembles C++ but is not directly compatible. Like C++, Java is object-oriented, providing classes and inheritance. It is a higher-level language than C++ since it presents object references, arrays, and strings instead of pointers. Java's automatic garbage collection frees the programmer from explicit memory management.

Java omits a number of C++'s more complicated features, such as operator overloading. While it can be a boon to readability (e.g., when performing operations on complex numbers), it can also be a powerful obfuscating force. Java also does not support C++'s complex multiple inheritance mechanism. Instead, it provides the notion of an interface—a set of methods provided by a class—that is equivalent to one of the most common uses of multiple inheritance. Templates were absent from the language for many years, although have now been added as "generics."

Java provides concurrent threads (Figure 5.24). Creating a thread involves extending the *Thread* class, creating instances of these objects, and calling their *start* methods to start a new thread of control that executes the objects' *run* methods.

Synchronizing a method or block uses a per-object lock to resolve contention when two or more threads attempt to access the same object simultaneously. A thread that attempts to gain a lock owned by another thread will block until the lock is released, which can be used to grant a thread exclusive access to a particular object.

For embedded systems, Java holds promise but also many caveats. On the positive side, it is a simple, powerful language that provides the programmer a more convenient set of abstractions than C or C++. For example, unlike C, Java provides true strings and variable-sized arrays. On the negative side, Java is a heavyweight language, even more so than C++. Its runtime system is large, consisting of either a bytecode interpreter or a just-in-time compiler, or perhaps both, and its libraries are absolutely vast. While work has been done on paring down these things, Java still requires far more memory than C.

Unpredictable runtimes are a more serious problem for Java. For time-critical embedded systems, Java's automatic garbage collector, bytecode interpreter, or just-in-time compiler make runtimes both unpredictable and variable, making it difficult to assess efficiency both beforehand and during simulation.

The real-time Java specification [14] attempts to address many of these concerns. It introduces mechanisms for more precise control over the scheduling policy for concurrent threads (the standard Java specification is deliberately vague on this point to improve portability), memory regions for which automatic garbage collection can be disabled, synchronization mechanisms for avoiding priority inversion, and various other real-time features such as timers. It remains to be seen, however, whether this specification addresses enough real-time concerns and is sufficiently efficient to be

```java
import java.io.*;
class Counter {
    int value = 0;
    boolean present = false;
    public synchronized void count() {
        try { while (present) wait(); } catch (InterruptedException e) {}
        value++; present = true; notifyAll();
    }
    public synchronized int read() {
        try { while (!present) wait(); } catch (InterruptedException e) {}
        present = false; notifyAll();
        return value;
    }
}
class Count extends Thread {
    Counter cnt;
    public Count(Counter c) { cnt = c; start(); }
    public void run() { for (;;) cnt.count(); }
}
class Mod5 {
    public static void main(String args[]) {
        Counter c = new Counter();
        Count count = new Count(c);
        int v;
        for (;;) if ( (v = c.read()) % 5 == 0 ) System.out.println(v);
    }
}
```

FIGURE 5.24 Contrived Java program that spawns a counting thread to print all numbers divisible by 5. The *main* method in the Mod5 class creates a new Counter and then a new Count object. The Count class extends the Thread class and spawns a new thread in its constructor by executing *start*. This invokes its *run* method, which calls the method *count*. Both *count* and *read* are synchronized, meaning at most one may run on a particular Count object at once, here guaranteeing the counter is either counting or waiting for its value to be read.

practical. For example, a naïve implementation of the memory management policies would be very inefficient.

5.4.5 Real-Time Operating Systems

Many embedded systems use a real-time operating system (RTOS) to simulate concurrency on a single processor. An RTOS manages multiple running processes, each written in a sequential language such as C. The processes perform the system's computation and the RTOS schedules them—attempts to meet deadlines by deciding which process runs when. Labrosse [45] describes the implementation of a particular RTOS (Figure 5.25).

Most RTOSs use fixed-priority preemptive scheduling in which each process is given a particular priority (a small integer) when the system is designed. At any time, the RTOS runs the highest-priority runnable process, which is expected to run for a short period of time before suspending itself to wait for more data. Priorities are usually assigned using rate-monotonic analysis [17] (due to Liu and Layland [50]), which assigns higher priorities to processes that must meet more frequent deadlines.

Priority inversion is a fundamental problem in fixed-priority preemptive scheduling that can lead to missed deadlines by enabling a lower-priority process to delay indefinitely the execution of a higher-priority one. Figure 5.26 illustrates the typical scenario: a low priority process L runs and acquires a resource. Shortly thereafter, a high-priority process H preempts L, attempts to acquire the same resource, and blocks waiting for L to release it. This can cause H to miss its deadline even though it is at a higher priority than L. Even worse, if a process M with priority between L and now starts,

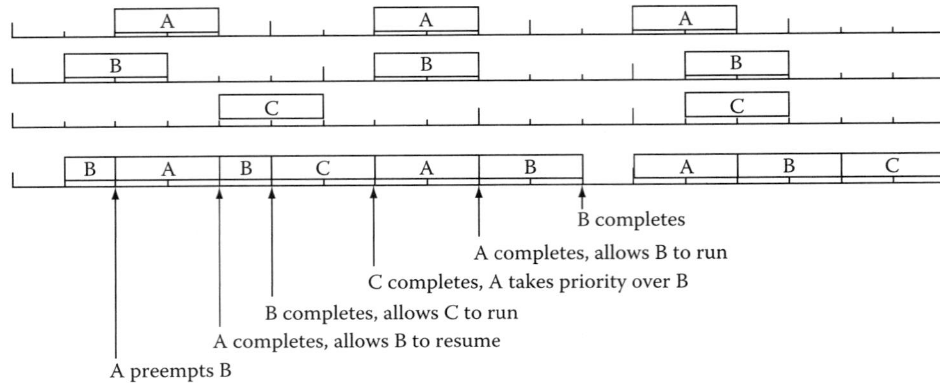

FIGURE 5.25   Behavior of an RTOS with fixed-priority preemptive scheduling. Rate-monotonic analysis gives process A the highest priority since it has the shortest period; C has the lowest.

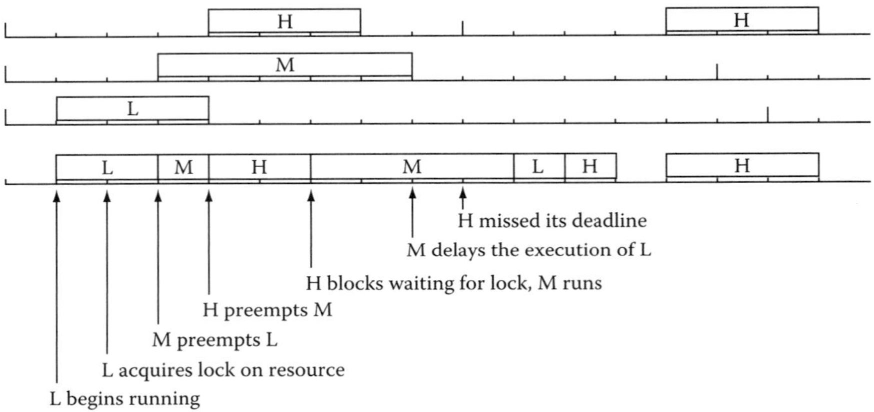

FIGURE 5.26   Priority inversion illustrated. When low-priority process L acquires a lock on a resource needed by process H, it effectively blocks process H, but then intermediate-priority process M preempts L, preventing it from running and releasing the resource needed by H. Priority inheritance, the common solution, temporarily raises the priority of L to that of H when H requests the resource held by L.

it can delay the execution of H indefinitely, even if it does not attempt to access the shared resource. Process M does not allow L to run since M is at a higher priority, so L cannot execute and release the lock and H will continue to block.

Priority inversion is usually solved with priority inheritance. When a process L acquires a lock, its priority is temporarily raised to a level where it will not be preempted by any other process that will also attempt to acquire the lock. Many RTOSs provide a mechanism for doing this automatically.

5.5   Domain-Specific Languages

The hardware and software languages described earlier have semantics very close to those of their implementations (e.g., as instructions on a sequential processor or as digital logic gates), which makes for efficient realizations, but some problems are better described using different models of computation.

Many embedded systems perform signal processing tasks such as reconstructing a compressed audio signal. While such tasks can be described and implemented using the hardware and software languages described earlier, signal processing tasks are more conveniently represented with systems of processes that communicate through queues. Although clumsy for general applications, dataflow languages are a perfect fit for signal-processing algorithms, which use vast quantities of arithmetic derived from linear system theory to decode, compress, or filter data streams that represent periodic samples of continuously changing values such as sound or video. Dataflow semantics are natural for expressing the block diagrams typically used to describe signal-processing algorithms, and their regularity makes dataflow implementations very efficient because otherwise costly run-time scheduling decisions can be made at compile time, even in systems containing multiple sampling rates.

The two other languages in this section are representative of those that use more novel models of computation than the hardware or software languages presented earlier. While such languages are more restrictive than general-purpose ones, they are much better suited for certain applications. Esterel excels at discrete control by blending software-like control flow with the synchrony and concurrency of hardware. Communication protocols are SDL's forte; it uses extended finite-state machines with single input queues.

5.5.1 Kahn Process Networks

Kahn Process Networks [43] form a formal basis for dataflow computation. Kahn's systems consist of processes that communicate exclusively through unbounded point-to-point first-in, first-out queues (Figure 5.27). Reading from a port makes a process wait until data is available, but writing to a port always completes immediately.

Deterministic behavior is the most unique aspect of Kahn's networks. Processes' blocking read behavior guarantees the overall system behavior (specifically, the sequence of data tokens that flow through each queue) is the same regardless of the relative execution rates of the processes, i.e., regardless of the scheduling policy. This is generally a very desirable property because it provides a guarantee about the behavior of the system, ensures that simulation and reality will match, and greatly simplifies the design task since a designer is not obligated to ensure this himself/herself.

Balancing processes' relative execution rates to avoid an unbounded accumulation of tokens is the challenge in scheduling a Kahn network. One general approach, proposed by Parks [58], places artificial limits on the size of each buffer. Any process that writes to a full buffer blocks until space is available, but if the system deadlocks because all buffers are full, the scheduler increases the capacity of the smallest buffer.

In practice, Kahn networks are rarely used in their pure form since they are fairly costly to schedule and their completely deterministic behavior is sometimes overly restrictive since they cannot easily handle sporadic events (e.g., an occasional change of volume level in a digital volume control) or server-like behavior where the environment may make requests in an unpredictable order. Nevertheless, Kahn's model still has useful properties and forms a starting point for other dataflow models.

5.5.2 Synchronous Dataflow

Lee and Messerschmitt's [46] Synchronous Dataflow (SDF) fixes the communication patterns of the blocks in a Kahn network (Figure 5.28 is an example after Bhattacharyya et al. [13]). Each time a block runs, it consumes and produces a fixed number of data tokens on each of its ports. Although more restrictive than Kahn networks, SDF's predictability allows it to be scheduled completely at compile time, producing very efficient code.

```
process f(in int u, in int v, out int w)
{
  int i; bool b = true;
  for (;;) {
    i = b ? wait(u) : wait(w);
    printf("%i\n", i);
    send(i, w);
    b = !b;
  }
}

process g(in int u, out int v, out int w)
{
  for (;;) {
    send(wait(u), v); send(wait(u), w);
  }
}

process h(in int u, out int v, int init)
{
  send(v, init);
  for(;;)
    send(wait(u), v);
}

channel int X, Y, Z, T1, T2;
f(Y, Z, X);
g(X, T1, T2);
h(T1, Y, 0);
h(T2, Z, 1);
```

FIGURE 5.27　　Kahn Process Network written in a C-like dialect. Here, processes are functions that run continuously, may be attached to communication channels, and may call *wait* to wait for data on a particular port and *send* to write data to a particular port. The f process alternately copies from its u and v ports to its w port; the g process does the opposite, copying its u port to alternately v and w; and h simply copies its input to its output.

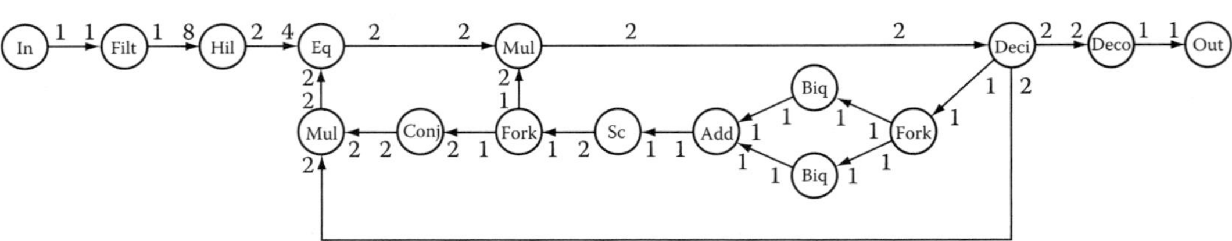

FIGURE 5.28　　Modem in SDF. Each node represents a process. The labels on each arc indicate the number of tokens sent or received by a process each time it fires.

Scheduling operates in two steps. First, the rate at which each block fires is established by considering the production and consumption rates of each block at the source and sink of each queue. For example, the arc between the Hil and Eq nodes in Figure 5.28 implies Hil runs twice as frequently. Once the rates are established, any algorithm that simulates the execution of the network without buffer underflow will produce a correct schedule if one exists. However, more sophisticated techniques reduce generated code and buffer sizes by better ordering the execution of the blocks (see Bhattacharyya et al. [12]).

SDF specifications are built by assembling blocks typically written in an imperative language such as C. The SDF block interface is specific enough to make it easy to create libraries of general-purpose blocks such as adders, multipliers, and even FIR filters.

While SDF is often used as a simulation language, it is also well suited to code generation. It enables a practical technique for generating code for DSPs, for which C compilers often cannot generate efficient code. Assembly code is handcrafted for each block in a library, and code synthesis consists of assembling these handwritten blocks, sometimes generating extra code that handles the interblock buffers. For large, specialized blocks such as fast Fourier transforms, this can be very effective because most of the generated code was carefully optimized by hand.

5.5.3 Esterel

Intended for specifying control-dominated reactive systems, Esterel [9] combines the control constructs of an imperative software language with concurrency, preemption, and a synchronous model of time like that used in synchronous digital circuits. In each clock cycle, the program awakens, reads its inputs, produces outputs, and suspends.

An Esterel program communicates through signals that are either present or absent each cycle. In each cycle, each signal is absent unless an emit statement for the signal runs and makes the signal present for that cycle only. Esterel guarantees determinism by requiring each emitter of a signal to run before any statement that tests the signal.

Esterel is strongest at specifying hierarchical state machines. In addition to sequentially composing statements (separated by a semicolon), it has the ability to compose arbitrary blocks of code in parallel (the double vertical bars) and abort or suspend a block of code when a condition is true. For example, the *every-do* construct in Figure 5.29 effectively wraps a reset statement around two state machines running in parallel.

```
module Example:

input   S, I;
output  O;

signal  R, A in
  every S do
    await I;
    weak abort
      sustain R
    when immediate A;
    emit O
  ||
    loop
      pause; pause;
      present R then emit A end;
    end
  end
end

end module
```

FIGURE 5.29 Esterel program modeling a shared resource. This implements two parallel threads (separated by | |), one that waits for an I signal, then asserts an R until it received an A from the other thread and emits an O. Meanwhile, the second thread emits an R in response to an A in alternate cycles.

5.5.4   SDL

SDL is a graphical specification language developed for describing telecommunication protocols defined by the ITU [42] (Ellsberger [26] provides a tutorial). A system consists of concurrently running FSMs, each with a single input queue, connected by channels that define which messages they carry. Each FSM consumes the most recent message in its queue, reacts to it by changing internal state or sending messages to other FSMs, changes to its next state, and repeats the process. Each FSM is deterministic, but because messages from other FSMs may arrive in any order because of varying execution speed and communication delays, an SDL system may behave nondeterministically.

In addition to a fairly standard textual format, SDL has a formalized graphical notation. There are three types of diagrams. Flowcharts define the behavior of state machines at the lowest level (Figure 5.30). Block diagrams illustrating the communication among state machines local to a single processor are at the next level up. Each communication channel is labeled with the set of messages that it conveys. The top level is another block diagram that depicts the communication among processors. The communication channels in these diagrams are also labeled with the signals they convey but are assumed to have significant delay unlike the channels among FSMs in a single processor.

The behavior of an SDL state machine is straightforward. At the beginning of each cycle, it gets the next signal in its input queue and sees if there is a "receive" block for that signal off the current state. If there is, the associated code is executed, possibly emitting other signals and moving to the next state. Otherwise, the signal is simply discarded and the cycle repeats from the same state. By itself, such semantics have a hard time dealing with signals that arrive out-of-order, but SDL has an additional construct for handling this condition. The save construct is like the receive construct, appearing immediately a state and matching a signal, but when it matches it stores the signal in a buffer that holds it until another state has a matching rule.

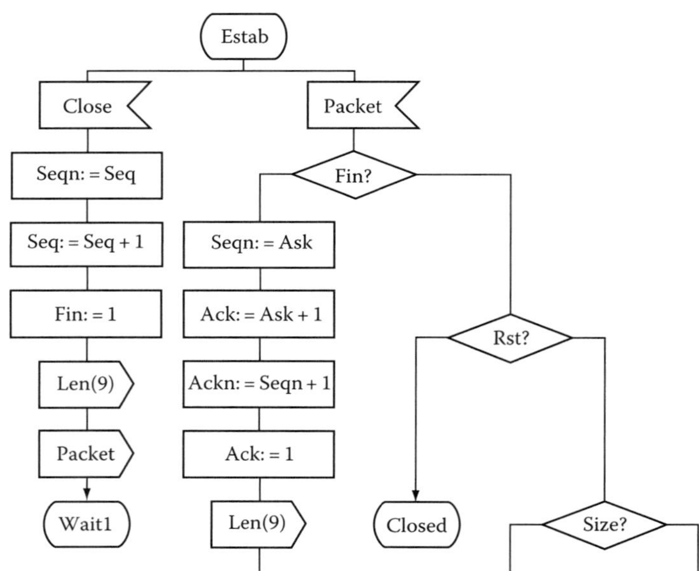

FIGURE 5.30   Fragment of an SDL flowchart specification for the TCP. The rounded boxes denote states (Estab, waitl, and Closed). Immediately below Estab are inward-pointing boxes that receive signals (Close, Packet). The square and diamond boxes below these are actions and decisions. The outward-pointed boxes (e.g., Packet) emit signals.

5.6 Summary

Currently, most embedded systems are programmed using C for software and Verilog, or possibly VHDL, for hardware components such as FPGAs or ASICs, but this will probably change. The increased complexity of such designs makes a compelling case for different, higher-level languages. Years ago, designers made the jump from assembly to C, and the higher-level constructs of Java are growing more attractive despite its performance loss.

Domain-specific languages, especially for signal-processing problems, already have a significant beachhead and will continue to make inroads. Most signal-processing algorithms are already prototyped using a higher-level language (Matlab®), but it remains to be seen whether synthesis from Matlab will ever be practical.

For hardware, the direction is less clear. While modeling languages such as SystemC will continue to grow in importance, there is currently no clear winner for the successor to VHDL and Verilog. Years ago, a different, high-level subset of VHDL and Verilog was proposed as the new "behavioral" synthesis subset but did not catch on because it was too limiting, largely because of restrictions placed on it by the synthesis algorithms. Additions such as SystemVerilog are incremental, if helpful, improvements, but will not provide the quantum leap forward that synthesis from the RTL subsets of Verilog and VHDL provided. Perhaps future hardware languages may contain constructs such as Esterel's.

References

1. Accelera. *SystemVerilog 3.1a Language Reference Manual: Accellera's Extensions to Verilog*, Napa, CA, May 2004.
2. Ken Arnold, James Gosling, and David Holmes. *The Java Programming Language*. Addison-Wesley, Reading, MA, Third edition, 2000.
3. Guido Arnout. SystemC standard. In *Proceedings of the Asia South Pacific Design Automation Conference (ASP-DAC)*, pp. 573–578, Yokohama, Japan, January 2000.
4. Peter J. Ashenden. *The Designer's Guide to VHDL*. Morgan Kaufmann, San Francisco, CA, 1996.
5. Peter J. Ashenden. *The Student's Guide to VHDL*. Morgan Kaufmann, San Francisco, CA, 1998.
6. Ilan Beer, Shoham Ben-David, Cindy Eisner, Dana Fisman, Anna Gringauze, and Yoav Rodeh. The temporal logic Sugar. In *Proceedings of the 13th International Conference on Computer-Aided Verification (CAV)*, volume 2102 of Lecture Notes in Computer Science, pp. 363–367, Paris, France, 2001. Springer-Verlag.
7. C. Gordon Bell and Allen Newell. *Computer Structures: Readings and Examples*. McGraw-Hill, New York, 1971.
8. Janick Bergeron. *Writing Testbenches: Function Verification of HDL Models*. Kluwer, Boston, MA, Second edition, 2003.
9. Gérard Berry and Georges Gonthier. The Esterel synchronous programming language: Design, semantics, implementation. *Science of Computer Programming*, 19(2):87–152, November 1992.
10. J. Bhasker. *A VHDL Synthesis Primer*. Star Galaxy Publishing, Allentown, PA, Second edition, 1998.
11. J. Bhasker. *A SystemC Primer*. Star Galaxy Publishing, Allentown, PA, Second edition, 2004.
12. Shuvra S. Bhattacharyya, Ranier Leupers, and Peter Marwedel. Software synthesis and code generation for signal processing systems. *IEEE Transactions on Circuits and Systems—II: Analog and Digital Signal Processing*, 47(9):849–875, September 2000.
13. Shuvra S. Bhattacharyya, Praveen K. Murthy, and Edward A. Lee. Synthesis of embedded software from synchronous dataflow specifications. *Journal of VLSI Signal Processing Systems*, 21(2):151–166, June 1999.

14. Greg Bollella, Ben Brosgol, Peter Dibble, Steve Furr, James Gosling, David Hardin, Mark Turnbull, Rudy Belliardi, Doug Locke, Scott Robbins, Pratik Solanki, and Dionisio de Niz. *The Real-Time Specification for Java*. Addison-Wesley, Reading, MA, 2000.

15. Dominique Borrione, Robert Piloty, Dwight Hill, Karl J. Lieberherr, and Philip Moorby. Three decades of HDLs: Part II, Colan through Verilog. *IEEE Design & Test of Computers*, 9(3):54–63, September 1992.

16. Robert K. Brayton, Gary D. Hachtel, and Alberto L. Sangiovanni-Vincentelli. Multilevel logic synthesis. *Proceedings of the IEEE*, 78(2):264–300, February 1990.

17. Loic P. Briand and Daniel M. Roy. *Meeting Deadlines in Hard Real-Time Systems: The Rate Monotonic Approach*. IEEE Computer Society Press, New York, 1999.

18. Yaohan Chu, Donald L. Dietmeyer, James R. Duley, Fredrick J. Hill, Mario R. Barbacci, Charles W. Rose, Greg Ordy, Bill Johnson, and Martin Roberts. Three decades of HDLs: Part I, CDL through TI-HDL. *IEEE Design & Test of Computers*, 9(2):69–81, June 1992.

19. Youhan Chu. An ALGOL-like computer design language. *Communications of the ACM*, 8(10):607–615, October 1965.

20. Edmund M. Clarke and E. Allen Emerson. Design and synthesis of synchronization skeletons using branching time temporal logic. In *Proceedings of the Workshop on Logic of Programs*, volume 131 of Lecture Notes in Computer Science, pp. 52–71, Yorktown Heights, New York, May 1981. Springer-Verlag.

21. Ben Cohen. *VHDL Coding Styles and Methodologies*. Kluwer, Boston, MA, Second edition, 1999.

22. Ben Cohen, Srinivasan Venkataramanan, and Ajeetha Kumari. *Using PSL/Sugar for Formal Verification*. VhdlCohen Publishing, Palos Verdes Peninsula, CA, 2004.

23. Al Dewey. VHSIC hardware description (VHDL) development program. In *Proceedings of the 20th Design Automation Conference*, pp. 625–628, Miami Beach, FL, June 1983.

24. Allen Dewey and Aart J. de Geus. VHDL: Toward a unified view of design. *IEEE Design & Test of Computers*, 9(2):8–17, April 1992.

25. Allen M. Dewey. *Analysis and Design of Digital Systems with VHDL*. Brooks/Cole Publishing (Formerly PWS), Pacific Grove, CA, 1997.

26. Jan Ellsberger, Dieter Hogrefe, and Amardeo Sarma. *SDL: Formal Object-Oriented Language for Communicating Systems*. Prentice Hall, Upper Saddle River, NJ, Second edition, 1997.

27. Peter L. Flake and Simon J. Davidmann. Superlog, a unified design language for system-on-chip. In *Proceedings of the Asia South Pacific Design Automation Conference (ASP-DAC)*, pp. 583–586, Yokohama, Japan, January 2000.

28. Peter L. Flake, Philip R. Moorby, and G. Musgrave. An algebra for logic strength simulation. In *Proceedings of the 20th Design Automation Conference*, pp. 615–618, Miami Beach, FL, June 1983.

29. Robert S. French, Monica S. Lam, Jeremy R. Levitt, and Kunle Olukotun. A general method for compiling event-driven simulations. In *Proceedings of the 32nd Design Automation Conference*, pp. 151–156, San Francisco, CA, June 1995.

30. Michael J. C. Gordon. The semantic challenge of Verilog HDL. In *Proceedings of the Tenth Annual IEEE Symposium on Logic in Computer Science (LICS)*, San Diego, CA, June 1995.

31. James Gosling, Bill Joy, Guy Steele, and Gilad Bracha. *The Java Language Specification*. Addison-Wesley, Reading, MA, Second edition, 2000.

32. Thorsten Grötker, Stan Liao, Grant Martin, and Stuart Swan. *System Design with SystemC*. Kluwer, Boston, MA, 2002.

33. Faisal Haque, Jonathan Michelson, and Khizar Khan. *The Art of Verification with Vera*. Verification Central, www.verificationcentral.com, 2001.

34. Randolph E. Harr and Alec G. Stanculescu, editors. *Applications of VHDL to Circuit Design*. Kluwer, Boston, MA, 1991.

35. IEEE Computer Society, New York. *IEEE Standard VHDL Language Reference Manual (1076–1993)*, 1994.

36. IEEE Computer Society, New York. *IEEE Standard Hardware Description Language Based on the Verilog Hardware Description Language (1364–1995)*, 1996.

37. IEEE Computer Society, New York. *IEEE Standard Verilog Hardware Description Language (1364–2001)*, September 2001.

38. Sasan Iman and Sunita Joshi. *The e Hardware Verification Langauge*. Kluwer, Boston, MA, 2004.

39. The Institute of Electrical and Electronics Engineers (IEEE), New York. *IEEE Standard VHDL Reference Manual (1076–1987)*, 1988.

40. The Institute of Electrical and Electronics Engineers (IEEE), New York. *IEEE Standard Multivalue Logic System for VHDL Model Interoperability (Std_logic_1164)*, 1993.

41. The Institute of Electrical and Electronics Engineers (IEEE), New York. *IEEE Standard VHDL Synthesis Packages (1076.3–1997)*, 1997.

42. International Telecommunication Union, Place des Nations, CH-1221, Geneva 20, Switzerland. *ITU-T Recommendation Z.100: Specification and Description Language*, 1999.

43. Gilles Kahn. The semantics of a simple language for parallel programming. In *Information Processing 74: Proceedings of IFIP Congress 74*, pp. 471–475, North-Holland Stockholm, Sweden, August 1974.

44. Brian W. Kernighan and Dennis M. Ritchie. *The C Programming Language*. Prentice Hall, Upper Saddle River, NJ, Second edition, 1988.

45. Jean Labrosse. *MicroC/OS-II*. CMP Books, Lawrence, KS, 1998.

46. Edward A. Lee and David G. Messerschmitt. Synchronous data flow. *Proceedings of the IEEE*, 75(9):1235–1245, September 1987.

47. Stan Liao, Steve Tjiang, and Rajesh Gupta. An efficient implementation of reactivity for modeling hardware in the Scenic design environment. In *Proceedings of the 34th Design Automation Conference*, pp. 70–75, Anaheim, CA, June 1997.

48. Tim Lindholm and Frank Yellin. *The Java Virtual Machine Specification*. Addison-Wesley, Reading, MA, 1999.

49. Roger Lipsett, Carl F. Schaefer, and Cary Ussery. *VHDL: Hardware Description and Design*. Kluwer, Boston, MA, 1989.

50. C. L. Liu and James W. Layland. Scheduling algorithms for multiprogramming in a hard real-time environment. *Journal of the Association for Computing Machinery*, 20(1):46–61, January 1973.

51. Carver Mead and Lynn Conway. *Introduction to VLSI Systems*. Addison-Wesley, Reading, MA, 1980.

52. Swapnajit Mittra. *Principles of Verilog PLI*. Kluwer, Boston, MA, 1999.

53. Wolfgang Muller, Wolfgang Rosenstiel, and Jurgen Ruf, editors. *SystemC: Methodologies and Applications*. Kluwer, Boston, MA, 2003.

54. Wayne E. Omohundro and James H. Tracey. Flowware—a flow charting procedure to describe digital networks. In *Proceedings of the First International Conference on Computer Architecture (ISCA)*, pp. 91–97, Gainesville, FL, December 1973.

55. Samir Palnitkar. *Verilog HDL: A Guide to Digital Design and Synthesis*. Prentice Hall, Upper Saddle River, NJ, 1996.

56. Samir Palnitkar. *Design Verification with e*. Prentice Hall, Upper Saddle River, NJ, 2003.

57. Federic I. Parke. An introduction to the N.mPc design environment. In *Proceedings of the 16th Design Automation Conference*, pp. 513–519, San Diego, CA, June 1979.

58. Thomas M. Parks. *Bounded Scheduling of Process Networks*. PhD thesis, University of California, Berkeley, CA, 1995. Available as UCB/ERL M95/105.

59. David A. Patterson and David R. Ditzel. The case for the reduced instruction set computer. *ACM SIGARCH Computer Architecture News*, 8(6):25–33, October 1980.

60. Douglas L. Perry. *VHDL*. McGraw-Hill, New York, Third edition, 1998.

61. Irving S. Reed. Symbolic synthesis of digital computers. In *Proceedings of the ACM National Meeting*, pp. 90–94, Toronto, Canada, September 1952.

62. Dennis M. Ritchie. The development of the C language. In *History of Programming Languages II*, Cambridge, MA, April 1993.

63. Douglas J. Smith. VHDL & Verilog compared & constrasted—plus modeled examples written in VHDL, Verilog, and C. In *Proceedings of the 33rd Design Automation Conference*, pp. 771–776, Las Vegas, NV, June 1996.

64. Bjarne Stroustrup. *The C++ Programming Language*. Addison-Wesley, Reading, MA, Third edition, 1997.

65. Stuart Sutherland. *The Verilog PLI Handbook*. Kluwer, Boston, MA, 1999.

66. Donald E. Thomas and Philip R. Moorby. *The Verilog Hardware Description Language*. Kluwer, Boston, MA, Fifth edition, 2002.

6

Synchronous Hypothesis and Polychronous Languages

Dumitru Potop-Butucaru
National Institute for Research in Computer Science and Control (INRIA)

Robert de Simone
National Institute for Research in Computer Science and Control (INRIA)

Jean-Pierre Talpin
National Institute for Research in Computer Science and Control (INRIA)

6.1 Introduction

Electronic embedded systems are not new, but their pervasive introduction in ordinary-life objects (cars, phones, home appliances) brought a new focus on design methods for such systems. New development techniques are needed to meet the challenges of productivity in a competitive environment. This book reports on a number of such innovative approaches to the matter. We concentrate here on synchronous reactive (S/R) languages [4,7,10,37].

S/R languages rely on the synchronous hypothesis, which lets computations and behaviors be divided into a discrete sequence of computation steps, which are equivalently called reactions or execution instants. In itself, this assumption is rather common in practical embedded system design. But the synchronous hypothesis adds to this the fact that, inside each instant, the behavioral propagation is well-behaved (causal), so that the status of every signal or variable is established and defined prior to being tested or used. This criterion, which may be seen at first as an isolated technical requirement, is in fact the key point of the approach. It ensures strong semantic soundness by allowing universally recognized mathematical models such as the Mealy machines and the digital circuits to be used as supporting foundations. In turn, these models give access to a large corpus of efficient optimization, compilation, and formal verification techniques. The synchronous hypothesis also guarantees full equivalence between various levels of representation, thereby avoiding altogether the pitfalls of non-synthesizability of other similar formalisms. In that sense, the synchronous hypothesis is, in our view, a major contribution to the goal of model-based design of embedded systems.

Structured languages have been introduced for the modeling and programming of S/R and polychronous applications. They are roughly classified into two families:

Imperative languages, such as Esterel [13,14,20] and SyncCharts [2], provide constructs to shape control-dominated programs as hierarchical synchronous automata, in the wake of the State-Charts formalism, but with a full-fledged treatment of simultaneity, priority, and absence notification of signals in a given reaction. Thanks to this, signals assume a consistent status for all parallel components in the system at any given instant.

Declarative languages, such as Lustre [38] and Signal [9], shape applications based on intensive data computation and data-flow organization, with the control flow part operating under the form of (internally generated) activation clocks. These clocks prescribe which data computation blocks are to be performed as part of the current reaction. Here again, the semantics of the languages deal with the issue of behavior consistency, so that every value needed in a computation is indeed available at that instant.

We describe here the synchronous hypothesis and its mathematical background, together with a range of design techniques empowered by the approach and a short comparison with neighboring formalisms; then, we introduce both classes of S/R languages, with their special features and a couple of programming examples; finally we comment on the benefits and shortcomings of S/R modeling, closing with a look at future perspectives and extensions. We shall mostly insist on the multiclock or polychronous extensions of the S/R model and languages.

A polychronous system is formed of synchronous subsystems whose executions only synchronize from time to time, or whose synchronization does not meet the requirements of the synchronous hypothesis (being done, for instance, through asynchronous message passing). This definition covers not only synchronous systems, but also asynchronous ones (pure asynchrony being modeled with independent clocks), and any combination of the two. Dealing with such systems requires specific support under the form of multiclock/polychronous languages and specific implementation techniques.

6.2 Synchronous Hypothesis

6.2.1 What For?

Program correctness (the process performs as intended) and program efficiency (it performs as fast as possible) are major concerns in all of computer science, but they are even more stringent in the embedded area, as no online debugging is feasible and time budgets are often imperative (for instance in multimedia applications).

Program correctness is sought by introducing appropriate syntactic constructs and dedicated languages, making programs more easily understandable by humans, as well as allowing high-level modeling and associated verification techniques. Provided semantic preservation is ensured down to actual implementation code, this provides reasonable guarantees on functional correctness. However, while this might sound obvious for traditional software compilation schemes, the hardware synthesis process is often not "seamless," as it includes manual rewriting.

Program efficiency is traditionally handled in the software world by algorithmic complexity analysis, and expressed in terms of individual operations. But in modern systems, due to a number of phenomena, this "high-level" complexity reflects rather imperfectly the "low-level" complexity in numbers of clock cycles spent. In the hardware domain, one considers various levels of modeling, corresponding to more abstract (or conversely more precise) timing account: transaction-level, cycle-accurate, time-accurate, etc.

One possible way (among many) to view synchronous languages is to take up the analogy of cycle-accurate programming to a more general setting, including (reactive) software as well. This analogy is supported by the fact that simulation environments in many domains (from scientific engineering to hardware description language (HDL) simulators) often use lockstep computation paradigms, very close to the synchronous cycle-based computation. In these settings, cycles represent logical steps, not physical time. Of course timing analysis is still possible afterward and in fact often simplified by the previous division into cycles.

The focus of synchronous languages is thus to allow modeling and programming of systems where cycle (computation step) precision is needed. The objective is to provide domain-specific structured languages for their description and to study matching techniques for efficient design, including compilation/synthesis, optimization, and analysis/verification. The strong condition ensuring the feasibility of these design activities is the synchronous hypothesis, described next.

6.2.2 Basic Notions

What has come to be known as the synchronous hypothesis, laying foundations for S/R systems, is really a collection of assumptions of a common nature, sometimes adapted to the framework considered. We shall avoid heavy mathematical formalization in this presentation and refer the interested reader to the existing literature, such as Refs. [4,7]. The basics are

Instants and reactions: Behavioral activities are divided according to (logical, abstract) discrete time. In other words, computations are divided according to a succession of nonoverlapping execution instants. In each instant, input signals possibly occur (for instance by being sampled); internal computations take place; and control and data are propagated until output values are computed and a new global system state is reached. This execution cycle is called the reaction of the system to the input signals. Although we used the word "time" just before, there is no real physical time involved, and instant durations need not be uniform (or even considered!). All that is required is that reactions converge and computations are entirely performed before the current execution instant ends and a new one begins. This empowers the obvious conceptual abstraction that computations are infinitely fast ("instantaneous," "zero-time") and take place only at discrete points in (physical) time, with no duration. When presented without sufficient explanations, this strong formulation of the synchronous hypothesis is often discarded by newcomers as unrealistic (while, again, it is only an abstraction, amply used in other domains where "all-or-nothing" transaction operations take place).

Signals: Broadcast signals are used to propagate information. At each execution instant, a signal can either be present or absent. If present, it also carries some value of a prescribed type ("pure" signals that carry only their presence status exist as well). The key rule is that a signal must be consistent (same present/absent status, same data) for all read operations during any given instant. In particular, reads from parallel components must be consistent, meaning that signals act as controlled shared variables.

Causality: The task of deciding whenever a signal can be declared absent is of utter importance in the theory of S/R systems and an important part of the theoretical body behind the synchronous hypothesis. This is of course true especially of local signals that are both generated and tested inside the system. The fundamental rule is that the present status and value of a signal should be defined before they are read (and tested). This requirement takes various practical forms depending on the actual language or formalism considered, and we shall come back to this later. Note that "before" refers here to causal dependency in the computation of the instant, and not to physical or even logical time between successive instants [12]. The synchronous hypothesis ensures that all possible schedules of operations amount to the same result (convergence); it also leads to the definition of "correct" programs, as opposed to ill-behaved ones where no causal scheduling can be found.

Activation conditions and clocks: Each signal can be seen as defining (or generating) a new clock, ticking when it occurs; in hardware design, this is called gated clocks. Clocks and subclocks, whether externally or internally generated, can be used as control entities to activate (or not) component blocks of the system. We shall also call them activation conditions.

6.2.3   Mathematical Models

If one temporarily forgets about data values and one accepts the duality of present/absent signals mapped to true/false values, then there is a natural interpretation of synchronous formalisms as synchronous digital circuits at schematic gate level, or "netlists" (roughly register transfer level [RTL] with only Boolean variables and registers). In turn, such circuits have a straightforward behavioral expansion into Mealy finite state machines (FSMs).

The two slight restrictions above are not essential: the adjunction of types and values into digital circuit models has been successfully attempted in a number of contexts, and S/R systems can also be seen as contributing to this goal. Meanwhile, the introduction of clocks and presence/absence signal status in S/R languages departs drastically from the prominent notion of sensitivity list generally used to define the simulation semantics of HDLs.

We now comment on the opportunities made available through the interpretation of S/R systems into Mealy machines or netlists.

Netlists: We consider here a simple form, as Boolean equation systems defining the values of wires and Boolean registers as a Boolean function of other wires and previous register values. Some wires represent input and output signals (with value `true` indicating signal presence); others are internal variables.

This type of representation is of special interest because it can provide exact dependency relations between variables and thus the good representation level to study causality issues with accurate analysis. Notions of "constructive" causality have been the subject of much attention here. They attempt to refine the usual crude criterion for synthesizability, which forbids cyclic dependencies between nonregister variables (so that a variable seems to depend upon itself in the same instant) but does not take into account the Boolean interpretation, nor the potentially reachable configurations. Consider the equation $x = y \vee z$, while it has been established that y is the constant `true`. Then x does not really depend on z, since its (constant) value is forced by y's. Constructive causality seeks for the best possible faithful notion of true combinatorial dependency taking the Boolean interpretation of functions into account. For details, see Shiple et al. [60].

Another equally important aspect of the mathematical model is that a number of combinatorial and sequential optimization techniques have been developed over the years, in the context of hardware synthesis approaches. The main ones are now embedded in the SIS and MVSIS optimization suites, from UC Berkeley [32,58]. They come as a great help in allowing programs written in high-level S/R formalisms to compile into efficient code, whether software or hardware targeted [59].

Mealy machines: Mealy machines are finite-state automata corresponding strictly to the synchronous assumption. In a given state, if a certain input valuation is provided (a subset of present signals), the machine reacts by immediately producing a set of output signals before entering a new state.

The Mealy machines can be generated from netlists (and by extension from any S/R system). The Mealy machine construction can then be seen as a symbolic expansion of all possible behaviors, computing the space of reachable states (RSS) on the way. But while the precise RSS is won, the precise causal dependencies relations are lost, which is why Mealy FSM and netlist models are both useful in the course of S/R design [65].

When the RSS is extracted, often in a symbolic binary decision diagram (BDD) form, it can be used in a number of ways. We already mentioned that constructive causality only considers dependencies inside the RSS; similarly, all activities of model-checking formal verification and test coverage analysis are strongly linked to the RSS construction [3,17,18,29].

The modeling style of netlists can be extrapolated to block-diagram networks, often used in multi-media digital signal processing, by adding more types and arithmetic operators, as well as activation conditions, to introduce some amount of control-flow. The declarative synchronous languages can be seen as attempts to provide structured programming to compose large systems modularly in this class of applications, as described in Section 6.4. Similarly, imperative languages provide ways to program in a structured way hierarchical systems of interacting Mealy FSMs, as described in Section 6.3.

6.2.3.1 Synchronous Hypothesis vs. Neighboring Models

Many quasi-synchronous formalisms exist in the fields of embedded system (co-)simulation: the simulation semantics of SystemC and regular HDLs at RTL level or the discrete-step Simulink/Stateflow simulation or the official StateCharts semantics for instance. Such formalisms generally employ a notion of physical time in order to establish when to start the next execution instant. Inside the current execution instant, however, delta-cycles allow zero-delay activity propagation, and potentially complex behaviors occur inside a given single reaction. The main difference here is that no causality analysis (based on the synchronous hypothesis) is performed at compilation time, so that an efficient ordering/scheduling cannot be precomputed before simulation. Instead, each variable change recursively triggers further recomputations of all depending variables in the same reaction.

6.2.4 Implementation Issues

The problem of implementing a synchronous specification mainly consists in defining the step reaction function that will implement the behavior of an instant, as shown in Figure 6.1. Then, the global behavior is computed by iterating this function for successive instants and successive input signal valuations. Following the basic mathematical interpretations, the compilation of an S/R program may either consist in the expansion into a flat Mealy FSM or in the translation into a flat netlist (with more types and arithmetic operators, but without activation conditions). The run-time implementation consists here in the execution of the resulting Mealy machine or netlist. In the first case, the automaton structure is implemented as a big top-level switch between states. In the second case, the netlist is totally ordered in a way compatible with causality, and all the equations in the ordered list are evaluated at each execution instant. These basic techniques are at the heart of the first compilers, and still of some industrial ones.

In the last decade fancier implementation schemes have been sought, relying on the use of activation conditions. During each reaction, execution starts by identifying the "truly useful" program blocks, which are marked as "active." Then only the actual execution of the active blocks is scheduled

```
reaction () {
    decode_state ; read_input ;
    compute ;
    write_output ; encode_state ;
}
```

FIGURE 6.1 Reaction function is called at each instant to perform the computation of the current step.

(a bit more dynamically) and performed in an order that respects the causality of the program. In the case of declarative languages, the activation conditions come in the form of a hierarchy of clock undersamplings—the clock tree, obtained through a "clock calculus" computation performed at compile time (see Section 6.4.3). In the case of imperative formalisms, activation conditions are based on the halting points (where the control flow can stop between execution instants) and on the signal-generated (sub-)clocks (see Section 6.3.3).

6.3 Imperative Style: Esterel and SyncCharts

For control-dominated systems, comprising a fair number of (sub-)modes and macrostates with activity swapping between them, it is natural to employ a description style that is algorithmic and imperative, describing the changes and progression of control in an explicit flow. In essence, one seeks to represent hierarchical (Mealy) FSMs, but with some data computation and communication treatment performed inside states and transitions. Esterel provides this in a textual fashion, while SyncCharts propose a graphical counterpart, with visual macrostates. It should be noted that systems here remain finite-state (at least their control structure).

6.3.1 Syntax and Structure

Esterel introduces a specific `pause` construct, used to divide behaviors into successive instants (reactions). The `pause` statement excepted, control is flowing through sequential, parallel, and if-then-else constructs, performing data operations and interprocess signaling. But it stops at `pause`, memorizing the activity of that location point for the next execution instant. This provides the needed atomicity mechanism, since the instant is over when all currently active parallel components reach a `pause` statement.

The full Esterel language contains a large number of constructs that facilitate modeling, but there exists a reduced kernel of primitive statements (corresponding to the natural structuring paradigms) from which all the other constructs can be derived. This is of special interest for model-based approaches, because only primitives need to be assigned semantics as transformations in the model space. The semantics of the primitives are then combined to obtain the semantics of composed statements. Figure 6.2 provides the list of primitive operators for the data-less subset of Esterel (also called Pure Esterel). A few comments are here in order:

- In p ; q the reaction where p terminates is the same as the reaction where q starts (control can be split into reactions only by `pause` statements inside p or q).

`[`p`]`	enforces precedence by parenthesis
`pause`	suspends the execution until next instant
p`;`q	executes p, then q as soon as p terminates
`loop` p `end`	iterates p forever in sequence
`[`p`|`q`]`	executes p and q in parallel, synchronously
`signal` S `in` `end`	declares local signal S in p
`emit` S	emits signal S
`present` S `then` p `else` q `end`	executes p or q upon S being present *or absent !*
`abort` p `when` S	executes p until S occurs (exclusive)
`weak abort` p `when` S	executes p until S occurs (inclusive)
`suspend` p `when` S	executes p unless S occurs
`trap` T `in` p `end`	declare/catch exception T in p
`exit` T	raise exception T

FIGURE 6.2 Pure Esterel statements.

- The `loop` constructs do not terminate, unless aborted from above. This abortion can be due to an external signal received by an `abort` statement or due to an internal exception raised through the `trap/exit` mechanism or due to any of the two (like for the `weak abort` statement). The body of a loop statement should not instantly terminate, or else the loop will unroll endlessly in the same instant, leading to divergence. This is checked by static analysis techniques. Finally, loops are the only means of defining iterating behaviors (there is no general recursion), so that the system remains finite-state.

- The `present` signal testing primitive allows an `else` part. This is essential to the expressive power of the language and has strong semantic implications pertaining to the synchronous hypothesis. It is enough to note here that, according to the synchronous hypothesis, signal absence can effectively be asserted.

- The difference between "abort p when S" and "weak abort p when S" is that in the first case signal S can only come from outside p and its occurrence prevents p from executing during the execution instant where S arrives. In the second case, S can also be emitted by p, and the preemption occurs only after p has completed its execution for the instant.

- Technically speaking, the `trap/exit` mechanisms can emulate the `abort` statements. But we feel that the ease of understanding makes the latter worth inclusion in the set of primitives. Similarly, we shall sometimes use "await S" as a shorthand for "abort loop pause end when S," and "sustain S" for "loop emit S end."

Most of the data-handling part of the language is deferred to a general-purpose host language (C, C++, Java, …). Esterel only declares type names, variables types, and function signatures (which are used as mere abstract instructions). The actual type specifications and function implementations must be provided and linked at later compilation time.

In addition to the structuring primitives of Figure 6.2, the language contains (and requires) interface declarations (for signals, most notably) and modular division with submodules invocation. Submodule instantiation allows signal renaming, i.e., transforming virtual name parameters into actual ones (again, mostly for signals). Rather than providing a full user manual for the language, we shall illustrate most of these features on an example.

The small example of Figure 6.3 has four input signals and one output signal. Meant to model a cyclic computation like a communication protocol, the core of our example is the loop, which awaits the input I, emits O, and then awaits J before instantly restarting. The local signal END signals the completion of loop cycles. When started, the `await` statement waits for the next clock cycle where its signal is present. The computation of all the other statements present in our example is performed during a single clock cycle, so that the `await` statements are the only places where control can be suspended between reactions (they preserve the state of the program between cycles). A direct consequence is that the signals I and J must come in different clock cycles in order not to be discarded.

The loop is preempted by the exception handling statement `trap` when "exit T" is executed. In this case, `trap` instantly terminates, control is given in sequence, and the program terminates. The preemption protocol is triggered by the input signal KILL, but the exception T is raised only when END is emitted. The program is suspended—no computation is performed and the state is kept unchanged—in clock cycles where the SUSP signal is received. A possible execution trace for our program is given in Figure 6.4.

6.3.2 Semantics

Esterel enjoys a full-fledged formal semantics, in the form of structural operational semantic (SOS) rules [12]. In fact, there are two main levels of such rules, with the coarser describing all potential logically consistent behaviors, while the more precise one only selects those that can be obtained in

```
module Example: input I,J,KILL,SUSP; output O;
suspend
  trap T in %exception handler, performs the preemption
    signal END in
      loop %basic computation loop
        await I;emit O;await J;emit END
      end
    ||
      %preemption protocol, triggered by KILL
      await KILL;await END;exit T
    end
  end;
when SUSP %suspend signal
end module
```

FIGURE 6.3 Simple Esterel program modeling a cyclic computation (like a communication protocol), which can be interrupted between cycles and which can be suspended.

Clock	Inputs	Outputs	Comments
0	any		all inputs discarded
1	I	O	
2	KILL		preemption protocol triggered
3			nothing happens
4	J,SUSP		suspend, J discarded
5	J		END emitted, T raised, program terminates

FIGURE 6.4 Possible execution trace for our example.

a constructive way (thereby discarding some programs as "unnatural" in this respect). This issue can be introduced with two small examples:

```
present S then emit S end            present S else emit S end
```

In the first case, the signal S can logically be assumed as either present or absent: if assumed present, it will be emitted, so it will become present; if assumed absent, it will not be emitted. In the second case, following a similar reasoning, the signal can be neither present nor absent. In both cases, anyhow, the analysis is done by "guessing" before branching to the potentially validating emissions. While more complex causality paradoxes can be built using the full language, these two examples already show that the problem stems from the existence of causality dependencies inside a reaction, prompted by instantaneous sequential control propagation and signal exchanges. The so-called constructive causality semantic of Esterel checks precisely that control and signal propagation are well-behaved, so that no "guess" is required. Programs that pass this requirement are deemed as "correct," and they provide deterministic behaviors for whatever input is presented to the program (which is a desirable feature in embedded system design).

6.3.3 Compilation and Compilers

Following the pattern presented in Section 6.2.4, the first compilers for Esterel were based on the translation of the source into (Mealy) finite automata or into digital synchronous circuits at netlist level. Then, the generated sequential code was a compiled automata or netlist simulator. The automata-based compilation [14] was used in the first Esterel compilers (known as Esterel V3). Automaton generation was done here by exhaustive expansion of all reachable states using symbolic execution (all data are kept uninterpreted). Execution time was then theoretically optimal, but code size could blow up (as the number of states), and huge code duplication was mandatory for actions that were performed in several different states. The netlist-based compilation (Esterel V5) is based on a quasi-linear, structural Esterel-to-circuits translation scheme [11] that ensures the tractability of compilation even for the largest examples. The drawback of the method is the reaction time (the simulation time for the generated netlist), which increases linearly with the size of the program.

Apart from these two previous compilation schemes, which have matured into full industrial-strength compilers, several attempts have been made to develop a more efficient, basically event-based type of compilation, which follows more readily the naive execution path and control propagation inside each reaction, and in particular executes "as much as possible" only the truly active parts of the program.* We mention here three such approaches: the Saxo compiler of Closse et al. [66], the EC compiler of Edwards [30], and the GRC2C compiler of Potop and de Simone [55]. All of them are structured around flowgraph-based intermediate representations that are easily translated into well-structured sequential code. The different intermediate representations also give the differences between approaches, by determining which Esterel programs can be represented and what optimization and code generation techniques can be applied.

We exemplify the GRC2C compiler [55], which is structured around the GRC intermediate form. The GRC representation of our example, given in Figure 6.5, uses two graph-based structures—a hierarchical state representation (HSR) and a concurrent control-flow graph (CCFG)—to preserve most of the structural information of the Esterel program while making the control flow explicit with few graph-building primitive nodes. The HSR is an abstraction of the syntax tree of the initial Esterel program. It can be seen as a structured data memory that preserves state information across reactions. During each instant, a set of activation conditions (clocks) is computed from this memory state, to drive the execution toward active instructions. The CCFG represents in an operational fashion the computation of an instant (the transition function). During each reaction, the dynamic CCFG operates on the static HSR by marking/unmarking component nodes (subtrees) with "active" tags as they are activated or deactivated by the semantics.

For instance, when we start our small example (Figures 6.3 and 6.5), the "program start (1)" and "program (0)" HSR nodes are active, while all the statements of the program (and the associated HSR nodes) are not. Like in any instant, control enters the CCFG by the topmost node and uses the first state decoding node (labeled 0) to read the state of the HSR and branch to the start behavior, which sets the "program start (1)" indicator to inactive (with "exit 1"), and activates "await I" and "await KILL" (with "enter 8" and "enter 11").

The HSR also serves as a repository for tags, which record redundancies between various activation clocks and are used by the optimization and code generation algorithms. Such a tag is #, which tells that at most one child of the tagged node can retain control between reactions at a time (the activation clocks of the branches are exclusive). Other tags (not figured here) are computed through complex static analysis of both the HSR and CCFG. The tags allow efficient optimization and sequential code generation.

---

* Recall that this is a real issue in Esterel, since programs may contain reaction to absence of signals, and determining this absence may require to check that no emission remains possible in the potential behaviors, whatever feasible test branches could be taken. To achieve this goal at a reasonable computational price, current compilers require in fact additional restrictions—in essence, the acyclicity of the dependency/causality graph at some representation level. Acyclicity ensures constructiveness, because any topological order of the operations in the graph gives an execution order, which is correct for all instants.

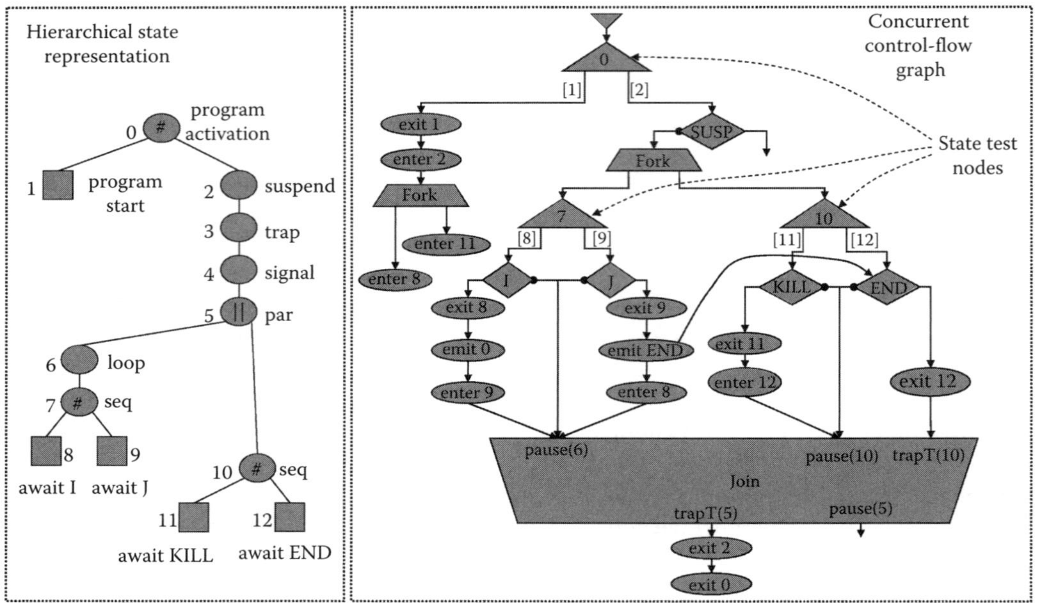

FIGURE 6.5   GRC intermediate representation for our Esterel example.

The CCFG is obtained by making the control flow of the Esterel program explicit (a structural, quasi-linear translation process).* Usually, it can be highly optimized using classical compiler techniques and some methods derived from circuit optimization, both driven by the HSR tags computed by static analysis. Code generation from a GRC representation is done by encoding the state on sequential variables and by scheduling the CCFG operators using classical compilation techniques [49].

The Saxo compiler of Closse et al. [66] uses a discrete-event interpretation of Esterel to generate a compiled event-driven simulator. The compiler flow is similar to that of VeriSUIF [31], but Esterel's synchronous semantics are used to highly simplify the approach. An event graph intermediate representation is used here to split the program into a list of guarded procedures. The guards intuitively correspond to events that trigger computation. At each clock cycle, the simulation engine traverses the list once, from the beginning to the end, and executes the procedures with an active guard. The execution of a procedure may modify the guards for the current cycle and for the next cycle. The resulting code is slower than its GRC2C-generated counterpart for two reasons. First, it does not exploit the hierarchy of exclusion relations determined by switching statements like the tests. Second, optimization is less effective because the program hierarchy is lost when the state is (very redundantly) encoded using guards.

The EC compiler of Edwards [30] treats Esterel as having control-flow semantics (in the spirit of Refs. [44,49]) in order to take advantage of the initial program hierarchy and produce efficient, well-structured C code. The Esterel program is first translated here into a CCFG representing the computation of a reaction. The translation makes the control flow explicit and encodes the state access operations using tests and assignments of integer variables. Its static scheduling algorithm

---

* Such a process is necessary, because most Esterel statements pack together two distinct, and often disjoint behaviors: one for the execution instants where they are started, and one for instants where control is resumed from inside.

takes advantage of the mutual exclusions between parts of the program and generates code that uses program counter variables instead of simple Boolean guards. The result is therefore faster than its Saxo-generated counterpart. However, it is usually slower than the GRC2C-generated code because the GRC representation preserves the state structure of the initial Esterel program and uses static analysis techniques to determine redundancies in the activation pattern. Thus, it is able to better simplify the final state representation and the CCFG.

6.3.4 Analysis/Verification/Test Generation: Benefits from Formal Approaches

We claimed that the introduction of well-chosen structuring primitives, endowed with formal mathematical semantics and interpretations as well-defined transformations in the realms of Mealy machines and synchronous circuits, was instrumental in allowing powerful analysis and synthesis techniques as part of the design of synchronous programs. What are they, and how do they appear in practice to enhance the confidence in the correctness of safety–critical embedded applications?

Maybe the most obvious is that synchronous formalisms can fully benefit from the model-checking and automatic verification usually associated with the netlist and Mealy machine representations, and now widely popular in the hardware design community with the PSL/SuGaR and assertion-based design approaches. Symbolic BDD- and SAT-based model-checking techniques are thus available on all S/R systems. Moreover,the structured syntax allows in many cases the introduction of modular approaches or guide abstraction techniques with the goal of reducing complexity of analysis.

The ability of formal methods akin to model-checking can also be used to automatically produce test sequences, which seek to reach the best possible coverage in terms of visited states or exercised transitions. Here again specific techniques are developed to match the S/R models.

Also, symbolic representations of the RSS (or abstracted over-approximations), which can effectively be produced and certified correct thanks to the formal semantics, can be used in the course of compilation and optimization. In particular for Esterel, the RSS computation allows more "correct" programs w.r.t. constructiveness: indeed causal dependencies may vary in direction depending on the state. If all dependencies are put together regardless of the states, then a causality cycle may appear, while not all components of the cycle may be active at the same instant, and so no real cycle exists (but it takes a dynamic analysis to establish this). Similarly, the RSS may exhibit combinatorial relations between registers encoding the local states, so that register elimination is possible to further simplify the state space structure.

Finally, the domain-specific structuring primitives empowering dedicated programming can also be seen as an important criterion. Readable, easily understandable programs are a big step toward correct programs. And when issues of correctness are not so plain and easy, for instance, when regarding proper scheduling of behaviors inside a reaction to respect causal effects, powerful abstract hypotheses are defined in the S/R domain, which define admissible orderings (and build them for correct programs). A graphical version of Esterel, named SyncCharts for Synchronous StateCharts, has been defined to provide a visual formalism with a truly synchronous semantics.

6.4 Declarative Style: Lustre and Signal

The presentation of declarative formalisms implementing the synchronous hypothesis as defined in Section 6.2 can be cast into a model of computation (proposed in Guernic et al. [35]) consisting of a domain of traces/behaviors and of a semilattice structure that renders the synchronous hypothesis using a timing equivalence relation: clock equivalence. Asynchrony can be superimposed on this model by considering a flow equivalence relation. Heterogeneous systems [6] can also be modeled by parameterizing the composition operator using arbitrary timing relations.

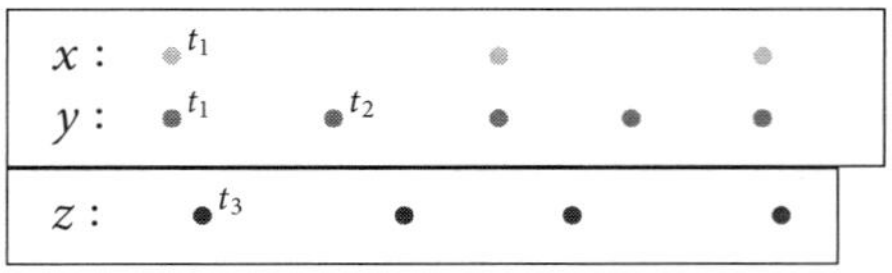

$$
\begin{array}{ll}
x: & \bullet^{t_1} \qquad\qquad\quad \bullet \qquad\quad \bullet \\
y: & \bullet^{t_1} \qquad \bullet^{t_2} \qquad \bullet \qquad \bullet \qquad \bullet \\
\end{array}
$$

$$
z: \quad \bullet^{t_3} \qquad \bullet \qquad \bullet \qquad \bullet
$$

FIGURE 6.6   A behavior (named b) over three signals (x, y, and z) belonging to two clock domains.

6.4.1 Synchronous Model of Computation

We consider a partially ordered set of tags, t, to denote instants (which are seen, in the sense of Section 6.2.2, as symbolic periods in time during which one reaction takes place). The relation $t_1 \leq t_2$ says that t_1 occurs before t_2. A minimum tag exists, denoted by 0. A totally ordered set of tags, C, is called a chain and denotes the sampling of a possibly continuous or dense signal over a countable series of causally related tags. Events, signals, behaviors, and processes are defined as follows:

- An event e is a pair consisting of a value v and a tag t
- A signal s is a function from a *chain* of tags to a set of values
- A behavior b is a function from a set of names x to signals
- A process p is a set of behaviors that have the same domain

In the remainder, we write tags(s) for the tags of a signal s, vars(b) for the domains of b, $b|_X$ for the projection of a behavior b on a set of names X, and b/X for its complementary. Figure 6.6 depicts a behavior (b) over three signals named x, y, and z. Two frames depict timing domains formalized by chains of tags. Signals x and y belong to the same timing domain: x is a downsampling of y. Its events are synchronous to odd occurrences of events along y and share the same tags, e.g., t_1. Even tags of y, e.g., t_2, are ordered along its chain, e.g., $t_1 < t_2$, but absent from x. Signal z belongs to a different timing domain. Its tags, e.g., t_3 are not ordered with respect to the chain of y, e.g., $t_1 \nleq t_3$ and $t_3 \nleq t_1$.

"The synchronous composition" of the processes p and q is denoted by $p\,||\,q$. It is defined by the union $b \cup c$ of all behaviors b (from p) and c (from q), which hold the same values at the same tags $b|_I = c|_I$ for all signal $x \in I = \text{vars}(b) \cap \text{vars}(c)$ they share. Figure 6.7 depicts the synchronous composition, right, of the behaviors b, left, and the behavior c, middle. The signal y, shared by b and c, carries the same tags and the same values in both b and c. Hence, $b \cup c$ defines the synchronous composition of b and c.

"A scheduling structure" is defined to schedule the occurrence of events along signals during an instant t. A scheduling $\rightarrow$ by a preorder relation between dates x_t where t represents the time and x the location of the event. Figure 6.8 depicts such a relation, superimposed to the signals x and y of Figure 6.6. The relation $y_{t_1} \rightarrow x_{t_1}$, for instance, requires y to be calculated before x at the instant t_1. Naturally, scheduling is contained in time: if $t < t'$ then $x_t \rightarrow^b x_{t'}$ for any x and b and if $x_t \rightarrow^b x_{t'}$ then $t' \nleq t$.

"A synchronous structure" is defined by a semilattice structure to denote behaviors that have the same timing structure. The intuition behind this relation (depicted in Figure 6.9) is to consider a signal as an elastic with ordered marks on it (tags). If the elastic is stretched, marks remain in the

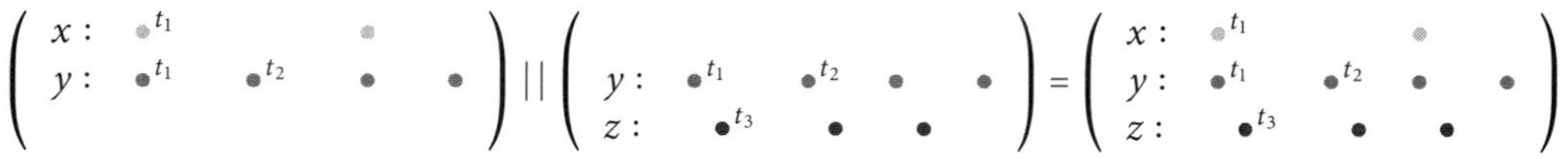

$$
\left(
\begin{array}{ll}
x: & \bullet^{t_1} \qquad\quad \bullet \\
y: & \bullet^{t_1} \quad \bullet^{t_2} \quad \bullet \quad \bullet
\end{array}
\right)
\;||\;
\left(
\begin{array}{ll}
y: & \bullet^{t_1} \quad \bullet^{t_2} \quad \bullet \quad \bullet \\
z: & \quad \bullet^{t_3} \quad \bullet \quad \bullet
\end{array}
\right)
=
\left(
\begin{array}{ll}
x: & \bullet^{t_1} \qquad\quad \bullet \\
y: & \bullet^{t_1} \quad \bullet^{t_2} \quad \bullet \quad \bullet \\
z: & \quad \bullet^{t_3} \quad \bullet \quad \bullet
\end{array}
\right)
$$

FIGURE 6.7   Synchronous composition of $b \in p$ and $c \in q$.

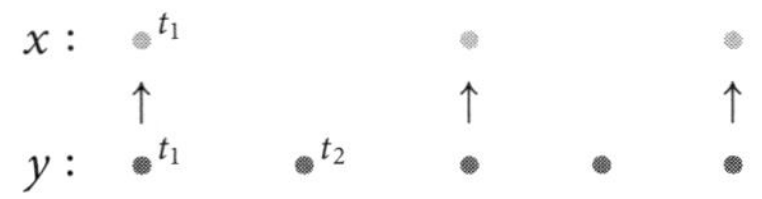

FIGURE 6.8 Scheduling relations between simultaneous events.

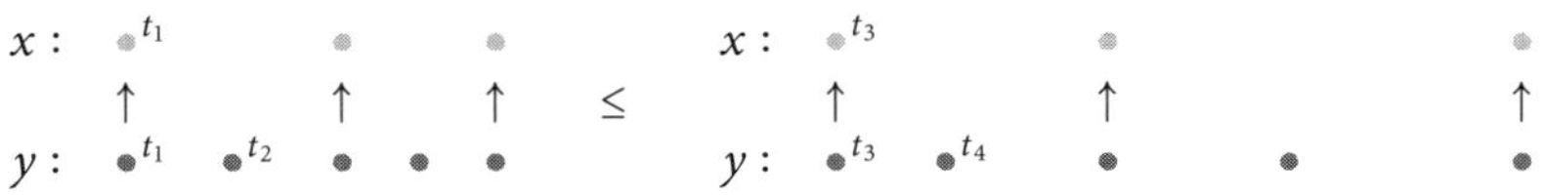

FIGURE 6.9 Relating synchronous behaviors by stretching.

$$P, Q ::= x = y\,f\,z \mid P/x \mid P\|Q$$

FIGURE 6.10 Common syntactic core for Lustre and Signal.

same relative and partial order but have more space (time) between each other. The same holds for a set of elastics: a behavior. If elastics are equally stretched, the order between marks is unchanged. In Figure 6.9, the time scale of x and y changes but the partial timing and scheduling relations are preserved. Stretching is a partial-order relation which defines clock equivalence. Formally, a behavior c is a stretching of b of same domain, written $b \leq c$, if there exists an increasing bijection on tags f that preserves the timing and scheduling relations. If so, c is the image of b by f. Finally, the behaviors b and c are said clock-equivalent, written $b \sim c$, *iff* there exists a behavior d s.t. $d \leq b$ and $d \leq c$.

6.4.2 Declarative Design Languages

The declarative design languages Lustre [38] and Signal [9] share the core syntax of Figure 6.10 and can both be expressed within the synchronous model of computation of Section 6.4.1. In both languages, a process P is an infinite loop that consists of the synchronous composition $P\|Q$ of simultaneous equations $x = y\,f\,z$ over signals named x, y, and z. Both Lustre and Signal support the restriction of a signal name x to a process P, noted P/x. The analogy stops here as Lustre and Signal differ in fundamental ways. Lustre is a single-clocked programming language, while Signal is a multiclocked (polychronous) specification formalism. This difference originates in the choice of different primitive combinators (named f in Figure 6.10) and results in orthogonal system design methodologies.

6.4.2.1 Combinators for Lustre

In a Lustre process, each equation processes the nth event of each input signal during the nth reaction (to possibly produce an output event). As it synchronizes upon availability of all inputs, the timing structure of a Lustre program is easily captured within a single clock domain: all input events are related to a master clock and the clock of the output signals is defined by sampling the master. There are three fundamental combinators in Lustre:

- **Delay:** "$x = \texttt{pre}\ y$" initially lets x undefined and then defines it by the previous value of y.
- **Followed-by:** "$x = y\ \texttt{->}\ z$" initially defines x by the value v, and then by z. The `pre` and `->` operators are usually used together, like in "$x = v\ \texttt{->}\ \texttt{pre}\,(y)$," to define a signal x initialized to v and defined by the previous value of y. Scade, the commercial version of Lustre, uses a 1-bit analysis to check that each signal defined by a `pre` is effectively initialized by an `->`.

- **Conditional:** "x = `if` b `then` y `else` z" defines x by y if b is true and by z if b is false. It can be used without alternative "x = `if` b `then` y" to sample y at the clock b, as shown in Figure 6.11.

Lustre programs are structured as data-flow functions, also called nodes. A node takes a number of input signals and defines a number of output signals upon the presence of an activation condition. If that condition matches an edge of the input signal clock, then the node is activated and possibly produces output. Otherwise, outputs are undetermined or defaulted. As an example, Figure 6.12 defines a resettable counter. It takes an input signal `tick` and returns the `count` of its occurrences. A Boolean `reset` signal can be triggered to reset the count to 0. We observe that the Boolean input signals `tick` and `reset` are synchronous to the output signal `count` and define a data-flow function.

6.4.2.2 Combinators for Signal

As opposed to nodes in Lustre, equations $x\ :=\ y\ f\ z$ in Signal more generally denote processes that define timing relations between input and output signals. There are three primitive combinators in Signal:

- **Delay:** "$x\ :=\ y\$1$ `init` v" initially defines the signal x by the value v and then by the previous value of the signal y. The signal y and its delayed copy "$x\ :=\ y\$1$ `init` v" are synchronous: they share the same set of tags $t_1, t_2, \ldots$. Initially (at t_1), the signal x takes the declared value v. At tag t_n, x takes the value of y at tag t_{n-1}. This is displayed in Figure 6.13.
- **Sampling:** "$x\ :=\ y$ `when` z" defines x by y when z is true (and both y and z are present); x is present with the value v_2 at t_2 only if y is present with v_2 at t_2 and if z is present at t_2 with the value true. When this is the case, one needs to schedule the calculation of y and z before x, as depicted by $y_{t_2} \rightarrow x_{t_2} \leftarrow z_{t_2}$.
- **Merge:** "$x\ =\ y$ `default` z" defines x by y when y is present and by z otherwise. If y is absent and z present with v_1 at t_1 then x holds (t_1, v_1). If y is present (at t_2 or t_3) then x holds its value whether z is present (at t_2) or not (at t_3). This is depicted in Figure 6.14.

FIGURE 6.11   The if-then-else conditional in Lustre.

```
node counter (tick, reset: bool) returns (count: int);
let
     count = if    true->reset
             then 0
             else if tick then pre count+1 else pre count;
```

FIGURE 6.12   Resettable counter in Lustre.

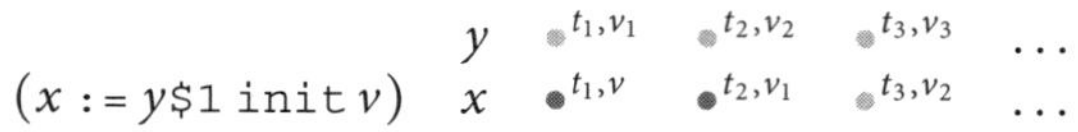

FIGURE 6.13   Delay operator in Signal.

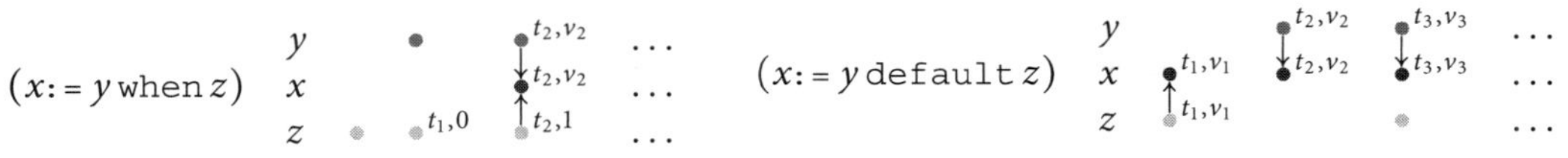

$$(x := y \text{ when } z)$$

$$(x := y \text{ default } z)$$

FIGURE 6.14   Merge operator in Signal.

```
process counter = (? event tick, reset ! integer value)
        (| value := (0 when reset)
            default ((value$ init 0 + 1) when tick)
            default (value$ init 0)
        |);
```

FIGURE 6.15   Resettable counter in Signal.

The structuring element of a Signal specification is a process. A process accepts input signals originating from possibly different clock domains to produce output signals when needed. Recalling the example of the resettable counter (Figure 6.12), this allows, for instance, to specify a counter (pictured in Figure 6.15) where the inputs `tick` and `reset` and the output `value` have independent clocks. The body of `counter` consists of one equation that defines the output signal `value`. Upon the event `reset`, it sets the count to 0. Otherwise, upon a `tick` event, it increments the count by referring to the previous value of `value` and adding 1 to it. Otherwise, if the count is solicited in the context of the counterprocess (meaning that its clock is active), the counter just returns the previous count without having to obtain a value from the `tick` and `reset` signals.

A Signal process is a structuring element akin to a hierarchical block diagram. A process may structurally contain subprocesses. A process is a generic structuring element that can be specialized to the timing context of its call. For instance, a definition of the Lustre counter (Figure 6.12) starting from the specification of Figure 6.15 consists of the refinement depicted in Figure 6.16. The input tick and reset clocks expected by the process `counter` are sampled from the Boolean input signals `tick` and `reset` by using the "when tick" and "when reset" expressions. The count is then synchronized to the inputs by the equation `reset ^= tick ^= count`.

6.4.3 Compilation of Declarative Formalisms

The analysis and code generation techniques of Lustre and Signal are necessarily different, tailored to handle the specific challenges determined by the different models of computation and programming paradigms.

6.4.3.1 Compilation of Signal

Sequential code generation starting from a Signal specification starts with an analysis of its implicit synchronization and scheduling relations. This analysis yields the control and data-flow graphs that define the class of sequentially executable specifications and allow to generate code.

```
process synccounter = (? boolean tick, reset ! integer value)
        (| value := counter (when tick, when reset)
         | reset ^= tick ^= value
        |);
```

FIGURE 6.16   Synchronization of the counterinterface.

$$e \quad ::= \quad \hat{\ }x \mid \text{when } x \mid \text{when not } x \mid e\,\hat{\ }{+}\,e' \mid e\,\hat{\ }{-}\,e' \mid e\,\hat{\ }{*}\,e' \mid 0 \quad \text{(clock expression)}$$
$$E \quad ::= \quad (\,) \mid e\,\hat{=}\,e' \mid e\,\hat{<}\,e' \mid x \to y \text{ when } e \mid E \,\|\, E' \mid E/x \quad \text{(clock relations)}$$

FIGURE 6.17   Syntax of clock expressions and clock relations (equations).

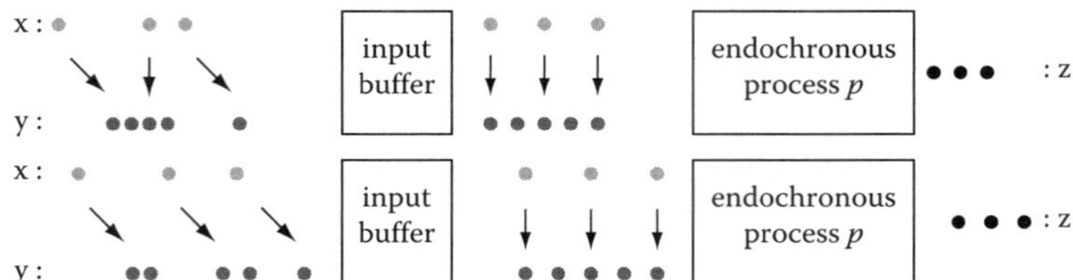

$$x := y\$1 \text{ init } v : \hat{\ }x\hat{\ }= \hat{\ }y$$
$$x := y \text{ when } z : \hat{\ }x\hat{\ }= \hat{\ }y \text{ when } z \,\|\, y \to x \text{ when } z$$
$$x := y \text{ default } z : \hat{\ }x\hat{\ }= \hat{\ }y\,\hat{\ }{+}\,\hat{\ }z \,\|\, y \to x \,\|\, z \to x \text{ when } (\hat{\ }z\,\hat{\ }{-}\,\hat{\ }y)$$

$$\frac{P : E \quad Q : E'}{P \,\|\, Q : E \,\|\, E'} \qquad \frac{P : E}{P/x : E/x}$$

FIGURE 6.18   Clock inference system of Signal.

FIGURE 6.19   Endochrony: from flow-equivalent inputs to clock-equivalent outputs.

Synchronization and scheduling analysis. In SIGNAL, the clock $\hat{\ }x$ of a signal x denotes the set of instants at which the signal x is present. It is represented by a signal, which is true when x is present and is absent otherwise. Clock expressions (see Figure 6.17) represent control. The clock "when x" (resp. "when not x") represents the time tags at which a Boolean signal x is present and true (resp. false). The empty clock is denoted by 0. Clock expressions are obtained using conjunction, disjunction, and symmetric differences over other clocks. Clock equations (also called clock relations) are Signal processes: the equation "$e \,\hat{=}\, e'$" synchronizes the clocks e and e' while "$e \,\hat{<}\, e'$" specifies the containment of e in e'. Explicit scheduling relations "$x \to y$ when e" allow the representation of causality in the computation of signals (e.g., x after y at the clock e).

A system of clock relations E can be easily associated (using the inference system $P : E$ of Figure 6.18) with any Signal process P, to represent its timing and scheduling structure.

Hierarchization. The clock and scheduling relations, E, of a process P define the control-flow and data-flow graphs that hold all necessary information to compile a Signal specification upon satisfaction of the property of endochrony, as illustrated in Figure 6.19. A process is said endochronous *iff* given a set of input signals (x and y in Figure 6.19) and flow-equivalent input behaviors (datagrams on the left of Figure 6.19); it has the capability to reconstruct a unique synchronous behavior up to clock-equivalence: the datagrams of the input signals in the middle of Figure 6.19 and of the output signal on the right of Figure 6.19 are ordered in clock-equivalent ways.

To determine the order $x \leq y$ in which signals are processed during the period of a reaction, clock relations E play an essential role. The process of determining this order is called hierarchization and consists of an insertion algorithm, which proceeds in three easy steps:

1. First, equivalence classes are defined between signals of same clock: if $E \Rightarrow \hat{\ }x\hat{\ }= \hat{\ }y$ then $x \leq y$ (we write $E \Rightarrow E'$ *iff* E implies E').

2. Second, elementary partial order relations are constructed between sampled signals: if $E \Rightarrow \hat{\ }x\hat{\ }= \text{when } y$ or $E \Rightarrow \hat{\ }x\hat{\ }= \text{when not } y$ then $y \leq x$.

3. Finally, assume a partial order of maximum z such that $E \Rightarrow {}^\wedge z = {}^\wedge y f {}^\wedge w$ (for some $f \in \{ {}^\wedge+, {}^\wedge*, {}^\wedge- \}$) and a signal x such that $y \le x \ge w$, and then insertion consists of attaching z to x by $x \le z$.

The insertion algorithm proposed in Amagbegnon et al. [1] yields a canonical representation of the partial order $\le$ by observing that there exists a unique minimum clock x below z such that rule 3 holds. Based on the order $\le$, one can decide whether E is hierarchical by checking that its clock relation $\le$ has a minimum, written $\min_\le E \in \text{vars}(E)$, so that $\forall x \in \text{vars}(E), \exists y \in \text{vars}(E), y \le x$. If E is furthermore acyclic (i.e. $E \Rightarrow x \to x$ when e implies $E \Rightarrow e^\wedge = 0$, for all $x \in \text{vars}(E)$) then the analyzed process is endochronous, as shown in Guernic et al. [35].

Example 6.1

The implications of hierarchization for code generation can be outlined by considering the specification of one-place buffer in Signal (Figure 6.20, left). Process buffer implements two functionalities. One is the process alternate that desynchronizes the signals i and o by synchronizing them to the true and false values of an alternating Boolean signal b. The other functionality is the process current. It defines a cell in which values are stored at the input clock ^i and loaded at the output clock ^o. cell is a predefined Signal operation defined by

$$x := y \; \texttt{cell} \; z \; \texttt{init} \; v =^{def} (m := x\$1 \; \texttt{init} \; v \,\|\, x := y \; \texttt{default} \; m \,\|\, {}^\wedge x^\wedge = {}^\wedge y {}^\wedge + {}^\wedge z) / m$$

Clock inference (Figure 6.20, middle) applies the clock inference system of Figure 6.18 to the process buffer to determine three synchronization classes. We observe that b, c_b, zb, zo are synchronous and define the master clock synchronization class of buffer. There are two other synchronization classes, c_i and c_o, which correspond to the true and false values of the Boolean flip-flop variable b, respectively :

$$\texttt{b} \diamond \texttt{c\_b} \diamond \texttt{zb} \diamond \texttt{zo} \text{ and } \texttt{b} \le \texttt{c\_i} \diamond \texttt{i} \text{ and } \texttt{b} \le \texttt{c\_o} \diamond \texttt{o}$$

This defines three nodes in the control-flow graph of the generated code (Figure 6.20, right). At the main clock c_b, b, and c_o are calculated from zb. At the subclock b, the input signal i is read. At the subclock c_o the output signal o is written. Finally, zb is determined. Notice that the sequence of instructions follows the scheduling relations determined during clock inference.

<table>
<tr><td valign="top">

```
process buffer = (? i ! o)
   (| alternate (i, o)
    | o := current (i)
    |) where
process alternate = (? i, o ! )
   (| zb := b$1 init true
    | b := not zb
    | o ^= when not b
    | i ^= when b
    |) / b, zb;
process current = (? i ! o)
   (| zo := i cell ^o init false
    | o   := zo when ^o
    |) / zo;
```

</td><td valign="top">

```
(| c_b ^= b
 | b    ^= zb
 | zb   ^= zo
 | c_i  := when b
 | c_i  ^= i
 | c_o  := when not b
 | c_o  ^= o
 | i -> zo when ^i
 | zb -> b
 | zo -> o when ^o
 |) / zb, zo, c_b,
        c_o, c_i, b;
```

</td><td valign="top">

```
buffer_iterate () {
   b = !zb;
   c_o = !b;
   if (b) {
     if (!r_buffer_i(&i))
       return FALSE;
   }
   if (c_o) {
     o = i;
     w_buffer_o(o);
   }
   zb = b;
   return TRUE;
}
```

</td></tr>
</table>

FIGURE 6.20   Specification, clock analysis, and code generation in Signal.

6.4.3.2 Compilation of Lustre

Whereas Signal uses a hierarchization algorithm to find a sequential execution path starting from a system of clock relations, Lustre leaves this task to engineers, which must provide a sound, fully synchronized program in the first place: well-synchronized Lustre programs correspond to hierarchized Signal specifications.

The classic compilation of Lustre starts with a static program analysis that checks the correct synchronization and cycle freedom of signals defined within the program. Then, it essentially partitions the program into elementary blocks activated upon Boolean conditions [38] and focuses on generating efficient code for high-level constructs, such as iterators for array processing [46].

Recently efforts have been made to enhance this compilation scheme by introducing effective activation clocks, whose soundness is checked by typing techniques. In particular, this was applied to the industrial SCADE version, with extensions [27,28].

6.4.3.3 Certification

The simplicity of the single-clocked model of Lustre eases program analysis and code generation. Therefore, its commercial implementation—Scade by Esterel Technologies—provides a certified C code generator. Its combination to Sildex (the commercial implementation of Signal by TNI-Valiosys) as a front-end for architecture mapping and early requirement specification is the methodology advocated in the IST project Safeair (URL: `http://www.safeair.org`). The formal validation and certification of synchronous program properties have been the subject of numerous studies. In Nowak et al. [50], a co-inductive axiomatization of Signal in the proof assistant Coq [33], based on the calculus of constructions [67], is proposed.

The application of this model is twofold. It allows, first of all, for the exhaustive verification of formal properties of infinite-state systems. Two case studies have developed. In Kerboeuf et al. [40], a faithful model of the steam-boiler problem was given in Signal and its properties proved with Signal's Coq model. In Kerboeuf et al. [41], it is applied to proving the correctness of real-time properties of a protocol for loosely time-triggered architectures, extending previous work proving the correctness of its finite-state approximation [8].

Another important application of modeling Signal in the proof assistant Coq is being explored: the development of a reference compiler translating Signal programs into Coq assertions. This translation allows to represent model transformations performed by the Signal compiler as correctness-preserving transformations of Coq assertions, yielding a costly yet correct-by-construction synthesis of the target code.

Other approaches to the certification of generated code have been investigated. In Pnueli et al. [52], validation is achieved by checking a model of the C code generated by the Signal compiler in the theorem prover PVS with respect to a model of its source specification (translation validation).

Related work on modeling Lustre has equally been numerous and started in Paulin-Mohring [51] with the verification of a sequential multiplier using a model of stream functions in Coq. In Canovas and Caspi [21], the verification of Lustre programs is considered under the concept of generating proof obligations and by using PVS. In Boulme and Hamon [19], a semantics of Lucid-Synchrone, an extension of Lustre with higher-order stream functions, is given in Coq.

6.5 Success Stories—A Viable Approach for System Design

S/R formalisms were originally defined and developed in the mid-1980s in an academic context, specially in France around INRIA, Ecole des Mines de Paris, and the Verimag CNRS laboratory. Research was also extensively contributed by German and US teams. It altogether provided the theoretical background for the current chapter, while the topic is still fairly active. Several large-scale cooperative IT R&D projects such as Syrf, Sacres, and Safeair were launched then.

S/R modeling and programming environments are today mainly marketed by two French software houses: Esterel Technologies for Esterel and SCADE/Lustre, and Geensys (formerly TNI-Valiosys) for Sildex/Signal. The influence of S/R systems tentatively pervaded to hardware CAD products such as Synopsys CoCentric Studio and Cadence VCC, despite the omnipotence of classical HDLs there. The Ptolemy cosimulation environment from UC Berkeley comprises an S/R domain based on the synchronous hypothesis.

There have been a number of industrial take-ups on S/R formalisms, most of them in the aeronautics industry. Airbus Industries is now using SCADE for the real design of parts of the new Airbus A-380 aircraft. S/R languages are also used by Dassault Aviation (for the next-generation Rafale fighter jet) and Snecma (Ref. [7] gives an in-depth coverage of these prominent collaborations). A highly praised feature of SCADE is that its formal basis is a current big enabler for certification of the design methodology in the (safety–critical) transportation domains. This has attracted considerable attention in the fields of avionics and trains, with possible extensions soon to car manufacturing. Phone and handheld manufacturers are also paying increasing attention to the design methods of S/R languages (for instance at Texas Instruments). A special subsidiary of Esterel Technologies, named Esterel-EDA, is dedicated to the use of synchronous and polychronous languages in the context of SoC ESL (Electronic-System Level design for Systems-on-Chip).

6.6 Into the Future: Perspectives and Extensions

Future advances in and around synchronous languages can be predicted in several directions:

Certified compilers. As already seen, this is the case for the basic SCADE compiler. But as the demand becomes higher, due to the critical–safety aspects of applications (in transportation fields notably), the impact of full-fledged operational semantics backing the actual compilers should increase.

Formal models and embedded code targets. Following the trend of exploiting formal models and semantic properties to help define efficient compilation and optimization techniques, one can consider the case of targeting distributed platforms (but still with a global reaction time). Then, the issues of spatial mapping and temporal scheduling of elementary operations composing the reaction inside a given interconnect topology become a fascinating (and NP-complete) problem. Heuristics for user guidance and semiautomatic approaches are the main topic of the SynDEx environment [34,43]. Of course this requires the estimation of the time budgets for the elementary operations and communications.

Loosely synchronized systems. In larger designs, the full global synchronous assumption is hard to maintain, especially if long propagation chains occur inside a single reaction (in hardware, for instance, the clock tree cannot be distributed to the whole chip). Several types of answers are currently being brought to this issue, trying to instill a looser coupling of synchronous modules into a desynchronized network (one then talks of "Globally Asynchronous Locally Synchronous" [GALS] systems). One such solution is given by the theory of latency-insensitive design, where each synchronous module of the network is supposed to be able to stall until the full information is synchronously available. The exact latency duration meant to recover a (slower) synchronous model is computed afterward, only after functional correctness on the more abstract level is achieved [22,57]. A broader presentation of the issues and solutions is given in Section 6.7.

Relations between transactional and cycle-accurate levels. If synchronous formalisms can be seen as a global attempt at transferring the notion of cycle-accurate modeling to the design of

SW/HW embedded systems, then the existing gap between these levels must also be reconsidered in the light of formal semantics and mathematical models. Currently, there exists virtually no automation for the synthesis of RTL at TLM levels. The previous item, with its well-defined relaxation of synchronous hypothesis at specification time, could be a definite step in this direction (of formally linking two distinct levels of modeling).

Relations between cycle-accurate and timed models. Physical timing is of course a big concern in synchronous formalisms, if only to validate the synchronous hypothesis and establish converging stabilization of all values across the system before the next clock tick. While in traditional software implementations one can decide that the instant is over when all treatments were effectively completed, in hardware or other real-time distributed settings a true compile-time timing analysis is in order. Several attempts have been made in this direction [26,45].

6.7 Loosely Synchronized Systems

The relations between synchronous and asynchronous models have long remained unclear, but investigations in this direction have recently received a boost due to demands coming from the engineering world. The problem is that many classes of embedded applications are best modeled, at least in part, under the cycle-based synchronous paradigm, while their desired implementation is not. This problem covers implementation classes that become increasingly popular (such as distributed software or even complex digital circuits like the Systems-on-a-Chip), hence the practical importance of the problem. Such implementations are formed of components that are only loosely connected through communication lines that are best modeled as asynchronous. At the same time, the existing synchronous tools for specification, verification, and synthesis are very efficient and popular, meaning that they should be used for most of the design process.

6.7.1 Asynchronous and Distributed Implementation of Synchronous Specifications

Much effort has been dedicated to the implementation of synchronous specifications onto loosely synchronized architectures. The difficulty is that of providing efficient simulations and implementations that still preserve in some formal sense the semantics of the specification. Three main classes of solutions have been proposed based on synchronous platforms, endochronous systems, and quasi-synchronous systems, which are presented in the following sections.

6.7.1.1 Synchronous Platforms

Various platforms that provide a system-wide notion of execution instant (a "simulated" global clock) have been defined. Provided that the system-wide synchronization overhead is acceptable, such platforms allow the direct implementation of the synchronous semantics.

In distributed software, the need for global synchronization mechanisms always existed. However, in order to be used in aerospace and automotive applications, an embedded system must also satisfy very high requirements in the areas of safety, availability, and fault tolerance. These needs prompted the development of integrated platforms, such as **TTA** [42], which offer higher-level, proven synchronization primitives, more adapted to specification, verification, and certification. The same correctness and safety goals are followed in a purely synchronous framework by two approaches: The AAA methodology and the **SynDEx** software of Sorel et al. [34] and the **Ocrep** tool of Girault et al. [24]. Both approaches take as input a synchronous specification, an architecture model, and some real-time and embedding constraints and produce a distributed implementation that satisfies the constraints and the synchrony hypothesis (supplementary signals simulate at run-time the global clock of the initial specification). The difference is that Ocrep is rather tailored for control-dominated synchronous programs, while SynDEx works best on data-flow specifications with simple control.

In the (synchronous) hardware world, problems appear when the clock speed and circuit size become large enough to make global synchrony unfeasible (or at least very expensive), most notably in what concerns the distribution of the clock and the transmission of data over long wires between functional components. The problem is to ensure that no communication error occurs due to the clock skew or due to the interconnect delay between the emitter and the receiver. Given the high cost (in area and power consumption) of precise clock distribution, it appears in fact that the only long-term solution is the division of large systems into several clocking domains, accompanied by the use of novel on-chip communication and synchronization techniques. When the multiple clocks are strongly correlated, we talk about mesochronous or plesiochronous systems [68], and communication between the clocking domains can be ensured without recourse to special devices [64]. This means that existing synchronous techniques and tools can be used without important modifications.

However, when the different clocks are unrelated (e.g., for power saving reasons), the resulting circuit is best modeled as a GALS system [25] where the synchronous domains are connected through asynchronous communication lines (e.g., FIFOs). Multiple problems occur here. Ensuring communication between clock domains so that data are not lost or duplicated has been addressed by both the asynchronous and synchronous hardware communities. On the asynchronous side, for instance, pausible clocking by Yun and Donohue [69] ensures reliable communication by synchronizing the clock of the receiver with incoming data to avoid metastability-related failures.

On the synchronous side, the theory of latency-insensitive design by Carloni and Sangiovanni-Vincentelli [22] investigates the case where a synchronous specification is implemented by a synchronous circuit, but the wires implementing the communication between major subsystems are too long for the given circuit technology, so that they must be segmented and transformed into FIFOs. Such FIFOs being dependent onto low-level technology and routing details are best modeled as unbounded asynchronous FIFOs. In the resulting GALS implementation model, the difficulty is that of defining and ensuring correctness with respect to the modular synchronous specification. The solution of Carloni and Sangiovanni-Vincentelli is based on the notion of stallable process, which identifies synchronous modules for which inputs on various input channels can be arbitrarily delayed without changing the outputs, modulo delays. Such stallable processes can be composed at will, and then a GALS implementation is easily derived by synchronizing at the input of each process the incoming inputs into synchronous events assigning a value to each input.

A more radical approach to the hardware implementation of a synchronous specification is desynchronization [16], where the clock subsystem is entirely removed and replaced with asynchronous handshake logic that simulates a system-wide notion of global clock. The advantages of such implementations are those of asynchronous logic: smaller power consumption, average-case performance, and smaller electromagnetic interference.

6.7.1.2 Endochronous Systems

We have seen earlier in this chapter that computation instants are well defined in a synchronous systems. Thus, the the absence of a signal is also well defined, allowing the modeling of inactive subsystems and communication lines. Reaction to absence is allowed, i.e., a change can be caused by the absence of a signal on a new clock tick. Since component inputs may become local signals in a larger concurrent system, absent values may have to be computed and propagated, to implement correctly the synchronous semantics.

When an asynchronous implementation is meant, where possibly distributed components communicate via message passing, signal/event absence in a reaction cannot be taken as granted because of communication latencies. A simple solution consists in systematically sending signal absence notifications. This corresponds to simulating the global clock, as it is done in the approaches of Section 6.7.1.1. However, such a solution may be unacceptable due to the communication overhead (in both time and communication resources). A natural question arises: when can one dispose of such "absent signal" communications?

The question is difficult, and its best formalization is due to Benveniste et al. [5], which noted that

- Some *absent* values are semantically needed to ensure that given asynchronous input always results in the same asynchronous output (asynchronous determinism). This is the case in programs involving priority constructs, such as the Esterel fragment:

```
module PRIORITY : input A,B ; output O ;
abort
   await A ; emit B
when C
end module
```

If both A and B are given to module PRIORITY, the output depends on the synchronous arrival order of A and B (O is emitted only when A arrives before B). This means that the absent values are needed to define the relative arrival order of A and B.

- Some absent values are semantically needed to ensure that the composition of synchronous modules through asynchronous FIFOs does not allow behaviors that are prohibited under synchronous composition rules. For instance, the synchronous composition of the following modules does not emit signal O, because A and B are emitted on different instants.

```
module SEND: output A,B ;
emit A ; pause ; emit B
end module

module RECV: input A,B ; output O ;
await [A and B] ; emit O
end module
```

However, the GALS implementation may allow A and B to arrive synchronously, resulting in the emission of O.

The solution proposed by Benveniste et al. involves the definition of two properties: Endochrony of a synchronous module ensures that signal absence is not needed on inputs to ensure asynchronous determinism. Isochrony of a synchronous composition ensures that signal absence is not needed to ensure correct synchronization in a GALS implementation.

Starting from the work on Signal language compilation [1] and the initial proposal of Benveniste, much effort has been dedicated to the understanding of endochrony [5,35,53,54], which is needed even for nondistributed implementations of synchronous specifications. It turned out that the fundamental property here is confluence (as coined by Milner [48]), which allows independent reactions (that share no common input) to be executed in any order (or synchronously) so that the first does not discard the second. This is also linked to the Kahn principles for networks [39], where only internal choice is allowed to ensure that overall lack of confluence cannot be caused by input signal speed variations. Similar reasoning in a hardware setting prompted the definition of the generalized latency-insensitive systems [61]. Isochrony has been more difficult to characterize. It turned out that it is akin to the absence of deadlocks in the execution of a distributed system [53].

Another line of work resulted in the finite flow-preservation criterion [62,63], which focuses on checking equivalence through finite desynchronization protocols.

6.7.1.3 Quasi-Synchronous Systems

The previous approaches to asynchronously implementing synchronous systems are based on the hypothesis that the behavior of the system, which must be preserved in the implementation, is given by the sequences of (nonabsent) values and meaningful synchronizations, which must be preserved.

However, in large classes of systems originating in the automatic control of physical systems, the behavior is fundamentally defined as a continuous-time process, the synchronous specification being only its quantized and time-discretized version. By consequence, we can exploit the continuity and robustness properties of the initial specification when implementing.

Consider now our problem of distributed implementation of a modular synchronous specification. The continuity and robustness of our system mean that the distributed implementation can lose some values and maybe read some other several times provided that the cyclic execution of each module respects the timing and accuracy prescribed by the control theory specification. In practice, this means that

- Communication can be done through a (nonsynchronizing) shared memory
- Ensuring the correctness of the implementation is choosing the periodic activation clocks of each module so that the aforementioned timing and accuracy constraints are met

This problem has been mainly investigated by Caspi et al. and resulted in a novel design methodology for distributed control systems [23,56].

Similar results are delivered by the loosely time-triggered architectures of Benveniste et al. [8], where sampling results are applied to ensure lossless communication between synchronous systems running on different clocks.

6.7.2 Modeling and Analysis of Polychronous Systems—Multiclock/Polychronous Languages

As explained by Milner [47], asynchronous systems can be modeled inside a synchronous framework. The essential ingredient of doing so is nondeterministic module activation, which is easily done using, for instance, additional inputs (oracles) used as activation conditions [36]. For instance, the following Esterel fragment can be used to model two asynchronously running modules M1 and M2 running on the activation conditions ORACLE1 and ORACLE2, which are assumed independent.

```
module ASYNC_MODEL:
input ORACLE1, ORACLE2;
[
  suspend
    run M1
  when not ORACLE1
||
  suspend
    run M2
  when not ORACLE2
]
end module
```

Of course, multiclock/polychronous systems can also be modeled using the same approach.

This approach is well adapted if the goal is simulation, verification, or monolithic implementation in the synchronous model. Problems appear when the synchronous specification is implemented in a globally asynchronous fashion. In such cases, the independence between various subsystems (which represents asynchrony) can be encoded with true asynchrony in the implementation, thus reducing the synchronization overhead. Doing this, however, requires that independence between clocks (activation conditions) be rediscovered in the globally synchronous specification, which comes down to the same techniques used to check endochrony. A native multiclock/polychronous modeling is therefore better suited in cases where the goal is globally asynchronous (distributed) implementation.

Two industrial-strength languages exist allowing the native modeling of multiclock/polychronous systems: Signal/Polychrony and Multiclock Esterel. The Signal language, presented earlier in this chapter, has been the first to adopt a polychronous model. As explained in Section 6.4, a Signal specification is a dataflow specification where each equation also defines implicit or explicit clock constraints. Clocks are associated with signals, and the clock of a signal is the same in its producer and in all consumers. This formalization naturally lead to globally asynchronous implementations based on lossless message passing and to the development of the successive notions of endochrony, which determine which synchronous programs have deterministic asynchronous implementations.

In Multiclock Esterel [15,64], the focus is on hardware implementations with multiple clock domains, the goal being to allow the modeling of large classes of such systems. Based on the purely synchronous Esterel language defined in Section 6.3, Multiclock Esterel enriches it with basic and derived clocks, clock domains (modules run on a given clock), signals that cross the clock domain barriers, and the possibility of defining more complex communication protocols, ensuring stronger synchronization properties. The signals that cross clock domains have semantics that are similar to the basic shared memory in distributed computing (only metastability issues are assumed solved). If two clocks are not related by the derivation process, then one cannot define complex clock synchronization properties like Signal. Timing constraints such as the ones used in the quasi-synchronous model cannot be modeled either, but such properties can be considered at a later design step. Multiclock Esterel is now part of the Esterel v7 language, implemented by the Esterel Studio environment and in the process of IEEE standardization.

References

1. Pascalin Amagbegnon, Loïc Besnard, and Paul Le Guernic. Implementation of the data-flow synchronous language signal. In *Conference on Programming Language Design and Implementation (PLDI'95)*. ACM Press, La Jolla, CA, 1995.
2. Charles André. Representation and analysis of reactive behavior: A synchronous approach. In *Computational Engineering in Systems Applications (CESA'96)*, pp. 19–29. IEEE-SMC, Lille, France, 1996.
3. Laurent Arditi, Hédi Boufaïed, Arnaud Cavanié, and Vincent Stehlé. Coverage-directed generation of system-level test cases for the validation of a DSP system. In Lecture Notes in Computer Science, vol. 2021. Springer-Verlag, 2001.
4. Albert Benveniste and Gérard Berry. The synchronous approach to reactive and real-time systems. *Proceedings of the IEEE*, 79(9):1270–1282, September 1991.
5. Albert Benveniste, Benoît Caillaud, and Paul Le Guernic. Compositionality in dataflow synchronous languages: Specification and distributed code generation. *Information and Computation*, 163:125–171, 2000.
6. Albert Benveniste, Paul Caspi, Luca Carloni, and Alberto Sangiovanni-Vincentelli. Heterogeneous reactive systems modeling and correct-by-construction deployment. In *Embedded Software Conference (EMSOFT'03)*. Springer-Verlag, Philadelphia, PA, October 2003.
7. Albert Benveniste, Paul Caspi, Stephen Edwards, Nicolas Halbwachs, Paul Le Guernic, and Robert de Simone. Synchronous languages twelve years later. *Proceedings of the IEEE*, 91(1):64–83, January 2003 (special issue on embedded systems).
8. Albert Benveniste, Paul Caspi, Paul Le Guernic, Hervé Marchand, Jean-Pierre Talpin, and Stavros Tripakis. A protocol for loosely time-triggered architectures. In *Embedded Software Conference (EMSOFT'02)*, vol. 2491 of Lecture Notes in Computer Science. Springer-Verlag, October 2002.

9. Albert Benveniste, Paul Le Guernic, and Christian Jacquemot. Synchronous programming with events and relations: The signal language and its semantics. *Science of Computer Programming*, 16:103–149, 1991.

10. Gérard Berry. Real-time programming: General-purpose or special-purpose languages. In G. Ritter, editor, *Information Processing 89*, pp. 11–17. Elsevier Science Publishers B.V., North Holland, 1989.

11. Gérard Berry. Esterel on hardware. *Philosophical Transactions of the Royal Society of London, Series A*, 19(2):87–152, 1992.

12. Gérard Berry. *The Constructive Semantics of Pure Esterel*. Esterel Technologies, electronic version available at http://www.esterel-technologies.com, 1999.

13. Gérard Berry and Laurent Cosserat. The synchronous programming language Esterel and its mathematical semantics. In Lecture Notes in Computer Science, vol. 197. Springer-Verlag, 1984.

14. Gérard Berry and Georges Gonthier. The Esterel synchronous programming language: Design, semantics, implementation. *Science of Computer Programming*, 19(2):87–152, 1992.

15. Gérard Berry and Ellen Sentovich. Multiclock Esterel. In *Proceedings CHARME'01*, vol. 2144 of Lecture Notes in Computer Science, 2001.

16. Ivan Blunno, Jordi Cortadella, Alex Kondratyev, Luciano Lavagno, Kelvin Lwin, and Christos Sotiriou. Handshake protocols for de-synchronization. In *Proceedings of the International Symposium on Asynchronous Circuits and Systems (ASYNC'04)*, Crete, Greece, 2004.

17. Amar Bouali. Xeve, an Esterel verification environment. In *Proceedings of the Tenth International Conference on Computer Aided Verification (CAV'98)*, vol. 1427 of Lecture Notes in Computer Science, UBC, Vancouver, Canada, June 1998.

18. Amar Bouali, Jean-Paul Marmorat, Robert de Simone, and Horia Toma. Verifying synchronous reactive systems programmed in Esterel. In *Proceedings FTRTFT'96*, vol. 1135 of Lecture Notes in Computer Science, pp. 463–466, 1996.

19. Sylvain Boulme and Grégoire Hamon. Certifying synchrony for free. In *Logic for Programming, Artificial Intelligence and Reasoning*, vol. 2250 of Lecture Notes in Artificial Intelligence. Springer-Verlag, 2001.

20. Frédéric Boussinot and Robert de Simone. The Esterel language. *Proceedings of the IEEE*, 79(9):1293–1304, September 1991.

21. Cécile Dumas Canovas and Paul Caspi. A PVS proof obligation generator for Lustre programs. In *International Conference on Logic for Programming and Reasonning*, vol. 1955 of Lecture Notes in Artificial Intelligence. Springer-Verlag, 2000.

22. Luca Carloni, Ken McMillan, and Alberto Sangiovanni-Vincentelli. The theory of latency-insensitive design. *IEEE Transactions on Computer-Aided Design of Integrated Circuits and Systems*, 20(9):1059–1076, 2001.

23. Paul Caspi. Embedded control: From asynchrony to synchrony and back. In *Proceedings EMSOFT'01*, Lake Tahoe, October 2001.

24. Paul Caspi, Alain Girault, and Daniel Pilaud. Automatic distribution of reactive systems for asynchronous networks of processors. *IEEE Transactions on Software Engineering*, 25(3):416–427, 1999.

25. Daniel Marcos Chapiro. Globally-asynchronous locally-synchronous systems. PhD thesis, Stanford University, Stanford, CA, October 1984.

26. Etienne Closse, Michel Poize, Jacques Pulou, Joseph Sifakis, Patrick Venier, Daniel Weil, and Sergio Yovine. TAXYS: A tool for the development and verification of real-time embedded systems. In *Proceedings CAV'01*, vol. 2102 of Lecture Notes in Computer Science, 2001.

27. Jean-Louis Colaço, Alain Girault, Grégoire Hamon, and Marc Pouzet. Towards a higher-order synchronous data-flow language. In *Proceedings EMSOFT'04*, Pisa, Italy, 2004.

28. Jean-Louis Colaço and Marc Pouzet. Clocks as first class abstract types. In *Proceedings EMSOFT'03*, Philadelphia, PA, 2003.

29. Robert de Simone and Annie Ressouche. Compositional semantics of Esterel and verification by compositional reductions. In *Proceedings CAV'94*, vol. 818 of Lecture Notes in Computer Science, 1994.

30. Stephen Edwards. An Esterel compiler for large control-dominated systems. *IEEE Transactions on Computer-Aided Design of Integrated Circuits and Systems*, 21(2):240–247, February 2002.

31. Robert French, Monica Lam, Jeremy Levitt, and Kunle Olukotun. A general method for compiling event-driven simulations. In *Proceedings of the 32nd Design Automation Conference (DAC'95)*, San Francisco, CA, 1995.

32. Minxi Gao, Jie-Hong Jiang, Yunjian Jiang, Yinghua Li, Subarna Sinha, and Robert Brayton. MVSIS. In *Proceedings of the International Workshop on Logic Synthesis (IWLS'01)*, Tahoe City, June 2001.

33. Eduardo Giménez. Un Calcul de Constructions Infinies et son Application à la Vérification des Systèmes Communicants. PhD thesis, Laboratoire de l'Informatique du Parallélisme, Ecole Normale Supérieure de Lyon, December 1996.

34. Thierry Grandpierre, Christophe Lavarenne, and Yves Sorel. Optimized rapid prototyping for real time embedded heterogeneous multiprocessors. In *Proceedings of the 7th International Workshop on Hardware/Software Co-Design (CODES'99)*, Rome, 1999.

35. Paul Le Guernic, Jean-Pierre Talpin, and Jean-Christophe Le Lann. Polychrony for system design. *Journal of Circuits, Systems and Computers*, 2000 (special issue on application-specific hardware design).

36. Nicolas Halbwachs and Louis Mandel. Simulation and verification of asynchronous systems by means of a synchronous model. In *Proceedings ACSD'06*, Turku, Finland, 2006.

37. Nicolas Halbwachs. Synchronous programming of reactive systems. In *Computer Aided Verification (CAV'98)*, Vancouver, Canada, pp. 1–16, 1998.

38. Nicolas Halbwachs, Paul Caspi, and Pascal Raymond. The synchronous data-flow programming language Lustre. *Proceedings of the IEEE*, 79(9):1305–1320, 1991.

39. Gilles Kahn. The semantics of a simple language for parallel programming. In J.L. Rosenfeld, editor, *Information Processing '74*, pp. 471–475. North Holland, Amsterdam, 1974.

40. Mickael Kerboeuf, David Nowak, and Jean-Pierre Talpin. Specification and verification of a steam-boiler with Signal-Coq. In *International Conference on Theorem Proving in Higher-Order Logics*, vol. 1869 of Lecture Notes in Computer Science. Springer-Verlag, 2000.

41. Mickael Kerboeuf, David Nowak, and Jean-Pierre Talpin. Formal proof of a polychronous protocol for loosely time-triggered architectures. In *International Conference on Formal Engineering Methods*, vol. 2885 of Lecture Notes in Computer Science. Springer-Verlag, 2003.

42. Hermann Kopetz. *Real-Time Systems: Design Principles for Distributed Embedded Applications*. Kluwer Academic Publishers, 1997.

43. Christophe Lavarenne, Omar Seghrouchni, Yves Sorel, and Michel Sorine. The SynDEx software environment for real-time distributed systems design and implementation. In *European Control Conference ECC'91*, Grenoble, France, 1991.

44. Jaejin Lee, David Padua, and Samuel Midkiff. Basic compiler algorithms for parallel programs. In *Proceedings of the 7th ACM SIGPLAN Symposium on Principles and Practice of Parallel Programming*, Atlanta, GA, 1999.

45. George Logothetis and Klaus Schneider. Exact high-level WCET analysis of synchronous programs by symbolic state space exploration. In *Proceedings DATE2003*, Munich, Germany, 2003.

46. Florence Maraninchi and Lionel Morel. Arrays and contracts for the specification and analysis of regular systems. In *International Conference on Applications of Concurrency to System Design (ACSD'04)*. IEEE Press, Hamilton, Ontario, Canada, 2004.

47. Robin Milner. Calculi for synchrony and asynchrony. *Theoretical Computer Science*, 25(3):267–310, July 1983.

48. Robin Milner. *Communication and Concurrency*. Prentice Hall, Upper Saddle River, NJ, 1989.

49. Steven Muchnick. *Advanced Compiler Design and Implementation*. Morgan Kaufmann Publishers, San Francisco, CA 1997.

50. David Nowak, Jean-Rene Beauvais, and Jean-Pierre Talpin. Co-inductive axiomatization of a synchronous language. In *International Conference on Theorem Proving in Higher-Order Logics*, vol. 1479 of Lecture Notes in Computer Science. Springer-Verlag, 1998.

51. Christine Paulin-Mohring. Circuits as streams in Coq: Verification of a sequential multiplier. In S. Berardi and M. Coppo, editors, *Types for Proofs and Programs, TYPES'95*, vol. 1158 of Lecture Notes in Computer Science, 1996.

52. Amir Pnueli, O. Shtrichman, and M. Siegel. Translation validation: From Signal to C. In *Correct System Design Recent Insights and Advance*, vol. 1710 of Lecture Notes in Computer Science. Springer-Verlag, 2000.

53. Dumitru Potop-Butucaru and Benoit Caillaud. Correct-by-construction asynchronous implementation of modular synchronous specifications. *Fundamenta Informaticae*, 78(1):131–159, 2007.

54. Dumitru Potop-Butucaru, Benoit Caillaud, and A. Benveniste. Concurrency in synchronous systems. *Formal Methods in System Design*, 28(2):111–130, March 2006.

55. Dumitru Potop-Butucaru and Robert de Simone. Optimizations for faster execution of Esterel programs. In Rajesh Gupta, Paul Le Guernic, Sandeep Shukla, and Jean-Pierre Talpin, editors, *Formal Methods and Models for System Design*, Kluwer, 2004.

56. The CRISYS ESPRIT project. The quasi-synchronous approach to distributed control systems. Technical report, CNRS, UJF, INPG, Grenoble, France, 2001. Available online at http://wwwverimag.imag.fr/ caspi/CRISYS/cooking.ps

57. Alberto Sangiovanni-Vincentelli, Luca Carloni, Fernando De Bernardinis, and Marco Sgroi. Benefits and challenges of platform-based design. In *Proceedings of the Design Automation Conference (DAC'04)*, San Diego, CA, 2004.

58. Ellen Sentovich, Kanwar Jit Singh, Luciano Lavagno, Cho Moon, Rajeev Murgai, Alexander Saldanha, Hamid Savoj, Paul Stephan, Robert Brayton, and Alberto Sagiovanni-Vincentelli. SIS: A system for sequential circuit synthesis. Memorandum UCB/ERL M92/41, UCB, ERL, 1992.

59. Ellen Sentovich, Horia Toma, and Gérard Berry. Latch optimization in circuits generated from high-level descriptions. In *Proceedings of the International Conference on Computer-Aided Design (ICCAD'96)*, San Jose, CA, 1996.

60. Tom Shiple, Gérard Berry, and Hervé Touati. Constructive analysis of cyclic circuits. In *Proceedings of the International Design and Testing Conference (ITDC)*, Paris, 1996.

61. Montek Singh and Michael Theobald. Generalized latency insensitive systems for GALS architectures. In *Proceedings FMGALS2003*, Pisa, Italy, 2003.

62. Jean-Pierre Talpin and Paul Le Guernic. Algebraic theory for behavioral type inference. *Formal Methods and Models for System Design (chap VIII)*. Kluwer Academic Press, Boston, MA, 2004.

63. Jean-Pierre Talpin, Paul Le Guernic, Sandeep Kumar Shukla, Frédéric Doucet, and Rajesh Gupta. Formal refinement checking in a system-level design methodology. *Fundamenta Informaticae*. IOS Press, Amsterdam, 2004.

64. Esterel Technologies. The Esterel v7 reference manual. version v7_30. initial IEEE standardization proposal. Online at http://www.esterel-eda.com/style-EDA/files/papers/Esterel-Language-v7-Ref-Man.pdf, November 2005.

65. Hervé Touati and Gérard Berry. Optimized controller synthesis using Esterel. In *Proceedings of the International Workshop on Logic Synthesis (IWLS'93)*, Lake Tahoe, 1993.

66. Daniel Weil, Valérie Bertin, Etienne Closse, Michel Poize, Patrick Vernier, and Jacques Pulou. Efficient compilation of Esterel for real-time embedded systems. In *Proceedings CASES'00*, San Jose, CA, 2000.

67. Benjamin Werner. Une Théorie des Constructions Inductives. PhD thesis, Université Paris VII, Mai. 1994.

68. Wade L. Williams, Philip E. Madrid, and Scott C. Johnson. Low latency clock domain transfer for simultaneously mesochronous, plesiochronous and heterochronous interfaces. In *Proceedings ASYNC'07*, CA, March 2007. University of California at Berkeley.

69. Kenneth Yun and Ryan Donohue. Pausible clocking: A first step toward heterogenous systems. In *International Conference on Computer Design (ICCD'96)*, Austin, TX, 1996.

7

Processor-Centric Architecture Description Languages

Steve Leibson
Tensilica, Inc.

Himanshu Sanghavi
Tensilica, Inc.

Nupur Andrews
Tensilica, Inc.

7.1 Introduction

In the twenty-first century, embedded systems have become ubiquitous and the turnover in products has become fierce. Product design-cycle times and product lifetimes are now measured in months. At the same time, board-level embedded microprocessor and on-chip processor core use have sky-rocketed. Multicore processor designs are becoming common. Consequently, design teams need a way to perform rapid design-space exploration (DSE) of programmable processor architectures to meet the pressures of shrinking time-to-market and ever-shrinking product lifetimes.

Design teams use architecture description languages (ADLs) to perform early exploration, synthesis, test generation, and validation of processor-based designs. Although various ADLs have been devised to develop and model software systems, hardware/software systems, and processors, they have gained the most traction in the specification, modeling, and validation of various processor architectures.

Processor-centric ADL specifications can be used to automatically generate a software toolkit including the compiler, assembler, instruction-set simulator (ISS), and debugger; a description of the target processor hardware composed using an hardware description language (HDL), such as Verilog or VHDL; and various simulation models and related tools for processor simulation and validation. The specification can also be used to generate device drivers for real-time operating systems (OSs).

Application programs can be compiled and simulated using the generated software tools and the feedback on program performance and memory footprint can then be used to modify the ADL specification with the goal of iterating to the best possible architecture for the given set of applications. The ADL specification can also be used for generating hardware prototypes from the generated HDL descriptions under design constraints such as area, power, and clock speed. Several researches have shown the usefulness of ADL-driven generation of functional test programs and test interfaces.

7.2 ADL Genesis

The term "architecture description language" refers to languages developed for designing both software and hardware architectures. Software ADLs represent and permit the rapid analysis of software architectures [1,2]. These sorts of ADLs capture the behavioral specifications of the software components and their interactions, which comprise the software architecture. Hardware ADLs capture hardware structure (hardware components and their connectivity) and the behavior (instruction set) of processor architectures. Processor-centric ADLs capture the essence of a processor's instruction-set architecture (ISA, the processor's instructions, registers, register files, and other state). The concept of using machine description (MDES) languages for specification of architectures has been around for a long time.

At least as far back as 1971, early ADLs such as Bell and Newell's instruction-set processor (ISP) notation [3] were used to classify and describe processors and whole computers. By the end of the 1970s, these ISP descriptions had progressed to the point where they could be used to simulate and evaluate proposed processor architectures [4]. Now, the ADLs that grew out of the tradition of ISP notations can be used to generate, simulate, and analyze processor architectures.

It is appropriate to quickly discuss rapid prototyping and evaluation of new processor architectures, which has suddenly grown in importance. The microprocessor age (and the age of embedded systems) started in 1971 with Intel's introduction of the 4 bit 4004 microprocessor, which was the first commercially available microprocessor chip. From 1971 to about 1995, most embedded designers purchased predesigned microprocessors, first in the form of manufactured and packaged chips and then, later, as intellectual property cores that could be incorporated into application-specific integrated circuit designs. ASICs incorporating processor cores became known as systems on chips (SOCs).

During this period, the majority of processors used were purchased from vendors who designed the processor hardware and developed the required software tool sets. Embedded systems designers used processors. Very few designed them. Processor design was relegated to the few companies that sold microprocessors and to the few companies that could not meet performance goals or achieve sufficient performance with commercial processor offerings.

All that changed with the advent of the SOC. As soon as the processor was freed from its package and placed on a silicon substrate along with other system blocks, it became possible—at least theoretically—to custom design a processor for each application. However, just because it was possible does not mean it was practical. A microprocessor's utility to system designers is not completely defined by the hardware. All microprocessors are surrounded by essential software components including development tools (compiler, assembler, debugger, and ISS) and by simulation tools needed to incorporate that processor into the SOC design flow and to create the software that will run on that processor. Tool development can represent just as big an R&D investment as the hardware design of the processor.

Processor-centric ADLs automate the process of generating processor hardware, development tools, and simulation aids from descriptions of the target processor. To be able to serve as the root of all these generated items, these descriptions must somehow capture the processor's structure and its behavior. Different processor-centric ADLs capture this information in different ways.

7.3 Classifying Processor-Centric ADLs

ADLs differ from modeling languages. Modeling languages such as unified modeling language (UML) are more concerned with the behaviors of whole systems rather than the parts. ADLs represent components and processor-centric ADLs represent processors. In practice, many designers have used modeling languages to represent systems of cooperating components and processor architectures. However, modeling languages' high level of abstraction makes it harder to create detailed descriptions of a processor's ISA. At the other extreme, HDLs such as VHDL and Verilog lack sufficient abstraction to describe processor architectures and explore them at the system level without gate-level simulation, which is very slow for reasonably complex architectures. In practice, some HDL variants have been made to work as ADLs for specific processor classes but the clear trend is to use purpose-built ADLs for processor DSE.

Figure 7.1, based on a system devised by Mishra and Dutt [5,6], classifies ADLs based on two aspects: content and objective. The content-oriented classification is based on the ADL's descriptive nature while the objective-oriented classification is based on the ADL's ultimate purpose. Mishra and Dutt divide contemporary ADLs into six categories based on the objective: simulation oriented, synthesis oriented, test oriented, compilation oriented, validation oriented, and OS oriented.

ADLs can also be classified into four categories based on the nature of the information: structural, behavioral, mixed, and partial. Structural ADLs capture processor structure in terms of architectural components and their connectivity. Behavioral ADLs capture ISA behavior. Mixed ADLs capture both structure and behavior of the architecture. Partial ADLs capture specific information about the architecture for some specific task. For example, a partial ADL used for interface synthesis need not describe a processor's internal structure or behavior—only interfaces need to be described.

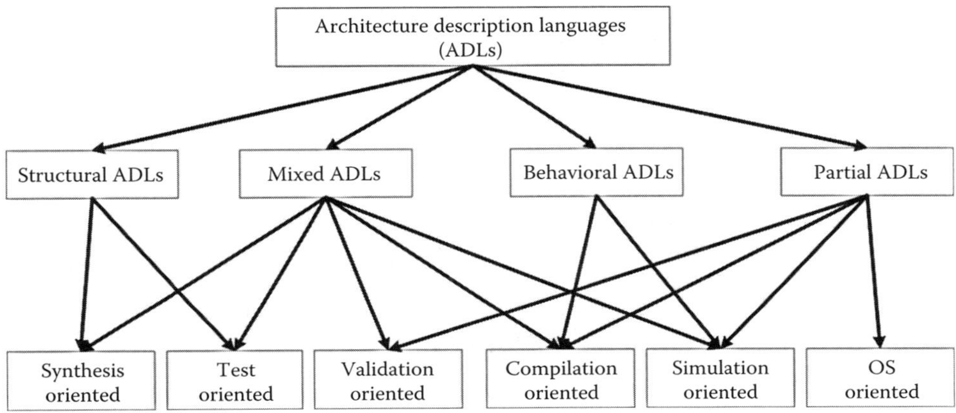

FIGURE 7.1   Mishra/Dutt ADL classifications.

7.3.1 Structural ADLs

Structural ADLs are suitable for synthesis and test generation. Structural ADLs often use the register transfer level (RT-level or simply RTL) abstraction level because this abstraction level is low enough to model detailed processor behavior yet it is high enough to hide gate-level implementation details, which cuts simulation time. Early ADLs were based on RTL descriptions. Structural ADLs include "machine independent microprogramming language" (MIMOLA developed at the University of Dortmund, Germany) and "unified design language for integrated circuits" (UDL/I developed at Kyushu University, Japan). MIMOLA and UDL/I are oriented toward logic synthesis.

7.3.2 Behavioral ADLs

Behavioral ADLs are suited for simulation and compilation. Therefore, behavioral ADLs specify a processor's instruction semantics and ignore the underlying hardware structure. Mishra and Dutt classify nML (developed at the Technical University of Berlin, Germany) as a behavioral ADL while the developers themselves classify the language as a mixed ADL [7]. Whatever its classification, nML is commercially available as the Chess/Checkers from Target Compiler Technologies. Another behavioral compiler is "instruction-set description language" (ISDL developed at MIT, Cambridge, MA), which was principally developed for the description of very long instruction word (VLIW) processors.

7.3.3 Mixed ADLs

Mixed ADLs capture both structural and behavioral architectural details. High-level machine description (HMDES), EXPRESSION, and language for instruction-set architecture (LISA) are three examples of mixed HDLs. An HMDES (developed at University of Illinois at Urbana-Champaign for the IMPACT research compiler) serves as the input to the MDES system of the Trimaran compiler, which contains IMPACT as well as the Elcor research compiler from HP Labs. The description is optimized and then translated into a low-level representation file. MDES captures both structure and behavior of target processors.

The EXPRESSION ADL (developed at University of California, Irvine, CA) describes a processor as a netlist of units and storage elements. Unlike MIMOLA, an EXPRESSION netlist is coarse-grained. It uses a higher level of abstraction similar to the block-diagram-level description in an architecture manual. EXPRESSION has been used by the EXPRESS retargetable compiler and SIMPRESS simulator.

LISA (developed at Aachen University of Technology, Germany) has a simulator-centric view. The language has since formed the basis of a company, LisaTek, now owned by CoWare. The LISA explicitly captures control paths. LISA's explicit modeling of both datapath and control permits cycle-accurate simulation.

7.3.4 Partial ADLs

Many ADLs only capture partial architectural information to perform specific tasks. Two such ADLs are AIDL (developed at University of Tsukuba, Japan) for the design of high-performance superscalar processors and PEAS-I (developed at Tsuruoka National College of Technology and Toyohashi University of Technology, Japan). AIDL does not aim for datapath optimization. Instead, it is targeted at validating pipeline behavior. AIDL does not support software toolkit generation but AIDL descriptions can be simulated using the AIDL simulator. PEAS-I is an instruction-set optimizer that takes a C program and a data set as input. The output of PEAS-I is an optimized instruction set that accelerates the execution of the input C program. Therefore, PEAS-I is not really an ADL but serves a similar purpose.

7.3.5 Some Specific ADL Overviews

Collectively, the following thumbnail sketches provide a good overview of the many types of processor-design problems addressed by various processor-centric ADLs. These sketches illustrate the many facets of processor design and elaborate the myriad details that a processor designer must address to produce a complete design.

7.3.5.1 MIMOLA

MIMOLA (developed at the University of Dortmund, Dortmund, North Rhine-Westphalia, Germany) is a high-level programming language (HLL), a register transfer language (RTL), and a computer hardware description language (CHDL) all rolled into one language [8]. MIMOLA describes a processor as a netlist of modules and a detailed interconnection scheme. MIMOLA can be used for high- or intermediate-level microprogramming.

MIMOLA allows the description of

- Digital computer hardware, modeled as a structure built from register transfer (RT) modules (e.g., adders, buses, memories) and gate-level modules (included as special-case RT modules)
- Behavior of digital hardware modules
- Information required for linking behavioral and structural domains
- Programs at a PASCAL-like level
- Programs at the level of RTs
- Initial simulation stimuli

MIMOLA 4.1 serves the common input language for a variety of CAD tools. These CAD tools have been designed to support essential VLSI design activities such as interactive synthesis, microcode generation, test generation, and simulation.

7.3.5.2 nML

nML is a high-level language developed at the Technical University of Berlin [7]. Descriptions written in nML express the connectivity, parallelism, and architectural details of embedded processors. The nML language allows the designer to specify the target processor architecture in a way that it parallels instruction-set descriptions found in a user's programming manual. In contrast to a description written in MIMOLA, an nML MDES contains behavioral as well as structural information. The first part of an nML description, called the "structural skeleton," declares storage, connectivity, and functional units. Storage and connections have a data type, defined as C++ classes in a user-extensible library. The second part of the description declares an instruction set defined by an attribute grammar. The grammar breaks down the instruction set into a hierarchical set of classes. *Or* rules in the grammar specify alternative choices while *and* rules specify concurrency. A modified version of the nML formalism is the foundation of patented compilation and hardware generation techniques used in the Chess/Checkers tool suite from Target Compiler Technologies, a spinoff from IMEC, the microelectronics research center in Belgium [9].

7.3.5.3 ISDL

ISDL was developed at MIT in Cambridge, MA [10]. ISDL's main focus is the description of VLIW processor architectures; however, the language also supports the description of standard microcontrollers and DSP cores with custom datapaths.

An ISDL description consists of six sections:

- Instruction word format
- Global definitions
- Storage resources
- Instruction set
- Constraints
- Optional architectural details

ISDL descriptions can express multiple functional units, different interconnect topologies, complex instructions, resource conflicts, pipelining idiosyncrasies, etc. Such architectures cannot be guaranteed to have clean instruction sets (i.e., instruction sets where every operation combination is valid) so ISDL supports explicit constraints that define which operation groupings are valid. The ISDL compiler can therefore avoid generating invalid instructions by ensuring that each instruction meets these constraints.

7.3.5.4   MDES and HMDES

The MDES language was jointly developed by the IMPACT group at the University of Illinois at Urbana-Champaign and the FAST group at HP Labs, responsible for the program in, chip out project [11]. The goal for MDES was a language that could describe a processor's resources and how the processor's instruction set uses these resources in sufficient detail so that a compiler could efficiently schedule instructions for that processor. The developers of MDES wanted the language to be "intuitive," to minimize the tedium of writing and modifying MDESs. They also wanted to make it easy for a compiler to efficiently load the information contained in an MDES without having to deal with syntax and typographic errors and they wanted the MDES language to be compiler independent. Consequently, the language's designers split the MDES into an HMDES (a high-level machine-description language) and an LMDES (a low-level machine-description language).

The HMDES allows a designer to write a processor description more intuitively using comments, text substitution, and flexible indentation and text formatting. For efficiency, the HMDES is then translated into a machine-readable LMDES file using preprocessing algorithms that check the description's grammar and syntax; it is essentially compiled.

HMDESs contain many sections:

- "Define" section of an HMDES specifies the number of predicate, destination, source, and synchronization operands supported by the processor's instruction set. It also specifies the processor type (superscalar or VLIW).
- Formally, the "Register_files" section was intended to describe the capacities (number of entries) and width of the processor's register files. Practically, the "Register_files" section is used to describe the allowed operand types in the processor's assembly language.
- "IO_set" section allows the designer to group register files into convenient sets.
- "IO_items" section gives names to legal operand groupings.
- "Resources" section names all resources available to model the processor.
- "ResTables" section describes how and when these resources can be used.
- "Latencies" section defines the cycle within an instruction's execution that is to be used for reading from registers with predicate, source, or incoming-synchronization operands and for writing to destination or outgoing-synchronization registers.

- Entries in the "Operation_Class" section describe instruction classes. All instructions within a class use the same operands, use the same processor resources, and have the same instruction latency. The "Operation" section then associates an operation's opcode with opcode flags, an assembly name, assembly flags, and an "Operation_class" designation or a direct specification of scheduling alternatives.

7.3.5.5 EXPRESSION

EXPRESSION (developed at the University of California, Irvine, CA) is a language targeting DSE for embedded SOC processor architectures and automatic generation of retargetable compiler/simulator toolkits [12]. The language and associated design methodology feature a mixed behavioral/structural representation supporting a "natural" architecture specification, explicit specification of the memory subsystem allowing novel memory organizations and hierarchies, clean syntax with easy modification to encourage architectural exploration, a single specification to simplify consistency and completeness checking, and efficient specification of architectural resource constraints allowing extraction of detailed reservation tables for compiler scheduling.

EXPRESSION allows the designer to specify the RTL netlist of the processor's datapath at an abstract level (a datapath netlist omits control signals). First, each RTL architectural component is specified. Then, pipeline paths and all valid data-transfer paths are specified. This information then guides the generation of both the netlist for a simulator and the reservation tables for a compiler. EXPRESSION's resource constraint specification scheme reduces specification complexity and eases consistency and completeness checking of the specification.

The EXPRESSION language employs a LISP-like syntax and descriptions written in EXPRESSION consist of two main sections:

- Behavior (or IS—the processor's instruction set)
- Structure

Each section of an EXPRESSION description is further subdivided into three subsections. The "Behavior" section contains the following subsections:

- Operations
- Instruction description
- Operation mappings

The "Operations" subsection describes the processor's instruction set. The "Instruction description" subsection captures the parallelism available in the architecture. The "Operation mappings" subsection specifies information needed for instruction selection and for architecture-specific compiler optimizations.

The "Structure" section contains the following subsections:

- Components
- Pipeline/Data-transfer paths
- Memory subsystem

The "Components" subsection describes each architectural RTL component including pipeline units, functional units, storage elements, ports, connections, and buses. The "Pipeline/Data-transfer paths" subsection describes the processor's netlist including the pipeline description, which specifies units in the processor's the pipeline stages and the data-transfer paths description, which specifies valid data transfers. The "Memory subsystem" subsection describes the types and attributes of various storage components (register files, SRAMs, DRAMs, caches, etc.).

7.3.5.6 LISA

The LISA ADL was developed at the Aachen University of Technology in Germany [13,14]. LISA was created to fill a perceived gap between

- Standard ISA models and ISDLs used in conjunction with compilers and ISSs
- Detailed behavioral/structural processor models and description languages used for hardware design

LISA descriptions express operation-level processor pipeline descriptions including descriptions of complex interlocking and bypassing techniques. Processor instructions described with LISA consist of multiple operations defined as RTs during a single control step, which can be resolved with instruction, clock-cycle, or clock-phase accuracy depending on the required modeling accuracy. LISA descriptions use modified Gantt charts (dubbed L-charts) to schedule operations and to specify the time and resource allocations for operations. Unlike classical reservation tables, LISA's L-charts permit modeling of data and control hazards and processor pipeline flushing. The LISA description produces a timed ISA model at the desired temporal accuracy. This model can then be used for several purposes including simulation and compilation.

LISA descriptions consist of "resources" and "operations." Declared "resources" represent hardware storage objects (registers, memories, pipelines) that hold system state. "Operations" express the designer's view of the processor's behavior, structure, and instruction set. A LISA MDES creates the following models:

- *The memory model*: A list of registers and system memories with their respective bit-widths, address ranges, and aliasing.
- *The resource model*: A description of the available hardware resources and the resource requirements for the processor's operations.
- *The instruction-set model*: A list of valid combinations of hardware operations and permissible operands expressed by a combination of assembly syntax, instruction-word coding, specification of legal operands, and addressing modes for each instruction.
- *The behavioral model*: An abstract of processor operations and resulting state changes used for simulation.
- *The timing model*: Specifies the activation sequences for hardware operations and hardware units.
- *The microarchitecture model*: Groups hardware operations to functional units and describes the microarchitecture implementation of structural blocks such as adders and multipliers.

LISA is now available in a product offered by CoWare Inc.

7.4 Purpose of ADLs

ADLs specify processor and memory architectures for three purposes:

1. Automated processor hardware generation
2. Automated generation of associated software tool suites
3. Processor validation

Two major approaches are used for synthesizable generation of processor HDL descriptions. The first is a parameterized approach: processor cores are based on a single processor template. In the simplest case, this approach produces configurable processors that can be modified through a click-box user interface, allowing the processor's architecture and development tools to be modified to a certain degree. Configuration options often include preconfigured execution units such as floating-point units and DSPs, the widths memory interfaces, inclusion of caches, and local memories. Processor configuration offers a useful but limited way to extend the performance reach of a processor's architecture. The second approach is based on the use of processor specification languages to define ISA extensions such as new instructions, registers, register files, I/O interfaces, etc.

7.5 Processor-Centric ADL Example: The Genesis of TIE

The remaining portion of this chapter describes the Tensilica Instruction Extension (TIE) ADL. TIE is a production-oriented ADL for system designers who want to quickly explore a design space using processor-centric ideas instead of being an ADL for processor designers. This distinction is not a subtle one. Processor users have very different needs than processor designers. Processor users are more concerned with the many aspects of system designs and they care relatively little about the arcane details of the processor design such as processor pipeline balancing, data hazards within a pipeline, or data forwarding. Yet embedded system designers are very concerned with results; they need software- and firmware-programmed processors that perform tasks more efficiently than off-the-shelf, general-purpose processors so they can achieve performance goals while reducing power dissipation and energy consumption.

Tensilica was founded with the idea of creating a configurable processor architecture specifically designed for the needs of embedded SOC designers. By the end of the 1990s, SOC designers were adopting available processor cores from such vendors as ARM and MIPS. Yet the processors from those vendors had not originally been designed as SOC cores. The ARM architecture arose in the early 1980s from a project that developed a then-new 32 bit RISC architecture for a British personal computer manufacturer named Acorn. (ARM originally meant Acorn RISC Machine.) The MIPS architecture grew out of the RISC research performed by Professor John Hennessy's team at Stanford, again during the early 1980s. Both of these architectures were designed to be fast and pipelined machines but not necessarily embedded cores intended for use on SOCs.

Tensilica's goal was to use those RISC concepts that were valuable for embedded SOC design (fast, pipelined, 32 bit architectures with a large address space) and to add features that became very important when placing a processor core onto a chip. The features deemed important especially for such on-chip, embedded applications included a memory-conserving instruction set using 16 and 24 bit instructions instead of the usual 32 bits employed by most RISC architectures.

Another feature deemed important for the embedded SOC market was processor configurability. Some processor features—such as multipliers, floating-point units, and MMUs—consume a lot of silicon and most on-chip applications cannot make use of all these features. So Tensilica's approach was to create a modular or configurable processor architecture that allowed system designers to tailor each instance of Tensilica's processor core to the assigned task(s) so that silicon would be conserved wherever possible.

It quickly became apparent that tailoring should extend to the ability to define new instructions not previously conceived of by the original processor designers. Click-box-configurable user interfaces were not sufficiently flexible to allow such an advanced ability so the notion of using a specialized ADL to permit the creation of new processor instructions and new processor state was born in the form of the TIE ADL.

TIE differs from almost all other ADLs in that it is not designed to allow the creation of an entire processor. That is because TIE assumes the existence of an underlying base processor architecture. In the case of TIE, the underlying architecture is Tensilica's 32 bit Xtensa RISC core, which was designed specifically for embedded SOC applications. A click-box user interface allows some amount of configuration such as the number and types of interrupts and timers, the inclusion of function units (multipliers, multiply-accumulates (MACs), floating-point units, and DSP architectural enhancements), the inclusion and sizes of instruction and data caches, and the number and sizes of local instruction and data memories.

TIE adds another level of configurability by permitting people who are not processor designers to add custom instructions, registers, register files, and specialized interfaces to the base Xtensa processor architecture. It does so by exploiting a key ADL characteristic: designers using TIE add these new architectural elements by describing only their function. Automation then handles the details of how these new elements are built in hardware. Automation also handles the modification of the associated software tools so that a new compiler, an assembler, an ISS, a debugger, and simulation models are created along with the HDL description of the new processor's hardware. All of these architectural components are automatically generated in about an hour once the descriptions have been written. Such speedy automation permits practical DSE by the SOC design team.

7.6 TIE: An ADL for Designing Application-Specific Instruction-Set Extensions

The TIE language describes ISA extensions for Tensilica's Xtensa family of 32 bit RISC processor cores. The language uses a simple and intuitive syntax to describe additional instructions, register files, data types, and interfaces to be added to the base Xtensa processor architecture [15,16]. Instructions described in TIE are not microcoded. They are implemented in hardware, taking a structural form that is very similar to the hardware implementation of the Xtensa processor's base instructions, which are themselves described in TIE that is been written by Tensilica. Further, these additional instructions are supported by the full software tool chain including a C/C++ compiler, an assembler, an ISS, and a debugger.

A tool called the TIE compiler automates the implementation of ISA extensions defined using the TIE language. The TIE compiler reads the ISA-extension descriptions defined by a designer and then generates both the HDL description of the resulting processor hardware and a software tool suite tailored the new customized processor. The process of generating the processor HDL and the software tools takes an hour or so, which makes the process of iterative DSE using these tools practical.

This section describes the TIE language from an embedded SOC designer's viewpoint. Here, TIE is used to extend an Xtensa processor's ISA with application specific instructions, registers, and interfaces. The base Xtensa core, with a typical RISC instruction set, already exists. This is the normal use for TIE by SOC designers. TIE's intent is not to create a new generation of processor designers. TIE's intent is to allow system designers to easily tailor preexisting microarchitectural processor designs for specific on-chip tasks, for the purpose of making these extended architectures more efficient at executing the target tasks. The result is a tailored processor that performs the task in fewer clock cycles, which cuts power dissipation, reduces energy consumption, and may also reduce the required memory footprint. All of these savings have a very positive effect on the SOC's and the system's manufacturing cost.

Note that the TIE language can also describe the Xtensa processor's base ISA. In fact, the Xtensa processor's entire instruction set, pipeline, and exception semantics are defined using the TIE language. Only a few of the processor's functional modules—such as the instruction-fetch, load/store, and bus-interface modules—are designed directly in RTL for gate-level efficiency. TIE is not designed to describe such microarchitectural features because it is optimized for processor users, not for processor designers.

7.6.1 TIE Design Methodology and Tools

ISA extensions accelerate data- and compute-intensive functions—often called "hot spots"—in application code. Programs heavily exercise these hot spots, so their implementation efficiency has a significant impact on the efficiency of the whole program. It is often possible to increase the performance of application code by a factor of 2, 5, or even an order of magnitude by adding a few carefully selected instructions that accelerate certain critical inner loops in the code. Such performance gains do cost additional hardware, but in most cases it is possible to achieve truly significant performance gains for literally a handful of additional gates.

Figure 7.2 is a flowchart showing a TIE-based design methodology for increasing application-code performance by creating application-specific ISA extensions. This flowchart highlights the important phases of application-specific ISA augmentation: identification of new instructions, functional description, verification, and hardware optimization.

TIE development starts with the selection of a base Xtensa processor. This is a key differentiating factor. Because system designers using Xtensa processor cores are expected to be processor users, not processor designers, starting with a validated, operational processor core is a significant project accelerator. The days when a designer should need to describe how to perform a 32 bit add or a left shift are long past and there is little advantage to be gained by going through this exercise again. Starting with a fully operational 32 bit processor core saves a significant amount of project-development time.

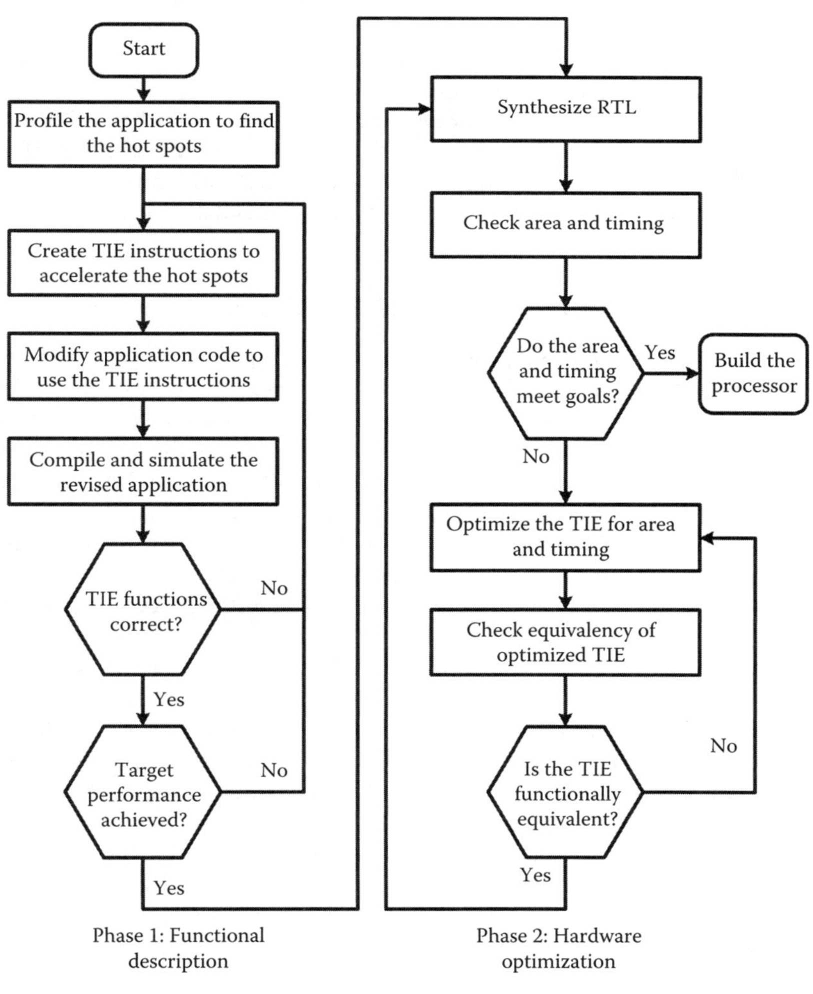

FIGURE 7.2 TIE development flowchart.

As illustrated in Figure 7.2, the application's execution profile is a good way to identify the application's hot spots. Next, the designer maps the behavior of these hot-spot functions to a set of TIE instructions by describing these new instructions in the TIE language. Note that the designer is not describing hardware structure, but only function. The mental focus remains on the task at hand, and not on the implementation details. Automation, in the form of the TIE compiler, handles the implementation details.

The TIE compiler generates a set of software tools that recognizes the new instructions. Software developers modify the HLL application code to replace the original hot-spot functions with TIE intrinsics. The modified application code is then run on the generated ISS to verify that the results from the modified application code match the results from running original code. Once the application is functionally verified to be correct, the revised code is profiled to evaluate the resulting performance. Because the TIE compiler runs very quickly, designers can iterate through this process several times, revising their TIE instructions or creating new ones until they achieve the desired performance. This process should be familiar to anyone who has tuned HLL application code by recoding hot spots using assembly code. The difference here is that hot-spot code sequences can often be replaced with one TIE instruction instead of a sequence of assembly-level RISC instructions.

Once the application-specific TIE instructions have been defined and verified, the next step is to optimize the hardware implementation of these new instructions. The TIE compiler generates the hardware implementation of these instructions as synthesizable Verilog, which can be directly synthesized to obtain the area and timing information.

Although Figure 7.2 shows a process for accelerating computation, data-bandwidth and memory-subsystem performance are often bigger bottlenecks in many systems. The use of TIE permits significant system performance gains through data-bandwidth optimization as well, using data interfaces that bypass the processor's bus and allow direct, high-speed data transfer between processors and other on-chip blocks.

7.6.2   SOC Design Automation with the TIE Compiler

Extending a processor's instruction set goes well beyond writing the RTL code of the new instruction hardware. Integrating new hardware into an existing processor requires in-depth knowledge of the processor pipeline and microarchitecture, to ensure that the hardware works correctly under all conditions such as data hazards, branches, and exceptions. Similarly, the task of modifying all the associated software tools to incorporate these ISA extensions is nontrivial. Yet the ability to modify the software tools quickly is crucial to designer productivity. It enables quick profiling of the ISA extensions' effect on application performance, thus allowing SOC designers to iteratively explore the design space using many different design options in a short amount of time.

The TIE compiler is a processor-synthesis tool that automates the process of tailoring an Xtensa processor with ISA extensions. The TIE developer simply describes the functional behavior of the new instructions, without regard to the processor's microarchitecture. The TIE compiler takes these descriptions, automatically generates the complete HDL hardware implementation of these instructions, and updates all the associated software tools to recognize these extensions. The TIE compiler generates the hardware implementation in synthesizable Verilog RTL, along with synthesis and place-and-route scripts, and test programs to verify the microarchitectural implementation of these instructions. It also updates the software tools so that the assembler can assemble these new instructions and the C/C++ compiler can recognize these new instructions as intrinsics, which allows it to automatically allocate registers properly for the new instructions and to schedule the new instructions efficiently in compiled code. The ISS can simulate these new instructions and the debugger is aware of any new state that has been added to the processor, so that the values of these new states can be examined during program debugging.

7.6.3　Basics of the TIE Language

This section introduces the fundamental concepts of the TIE language starting with a simple example that illustrates a basic TIE instruction. Note that these new instructions are added to a predefined core Xtensa ISA, which includes a 32 bit, 16-entry register file, and a RISC instruction set consisting of load/store, arithmetic logic unit (ALU), shift, and branch instructions.

Consider an example where the application code calculates a dot product from two arrays of unsigned 16 bit data. The product is shifted right by 8 bits before accumulating it into an unsigned, 32 bit accumulator.

```
unsigned int i, acc;
short *sample, *coeff;
for (i=0; i<N; i++) {
   acc += (sample[i] * coeff[i]) >> 8;
}
```

An Xtensa processor with a base ISA and a single-cycle multiplier needs at least three cycles to execute each loop iteration, excluding the respective loads and stores of data and result and updating the memory pointers. A fused TIE instruction can accelerate this computation. The fused operation combines the multiplication, shifting, and accumulation:

```
operation dotprod {inout AR acc, in AR sample, in AR coeff} {} {
assign acc = acc + ((sample[15:0] * coeff[15:0]) >> 8);
}
```

In the above example TIE code, "operation" is a TIE language keyword indicating the definition of a new instruction. The name or mnemonic of the instruction is `dotprod`. This instruction uses three operands in the Xtensa processor's 32 bit *AR* general-purpose register file. The operation reads two input operands, `sample`, and `coeff`, from the register file. It reads and writes the third operand, `acc`.

The TIE instruction `dotprod` can be used in C code as illustrated below:

```
#include <xtensa/tie/dotprod.h>
unsigned int i, acc;
short *sample, *coeff;
for (i=0; i<N; i++) {
   dotprod(acc, sample[i], coeff[i]);
}
```

The header file `dotprod.h` generated by the TIE compiler defines the prototype of the intrinsic `dotprod` as

```
void dotprod(unsigned int *acc, unsigned int sample,
unsigned int coeff);
```

This C code can be compiled by the automatically generated Xtensa C compiler and simulated by the associated ISS. The computation's inner loop shrinks from three cycles to one because of the new TIE instruction. If this inner loop sits in a critical section of an application, the resulting acceleration will significantly improve performance.

7.6.4 Defining a Basic TIE Instruction

The most fundamental TIE construct is "operation." It specifies the name, arguments, and the behavior of a TIE instruction. The syntax of the "operation" construct is

```
operation <name> {<argument list>} {<state interface list>}
          {<datapath description>}
```

The "operation name" is the instruction mnemonic, and its definition is followed by the "argument list," which takes the form of a comma-separated list of the register-file entries and immediate operands used by the instruction. Each argument in this list has three components. The first is a direction identifier, which can be "in," "out," or "inout." This direction identifier specifies whether the instruction reads, writes, or reads/modifies/writes the operand. The second component specifies the operand type—either the name of the register file referenced by this operand or the identifier for an immediate operand. The third component is the operand name, which is used in a similar manner as argument variable names in C functions. These names are used in the operation description. Operands listed in the "argument list" are encoded in the instruction word and appear in the assembly language syntax of the instruction.

The "argument list" is followed by the "state interface list," which lists the implicit instruction operands. Implicit operands are not encoded in identifiable fields in the instruction word and they are not listed in the assembly language syntax of the instruction. They are actually embedded into the instruction.

The "state interface list" is followed by the "datapath description," which is the main body of the "operation" definition. It describes the computation performed by the instruction using a list of one or more assignments.

Immediate instruction operands are defined using the "immediate_range" TIE construct, which defines a range of legal operand values using the following syntax:

```
immediate_range <name> <low value> <high value> <step size>
```

"name" is the identifier used to reference the immediate range in an "operation" definition; "low value" and "high value" are signed integers representing the lowest and highest value the operand can take; and "step size" is the amount by which the successive values in this range increment.

The `dotprod` operation as described above does not explicitly specify the encoding of the instruction. This is an implementation detail that should not ordinarily be of concern to a system designer. The TIE compiler can automatically generate encodings for instructions; this is just another aspect of the automation it provides. It is however possible for the designer to provide the encoding if desired.

For example, the `dotprod` instruction has three operands that reference the Xtensa processor's 16-entry *AR* register file. Each operand therefore requires 4 bits in the instruction word. The remaining bits in the instruction word can be used to specify the instruction's opcode.

The base Xtensa processor has a 24 bit instruction word, referred to by the TIE keyword "Inst." The TIE language allows the designer to create bit fields in this instruction word and to specify operand and opcode definitions using these bit fields. The following example illustrates how the TIE language constructs "field," "operand," and "opcode" can specify an encoding for the `dotprod` instruction.

```
// Define 4-bit bit-fields of Inst
field r Inst[ 3: 0]
field s Inst[ 7: 4]
field t Inst[11: 8]
```

```
// Create operands that reference the AR register file
operand ars s { AR[s] }
operand arr r { AR[r] }
operand art t { AR[t] }

// Specify the opcode encoding for the instruction
field opc Inst[23:12]
opcode dotprod opc=12'h0A2

// Alternative syntax for instruction definition, when using
// explicit operand definitions
iclass dp { dotprod } { inout arr, in ars, in art }
reference dotprod {
   assign arr = arr + ((ars[15:0]*art[15:0]) >> 8);
}
```

In the above description, "field" is a TIE keyword that defines 3 bit fields named r, s, and t. Each of these fields is 4 bits wide, and these fields are used to create the operands arr, ars, and art with the TIE keyword "operand." These operands reference the *AR* register file, and the "operand" syntax indicates that the "field" provides the register file address or index. Together, these three operands consume 12 bits out of the 24 bit instruction word. The remaining 12 bits are used to define another field, opc, and the opcode encoding of the dotprod instruction (using the TIE keyword "opcode"). The value of the opcode field is set to the hex value 0x0A2. The "iclass" definition specifies the instruction's input and output operands and "reference" describes the instruction's behavior, also called the datapath.

The "operation" definition describes the TIE instruction's datapath using a list of assignment statements with the following syntax:

```
assign <name> = <expression>;
```

"name" is the name of an "out" or "inout" operand of the "operation" or an intermediate "wire" that has previously been declared in the computation. ("Wires" are simply intermediate calculation values used to create TIE definitions.) Expressions in the TIE language follow combinational Verilog syntax and can be any expression using TIE operators and the names of "in" or "inout" operands or intermediate "wires." A list of TIE operators appears in Table 7.1. These operators are essentially identical to Verilog-95 HDL and follow the same precedence rules. All operators treat the data as unsigned. Thus the rules governing the computed width of expressions in TIE are the same as Verilog rules for unsigned data types.

TABLE 7.1   TIE Operators

Operator Type	Operator Symbol
Subrange	$a[n{:}m]$, where $n >= m$
Arithmetic	$+, -, {}^{*}$
Logical	!, &&, ||
Relational	>, <, >=, <=, ==,!=
Bitwise	~,&,|,^, ^~, ~^
Reduction	&, |,^, ^~, ~^
Shift	>>, <<
Concatenation	{ }
Replication	{n{ }}
Conditional	?:

7.6.4.1 TIE Wires

For all but the simplest of TIE instructions, it becomes impractical to describe the entire datapath using only the input and output operands of the instruction. Wires are temporary "scratchpad" variables and are assigned the intermediate results of a computation. The syntax for declaring a wire is

```
wire [n:m] <name> [= <expression>];
```

where "[*n:m*]" specifies the wire's width. By default a wire is 1 bit wide if the width is omitted. The assignment to the wire can be specified in a separate assign statement or in the declaration itself. Using wires, the `dotprod` operation can be rewritten as

```
operation dotprod {inout AR acc, in AR sample, in AR coeff} {} {
wire [15:0] s = sample[15:0];
wire [15:0] c = coeff[15:0];
assign acc = acc + ((s * c) >> 8);
}
```

Note that the lexical ordering of the assign statements in an operation's description may not bear any relationship with the actual order in which the statements are executed. All TIE assignments are evaluated for dependencies and computed in dependency order. All input arguments in a statement are read before any output arguments are computed. For an "inout" operand, the input value is read before any computation is performed and the output value written after all computations are performed.

7.6.4.2 Conditional Writes to an Output Operand

The TIE language provides a mechanism to predicate the result of a TIE operation by specifying a condition that can suppress or kill the write to an output operand. For every output operand "<name>," the TIE compiler automatically generates a signal "<name>_kill," which can be assigned in the TIE operation. Assertion of this signal kills the write. An example of a conditional move operation using this feature appears below:

```
operation MOVCOND { out AR d, in AR a } {} {
  assign d = a;
  assign d_kill = (a == 32'b0);
}
```

In the above example, the move is suppressed when the input is zero. Note that although the operand is an output of the move, the Xtensa compiler will be aware of the predication.

7.6.4.3 Built-In TIE Computation Modules

The TIE language contains a comprehensive set of built-in computation modules that can be used in expressions. These modules help specify computations that are otherwise cumbersome to describe. The hardware implementation of these modules has been preoptimized for area and timing. Therefore, using these computation models results in better-optimized hardware built with standard cell libraries. Table 7.2 includes a list of these modules.

TABLE 7.2   TIE Built-In Modules

Format	Description	Result Definition
$\text{TIEadd}(a, b, c_{\text{in}})$	Add with carry-in	$a + b + c_{\text{in}}$
$\text{TIEaddn}(a_0, a_1, \ldots, a_n)$	N number addition	$a_0 + a_1 + \ldots + a_n$
$\text{TIEcmp}(a, b, \text{sign})$	Signed and unsigned comparison	$\{a < b,\ a <= b,\ a == b,\ a >= b,\ a > b\}$
$\text{TIEcsa}(a, b, c)$	Carry save adder	$\{\text{carry, sum}\} = \{a\ \&\ b | a\ \&\ c | b\ \&\ c,\ a^\wedge b^\wedge c\}$
$\text{TIEmac}(a, b, c, \text{sign}, \text{negate})$	MAC, sign specifies how a and b are sign extended	$\text{negate}\ ?\ c - a^*b : c + a^*b$
$\text{TIEmul}(a, b, \text{sign})$	Signed and unsigned multiplication	$\{\{m\{a[n - 1]\ \&\ \text{sign}\}\}, a\}^* \{\{n\{b[m - 1]\ \&\ \text{sign}\}\}, b\}$, where n is the size of a and m is the size of b
$\text{TIEmulpp}(a, b, \text{sign}, \text{negate})$	Partial product multiply	$\{p0, p1\} = \text{result}$ $p0 + p1 = \text{negate}\ ? - (a^*b) : (a^*b)$
$\text{TIEmux}(s, d_0, d_1, \ldots, d_{n-1})$	n-way multiplexer	$s == 0\ ?\ d_0 : s == 1\ ?\ d_1 : \ldots s == n - 2 : d_{n-2} : d_{n-1}$
$\text{TIEpsel}(s_0, d_0, s_1, d_1, \ldots s_{n-1}, d_{n-1})$	n-way priority selector	$s_0\ ?\ d_0 : s_1\ ?\ d_1 : \ldots s_{n-1}\ ?\ d_{n-1} : 0$
$\text{TIEsel}(s_0, d_0, s_1, d_1, \ldots s_{n-1}, d_{n-1})$	n-way 1 hot selector	$(\text{size}\{s_0\}\ \&\ d_0) | (\text{size}\{s_1\}\ \&\ d_1) \ldots | (\text{size}\{s_{n-1}\}\ \&\ d_{n-1})$

7.6.4.4   Adding Processor State

The TIE extensions described so far read and write the Xtensa processor's predefined 32 bit *AR* register file. However, many SOC designs could benefit from operands with a customized data size. For example, a designer might want to use a 24 bit data type for fixed-point audio processing or a 128 bit data type to represent eight 16 bit values for SIMD vector operations. The TIE language provides constructs to add new processor states and register files that can then be used as operands of TIE instructions.

TIE language uses the term "state" to refer to a single-entry register file. TIE states are useful for accelerating certain processing operations by keeping frequently accessed data within the processor instead of memory. TIE states can be of arbitrary width. Once defined, they can be used as an operand of any TIE instruction.

The syntax for defining a TIE state is

```
state <name> <width> [<reset value>] [<add_read_write>] [<export>]
```

"name" is the unique identifier for referencing the state and "width" is the bit-width of the state. Optional parameters of a state definition include a "reset value" that specifies the state's value when the processor comes out of reset. The "add_read_write" keyword directs the TIE compiler to automatically create instructions that transfer data between the TIE state and the *AR* register file. Finally, the "export" keyword specifies that the state's value should be visible outside the processor as a new top-level interface.

Consider a modified version of the `dotprod` instruction in which a 40 bit accumulator is used. The accumulator is too wide to be stored in one 32 bit *AR* register-file entry. The following example illustrates the use of TIE state to create such an instruction. It also uses the built-in module "TIEmac" to perform the multiply and accumulate operation.

```
state ACC 40 add_read_write
operation dotprod{in AR sample, in AR coeff}{inout ACC} {
   assign ACC = TIEmac(sample[15:0], coeff[15:0], ACC, 1'b0, 1'b1);
}
```

In addition to allowing the use of TIE state as an operand for any arbitrary TIE instruction, it is useful to have direct read/write access to TIE state. This mechanism also allows a debugger to view the value of the state, or an OS to save and restore the value of the state across a context switch. The TIE compiler can automatically generate a *RUR* (read user register) instruction that reads the value of the TIE state into an *AR* register and a *WUR* (write user register) instruction that writes the value of an *AR* register to a TIE state. If the state is wider than 32 bits, multiple *RUR* and *WUR* instructions are generated, each with a numbered suffix that begins with 0 for the least significant word of the state.

The automatic generation of these instructions is enabled by the use of the "add_read_write" keyword in the state declaration. In the TIE example above, instructions RUR.ACC_0 and WUR.ACC_0 are generated to read and write bits [31:0] of the 40 bit state ACC, respectively, while instructions RUR.ACC_1 and WUR.ACC_1 are generated to access bits [39:32] of the state.

7.6.4.5 Defining a Register File

The TIE state is well suited for storing single variables but register files are for more general purposes. TIE register files are custom sets of addressable registers used by TIE instructions. While most microprocessors have only one general-purpose register file, many tasks and algorithms can benefit from multiple, custom register files to reduce memory accesses. Further, many algorithms operate on data types that are wider than 32 bits and wedging such algorithms into 32 bit datapaths is both cumbersome and inefficient. Custom register files that match the natural size of the data types are much more efficient. The TIE construct for defining a register file is

```
regfile <name> <width> <depth> <short name>
```

The "width" is the width of each register, while "depth" indicates the number of registers in the register file. The assembler and the debugger use the short name to reference the register file. When a register file is defined, its name can be used as an operand type in a TIE operation. An example of a 64 bit wide, 32-entry register file, and an XOR operation that operates on this register file is shown below:

```
regfile myreg 64 32 mr
operation widexor {out myreg o, in myreg i0, in myreg i1}{} {
   assign o = i0 ^ i1;
}
```

7.6.4.6 Load/Store Operations and Memory Interfaces

Every TIE register file definition is accompanied by automatically generated load, store, and move instructions for that register file. These basic instructions are typically generated automatically by the TIE compiler unless specified by the designer. An example of these instructions for the myreg register file is shown below:

```
immediate_range imm4 0 120 8
operation ld.myreg {out myreg d, in myreg *addr, in imm4 offset}
   {out VAddr, in MemDataIn64} {
   assign VAddr = addr + offset;
   assign d = MemDataIn64;
}
operation st.myreg {in myreg d, in myreg *addr, in imm4 offset}
   {out VAddr, out MemDataOut64} {
   assign VAddr = addr + offset;
   assign MemDataOut64 = d;
}
operation mv.myreg {out myreg b, in myreg b} {} {
   assign b = a;
}
```

TABLE 7.3 Load Store Memory Interface Signals

Name	Width	Direction[a]	Purpose
Vaddr	32	Out	Load/store address
MemDataIn {128,64,32,16,8}	128, 64,32,16,8	In	Load data
MemDataOut {128,64,32,16,8}	128,64,32,16,8	Out	Store data
LoadByteDisable	16	Out	Byte disable signal
StoreByteDisable	16	Out	Byte disable signal

[a] "In" signals go from Xtensa core to TIE logic; "out" signals go from TIE logic to Xtensa core.

The instruction `ld.myreg` performs a 64 bit load from memory with the effective virtual address provided by the pointer operand `addr`, plus an immediate `offset`. The operand `addr` is specified as a pointer. The `*` tells the compiler to expect a pointer to a data type that resides in the register file `myreg`. Because the 64 bit load operation requires the address to be aligned to a 64 bit boundary, the step size of the offset is 4 (bytes). The load/store operations send the virtual address to the load/store unit of the Xtensa processor in the processor pipeline's execution stage and the load data is received from the load/store unit in the pipeline's memory stage. The store data is sent in the memory stage as well. All memory transactions are performed using a set of standard interfaces to the processor's load/store unit(s). A list of these memory interfaces appears in Table 7.3.

The designer can also write custom load/store instructions with a variety of addressing modes. For example, auto-incrementing and auto-decrementing load instructions are useful to efficiently access an array of data values in a loop. Similarly, bit-reversed addressing is useful for DSP applications.

7.6.5 TIE Data Types and Compiler Support

The TIE compiler automatically creates new data types for every TIE register file. Each data type has the same name as the associated register file and can be used as the variable "type" in C/C++ programs. The Xtensa C/C++ compiler understands that variables of certain data types reside in the associated custom register files and the compiler also performs register allocation for these variables. The C/C++ compiler uses the register-file-specific load/store/move instructions described above to save and restore register values when performing register allocation and during a context change for multithreaded applications.

The C programming language does not support constant values wider than 32 bits. Thus initialization of data types wider than 32 bits is done indirectly, as illustrated in the example below for the `myreg` data type generated for the same register file:

```
#define align_by_8 __attribute__ ((aligned)8)
unsigned int data[4] align_by_8 = { 0x0, 0xffff, 0x0, 0xabcd };
myreg i1, *p1, i2, *p2, op;
p1 = (myreg *)&data[0]; p2 = (myreg *)&data[2];
i1 = *p1; i2 = *p2;
op = widexor(i1, i2);
```

In this example, variables `i1` and `i2` are of type `myreg` and are initialized by the pointer assignments to a memory location. The compiler uses the appropriate load/store instructions corresponding to the associated register file when initializing variables. Note that data values should be aligned to an 8 byte boundary in memory for the 64 bit load/store operations to function correctly, as specified by the attribute pragma in the code.

7.6.6 Multiple TIE Data Types

The TIE language provides constructs to define multiple data types that reside in a single register file and to perform type conversions between these various data types. The `myreg` register file described above holds a 40 bit data type and can also be configured to hold a 64 bit type using the "ctype" TIE construct as illustrated below:

```
ctype myreg 64 64 myreg default
ctype my40 40 64 myreg
```

The syntax of the "ctype" declaration provides the data width and memory alignment and specifies the register file it resides in. In the above description, the second data type `my40` has a width of 40 bits. Both data types are aligned to a 64 bit boundary in memory. The keyword "default" in a "ctype" declaration indicates the default data type to be used by any instruction that references the register file unless otherwise specified.

The Xtensa C/C++ compiler requires special load/store/move instructions that correspond to each "ctype" of a register file. The TIE language's "proto" construct tells the Xtensa C/C++ compiler which load/store/move instruction corresponds to each "ctype" as illustrated below:

```
proto loadi_myreg  {out myreg d, in myreg *p, in immediate o} {}
   {ld.myreg(d,p,o);}
proto storei_myreg {in myreg d, in myreg *p, in immediate o} {}
   {st.myreg(d,p,o);}
proto move_myreg   {out myreg d, in myreg a} {} {mv.myreg(d,a);}

proto loadi_my40   {out my40 d, in my40 *p, in immediate o} {}
   {ld.my40(d,p,o);}
proto storei_my40  {in my40 d, in my40 *p, in immediate o} {}
   {st.my40(d,p,o);}
proto move_my40    {out my40 d, in my40 a} {} {mv.my40(d,a);}
```

The "proto" construct uses stylized names of the form "loadi_<ctype>" and "storei_<ctype>" to define the instruction sequence for loading from and storing variables of type "<ctype>" to memory. The proto "move_<ctype>" defines a register-to-register move. The load/store instructions define the "proto" for the "ctype" `myreg`. Similar instructions for the 40 bit type `my40` can be defined using only the lower 40 bits of the memory interfaces "MemDataIn64" and "MemDataOut64." In some cases, the "proto" may need multiple instructions to perform operations such as loading a register file whose width is greater than the Xtensa processor's maximum allowable data-memory width, which is 128 bits.

The "proto" construct can also be used to specify type conversion from one "ctype" to another. For example, conversion from the fixed-point data type `my40` to `myreg` involves sign extension, while the reverse conversion involves truncation with saturation as shown below:

```
operation mr40to64 {out myreg o, in myreg i} {} {
   assign o = {{24{i[39]}}, i[39:0]};
}
operation mr64to40 {out myreg o, in myreg i} {} {
   assign o = (i[63:40] == {24{i[39]}}) ? i[39:0] :
   {i[63], {39{~i[63]}}}};
}
proto my40_rtor_myreg {out myreg o, in my40 i} {} {
   mr40to64 o, i; }
```

```
proto myreg_rtor_my40 {out my40 o, in myreg i} {} {
   mr64to40 o, i; }
```

The "proto" definition follows a stylized name of the type "<ctype1>_rtor_<ctype2>" and gives the instruction sequence required to convert a `ctype1` variable into a `ctype2` variable. The C/C++ compiler uses these "protos" when it assigns a variable of one data type to another.

The C intrinsic for all operations referencing the register file `myreg` will automatically use the default 64 bit "ctype" `myreg` because it is the default "ctype." If an operation uses the 40 bit data type `my40`, this can be specified by writing a proto as shown below:

```
operation add40 { out myreg o, in myreg i1, in myreg i2 } {
   assign o = { 24'h0, TIEadd(i1[39:0], i2[39:0], 1'b1) };
}
proto add40 { out my40 o, in my40 d1, in my40 d2}{} {
   add40 o, d1, d2; }
```

The "proto" `add40` specifies that the intrinsic for the operation `add40` uses the `my40` data type.

7.6.7 Data Parallelism, SIMD, and Performance Acceleration

The ability to use custom register files allows the designer to create new machines targeted for a wide variety of data-processing tasks. For example, the TIE language has been used to create a set of floating-point extensions for the Xtensa processor core. Many DSP algorithms that demand a high performance share common characteristics—in other words, the same sequence of operations is performed repetitively on a stream of data operands. Applications such as audio processing, video compression, and error correction and coding fit this computation model. These algorithms can see large performance benefits from the use of single-instruction, multiple-data (SIMD) processing, which is easy to design with custom instructions and register files.

The example below computes the average of two arrays. In one loop iteration, two short values are added and the result shifted by 1 bit, which requires two Xtensa instructions as well as load/store instructions:

```
unsigned short *a, *b, *c;
for ( i=0; i<N; i++ ) {
   c[i] = (a[i] + b[i]) » 1;
}
```

As discussed previously, creating a TIE fusion operation to combine the add and shift computations will accelerate the performance of this loop. However, the same computation is performed in every iteration of the loop so this code is ideally suited for a SIMD or vector instruction. A SIMD instruction that performs four averages in parallel can be written as shown below:

```
regfile VEC 64 8 v
operation VAVERAGE {out VEC res, in VEC input0, in VEC input1} {} {
   wire [16:0] res0 = input0[15: 0] + input1[15: 0];
   wire [16:0] res1 = input0[31:16] + input1[31:16];
   wire [16:0] res2 = input0[47:32] + input1[47:32];
   wire [16:0] res3 = input0[63:48] + input1[63:48];
   assign res = {res3[16:1], res2[16:1], res1[16:1], res0[16:1]};
}
```

Compared to a scalar fusion of the add and shift instructions, which requires one 16 bit adder, the SIMD averaging instruction requires four 16 bit adders. Also, a new 64 bit register file is needed because the SIMD vector data fill 64 bits (four 16 bit vectors).

7.6.8 TIE and VLIW Machine Design

VLIW processors are well known for speeding application performance by exploiting available instruction-level parallelism. The TIE language supports this same approach—combining multiple independent operations into one long instruction word so that the operations can be executed simultaneously—using a similar approach called flexible-length instruction extension (FLIX).

Although the base Xtensa processor has a native 24 bit instruction word, the TIE language can create 32 and 64 bit instructions as well. Instructions of different sizes can be freely and modelessly intermixed in Xtensa machine code because the processor identifies, decodes, and executes instruction words of any size in the instruction stream. The 32 and 64 bit instructions can be divided into multiple instruction issue slots, with independent operations placed in each slot. A VLIW instruction consists of any combination of operations allowed in each issue slot. The Xtensa C/C++ compiler recognizes the parallelism available through the definition of the VLIW formats and automatically reorders and bundles the operations into FLIX instructions to accelerate the compiled code.

7.6.8.1 Defining a TIE FLIX/VLIW Instruction

Defining a FLIX/VLIW instruction requires two TIE constructs. The first construct is "format," which defines the instruction length and the number of issue (operation) slots in the instruction. The second construct is "slot_opcodes," which defines all the possible operations that can be used in each slot of a FLIX-format instruction. The syntax for these constructs appears below:

```
format <name> <length> <list of slots>
slot_opcodes <slot_name> <list of operations>
```

"name" is a unique identifier for the "format." "length" defines the instruction length (either 32 or 64 bits). The slot list contains slot names in the order they appear in the instruction word. For each slot name defined in the slot list, there must be a corresponding "slot_opcodes" statement that declares the list of operations available in that slot. This list can include any of the base Xtensa ISA instructions and any designer-defined TIE instructions.

The size of each slot is automatically determined by the size of the operations defined in that slot. Slots within a FLIX-format instruction can be of different sizes. Further, the TIE compiler automatically determines the encoding of each operation in each FLIX slot. The designer can also explicitly specify the instruction bits corresponding to each slot and the encoding of the instructions within each slot.

Consider the inner loop in the SIMD application code discussed above. An Xtensa processor with just the base ISA uses the 32 bit add (ADD) and shift-right (SRAI) instructions to compute the average, two 16 bit load (L16I) instructions and one 16 bit store (S16I) instruction to load the input data and store the result, and three add immediate (ADDI) instructions to update the memory pointers. Using the processor's base ISA, the execution of this inner loop takes eight cycles. Four lines of TIE code can create a simple three-issue FLIX/VLIW processor to accelerate this code:

```
format vliw3 64 {slot0, slot1, slot2}
slot_opcodes slot0 {L16I, S16I}
slot_opcodes slot1 { ADDI }
slot_opcodes slot2 {ADD, SRAI}
```

Each 64 bit instruction of the `vliw3` format consists of three issue slots. Each issue slot can issue any instructions from its "slot_opcodes" list. `slot0` can issue a 16 bit load or store instruction, `slot1` can issue the "add-with-immediate" instruction, and `slot2` can issue either a 32 bit add or a "shift-right-arithmetic" instruction. All the operations in this example are from the base Xtensa ISA, so the description of these operations is not required in the TIE code and the resulting description of a VLIW processor is surprisingly small. With this simple description, the inner loop of this code reduces to three instructions.

7.6.8.2 Hardware Cost of FLIX/VLIW Instructions

One very important aspect of VLIW design is estimating and optimizing the hardware cost. To replicate operations in multiple slots, multiple copies of the hardware execution units must be instantiated, which substantially increases a gate count. Hardware for each slot is grouped by the slot index, where the index refers to the slot position in a format, starting at 0 for the least significant bit. Slots with the same slot index share the execution units. The single 24 bit slot "Inst" of the base processor has an index 0. This is an important consideration for creating VLIW instructions with base processor operations.

For the FLIX/VLIW format described above, `slot0`, `slot1`, and `slot2` have indices of 0, 1, and 2, respectively. The load/store operations in `slot0` therefore have no incremental hardware cost because they share the base processor's existing hardware for the base ISA's load/store operations. The `ADD` and `ADDI` instructions in `slot1` and `slot2` require the replication of hardware for the additional arithmetic units. (Note that load/store instructions can only be replicated if the base processor is configured with multiple load/store units.)

The general-purpose and custom register files' size and complexity increase with the addition of FLIX/VLIW instructions because the input and output operands of an operation are automatically mapped to a register file's read and write ports. Consequently, when an operation is replicated in multiple FLIX slots, the associated register-file read and write ports must also be replicated, which causes the register-file hardware to grow in complexity and size.

7.6.9 TIE Language Constructs for Efficient Hardware Implementation

Designers of TIE extensions must trade off improved application performance and hardware cost. The TIE language includes several constructs to improve hardware utilization and thus minimize the hardware cost of TIE extensions.

Many situations permit hardware execution units to be shared between two or more instructions. For example, the same adder can be used to compute the sum and difference of two operands. Multiply-add and multiply-subtract instructions can share a multiplier. The TIE language includes a construct called "semantic" that describes the hardware implementation of one or more previously defined instructions. The following example illustrates a "semantic" that implements 32 bit add and subtract instructions using a shared adder:

```
operation ADD {out AR res, in AR in0, in AR in1} {
  assign res = in0 + in1;
}
operation SUB {out AR res, in AR in0, in AR in1} {
  assign res = in0 - in1;
}

semantic addsub {ADD, SUB} {
  wire carry = SUB;
  wire [31:0] tmp = ADD ? in1 : ~in1;
  assign res = TIEadd(in0, tmp, carry);
}
```

The semantic `addsub` implements the instructions ADD and SUB. It uses the simple principle of 2's complement arithmetic that $(a - b)$ is equal to $(a + {\sim}b + 1)$. For every instruction implemented in a TIE semantic, a one-hot signal with the same name as the instruction is available in the semantic. This signal is true when the corresponding instruction is being executed and false otherwise. In the above example, the one-hot signal for the subtract instruction is used to set the value of the 1 bit "carry" and the one-hot signal for the add instruction is used to select between the original value and the bit inverted value of the second operand.

In the absence of the `addsub` semantic, the ADD and SUB instructions require separate hardware adders. The `addsub` semantic implements both instructions using a single 32 bit adder with carry and a 32 bit multiplexer.

7.6.10   TIE Functions

TIE functions resemble Verilog functions. They encapsulate a computation that can then be instantiated in other TIE descriptions. TIE functions can be used to modularize TIE descriptions and to optionally share hardware between multiple instantiations of a function. A TIE function has the following syntax:

```
function [n:m] <name> ([input_list]) [<shared/slot_shared>]
{<function statements>}
```

A function's "input_list" lists the function's input arguments. Each function has a single output with the same name as the function. The width of this output is "$[n{:}m]$" bits. An optional argument declares the function to be either "shared" or "slot_shared." A shared function has exactly one instance of the required hardware among all the instructions that use it. A "slot_shared" function has one hardware instance per FLIX/VLIW slot index and all the instructions within a slot index share this hardware but this hardware is not shared with instructions in other slots. If neither of these attributes is specified, the function is not shared and the function's purpose is simply to encapsulate the computation for modular code development.

The following example demonstrates the use of shared functions for the ADD and SUB instructions discussed above:

```
function [31:0] addsub([31:0]in0, [31:0]in1, cin) shared {
   assign addsub = TIEadd(in0, in1, cin);
}
operation ADD {out AR res, in AR input0, in AR input1 } {
   assign res = addsub(input0, input1, 1'b0);
}
operation SUB {out AR res, in AR input0, in AR input1 } {
     assign res = addsub(input0, ~input1, 1'b1);
}
```

Instead of writing a "semantic" for the two instructions to share the 32 bit adder, a shared function encapsulates the adder. The TIE compiler automatically generates the control logic required to multiplex the inputs from the operations that share the hardware into the single copy of the hardware for function `addsub`. The TIE compiler also resolves resource hazards between instructions using the same function.

7.6.11 Defining Multicycle TIE Instructions

The most basic Xtensa processor has a five-stage pipeline. The pipeline stages are instruction fetch (I), instruction decode and register read (R), execution (E), memory access (M), and write back (W). True to the RISC form, base Xtensa instruction computations execute in one clock cycle. By default, all TIE operations also perform the computation phase in one clock cycle, during the execution (E) stage of the pipeline. Output operands are written to a register file at the end of the execution cycle in which the input operands are read.

Depending on the nature of the computation associated with a TIE instruction and the target frequency of the processor, it may not be possible to finish a computation in one clock cycle. For example, a MAC hardware block may not be able to produce a result in the allotted time at a high target clock rate. Such a computation should be distributed over multiple clock cycles to allow the processor to sustain the high target clock rate.

The TIE language provides the "schedule" construct to define multicycle instructions and, more generally, to allow the designer more control over the pipeline stage in which input operands are read and output operands are computed.

The syntax of the TIE "schedule" construct is as shown below:

```
schedule <name> <instruction list> {
use <operand> <stage>; . . . use <operand> <stage>;
def <operand> <stage>; . . . def <operand> <stage>;
}
```

The "instruction list" is a list of all instructions to which the stated schedule applies. The "use" schedule defines the pipeline stage at the beginning of which the input operand is available to the execution unit. The "def" schedule defines the pipeline stage at the end of which the output operand is expected to be available from the execution unit. A "use" schedule can be applied to one or more input operands of an instruction, while a "def" schedule can be applied to output operands as well as the intermediate wires of a computation. The stage parameter of the schedule is defined in terms of the processor's pipeline stages, such as "Estage," "Mstage," "Wstage," and so on.

The following example illustrates the use of the schedule construct to define a multicycle MUL32 TIE instruction:

```
state PROD 64 add_read_write
operation MUL32 {in AR m0, in AR m1} {out PROD} {
  assign PROD = m0 * m1;
}
schedule mul {MUL32} {
use m0 Estage; use m1 Estage; def PROD Mstage;
}
```

The MUL32 instruction multiplies two 32 bit numbers to generate a 64 bit product. The schedule construct specifies that the input operands are read in the "Estage" and the result is available in the "Mstage," spreading the multiply computation over two clock cycles.

Multicycle TIE instructions are always implemented with hardware pipelining so that the processor can issue a new instruction every clock cycle even though the instruction takes more than one cycle to complete. The increased instruction latency does not affect performance unless the computation result is immediately required by the following instruction.

The TIE compiler automatically generates the control logic for detecting data hazards in the processor pipeline. This control logic inserts pipeline bubbles by stalling subsequent instructions in the pipeline, which prevents data hazards from occurring. The TIE compiler also creates data-forwarding logic so that an instruction that is stalled in the pipeline awaiting a result can proceed as soon as the earlier instruction completes its computation, without waiting for that result to be written to the register file. The Xtensa C/C++ compiler is aware of the TIE instruction schedules and performs intelligent instruction scheduling to avoid data-hazard stalls in hardware as far as possible.

7.6.12 Iterative Use of Shared TIE Hardware

The TIE language allows designers to create iterative instructions, which are instructions that use the same piece of hardware multiple times but in different clock cycles. This sort of time multiplexing can be viewed as a way of implementing nonpipelined, multicycle instructions. Reduced hardware cost is the primary advantage of this implementation approach. The disadvantage of this approach is that the processor cannot issue the same iterative instruction every cycle because the required hardware is busy.

The following example illustrates a vector dot product computation implemented using a single multiplier:

```
function [31:0] mul16 ([15:0] a, [15:0] b, signed) shared {
  assign mul16 = TIEmul(a,b,signed);
}
operation VDOTPROD {out AR acc, in AR m0, in AR m1} {
  wire [31:0] prod0 = mul16(m0[15: 0], m1[15: 0], 1'b1);
  wire [31:0] prod1 = mul16(m0[31:16], m1[31:16], 1'b1);
  assign acc = prod0 + prod1;
}
schedule dotprod {VDOTPROD} { def acc Estage + 2; }
```

In the above example, the VDOTPROD instruction performs two 16×16 multiplication operations using the shared function mul16. Because there is only one copy of the mul16 execution hardware, this instruction requires at least two cycles to execute, as specified by the schedule.

7.6.13 Creating Custom Data Interfaces with TIE

Over the last several years, processors have become significantly faster and more powerful but memory subsystems have not kept pace. Data bandwidth in and out of the processor has become the bottleneck in many systems [17]. For such systems, reducing the latency and increasing the bandwidth of the connection between the processor and the rest of the system is the key to improving system performance. Modern processors communicate with system memory and other peripheral devices through a main bus that requires several cycles to perform one data transfer. The bus becomes a bottleneck in systems where large amounts of data must be transferred between the processor and external devices very quickly.

The TIE language provides mechanisms to create new types of interfaces between the Xtensa processor and external devices in addition to the processor's main bus. These interfaces allow direct, point-to-point connections between the Xtensa processor and other SOC blocks, which can include other Xtensa processors. A very large number of such interfaces can be defined and they can be of arbitrary bit-width up to 1024 bits. This scheme permits orders of magnitude improvement in the processor's I/O bandwidth without increasing the clock rate. These new interfaces allow high-bandwidth, low-latency communication between the processor and external devices without using the processor's memory subsystem or main bus.

7.6.13.1 TIE Export State and Import Wire

The TIE export state and import wire constructs define status and control interfaces between an Xtensa processor and external devices. The value of any TIE state inside of the processor can be made visible outside the processor via a top-level port using an exported state. A state is exported by adding a keyword "export" to the state declaration. The architectural copy of an exported state is available as a primary output port of the Xtensa processor. Import wires allow TIE instructions to read inputs from the Xtensa processor's top-level ports. External devices drive this port to generate the data value held by the declared processor state. The export state of one processor can be connected to the import wire of another processor.

An import wire is defined in the TIE language as

```
import_wire <name> <width>
```

Import wires can be of arbitrary width (up to 1024 bits) and can be used as an input operand for any TIE instruction. The following example illustrates a 20 bit value being read from an import wire, sign-extended to 32 bits, and then the result being assigned to an *AR* register:

```
import_wire IMP_WIRE 20
operation READ_WIRE {out AR reg} {in IMP_WIRE} {
  assign reg = {{12{IMP_WIRE[19]}}, IMP_WIRE};
}
```

The processor samples the value of `IMP_WIRE` when the `READ_WIRE` instruction executes. This sampling occurs before the instruction commits. The read of the import wire interface is thus speculative in nature. The data sampled may not be consumed if the instruction does not commit due to a branch, exception, or interrupt. The speculative nature of executed instructions is an inherent property of all pipelined processors, which spread the various portions of instruction execution across multiple pipeline stages and clock cycles. The read must be free of any side effect (the value must not be affected by the fact that the processor reads the value) because an import wire has no handshake with the external device. Because of this property, import wires are suited for passing status or control information between devices but are less useful for data transfers.

7.6.13.2 TIE Queue Interfaces

Designers often use dual-ported FIFO buffers to create a point-to-point buffered data channel between two on-chip blocks. This sort of hardware design fits the common producer–consumer scenario quite well. In conventional processor-based systems, the FIFO channels are often implemented using a shared, bus-attached, single-ported memory to save cost [18]. However, this implementation approach suffers from data-bandwidth and data-latency limitations imposed by the shared bus and memory. This approach is also burdened with synchronization overhead. Consequently, this approach suffers from substantial inefficiency that is unwarranted in on-chip SOC system designs.

The TIE language provides the "queue" construct to add one or more FIFO-type interfaces to the Xtensa processor. These interfaces employ a handshake mechanism that directly connects to a two-ported FIFO queue. The interfaces can each be as wide as 1024 bits and designers can add hundreds of such interfaces to an Xtensa processor. Queue-oriented TIE instructions read data from an input FIFO or write data to an output FIFO using the "queue" interface. Synchronization is achieved automatically through the queue interface's handshake mechanism, which stalls the processor if it tries to write to a full output FIFO or read from an empty input FIFO. Thus TIE queues provide a natural and efficient way to map data-flow applications to Xtensa processor-based systems.

The syntax for defining a queue interface is

```
queue <name> <width> <in|out>
```

The width of the "queue" interface can be arbitrary (up to 1024 bits). The keyword "in" denotes an input queue interface from which data can be read while the keyword "out" denotes an output queue interface to which data can be written. Once a TIE queue is defined, it can be used as an operand in any TIE instruction. This is illustrated in the following example:

```
queue IPQ0 40 in
queue IPQ1 40 in
queue OPQ 40 out
operation QADD {} {in IPQ0, in IPQ1, out OPQ} {
   assign OPQ = IPQ0 + IPQ1;
}
```

The above example defines two 40 bit input queues and one 40 bit output queue. The instruction QADD reads from the two input queues, computes the sum of their data values, and writes the result to the output queue. The QADD instruction stalls if either of the input queues is empty, or the output queue is full. No further synchronization in software is needed.

For each queue defined in the TIE description, the TIE compiler generates the port interfaces necessary to connect the Xtensa processor to an external synchronous FIFO as illustrated in Figure 7.3. Hardware to read and write the FIFO using the pop/push interface and to stall an empty/full FIFO is also generated automatically.

When an instruction reads data from an input queue, the queue-interface hardware pops data from the external queue before the instruction commits. The read is thus speculative (i.e., the data popped from the external queue may not be consumed if the instruction does not commit due to an interrupt or exception). Because FIFOs do not tolerate speculation (if you remove a word from a FIFO, that word is gone and the FIFO cannot regenerate it), the processor's input-queue interface buffers this data internally so that the retried data is available when the instruction is re-executed and the external FIFO queue is not popped the second time erroneously. All RISC processors execute

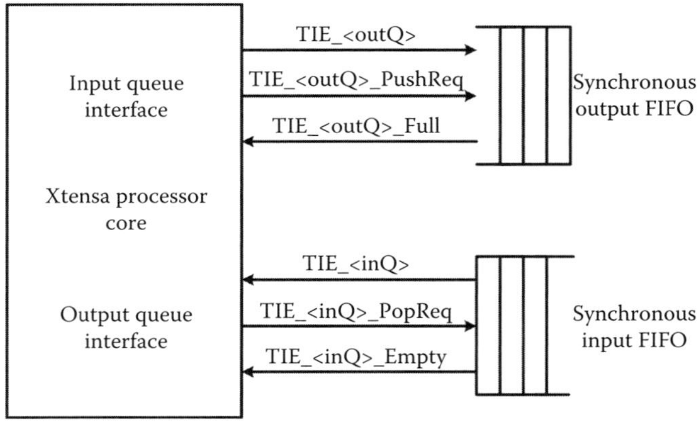

FIGURE 7.3   Top-level pins for TIE queue interfaces.

instructions speculatively and the Xtensa/TIE FIFO interface mechanism is an elegant way of hiding the processor's I/O speculation at very low cost in hardware.

Similarly, an output queue interface only writes to an output FIFO when the instruction performing the write commits. This sequence is achieved by buffering the write data inside the processor's output-FIFO interface hardware until the instruction commits. Thus writes to output queues are not speculative from the perspective of the attached FIFO.

7.6.13.3 Nonblocking Queue Access

The operation QADD discussed above stalls the processor if either of the input queues is empty or if the output queue is full. In some cases, it is more efficient if the software can check the status of a queue before issuing the instruction, preventing a stall. The TIE compiler automatically generates a signal for each queue called "<queue_name>_NOTRDY." When asserted, this signal indicates that the next access to the associated queue interface will stall. The signal is based on the state of the queue interface's internal buffer as well as the state of the attached, external FIFO queue. This signal can be read as an input operand of a TIE operation, which permits the creation of "test-and-branch" instruction sequences that guarantee successful I/O transactions when the processor finally decides to perform them.

The TIE compiler also creates an interface signal called "<queue_name>_KILL" for each TIE queue. This signal can serve as an output operand of a TIE instruction that reads or writes the queue. If this signal is asserted, the queue access is cancelled or killed. If the processor is stalled due to a full output queue or an empty input queue, asserting this signal releases the stall and the instruction completes. The "KILL" signal can be assigned based on a run-time condition, and used to programmatically access or not access a certain queue interface.

Using the two interfaces "NOTRDY" and "KILL," a nonblocking version of the QADD instruction discussed above can be defined as

```
operation QADD_NOBLOCK {} {in IPQ0, in IPQ1, out OPQ, out BR b} {
   wire kill = IPQ0_NOTRDY | IPQ1_NOTRDY | OPQ_NOTRDY;
   assign OPQ = IPQ0 + IPQ1;
   assign IPQ0_KILL = kill; assign IPQ1_KILL = kill;
   assign OPQ_KILL = kill; assign b = ~kill;
}
```

If either of the input queues is empty or if the output queue is full, the instruction QADD_NOBLOCK does not stall. Instead, the read and write operations are cancelled and the instruction completes after setting the status bit to 0, indicating an unsuccessful operation.

7.6.13.4 TIE Lookup Interface

A TIE lookup interface is designed to send a request or an address to an external device. The device then returns a response or data one or more cycles later. Designers can use the TIE lookup interface for a variety of tasks such as interfaces to external lookup tables stored in ROM or to set up a request–response connection with hardware blocks external to the processor. The external device can stall the processor if it is not ready, thus making arbitration between processors and other devices possible.

The lookup request can be issued at any pipeline stage at or beyond the execution stage. The response can be received a cycle or more later. The syntax for defining a lookup is

```
lookup <name> {<out width>, <out stage>}
   {<in width>, <in stage>} [rdy]
```

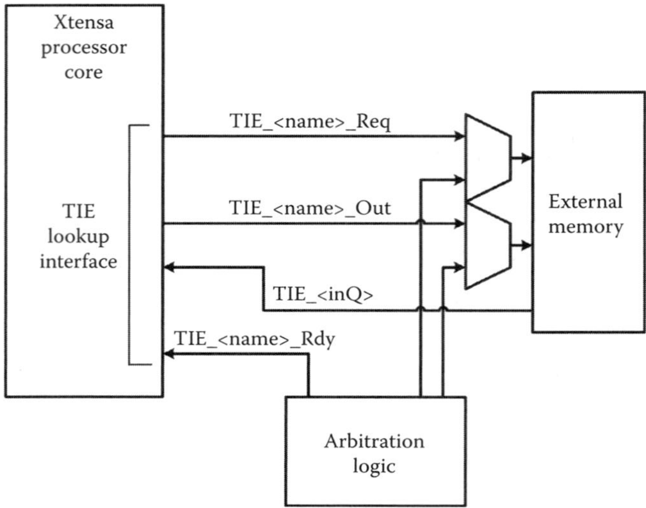

FIGURE 7.4   Top-level connections for a TIE lookup interface.

The width and pipeline stage in which the address/request is sent out are specified using the parameters "out width" and "out stage," respectively. The width and pipeline stage in which the data/response is read back are specified using the parameters "in width" and "in stage," respectively. Thus the latency of the external memory or device is ("<in stage>-<out stage>"). An optional argument "rdy" in the lookup definition indicates that the device should be able to stall the processor when it is not ready to accept a request. Note that once the device has accepted a request, the response must be guaranteed after the specified latency.

The TIE compiler automatically generates the hardware for the request–response protocol of a TIE lookup interface. Figure 7.4 shows the top-level ports generated for a TIE lookup interface. "TIE_<name>_Out" is the output or address of the lookup and can be connected to the address input of the memory. "TIE_<name>_In" is the input data for the lookup and can be connected to the data port of the memory. The processor asserts the 1 bit output port "TIE_<name>_Req" for one clock cycle when making a TIE lookup request. If the TIE lookup interface is configured with "rdy," the interface will have a signal called "TIE_<name>_Rdy," which is a 1 bit input that an external device can use to signal that it is not able to accept requests. If this input signal indicates that the external device is not ready, the processor's execution pipeline will stall and the TIE lookup interface port will present the request until the external device accepts it.

Once a TIE Lookup is defined, it can be used as an operand in any TIE instruction. The following example illustrates the use of a TIE lookup to access a coefficient table that is external to the processor:

```
lookup coeff {8, Estage} {256, Mstage}
state ROM_DATA 256
operation ROM_LKUP {in AR index}
                   {out coeff_Out, in coeff_In, out ROM_DATA } {
   assign coeff_Out = index;
   assign ROM_DATA = coeff_In;
}
```

7.6.14　Hardware Verification Using TIE

The TIE language allows the designer to describe instruction extensions at a high abstraction level. The designer typically specifies only the operands of the instruction and the combinational relationship between the output and the input. The designer can optionally specify the instruction encoding and can specify the pipeline stages in which input operands are read and output operands are written. The TIE compiler generates the RTL for implementing the datapath and the control logic for integrating the new instructions into the Xtensa processor's pipeline. This, then, is the higher level of abstraction: the designer specifies function and behavior but not hardware structure. Structural details are automatically generated.

The designer is responsible for verifying that the combinational relationship between the input and the output operands of the instruction is correct. This verification can be done by writing self-checking test programs for each TIE instruction. The general format of such test programs is to initialize the input operands of the instructions to a set of values, execute the instruction, and then ensure that the output value matches the expected value of the result. This checking should be performed over a number of test vectors that cover all the corner case values and several typical values. Simulation of the application code using the TIE instrinsics is also part of this verification.

The control logic that integrates the new instructions into the Xtensa processor is automatically generated by the TIE compiler. This hardware includes instruction-decode logic, pipelining controls such as data-hazard prevention and data forwarding, and appropriate interfaces to the processor's instruction-fetch and load/store units. The designer is not responsible for verifying these aspects of the instruction implementation because this logic is generated using "correct-by-construction" techniques that have been verified using literally trillions of simulations over a period of many years. Microarchitectural test programs that verify the control logic are automatically generated by the TIE compiler and associated tools.

7.6.14.1　Microarchitectural TIE Verification

A rich set of microarchitectural diagnostics is automatically generated for all TIE instructions to test the control logic of the TIE extensions. These diagnostics have no knowledge of the instruction datapath. They test specific microarchitectural features. For example, diagnostics are generated to test pipeline stalls caused by data dependencies between TIE operations. These tests depend entirely on the read and write stages of the operations described in TIE. The algorithm that generates an instruction sequence for such a diagnostic is given below:

```
foreach reg in ( designer defined register_files ) {
   foreach wstage in ( write/def stages of reg ) {
      foreach rstage in ( read/use stages of reg ) {
            1) Find instruction I1 that writes reg in wstage
            2) Find instruction I2 that reads reg in rstage
            3) Allocate registers or select operands of I1 and I2
               with the constraint to generate data dependency,
               i.e. read operand of I2 = write operand of I1
            4) Calculate stall cycles = (rstgae - wstage) if
               (rstage > wstage)
            5) Generate instruction sequence as
               a. Read cycle count register into AR register a1
               b. Print instruction I1
               c. Print instruction I2
               d. Read cycle count register into AR register a2
               e. Execution cycles = a2 - a1
```

```
        f. Compare execution cycles with expected value
           and generate error if not correct.
    }
  }
}
```

The algorithm described above can generate a self-checking diagnostic test program to check that the control logic appropriately handles read-after-write data hazards. This methodology can be used to generate an exhaustive set of diagnostics to verify specific TIE extension characteristics [19].

7.7 Case Study: Designing an Audio DSP Using an ADL

Tensilica, its customers, and research institutions have used the TIE language to design several complex extensions to the Xtensa processor core [20,21]. This section presents a case study of an audio DSP called the HiFi2 Audio Engine, which Tensilica designed using the TIE ADL. The HiFi2 Audio Engine is a programmable, 24 bit, fixed-point audio DSP designed to run a wide variety of present and future digital audio codecs.

7.7.1 HiFi2 Audio Engine Architecture and ISA

The HiFi2 Audio Engine is a VLIW-SIMD design that exploits both the instruction and the data parallelism that is typically found in audio-processing applications. In addition to the native 16 and 24 bit instruction formats of the Xtensa processor, the HiFi2 Audio Engine supports a 64 bit, two-slot VLIW instruction set defined using the TIE language's FLIX features. While most of the HiFi2 Audio Engine instructions are available in the VLIW instruction format, all the instructions in the first slot of the VLIW format are also available in the 24 bit format, which results in better code density.

Figure 7.5 shows the datapath of the HiFi2 design. In addition to the Xtensa core's general-purpose *AR* register file, the HiFi2 Audio Engine has two additional register files called *P* and *Q*. The *P* register

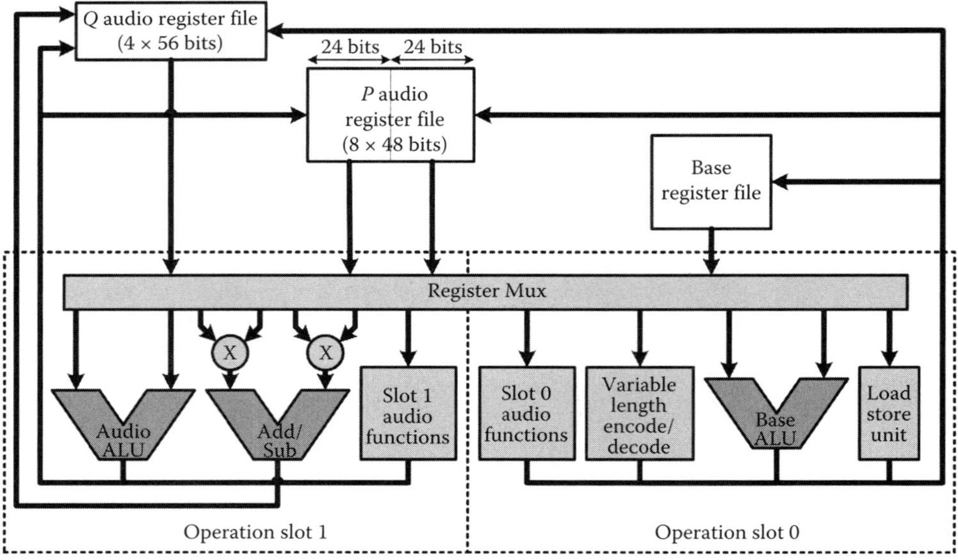

FIGURE 7.5 HiFi2 Audio Engine datapath.

TABLE 7.4 HiFi2 ISA Summary

Operation Type	Slot	Count	Description
Load/store	0	28	Load with sign extension, store with saturation to/from the P and Q register files. SIMD and scalar loads supported for P register file. Addressing modes include immediate and index offset, with or without update.
Bit manipulation	0	17	Bit-extraction and variable-length encode/decode operations.
Multiply and MAC	1	175	24×24 to 56 bit signed single and dual MAC. 24×24 to 56 bit signed single MAC with saturation. 32×16 to 56 bit single and dual MAC. 16×16 to 32 bit signed single MAC with saturation. Different saturation and accumulation modes.
Arithmetic	1	22	Add, subtract, negate, and absolute value on P (element wise) and Q registers, with and without saturation. Minimum and maximum value computations on P and Q registers.
Logic	1	8	Bitwise AND, NAND, OR, XOR on P and Q registers.
Shift	1/0	20	Arithmetic and logical right shift on P (element wise) and Q registers. Left shift with and without saturation. Immediate or variable shift amount (using special shift amount register).
Miscellaneous	1/0	42	Rounding, normalization, truncation, saturation, conditional and unconditional moves.

file is an eight-entry, 48 bit register file. Each 48 bit entry in the P register file can be operated upon as two 24 bit values (stereo pairs), making the HiFi2 Audio Engine a two-way SIMD engine. The Q register file is a four-entry, 56 bit register file that serves as the accumulator for the audio MAC operations. The HiFi2 Audio Engine architecture also defines a few special-purpose registers such as an overflow flag bit, a shift amount register, and a few registers for implementing efficient bit-extraction and bit-manipulation hardware. The HiFi2 Audio Engine's computation datapath features SIMD MAC, ALU, and shift units that operate on the two elements contained in each P register file entry. A variety of audio-oriented scalar operations on the Q register file and on the AR register file of the Xtensa core are also defined with TIE.

Table 7.4 provides a high-level summary of the HiFi2 Audio Engine's ISA, along with an indication of which slot of the VLIW format the instruction group is available in. The table also lists the approximate number of instructions belonging to each group. There are more than 300 operations in the two slots. In addition to the load/store, MAC, ALU, and shift instructions, the HiFi2 ISA supports several bit-manipulation instructions that enable efficient parsing of bit-stream data and variable-length encode and decode operations common to the manipulation of digital-audio bit streams.

While it is possible to program the HiFi2 Audio Engine in an assembly language, it is not necessary to do so. All of the HiFi2 Audio Engine's instructions are supported as C/C++ intrinsics and variables stored in the P and Q register files can be declared using custom data types. The Xtensa C/C++ compiler allocates registers for these variables and appropriately schedules the intrinsics. As a result, all digital-audio codecs developed by Tensilica are written in C and the performance numbers quoted in the next section are achieved without assembly programming.

7.7.2 HiFi2 Audio Engine Implementation and Performance

The HiFi2 Audio Engine is available as synthesizable RTL, along with synthesis and place-and-route scripts that allow the processor to be implemented in any modern digital-IC process technology. The processor's synthesized netlist corresponds to about 67K gates, of which approximately 45K gates correspond to the TIE extensions and 22K gates are for the base Xtensa processor core. In TSMC 90 nm process technology ("G" process), these numbers translate to a die area of $0.84\,\text{mm}^2$, a 350 MHz maximum clock frequency, and a power dissipation of 41.3 mW at 350 MHz.

While the HiFi2 Audio Engine design can achieve a maximum clock frequency of 350 MHz, the actual clock frequency needed to implement the digital-audio codecs is significantly lower. Thus, there is ample headroom for other computations to be performed on the processor. More likely, the SOC design team will run the HiFi2 Audio Engine at a much lower clock frequency to dissipate much less power for better battery life in mobile applications. More than 20 digital-audio codecs run on the HiFi2 Audio Engine and Table 7.5 lists the performance of a few of them. The table lists the millions

TABLE 7.5 HiFi2 Codec Performance

Codec	Clock Rate (MHz)	ROM Code Size (kB)	ROM Table Size (kB)	RAM Size (kB)	I/O Buffer RAM Size (kB)
Dolby Digital AC-3 Decoder, 5.1 ch (384 and 448 kbps/48 kHz)	27.3	19.6	7.9	17	9.8
Dolby Digital AC-3 Consumer Encoder (DDCE), stereo	19	39	7.7	49	8.5
Dolby Digital Consumer Encoder, 5.1 ch (640 kbps/48 kHz)	60	39	7.7	95	21
Dolby Digital Compatible Output Encoder, 5.1 ch (640 kbps/48 Hz)	60	39	7.7	95	21
Dolby Digital Plus Consumer Decoder, 7.1 channels	63	45	22	152	52
Dolby Digital Plus Decoder-Converter, 5.1 channels	43 (Decoder) 52 (Converter) 70 (Both)	54.7	23.1	97	36 (PCM) 4 (AC-3)
Dolby TrueHD Decoder, 7.1 ch (9.6 Mbps/48 kHz)	48	13	5.4	16	8.9
MP3 Stereo Decoder (128 kbps/44.1 kHz)	5.7	20	15	20	6.5
MP3 Stereo Decoder (320 kbps/44.1 kHz)	6.9	20	15	20	6.5
MP3 Stereo Encoder (128 kbps/44.1 kHz)	26	47	14	40	6.5
MP3 Stereo Encoder (320 kbps/44.1 kHz)	30.2	47	14	40	6.5
MPEG-4 aacPlus v2 Stereo Decoder (48 kbps/44.1 kHz)	19	71	33	52	18
MPEG-4 aacPlus v2 Stereo Decoder (64 kbps/44.1 kHz)	20	71	33	52	18
MPEG-4 aacPlus v2 Stereo Encoder (48 kbps/48 kHz)	33.6	128	41	124	9.5
MPEG-4 aacPlus v2 Stereo Encoder (64 kbps/48 kHz)	34.2	128	41	124	9.5
MPEG-4 aacPlus v1 Stereo Decoder (64 kbps/44.1 kHz)	19	63	31	47	18
MPEG-4 aacPlus v1 Stereo Decoder (128 kbps/44.1 kHz)	20	63	31	47	18
MPEG-4 aacPlus v1 Decoder, 7.1 ch (192 kbps/48 kHz)	72	63	31	141	70
MPEG-4 aacPlus v1 Stereo Encoder (64 kbps/48 kHz)	51.1	123	41	100	9.5
MPEG-4 aacPlus v1 Stereo Encoder (128 kbps/48 kHz)	52.8	123	41	100	9.5
MPEG 2/4 AAC LC Stereo Decoder (128 kbps/44.1 kHz)	8.1	34	20	18	5.5
MPEG 2/4 AAC LC Stereo Decoder (320 kbps/44.1 kHz)	11	34	20	18	5.5
MPEG-4 AAC-LC Decoder, 7.1 ch (576 kbps/48 kHz)	34	34	20	66	38
MPEG 2/4 AAC LC Stereo Encoder (128 kbps/48 kHz)	37.8	74	27	50	9.5
MPEG 2/4 AAC LC Stereo Encoder (320 kbps/48 kHz)	43.9	74	27	50	9.5
MPEG-4 BSAC Stereo Decoder (64 kbps/44.1 kHz)	25	33	24	33	10
MPEG-4 BSAC Stereo Decoder (128 kbps/48 kHz)	45	33	24	33	10
Ogg Vorbis Stereo Decoder (128 kbps/44.1 kHz)	12.3	28.1	19	86	16
Ogg Vorbis Stereo Decoder (320 kbps/44.1 kHz)	16.5	28.1	19	86	16
WMA Stereo Decoder (22 kbps/22 kHz)	12.4	32	52	52	8
WMA Stereo Decoder (128 kbps/44.1 kHz)	10.5	32	52	52	8
WMA Stereo Decoder (320 kbps/48 kHz)	14.9	32	52	52	8
WMA Stereo Encoder (128 kbps/44.1 kHz)	49	125	45	164	111
AMR Narrowband Speech Codec (5.15 kbps)	17.9	63	31	13	3.6
AMR Wideband Speech Codec (8.85 kbps)	37.2	61	27	18	3.0
G.729AB Speech Codec (8 kbps)	13.7	36	5.8	7	0.4

of clocks per second required for real-time audio encode/decode and the amount of program and data memory used by the codecs. The performance numbers illustrate the versatility and efficiency of the HiFi2 Audio Engine architecture in handling a wide variety of audio algorithms.

7.8 Conclusions

Processor-centric ADLs arose from the desire to explore processor architectures and their ability to solve a variety of problems. Initially, ADLs were of chief interest to processor designers who were concerned with fine architectural details such as the minutiae of pipeline operation. ADLs developed for these purposes did not provide the high abstraction level needed to make the ADLs useful to the broader group of embedded designers—people who were more concerned with using processors than designing them.

The phrase "design productivity gap" has been frequently used to refer to the imbalance between the number of available transistors on a piece of silicon and the ability of system designers to make good use of these transistors. Designing SOCs at a higher level of abstraction has often been proposed as a potential way to close this gap. Over the past few years, system-friendly ADLs like Tensilica's TIE have been introduced to help close this design productivity gap by providing SOC designers with a way to quickly explore new processor architectures that might ease the job of developing programmable, application-specific blocks that meet system design goals of performance, power, and cost.

References

1. P. C. Clements, A survey of architecture description languages, *Proceedings of the International Workshop on Software Specification and Design (IWSSD)*, pp. 16–25, Schloss Velen, Germany, 1996.
2. N. Medvidovic and R. Taylor, A framework for classifying and comparing architecture description languages, *Proceedings of the European Software Engineering Conference (ESEC)*, Springer-Verlag, pp. 60–76, Zurich, Switzerland, 1997.
3. C. G. Bell and A. Newell, *Computer Structures: Readings and Examples*, McGraw-Hill Book Company, New York, 1971.
4. H. Tomiyama, A. Halambi, P. Grun, N. Dutt, and A. Nicolau, Architecture description languages for system-on-chip design, *Proceedings of the APCHDL*, Fukuoka, Japan, October 1999.
5. P. Mishra and N. Dutt, Architecture description languages for programmable embedded systems, *IEE Proceedings on Computers and Digital Techniques (CDT)*, special issue on Embedded Microelectronic Systems: Status and Trends, 152(3):285–297, May 2005.
6. P. Mishra and N. Dutt, Architecture description languages, in *Customizable and Configurable Embedded Processors*, P. Ienne and R. Leupers, Editors, Morgan Kaufmann Publishers, San Francisco, CA, 2006.
7. A. Fauth, J. V. Praet, and M. Freericks, Describing instruction set processors using nML, *Proceedings of the European Design and Test Conference*, pp. 503–507, Brighton, England, UK, 1995. http://citeseer.ist.psu.edu/fauth95describing.html
8. S. Bashford, U. Bieker, B. Harking, R. Leupers, P. Marwedel, A. Neumann, and D. Voggenauer. The MIMOLA Language—Version 4.1. Technical Report, Computer Science Department, University of Dortmund, September 1994. http://citeseer.ist.psu.edu/bashford94mimola.html
9. G. Goossens, D. Lanneer, W. Geurts, and J. Van Praet, Design of ASIPs in multi-processor SoCs using the Chess/Checkers retargetable tool suite, *Proceedings of the International Symposium on System-on-Chip (SoC 2006)*, Tampere, November 2006.

10. G. Hadjiyiannis, S. Hanono, and S. Devadas, ISDL: An instruction set description language for retargetability, *Proceedings of the 34th Design Automation Conference*, pp. 299–302, Anaheim, CA, June 1997.

11. J. Gyllenhaal, B. Rau, and W. Hwu. HMDES Version 2.0 specification. Technical Report IMPACT-96-3, IMPACT Research Group, University of Illinois, Urbana, IL, 1996.

12. A. Halambi, P. Grun, V. Ganesh, A. Khare, N. Dutt, and A. Nicolau, EXPRESSION: A language for architecture exploration through compiler/simulator retargetability, *Proceedings of Design Automation and Test in Europe (DATE)*, pp. 485–490, Munich, Germany, 1999.

13. V. Zivojnovic, S. Pees, and H. Meyr, LISA—Machine description language and generic machine model for hw/sw co-design, W. Burleson, K. Konstantinides, and T. Meng, Editors, *VLSI Signal Processing IX*, pp. 127–136, San Francisco, CA, 1996.

14. A. Hoffmann, A. Nohl, and G. Braun, A novel methodology for the design of application-specific instruction-set processors, in *Embedded Systems Handbook*, R. Zurawski, Editor, pp. 19-1–19-9, CRC Press, Taylor & Francis Group, Boca Raton, FL, 2006.

15. Tensilica Instruction Extension (TIE) Language Reference Manual, Issue Date 11/2007, Tensilica, Inc., Santa Clara, CA.

16. Tensilica Instruction Extension (TIE) Language User's Guide, Issue Date 11/2007, Tensilica Inc., Santa Clara, CA.

17. D. Burger, J. Goodman, and A. Kagi, Limited bandwidth to affect processor design, *IEEE Micro*, 17(6):55–62, November/December 1997.

18. M. Rutten et al., A heterogeneous multiprocessor architecture for flexible media processing, *IEEE Design and Test of Computers*, 19(4):39–50, July–August 2002.

19. N. Bhattacharyya and A. Wang, Automatic test generation for micro-architectural verification of configurable microprocessor cores with user extensions, *High-Level Design Validation and Test Workshop*, pp. 14–15, Monterey, CA, November 2001.

20. M. Carchia and A. Wang, Rapid application optimization using extensible processors, *Hot Chips*, 13, Palo Alto, CA, 2001.

21. N. Cheung, J. Henkel, and S. Parameswaran, Rapid configuration and instruction selection for an ASIP: A case study, *Proceedings of the Conference on Design, Automation and Test in Europe*, pp. 802–807, Munich, Germany, March 2003.

8

Network-Ready, Open-Source Operating Systems for Embedded Real-Time Applications

Ivan Cibrario Bertolotti
National Research Council

8.1 Introduction

More often than not, modern embedded systems must provide some form of real-time execution capability and, most importantly, must be connected to a network.

At the same time, open-source operating systems have steadily gained popularity in recent years for embedded applications due to absence of licensing fees and royalties, a feature that promises to easily cut down the cost of their deployment. Moreover, many open-source operating systems have nowadays reached an excellent level of maturity and stability, comply with international standards and are able to support even demanding, hard real-time applications.

The goal of this chapter is to give an overview of the architectural choices for real-time and networking support adopted by many contemporary operating systems, within the framework of the IEEE 1003.1-2004 international standard.

In particular, Section 8.2 gives an overview of several widespread architectural choices for real-time support at the operating system level and especially describes the real-time application interface (RTAI) [27] approach. Then, Section 8.3 summarizes the real-time and networking support specified by the IEEE 1003.1-2004 international standard [12].

Finally, Section 8.4 describes the internal structure of a commonly used, open-source network protocol stack, in order to show how it can be extended to handle other protocols, besides the TCP/IP suite it was originally designed for. In this way, it becomes possible to seamlessly support communication media and protocols more closely tied to the real-time domain such as the Controller Area Network (CAN). A comprehensive set of bibliographic references for further reading concludes this chapter.

8.2 Embedded Operating System Architecture

The main goal of this section is to briefly recall the most widespread internal architectures of operating systems suitable for embedded applications and, at the same time, to describe several possible ways being used in practice to support the orderly coexistence between general-purpose and real-time applications on the same machine. The discussion being focused mostly on open-source operating systems, the RTAI [27] approach will be presented in more detail. Moreover, the kind and level of support for networking nowadays offered by popular, open-source operating systems is also be outlined.

Other books such as [32 and 36] give more general and thorough information about general-purpose operating systems, whereas [21] contains an in-depth discussion on the internal architecture of the influential Berkeley Software Distribution (BSD) operating system, also known as "Berkeley Unix."

In the following discussion, a rather broad definition of embedded system is adopted, from the assumption that, in general, an embedded system is a special-purpose computer system built into a larger device and is usually not programmable by the end user. It should also be noted that a common requirement for an embedded systems is some kind of real-time behavior. The strictness of this requirement varies with the application, but it is so common that some operating system vendors often use the two terms interchangeably and refer to their products either as "embedded operating systems" or as "real-time operating systems for embedded applications."

8.2.1 Overall System Architecture

An operating system can be built around several different architectural designs, depending on its characteristics and application domain. Some widespread designs are as follows.

8.2.1.1 Monolithic and Layered Systems

Even if this is the oldest design from the historical point of view, it is effective and still very popular for small real-time executives intended for embedded applications, due to its simplicity and very low processor and memory overhead. The same features make this approach attractive for the real-time portion of more complex systems as well.

In monolithic and layered systems, only the internal structure is usually induced by the way operating system services are invoked and implemented, and it mainly includes organizing the operating system as a hierarchy of layers at system design time. Each layer is built upon the services offered by the one below it and, in turn, offers a well-defined and usually richer set of services to the layer above it.

Better structure and modularity make maintenance easier, both because the operating system code is easier to read and understand and because the inner contents of a layer can be changed at will without interfering with other layers, provided the interface between layers does not change.

Moreover, the modular structure of the operating system enables the fine-grained configuration of its capabilities, to tailor the operating system itself to its target platform and avoid wasting valuable memory space for operating system functions that are never used by the application. As a consequence, it is possible to enrich the operating system with many capabilities, for example, network support, without sacrificing its ability to run on very small platforms when these features are not needed.

A number of contemporary operating systems, both commercial and open sources, for example [7,15,20,22,25,29, and 30], conform to this general design approach. They offer sophisticated build or link-time configuration tools, in order to tightly control what and how much code is actually put into the operating system executable image. Besides static configuration, some of them also have the capability of linking and loading additional modules into the kernel dynamically, that is, while the operating system is running.

In this kind of operating system, the operating system as a whole runs in privileged mode and the application software is confined to execute in user mode. It executes a special trapping instruction, usually known as the system call instruction, in order to request an operating system service by bringing the processor into privileged mode and transferring control to the operating system dispatcher.

Interrupt handling is done directly within the kernel for the most part and interrupt handlers are not full-fledged processes or tasks. As a consequence, the interrupt handling overhead is very small because there is no full task switching at interrupt arrival, but the interrupt handling code cannot invoke most system services, notably blocking synchronization primitives.

Moreover, the operating system scheduler is disabled while interrupt handling is in progress, and only the hardware-enforced prioritization of interrupt requests is in effect, and hence the interrupt handling code is implicitly executed at a priority higher than the priority of all other tasks in the system.

In order to alleviate these issues, which are especially relevant from the real-time execution point of view, some operating systems, for example [7], partition interrupt handling into two levels. The first-level interrupt handler runs with interrupts partially disabled and may schedule for deferred execution a second-level handler, which will run outside the interrupt service processor mode, with interrupts fully enabled, and will therefore be subject to less restrictions in its interaction with the operating system facilities.

In this way, the overall interrupt handling code can be split between these two levels, to achieve an optimal balance between a quick reaction to interrupt requests and an acceptable overall interrupt handling latency, which would be undermined by keeping interrupts disabled for a long time.

To further reduce processor overhead on small systems, it is also possible to run the application as a whole in supervisor mode. In this case, the application code can be bound with the operating system at link time and system calls become regular function calls. The interface between application code and operating system becomes much faster, because no user-mode state must be saved on system call invocation and no trap handling is needed.

On the other hand, the overall control that the operating system can exercise on bad application behavior is greatly reduced and debugging may become harder.

8.2.1.2 Client–Server Systems

This design moves most operating system functions from the kernel up into a set of operating system processes or tasks running in user mode, leaving a minimal microkernel and reducing to an absolute minimum the amount of privileged operating system code.

With this approach, the main function still allocated to the kernel is to handle interprocess communication, both among system tasks and between system tasks and applications, according to a message passing paradigm. As a consequence, applications request operating system services by sending a request message to the appropriate operating system server, and then waiting for a reply.

For what concerns interrupt requests, they are also transformed into messages as soon as possible: the interrupt handler proper runs in interrupt service mode and performs the minimum amount of work strictly required by the hardware, and then synthesizes a message and sends it to an appropriate interrupt service task, which is itself part of the operating system. In turn, the interrupt service task concludes interrupt handling running in user mode.

Being an ordinary task, the interrupt service task can, at least in principle, invoke the full range of operating system services, including blocking synchronization primitives and must not concern itself with excessive usage of the interrupt service processor mode. On the other hand, the overhead related to interrupt handling increases, because the activation of the interrupt service task requires a full task switch.

Besides this one, the other functions of the microkernel are to enforce an appropriate security policy on communications and to perform some critical operating system functions, such as accessing input/output (I/O) device registers, that would be impractical, or inefficient, to do from a user-mode task. An alternative approach allows some critical system tasks to run in privileged mode for the same reason.

This kind of design makes the operating system easier to manage and maintain. Also, the message passing interface between user tasks and operating system components encourages modularity and enforces a clear and well-understood structure on operating system components.

Moreover, the reliability of the operating system is increased: since the operating system tasks run in user mode, if one of them fails some operating system functions will no longer be available, but the system will not crash. Moreover, the failed component can be restarted and replaced without shutting down the whole system. For these reasons, this architecture has been chosen by several popular operating systems [19,28,39].

By contrast, making the message passing communication mechanism efficient has been a critical issue in the past, and the system call invocation mechanism often induced more overhead than in monolithic and layered systems. However, starting from the seminal work of Liedtke [18], improving message passing within a microkernel has been a fruitful research topic in recent years, leading to a considerable reduction of the associated overheads.

8.2.1.3 Virtual Machines

The internal architecture of operating systems based on virtual machines revolves around the basic observation that an operating system must perform two essential functions: multiprogramming and system services.

Accordingly, these operating systems fully separate the two functions and implement them as distinct operating system components:

1. A "virtual machine monitor" that runs in privileged mode, implements multiprogramming, and provides many virtual processors. In addition, it provides basic synchronization and communication mechanisms between virtual machines and partitions system resources among them, thus giving to each virtual machine its own set of (possibly virtualized) I/O devices.

2. A "guest operating system" that runs on each virtual machine and implements system services to support the execution of a set of applications within the virtual machine itself. Different virtual machines can run different operating systems.

In this way, it becomes possible to support, for example, the concurrent execution of both a real-time system and a general-purpose operating system, each one hosted by its own virtual machine.

The most interesting property of virtual machines is that, at least according to their original definition, also known as full or perfect virtualization [10], they are identical in all respects to the physical machine they are implemented on, barring instruction timings. As a consequence, no modifications to the operating systems hosted by the virtual machines are required, except the addition of a special-purpose device driver to handle intermachine communication, if required.

With this approach, guest operating systems are given the illusion of running in privileged mode but are instead constrained to operate in user mode; in this way, the virtual machine monitor is able to intercept all privileged instructions issued by the guest operating systems, check them against the security policy of the system, and then perform them on behalf of the guest operating system itself.

Interrupt handling is implemented in a similar way: the virtual machine monitor catches all interrupt requests and then redirects them to the appropriate guest operating system handler, reverting to user mode in the process; thus, the virtual machine monitor can intercept all privileged instructions issued by the guest interrupt handler and again check and perform them as appropriate.

The full separation of roles and the presence of a relatively small, centralized arbiter of all interactions between virtual machines has the advantage of making the enforcement of security policies easier. The isolation of virtual machines from each other also enhances reliability because, even if one virtual machine fails, it does not bring down the system as a whole. In addition, it is possible to run a distinct operating system in each virtual machine thus supporting, for example, the orderly coexistence between a real-time system and a general-purpose operating system.

More recently, in order to achieve improved efficiency, the guest operating systems and the virtual machine monitor were made capable of communicating and cooperating, to provide the latter with a better notion of the ongoing operating system activities. This approach is referred to as paravirtualization [3] and its implementation implies modifying and recompiling parts of the guest operating system. On the hardware side, several processor families nowadays include support for a more efficient processor virtualization [2,24]. In some cases, like [15], hardware support streamlines the virtualization of I/O devices as well.

As a consequence, beyond the early success of [23] for time-sharing systems, this kind of approach is nowadays becoming attractive for embedded systems, too [38].

8.2.2 "Double Kernel" Approach

The raw computing power of the processors commonly used to implement many kinds of embedded systems has increased steadily in recent years. As a consequence, it is nowadays possible to host on them a sophisticated system software, which offers the opportunity to tightly integrate real-time control tasks with a general-purpose operating system and its applications.

In an industrial environment this is especially appealing because it supports, for example, the orderly coexistence—on the same hardware—of both time-critical industrial control functions and application software that, for example, connects the system to the higher layers of the factory automation hierarchy, gives the system a friendly man–machine interface, and provides for online browsing of the system documentation.

One option to do this is to adopt a virtualization technique, as described in Section 8.2.1. An alternative approach consists of enhancing the inner components of an existing, general-purpose operating system in order to "nest" a real-time kernel inside it.

The ever-increasing viability and interest of this approach are corroborated by the importance and reputation of its supporters. For example, Refs. [4 and 31] provide solutions for Windows, whereas Refs. [9,27,40] do the same for Linux and other operating systems. Moreover, Ref. [33] provides a development kit and run-time systems for real-time execution on both Linux and Windows.

The RTAI [27] approach to provide real-time execution support is depicted in Figure 8.1. It has been chosen as an example because, being licensed under the open-source GNU General Public License (GPL), its internal architecture is well known. Moreover, most other open-source and commercial products mentioned above use comparable techniques.

A software component placed immediately above the hardware, called "Adeos" [41], enables the controlled sharing of hardware resources among multiple operating systems. In particular, it allows each operating system to safely handle and keep control of its own interrupt sources without hindering the other ones.

In order to do this, each operating system is encompassed in its own domain, over which it has total control. The parceling of hardware resources (such as main memory and I/O devices) among domains is a task left to the system designer. If the operating systems are aware of the Adeos's presence, they can

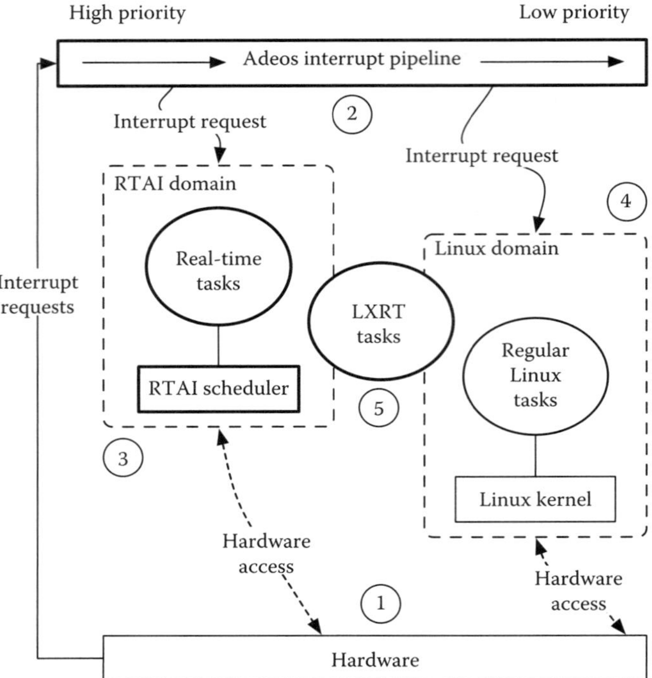

FIGURE 8.1　Adeos/RTAI approach to real-time execution.

also share part of these resources or access the resources of another domain. In any case (Figure 8.1, Ref. [1]), each operating system is free to access the hardware elements that have been assigned to it directly, without any interposition on the Adeos part, that is, without any performance penalty.

On the other hand, Adeos takes full control of interrupt handling, through a mechanism called interrupt pipeline (Figure 8.1, Ref. [2]). This is done for two related reasons:

1. If a domain were granted the ability to disable and enable interrupts at the hardware level, it would also be able to hinder the interrupt handling capabilities of any other domain. In turn, this would make interrupt latencies completely unpredictable.

2. By taking control of interrupt handling, it becomes possible to give to each domain its own interrupt handling priority and enforce it. This is important especially when a domain hosts a real-time operating system, which needs to be the first to be presented with interrupt requests, for reasons related to interrupt response determinism, latency, and performance.

The interrupt handling pipeline is made of a sequence of stages, one for each domain; the position of a domain within the pipeline implicitly determines its interrupt handling priority. Moreover, by interacting with Adeos, domains can declare whether or not they are willing to accept interrupts.

Upon arrival of an interrupt request, the interrupt pipeline is scanned, starting from the highest-priority stage, to locate a domain willing to accept it. When it is found, its corresponding interrupt handling facility is invoked, after setting the execution environment as if an actual hardware interrupt were being delivered to the operating system hosted by the domain, and the domain is then allowed to run until it is done. The latter event can be recognized either by an explicit interaction with Adeos, if the domain's operating system is aware of its presence, or by detecting when the domain's operating system schedules its idle task by means of a hook deliberately inserted into it at operating system initialization time.

At this point, pipeline scanning is resumed to look for other domains willing to handle the interrupt request, unless the domain just processed elected to terminate the interrupt itself. In this case, the interrupt is no longer propagated to the remaining pipeline stages. It is also possible for a domain to discard an interrupt request, that is, to ask Adeos to immediately pass the interrupt request along the pipeline to the other stages. In any case, when control is transferred from one domain to another, the execution state of the domain being abandoned must be saved, so that it can be restored at a later time.

When a domain wants to disable interrupts, it asks Adeos to stall the corresponding pipeline stage. When a stalled stage is encountered during the pipeline scan, interrupt requests are stalled within the stage as well and go no further in the pipeline. The same also happens if further interrupt requests are injected into the pipeline at a later time: all of them will remain at the stalled stage and will be delivered to that stage once the stall is removed, that is, when the associated domain wants to enable interrupts again. At this point, all the interrupts that were stalled resume their way through the pipeline, starting with the stage just unstalled, and their processing is resumed as usual.

Overall, the interrupt pipeline provides a convenient and efficient mechanism to allow each domain to keep control of its own interrupt delivery, and yet to prevent them from interfering with other domains in this respect. Hence, a low-priority domain can disable its own interrupts by stalling its own pipeline stage, without compromising the interrupt handling latency of the higher-priority domains, because the effects of the stall only affect the downstream pipeline stages.

The mapping between the interrupt disable/enable requests issued by the operating system hosted by a certain domain and the state of the corresponding pipeline stage can be based upon two distinct mechanisms:

- If the operating system is willing to cooperate with Adeos, it can explicitly and directly invoke the Adeos pipeline handling functions. With this approach, the overhead is minimal.
- Otherwise, any attempt to manipulate the interrupt handling state of the system made by the operating system must be trapped, by leveraging a suitable hardware mechanism. The occurrence of any trap of this kind is handled by a special domain provided by Adeos itself in order to perform the appropriate mapping.

As an example of the second approach, Ref. [41] shows how it is possible, on an `x86` platform, to constrain an uncooperative operating system to run in privilege ring 1 instead of ring 0 in order to trap the `cli` and `sti` instructions that, on this architecture, disable and enable interrupts, respectively. In this way, on the one hand, the operating system cannot take control of interrupt handling and, on the other hand, these attempts to disable and enable interrupts can be transformed into the appropriate interrupt pipeline handling commands.

The price to be paid is that many other instructions besides `cli` and `sti` and unrelated to interrupt handling are trapped as well. Even if the issue can be circumvented for several classes of instructions, for example, I/O instructions, the others must still be either emulated or executed by putting the processor in single-stepping mode.

Hence, even if the problem is somewhat alleviated in this case by the fact that Adeos trusts the code running within its domains and does not attempt to provide virtual I/O devices to them, the general technique bears some similarities with the virtualization technique discussed in Section 8.2.1 and inherits some of its shortcomings. Moreover, the inner implementation details are highly dependent on the target hardware architecture and are not easily ported from one architecture to another.

After each traversal of the interrupt pipeline, due to the arrival of an interrupt request, the last function of Adeos is to resume the domain whose execution was interrupted by the request itself, if any. To this purpose, Adeos checks whether all domains are idle: if this is the case, it invokes its own idle task, or else it restores the processor to the state it had when the interrupt request arrived. This has the effect of resuming the execution of the domain that was formerly interrupted, if any.

On top of Adeos the RTAI real-time scheduler, running in its own domain, supervises the real-time tasks (Figure 8.1, Ref. [3]). Another domain, with a lower-priority location in the interrupt pipeline, hosts Linux and its regular tasks (Figure 8.1, Ref. [4]). With this arrangement, whenever an interrupt request destined to RTAI arrives and RTAI has not stalled its interrupt pipeline stage, Adeos immediately gives control to RTAI itself and lets it run until it becomes idle. As a consequence, control is given back to Linux only when all real-time tasks are idle.

Since Linux runs only when no real-time tasks are ready for execution and is immediately preempted whenever a real-time interrupt request arrives, real-time activities always have an execution priority higher than any other task in the system. It should also be noted that this happens even if Linux disables its own interrupts, because this only affects the pipeline stages that follow Linux in the pipeline, and not the RTAI stage.

However, even if the interrupt pipeline mechanism discussed above is effective to prevent any interference between Linux and RTAI from the interrupt handling point of view, it is still the programmer's responsibility to ensure that any activity initiated within Linux cannot hinder the timing behavior of the real-time application in other ways.

For example, it is advisable to avoid heavy use of the bus mastering capabilities built in several kinds of peripherals, like accelerated graphics cards and USB controllers, because the delay they introduce in the execution of real-time code due to bus contention can be hard to predict and cannot be avoided. For the same reason, any technique that dynamically trades off processing performance for reduced power consumption, such as advanced configuration and power interface (ACPI) power management and CPU frequency scaling, should also be avoided.

At the programmer's choice, real-time applications can either be compiled as a kernel module and run in privileged mode, or they can be implemented as regular Linux tasks and run in unprivileged mode with the assistance of an RTAI component known as LXRT (Figure 8.1, Ref. [5]).

In the latter case, the real-time tasks are easier to develop, are protected from each other faults by the Linux memory management mechanisms, and have a wider set of interprocess communication facilities at their disposal. On the other hand, the price to be paid is a slightly greater overhead incurred when performing a context switch between real-time tasks. In either case, the real-time execution properties are guaranteed, because task scheduling is nevertheless under the control of RTAI.

8.2.3 Open-Source Networking Support

Most contemporary, open-source operating systems offer networking support, even if the details of the implementation can be slightly different from one system to another. In most cases, the application programming interface is conforming to the IEEE 1003.1-2004 international standard [12] which is discussed in Section 8.3.

Several open-source, real-time operating systems, for example Refs. [7,25], take the simplest route and offer networking support by means of an adaptation of the "Berkeley sockets" [21]. Whereas the application programming interface is kept unchanged so that standard conformance is ensured, the adaptation often includes enhancements to the determinism of the protocol stack, thus making it more adequate for use in real-time applications.

In the case of RTAI, instead, the networking support is more elaborate and encompasses two distinct protocol stacks:

1. The traditional protocol stack provided by Linux, and founded on a code base developed by the Swansea University Computer Society, is still available to non-real-time applications. This protocol stack has not been modified in any way, and hence it is well proved but not particularly suited for real-time execution.

2. An additional protocol stack called RTnet [17] provides an extensible framework for hard real-time communication over Ethernet and other communication media. The application programming interface is still widely based on IEEE 1003.1-2004, so that software portability is not a concern, with suitable extensions to cover the additional features of RTnet, mainly related to real-time medium access control.

Albeit the network software is often bundled with the operating system and provided by the same vendor, one last option is to get it from third-party software houses. For example, Refs. [16 and 37] are TCP/IP protocol stacks that run on a wide variety of hardware platforms and operating systems. Often, these products come with source code; hence, it is also possible to port them to custom operating systems developed in-house, and they can be extended and enhanced by the end user.

With respect to the other options, the main advantage of a third-party protocol stack is the possibility of integrating it with very small operating systems that do not have networking support and, in some cases, the ability to run the protocol stack even without an underlying operating system. This is a useful technique, for example, on very small embedded systems and on platforms with severe limits on execution resources.

8.3   IEEE 1003.1 Standard and Networking

The original version of the Portable Operating System Interface for Computing Environments, better known as "the POSIX standard," was first published between 1988 and 1990 and defines a standard way for applications to interface with the operating system.

The standard has since been constantly evolving and growing; the latest developments have been crafted by a joint working group of members of the IEEE Portable Applications Standards Committee, members of The Open Group, and members of ISO/IEC Joint Technical Committee 1. The joint working group is known as the Austin Group named after the location of the inaugural meeting held at the IBM facility in Austin, TX, in September 1998.

The overall set of documents now includes over 30 individual standards and covers a wide range of topics, from the definition of basic operating system services, such as process management, to specifications for testing the conformance of an operating system to the standard itself.

Among these, of particular interest is the System Interfaces (XSH) Volume of IEEE Std 1003.1-2004 [12], which defines a standard operating system interface and environment, including real-time extensions. The standard contains definitions for system service functions and subroutines, language-specific system services for the C programming language, and notes on portability, error handling, and error recovery.

Moreover, since embedded systems can have serious resource limitations, the IEEE Std 1003.13-2003 [13] profile standard groups functions from the standards mentioned above into units of functionality. Implementations can then choose the profile most suited to their needs and to the computing resources of their target platforms.

Table 8.1 summarizes the functional groups of IEEE Std 1003.1-2004 related to real-time execution that will be briefly discussed in Section 8.3.1. Assuming that the functions common to both the IEEE Std 1003.1-2004 and the ISO C [14] standards are well known to readers, we will not further describe them. On the other hand, the networking support will be more thoroughly described in Section 8.3.2.

TABLE 8.1 Basic Functional Groups of IEEE Std 1003.1-2004

Functional Group	Main Functions
Multiple threads	`pthread_create, pthread_exit, pthread_join, pthread_detach`
Process and thread scheduling	`sched_setscheduler, sched_getscheduler, sched_setparam, sched_getparam, pthread_setschedparam, pthread_getschedparam, pthread_setschedprio, pthread_attr_setschedpolicy, pthread_attr_getschedpolicy, pthread_attr_setschedparam, pthread_attr_getschedparam, sched_get_priority_max, sched_get_priority_min`
Real-time signals	`sigqueue, pthread_kill, sigaction, pthread_sigmask, sigemptyset, sigfillset, sigaddset, sigdelset, sigismember, sigwait, sigwaitinfo, sigtimedwait`
Interprocess synchronization and communication	`mq_open, mq_close, mq_unlink, mq_send, mq_receive, mq_timedsend, mq_timedreceive, mq_notify, mq_getattr, mq_setattr, sem_init, sem_destroy, sem_open, sem_close, sem_unlink, sem_wait, sem_trywait, sem_timedwait, sem_post, sem_getvalue, pthread_mutex_destroy, pthread_mutex_init, pthread_mutex_lock, pthread_mutex_trylock, pthread_mutex_timedlock, pthread_mutex_unlock, pthread_mutex_getprioceiling, pthread_mutex_setprioceiling, pthread_cond_init, pthread_cond_destroy, pthread_cond_wait, pthread_cond_timedwait, pthread_cond_signal, pthread_cond_broadcast, shm_open, close, shm_unlink, mmap, munmap`
Thread-specific data	`pthread_key_create, pthread_getspecific, pthread_setspecific, pthread_key_delete`
Mem. management	`mlock, mlockall, munlock, munlockall, mprotect`
Asynchronous I/O	`aio_read, aio_write, aio_error, aio_return, aio_fsync, aio_suspend, aio_cancel`
Clocks and timers	`clock_gettime, clock_settime, clock_getres, timer_create, timer_delete, timer_getoverrun, timer_gettime, timer_settime`
Cancellation	`pthread_cancel, pthread_setcancelstate, pthread_setcanceltype, pthread_testcancel, pthread_cleanup_push, pthread_cleanup_pop`

8.3.1 Overview of the Standard

8.3.1.1 Multithreading

The multithreading capability specified by the IEEE Std 1003.1-2004 standard includes functions to populate a process with new threads. In particular, the `pthread_create` function creates a new thread within a process and sets up a thread identifier for it, to be used to operate on the thread in the future.

After creation, the thread immediately starts executing a function passed to `pthread_create` as an argument; moreover, it is also possible to pass an argument to the function in order to share the same function among multiple threads and nevertheless be able to distinguish them.

The `pthread_create` function also takes an optional reference to an "attribute object" as argument. The attributes of a thread determine several of its characteristics such as its scheduling parameters.

A thread may terminate its execution in several different ways, either voluntarily (by returning from its main function or calling the `pthread_exit` function) or involuntarily (by accepting a cancellation request from another thread). In any case, the `pthread_join` function allows the calling thread to wait for the termination of another thread. When the thread finally terminates, this function also returns to the caller a summary information about the reason of the termination. For example, if the target thread terminated itself by means of the `pthread_exit` function, `pthread_join` returns the status code passed to `pthread_exit` in the first place.

If this information is not desired, it is possible and advisable to save system resources by detaching a thread, either dynamically by means of the `pthread_detach` function, orstatically by means of a thread's attribute. In this way, the storage associated with that thread can be immediately reclaimed when the thread terminates.

8.3.1.2 Process and Thread Scheduling

Functions in this group allow the application to select a specific policy that the operating system must follow to schedule a particular process or thread and to get and set the scheduling parameters associated with that process or thread.

In particular, the `sched_setscheduler` function sets both the scheduling policy and parameters associated with a process, and `sched_getscheduler` reads them back for examination. The simpler functions `sched_setparam` and `sched_getparam` set and get the scheduling parameters but not the policy. All functions take a process identifier as argument, to uniquely identify a process.

For threads, the `pthread_setschedparam` and `pthread_getschedparam` functions set and get the scheduling policy and parameters associated with a thread; `pthread_setschedprio` directly sets the scheduling priority of the given thread. All these functions take a thread identifier as an argument and can be used when the thread already exists in the system.

Otherwise, the scheduling policy and parameters can also be set indirectly through an attribute object, by means of the functions `pthread_attr_setschedpolicy`, `pthread_attr_getschedpolicy`, `pthread_attr_setschedparam`, and `pthread_attr_getschedparam`. This attribute object can subsequently be used to create one or more threads with the specified scheduling policy and attributes.

In order to support the orderly coexistence of multiple scheduling policies, the conceptual scheduling model defined by the standard assigns a global priority to all threads in the system and contains one ordered thread list for each priority; any runnable thread will be on the thread list for that thread's priority.

When appropriate, the scheduler shall select the thread at the head of the highest-priority, nonempty thread list to become a running thread, regardless of its associated policy; this thread is then removed from its thread list. When a running thread yields the CPU, either voluntarily or by preemption, it is returned to the thread list it belongs to.

The purpose of a scheduling policy is then to determine how the operating system scheduler manages the thread lists, that is, how threads are moved between and within lists when they gain or lose access to the CPU. Associated with each scheduling policy is a priority range, into which all threads scheduled according to that policy must lie. This range can be retrieved by means of the `sched_get_priority_min` and `sched_get_priority_max` functions.

The mapping between the multiple local priority ranges, one for each scheduling policy active in the system, and the single global priority range is usually performed by a simple relocation and is either fixed or programmable at system configuration time, depending on the operating system. In addition, operating systems may reserve some global priority levels, usually the higher ones, for interrupt handling.

The standard defines three scheduling policies: first in, first out (`SCHED_FIFO`), round robin (`SCHED_RR`), and, optionally, a variant of the sporadic server scheduler [35] (`SCHED_SPORADIC`). A fourth scheduling policy, `SCHED_OTHER`, can be selected to denote that a thread no longer needs a specific real-time scheduling policy: general-purpose operating systems with real-time extensions usually revert to the default, non-real-time scheduler when this scheduling policy is selected.

In addition, each implementation is free to redefine the exact meaning of the `SCHED_OTHER` policy and can provide additional scheduling policies besides those required by the standard, but any application using them will no longer be fully portable.

8.3.1.3 Real-Time Signals and Asynchronous Events

Signals are a facility specified by the ISO C standard and are widely available on most operating systems; they provide a mechanism to convey information to a process or thread when it is not necessarily waiting for input. The IEEE Std 1003.1-2004 further extends the signal mechanism to make it suitable for real-time handling of exceptional conditions and events that may occur asynchronously with respect to the notified process.

The signal mechanism owes most of its complexity to the need of maintaining compatibility with the historical implementations of the mechanism made, for example, by the various flavors of the influential Unix operating systems; however, in this section the compatibility interfaces will not be discussed for the sake of clarity and conciseness.

With respect to the ISO C signal behavior, the IEEE Std 1003.1-2004 specifies two main enhancements of interest to real-time programmers:

1. In the ISO C standard, the various kinds of signals are identified by an integer number (often denoted by a symbolic constant in application code) and, when multiple signals of different kind are pending, they are serviced in an unspecified order. The IEEE Std 1003.1-2004 continues to use signal numbers but specifies that for a subset of their allowable range, between `SIGRTMIN` and `SIGRTMAX`, a priority hierarchy among signals is in effect, so that the lowest-numbered signal has the highest priority of service.

2. In the ISO C standard, there is no provision for signal queues; hence, when multiple signals of the same kind are raised before the target process had a chance of handling them, all signals but the first are lost. Instead, the IEEE Std 1003.1-2004 specifies that the system must be able to keep track of multiple signals with the same number by enqueuing and servicing them in order. Moreover, it also adds the capability of conveying a limited amount of information with each signal request, so that multiple signals with the same signal number can be distinguished from each other. The queueing policy is always FIFO and cannot be changed by the user.

Figure 8.2 depicts the life of a signal from its generation up to its delivery. Depending on their kind and source, signals may be directed to either a specific thread in the process, or to the process as a whole; in the latter case, every thread belonging to the process is a candidate for the delivery of the signal, by the rules described later.

It should also be noted that for some kinds of events, the IEEE Std 1003.1-2004 standard specifies that the notification can also be carried out by the execution of a handling function in a separate thread, if the application so chooses; this mechanism is simpler and clearer than the signal-based notification but requires multithreading support on the system side.

8.3.1.3.1 *Generation of a Signal*

As outlined above, most signals are generated by the system rather than by an explicit action performed by a process. For these, the IEEE Std 1003.1-2004 standard specifies that the decision of whether the signal must be directed to the process as a whole or to a specific thread within a process must be carried out at the time of generation and must represent the source of the signal as closely as possible.

In particular, if a signal is attributable to an action carried out by a specific thread, for example, a memory access violation, the signal shall be directed to that thread and not to the process. If such an attribution is either not possible or not meaningful as in the case of the power failure signal, the signal shall be directed to the process.

Besides various error conditions, an important source of signals generated by the system relate to asynchronous event notification (for example, the completion of an asynchronous I/O operation or the availability of data on a communication endpoint) and are always directed to the process.

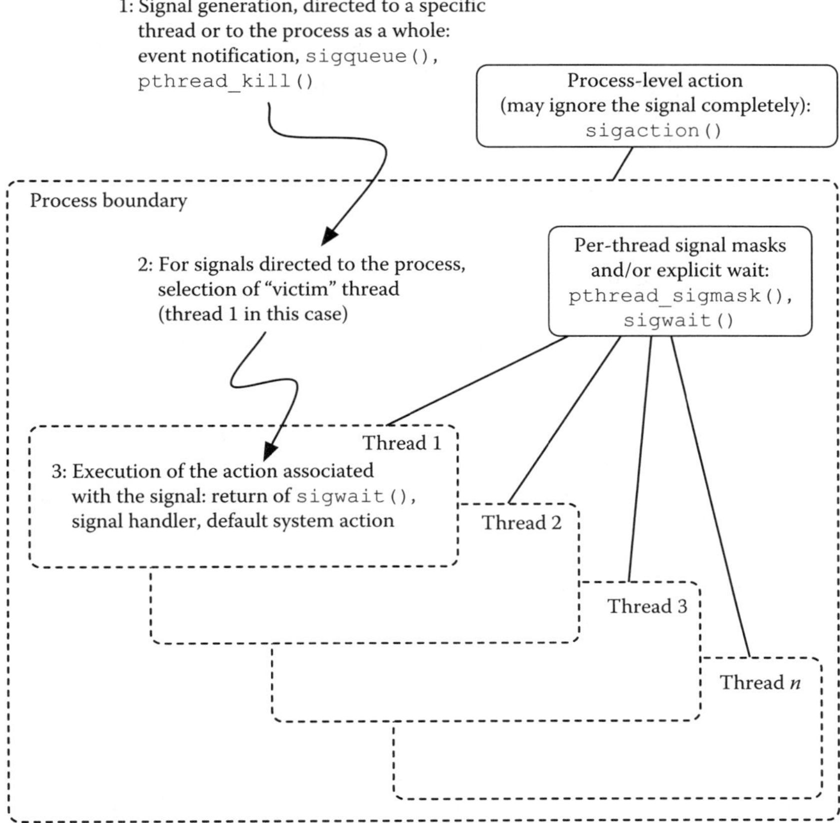

FIGURE 8.2 Simplified view of signal generation and delivery in the IEEE Std 1003.1-2004.

On the other hand, processes have the ability to synthesize signals by means of two main interfaces, depending on the target of the signal:

- The sigqueue function, given a process identifier and a signal number, generates a signal directed to that process. An additional argument allows the caller to associate a limited amount of information with the signal.
- The pthread_kill function generates a signal directed to a specific thread within the calling process and identified by its thread identifier.

8.3.1.3.2 *Process-Level Action*

For each kind of signal defined in the system, that is, for each valid signal number, processes may set up an action by means of the sigaction function; the action may consist of ignore the signal completely, perform a default action (for example, terminate the process), or execute a signal handling function specified by the programmer.

In addition, the same function allows the caller to set zero or more "flags" associated with the signal number. Each flag requests a variation on the default reaction of the system to the signal. For example, the SA_RESTART flag, when set, enables the automatic, transparent restart of interruptible system calls when the system call is interrupted by the signal. If this flag is clear, system calls that were interrupted by a signal fail with an error indication and must be explicitly restarted by the application, if appropriate.

It should be noted that the setting of the action associated with each kind of signal takes place at the process level, that is, all threads within a process share the same set of actions; hence, for example, it is impossible to set two different signal handling functions (for two different threads) to be executed in response to the same kind of signal.

Immediately after generation, the system checks the process-level action associated with the signal in the target process and immediately discards the signal if that action is set to ignore it; otherwise, it proceeds to check whether the signal can be acted on immediately.

8.3.1.3.3 *Signal Delivery and Acceptance*

Provided that the action associated with the signal at the process level does not specify to ignore the signal in the first place, a signal can be either "delivered to" or "accepted by" a thread within the process.

Unlike the action associated with each kind of signal discussed above, each thread has its own "signal mask"; by means of the signal mask, each thread can selectively block some kinds of signals from being delivered to it, depending on their signal number. The `pthread_sigmask` function allows the calling thread to examine or change (or both) its signal mask. A separate group of functions (namely, `sigemptyset`, `sigfillset`, `sigaddset`, `sigdelset`, and `sigismember`) allows the programmer to set up and manipulate a signal mask.

A signal can be delivered to a thread if, and only if, that thread does not block the signal; when a signal is successfully delivered to a thread, that thread executes the process-level action associated with the signal.

On the other hand, a thread may perform an explicit wait for one or more kinds of signal, by means of the `sigwait` function; that function stops the execution of the calling thread until one of the signals passed as an argument to `sigwait` is conveyed to the thread. When this occurs, the thread accepts the signal and continues past the `sigwait` function. Since the standard specifies that signals in the range from `SIGRTMIN` to `SIGRTMAX` are subject to a priority hierarchy, when multiple signals in this range are pending, the `sigwait` shall consume the lowest-numbered one.

It should also be noted that for this mechanism to work correctly, the thread must block the signals that it wishes to accept by means of `sigwait` (through its signal mask); otherwise, signal delivery takes precedence.

Two, more powerful, variants of the `sigwait` function exist: `sigwaitinfo` has an additional argument used to return additional information about the signal just accepted, including the information associated with the signal when it was first generated; furthermore, `sigtimedwait` also allows the caller to specify the maximum amount of time that shall be spent waiting for a signal to arrive.

The way in which the system selects a thread within a process to convey a signal depends on where the signal is directed:

- If the signal is directed toward a specific thread, only that thread is a candidate for delivery or acceptance.
- If the signal is directed to a process as a whole, any thread belonging to that process is a candidate to receive the signal; hence, the system selects exactly one thread within the process with the appropriate signal mask (for delivery), or performing a suitable `sigwait` (for acceptance).

If there is no suitable thread to convey the signal when it is first generated, the signal remains pending until either its delivery or acceptance becomes possible, by following the same rules outlined above, or the process-level action associated with that kind of signal is changed and set to ignore it. In the latter case, the system forgets everything about the signal and all other signals of the same kind.

8.3.1.4 Interprocess Synchronization and Communication

The main interprocess synchronization and communication mechanisms offered by the standard are the semaphore and the message queue. The blocking synchronization primitives have a nonblocking and a timed counterpart, to make them more flexible in a real-time execution environment. Moreover, multithreading support also adds support for mutual exclusion devices, condition variables, and other synchronization mechanisms. The scope of these mechanisms can be limited to threads belonging to the same process to enhance their performance.

8.3.1.4.1 *Message Queues*

The `mq_open` function either creates or opens a message queue and connects it with the calling process; in the system, each message queue is uniquely identified by a "name," like a file. This function returns a message queue "descriptor" that refers to and uniquely identifies the message queue; the descriptor must be passed to all other functions that operate on the message queue.

Conversely, `mq_close` removes the association between the message queue descriptor and its message queue. As a result, the message queue descriptor is no longer valid after successful return from this function. Finally, the `mq_unlink` function removes a message queue, provided no other processes reference it; if this is not the case, the removal is postponed until the reference count drops to zero.

The number of elements that a message queue is able to buffer, and their maximum size, are constant for the lifetime of the message queue and are set when the message queue is first created.

The `mq_send` and `mq_receive` functions send and receive a message to and from a message queue, respectively. If the message cannot be immediately stored or retrieved (e.g., when `mq_send` is executed on a full message queue) these functions block as long as appropriate, unless the message queue was opened with the nonblocking option set. If this is the case, these functions return immediately if they are unable to perform their job.

The `mq_timedsend` and `mq_timedreceive` functions have the same behavior but allow the caller to place an upper bound on the amount of time they may spend waiting.

The standard allows to associate a priority with each message, and specifies that the queueing policy of message queues must obey the priority so that `mq_receive` retrieves the highest-priority message that is currently stored in the queue.

The `mq_notify` function allows the caller to arrange for the asynchronous notification of message arrival at an empty message queue, when the status of the queue transitions from empty to nonempty, according to the mechanism described in Section 8.3.1.3. The same function also allows the caller to remove a notification request it made previously.

At any time, only a single process may be registered for notification by a message queue. The registration is removed implicitly when a notification is sent to the registered process, or when the process owning the registration explicitly removes it; in both cases, the message queue becomes available for a new registration.

If both a notification request and a `mq_receive` call are pending on a given message queue, the latter takes precedence, that is, when a message arrives at the queue, it satisfies the `mq_receive` and no notification is sent.

Finally, the `mq_getattr` and `mq_setattr` functions allow the caller to get and set, respectively, some attributes of the message queue dynamically after creation; these attributes include the nonblocking flag just described and may also include additional, implementation-specific flags.

8.3.1.4.2 *Semaphores and Mutexes*

Semaphores come in two flavors: unnamed and named. Unnamed semaphores are created by the `sem_init` function and must be shared among processes by means of the usual memory sharing mechanisms provided by the system. On the other hand, named semaphores created and accessed by the sem_open function exist as named objects in the system, like the message queues described above, and can therefore be accessed by name. Both functions, when successful, associate the calling process with the semaphore and return a descriptor for it.

Depending on the kind of semaphore, either the sem_destroy (for unnamed semaphores) or the sem_close function (for named semaphores) must be used to remove the association between the calling process and a semaphore.

For unnamed semaphores, the sem_destroy function also destroys the semaphore; but, named semaphores must be removed from the system with a separate function, sem_unlink.

For both kinds of semaphore, a set of functions implements the classic P and V primitives, namely,

- sem_wait function performs a P operation on the semaphore; the sem_trywait and sem_timedwait functions perform the same function in polling mode and with a user-specified timeout, respectively.
- sem_post function performs a V operation on the semaphore.
- sem_getvalue function has no counterpart in the definition of semaphore found in literature and returns the current value of a semaphore.

A "mutex" is a very specialized binary semaphore that can only be used to ensure the mutual exclusion among multiple threads; it is therefore simpler and more efficient than a full-fledged semaphore. Optionally, it is possible to associate with each mutex a protocol to deal with priority inversion.

The pthread_mutex_init function initializes a mutex and prepares it for use. It takes an attribute object as an argument, useful to better specify several characteristics of the mutex like, for example, which priority inversion protocol it must use. The pthread_mutex_destroy function destroys a mutex.

The following main functions operate on the mutex after creation:

- The pthread_mutex_lock function locks the mutex if it is free; otherwise, it blocks until the mutex becomes available and then locks it. The pthread_mutex_trylock function does the same but returns to the caller without blocking if the lock cannot be acquired immediately. The pthread_mutex_timedlock function allows the caller to specify a maximum amount of time to be spent waiting for the lock to become available.
- The pthread_mutex_unlock function unlocks a mutex.

Additional functions are defined for particular flavors of mutexes; for example, the pthread_mutex_getprioceiling and pthread_mutex_setprioceiling functions allow the caller to get and set, respectively, the priority ceiling of a mutex and make sense only if the priority ceiling protocol has been selected for the mutex, by means of a suitable setting of its attributes.

8.3.1.4.3 Condition Variables

A set of condition variables, in concert with a mutex, can be used to implement a synchronization mechanism similar to the monitor, without requiring the notion of monitor to be known at the programming language level.

A condition variable must be initialized before use by means of the pthread_cond_init function. Like pthread_mutex_init, this function also takes an attribute object as an argument. When default attributes are appropriate, the macro PTHREAD_COND_INITIALIZER is available to initialize a condition variable that the application has statically allocated.

Then, the mutex and the condition variables can be used as follows:

- Each procedure belonging to the monitor must be explicitly bracketed with a mutex lock at the beginning, and a mutex unlock at the end.

- To block on a condition variable, a thread must call the `pthread_cond_wait` function giving both the condition variable and the mutex used to protect the procedures of the monitor as arguments. This function atomically unlocks the mutex and blocks the caller on the condition variable; the mutex will be reacquired when the thread is unblocked, and before returning from `pthread_cond_wait`. To avoid blocking for a (potentially) unbound time, the `pthread_cond_timedwait` function allows the caller to specify the maximum amount of time that may be spent waiting for the condition variable to be signaled.

- Inside a procedure belonging to the monitor, the `pthread_cond_signal` function, taking a condition variable as an argument, can be called to unblock at least one of the threads that are blocked on the specified condition variable; the call has no effect if no threads are blocked on the condition variable.

- A variant of `pthread_cond_signal`, called `pthread_cond_broadcast`, is available to unblock all threads that are currently waiting on a condition variable. As before, this function has no effect if no threads are waiting on the condition variable.

When no longer needed, condition variables shall be destroyed by means of the `pthread_cond_destroy` function, to save system resources.

8.3.1.4.4 *Shared Memory*

Except message queues, all IPC mechanisms described so far only provide synchronization among threads and processes, and not data sharing.

Moreover, while all threads belonging to the same process share the same address space, so that they implicitly and inherently share all their global data, the same is not true for different processes; therefore, the IEEE Std 1003.1-2004 standard specifies an interface to explicitly set up a shared memory object among multiple processes.

The `shm_open` function either creates or opens a new shared memory object and associates it with a "file descriptor," which is then returned to the caller. In the system, each shared memory object is uniquely identified by a "name," like a file. After creation, the state of a shared memory object, in particular all data it contains, persists until the shared memory object is unlinked and all active references to it are removed. Instead, the standard does not specify whether or not a share memory object remains valid after a reboot of the system.

Conversely, `close` removes the association between a file descriptor and the corresponding shared memory object. As a result, the file descriptor is no longer valid after successful return from this function. Finally, the `shm_unlink` function removes a shared memory object, provided no other processes reference it; if this is not the case, the removal is postponed until the reference count drops to zero.

It should be noted that the association between a shared memory object and a file descriptor belonging to the calling process, performed by `shm_open`, does not map the shared memory into the address space of the process. In other words, merely opening a shared memory object does not make the shared data accessible to the process.

In order to perform the mapping, the `mmap` function must be called; since the exact details of the address space structure may be unknown to, and uninteresting for the programmer, the same function also provides the capability of choosing a suitable portion of the caller's address space to place the mapping automatically. The function `munmap` removes mapping.

8.3.1.5 Thread-Specific Data

All threads belonging to the same process implicitly share the same address space, so that they have shared access to all their global data. As a consequence, only the information allocated on the thread's stack, such as function arguments and local variables, is private to each thread. On the other hand, it is often useful in practice to have data structures that are private to a single thread but can be accessed globally by the code of that thread. The IEEE Std 1003.1-2004 standard responds to this need by defining the concept of "thread-specific data."

The `pthread_key_create` function creates a thread-specific data key visible to, and shared by, all threads in the process. The key values provided by this function are opaque objects used to access thread-specific data.

In particular, the pair of functions `pthread_getspecific` and `pthread_setspecific` take a key as argument and allow the caller to get and set, respectively, a pointer uniquely bound with the given key, and "private" to the calling thread. The pointer bound to the key by `pthread_setspecific` persists with the life of the calling thread, unless it is replaced by a subsequent call to `pthread_setspecific`.

An optional "destructor" function may be associated with each key when the key itself is created. When a thread exits, if a given key has a valid destructor, and the thread has a valid (i.e., not NULL) pointer associated with that key, the pointer is disassociated and set to NULL, and then the destructor is called with the previously associated pointer as an argument.

When it is no longer needed, a thread-specific data key should be deleted by invoking the `pthread_key_delete` function on it. It should be noted that, unlike in the previous case, this function does not invoke the destructor function associated with the key, so that it is the responsibility of the application to perform any cleanup actions for data structures related to the key being deleted.

8.3.1.6 Memory Management

The standard allows processes to lock parts or all of their address space in main memory by means of the `mlock` and `mlockall` functions; in addition, `mlockall` also allows the caller to demand that all of the pages that will become mapped into the address space of the process in the future must be implicitly locked.

The lock operation both forces the memory residence of the virtual memory pages involved and prevents them from being paged out in the future. This is vital in operating systems that support demand paging and must nevertheless support any real-time processing, because the paging activity could introduce undue and highly unpredictable delays when a real-time process attempts to access a page that is currently not in the main memory and must therefore be retrieved from secondary storage.

When the lock is no longer needed, the process can invoke either the `munlock` or the `munlockall` function to release it and enable demand paging again.

Finally, it is possible for a process to change the access protections of portions of its address space by means of the `mprotect` function; in this case, it is assumed that protections will be enforced by the hardware. For example, to prevent inadvertent data corruption due to a software bug, one could protect critical data intended for read-only usage against write access.

8.3.1.7 Asynchronous Input and Output

Many operating systems carry out I/O operations synchronously with respect to the process requesting them. Thus, for example, if a process invokes a file read operation, it stays blocked until the operating system has finished it, whether successfully or unsuccessfully. As a side effect, any process can have at most one pending I/O operation at any given time.

While this programming model is intuitive and adequate for a general-purpose system, in a real-time environment it may not be wise to suspend the execution of a process until the I/O operation completes, because this would introduce a source of unpredictability in the system. It may also be desirable, for example to enhance system performance by exploiting I/O hardware parallelism and to start more than one I/O operation simultaneously, under the control of a single process.

To satisfy these requirements, the standard defines a set of functions to start one or more I/O requests, to be carried out in parallel with process execution, and whose completion status can be retrieved asynchronously by the requesting process.

Asynchronous and list-directed I/O functions revolve around the concept of asynchronous I/O control block, `struct aiocb`, a structure that contains all the information needed to describe an I/O operation. For instance, it contains members to specify the operation to be performed (read or write), identify the target file, indicate what portion of the file will be affected by the operation (by means of a file offset and a transfer length), locate a data buffer in memory, and give a priority classification to the operation.

In addition, it is possible to request the asynchronous notification of the completion of the operation, either by a signal or by the asynchronous execution of a function, as described in Section 8.3.1.3.

Then, the following functions are available:

- `aio_read` and `aio_write` functions take an I/O control block as an argument and schedule a read or a write operation, respectively; both return to the caller as soon as the request has been queued for execution.

- `aio_error` and `aio_return` functions allow the caller to retrieve the error and status information associated with an I/O control block, after the corresponding I/O operation has been completed.

- `aio_fsync` function asynchronously forces all I/O operations associated with the file indicated by the I/O control block passed as an argument and currently queued to the synchronized I/O completion state.

- `aio_suspend` function can be used to block the calling thread until at least one of the I/O operations associated with a set of I/O control blocks passed as argument completes, or up to a maximum amount of time.

- `aio_cancel` function cancels an I/O operation that has not yet been completed.

8.3.1.8 Clocks and Timers

Real-time applications very often rely on timing information to operate correctly; the IEEE Std 1003.1-2004 standard specifies support for one or more timing bases, called clocks, of known resolution and whose value can be retrieved at will. In the system, each clock has its own unique identifier. The `clock_gettime` and `clock_settime` functions get and set the value of a clock, respectively, while the `clock_getres` function returns the resolution of a clock. Clock resolutions are implementation-defined and cannot be set by a process; some operating systems allow the clock resolution to be set at system generation or configuration time.

In addition, applications can set one or more perprocess timers, using a specified clock as a timing base, by means of the `timer_create` function. Each timer has a current value and, optionally, a reload value associated with it. The operating system decrements the current value of timers according to their clock and, when a timer expires, it notifies the owning process with an asynchronous notification of timer expiration. As described in Section 8.3.1.3, the notification can be carried out either by a signal or by awakening a thread belonging to the process. On timer expiration, the operating system also reloads the timer with its reload value, if it has been set, thus possibly realizing a repetitive timer.

When a timer is no longer needed, it shall be removed by means of the `timer_delete` function, that both stops the timer and frees all resources allocated to it.

Since, due to scheduling or processor load constraints, a process could lose one or more notifications of expiration, the standard also specifies a way for applications to retrieve, by means of the `timer_getoverrun` function, the number of "missed" notifications, that is, the number of extra timer expirations that occurred between the time at which a given timer expired and when the notification associated with the expiration was eventually delivered to, or accepted by, the process.

At any time, it is also possible to store a new value into, or retrieve the current value of, a timer by means of the `timer_settime` and `timer_gettime` functions, respectively.

8.3.1.9 Cancellation

Any thread may request the "cancellation" of another thread in the same process by means of the `pthread_cancel` function. Then, the target thread's cancelability state and type determine whether and when the cancellation takes effect. When the cancellation takes effect, the target thread is terminated.

Each thread can atomically get and set its own way to react to a cancellation request by means of the `pthread_setcancelstate` and `pthread_setcanceltype` functions. In particular, three different settings are possible:

- Thread can ignore cancellation requests completely.
- Thread can accept the cancellation request immediately.
- Thread can be willing to accept the cancellation requests only when its execution flow crosses a "cancellation point." A cancellation point can be explicitly placed in the code by calling the `pthread_testcancel` function. Also, it should be remembered that many functions specified by the IEEE Std 1003.1-2004 standard act as implicit cancellation points.

The choice of the most appropriate response to cancellation requests depends on the application and is a trade-off between the desirable feature of really being able to cancel a thread and the necessity of avoiding the cancellation of a thread while it is executing in a critical section of code, both to keep the guarded data structures consistent and to ensure that any IPC object associated with the critical section, such as a mutex, is released appropriately; otherwise, the critical region would stay locked forever, likely inducing a deadlock in the system.

As an aid to do this, the IEEE Std 1003.1-2004 standard also specifies a mechanism that allows any thread to register a set of "cleanup handlers" on a stack to be executed, in LIFO order, when the thread either exits voluntarily or accepts a cancellation request. The `pthread_cleanup_push` and `pthread_cleanup_pop` functions push and pop a cleanup handler into and from the handler stack; the latter function also has the ability to execute the handler it is about to remove.

8.3.2 Networking Support

8.3.2.1 Design Guidelines

There are two basic approaches to implement network support in a real-time operating system and to offer it to applications:

- The IEEE Std 1003.1-2004 standard [12] specifies the "socket" paradigm, and most real-time operating systems conforming to the standard provide it. Sockets, fully described in Ref. [21], were first introduced in the "Berkeley Unix" operating system and are now available on virtually all general-purpose operating systems; as a consequence, most programmers are likely to be proficient with them.

The main advantage of sockets is that they support in a uniform way any kind of communication network, protocol, naming conventions, hardware, and so on. Semantics of communication and naming are captured by communication domains and socket types, both specified upon socket creation. For example, communication domains are used to distinguish between IPv4 and X25 network environments, whereas the socket type determines whether communication will be stream based or datagram based and also implicitly selects the network protocol a socket will use.

Additional socket characteristics can be set up after creation through abstract socket options; for example, socket options provide a uniform, implementation-independent way to set the amount of receive buffer space associated with a socket.

- Other operating system specifications, mostly focused on a specific class of embedded applications, offer network support through a less general, but more rich and efficient, application programming interface. For example, Ref. [26] is an operating system specification oriented to automotive applications; it specifies a communication environment (OSEK/VDX COM) less general than sockets and oriented to real-time message-passing networks, such as the CAN.

  In this case, for example, the application programming interface allows applications to easily set message filters and perform out-of-order receives, thus enhancing their timing behavior; both these functions are not completely straightforward to implement with sockets, because they do not fit very well within the general socket paradigm.

In both cases, network device drivers are usually supplied by third-party hardware vendors and conform to a well-defined interface defined by the operating system vendor.

The socket facility and its application programming interface were initially designed to enhance the interprocess communication capabilities of the Berkeley 4.2BSD operating system, a predecessor of 4.4BSD [21]. Before that release, Unix systems were generally weak in this area, leading to the offspring of several, incompatible experimental facilities which did not enjoy widespread adoption.

The interprocess-communication facility of 4.2BSD was developed with several goals in mind, of which the most important one was to provide access to communication networks, such as the Internet that was just born at that time [6]; hence, the interprocess-communication and network-communication subsystems were tightly intertwined from the very beginning.

Another important goal was to overcome many of the limitations of the existing pipe mechanism, in order to allow multiprocess programs—such as distributed databases—to be implemented in an efficient and straightforward way. In order to do this it was necessary to grant any pair of processes the ability to communicate between themselves, even if they did not share a common ancestor. In summary, the socket facility was designed to support:

- *Transparency*: The communication among processes should not depend on the physical location of the communicating processes (on a single host or on multiple hosts) and should be as much independent as possible from the communication protocols being used.
- *Efficiency*: In order to obtain higher performance, it was decided to layer interprocess communication on top of network communication and not vice-versa, even if the latter option would have been more modular, because this would have implied an indirect (and less efficient) access to the network communication services by means of a network server accessed by means of the interprocess-communication facility, thus involving multiple context switches during each client–server interaction.

TABLE 8.2   Main Functions of the IEEE Std 1003.1-2004 Application Programming Interface for Sockets

Functional Group	Main Functions
Communication endpoint management	`socket, close, shutdown, getsockopt, setsockopt`
Local socket address	`bind`
Connection establishment	`connect, listen, accept`
Data transfer	`send, sendto, sendmsg, recv, recvfrom, recvmsg, read, write`
Socket polling and selection	`fcntl, select, pselect, poll, FD_CLR, FD_ISSET, FD_SET, FD_ZERO`

- *Compatibility*: The new communication facility should not depart significantly from the traditional standard input and standard output interfaces commonly used by Unix programs, so that naive processes using it should still be usable with either no or minimal modifications in a distributed environment.

8.3.2.2 Communication Endpoint Management

Table 8.2 summarizes the main functions that compose the IEEE Std 1003.1-2004 application programming interface for sockets and will be described in this and the following sections.

In order to use the interprocess-communication facility, a process must first of all create one or more communication endpoints, known as "sockets." This is accomplished through the invocation of the `socket` function with three arguments:

1. A "protocol family" identifier, which uniquely identifies the network communication domain the socket belongs to and operates within. A communication domain is an abstraction introduced to group together sockets with common communication properties, such as their endpoint addressing scheme, and also implicitly determines a communication boundary because data exchange can take place only among sockets belonging to the same domain. For example, the `PF_INET` domain identifies the Internet and `PF_ISO` identifies the ISO/OSI communication domain. Another commonly used domain is `PF_UNIX`; it identifies a communication domain local to a single host.

2. A "socket type" identifier that specifies which communication model will be obeyed by the socket and, consequently, determines which communication properties will be visible and available to its user.

3. A "protocol identifier" that specifies protocol stack—among those suitable for the given protocol family and socket type—the socket will use. In other words, the communication domain and the socket type are orthogonal to each other and together determine a (possibly empty) set of communication protocols that belongs to the domain and obeys the communication model the socket type calls for. Then, the protocol identifier can be used to narrow the choice down to a specific protocol in the set.

   The special identifier 0 (zero) specifies that a default protocol, selected by the underlying socket implementation, shall be used. It should also be noted that, in most cases, this is not a source of ambiguity, because most protocol families support exactly one protocol for each socket type.

   For example, the identifiers `IPPROTO_TCP` and `IPPROTO_ICMP` select the well-known transmission control protocol (TCP) and Internet control message protocol (ICMP), respectively. Both protocols are defined in the Internet communication domain, so they shall be used only in concert with the `PF_INET` protocol family.

As a result, the `socket` function returns to the caller either a small integer, known as socket descriptor, which represents the socket just created, or a failure indication. The socket descriptor shall be passed to all other socket-related functions in order to reference the socket itself. Like most other IEEE Std 1003.1-2004 functions, in case of failure the "errno" variable conveys to the caller additional information about the reason of the failure.

The standard currently specifies three different socket types, albeit implementations are free to furnish additional ones:

1. The `SOCK_STREAM` socket type provides a connection-oriented, bidirectional, sequenced, reliable transfer of a byte stream, without any notion of message boundaries. It is hence possible for a message sent as a single unit to be received as two or more separate pieces, or for multiple messages to be grouped together at the receiving side.

2. Sockets of type `SOCK_SEQPACKET` behave like stream sockets, but they also keep track of message boundaries.

3. On the other hand, the `SOCK_DGRAM` socket type still supports a bidirectional data flow with message boundaries but does not provide any guarantee of sequenced or reliable delivery. In particular, the messages sent through a datagram socket may be duplicated, lost completely, or received in an order different from the transmission order, with no indication about these facts being conveyed to the user.

The semantics of the `close` function, already defined to close a file descriptor, has also been overloaded to destroy a socket, given a socket descriptor. It may (and should) be used to recover system resources when a socket is no longer in use.

If the `SO_LINGER` option has not been set for the socket, the `close` call is handled in a way that allows the calling thread to proceed as soon as possible, possibly discarding part or all of the data still queued for transmission. Instead, if the `SO_LINGER` option has been set, `close` blocks the calling thread until any unsent data is successfully transmitted or a user-defined timeout, also called lingering period, expires.

Regardless of the setting of the `SO_LINGER` option it is also possible to disable further send and/or receive operations on a socket by means of the `shutdown` function.

Socket options can be retrieved and set by means of a pair of generic functions, `getsockopt` and `setsockopt`. The way of specifying options to these functions is modeled after the typical layered structure of the underlying communication protocols and software. In particular, each option is uniquely specified by a (`level`, `name`) pair, in which

- `level` indicates the protocol level at which the option is defined. In addition, a separate level identifier (`SOL_SOCKET`) is reserved for the upper layer, that is, the socket level itself, which does not have a direct correspondence with any protocol.

- `name` determines the option to be set or retrieved within the level and, implicitly, the additional arguments of the functions.

As outlined above, additional arguments allow the caller to pass to the functions the location and size of a memory buffer used to store or retrieve the value of the option, depending on the function being invoked. Continuing the previous example, the `SO_LINGER` option is defined at the socket level. The memory buffer associated with it shall contain a data structure of type `struct linger`, which contains two fields: the first one is a flag that specifies whether the option is active or not, whereas the second one is an integer value that holds the lingering timeout expressed in seconds.

8.3.2.3 Local Socket Address

A socket has no address when it is initially created by means of the `socket` function. On the other hand, a socket must have a unique local address to be actively engaged in data reception, because communicating sockets are bound by associating them and, to make an association, their addresses must be known.

The exact address format and its interpretation may vary depending on the communication domain. For example, within the Internet communication domain, addresses contain a 4 byte IP address and a 16 bit port number (assuming that IPv4 is in use). Other domains, for example `PF_UNIX`, use character strings of variable length, formatted as path names, as addresses.

A local address is bound to a socket in two distinct ways, depending on how the application intends to use it:

- The `bind` function "explicitly" gives a certain local address, specified as a function argument, to a socket. This course of action gives to the caller full control on what address will be assigned to the socket.
- Other functions, `connect` for instance, automatically and "implicitly" bind an appropriate, unique address when invoked on an unbound socket. In this case, the caller is not concerned at all with address selection but, on the other hand, has also no control on local address assignment.

8.3.2.4   Connection Establishment

The `connect` has a local socket descriptor and a target socket address as arguments. When invoked on a connection-oriented socket, it sets out a connection request directed toward the target address specified in the call. Moreover, if the local socket is currently unbound, the system also selects and binds an appropriate local address to it before attempting the connection.

If the function succeeds, it associates the local and the target sockets and data transfer can begin. Otherwise, it returns to the caller an error indication.

In order to be a valid target for a `connect`, a socket must first of all have a well-known address (because the process willing to connect has to specify it is in the `connect` call); then, it must be marked as willing to accept connection requests.

As described in the previous section, address assignment can be performed by means of the `bind` function. On the other hand, the willingness to listen to incoming connection requests is expressed by calling the `listen` function.

The first argument of this function is, as usual, a socket descriptor to be acted upon. The second argument is an integer that specifies the maximum number of outstanding connection requests that may be waiting acceptance by the process using the socket, known as "backlog." It should be noted that the user-specified value is handled as a hint by the socket implementation, which is free to reduce it if necessary.

If a new connection request is attempted while the queue of outstanding requests is full, the connection can either be refused immediately or, if the underlying protocol implementation supports this feature, the request can be retried at a later time.

After a successful execution of `listen`, the `accept` function can then be used to wait for the arrival of a connection request on a given socket. The function blocks the caller until a connection request arrives and then accepts it and clones the original socket so that the new socket is connected to the originator of the connection request, and the old one is still available to wait for further connection requests.

The descriptor of the new socket is returned to the caller to be used for subsequent data transfer. Moreover, the `accept` function has also the ability of providing to the caller the address of the socket that originated the connection request.

Figure 8.3 summarizes the steps to be performed by the communicating processes in order to prepare a pair of connection-oriented sockets for data transfer.

8.3.2.5   Connectionless Sockets

Connectionless interactions are typical of datagram sockets and do not require any form of connection negotiation or establishment before data transfer can take place.

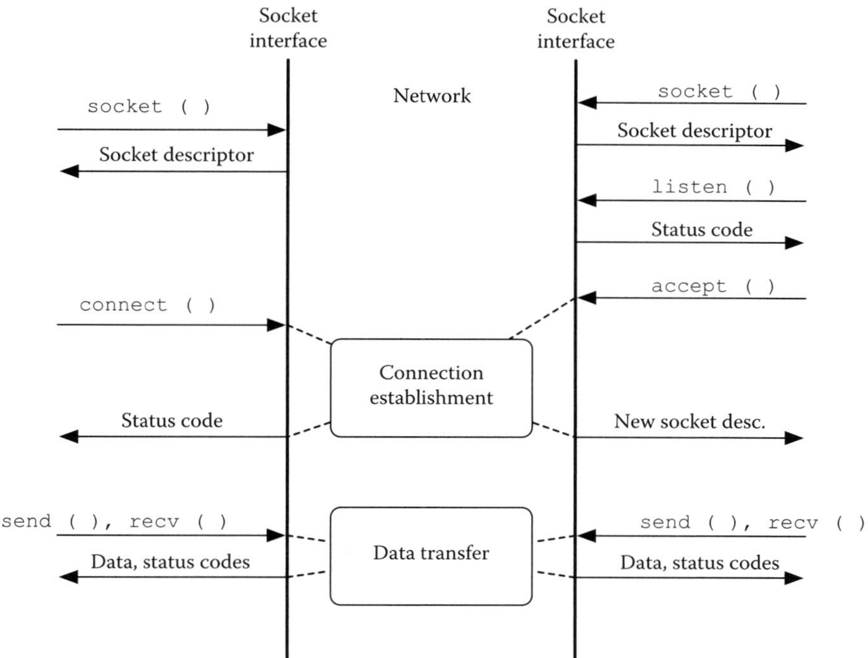

FIGURE 8.3 Socket creation and connection establishment in the IEEE Std 1003.1-2004.

Socket creation proceeds as for connection-oriented sockets, and `bind` can be used to assign a specific local address to a socket. Moreover, if a send operation is invoked on an unbound socket, the socket is implicitly bound to an appropriate local address before transmission.

Due to the lack of need (and support) for connection establishment, `listen` and `accept` cannot be used on a connectionless socket. On the other hand, `connect` simply associates a destination address with the socket so that, in the future, it will be possible to use it with data transmission functions which do not explicitly indicate the destination address, like `send`. Moreover, after a successful `connect`, only data received from that remote address will be delivered to the user.

The `connect` function can be used multiple times on the same socket, but only the last address specified remains effective. Unlike for connection-oriented sockets, in which `connect` implies a certain amount of network activity, connect requests on connectionless sockets return much faster to the caller, because they simply result in the system recording the remote address locally.

If `connect` has not been used, the only way to send data through a connectionless socket is by means of a function that allows the caller to specify the destination address for each message to be sent such as `sendto` and `sendmsg`.

8.3.2.6 Data Transfer

The functions `send`, `sendto`, and `sendmsg` allow the caller to send data through a socket, with different trade-offs between expressive power and interface complexity:

- The `send` function is the simplest one and assumes that the destination address is already known to the system. Hence, it does not support the explicit indication of this address and cannot be used on connectionless sockets without a former `connect`.

  Instead, its four arguments specify the socket to be used, the position and size of a memory buffer containing the data to be sent, and a set of flags that may alter the semantics of the function.

- Compared to the previous one, the `sendto` function is more powerful, because it allows the caller to explicitly specify a destination address, making it most useful for connectionless sockets.
- The `sendmsg` function is the most powerful and adds the capability of

  - Gathering the data to be transmitted as a single unit from a sequence of distinct buffers in memory instead of a single one

  - Specifying additional data related to protocol management or other miscellaneous ancillary data

  In order to keep the total argument count reasonably low, `sendmsg` takes as argument a single data structure of type `struct msghdr` that holds in its fields most of the information described above.

Conversely, the `recv`, `recvfrom`, and `recvmsg` functions allow a process to wait for and retrieve incoming data from a socket. Like their transmitting-side counterparts, they have different levels of expressive power:

- The `recv` function waits for the arrival of a message from a socket, stores it into a data buffer in memory, and returns to the caller the length of the data just received. It also accepts as argument a set of flags that may alter the semantics of the function.
- In addition, the `recvfrom` function allows the caller to retrieve the address of the sending socket, making it useful for connectionless sockets, in which the communication endpoints may not be permanently paired.
- Finally, the `recvmsg` function allows the caller to scatter the received data into a set of distinct buffers in memory instead of a single one. It also allows the caller to retrieve additional data related to protocol management or other miscellaneous ancillary data. Like `sendmsg`, `recvmsg` also groups most of its arguments into a single data structure of type `struct msghdr` to keep the total argument count low.

Besides these specialized functions, very simple applications can also use the normal `read` and `write` functions, originally devised to operate on file descriptors, with connection-oriented sockets, after a connection has been successfully established.

8.3.2.7 Socket Polling and Selection

The default semantics of the socket functions described so far is to block the caller until they can proceed: for example, the `recv` function blocks the caller until there is some data available to be retrieved from the socket or an error occurs.

Even if this behavior is quite useful in some situations because it allows the software to be written in a simple and intuitive way, it may become a disadvantage in other, more complex situations. If we consider, for example, a network server, it will probably be connected to a number of clients at a time and will not know in advance from which socket the next request message will arrive. In this case, if the server performs a `recv` on a specific socket, it runs into the risk of ignoring messages from other sockets for an indeterminate amount of time.

The standard provides two distinct and independent ways to avoid this issue. They can be used either alone or in combination, even in the same program.

1. By means of the `fcntl` function, it is possible to change the socket I/O mode by setting the O_NONBLOCK flag. When this flag is set, all socket functions that would normally block until completion either return a special error code if they cannot immediately finish their operation or conclude the operation asynchronously with respect to the execution of

the caller. In this way, it becomes possible to perform a periodic polling on each member of a set of sockets.

For example, the `recv` function used for data transfer immediately returns to the caller an "error" indication when it is invoked on a `O_NONBLOCK` socket and no data is immediately available to be retrieved. The caller can then distinguish the mere unavailability of data from other errors by consulting the global variable "errno" as usual.

In the same situation, the `connect` function shall initiate a connection but, instead of waiting for its completion, shall immediately return to the caller with an error indication. Again, this does not mean that a true error condition has been encountered, but only that the connection (albeit successfully initiated) has not been completed yet.

2. If polling is not an option for a given application, because its overhead would be unacceptable due to the large number of sockets to be handled, it is also possible to use synchronous multiplexing and block a process until certain I/O operations are possible on any socket in a set. The standard specifies three main functions for this:

`select` takes as input three, possibly overlapping, sets of file or socket descriptors and a timeout value. It examines the descriptors belonging to each set in order to check whether at least one of them is ready for reading, ready for writing, or has an exceptional condition pending, respectively.

The function blocks the calling process until the timeout expires, a signal is caught, or at least one of the events being watched occurs. In the latter case, the function also updates its arguments to inform the caller about which descriptors were involved in the events just detected. A more sophisticated function, `pselect`, allows the caller to specify the timeout with a higher resolution and to alter the signal mask in effect for the calling thread during the wait.

A set of descriptors can be manipulated by means of the macros `FD_ZERO` (to initialize a set to be empty), `FD_SET` and `FD_CLR` (to insert and remove a descriptor into/from the set, respectively), and `FD_ISSET` (to check whether or not a certain descriptor belongs to the set.

`poll` takes a different approach and, instead of partitioning the descriptors being watched into three broad categories, it supports a much wider and more specific set of conditions to be watched for each descriptor. In order to do this, the function takes as input a set of data structures, one for each descriptor to be watched, and each structure contains the set of interest for the corresponding descriptor.

The main disadvantage of `poll` is that (unlike `select` and `pselect`) it only accepts timeout values with a millisecond resolution.

8.4   Extending the Berkeley Sockets

The implementation of sockets known as Berkeley sockets found in the BSD operating system and thoroughly described in Ref. [21] has been much influential because it was one of the very first implementations of the socket concept. For this reason, it was used as a starting point by most other socket implementations widespread nowadays and, in particular, it has been adopted by several popular open-source, real-time operating systems [7,25].

Then, looking at how Berkeley sockets can be extended to handle other protocols, besides the TCP/IP suite for which they were originally designed, is of particular interest because, in this way, it becomes possible to seamlessly support communication media and protocols more closely tied to the real-time domain such as the CAN [5].

Moreover, similar conclusions can also be drawn with few modifications for a wide range of other operating systems and communication protocols. For example, in the open-source community a similar goal, albeit more focused on the Linux operating system (either with or without real-time support), is being pursued by the Socket-CAN project [34].

8.4.1 Main Data Structures

The Berkeley sockets implementation is based on, and revolves around, several data structures:

- The `domain` structure holds all information about a communication domain; in particular, it contains the symbolic protocol family identifier assigned to the communication domain (for example, `PF_INET` for the Internet), its human-readable name, and a pointer to an array of protocol switch structures, one for each protocol supported by the communication domain, to be described later.

  In addition, it contains a set of pointers to domain-specific routines for the management and transfer of access rights and for routing initialization.

  The socket implementation maintains a globally accessible table of domain structures, one for each communication domain known to the system.

- The `protosw` (protocol switch) data structure describes a protocol module. Among other fields, it contains some information to identify the protocol (i.e., the type of socket it supports, the domain which it belongs to, and its protocol number) and pointers to the set of externally accessible entry points of the protocol module. The former information is used to choose the right protocol when creating a new socket, whereas the latter is used as a uniform interface to access the protocol module.

  The main interface between layered protocol modules and between the topmost protocol module and the socket level, is the `pr_usrreq` entry point. It is invoked by the upper levels of the protocol stack, with an appropriate request code and additional arguments, whenever the protocol module must perform an action. For example, `pr_usrreq` is invoked with the `PRU_SEND` request code when a `send` operation is invoked at the socket level.

- The `socket` data structure represents a communication endpoint and contains information about the type of socket it supports and its state. In addition, it provides buffer space for data coming from, and directed to, the process that owns the socket and may hold a pointer to a chain of protocol state information.

  Upon creation of a new socket, the table of domain structures and the table of protocol switch structures associated with its elements are scanned, looking for a protocol switch entry that matches the arguments passed to the socket creation function. That entry is then linked to the socket data structure through the `so_proto` pointer and is used as the only interface point between the top-level socket layer and the communication protocol.

  Within the socket data structure there are two data queues, one for transmission and the other for reception. These queues are manipulated through a uniform set of utility functions. For example, the `sbappend` function appends a chunk of data to a queue and is therefore invoked whenever a new data message is received from the lower levels.

- The `ifnet` (network interface) data structure represents a network interface module, with which a hardware device is usually associated, provides a uniform interface to all network devices that may be present on a host, and insulates the upper layers of software from the implementation details of each device. The main purpose of a network interface module is to interact with the corresponding hardware device, in order to send and receive data-link level packets. A list of `ifaddr` data structures, each representing an interface address in possibly different communication domains, is linked to the main `ifnet` structure.

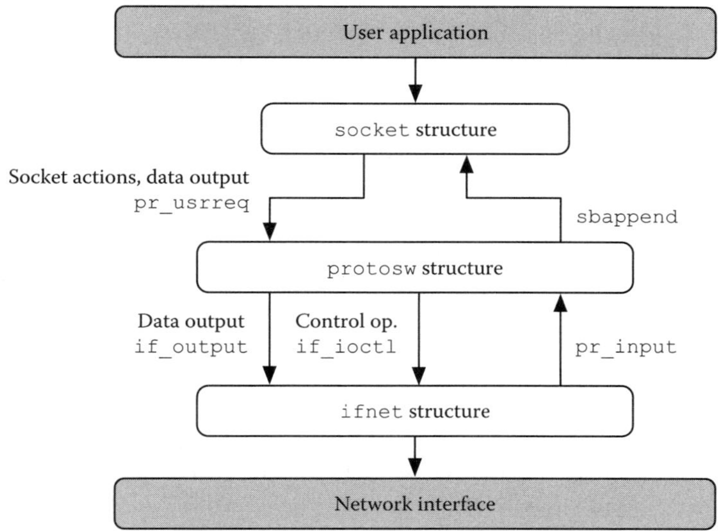

FIGURE 8.4   Main Berkeley sockets data structures and their relationships.

At this level, the main entry points are

`if_output,` which is responsible for data output through the interface

`if_ioctl,` which performs all control operations on the interface

In addition, another data structure, the `mbuf`, is used ubiquitously when dynamic storage allocation is needed. Its implementation makes it particularly suitable to prepend and append further data to an existing buffer, an operation frequently used in communication protocols for encapsulation and deencapsulation.

Figure 8.4 summarizes the relationships among some of the data structures just mentioned for the CAN data-link communication domain. It should be noted that, although in principle there could be multiple protocol modules stacked between the socket and the network interface layers, for CAN only one module is needed, because the protocols defined within the CAN data-link communication domain have the purpose of giving direct access to the CAN data-link layer, and are not built one on top of another. On the other hand, multiple protocols can still be linked to the same interface data structure to support different interface access paradigms.

At the socket level, the association between a socket descriptor and the corresponding socket structure is carried out by means of a table that, in general-purpose operating systems, often coincides with the file table. On simpler systems, it can simply be a per-process array of pointers to socket structures, in which the socket descriptor (a small integer) is used as an index.

Between the socket structure and the protocol switch structure, there is a direct link held in the `so_proto` field of the socket structure. For CAN sockets, like most other kinds of sockets, this link is initialized once and for all at socket creation time, depending on the protocol argument passed to `socket`. On network infrastructures that provide for network-layer routing, for example, the Internet Protocol (IP), the selection of the right output interface is usually carried out by the local routing algorithm, but this possibility has not been further considered because no routing facilities have been introduced for the CAN communication domain.

On the other hand, since the application still can, at least in principle, access different network interfaces from the same socket, it is impossible to establish a static link between the protocol switch structure and a particular interface structure for CAN sockets. Instead, the link is formed on a frame-by-frame basis from the information found in the destination address that explicitly holds, among other information, the destination interface number.

For input messages a similar, dynamic association between the input interface, the protocol switch, and the socket data structure must be established, in order to transfer the messages up to the user level and make them available at the right service endpoint. For CAN sockets, both associations can easily be carried out by means of the message identifier, possibly integrated by a filtering mechanism based on the message identifier itself and a per-socket mask. In this respect, the CAN message identifier is used in the same way as the port number of more sophisticated communication protocols, for example, TCP.

8.4.2   Interrupt Handling

The main adaptation to make the Berkeley sockets implementation more suitable for a real-time execution environment is related to the mutual exclusion and synchronization methods used within the communication modules themselves. In fact, for the sake of maximum efficiency and optimal (mean) latency, in the original implementation the network communication functions are executed at three different interrupt priority levels (IPLs):

1. The highest level, traditionally called `splimp`, is used to execute the interrupt handlers of the network interfaces.
2. An intermediate level, `splnet`, is used by a software interrupt handler to carry out the bulk of network protocol processing.
3. All high-level socket operations discussed in Section 8.3.2 are carried out with interrupts fully enabled.

To enforce the mutual exclusion between these concurrent activities, the processor IPL is temporarily raised to either `splimp` or `splnet` when necessary. For example, in order to retrieve data from the network interface buffers, the IPL is temporarily raised to `splimp`, to prevent the receive interrupt handler of the interface from performing a concurrent store into the same buffers, which would lead to a race condition.

For synchronization, the Berkeley sockets implementation makes widespread use of "wait channels," the traditional synchronization mechanism of the BSD operating system. A process can perform a passive wait on a channel by means of the `tsleep` primitive, whereas `wakeup` awakens all processes sleeping on a given channel.

On the one hand, the mutual exclusion mechanism just described is not particularly suitable for a real-time system, because it keeps interrupts partially disabled for an amount of time which is difficult to predict and may be long. On the other hand, wait channels are unavailable on most contemporary operating systems.

As a consequence, both facilities must be emulated by means of an adaptation layer and implemented in terms of the mutual exclusion and synchronization mechanisms provided by the operating system. For example, in Ref. [7] a mutual exclusion lock is used to emulate IPL settings, synchronization is ensured by means of a semaphore for each wait channel, and the critical regions within the adaptation layer itself are implemented by temporarily locking out the scheduler.

With this approach, hardware interrupts are never disabled except for the tiny and bounded amount of time needed to execute the basic synchronization primitives just mentioned. In addition, it is still possible to develop a network device driver by reusing the same, well-known coding structure of the corresponding BSD driver, because the emulation layer is available also in that case.

For what concerns the CAN protocol, most controllers take a very simple approach to interrupt handling, and hence the standard interrupt handling framework of Berkeley sockets is more than adequate to accommodate them. When working with real-time operating systems that, as described in Section 8.2.1, support two-level interrupt handling, the main design choice to be taken is how to partition the overall interrupt handling code into the two levels, to achieve an optimal balance between a quick reaction to interrupt requests coming from the CAN controller (needed, for example, to avoid overflowing the on-chip receive buffer) and an acceptable overall interrupt handling latency (undermined by keeping interrupts disabled for a long time).

However, in the case described by Cena et al. [5], experimental evidence showed that the total time spent in the interrupt handler is dominated by the data transfer operations associated with the movement of a CAN message to/from the controller and is very short due to the high speed of the controller itself and to the very limited quantity of data to be transferred. Hence, for the sake of simplicity, an acceptable choice is to place the interrupt handling code as a whole within the first-level interrupt handler and keep the second level completely empty.

8.4.3 Interface-Level Resources

Unlike other communication protocols the CAN data-link, when implemented on a network interface that follows the FullCAN design principle [8], requires the allocation of interface-level resources private to each socket opened on the interface, for example, message buffers. The allocation happens in different phases of the lifetime of a socket, depending on the protocol in use and on the user choice. However, this aspect does not pose significant implementation problems because most socket-level actions either explicitly requested by a user or implicitly carried out by the socket are propagated to the protocol-switch level through the `pr_usrreq` entry point of the protocol switch structure. From there, they can be acted upon and/or propagated to the interface level through the interface `if_ioctl` (I/O control) entry point, to trigger the allocation of interface-level resources.

For example, the creation of a new socket with a given protocol results in the assignment of the `so_proto` field of the socket to point to the right protocol switch data structure, and in the activation of the `pr_usrreq` entry point of the protocol switch with the request code `PRU_ATTACH`, to denote that a fresh socket is being attached to the protocol.

Similarly, the invocation of the `bind` function on a socket results in the activation of the same entry point with the request code `PRU_BIND` and may trigger the allocation of a receive message buffer in a FullCAN controller. The controlled release of the protocol and interface-level resource is carried out in the same manner, for example, when the socket is closed.

8.4.4 Data Transfer

The transmission of a data frame is triggered by the invocation of the `send` function and follows the usual path inside Berkeley sockets. In the downward path, the main concern of the various functions involved is to validate the transmission request. With respect to data reception, three different possibilities must be accounted for

1. Sometimes, the received data must be passed to the user level immediately and without further actions, to satisfy a pending `recv`. This is the simplest case, and follows the same course of actions as, for example, the reception of a UDP datagram. For UDP datagrams, the destination socket structure is determined from the UDP port number found in the datagram. Instead, in this case, the destination socket is determined by applying the filtering mechanism to match the incoming message with one of the open sockets in the system.

2. When remotely requested transmissions are enabled, a mechanism that has no exact counterpart in any of the protocols originally supported by Berkeley sockets, the reception of an RTR frame must be handled specially, because it may require both an immediate reaction from the protocol layer, namely, the transmission of the requested

data frame and an optional notification of the user layer when it is waiting on a `recv` focused on RTR frames. Albeit slightly more complex than the previous case, this execution flow is well supported by the Berkeley sockets framework, too. In fact it is typical of TCP, where the reception of a segment may trigger both the transmission of an acknowledge and the propagation of the segment contents to the user layer.

3. Finally, one must also consider the activities related to nonattributable error conditions, that is, errors that are not directly attributable to a specific socket, for example, a cyclic redundancy check (CRC) error in a received message (its message identifier could be invalid) or the CAN interface entering the bus off state due to an excessive error rate.

   Also in this case, there is no direct relationship with other protocols already available in the Berkeley sockets, but they can nonetheless be implemented in a straightforward way, because they can be handled like a normal data reception. In fact, the events that trigger these activities have origin from an interrupt request made by the network interface, like data reception, and they must be conveyed to a particular socket, solely devoted to gather and propagate to a dedicated task error indications of this kind. Hence, this is an activity that can be seen as a (peculiar) form of filtering and treated in the same way.

8.4.5 Real-Time Properties

The original Berkeley sockets interface and implementation were designed for a general-purpose operating system and did not take into account any kind of real-time execution constraint. Also, most underlying communication protocols to which the design was targeted, for example, TCP, were not engineered to provide any real-time guarantee. However, this has not impaired the integration of Berkeley sockets into several popular real-time operating systems [7,25]. Furthermore, the ability of the socket framework to accommodate specialized protocols for real-time communications was shown several times in literature.

As a consequence, a socket implementation including the CAN communication protocol, when accompanied by an appropriate support at the operating system level (e.g., for real-time scheduling and time measurement and synchronization) could be used not only for device management activities, whose real-time execution constraints are usually quite relaxed, but for real-time data exchange as well.

References

1. Advanced Micro Devices Inc. *AMD I/O Virtualization Technology (IOMMU) Specification*, February 2007. Available online, at http://www.amd.com/
2. ARM Ltd. *ARM Architecture Reference Manual*, July 2005. Available online, at http://www.arm.com/
3. P. Barham, B. Dragovic, K. Fraser, S. Hand, T. Harris, A. Ho, R. Neugerbauer, I. Pratt, and A. Warfield. Xen and the art of virtualization. In *Proceedings of the 19th ACM Symposium on Operating System Principles*, Bolton Landing, NY, October 2003.
4. Beckhoff Automation GmbH. *TwinCAT System Overview*. Available online, at http://www.beckhoff.com/.
5. G. Cena, I. Cibrario Bertolotti, A. Valenzano. A socket interface for CAN devices. *Elsevier Computer Standards & Interfaces*, 29(6), 662–673, 2007, doi 10.1016/j.csi.2007.07.004.
6. V. Cerf. The catenet model for internetworking. Tech. Rep. IEN 48, SRI Network Information Center, 1978.
7. eCosCentric Ltd. *eCos User Guide*. Available online, at http://ecos.sourceware.org/

8. K. Etschberger. *Controller Area Network Basics, Protocols, Chips and Applications*. IXXAT Press, Weingarten, Germany, 2001.

9. FSMLabs Inc. *Real-Time Management Systems (RTMS) Overview*. Available online, at http://www.fsmlabs.com/.

10. R. Goldberg, Architectural principles for virtual computer systems. PhD thesis, Harvard University, 1972.

11. Green Hills Software Inc. *velOSity Real-Time Operating System*. Available online, at http://www.ghs.com/

12. IEEE Std 1003.1-2004. *Standard for Information Technology—Portable Operating System Interface (POSIX)—System Interfaces*. The IEEE and The Open Group, 2004.
    Also available online, at http://www.opengroup.org/

13. IEEE Std 1003.13-2003. *Standard for Information Technology—Standardized Application Environment Profile (AEP)—POSIX Realtime and Embedded Application Support*. The IEEE, New York, 2004.

14. ISO/IEC 9899:1999. *Programming Languages—C*. International Standards Organization, Geneva, 1999.

15. KADAK Products Ltd. *AMX User's Guide and Programming Guide*. Available online, at http://www.kadak.com/

16. KADAK Products Ltd. *KwikNet TCP/IP Stack Reference Manual*. Available online, at http://www.kadak.com/

17. J. Kiszka, B. Wagner. RTnet—A flexible hard real-time networking framework. In *Proceedings of the 10th IEEE Conference on Emerging Technologies and Factory Automation (ETFA)*, Catania, Italy; vol. 1, pp. 449–456, September 2005.

18. J. Liedtke. On μ-kernel construction. In *Proceedings of the 15th ACM Symposium on Operating System Principles (SOSP)*, Copper Mountain Resort, CO, December 1995.

19. J. Liedtke. *L4 Reference Manual (486, Pentium, Pentium Pro)*. Arbeitspapier 1021 of GMD—German National Research Center for Information Technology, September 1996.

20. LynuxWorks Inc. *LynxOS Product Information*. Available online, at http://www.lynuxworks.com/

21. M. K. McKusick, K. Bostic, M. J. Karels, and J. S. Quarterman. *The Design and Implementation of the 4.4BSD Operating System*. Addison-Wesley, Reading, MA, 1996.

22. Mentor Graphics, Inc. *Nucleus OS Brochure*. Available online, at http://www.mentor.com/.

23. R. A. Meyer and L. H. Seawright. A virtual machine time-sharing system. *IBM Systems Journal*, 9(3), 199–218, 1970.

24. G. Neiger, A. Santoni, F. Leung, D. Rodgers, and R. Uhlig. Intel virtualization technology: Hardware support for efficient processor virtualization. *Intel Technology Journal*, 10(3), 167–177, 2006. Available online, at http://www.intel.com/technology/itj/

25. On-Line Applications Research Corp. *RTEMS Documentation*. Available online, at http://www.rtems.com/

26. OSEK/VDX. *OSEK/VDX Operating System Specification*. Available online, at http://www.osek-vdx.org/

27. Politecnico di Milano, Dip. di Ingegneria Aerospaziale. *RTAI 3.4 User Manual*. Available online, at https://www.rtai.org/

28. QNX Software Systems Ltd. *QNX Neutrino RTOS Data Sheet*. Available online, at http://www.qnx.com/

29. RadiSys Corp. *OS-9 Product Data Sheet*. Available online, at http://www.radisys.com/

30. RTMX Inc. *RTMX O/S Data Sheet*. Available online, at http://www.rtmx.com/

31. Siemens AG. *Simatic WinAC RTX Product Information*. Available online, at http://www.siemens.com/

32. A. Silberschatz, P. B. Galvin, and G. Gagne. *Operating Systems Concepts*, 7th edn., John Wiley & Sons, Hoboken, NJ, 2005.

33. 3S - Smart Software Solutions GmbH. *CoDeSys Product Tour*. Available online, at http://www.3s-software.com/

34. Socket-CAN. *The Socket-CAN Project.* Available online, at http://socketcan.berlios.de/

35. B. Sprunt, L. Sha, and J. P. Lehoczky. A periodic task scheduling for hard real-time systems. *The Journal of Real-Time Systems,* 1(1), 27–60, 1989.

36. A. S. Tanenbaum. *Modern Operating Systems,* 2nd edn., Prentice Hall, Upper Saddle River, NJ, 2001.

37. Unicoi Systems Inc. *Fusion Net Overview.* Available online, at http://www.unicoi.com/

38. VirtualLogix Inc. *VirtualLogix VLX Product Information.* Available online, at http://www.virtuallogix.com/

39. Wind River Systems Inc. *VxWorks Datasheet.* Available online, at http://www.windriver.com/

40. Xenomai Development Team. *Xenomai: Real-Time Framework for Linux—Wiki.* Available online, at http://www.xenomai.org/.

41. K. Yaghmour. *Adaptive Domain Environment for Operating Systems,* 2001. Available online, at http://www.opersys.com/ftp/pub/Adeos/adeos.pdf.

9

Determining Bounds on Execution Times

Reinhard Wilhelm
University of Saarland

9.1   Introduction

Hard real-time systems are subject to stringent timing constraints which are dictated by the surrounding physical environment. We assume that a real-time system consists of a number of tasks, which realize the required functionality. A schedulability analysis for this set of tasks and a given hardware has to be performed in order to guarantee that all the timing constraints of these tasks will be met ("timing validation"). Existing techniques for schedulability analysis require upper bounds for the execution times of all the system's tasks to be known. These upper bounds are commonly called the worst-case execution times (WCETs), a misnomer that causes a lot of confusion and will therefore not be adopted in this presentation. In analogy, lower bounds on the execution time have been named best-case execution times, (BCET). These upper bounds (and lower bounds) have to be

---

This is an extended and updated version of the material published in 2005 in the Embedded Systems Handbook.

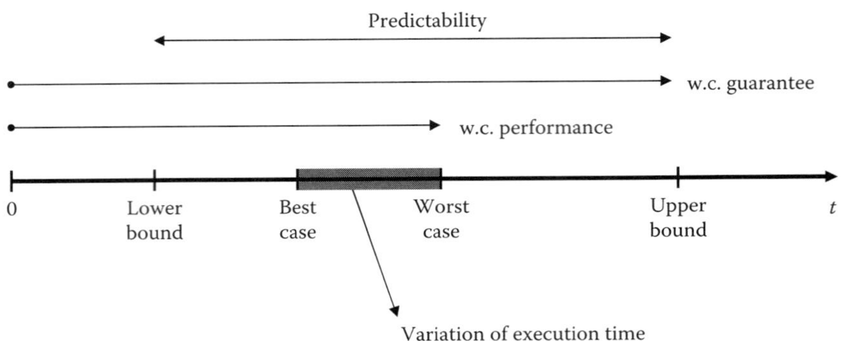

FIGURE 9.1 Basic notions concerning timing analysis of systems.

"safe," i.e., they must never underestimate (overestimate) the real execution time. Furthermore, they should be *tight*, i.e., the overestimation (underestimation) should be as small as possible.

Figure 9.1 depicts the most important concepts of our domain. The system shows a variation of execution times depending on the input data, the initial execution state, and different behavior of the environment. In general, the state space of input data, initial state, and potential interferences is too large to exhaustively explore all possible executions and so determine the exact worst-case and best-case execution times. Some abstraction of the system is necessary to make a timing analysis of the system feasible. These abstractions lose information, and thus are responsible for the distance between WCETs and upper bounds and between BCETs and lower bounds. How much is lost depends both on the methods used for timing analysis and on system properties, such as the hardware architecture and the cleanness of the software. So, the two distances mentioned above, termed "upper predictability" and "lower predictability," can be seen as a measure for the timing predictability of the system. Experience has shown that the two predictabilities can be quite different, cf. [HLTW03]. The methods used to determine upper bounds and lower bounds are the same. We will concentrate on the determination of upper bounds unless otherwise stated.

Methods to compute sharp bounds [PK89,PS91] for processors with fixed execution times for each instruction have long been established. However, in modern microprocessor architectures caches, pipelines, and all kinds of speculation are key features for improving (average-case) performance. Caches are used to bridge the gap between processor speed and the access time of main memory. Pipelines enable acceleration by overlapping the executions of different instructions. The consequence is that the execution time of individual instructions, and thus the contribution of one execution of an instruction to the program's execution time can vary widely. The interval of execution times for one instruction is bounded by the execution times of the following two cases:

- Instruction goes "smoothly" through the pipeline; all loads hit the cache, no pipeline hazard happens, i.e., all operands are ready, no resource conflicts with other currently executing instructions exist.
- "Everything goes wrong," i.e., instruction and/or operand fetches miss the cache, resources needed by the instruction are occupied, etc.

Figure 9.2 shows the different paths through a multiply instruction of a PowerPC processor. The instruction-fetch phase may find the instruction in the cache ("cache hit"), in which case it takes 1 cycle to load it. In the case of a cache miss, it may take 30 cycles (or more, depending on the memory subsystem) to load the memory block containing the instruction into the cache. The instruction needs an arithmetic unit, which may be occupied by a preceding instruction. Waiting for the unit to become free may take up to 19 cycles. This latency would not occur, if the instruction fetch had

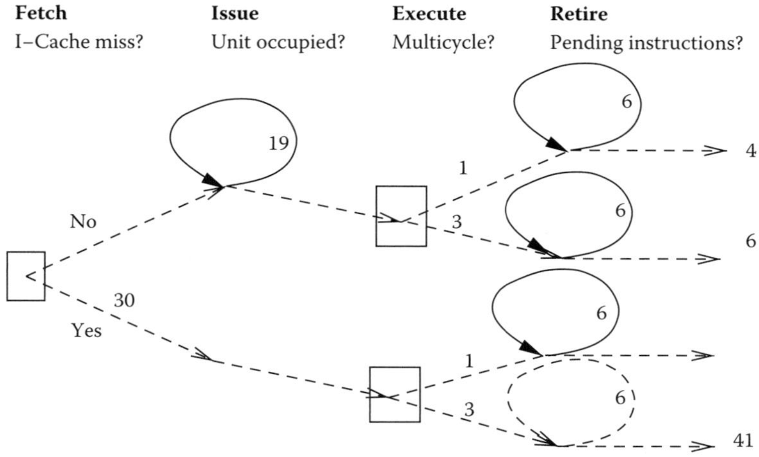

FIGURE 9.2 Different paths through the execution of a multiply instruction. Unlabeled transitions take 1 cycle.

missed the cache, because the cache-miss penalty of 30 cycles has allowed any preceding instruction to terminate its arithmetic operation. The time it takes to multiply two operands depends on the size of the operands; for small operands, one cycle is enough, for larger, three are needed. When the operation has finished, it has to be retired in the order it appeared in the instruction stream. The processor keeps a queue for instructions waiting to be retired. Waiting for a place in this queue may take up to 6 cycles. On the dashed path, where the execution always takes the fast way, its overall execution time is 4 cycles. However, on the dotted path, where it always takes the slowest way, the overall execution time is 41 cycles.

We call any increase in execution time during an instruction's execution a "timing accident" and the number of cycles by which it increases the "timing penalty" of this accident. Timing penalties for an instruction can add up to several hundred processor cycles. Whether or not the execution of an instruction encounters a timing accident depends on the execution state, e.g., the contents of the cache(s), the occupancy of other resources, and thus on the execution history. It is therefore obvious that the attempt to predict or exclude timing accidents needs information about the execution history.

For certain classes of architectures, namely those without timing anomalies, excluding timing accidents means decreasing the upper bounds. However, for those with timing anomalies, this assumption is not true.

9.1.1 Tool Architecture and Algorithm

A more or less standard architecture for timing-analysis tools has emerged [HWH95,TFW00, Erm03]. Figure 9.3 shows one instance of this architecture. The first phase, depicted on the left, predicts the behavior of processor components for the instructions of the program. It usually consists of a sequence of static program analyses of the program. They altogether allow to derive safe upper bounds for the execution times of basic blocks. The second phase, the column on the right, computes an upper bound on the execution times over all possible paths of the program. This is realized by mapping the control flow of the program to an Integer Linear Program and solving this by appropriate methods. This architecture has been successfully used to determine precise upper bounds on the execution times of real-time programs running on processors used in embedded systems [AFMW96,FMW99,FHL+01,TSH+03,HLTW03]. A commercially available tool, aiT by AbsInt, cf.

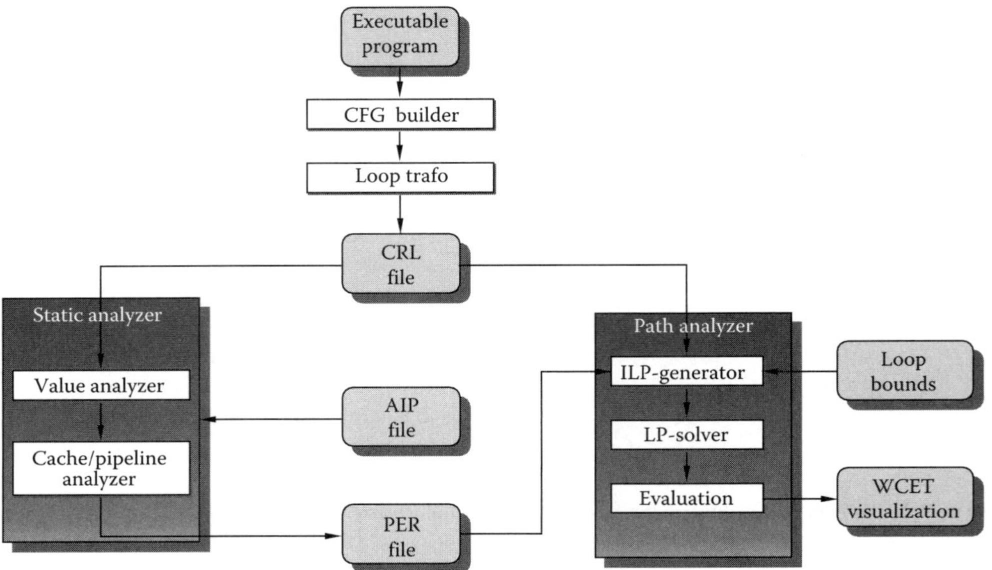

FIGURE 9.3   The architecture of the `aiT` timing-analysis tool.

`http://www.absint.de/wcet.htm`, was implemented and is used in the aeronautics and automotive industries.

The structure of the first phase, "processor-behavior prediction," often called "microarchitecture analysis," may vary depending on the complexity of the processor architecture. The first, modular approach would be the following:

1. Cache-behavior prediction determines statically and approximately the contents of caches at each program point. For each access to a memory block, it is checked, whether the analysis can safely predict a cache hit.

   Information about cache contents can be forgotten after the cache analysis. Only the miss/hit information is needed by the pipeline analysis.

2. Pipeline-behavior prediction analyzes how instructions pass through the pipeline taking cache-hit or miss information into account. The cache-miss penalty is assumed for all cases, where a cache hit cannot be guaranteed.

   At the end of simulating one instruction, the pipeline analysis continues with only those states that show the locally maximal execution times. All others can be forgotten.

9.1.2 Timing Anomalies

Unfortunately, this approach is not safe for many processor architectures. Most powerful microprocessors have so-called timing anomalies. Timing anomalies are contra-intuitive influences of the (local) execution time of one instruction on the (global) execution time of the whole program. The interaction of several processor features can interact in such a way that a locally faster execution of an instruction can lead to a globally longer execution time of the whole program.

For example, a cache miss contributes the cache-miss penalty to the execution time of a program. It was, however, observed for the MCF 5307 [RSW02] that a cache miss may actually speed up program execution. Since the MCF 5307 has a unified cache and the fetch and execute pipelines are independent, the following can happen: A data access that is a cache hit is served directly from the cache.

At the same time, the fetch pipeline fetches another instruction block from main memory, performing branch prediction and replacing two lines of data in the cache. These may be reused later on and cause two misses. If the data access was a cache miss, the instruction fetch pipeline may not have fetched those two lines, because the execution pipeline may have resolved a misprediction before those lines were fetched.

The general case of a timing anomaly is the following. Different assumption about the processor's execution state, e.g., the fact that the instruction is or is not in the instruction cache, will result in a difference, $\Delta T_{\mathrm{local}}$, of the execution time of the instruction between these two cases. Either assumption may lead to a difference ΔT of the global execution time compared to the other one. We say that a "timing anomaly" occurs if either

$\Delta T_{\mathrm{local}} < 0$, i.e., the instruction executes faster, and

> $\Delta T < \Delta T_{\mathrm{local}}$, the overall execution is accelerated by more than the acceleration of the instruction, or

> $\Delta T > 0$, the program runs longer than before.

$\Delta T_{\mathrm{local}} > 0$, i.e., the instruction takes longer to execute, and

> $\Delta T > \Delta T_{\mathrm{local}}$, i.e., the overall execution is extended by more than the delay of the instruction, or

> $\Delta T < 0$, i.e., the overall execution of the program takes less time to execute than before.

The case $\Delta T_{\mathrm{local}} < 0$ and $\Delta T > 0$ is a critical case for our timing analysis. It makes it impossible to use local worst cases for the calculation of the program's execution time. The analysis has to follow all possible paths as is explained in Section 9.3.

9.1.2.1 Open Questions

Timing anomalies complicate timing analysis enormously. They threaten the correctness of many simplifying assumptions and efficient methods based on them. Pipeline analysis could be made very efficient if always the local worst-case transition could be taken. This, however, would not be safe for processors with timing anomalies as has been said above. Instead, all transitions and thus quite large state spaces have to be explored. It would be quite helpful if an analysis of the abstract processor model could identify "anomaly-free zones" in these state spaces, more precisely could compute a predicate on the set of execution states indicating whether a state could be the start of several execution paths exhibiting timing anomalies. If this were not the case, only the local worst case transition needed to be followed.

The phenomen timing anomaly is still awaiting a final characterization. An attempt has been made in [RWT+06]. It covers timing anomalies that are instances of the well-known scheduling anomalies [Gra69] as well as speculation anomalies, which have a different character. Scheduling anomalies could be seen as resulting from the execution of the same set of tasks, albeit in different schedules, while speculation anomalies result from executing tasks whose execution may not be needed.

9.1.3 Contexts

The contribution of an individual instruction to the total execution time of a program may vary widely depending on the execution history. For example, the first iteration of a loop typically loads the caches, and later iterations profit from the loaded memory blocks being in the caches. In this case, the execution of an instruction in the first iteration encounters one or more cache misses and pays with the cache-miss penalty. Later executions, however, will execute much faster because they hit the cache. A similar observation holds for dynamic branch predictors. They may need a few iterations until they stabilize and predict correctly.

Therefore, precision is increased if instructions are considered in their control–flow context, i.e., the way control reached them. Contexts are associated with basic blocks, i.e., maximally long straight-line code sequences that can be entered only at the first instruction and left at the last. They indicate through which sequence of function calls and loop iterations control arrived at the basic block. Thus, when analyzing the cache behavior of a loop, precision can be increased considering the first iteration of the loop and all other iterations separately, more precisely, to unroll the loop once and then analyze the resulting code.[*]

DEFINITION 9.1   *Let p be a program with set of functions $P = \{p_1, p_2, \ldots, p_n\}$ and set of loops $L = \{l_1, l_2, \ldots, l_n\}$. A word c over the alphabet $P \cup L \times \mathbb{N}$ is called a* context *for a basic block b, if b can be reached by calling the functions and iterating through the loops in the order given in c.*

Even if all loops have static loop bounds and recursion is also bounded, there are in general too many contexts to consider them exhaustively. A heuristics is used to keep relevant contexts apart and summarize the rest conservatively, if their influence on the behavior of instructions does not significantly differ. Experience has shown [TSH⁺03] that a few first iterations and recursive calls are sufficient to "stabilize" the behavior information, as the above example indicates, and that the right differentiation of contexts is decisive for the precision of the prediction [MAWF98].

A particular choice of contexts transforms the call and the control flow graph into a context-extended control-flow graph by virtually unrolling the loops and virtually inlining the functions as indicated by the contexts. The formal treatment of this concept is quite involved and shall not be given here. It can be found in [The02].

9.2   Cache-Behavior Prediction

Abstract Interpretation [CC77] is used to compute invariants about cache contents. How the behavior of programs on processor pipelines is predicted follows in Section 9.3.

9.2.1   Cache Memories

A cache can be characterized by three major parameters:

- Capacity is the number of bytes it may contain.
- Line size (also called block size) is the number of contiguous bytes that are transferred from memory on a cache miss. The cache can hold at most $n = capacity/line\ size$ blocks.
- Associativity is the number of cache locations where a particular block may reside. $n/associativity$ is the number of sets of a cache.

If a block can reside in any cache location, then the cache is called fully associative. If a block can reside in exactly one location, then it is called direct mapped. If a block can reside in exactly A locations, then the cache is called A-way set associative. The fully associative and the direct mapped caches are special cases of the A-way set associative cache where $A = n$ and $A = 1$, respectively.

---

[*] Actually, this unrolling transformation need not be really performed but can be incorporated into the iteration strategy of the analyzer. So, we talk of virtual unrolling the loops.

In the case of an associative cache, a cache line has to be selected for replacement when the cache is full and the processor requests further data. This is done according to a "replacement strategy." Common strategies are *LRU* (least recently used), *FIFO* (first in first out), and "random."

The set where a memory block may reside in the cache is uniquely determined by the address of the memory block, i.e., the behavior of the sets is independent of each other. The behavior of an A-way set associative cache is completely described by the behavior of its n/A fully associative sets. This holds also for direct mapped caches where $A = 1$.

For the sake of space, we restrict our description to the semantics of fully associative caches with LRU replacement strategy. More complete descriptions that explicitly describe direct mapped and A-way set associative caches can be found in [Fer97,FMW99].

9.2.2 Cache Semantics

In the following, we consider a (fully associative) cache as a set of cache lines $L = \{l_1, \ldots, l_n\}$ and the store as a set of memory blocks $S = \{s_1, \ldots, s_m\}$.

To indicate the absence of any memory block in a cache line, we introduce a new element I; $S' = S \cup \{I\}$.

DEFINITION 9.2 (Concrete Cache State)   *A (concrete) cache state is a function $c : L \to S'$. C_c denotes the set of all concrete cache states. The initial cache state c_I maps all cache lines to I.*

If $c(l_i) = s_y$ for a concrete cache state c, then i is the relative age of the memory block according to the LRU replacement strategy and not necessarily the physical position in the cache hardware.

The update function describes the effect on the cache of referencing a block in memory. The referenced memory block s_x moves into l_1 if it was in the cache already. All memory blocks in the cache that had been used more recently than s_x increase their relative age by one, i.e., they are shifted by one position to the next cache line. If the referenced memory block was not yet in the cache, it is loaded into l_1 after all memory blocks in the cache have been shifted and the "oldest," i.e., LRU memory block, has been removed from the cache if the cache was full.

DEFINITION 9.3 (Cache Update)   *A cache update function $\mathcal{U} : C_c \times S \to C_c$ determines the new cache state for a given cache state and a referenced memory block.*

Updates of fully associative caches with LRU replacement strategy are pictured as in Figure 9.4.

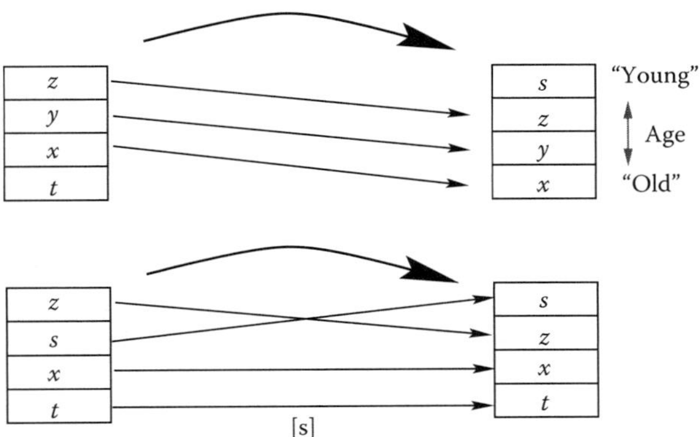

FIGURE 9.4   Update of a concrete fully associative (sub-) cache.

Control flow representation: We represent programs by control flow graphs consisting of nodes and typed edges. The nodes represent basic blocks. A basic block is a sequence (of fragments) of instructions in which control flow enters at the beginning and leaves at the end without halt or possibility of branching except at the end. For cache analysis, it is most convenient to have one memory reference per control flow node. Therefore, the nodes may represent the different fragments of machine instructions that access memory. For nonprecisely determined addresses of data references, one can use a set of possibly referenced memory blocks. We assume that for each basic block, the sequence of references to memory is known (This is appropriate for instruction caches and can be too restricted for data caches and combined caches. See [Fer97,AFMW96] for weaker restrictions.), i.e., there exists a mapping from control flow nodes to sequences of memory blocks: $\mathcal{L}: V \to S^*$.

We can describe the effect of such a sequence on a cache with the help of the update function $\mathcal{U}$. Therefore, we extend $\mathcal{U}$ to sequences of memory references by sequential composition: $\mathcal{U}(c, \langle s_{x_1}, \ldots, s_{x_y} \rangle) = \mathcal{U}(\ldots (\mathcal{U}(c, s_{x_1})) \ldots, s_{x_y})$.

The cache state for a path $(k_1, \ldots, k_p)$ in the control flow graph is given by applying $\mathcal{U}$ to the initial cache state c_I and the concatenation of all sequences of memory references along the path: $\mathcal{U}(c_I, \mathcal{L}(k_1), \ldots, \mathcal{L}(k_p))$.

The *Collecting Semantics* of a program gathers at each program point the set of all execution states, which the program may encounter at this point during some execution. A semantics on which to base a cache analysis has to model cache contents as part of the execution state. One could thus compute the collecting semantics and project the execution states onto their cache components to obtain the set of all possible cache contents for a given program point. However, the collecting semantics is in general not computable.

Instead, one restricts the standard semantics to only those program constructs, which involve the cache, i.e., memory references. Only they have an effect on the cache modeled by the cache update function, $\mathcal{U}$. This coarser semantics may execute program paths which are not executable in the start semantics. Therefore, the collecting cache semantics of a program computes a superset of the set of all concrete cache states occurring at each program point.

DEFINITION 9.4 (Collecting Cache Semantics)   *The* Collecting Cache Semantics *of a program is*

$$C_{coll}(p) = \{\mathcal{U}(c_I, \mathcal{L}(k_1), \ldots, \mathcal{L}(k_n)) \mid (k_1, \ldots, k_n) \text{ path in the CFG leading to } p\}$$

This collecting semantics would be computable, although often of enormous size. Therefore, another step abstracts it into a compact representation, so-called abstract cache states. Note that every information drawn from the abstract cache states allows to safely deduce information about sets of concrete cache states, i.e., only precision may be reduced in this two step process. Correctness is guaranteed.

Abstract semantics: The specification of a program analysis consists of the specification of an abstract domain and of the abstract semantic functions, mostly called "transfer functions." The least upper bound operator of the domain combines information when control flow merges.

We present two analyses. The "must analysis" determines a set of memory blocks that are in the cache at a given program point whenever execution reaches this point. The "may analysis" determines all memory blocks that may be in the cache at a given program point. The latter analysis is used to determine the absence of a memory block in the cache.

The analyses are used to compute a categorization for each memory reference describing its cache behavior. The categories are described in Table 9.1.

TABLE 9.1   Categorizations of Memory References and Memory Blocks

Category	Abb.	Meaning
always hit	ah	The memory reference will always result in a cache hit.
always miss	am	The memory reference will always result in a cache miss.
not classified	nc	The memory reference could neither be classified as ah nor am.

The domains for our abstract interpretations consist of *abstract cache states*:

DEFINITION 9.5 (Abstract Cache State)   An *abstract cache state* $\hat{c} : L \to 2^S$ maps cache lines to sets of memory blocks. $\hat{C}$ denotes the set of all abstract cache states.

The position of a line in an abstract cache will, as in the case of concrete caches, denote the relative age of the corresponding memory blocks. Note, however, that the domains of abstract cache states will have different partial orders and that the interpretation of abstract cache states will be different in the different analyses.

The following functions relate concrete and abstract domains. An "extraction function," *extr*, maps a concrete cache state to an abstract cache state. The "abstraction function," *abstr*, maps sets of concrete cache states to their best representation in the domain of abstract cache states. It is induced by the extraction function. The "concretization function," *concr*, maps an abstract cache state to the set of all concrete cache states represented by it. It allows to interpret abstract cache states. It is often induced by the abstraction function, cf. [NNH99].

DEFINITION 9.6 (Extraction, Abstraction, Concretization Functions)   *The* extraction function *extr* : $C_c \to \hat{C}$ *forms singleton sets from the images of the concrete cache states it is applied to, i.e.,* $extr(c)(l_i) = \{s_x\}$ *if* $c(l_i) = s_x$.

The abstraction function *abstr* : $2^{C_c} \to \hat{C}$ *is defined by* $abstr(C) = \bigsqcup \{extr(c) \mid c \in C\}$
The concretization function *concr* : $\hat{C} \to 2^{C_c}$ *is defined by* $concr(\hat{c}) = \{c \mid extr(c) \sqsubseteq \hat{c}\}$.

So much of commonalities of all the domains are to be designed. Note that all the constructions are parameterized in $\sqcup$ and $\sqsubseteq$.

The transfer functions, the "abstract cache update" functions, all denoted $\hat{\mathcal{U}}$, will describe the effects of a control flow node on an element of the abstract domain. They will be composed of two parts,

1. "Refreshing" the accessed memory block, i.e., inserting it into the youngest cache line
2. "Aging" some other memory blocks already in the abstract cache

Termination of the analyses: There are only a finite number of cache lines and for each program a finite number of memory blocks. This means that the domain of abstract cache states $\hat{c} : L \to 2^S$ is finite. Hence, every ascending chain is finite. Additionally, the abstract cache update functions, $\hat{\mathcal{U}}$, are monotonic. This guarantees that all the analyses will terminate.

Must analysis: As explained above, the must analysis determines a set of memory blocks that are in the cache at a given program point whenever execution reaches this point. Good information, in the sense of valuable for the prediction of cache hits, is the knowledge that a memory block is in this set. The bigger the set, the better. As we will see, additional information will even tell how long it will at least stay in the cache. This is connected to the "age" of a memory block. Therefore, the partial order on the must–domain is as follows. Take an abstract cache state $\hat{c}$. Above $\hat{c}$ in the domain, i.e., less precise, are states where memory blocks from $\hat{c}$ are either missing or are older than in $\hat{c}$. Therefore, the $\sqcup$-operator applied to two abstract cache states $\hat{c}_1$ and $\hat{c}_2$ will produce a state $\hat{c}$ containing only those memory blocks contained in both, and will give them the maximum of their ages in $\hat{c}_1$ and $\hat{c}_2$

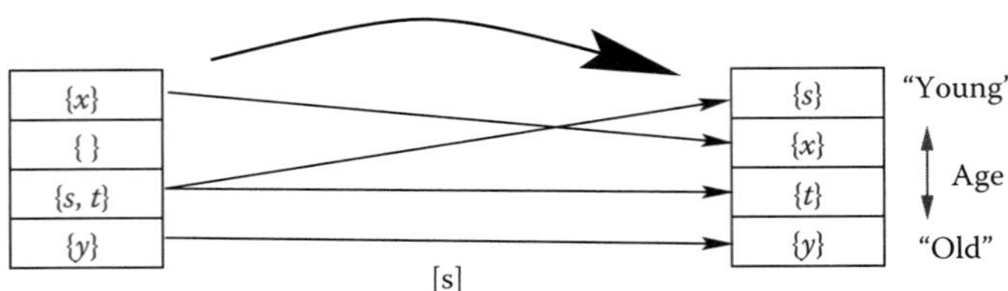

FIGURE 9.5   Update of an abstract fully associative (sub-) cache.

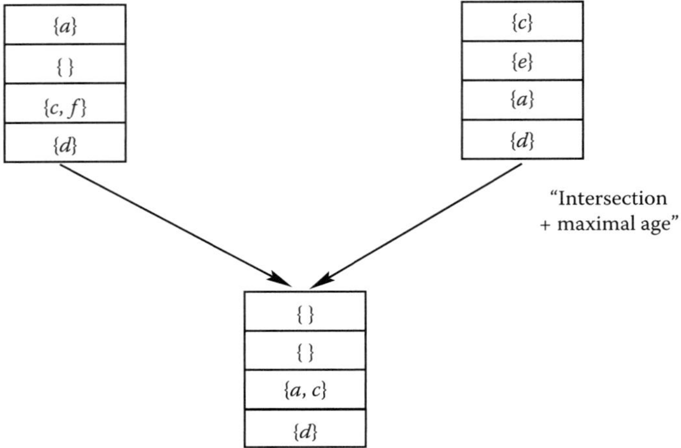

FIGURE 9.6   Combination for must analysis.

(see Figure 9.5). The positions of the memory blocks in the abstract cache state are thus the upper bounds of the "ages" of the memory blocks in the concrete caches occurring in the collecting cache semantics. Concretization of an abstract cache state, $\hat{c}$, produces the set of all concrete cache states, which contain all the memory blocks contained in $\hat{c}$ with ages not older than in $\hat{c}$. Cache lines not filled by these are filled by other memory blocks.

We use the abstract cache update function depicted in Figure 9.6. Let us argue the correctness of this update function. The following theorem formulates the soundness of the must-cache analysis.

THEOREM 9.1   *Let n be a program point, $\hat{c}_{in}$ the abstract cache state at the entry to n, s a memory line in $\hat{c}_{in}$ with age k.*

(i) For each $1 \le k \le A$ there are at most k memory lines in lines $1, 2, \ldots, k$
(ii) On all paths to n, s is in cache with age at most k.

The solution of the must analysis problem is interpreted as follows: Let $\hat{c}$ be an abstract cache state at some program point. If $s_x \in \hat{c}(l_i)$ for a cache line l_i, then s_x will definitely be in the cache whenever execution reaches this program point. A reference to s_x is categorized as "always hit" (ah). There is even a stronger interpretation of the fact that $s_x \in \hat{c}(l_i)$. s_x will stay in the cache at least for the next $n - i$ references to memory blocks that are not in the cache or are *older* than the memory blocks in $\hat{c}$, whereby s_a is older than s_b means: $\exists l_i, l_j : s_a \in \hat{c}(l_i), s_b \in \hat{c}(l_j), i > j$.

May analysis: To determine, if a memory block s_x will never be in the cache, we compute the complimentary information, i.e., sets of memory blocks that may be in the cache. "Good" information is that

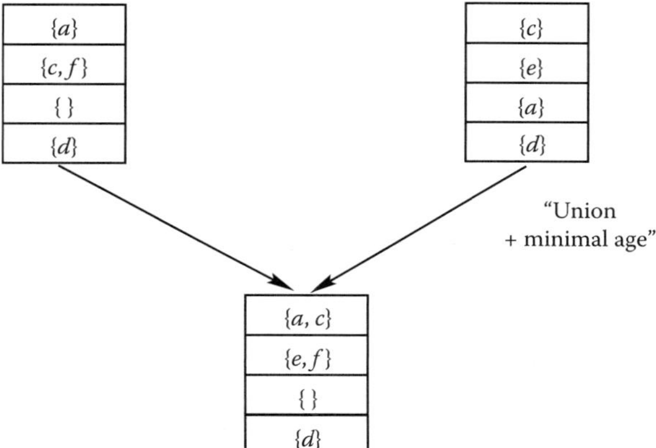

FIGURE 9.7   Combination for may analysis.

a memory block is not in this set, because this memory block can be classified as definitely not in the cache whenever execution reaches the given program point. Thus, the smaller the sets are, the better. Additionally, the older blocks will reach the desired situation to be removed from the cache faster than the younger ones. Therefore, the partial order on this domain is as follows. Take some abstract cache state $\hat{c}$. Above $\hat{c}$ in the domain, i.e., less precise, are those states which contain additional memory blocks or where memory blocks from $\hat{c}$ are younger than in $\hat{c}$. Therefore, the $\sqcup$-operator applied to two abstract cache states $\hat{c}_1$ and $\hat{c}_2$ will produce a state $\hat{c}$ containing those memory blocks contained in $\hat{c}_1$ or $\hat{c}_2$ and will give them the minimum of their ages in $\hat{c}_1$ and $\hat{c}_2$ (Figure 9.7).

The positions of the memory blocks in the abstract cache state are thus the lower bounds of the *ages* of the memory blocks in the concrete caches occurring in the collecting cache semantics.

The solution of the may analysis problem is interpreted as follows: The fact that s_x is in the abstract cache $\hat{c}$ means that s_x may be in the cache during some execution when the program point is reached. If s_x is not in $\hat{c}(l_i)$ for any l_i, then it will definitely be not in the cache on any execution. A reference to s_x is categorized as "always miss" (am).

9.3 Pipeline Analysis

Pipeline analysis attempts to find out how instructions move through the pipeline. In particular, it determines how many cycles they spend in the pipeline. This largely depends on the timing accidents the instructions suffer. Timing accidents during pipelined executions can be of several kinds. Cache misses during instruction or data load stall the pipeline for as many cycles as the cache miss penalty indicates. Functional units that an instruction needs may be occupied. Queues into which the instruction may have to be moved may be full, and prefetch queues, from which instructions have to be loaded, may be empty. The bus needed for a pipeline phase may be occupied by a different phase of another instruction. Again, for an architecture without timing anomalies, we can use a simplified picture, in which the task is to find out which timing accidents can be safely excluded, because each excluded accident allows to decrease the bound for the execution time. Accidents that cannot be safely excluded are assumed to happen.

A cache analysis as described in Section 9.2 has annotated the instructions with cache-hit information. This information is used to exclude pipeline stalls at instruction or data fetches.

We will explain pipeline analysis in a number of steps starting with "concrete-pipeline execution." A pipeline goes through a number of pipeline phases and consumes a number of cycles when it executes a sequence of instructions; in general, a different number of cycles for different initial execution states. The execution of the instructions in the sequence overlaps in the instruction pipeline as far as the data dependences between instructions permit it and if the pipeline conditions are satisfied. Each execution of a sequence of instructions starting in some initial state produces one "trace," i.e., sequence of execution states. The length of the trace is the number of cycles this execution takes.

Thus, concrete execution can be viewed as applying a function

function $exec$ (b : **basic block**, s : **pipeline state**) t : **trace**

that executes the instruction sequence of basic block b starting in concrete pipeline state s producing a trace t of concrete states. $last(t)$ is the final state when executing b. It is the initial state for the successor block to be executed next.

So far, we talked about concrete execution on a concrete pipeline. Pipeline analysis regards abstract execution of sequences of instructions on abstract (models of) pipelines. The execution of programs on abstract pipelines produces abstract traces, i.e., sequences of abstract states, where some information contained in the concrete states may be missing. There are several types of missing information.

- The cache analysis in general has incomplete information about cache contents.
- The latency of an arithmetic operation, if it depends on the operand sizes, may be unknown. It influences the occupancy of pipeline units.
- The state of a dynamic branch predictor changes over iterations of a loop and may be unknown for a particular iteration.
- Data dependences cannot safely be excluded because effective addresses of operands are not always statically known.

9.3.1 Simple Architectures without Timing Anomalies

In the first step, we assume a simple processor architecture, with in-order execution and without "timing anomalies," i.e., architectures, where local worst cases contribute to the program's global execution time, cf. Section 9.1.2. Also, it is safe to assume the local worst cases for unknown information. For both of them, the corresponding timing penalties are added. For example, the cache miss penalty has to be added for instruction fetch of an instruction in the two cases, that a cache miss is predicted or that neither a cache miss nor a cache hit can be predicted.

The result of the abstract execution of an instruction sequence for a given initial abstract state is again one trace; however, possibly of a greater length and thus an upper bound properly bounding the execution time from above. Because worst cases were assumed for all uncertainties, this number of cycles is a safe upper bound for all executions of the basic block starting in concrete states represented by this initial abstract state.

The Algorithm for pipeline analysis is quite simple. It uses a function

function $\hat{exec}$ (b : **cache-annotated basic block**, $\hat{s}$: **abstract pipeline state**)
         $\hat{t}$: **abstract trace**

that executes the instruction sequence of basic block b, annotated with cache information, starting in the abstract pipeline state $\hat{s}$ and producing a trace $\hat{t}$ of abstract states.

This function is applied to each basic block b in each of its contexts and the empty pipeline state $\hat{s}_0$ corresponding to a flushed pipeline. Therefore, a linear traversal of the cache-annotated context-extended Basic-Block Graph suffices. The result is a trace for the instruction sequence of the block, whose length is an upper bound for the execution time of the block in this context. Note that it still makes sense to analyze a basic block in several contexts because the cache information for them may be quite different.

Note that this algorithm is simple and efficient, but not necessarily very precise. Starting with a flushed pipeline at the beginning of the basic block is safe, but it ignores the potential overlap between consecutive basic blocks.

A more precise algorithm is possible. The problem is with basic blocks having several predecessor blocks. Which of their final states should be selected as initial state of the successor block? First solution involves working with sets of states for each pair of basic block and context. Then, one analysis of each basic block and context would be performed for each of the initial states. The resulting set of final states would be passed on to successor blocks, and the maximum of the trace lengths would be taken as upper bound for this basic block in this context.

Second solution would work with a single state per basic block and context and would combine the set of predecessor final states conservatively to the initial state for the successor.

9.3.2 Processors with Timing Anomalies

In the next step, we assume more complex processors, including those with out-of-order execution. They typically have timing anomalies. Our assumption above, i.e., local worst cases contribute worst-case times to the global execution times, is no more valid. This forces us to consider several paths, wherever uncertainty in the abstract execution state does not allow to take a decision between several successor states. Note that the absence of information leads from the deterministic concrete pipeline to an abstract pipeline that is nondeterministic. This situation is depicted in Figure 9.8. It demonstrates two cases of missing information in the abstract state. First, the abstract state lacks the information whether the instruction is in the I-cache. Pipeline analysis has to follow both cases in case of instruction fetch, because it could turn out that the I-cache miss, in fact, is not the global worst case. Second, the abstract state does not contain information about the size of the operands. We also have to follow both paths. The dashed paths have to be explored to obtain the execution times for this instruction. Depending on the architecture, we may be able to conservatively assume the case of large operands and surpress some paths.

The algorithm has to combine cache and pipeline analysis because of the interference between both, which actually is the reason for the existence of the timing anomalies. For the cache analysis, it uses the abstract cache states discussed in Section 9.2. For the pipeline part, it uses "analysis states," which are sets of abstract pipeline states, i.e., sets of states of the abstract pipeline. The question arises

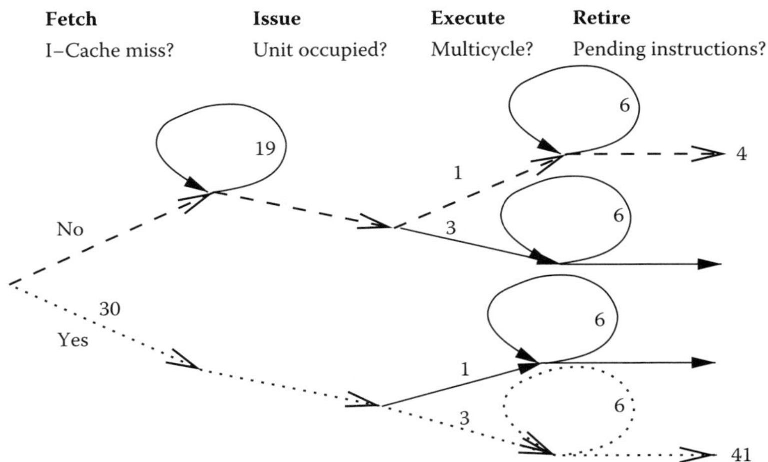

FIGURE 9.8 Different paths through the execution of a multiply instruction. Decisions inside the boxes cannot be deterministically taken based on the abstract execution state because of missing information.

whether an abstract cache state is to be combined with an analysis state $\hat{s}s$ or an individual one with each of the abstract pipeline states in $\hat{s}s$. So, there could be one abstract cache state for $\hat{s}s$ representing the concrete cache contents for all abstract pipeline states in $\hat{s}s$, or there could be one abstract cache state per abstract pipeline state in $\hat{s}s$. The first choice saves memory during the analysis but loses precision. This is because different pipeline states may cause different memory accesses and thus cache contents, which have to be merged into the one abstract state thereby losing information. The second choice is more precise but requires more memory during the analysis. We choose the second alternative and thus define a new domain of "analysis states" $\hat{A}$ of the following type:

$$\hat{A} = 2^{\hat{S} \times \hat{C}} \tag{9.1}$$

$$\hat{S} = \text{set of abstract pipeline states} \tag{9.2}$$

$$\hat{C} = \text{set of abstract cache states} \tag{9.3}$$

The Algorithm again uses a new function $\hat{exec}_c$.

function $\hat{exec}_c$ (b : **basic block**, $\hat{a}$: **analysis state**) $\hat{T}$: **set of abstract trace,**

which analyzes a basic block b starting in an analysis state $\hat{a}$ consisting of pairs of abstract pipeline states and abstract cache states. As a result, it will produce a set of abstract traces.

The algorithm is as follows:

Algorithm Pipeline-Analysis

Perform fixpoint iteration over the context-extended Basic-Block Graph:
For each basic block b in each of its contexts c, and for the initial analysis state $\hat{a}$,
compute $\hat{exec}_c(b, \hat{a})$ yielding a set of traces $\{\hat{t}_1, \hat{t}_2, \ldots, \hat{t}_m\}$.
$\max(\{|\hat{t}_1|, |\hat{t}_2|, \ldots, |\hat{t}_m|\}$ is the bound for this basic block in this context.

The set of output states $\{last(\hat{t}_1), last(\hat{t}_2), \ldots, last(\hat{t}_m)\}$ will be passed on to the successor block(s) in context c as initial states.

Basic blocks (in some context) having more than one predecessor receive the union of the set of output states as initial states.

The abstraction we use as analysis states is a *set* of abstract pipeline states, since the number of possible pipeline states for one instruction is not too big. Hence, our abstraction computes an upper bound to the collecting semantics. The abstract update for an analysis state $\hat{a}$ is thus the application of the concrete update on each abstract pipeline state in $\hat{a}$ extended with the possibility of multiple successor states in case of uncertainties.

Figure 9.9 shows the possible pipeline states for a basic block in this example. Such pictures are shown by **aiT** tool upon special demand. The large dark grey boxes correspond to the instructions of the basic block, and the smaller rectangles in them stand for individual pipeline states. Their cyclewise

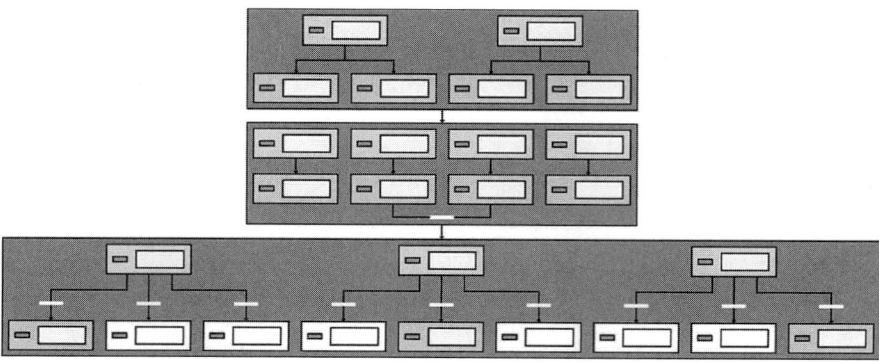

FIGURE 9.9   Possible pipeline states in a basic block.

evolution is indicated by the strokes connecting them. Each layer in the trees corresponds to one CPU cycle. Branches in the trees are caused by conditions that could not be statically evaluated, e.g., a memory access with unknown address in presence of memory areas with different access times. On the other hand, two pipeline states fall together when details they differ in leave the pipeline. This happened, for instance, at the end of the second instruction, reducing the number of states from four to three.

The update function belonging to an edge (v, v') of the control-flow graph updates each abstract pipeline state separately. When the bus unit is updated, the pipeline state may split into several successor states with different cache states. The initial analysis state is a set of empty pipeline states plus a cache that represents a cache with unknown content. There can be multiple concrete pipeline states in the initial states, since the adjustment of internal to external clock of the processor is not known in the beginning and every possibility (aligned, one cycle apart, etc.) has to be considered. Thus prefetching must start from scratch, but pending bus requests are ignored. To obtain correct results, they must be taken into account by adding a fixed penalty to the calculated upper bounds.

9.3.3 Pipeline Modeling

The basis for pipeline analysis is a model of an abstract version of the processor pipeline, which is conservative with respect to the timing behavior, i.e., times predicted by the abstract pipeline must never be lower than those observed in concrete executions. Some terminology is needed to avoid confusion. Processors have "concrete" pipelines, which may be described in some formal language, e.g., VHDL. If this is the case, there exists a "formal model" of the pipeline. Our abstraction step, by which we eliminate many components of a concrete pipeline that are not relevant for the timing behavior lead us to an "abstract pipeline." This may again be described in a formal language, e.g., VHDL, and thus have a formal model. Deriving an abstract pipeline is a complex task. It is demonstrated for the Motorola ColdFire processor, a processor quite popular in the aeronautics and the submarine industry. The presentation follows closely that of [LTH02].*

9.3.3.1 The ColdFire MCF 5307 Pipeline

The pipeline of the ColdFire MCF 5307 consists of a "fetch pipeline" that fetches instructions from memory (or the cache), and an "execution pipeline" that executes instructions, cf. Figure 9.10. Fetch and execution pipelines are connected and as far as speed is concerned decoupled by a FIFO instruction buffer that can hold at most eight instructions.

The MCF 5307 accesses memory through a bus hierarchy. The fast pipelined K-bus connects the cache and an internal 4KB SRAM area to the pipeline. Accesses to this bus are performed by the IC1/IC2 and the AGEX and DSOC stages of the pipeline. On the next level, the M-Bus connects the K-Bus to the internal peripherals. This bus runs at the external bus frequency, while the K-Bus is clocked with the faster internal core clock. The M-Bus connects to the external bus, which accesses off-chip peripherals and memory.

The "fetch pipeline" performs branch prediction in the IED stage, redirecting fetching long before the branch reaches the execution stages. The fetch pipeline is stalled if the instruction buffer is full, or if the execution pipeline needs the bus for a memory access. All these stalls cause the pipeline to wait for 1 cycle. After that, the stall condition is checked again.

The fetch pipeline is also stalled if the memory block to be fetched is not in the cache (cache miss). The pipeline must wait until the memory block is loaded into the cache and forwarded to

---

* The model of the abstract pipeline of the MCF 5307 has been derived by hand. A computer-supported derivation would have been preferable. Ways to develop this are subject of actual research.

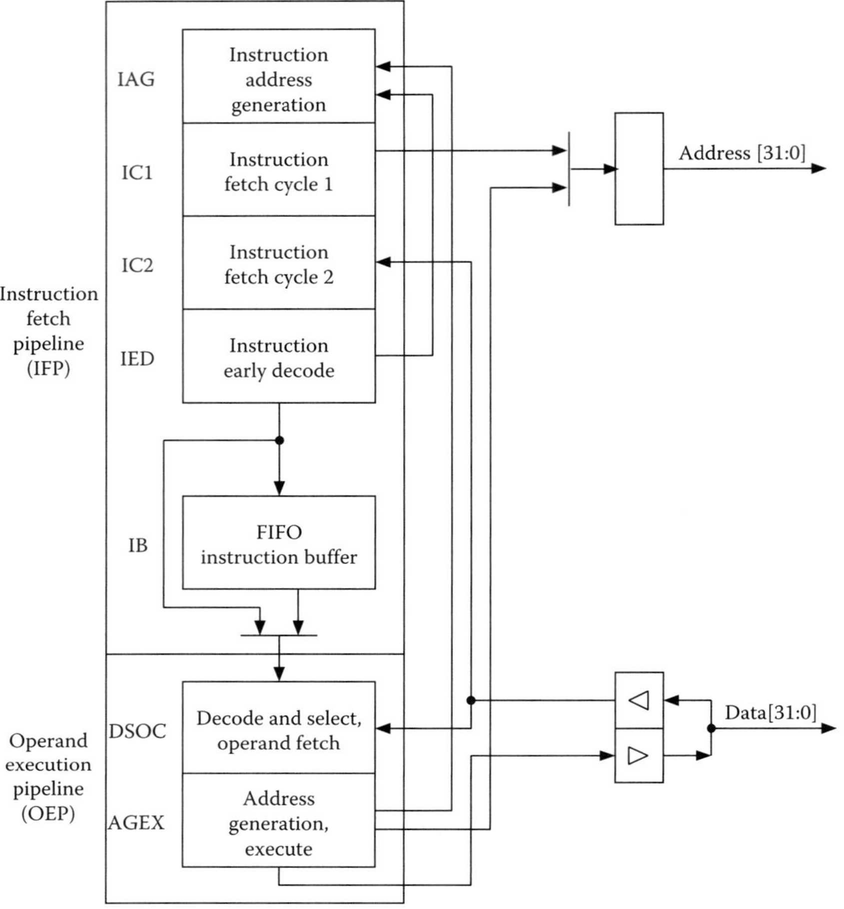

FIGURE 9.10   Pipeline of the Motorola ColdFire 5307 processor.

the pipeline. The instructions that are already in the later stages of the fetch pipeline are forwarded to the instruction buffer.

The "execution pipeline" finishes the decoding of instructions, evaluates their operands, and executes the instructions. Each kind of operation follows a fixed schedule. This schedule determines how many cycles the operation needs and in which cycles memory is accessed.[*] The execution time varies between 2 cycles and several dozen cycles. Pipelining admits a maximum overlap of 1 cycle between consecutive instructions: the last cycle of each instruction may overlap with the first of the next one. In this first cycle, no memory access and no control-flow alteration happen. Thus, cache and pipeline cannot be affected by two different instructions in the same cycle. The execution of an instruction is delayed if memory accesses lead to cache misses. Misaligned accesses lead to small time penalties of 1–3 cycles. Store operations are delayed if the distance to the previous store operation is less than 2 cycles. (This does not hold if the previous store operation was issued by a **MOVEM** instruction.) The start of the next instruction is delayed if the instruction buffer is empty.

---

[*] In fact, there are some instructions like **MOVEM** whose execution schedule depends on the value of an argument as immediate constant. These instructions can be taken into account by special means.

9.3.4　Formal Models of Abstract Pipelines

An abstract pipeline can be seen as a big finite state machine, which makes a transition on every clock cycle. The states of the abstract pipeline, although greatly simplified, still contain all timing relevant information of the processor. The number of transitions it takes from the beginning of the execution of an instruction until its end gives the execution time of that instruction.

The abstract pipeline although greatly reduced by leaving out irrelevant components still is a really big finite state machine, but it has structure. Its states can be naturally decomposed into components according to the architecture. This makes it easier to specify, verify, and implement a model of an abstract pipeline. In the formal approach presented here, an abstract pipeline state consists of several "units" with inner "states" that communicate with one another and the memory via "signals," and evolve cycle-wise according to their inner state and the signals received. Thus, the means of decomposition are units and signals.

Signals may be "instantaneous," meaning that they are received in the same cycle as they are sent, or "delayed," meaning that they are received 1 cycle after they have been sent. Signals may carry data, e.g., a fetch address. Note that these signals are only part of the formal pipeline model. They may or may not correspond to real hardware signals. The instantaneous signals between units are used to transport information between the units. The state transitions are coded in the evolution rules local to each unit.

Figure 9.11 shows the formal pipeline model for the ColdFire MCF 5307. It consists of the following units: IAG (instruction address generation), IC1 (instruction fetch cycle 1), IC2 (instruction fetch cycle 2), IED (instruction early decode), IB (instruction buffer), EX (execution unit), and SST (store stall timer). In addition, there is a "bus unit" modeling the busses that connect the CPU, the static RAM, the cache, and the main memory. The signals between these units are shown as arrows. Most units directly correspond to a stage in the real pipeline. However, the SST unit is used to model the fact that two stores must be separated by at least two clock cycles. It is implemented as a (virtual) counter. The two stages of the execution pipeline are modeled by a single stage, EX, because instructions can only overlap by 1 cycle.

The inner states and emitted signals of the units evolve in each cycle. The complexity of this state update varies from unit to unit. It can be as simple as a small table, mapping pending signals and inner state to a new state and signals to be emitted, e.g., for the IAG unit and the IC1 unit. It can be much more complicated, if multiple dependencies have to be considered, e.g., the instruction reconstruction and branch prediction in the IED stage. In this case, the evolution is formulated in pseudo code. Full details on the model can be found in [The04].

9.3.5　Pipeline States

Abstract Pipeline States are formed by combining the inner states of IAG, IC1, IC2, IED, IB, EX, SST, and bus unit plus additional entries for pending signals into one overall state. This overall state evolves from 1 cycle to the next. Practically, the evolution of the overall pipeline state can be implemented by updating the functional units one by one in an order that respects the dependencies introduced by input signals and the generation of these signals.

9.3.5.1　Update Function for Pipeline States.

For pipeline modeling, one needs a function that describes the evolution of the concrete pipeline state while traveling along an edge (v, v') of the control-flow graph. This function can be obtained by iterating the cycle-wise update function of the previous paragraph.

An initial concrete pipeline state at v has an empty execution unit EX. It is updated until an instruction is sent from IB to EX. Updating of the concrete pipeline state continues using the knowledge that the successor instruction is v' until EX has become empty again. The number of cycles needed from

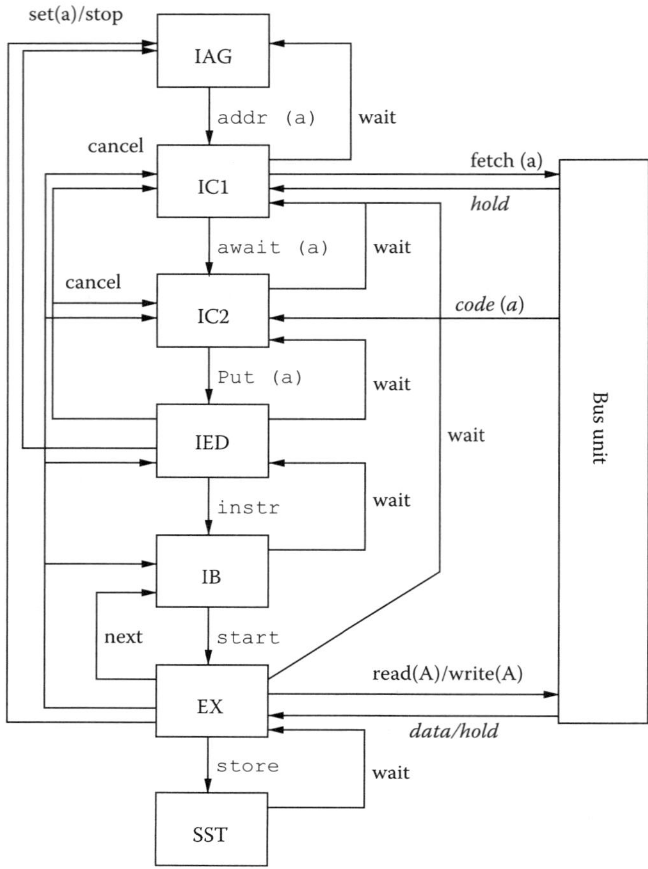

FIGURE 9.11   Abstract model of the Motorola ColdFire 5307 processor.

the beginning until this point can be taken as the time needed for the transition from v to v' for this concrete pipeline state.

9.3.6   Modeling the Periphery

System performance of real-time control applications is dominated by the performance of peripherals, especially the memory access times. The system controller's timing behavior has thus a huge influence on the overall performance. Modeling just the processor puts the emphasis on the wrong spot. [The06] describes the systematic derivation of a timing model for a complex system controller. This controller connects the CPU to main memory and several busses (PCI, etc.). A timing model for this controller was derived from a VHDL description provided by EADS Airbus. The resulting model quite accurately captured the controller's behavior.

9.3.7   Support for the Derivation of Timing Models

The VHDL model of the controller mentioned above is quite large, 18000 lines of VHDL. This is small compared to the full specification of a modern processor. The Leon 2 processor, also called the ESA SPARC, has a VHDL-Specification of 68000 lines [Gai]. Deriving timing models from so large specifications by hand is next to impossible. A computer-supported semiautomatic process is currently being developed to support the designer of a timing model [SP07]. It reflects the manual process described in [The06]. The timing model is obtained by a series of static analyses and transformations, all on VHDL representations. Arithmetic is factored out; whatever is relevant for the timing behavior in the computations is approximately determined by a Value Analysis, cf. Section 9.5.1. The specification of the processor's caches is similarly factored out. A cache analysis is imported or designed separately, as described in Section 9.2, and then integrated with the pipeline analysis. Asynchronous events such as SDRAM refreshes or DMA transfers are also eliminated. They have to deal with by other means. Slicing is the most profitable static analysis to cut down the size of the remaining specification. A backward slice starting with timing signals contains all the logic influencing the timing behavior. The logic not contained in this backward slice can be eliminated. Generic constructs can be instantiated by a constant propagation as soon as the actual parameters are known. Some of the logic then is unreachable and can be eliminated. This way, the specification of the concrete processor can be reduced to a timing model in a series of analyses and transformations.

9.4   Path Analysis Using Integer Linear Programming

The structure of a program and the set of program paths can be mapped to an ILP in a very natural way. A set of constraints describes the control flow of the program. Solving these constraints yields very precise results [TFW00]. However, requirements for precision of the results demand analyzing basic blocks in different contexts, i.e., in different ways, how control reached them. This makes the control quite complex, so that the mapping to an ILP may be very complex [The02].

A problem formulated in an ILP consists of two parts: the cost function and constraints on the variables used in the cost function. The cost function represents the number of CPU cycles. Correspondingly, it has to be maximized. Each variable in the cost function represents the execution count of one basic block of the program and is weighted by the execution time of that basic block. Additionally, variables are used corresponding to the traversal counts of the edges in the control flow graph, see Figure 9.12.

The integer constraints describing how often basic blocks are executed relative to each other can be automatically generated from the control flow graph (Figure 9.13). However, additional information about the program provided by the user is usually needed, as the problem of finding the worst case program path is unsolvable in the general case. Loop and recursion bounds cannot always be inferred automatically and must therefore be provided by the user.

The ILP approach for program path analysis has the advantage that users are able to describe in precise terms virtually anything they know about the program by adding integer constraints. The system first generates the obvious constraints automatically and then adds user supplied constraints to tighten the WCET bounds.

9.5   Other Ingredients

9.5.1   Value Analysis

A static method for data-cache behavior prediction needs to know effective memory addresses of data, in order to determine where a memory access goes. However, effective addresses are only available at run time. Interval analysis as described by Patrick and Radhia Cousot [CC76] can help here. It can compute intervals for address-valued objects like registers and variables. An interval computed for such an object at some program point bounds the set of potential values the object may have when

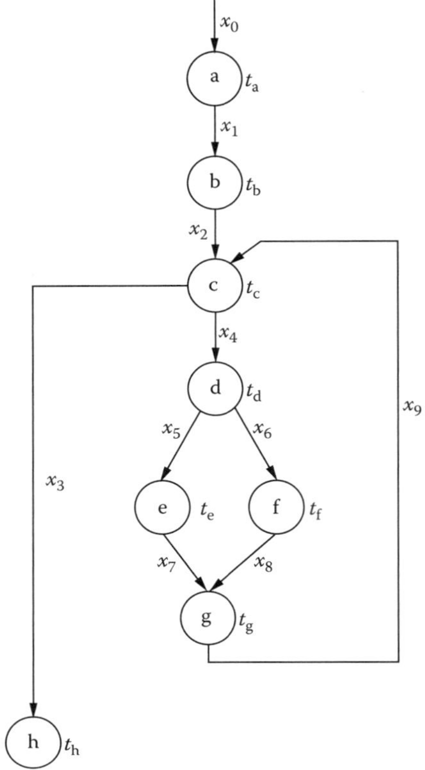

FIGURE 9.12　Program snippet, the corresponding control flow graph, and the ILP variables generated.

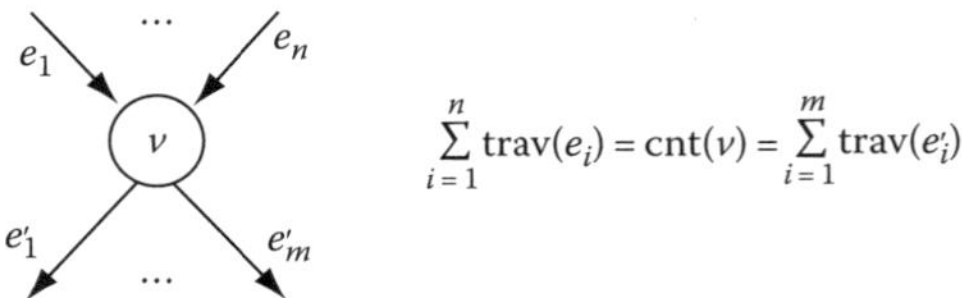

$$\sum_{i=1}^{n} \text{trav}(e_i) = \text{cnt}(v) = \sum_{i=1}^{m} \text{trav}(e_i')$$

FIGURE 9.13　Control flow joins and splits and flow-preservation laws.

program execution reaches this program point. Such an analysis, in `aiT` called "value analysis" has shown to be able to determine many effective addresses in disciplined code statically [TSH⁺03].

9.5.2　Control-Flow Specification and Analysis

Any information about the possible flow of control of the program may increase the precision of the subsequent analyses. Control-flow analysis may attempt to exclude infeasible paths, determine execution frequencies of paths or the relation between execution frequencies of different paths or subpaths, etc.

The purpose of control flow analysis is to determine the dynamic behavior of the program. This includes information about what functions are called and with which arguments, how many times loops iterate, if there are dependencies between successive `if`-statements, etc. The main focus of flow analysis has been the determination of loop bounds, since the bounding of loops is a necessary step in order to find an execution time bound for a program.

Control-flow analysis can be performed manually or automatically. Automatic analyses have been based on various techniques, like symbolic execution, abstract interpretation, and pattern recognition on parse trees. The best precision is achieved by using interprocedural analysis techniques, but this has to be traded off with the extra computation time and memory required. All automatic techniques allow a user to complement the results and guide the analysis using manual annotations, since this is sometimes necessary in order to obtain reasonable results.

Since the flow analysis in general is performed separately from the path analysis, it does not know the execution times of individual program statements, and must thus generate a safe (over) approximation including *all* possible program executions. The path analysis will later select the path from the set of possible program paths that corresponds to the upper bound using the time information computed by processor behavior prediction.

Control-flow specification is preferrably done on the source level. Concepts based on source-level constructs are used in [EG97,Erm03].

9.5.3 Frontends for Executables

Any reasonably precise timing analysis takes fully linked executable programs as input. Source programs do not contain information about program and data allocation, which is essential for the described methods to predict the cache behavior.

Executables must be analyzed to reconstruct the original control flow of the program. This may be a difficult task depending on the instruction set of the processor and the code generation of the used compiler. A generic approach to this problem is described in [The00,The01,The02].

9.6 Related Work

It is not possible in general to obtain upper bounds on running times for programs. Otherwise, one could solve the halting problem. However, real-time systems use only a restricted form of programming, which guarantees that programs always terminate. That is, recursion is not allowed (or explicitly bounded) and the maximal iteration counts of loops are known in advance.

A worst-case running time of a program could easily be determined if the worst-case input for the program and the worst-case initial state of the processor were known. This is in general not the case. The alternative, to execute the program with all possible inputs starting in all possible processor states, is prohibitively expensive. As a consequence, approximations for the worst-case execution time are determined. Two classes of methods to obtain bounds can be distinguished:

- *Dynamic* methods execute the program to obtain execution times. These may be end-to-end executions of the whole program or piecewise executions, e.g., of sequences of basic blocks. Measurement-based methods are in general "unsafe" as they only compute the maximum of a subset of all executions.

- *Static* methods only need the program itself, extended with some additional information about the program like loop bounds and information about the execution platform like access characteristics for the memory areas the program is using, bus frequencies, and CPU cycle lengths.

9.6.1 (Partly) Dynamic Method

A traditional method, still used in industry, combines measuring and static methods. Here, small snippets of code are measured for their execution time, then a "safety margin" is applied and the results for code pieces are combined according to the structure of the whole task. For example, if a task first executes snippet A and then snippet B, the resulting time is that measured for A, t_A, added to that measured for B, t_B: $t = t_A + t_B$. This reduces the amount of measurements that have to be made, as code snippets tend to be reused a lot in control software and only the different snippets need to be measured. It adds, however, the need for an argumentation about the correctness of the composition step of the measured snippet times. This typically relies on certain implicit assumptions about the worst-case initial execution state for these measurements. For example, the snippets are measured with an empty cache at the beginning of the measurement under the assumption that this is the worst-case cache state. In [The04], it is shown that this assumption can be wrong. The problem of unknown worst-case input exists for this method as well, and it is still infeasible to measure execution times for all input values.

The approaches using piecewise measurement claim to add a conservative overhead in order to compensate for choosing the "wrong" initial state. Typically, they start execution with an empty cache, which for most replacement strategies is not the worst case.

Recently, it has been shown that non-LRU caches are very sensitive to the initial cache state, e.g., for a PLRU cache, the observed cache hit rate when starting execution in one state gives no clue about the hit rate for an execution starting in another state. The deviation is only bounded by the number of memory accesses [RG08].

The next problem is how to combine the results of piecewise executions plus the assumed conservative overhead. A pessimistic combination ignoring the sequencing through consecutive blocks of the program may end up with larger over-estimations than a safe static approach, even though it starts with under-estimated execution-time bounds for program pieces.

9.6.2 Purely Static Methods

9.6.2.1 Timing Schema Approach

In the timing-schemata approach [Sha89], bounds for the execution times of a composed statement are computed from the bounds of the constituents. One timing schema is given for each type of statement. Bases are known times of the atomic statements. These are assumed to be constant and available from a manual or are assumed to be computed in a preceding phase. A bound for the whole program is obtained by combining results according to the structure of the program.

The precision can be very bad when applied for modern architectures with high variability of execution times. Ignoring the control-flow context of program pieces forces one to combine worst-case bounds if one wants to be on the safe side. Worst-case bounds that are independent of the execution history are in general unrealistic.

9.6.2.2 Symbolic Simulation

Another static method simulates the execution of the program on an abstract model of the processor. The simulation is performed without input; the simulator thus has to be capable to deal with partly unkown execution states. This method combines flow analysis, processor-behavior prediction, and path analysis in one integrated phase [LS99,Lun02]. One problem with this approach is that analysis time is proportional to the actual execution time of the program with a usually large factor for doing a simulation.

9.6.2.3 WCET Determination by ILP

Li, Malik, and Wolfe proposed an ILP-based approach to WCET determination [LM95,LMW95a, LMW95b,LMW96]. Cache and pipeline behavior prediction are formulated as a single linear program. The i960KB, a 32-bit microprocessor with a 512 byte direct mapped instruction cache and a fairly simple pipeline is investigated. Only structural hazards need to be modeled, thus keeping the complexity of the integer linear program moderate compared to the expected complexity of a model for a modern microprocessor. Variable execution times, branch prediction, and instruction prefetching are not considered at all. Using this approach for super-scalar pipelines does not seem very promising, considering the analysis times reported in one of the articles.

One of the severe problems is the exponential increase of the size of the ILP in the number of competing l-blocks. l-blocks are maximally long contiguous sequences of instructions in a basic block mapped to the same cache set. Two l-blocks mapped to the same cache set "compete" if they do not have the same address tag. For a fixed cache architecture, the number of competing l-blocks grows linearly with the size of the program. Differentiation by contexts, absolutely necessary to achieve precision, increases this number additionally. Thus, the size of the ILP is exponential in the size of the program. Even though the problem is claimed to be a network-flow problem, the size of the ILP is killing the approach. Growing associativity of the cache increases the number of competing l-blocks. Thus, increasing cache-architecture complexity also plays against this approach.

Nonetheless, their method of modeling the control flow as an ILP, the so-called Implicit Path Enumeration, is elegant and can be efficient if the size of the ILP is kept small. It has been adopted by many groups working in this area.

9.6.2.4   Timing Analysis by Static Program Analysis

The method described in this chapter uses a sequence of static program analyses for determining the program's control flow and its data accesses and for predicting the processor's behavior for the given program.

An early approach to timing analysis using data-flow analysis methods can be found in [AMWH94, MWH94]. Jakob Engblom showed how to precompute parts of a timing analyzer to speed up the actual timing analysis for architectures without timing anomalies [Eng02].

[WEE+08] gives an overview of existing tools for timing analysis, both commercially available tools and academic prototypes.

9.7   State of the Art and Future Extensions

The timing-analysis technology described in this chapter is realized in the **aiT** tool of AbsInt Angewandte Informatik, Saarbrücken [Inf]. **aiT** is used in the aeronautics and automotive industries. The European Airworthiness Authorities have admitted the tool for the certification of several time-critical systems of the Airbus A380 plane and have attributed to it the status of a "validated tool" for these airplane functions.

There are a number of published benchmark results about the precision obtained by timing-analysis tools, and there are the results of the WCET Tool Challenge 2006 organized by the European Network of Excellence ARTIST2. Figure 9.14 presents results. [LBJ+95] is a study done by the authors of a method carefully explaining the reasons for overestimation. [TSH+03,SlPH+05] report experiences made by developers. The developers are experienced, and the tool is integrated into the development process. The figure contains two curves, one showing the degree of overestimation observed in the experiments, the other the assumed cache-miss penalty. The latter curve reflects the development of processor architectures and in particular the divergence of processor and memory speeds. Both have increased the timing variability and thus the penalty for imprecision of the analysis. This has made the challenge of timing analysis harder all the time.

In [LBJ+95], published in 1995, a cache-miss penalty of 4 cycles was assumed. In [TSH+03], a cache-miss penalty of 25 was given, and finally in the setting described in [SlPH+05], the cache-miss penalty was 60 internal cycles for a worst-case access to an instruction in SDRAM and roughly 200 internal cycles for an access to data over the PCI bus. [SlPH+05] reports overestimations between

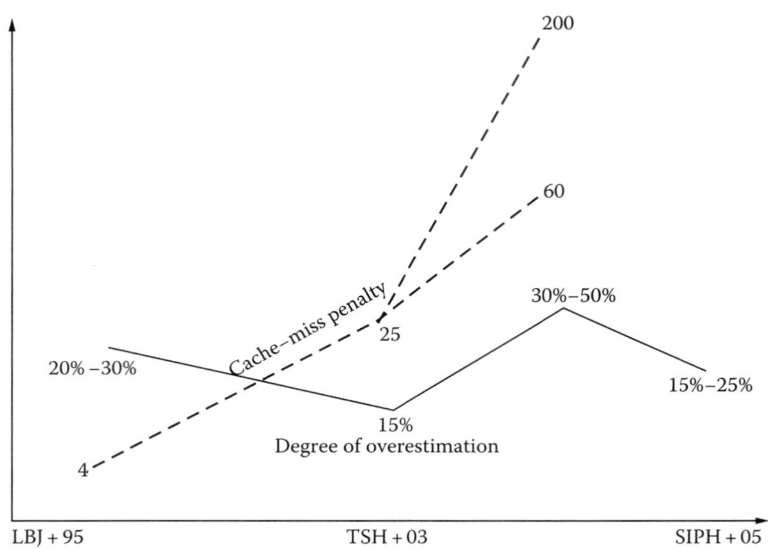

FIGURE 9.14   Several benchmarks with their degrees of overestimation and the development of the cache-miss penalty.

30% and 50% for some Airbus applications. Improvements of the used tool, **aiT**, in particular the strengthening of the analysis across basic-block boundaries has reduced the overestimations for these applications to between 15% and 25%.

The figure says that the significant methodological progress made in the last 12 years has just sufficed to keep the degree of overestimation roughly constant. An overestimation of 30% in 2005 as reported in [SlPH$^+$05] means a huge progress in method and tool performance compared to an overestimation of 30% reported in 1995 [LBJ$^+$95]!

The results of the ARTIST2 WCET Tool Challenge have shown overestimations of 8% for **aiT**, which turned out as the dominating tool [Tan06]. However, the target platforms chosen for the challenge were simple architectures without caches.

The computational effort is high, but acceptable. Future optimizations will reduce this effort. As often in static program analysis, there is a trade-off between precision and effort. Precision can be reduced if the effort is intolerable.

The only really drawback of the described technology is the huge effort for producing abstract processor models. As described in Section 9.3, work is under way to support this activity through transformations on the VHDL level [SP07].

9.8   Timing Predictability

Experience has shown that several factors influence the achievable precision of the execution-time bounds and the necessary efforts to determine them [HLTW03,TW04]:

- The architecture of the execution platform
- Characteristics of the software

The timing predictability of a system is a measure for the possibility of determining tight bounds on execution times. The difference between the best determinable upper bound and the worst-case execution time and the difference between the best-case execution time and the best determinable lower bound correspond to this notion. They could be called "worst-case predictability" and "best-case predictability," respectively. This timing predictability is then composed of the remaining uncertainty after employing the strongest static analyses and the associated penalties to be paid for this uncertainty. Uncertainty comprises timing accidents that cannot be excluded statically, but they never actually happen during execution. High penalties do not automatically make a system unpredictable: They are no problem if a system can be analyzed without remaining uncertainty. On the other hand, high levels of uncertainty become harmful to timing predictability if the associated penalties are large.

[RGBW07] describes predictability properties of cache architectures. How would one define the predictability of a cache architecture? The notion should express what the strongest methods can find out about cache contents at program points. It is highly unlikely that a cache analysis can find out perfect information. This would require the knowledge of the initial cache contents, would only allow straight-line code, and would require the complete static knowledge of effective addresses. Thus, we can safely assume that there are program points with a certain amount of uncertainty about the cache contents. Therefore, it makes sense, as it is done in [RGBW07] to define the predictability of a cache architecture as the speed of recovery from unknown information. This section introduces two metrics, termed "evict" and "fill," which express how quickly cache contents become known by accessing sequences of memory blocks starting with a completely unknown cache state. These metrics are functions in the associativity, k, of the cache. "evict" tells after how many distinct memory accesses a cache analysis can safely predict that some memory blocks are no more in the cache. This is relevant for the prediction of cache misses. "fill" tells after how many distinct memory accesses a cache analysis has full information about the cache contents. This is relevant for the prediction of cache hits.

The cache replacement-strategy has the strongest influence on the predictability. It comes as no surprise that caches with an LRU replacement strategy are the most predictable; full information about what is in the cache is obtained after k distinct memory accesses, namely what has been accessed in these last k accesses. A cache with a FIFO replacement strategy needs $2k - 1$ distinct accesses to perfectly predict cache misses and $3k - 1$ to perfectly predict cache hits. For caches with a PLRU replacement strategy such as PowerPCs, these values are $k/2 \, \log_2 \, k + 1$ and $k/2 \, \log_2 \, k + k - 1$, respectively and that caches with FIFO or PLRU replacement strategies are significantly less predictable. Thus, LRU caches are preferrable for embedded systems with rigid timing constraints.

9.9   Acknowledgments

Many former students have worked on different parts of the method presented in this chapter and have together built a timing-analysis tool satisfying industrial requirements. Christian Ferdinand studied cache analysis and showed that precise information about cache contents can be obtained. Stephan Thesing together with Reinhold Heckmann and Marc Langenbach developed methods to model abstract processors. Stephan went through the pains of implementing several abstract models for real-life processors such as the Motorola ColdFire MCF 5307 and the Motorola PPC 755. I owe him my thanks for help with the presentation of pipeline analysis; Henrik Theiling contributed the preprocessor technology for the analysis of executables and the translation of complex control flow to integer linear programs. Many thanks to him for his contribution to the path analysis section. Michael Schmidt implemented several powerful value analyses. Reinhold Heckmann managed to model even very complex cache architectures. Jan Reineke and Daniel Grund developed the missing theory about the predictability of caches, which allowed us to see the general picture behind individual observations. Florian Martin implemented the program-analysis generator, PAG, which is the basis for many of the program analyses.

Over the years, the work of my group was supported by the European IST Project DAEDALUS, Validation of Critical Software by Static Analysis and Abstract Testing, the German Transregional Collaborative Research Centre AVACS (Automatic Verification and Analysis of Complex Systems) of the Deutsche Forschungsgemeinschaft, the European Network of Excellence ARTIST2, and the European ICT project PREDATOR, Reconciling Performance with Predictability. I owe thanks to the members of the cluster on Compilation and Timing Analysis of ARTIST2 for many interesting discussions and the collaboration in writing a survey on Timing Analysis [WEE+08].

References

AFMW96. Martin Alt, Christian Ferdinand, Florian Martin, and Reinhard Wilhelm. Cache behavior prediction by abstract interpretation. In *Proceedings of SAS'96, Static Analysis Symposium*, volume 1145 of Lecture Notes in Computer Science, pages 52–66. Springer-Verlag, September 1996.

AMWH94. Robert Arnold, Frank Mueller, David B. Whalley, and Marion Harmon. Bounding worst-case instruction cache performance. In *Proc. of the IEEE Real-Time Systems Symposium*, pages 172–181, Puerto Rico, December 1994.

CC76. P. Cousot and R. Cousot. Static determination of dynamic properties of programs. In *Proceedings of the Second International Symposium on Programming*, pages 106–130. Dunod, Paris, France, 1976.

CC77. Patrick Cousot and Radhia Cousot. Abstract interpretation: A unified lattice model for static analysis of programs by construction or approximation of fixpoints. In *Proceedings of the 4th ACM Symposium on Principles of Programming Languages*, pages 238–252, Los Angeles, California, 1977.

EG97. Andreas Ermedahl and Jan Gustafsson. Deriving annotations for tight calculation of execution time. In Christian Lengauer, Martin Griebl, and Sergei Gorlatch (Eds.), *Proceedings of the Third International Euro-Par Conference on Parallel Processing, Euro-Par '97*, pages 1298–1307, Passau, Germany, August 26–29, 1997, volume 1300 of Lecture Notes in Computer Science, Springer, 1997.

Eng02. Jakob Engblom. Processor pipelines and static worst-case execution time analysis. PhD thesis, Uppsala University, 2002.

Erm03. Andreas Ermedahl. A modular tool architecture for worst-case execution time analysis. PhD thesis, Uppsala University, 2003.

Fer97. Christian Ferdinand. Cache behavior prediction for real-time systems. PhD Thesis, Universität des Saarlandes, September 1997.

FHL+01. C. Ferdinand, R. Heckmann, M. Langenbach, F. Martin, M. Schmidt, H. Theiling, S. Thesing, and R. Wilhelm. Reliable and precise WCET determination for a real-life processor. In *EMSOFT*, volume 2211 of LNCS, pages 469–485, 2001.

FMW99. Christian Ferdinand, Florian Martin, and Reinhard Wilhelm. Cache behavior prediction by abstract interpretation. *Science of Computer Programming*, 35:163–189, 1999.

Gai. Aeroflex Gaisler. http:www.gaisler.com

Gra69. Ronald L. Graham. Bounds on multiprocessing anomalies. *SIAM Journal of Applied Mathematics*, 17(2):416–429, 1969.

HLTW03. Reinhold Heckmann, Marc Langenbach, Stephan Thesing, and Reinhard Wilhelm. The influence of processor architecture an the design and the results of WCET tools. *IEEE Proceedings on Real-Time Systems*, 91(7):1038–1054, 2003.

HWH95. Christopher A. Healy, David B. Whalley, and Marion G. Harmon. Integrating the timing analysis of pipelining and instruction caching. In *Proceedings of the IEEE Real-Time Systems Symposium*, pages 288–297, December 1995.

Inf. AbsInt Angewandte Informatik.

LBJ$^+$95.   Sung-Soo Lim, Young Hyun Bae, Gye Tae Jang, Byung-Do Rhee, Sang Lyul Min, Chang Yun Park, Heonshik Shin, Kunsoo Park, Soo-Mook Moon, and Chong Sang Kim. An accurate worst case timing analysis for RISC processors. *IEEE Transactions on Software Engineering*, 21(7):593–604, July 1995.

LM95.   Yau-Tsun Steven Li and Sharad Malik. Performance analysis of embedded software using implicit path enumeration. In *Proceedings of the 32nd ACM/IEEE Design Automation Conference*, pages 456–461, June 1995.

LMW95a.   Yau-Tsun Steven Li, Sharad Malik, and Andrew Wolfe. Efficient microarchitecture modeling and path analysis for real-time software. In *Proceedings of the IEEE Real-Time Systems Symposium*, pages 298–307, December 1995.

LMW95b.   Yau-Tsun Steven Li, Sharad Malik, and Andrew Wolfe. Performance estimation of embedded software with instruction cache modeling. In *Proceedings of the IEEE/ACM International Conference on Computer-Aided Design*, pages 380–387, November 1995.

LMW96.   Yau-Tsun Steven Li, Sharad Malik, and Andrew Wolfe. Cache modeling for real-time software: Beyond direct mapped instruction caches. In *Proceedings of the IEEE Real-Time Systems Symposium*, December 1996.

LS99.   Thomas Lundqvist and Per Stenström. An integrated path and timing analysis method based on cycle-level symbolic execution. *In Real-Time Systems*, 17((2/3)), November 1999.

LTH02.   Marc Langenbach, Stephan Thesing, and Reinhold Heckmann. Pipeline modelling for timing analysis. In Manuel V. Hermenegildo and German Puebla, editors, *Static Analysis Symposium SAS 2002*, volume 2477 of Lecture Notes in Computer Science, pages 294–309. Springer-Verlag, 2002.

Lun02.   Thomas Lundqvist. A WCET analysis method for pipelined microprocessors with cache memories. PhD thesis, Dept. of Computer Engineering, Chalmers University of Technology, Sweden, June 2002.

MAWF98.   Florian Martin, Martin Alt, Reinhard Wilhelm, and Christian Ferdinand. Analysis of loops. In *Proceedings of the International Conference on Compiler Construction (CC'98)*, volume 1383 of Lecture Notes in Computer Science, pages 80–94. Springer-Verlag, 1998.

MWH94.   Frank Mueller, David B. Whalley, and Marion Harmon. Predicting instruction cache behavior. In *Proceedings of the ACM SIGPLAN Workshop on Language, Compiler and Tool Support for Real-Time Systems*, Orlando, FL, 1994.

NNH99.   Flemming Nielson, Hanne Riis Nielson, and Chris Hankin. *Principles of Program Analysis*. Springer-Verlag, 1999.

PK89.   Peter Puschner and Christian Koza. Calculating the maximum execution time of real-time programs. *Real-Time Systems*, 1:159–176, 1989.

PS91.   Chang Yun Park and Alan C. Shaw. Experiments with a program timing tool based on source-level timing schema. *IEEE Computer*, 24(5):48–57, May 1991.

RG08.   Jan Reineke and Daniel Grund. Sensitivity of cache replacement policies. Reports of SFB/TR 14 AVACS 36, SFB/TR 14 AVACS, March 2008. ISSN: 1860-9821, http://www.avacs.org.

RGBW07.   Jan Reineke, Daniel Grund, Christoph Berg, and Reinhard Wilhelm. Timing predictability of cache replacement policies. *Real-Time Systems*, 37(2):99–122, November 2007.

RSW02.   T. Reps, M. Sagiv, and R. Wilhelm. Shape analysis and applications. In Y N Srikant and Priti Shankar, editors, *The Compiler Design Handbook: Optimizations and Machine Code Generation*, pages 175–217. CRC Press, 2002.

RWT$^+$06.   Jan Reineke, Bjoern Wachter, Stephan Thesing, Reinhard Wilhelm, Ilia Polian, Jochen Eisinger, and Bernd Becker. A definition and classification of timing anomalies. In *Proceedings of 6th International Workshop on Worst-Case Execution Time (WCET) Analysis*, Dresden, July 2006.

Sha89.   Alan C. Shaw. Reasoning about time in higher-level language software. *IEEE Transactions on Software Engineering*, 15(7):875–889, 1989.

SlPH+05.   Jean Souyris, Erwan le Pavec, Guillaume Himbert, Victor Jégu, Guillaume Borios, and Reinhold Heckmann. Computing the WCET of an avionics program by abstract interpretation. In *Proceedings of 5th International Workshop on Worst-Case Execution Time (WCET) Analysis*, pages 15–18, 2005.

SP07.   Marc Schlickling and Markus Pister. A framework for static analysis of VHDL code. In Christine Rochange, editor, *Proceedings of 7th International Workshop on Worst-Case Execution Time (WCET) Analysis*, Pisa, Italy, July 2007.

Tan06.   Lili Tan. The worst case execution time tool challenge 2006: The external test. In Tiziana Margaris, Anna Philippou, and Bernhard Steffen, editors, *2nd International Symposium on Leveraging Applications of Formal Methods (ISOLA'06)*, Paphos, Cyprus, November 2006.

TFW00.   Henrik Theiling, Christian Ferdinand, and Reinhard Wilhelm. Fast and precise WCET prediction by separated cache and path analyses. *Real-Time Systems*, 18(2/3):157–179, May 2000.

The00.   Henrik Theiling. Extracting safe and precise control flow from binaries. In *Proceedings of the 7th International Conference on Real-Time Systems and Applications (RTCSA'00)*, pages 23–30, 2000.

The01.   Henrik Theiling. Generating decision trees for decoding binaries. In *Proceedings of the Workshop on Languages, Compilers, and Tools for Embedded Systems (LCTES 2001)*, Snowbird, Utah, pages 112–120, 2001.

The02.   Henrik Theiling. Control flow graphs for real-time systems analysis. PhD thesis, Universität des Saarlandes, Saarbrücken, Germany, 2002.

The04.   Stephan Thesing. Safe and precise WCET determination by abstract interpretation of pipeline models. PhD thesis, Saarland University, 2004.

The06.   Stephan Thesing. Modeling a system controller for timing analysis. In *EMSOFT '06: Proceedings of the 6th ACM & IEEE International Conference on Embedded Software*, pages 292–300, New York, NY, USA, 2006. ACM.

TSH+03.   Stephan Thesing, Jean Souyris, Reinhold Heckmann, Famantanantsoa Randimbivololona, Marc Langenbach, Reinhard Wilhelm, and Christian Ferdinand. An abstract interpretation-based timing validation of hard real-time avionics software systems. In *Proceedings of the 2003 International Conference on Dependable Systems and Networks (DSN 2003)*, pages 625–632. IEEE Computer Society, June 2003.

TW04.   Lothar Thiele and Reinhard Wilhelm. Design for timing predictability. *Real-Time Systems*, 28:157–177, 2004.

WEE+08.   Reinhard Wilhelm, Jakob Engblom, Andreas Ermedahl, Niklas Holsti, Stephan Thesing, David Whalley, Guillem Bernat, Christian Ferdinand, Reinhold Heckmann, Frank Mueller, Isabelle Puaut, Peter Puschner, Jan Staschulat, and Per Stenström. The determination of worst-case execution times—Overview of the methods and survey of tools. *ACM Transactions on Embedded Computing Systems (TECS)*, 2008.

10

Performance Analysis of Distributed Embedded Systems

Lothar Thiele
Swiss Federal Institute of Technology

Ernesto Wandeler
Swiss Federal Institute of Technology

Wolfgang Haid
Swiss Federal Institute of Technology

10.1 Performance Analysis

10.1.1 Distributed Embedded Systems

An embedded system is a special-purpose information processing system that is closely integrated into its environment. It is usually dedicated to a certain application domain, and knowledge about the system behavior at design time can be used to minimize resources while maximizing predictability.

The embedding into a technical environment and the constraints imposed by a particular application domain very often lead to heterogeneous and distributed implementations. In this case, systems are composed of hardware components that communicate via some interconnection network. The functional and nonfunctional properties of the whole system do not only depend on the computations inside the various nodes but also on the interaction of the various data streams on the common communication media. In contrast to multiprocessor or parallel computing platforms, the individual computing nodes have a high degree of independence and usually communicate via message passing. It is particularly difficult to maintain global state and workload information as the local processing nodes usually make independent scheduling and resource access decisions.

In addition, the dedication to an application domain very often leads to heterogeneous distributed implementations, where each node is specialized to its local environment and/or its functionality. For example, in an automotive application one may find nodes (usually called embedded control units) that contain a communication controller, a CPU, memory, and I/O interfaces. But depending on the particular task of a node, it may contain additional digital signal processors (DSPs), different kinds of CPUs and interfaces, and different memory capacities.

The same observation also holds for interconnection networks. They may be composed of several interconnected smaller subnetworks, each with its own communication protocol and topology.

For example, in automotive applications we may find controller area networks (CAN), time-triggered protocols (TTP) like in TTCAN, or hybrid protocols like in FlexRay. The complexity of a design is particularly high if the computation nodes responsible for a single application are distributed across several networks. In this case, critical information may flow through several subnetworks and connecting gateways before it reaches its destination.

The above described architectural concepts of heterogeneity, distributivity, and parallelism can be observed on several layers of granularity. For example, heterogeneous multiprocessor systems-on-chip (MPSoC) integrating heterogeneous computing resources, such as CPUs, controllers, and DSPs, are becoming more and more popular in embedded systems. These individual components are connected using "networks-on-chip" that can be regarded as dedicated interconnection networks involving adapted protocols, bridges, or gateways. Typical applications executing on heterogeneous MPSoCs are real-time applications, such as audio- and video-streaming applications, signal processing applications, or packet processing applications with quality-of-service constraints.

The challenge is to exploit the performance potential of these architectures while maintaining a predictable execution in real time. One reason for requesting timing predictability is the fact that embedded systems are frequently connected to a physical environment through sensors and actuators. Typically, embedded systems are reactive systems that are in continuous interaction with their environment and they must execute at a pace determined by that environment. Examples are automatic control tasks, manufacturing systems, mechatronic systems, automotive/air/space applications, radio receivers and transmitters, and signal processing tasks in general. Also, in the case of multimedia and content production, missing audio or video samples need to be avoided under all circumstances. As a result, many embedded systems must meet real-time constraints, i.e., they must react to stimuli within the time interval dictated by the environment. A real-time constraint is called hard, if not meeting that constraint could result in a catastrophic failure of the system, and it is called soft otherwise. As a consequence, time-predictability in the strong sense cannot be guaranteed using statistical arguments.

To achieve these goals, suitable methods to estimate the timing properties of a design point are required during design space exploration as well as for the verification of the final design. Due to the distributed, heterogeneous architectures, this is not an easy task. Effects that need to be taken into account are, among others, the execution time and scheduling of tasks on processors, the arbitration of shared interconnection networks, the access to shared memories, and the overhead of the software run-time environment. In addition to analyzing the properties of a specific design point, more general questions regarding the analysis or optimization might arise during design space exploration, such as: How does the system behavior change when a certain task requires more time to execute? Which processing capabilities are needed to process an input data stream with a certain workload? What is an optimal binding of tasks to computing resources? What is an optimal slot length for a time-slotted bus?

Finally, let us give an example that shows part of the complexity in the performance and timing analysis of distributed embedded systems. The example adapted from [16] is particularly simple in order to point out one source of difficulties, namely, the interaction of event streams on a communication resource (Figure 10.1).

The application $A1$ consists of a sensor that periodically sends bursts of data to the CPU, which stores them in the memory using a task, $P1$. These data are processed by the CPU using a task, $P2$, with a worst-case execution time (WCET) and a best-case execution time (BCET). The processed data are transmitted via the shared bus to a hardware input/output device that is running task $P3$. We suppose that the CPU uses a preemptive fixed-priority scheduling policy, where $P1$ has the highest priority. The maximal workload on the CPU is obtained when $P2$ continuously uses the WCET and when the sensor simultaneously submits data. There is a second streaming application, $A2$, that receives real-time data in equidistant packets via the input interface. The input interface is running task $P4$ to send the data to a DSP for processing with task $P5$. The processed packets are then transferred to a playout buffer and task $P6$ periodically removes packets from the buffer, for example,

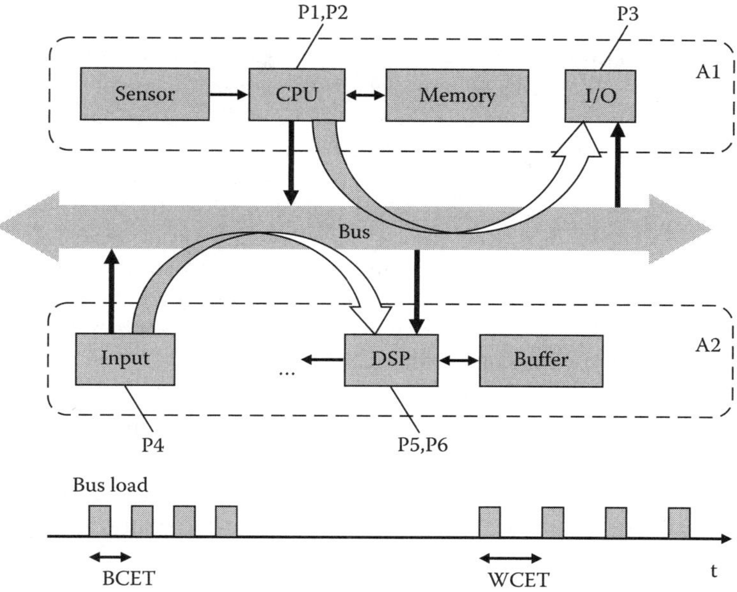

FIGURE 10.1 Interference of two applications on a shared communication resource. (From Thiele, L. and Wandeler, E., Performance analysis of distributed embedded systems, in Zurawski, R. (Ed.), *Embedded Systems Handbook*, CRC Press, Boca Raton, FL, 2005, 15-1–15-18. With permission.)

for playback. We suppose that the bus uses a FCFS (first come, first serve) scheme for arbitration. As the bus transactions from applications *A1* and *A2* interfere on the common bus, there will be a jitter in the packet stream received by the DSP that eventually may lead to an undesirable buffer overflow or underflow. It is now interesting to note that the worst-case situation in terms of jitter occurs if the processing in *A1* uses its BCET, as this leads to a blocking of the bus for a long time period. Therefore, the worst-case situation for the CPU load leads to a best case for the bus, and vice versa.

In more realistic situations, there will be simultaneous resource sharing on the computing and communication resources, there may be different protocols and scheduling policies on these resources, there may be a distributed architecture using interconnected subnetworks, and there may be additional nondeterminism caused by unknown input patterns and data. It is the purpose of performance analysis to determine the timing and memory properties of such systems.

10.1.2 Basic Terms

As a starting point to the analysis of timing and performance of embedded systems, it is very useful to clarify a few basic terms. Very often, the timing behavior of an embedded system can be described by the time interval between a specified pair of events. For example, the instantiation of a task, the occurrence of a sensor input, or the arrival of a packet could be a start event. Such events will be denoted as *arrival events*. Similarly, the finishing of an application or a part of it can again be modeled as an event, denoted as a *finishing event*. In the case of a distributed system, the physical location of the finishing event may not be equal to that of the corresponding arrival event and it may require the processing of a sequence or set of tasks, as well as the use of distributed computing and communication resources. In this case, we talk about end-to-end timing constraints. Note that not all pairs of events in a system are necessarily critical, i.e., have deadline requirements.

An embedded system processes the data associated with arrival events. The timing of computations and communications within the embedded system may depend on the input data (because of the data-dependent behavior of the tasks) and on the arrival pattern. In case of a conservative resource sharing strategy, such as a time-triggered architecture, the interference between these tasks is resolved by applying a static sharing strategy. If the use of shared resources is controlled by dynamic policies, all activities may interact with each other and the timing properties influence each other. As has been shown in the previous section, it is necessary to distinguish between the following terms:

- Worst case and best case: The worst case and the best case are the maximal and minimal time intervals between the arrival and finishing events under all admissible system and environment states. The execution time may vary largely, due to different input data and interference between concurrent system activities.
- Upper and lower bounds: Upper and lower bounds are quantities that bound the worst-case and best-case behaviors. These quantities are usually computed off-line, i.e., not during the run-time of the system.
- Statistical measures: Instead of computing bounds on the worst-case and best-case behaviors, one may also determine a statistical characterization of the run-time behavior of the system, for example, expected values, variances, and quantiles.

In the case of real-time systems, we are particularly interested in upper and lower bounds. They are used in order to statically verify whether the system meets its timing requirements, for example, deadlines. In contrast to end-to-end timing properties, the term *performance* is less well defined. Usually, it refers to a mixture of the achievable deadline, the delay of events or packets, and the number of events that can be processed per time unit (throughput).

Several methods do exist, such as analysis, simulation, emulation, and implementation, in order to determine or to approximate the above quantities. Besides analytic methods based on formal models, one may also consider simulation, emulation, or even implementation. All the latter possibilities should be used with care as only a finite set of initial states, environment behaviors, and execution traces can be considered. As is well known, the corner cases that lead to a WCET or BCET are usually not known, and thus incorrect results may be obtained. The huge state space of realistic system architectures makes it highly improbable that the critical instances of the execution can be determined without the help of analytical methods.

In order to understand the requirements for performance analysis methods in distributed embedded systems, we will classify the possible causes for a large difference between the worst-case and best-case or between the upper and lower bounds.

- Nondeterminism and interference: Let us suppose that there is only limited knowledge about the environment of the embedded system, for example, about the time when external events arrive or about their input data. In addition, there is interference of computation and communication on shared resources such as the CPU, memory, bus, or network. Then, we will say that the timing properties are nondeterministic with respect to the available information. Therefore, there will be a difference between the worst-case and the best-case behaviors as well as between the associated bounds. An example may be that the execution time of a task may depend on its input data. Another example is the communication of data packets on a bus in case of an unknown interference.
- Limited analyzability: If there is complete knowledge about the whole system, then the behavior of the system can be determined. Nevertheless, it may be that because of the system complexity, i.e., its complex state space as well as long-range dependencies, there is no feasible way of determining close upper and lower bounds on the worst-case and best-case timing, respectively.

As a result of this discussion, we understand that methods to analyze the performance of distributed embedded systems must be (a) correct in that they determine valid upper and lower bounds and (b) accurate in that the determined bounds are close to the actual worst case and best case.

10.1.3 Role in the Design Process

One of the major challenges in the design process of embedded systems is to the estimate essential characteristics of the final implementation early in the design. This can help in making important design decisions before investing too much time in detailed implementations. Typical questions faced by a designer during a system-level design process are: Which functions should be implemented in hardware and which in software (partitioning)? Which hardware components should be chosen (allocation)? How should the different functions be mapped onto the chosen hardware (binding)? Do the system-level timing properties meet the design requirements? What are the different bus utilizations and which bus or processor acts as a bottleneck? Then there are also questions related to the on-chip memory requirements and off-chip memory bandwidth.

Typically, the performance analysis or estimation is part of the design space exploration, where different implementation choices are investigated in order to determine the appropriate design trade-offs between the different conflicting objectives. Following Figure 10.2, the estimation of system properties in an early design phase is an essential part of the design space exploration. Different choices of the underlying system architecture, the mapping of the applications onto this architecture, and the chosen scheduling and arbitration schemes will need to be evaluated in terms of the different quality criteria.

In order to achieve acceptable design times, though, there is a need for automatic or semiautomatic (interactive) exploration methods. As a result, there are additional requirements for performance analysis if used for design space exploration, namely, (a) the simple reconfigurability with respect to architecture, mapping, and resource sharing policies; (b) a short analysis time in order to be able to test many different choices in a reasonable time frame; and (c) the possibility of coping with incomplete design information, as typically the lower layers are not designed or implemented (yet).

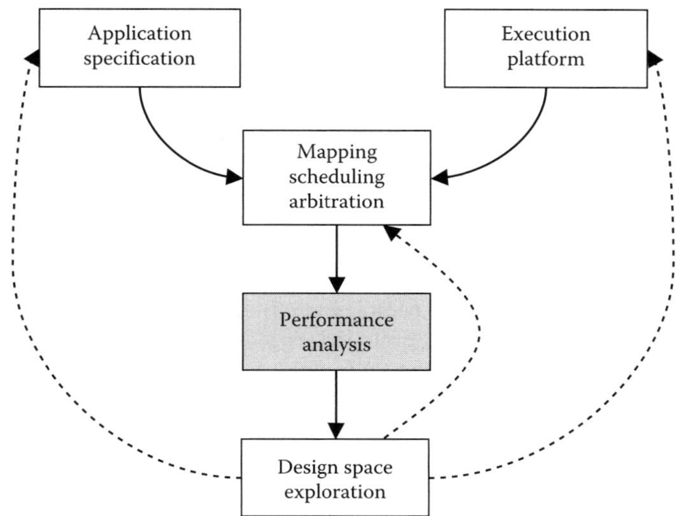

FIGURE 10.2 Relation between design space exploration and performance analysis. (From Thiele, L. and Wandeler, E., Performance analysis systems, in Zurawski, R. (Ed.), *Embedded Systems Handbook*, CRC Press, Boca Raton, FL, 15-1–15-18. With permission.)

Even if the design space exploration as described above is not a part of the chosen design methodology, the performance analysis is often part of the development process of software and hardware. In embedded system design, the functional correctness is validated after each major design step using simulation or formal methods. If there are nonfunctional constraints such as deadline or throughput requirements, they need to be validated as well, and all aspects of the design representation related to performance become "first-class citizens."

Finally, performance analysis of the whole embedded system may be done after completion of the design, in particular if the system is operated under hard real-time conditions where timing failures lead to a catastrophic situation. As has been mentioned above, performance simulation is not appropriate in this case because the critical instances and test patterns are not known in general.

10.1.4　Requirements

Based on the discussion above, one can list some of the requirements that a methodology for performance analysis of distributed embedded systems must satisfy.

- Correctness: The results of the analysis should be correct, i.e., there exist no reachable system states and feasible reactions of the system environment such that the calculated bounds are violated.
- Accuracy: The lower and upper bounds determined by the performance analysis should be close to the actual worst-case and best-case timing properties.
- Embedding into the design process: The underlying performance model should be sufficiently general to allow the representation of the application (which possibly uses different specification mechanisms), of the environment (periodic, aperiodic, bursty, different event types), of the mapping including the resource sharing strategies (preemption, priorities, time triggered), and of the hardware platform. The method should seamlessly integrate into the functional specification and design methodology.
- Short analysis time: Especially, if the performance analysis is part of a design space exploration, a short analysis time is important. In addition, the underlying model should allow for reconfigurability in terms of application, mapping, and hardware platform.

As distributed systems are heterogeneous in terms of the underlying execution platform, the diverse concurrently running applications, and the different scheduling and arbitration policies used, modularity is a key requirement for any performance analysis method. We can distinguish between several composition properties:

- Process Composition: Often, events need to be processed by several consecutive application tasks. In this case, the performance analysis method should be modular in terms of this functional composition.
- Scheduling Composition: Within one implementation, different scheduling methods can be combined, even within one computing resource (hierarchal scheduling); the same property holds for the scheduling and arbitration of communication resources.
- Resource Composition: A system implementation can consist of different heterogeneous computing and communication resources. It should be possible to compose them in a similar way as processes and scheduling methods.
- Building Components: Combinations of processes, associated scheduling methods, and architecture elements should be combined into components. This way, one could associate a performance component to a combined hardware/OS/software module of the implementation that exposes the performance requirements but hides internal implementation details.

It should be mentioned that none of the approaches known to date are able to satisfy all of the above mentioned criteria. On the other hand, depending on the application domain and the chosen design approach, not all of the requirements are equally important. The next section summarizes some of the available methods.

10.2 Approaches to Performance Analysis

In this survey, we select just a few representative and promising approaches that have been proposed for the performance analysis of distributed embedded systems. More detailed comparisons are contained in [7,11].

10.2.1 Simulation-Based Methods

Simulators have become an integral part of the embedded system design process. Currently, the performance estimation of embedded systems is mainly based on simulation, either for the performance estimation of the complete system or for the performance estimation of single system components. In simulation-based methods, many dynamic and complex interactions can be taken into account whereas analytic methods usually have to stick to a restrictive underlying model and suffer from limited scope. In addition, it is possible to match the level of abstraction in the representation of time to the required degree of accuracy. A wide range of simulators at different levels of abstractions is available today. The most accurate but also the slowest simulators are cycle-accurate simulators that are mainly used for the simulation of single processors [1]. Instruction-accurate simulators (also referred to as instruction-set simulators or virtual platforms) provide a good trade-off between speed and accuracy that allows one to model and simulate entire distributed systems. SystemC [17] has evolved as a widely used standard for implementing simulators at this level. Especially during design space exploration, simulations at even higher levels of abstraction are used such as trace-based simulation and different kinds of discrete event simulators.

Independent of the level of abstraction, a simulation framework not only has to consider the functional behavior of an application but also requires a concept of time in order to determine the performance of an embedded system. Additionally, properties of the execution platform, of the mapping between functional computation and communication processes to the elements of the underlying hardware, and of resource sharing policies (as usually implemented in the operating system or directly in hardware) need to be taken into account. This additional complexity leads to higher computation times, and performance estimation quickly becomes a bottleneck in the design even when advanced techniques are used such as mixed-level simulation or sampling-based simulation. Besides, there is often a substantial setup effort necessary if the mapping of the application to the underlying hardware platform changes.

The fundamental problem of simulation-based approaches to performance estimation is the insufficient corner case coverage. As shown in the example in Figure 10.1, the subsystem corner case (high computation time of $A1$) does not lead to the system corner case (small computation time of $A1$). Designers must provide a set of appropriate simulation stimuli in order to cover all the corner cases that exist in the distributed embedded system. Failures of embedded systems very often relate to timing anomalies that happen infrequently and, therefore, are almost impossible to discover by simulation. In general, simulation provides estimates of the average system performance but does not yield worst-case results and cannot determine whether the system satisfies the required timing constraints.

In the design process of distributed embedded systems, often a combination of different simulation approaches is used. An example is the approach taken in the Sesame framework [12] for exploring the design space of heterogeneous MPSoC architectures. It employs trace-based simulation that is intended to bridge the gap between pure simulation, which might be too slow to be used in a design space exploration cycle, and analytic methods, which are often too restricted in scope and not accurate enough. The performance estimation is split into several stages:

- Stage 1: A pure functional simulation is executed. During this simulation, event traces are extracted that represent the behavior of the application in terms of computation and communication events but does not contain data information anymore. Here, we do not take into account resource sharing such as different arbitration schemes and access conflicts. The output of this step is a timing inaccurate system execution trace.
- Stage 2: In the calibration phase, low-level simulation is used to determine the timing quantities of the tasks that are associated to the events. If tasks can be executed on different hardware resources, the corresponding quantities need to be determined for each resource.
- Stage 3: An architecture is chosen, a binding of the application onto the allocated resources is specified, and, finally, the scheduling and arbitration policies for processors and the interconnection networks is defined.
- Stage 4: Using trace-based simulation, the trace obtained by functional simulation is transformed and refined according to the decisions made in stage 3, taking into account the quantities obtained in the calibration phase. Thereby, the simulation captures the computation, communication, and synchronization as seen on the target system. The resulting simulation can then be used to estimate the system performance, determine critical paths, and collect various statistics about the computation and communication components.

The above approach still suffers from several disadvantages. All traces are the result of a simulation, and the coverage of corner cases is still limited. The underlying representation is a complete execution of the application in the form of a graph that may be of prohibitive size. The effects of the transformations applied in order to incorporate the concrete architecture are not formally specified. Therefore, it is not clear what the final analysis results represent. Finally, because of the separation between the functional simulation and the nonfunctional trace-based simulation, no feedback is possible. For example, a buffer overflow because of a sporadic communication overload situation may lead to a different functional behavior. Nevertheless, the described approach shows how simulations at different levels of abstraction can be used for performance estimation that is fast enough to be used during design space exploration.

10.2.2   Holistic Scheduling Analysis

There is a large body of formal methods available for analyzing the scheduling of shared computing resources, for example, fixed-priority, rate-monotonic, earliest deadline first scheduling, time-triggered policies like TDMA or round-robin, and static cyclic scheduling. From the WCET of individual tasks, the arrival pattern of activation, and the particular scheduling strategy, one can analyze in many cases the schedulability and worst-case response times (see for example, [3]). Many different application models and event patterns have been investigated such as sporadic, periodic, jitter, and bursts. There exists a large number of commercial tools that enable the analysis of quantities like resource load and response times. In a similar way, network protocols are increasingly supported by analysis and optimization tools.

The classical scheduling theory has been extended towards distributed systems where the application is executed on several computing nodes and the timing properties of the communication between these nodes cannot be neglected. The seminal work of Tindell and Clark [20] combined fixed priority preemptive scheduling at computation nodes with TDMA scheduling on the interconnecting bus. These results are based on two major achievements:

- Communication system (in this case, the bus) was handled in a similar way to the computing nodes. Because of this integration of process and communication scheduling, the method was called a holistic approach to the performance analysis of distributed real-time systems.
- Second contribution was the analysis of the influence of the release jitter on the response time, where the release jitter denotes the worst-case time difference between the arrival (or activation) of a process and its release (making it available to the processor). Finally, the release jitter has been linked to the message delay induced by the communication system.

This work was improved in terms of accuracy by Wolf [26] by taking into account correlations between arrivals of triggering events. In the meantime, many extensions and applications have been published based on the same line of thoughts. Other combinations of scheduling and arbitration policies have been investigated, such as CAN [19], and, more recently, the FlexRay protocol [13]. The latter extension opens the holistic scheduling methodology to mixed event-triggered and time-triggered systems where the processing and communication are driven by the occurrence of events or the advance of time, respectively. Several holistic analysis techniques are aggregated and implemented in the Modeling and Analysis Suite for Real-Time Applications (MAST) [9].

Nevertheless, it must be noted that the holistic approach does not scale to general distributed architectures in that for every new kind of application structure, sharing of resources, and combinations thereof, a new analysis needs to be developed. In general, the model complexity grows with the size of the system and the number of different scheduling policies. In addition, the method is restricted to the classical models of task arrival patterns such as periodic, or periodic with jitter.

10.2.3 Compositional Methods

Three main problems arise in the case of complex distributed embedded systems: Firstly, the architecture of such systems, as already mentioned, is highly heterogeneous—the different architectural components are designed assuming different input event models and use different arbitration and resource sharing strategies. This makes any kind of *compositional performance analysis* difficult. Secondly, applications very often rely on a high degree of concurrency. Therefore, there are multiple control threads, which additionally complicates timing analysis. And thirdly, we cannot expect that an embedded system only needs to process periodic events where each event is associated to a fixed number of bytes. If, for example, the event stream represents a sampled voice signal, then after several coding, processing, and communication steps, the amount of data per event as well as the timing may have changed substantially. In addition, often stream-based systems also have to process other event streams that are sporadic or bursty, for example, they have to react to external events or deal with best-effort traffic for coding, transcription, or encryption. There are only a few approaches available that can handle such complex interactions.

One prominent approach named modular performance analysis (MPA) [4] is based on a unifying model of different event patterns in the form of arrival curves as known from the networking domain [6]. The proposed real-time calculus (RTC) [18] represents the resources and their processing or communication capabilities in a compatible manner, and therefore allows for a modular hierarchical scheduling and arbitration for distributed embedded systems. The approach will be explained in Section 10.3 in some more detail.

Richter et al. proposed in [10,15] a method that is based on classical real-time scheduling results. They combine different well-known abstractions of event task arrival patterns and provide additional interfaces between them. The approach is based on the following principles:

- Main goal is to make use of the very successful results in real-time scheduling, in particular for sharing a single processor or a single communication link (see for example, [3]). For a large class of scheduling and arbitration policies and a set of arrival patterns (periodic, periodic with jitter, sporadic, and bursty), upper and lower bounds on the response time can be determined, i.e., the time difference between the arrival of a task and its finishing time. Therefore, the abstraction of a task of the application consists of a triggering event stream with a certain arrival pattern, the WCET, and the BCET on the resource. Several tasks can be mapped onto a single resource. Together with the scheduling policy, one can obtain for each task the associated lower and upper bounds of the response time. Communication and shared busses can be handled in a similar way.

- Application model is a simple concatenation of several tasks. The end-to-end delay can now be obtained by adding the individual contributions of the tasks; the necessary buffer memory can simply be computed by taking into account the initial arrival pattern.

- Obviously, the approach is feasible only if the arrival patterns fit the few basic models for which results on computing bounds on the response time are available. In order to overcome this limitation, two types of interfaces are defined:

  - EMIF: Event model interfaces are used in the performance analysis only. They perform a type of conversion between certain arrival patterns, i.e., they change the mathematical representation of event streams.

  - EAF: Event adaptation functions need to be used in cases where there exists no EMIF. In this case, the hardware/software implementation must be changed in order to make the system analyzable, for example, by adding playout buffers at appropriate locations.

  In addition, a new set of six arrival patterns was defined [16] that is more suitable for the proposed type of conversion using EMIF and EAF (see Figure 10.3).

Several extensions have been worked out, for example, in order to deal with cyclic nonfunctional dependencies and to generalize the application model (see [10]). In addition, there have been notable extensions towards robustness and sensitivity analysis (see [14]).

Nevertheless, when referring to the requirements for a MPA, the approach has some inherent drawbacks. EAFs are caused by the limited class of supported event models and the available analysis methods. The analysis method enforces a change in the implementation. Furthermore, the approach is not modular in terms of the resources, as their service is not modeled explicitly. For example, when several scheduling policies need to be combined in one resource (hierarchical scheduling), then for each new combination an appropriate analysis method must be developed. In this way, the approach suffers from the same problem as the "holistic approach" described earlier. In addition, one is bound to the classical arrival patterns that are not sufficient in cases of stream processing applications. Other event models need to be converted with loss in accuracy (EMIF) or the implementation must be changed (EAF).

10.3 Modular Performance Analysis

This section describes an approach to the performance analysis of embedded systems that is influenced by the worst-case analysis of communication networks. The network calculus as described

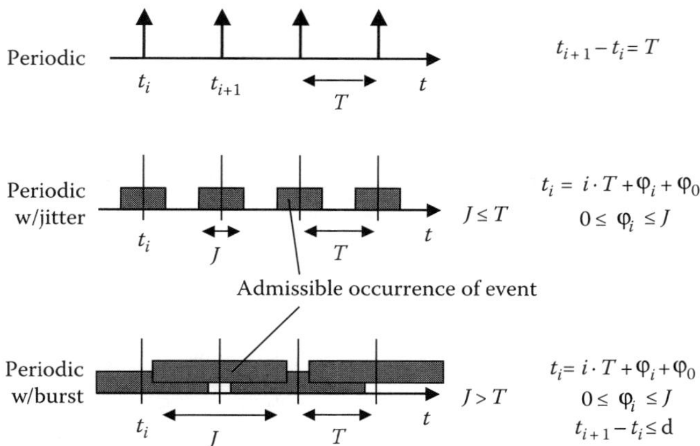

FIGURE 10.3 Some arrival patterns of tasks that can be used to characterize properties of event streams in [16]. T, J, and d denote the period, jitter, and minimal interarrival time, respectively. ϕ_0 denotes a constant phase shift. (From Thiele, L. and Wandeler, E., Performance analysis of distributed embedded systems, in Zurawski, R. (Ed.), *Embedded Systems Handbook*, CRC Press, Boca Raton, FL, 15-1–15-18. With permission.)

in [2] is based on [6] and uses (max,+) algebra to formulate the necessary operations. The network calculus is a promising analysis methodology as it is designed to be modular in various respects and as the representation of event (or packet) streams is not restricted to the few classes mentioned in the previous section. In [4,18], the method has been extended to RTC in order to deal with distributed embedded systems by combining computation and communication. Because of the detailed modeling of the capability of the shared computing and communication resources as well as the event streams, a high accuracy can be achieved, see [5]. The following sections serve to explain the basic approach.

10.3.1 Performance Network

In functional specification and verification, the given application is usually decomposed into components that communicate via event interfaces. The properties of the whole system are investigated by combining the behavior of the components. This kind of representation is common in the design of complex embedded systems and is supported by many tools and standards, for example, UML. It would be highly desirable if the performance analysis follows the same line of thinking as it could be easily integrated into the usual design methodology. Considering the discussion in the previous sections, we can identify two major additions that are necessary:

- Abstraction: Performance analysis is aimed at making statements about the timing behavior not just for one specific input characterization but for a larger class of possible environments. Therefore, the concrete event streams that flow between the components must be represented in an abstract way. As an example, we have seen them characterized as "periodic" and "sporadic with jitter." In the same way, the nonfunctional properties of the application and the resource sharing mechanisms must be modeled appropriately.

- Resource Modeling: In comparison to functional validation, we need to model the resource capabilities and how they are changed by the workload of tasks or communication. Therefore, contrary to the approaches described before, we will model the resources explicitly as "first-class citizens" of the approach.

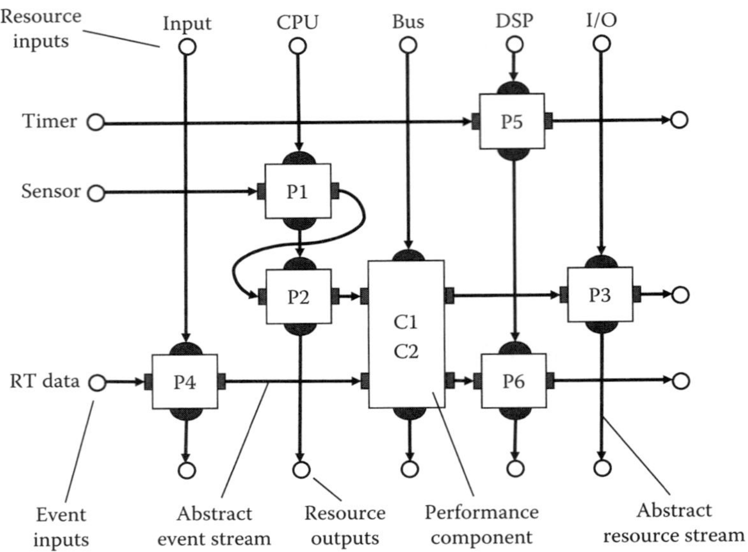

FIGURE 10.4 Simple performance network related to the example in Figure 10.1. (From Thiele, L. and Wandeler, E., Performance analysis of distributed embedded systems, in Zurawski, R. (Ed.), *Embedded Systems Handbook*, CRC Press, Boca Raton, FL, 15-1–15-18. With permission.)

As an example of a performance network, let us look again at the simple example from Figure 10.1. In Figure 10.4, we see a corresponding performance network. Because of the simplicity of the example, not all the modeling possibilities can be shown.

On the left-hand side of the figure, we see the *abstract event inputs* which model the sources of the event streams that trigger the tasks of the applications: "Timer" represents the periodic instantiation of the task that reads out the buffer for playback, "Sensor" models the periodic bursty events from the sensor, and "RT data" denotes the real-time data in equidistant packets via the input interface. The associated *abstract event streams* are transformed by the *performance components*. On the top, one can see the *resource inputs* that model the service of the shared resources, for example, the input, CPU, bus, CPU, and I/O component. The *abstract resource streams* (vertical direction) interact with the event streams on the performance components. The *resource outputs* at the bottom represent the remaining resource service that is available to other applications that may execute on the hardware platform.

The *performance components* represent (a) how the timing properties of input event streams are transformed to timing properties of output event streams and (b) the transformation of the resources. Of course, these components can be hierarchically grouped into larger components. How the performance components are grouped and interconnected reflects the resource sharing strategy. For example, *P1* and *P2* are connected serially in terms of the resource stream, and, therefore, they model a fixed priority scheme with the high priority assigned to task *P1*. If the bus implements the FCFS strategy or a TTP, the transfer function of *C1/C2* needs to be determined such that the abstract representations of the event and resource stream are correctly transformed.

10.3.2 Abstractions

The timing characterization of event and resource streams is based on variability characterization curves (VCCs), which substantially generalize the classical representations such as sporadic or periodic. As the event streams propagate through the distributed architecture, their timing properties get increasingly complex and the standard patterns cannot model them with appropriate accuracy.

The event streams are described using arrival curves, $\overline{\alpha}^u(\Delta)$, $\overline{\alpha}^l(\Delta) \in \mathbb{R}^{\geq 0}$, $\Delta \in \mathbb{R}^{\geq 0}$, which provide upper and lower bounds on the number of events in *any* time interval of length Δ. In particular, there are at most $\overline{\alpha}^u(\Delta)$ and at least $\overline{\alpha}^l(\Delta)$ events within the time interval $[t, t + \Delta)$ for all t.

In a similar way, the resource streams are characterized using service functions, $\beta^u(\Delta)$, $\beta^l(\Delta) \in \mathbb{R}^{\geq 0}$, $\Delta \in \mathbb{R}^{\geq 0}$, which provide upper and lower bounds on the available service in any time interval of length Δ. The unit of service depends on the kind of the shared resource, for example, instructions (computation) or bytes (communication).

Note that, as defined above, the VCC's $\overline{\alpha}^u(\Delta)$ and $\overline{\alpha}^l(\Delta)$ are expressed in terms of events (this is marked by a bar on their symbol), while the VCC's $\beta^u(\Delta)$ and $\beta^l(\Delta)$ are expressed in terms of workload/service. A method to transform event-based VCCs, to workload/resource-based VCCs, and vice versa is presented in [21]. A more fine-grained modeling of an application is possible too, for example, by taking into account different event types in event streams (see [23]). By the same approach, it is also possible to model more complex task models, for example, a task with state-dependent behavior.

Figure 10.5 shows arrival curves that specify the basic classical models shown in Figure 10.3. Note that in the case of sporadic patterns, the lower arrival curves are 0. In a similar way, Figure 10.6 shows a service curve of a simple TDMA bus access with period T, bandwidth b, and slot interval τ.

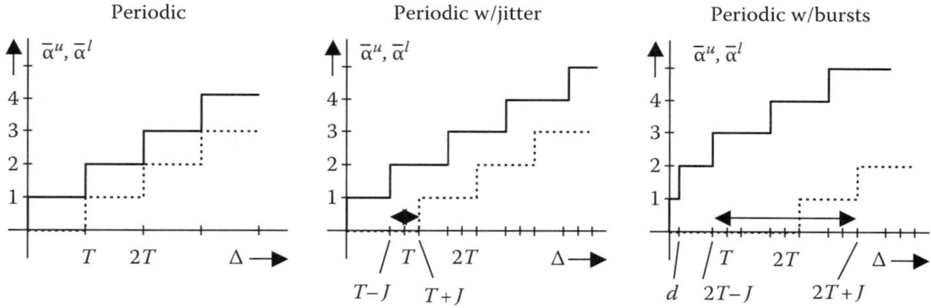

FIGURE 10.5 Basic arrival curves related to the patterns described in Figure 10.3. (From Thiele, L. and Wandeler, E., Performance analysis of distributed embedded systems, in Zurawski, R. (Ed.), *Embedded Systems Handbook*, CRC Press, Boca Raton, FL, 15-1–15-18. With permission.)

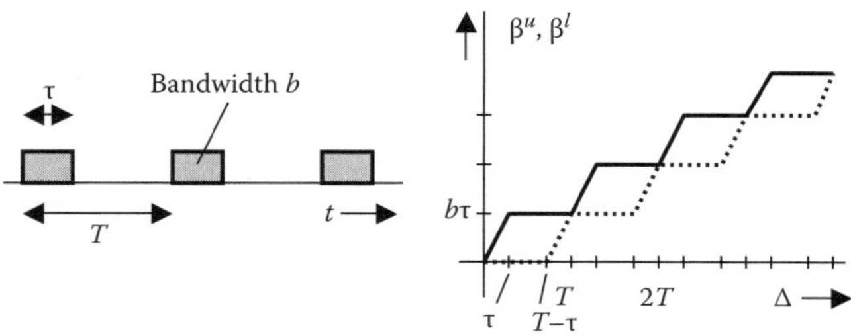

FIGURE 10.6 Example of a service curve that describes a simple TDMA protocol. (From Thiele, L. and Wandeler, E., Performance analysis of distributed embedded systems, in Zurawski, R. (Ed.), *Embedded Systems Handbook*, CRC Press, Boca Raton, FL, 15-1–15-18. With permission.)

Note that arrival curves can be approximated using linear approximations, i.e., piecewise linear functions. Moreover, there are of course finite representations of the arrival and service curves, for example, by decomposing them into an irregular initial part and a periodic part (see also [24]).

Where do we get the arrival and service functions from, for example, those characterizing a processor (CPU in Figure 10.4), or an abstract input (Sensor in Figure 10.4)?

- Pattern: In some cases, the patterns of the event or resource stream are known, for example, bursty, periodic, sporadic, and TDMA. In this case, the functions can be constructed analytically (see for example, Figures 10.5 and 10.6).
- Trace: In case of unknown arrival or service patterns, one may use a set of traces and compute the envelope. This can be done easily by using a sliding window of size Δ and determining the maximum and minimum number of events (or service) within the window. Such a trace could also be obtained through simulation.
- Data Sheets: In other cases, one can derive the curves by deriving the bounds from the characteristic of the generating device (in the case of the arrival curve) or the hardware component (in the case of the service curve).

In order to construct a scheduling network according to Figure 10.4, we still need to take into account the resource sharing strategy.

10.3.3 Resource Sharing and Analysis

We still need to describe how single event streams and resource streams interact on the available resources. The underlying model and analysis very much depends on the underlying execution platform. As an example, we suppose that the events (or data packets) corresponding to a single stream are stored in a queue before being processed (see Figure 10.7). The same model is used for computation as well as for communication resources. It matches well the usual structure of operating systems where ready tasks are lined up until the processor is assigned to one of them. Events belonging to one stream are processed in an FCFS manner whereas the order between different streams depends on the particular resource sharing strategy.

The resource sharing strategy determines the interconnection between performance components (see Figure 10.4). For example, the performance components associated to tasks $P1$ and $P2$ are connected serially. This way, we can model a preemptive fixed priority resource sharing strategy as $P2$ only gets the CPU resource that is left after the workload of $P1$ has been served. Other resource sharing

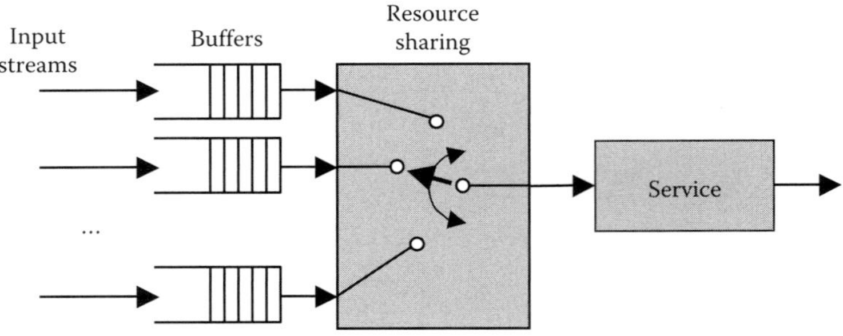

FIGURE 10.7   Functional model of resource sharing on computation and communication resources. (From Thiele, L. and Wandeler E., Performance analysis of distributed embedded systems, in Zurawski, R. (Ed.), *Embedded Systems Handbook*, CRC Press, Boca Raton, FL, 15-1–15-18. With permission.)

strategies can be modeled as well, such as TDMA, nonpreemptive fixed-priority scheduling [8], hierarchical scheduling, servers, and EDF [22].

The specification of a particular service model, i.e., the interaction of a single event stream on a single resource, leads to the equations that describe the functional transformation of arrival and service curves by a single performance component (see for example, [4]). Using these equations, the workload curves, and the characterization of input event and resource streams, we can now determine the characterizations of all event and resource streams in a performance network such as in Figure 10.4. From the computed arrival curves and service curves, we can compute all the relevant information such as the average resource loads, the end-to-end delays and the necessary buffer spaces on the event and packet queues (see Figure 10.7). A more elaborated example is described in [25].

All the above computations can be implemented efficiently, if appropriate representations for the VCCs are used, for example, piecewise linear, discrete points, or periodic (see [24]).

10.3.4 Concluding Remarks

Because of the modularity of the performance network, one can easily analyze a large number of different mapping and resource sharing strategies for design space exploration. Applications can be extended by adding tasks and the corresponding performance components. Moreover, different subsystems can use different kinds of resource sharing without sacrificing the performance analysis.

Of particular interest is the possibility building a performance component for a combined hardware–software system that describes the performance properties of a whole subsystem. This way, a subcontractor can deliver an HW/SW/OS module that already contains part of the application. The system house can now integrate the performance components of the subsystems in order to validate the performance of the whole system. To this end, he does not need to know the details of the subsystem implementations. In addition, a system house can also add an application to the subsystems. Using the resource interfaces that characterize the remaining available service from the subsystems, its timing correctness can easily be verified.

The performance network approach is correct in the sense that it yields upper and lower bounds on quantities like end-to-end delay and buffer space. On the other hand, it is a worst-case approach that covers all possible corner cases independent of their probability. Even if the deviations from simulation results can be small, see for example, [4], in many cases one is interested in the average case behavior of distributed embedded systems as well. Therefore, performance analysis methods such as those described in this chapter can be considered to be complementary to the existing simulation-based validation methods.

Furthermore, any automated or semi-automated exploration of different design alternatives (design space exploration) could be separated into multiple stages, with each stage having a different level of abstraction. It would then be appropriate to use an analytical performance evaluation framework, such as one of those described in this chapter, during the initial stages and resort to simulation only when a relatively small set of potential architectures is identified.

References

1. T. Austin, E. Larson, and D. Ernst, SimpleScalar: An infrastructure for computer system modeling, *Computer* 35(2), (2002), 59–67.
2. J.-Y. Le Boudec and P. Thiran, *Network Calculus — A Theory of Deterministic Queuing Systems for the Internet*, Lecture Notes in Computer Science 2050, Springer Verlag, 2001.
3. G.C. Buttazzo, *Hard Real-Time Computing Systems: Predictable Scheduling Algorithms and Applications*, Kluwer Academic Publishers, Boston, MA, 1997.

4. S. Chakraborty, S. Künzli, and L. Thiele, A general framework for analysing system properties in platform-based embedded system designs, *Proceedings of the 6th Design, Automation and Test in Europe (DATE)* (Munich, Germany), March 2003, pp. 190–195.

5. S. Chakraborty, S. Künzli, L. Thiele, A. Herkersdorf, and P. Sagmeister, Performance evaluation of network processor architectures: Combining simulation with analytical estimation, *Computer Networks* (51), (2003), 641–665.

6. R.L. Cruz, A calculus for network delay, Part I: Network elements in isolation, *IEEE Transactions on Information Theory* 37(1), (1991), 114–131.

7. M. Gries, Methods for evaluating and covering the design space during early design development, *Integration, the VLSI Journal* 38(2), (2004), 131–183.

8. W. Haid and L. Thiele, Complex task activation schemes in system level performance analysis, *Proceedings of the 5th International Conference on Hardware/Software Codesign and System Synthesis (CODES+ISSS)* (Salzburg, Austria), October 2007, pp. 173–178.

9. M. González Harbour, J. J. Gutiérrez García, J.C. Palencia Gutiérrez, and J.M. Drake Moyano, MAST: Modeling and analysis suite for real time applications, *Proceedings of the 13th Euromicro Conference on Real-Time Systems (ECRTS)* (Delft, The Netherlands), June 2001, pp. 125–134.

10. R. Henia, A. Hamann, M. Jersak, R. Racu, K. Richter, and R. Ernst, System level performance analysis — The SymTA/S approach, *Computers and Digital Techniques, IEE Proceedings* 152(2), (2005), 148–166.

11. S. Perathoner, E. Wandeler, L. Thiele, A. Hamann, S. Schliecker, R. Henia, R. Racu, R. Ernst, and M. González Harbour, Influence of different system abstractions on the performance analysis of distributed real-time systems, *Proceedings of the 7th International Conference on Embedded Software (EMSOFT)* (Salzburg, Austria), October 2007, pp. 193–202.

12. A.D. Pimentel, C. Erbas, and S. Polstra, A systematic approach to exploring embedded system architectures at multiple abstraction levels, *IEEE Transactions on Computers* 55(2), (2006), 99–112.

13. T. Pop, P. Eles, and Z. Peng, Holistic scheduling and analysis of mixed time/event triggered distributed embedded systems, *Proceedings of the 10th International Symposium on Hardware-Software Codesign (CODES)* (Estes Park, CO), May 2002, pp. 187–192.

14. R. Racu, M. Jersak, and R. Ernst, Applying sensitivity analysis in real-time distributed systems, *Proceedings of the 11th Real Time and Embedded Technology and Applications Symposium (RTAS)* (San Francisco, CA), March 2005, pp. 160–169.

15. K. Richter and R. Ernst, Model interfaces for heterogeneous system analysis, *Proceedings of the 6th Design, Automation and Test in Europe (DATE)* (Munich, Germany), March 2002, pp. 506–513.

16. K. Richter, D. Ziegenbein, M. Jersak, and R. Ernst, Model composition for scheduling analysis in platform design, *Proceedings of the 39th Design Automation Conference (DAC)* (New Orleans, LA), June 2002, pp. 287–292.

17. *Open SystemC Initiative*, http://www.systemc.org.

18. L. Thiele, S. Chakraborty, and M. Naedele, Real-time calculus for scheduling hard real-time systems, *Proceedings of the IEEE International Symposium on Circuits and Systems (ISCAS)* (Geneva, Switzerland), May 2000, pp. 101–104.

19. K. Tindell, A. Burns, and A.J. Wellings, Calculating controller area networks (CAN) message response times, *Control Engineering Practice* 3(8), (1995), 1163–1169.

20. K. Tindell and J. Clark, Holistic schedulability analysis for distributed hard real-time systems, *Microprocessing and Microprogramming—Euromicro Journal* (Special Issue on Parallel Embedded Real-Time Systems) 40, (1994), 117–134.

21. E. Wandeler, A. Maxiaguine, and L. Thiele, Quantitative characterization of event streams in analysis of hard real-time applications, *Real-Time Systems* 29(2) (2005), 205–225.

22. E. Wandeler and L. Thiele, Interface-based design of real-time systems with hierarchical scheduling, *Proceedings of the 12th IEEE Real-Time and Embedded Technology and Applications Symposium (RTAS)* (San Jose, CA), April 2006, pp. 243–252.

23. E. Wandeler and L. Thiele, Workload correlations in multi-processor hard real-time systems, *Journal of Computer and System Sciences* 73(2), (2007), 207–224.
24. E. Wandeler and L. Thiele, *Real-Time Calculus (RTC) Toolbox*, http://www.mpa.ethz.ch/Rtctoolbox, 2008.
25. E. Wandeler, L. Thiele, M. Verhoef, and P. Lieverse, System architecture evaluation using modular performance analysis—a case study, *Software Tools for Technology Transfer (STTT)* 8(6), (2006), 649–667.
26. T. Yen and W. Wolf, Performance estimation for real-time distributed embedded systems, *IEEE Transactions on Parallel and Distributed Systems* 9(11), (1998), 1125–1136.
27. L. Thiele and E. Wandeler, Performance analysis of distributed embedded systems, in R. Zurawski (Ed.), *Embedded Systems Handbook*, CRC Press, Boca Raton, FL, pp. 15-1–15-18, 2005.

11

Power-Aware Embedded Computing

Margarida F. Jacome
University of Texas at Austin

Anand Ramachandran
University of Texas at Austin

11.1 Introduction

Embedded systems are pervasive in modern life. State-of-the-art embedded technology drives the ongoing revolution in consumer and communication electronics, and it is on the basis of substantial innovation in many other domains, including medical instrumentation, process control, etc. [MRW02]. The impact of embedded systems in well-established "traditional" industrial sectors, e.g., automotive industry, is also increasing at a fast pace [MRW02,BGJ$^+$97].

Unfortunately, as CMOS technology rapidly scales, enabling the fabrication of ever faster and denser integrated circuits (ICs), the challenges that must be overcome to deliver each new generation of electronic products multiply. In the last few years, power dissipation has emerged as a major concern. In fact, projections on power density increases due to CMOS scaling clearly indicate that this is one of the fundamental problems that will ultimately preclude further scaling [Bor99,ITR]. Although the power challenge is indeed considerable, much can be done to mitigate the deleterious effects of power dissipation, thus enabling performance and device density to be taken to truly unprecedented levels by the semiconductor industry throughout the next 10–15 years.

Power density has a direct impact on packaging and cooling costs and can also affect system reliability, due to electromigration and hot-electron degradation effects. Thus, the ability to decrease power density, while offering similar performance and functionality, critically enhances the competitiveness of a product. Moreover, for battery-operated portable systems, maximizing battery lifetime translates into maximizing duration of service, an objective of paramount importance for this class of products. Power is thus a "primary figure of merit" in contemporaneous embedded system design.

Digital CMOS circuits have two main types of power dissipation: dynamic and static. Dynamic power is dissipated when the circuit performs the function(s) it was designed for, e.g., logic and arithmetic operations (computation), data retrieval, storage and transport, etc. Ultimately, all of this activity translates into switching of the logic states held on circuit nodes. Dynamic power dissipation is thus proportional to $C \cdot V_{DD}^2 \cdot f \cdot r$, where C denotes the total circuit capacitance; V_{DD} and f denote the circuit supply voltage and clock frequency, respectively; and r denotes the fraction of transistors expected to switch at each clock cycle [PR02,GGH97]. In other words, dynamic power dissipation is impacted to first order by circuit size/complexity, speed/rate, and switching activity. By contrast, static power dissipation is associated with preserving the logic state of circuit nodes between such switching activities and is caused by subthreshold leakage mechanisms. Unfortunately, as device sizes shrink, the severity of leakage power increases at an alarming pace [Bor99].

Clearly, the power problem must be addressed at all levels of the design hierarchy, from system to circuit, as well as through innovations on CMOS device technology [CSB92,PR02,JYWV03]. In this survey we provide a snapshot on the state of the art on system- and architecture-level design techniques and methodologies aimed at reducing both static and dynamic power dissipation. Since such techniques focus on the highest level of the design hierarchy, their potential benefits are immense. In particular, at this high level of abstraction, the specifics of each particular class of embedded applications can be considered as a whole and, as it will be shown in our survey, such an ability is critical to designing power/energy-efficient systems, that is, systems that spend energy strictly when and where it is needed. Broadly speaking, this requires a proper design and allocation of system resources, geared toward addressing critical performance bottlenecks in a power-efficient way. Substantial power/energy savings can also be achieved through the implementation of adequate dynamic power management policies, e.g., tracking instantaneous workloads (or levels of resource utilization) and "shutting-down" idling/unused resources, so as to reduce leakage power, or "slowing down" underutilized resources, so as to decrease dynamic power dissipation. These are clearly system-level decisions/policies, in that their implementation typically impacts several architectural subsystems. Moreover, different decisions/policies may interfere or conflict with each other and, thus, assessing their overall effectiveness requires a system-level (i.e., global) view of the problem.

A typical embedded system architecture consists of a processing subsystem (including one or more processor cores, hardware accelerators, etc.), a memory subsystem, peripherals, and global and local interconnect structures (buses, bridges, crossbars, etc.). Figure 11.1 shows an abstract view of two such architecture instances. Broadly speaking, a system-level design consists of defining the specific embedded system architecture to be used for a particular product, as well as defining how the target-embedded application (implementing the required functionality/services) is to be mapped into that architecture.

Embedded systems come in many varieties and with many distinct design optimization goals and requirements. Even when two products provide the same basic functionality, say, video encoding/decoding, they may have fundamentally different characteristics, namely, different performance and quality-of-service requirements, one may be battery operated and the other may not, etc. The implications of such product differentiation are of paramount importance when power/energy is considered. Clearly, the higher the system's required performance/speed (defined by metrics such as throughput, latency, bandwidth, response time, etc.), the higher its power dissipation will be. The key objective is thus to minimize the power dissipated to deliver the required level of performance [CB95a,PR02]. The trade-offs, techniques, and optimizations required to develop such power-aware or power-efficient designs vary widely across the vast spectrum of embedded systems

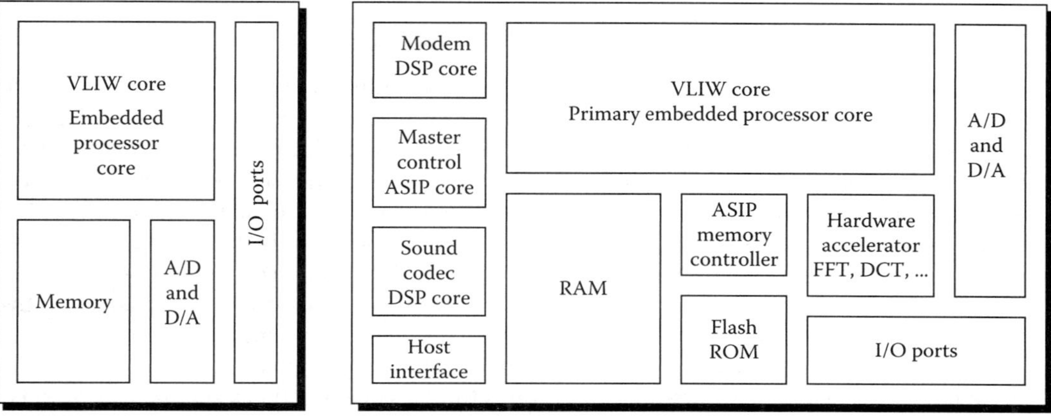

FIGURE 11.1 Illustrative examples of a simple and a more complex embedded system architecture.

available in today's market, encompassing many complex decisions driven by system requirements as well as intrinsic characteristics of the target applications [MSS⁺03].

Consider, for example, the design task of deciding on the number and type of processing elements to be instantiated on an embedded architecture, that is, defining its processing subsystem. Power, performance, cost, and time-to-market considerations dictate if one should rely entirely on readily available processors (i.e., off-the-shelf microcontrollers, DSPs and/or general-purpose RISC cores), or should also consider custom execution engines, namely, application-specific instruction set processors (ASIPs), possibly reconfigurable, and/or hardware accelerators (see Figure 11.1). Hardware/ software partitioning is a critical step in this process [MRW02]. It consists of deciding which of an application's segments/functions should be implemented in software (i.e., run on a processor core) and which (if any) should be implemented in hardware (i.e., execute on high performance, highly power-efficient custom hardware accelerators). Naturally, hardware/software partitioning decisions should reflect the power/performance criticality of each such segment/function. Clearly, this is a complex multiobjective optimization problem defined on a huge design space that encompasses, both, hardware- and software-related decisions. To compound the problem, the performance and energy-efficiency of an architecture's processing subsystem cannot be evaluated in isolation, since its effectiveness can be substantially impacted by the memory subsystem (i.e., the adopted memory hierarchy/organization) and the interconnect structures supporting communication/data transfers between processing components and to/from the environment in which the system is embedded. Thus, decisions with respect to these other subsystems and components must be concurrently made and jointly assessed.

Targeting up-front a specific embedded system platform, that is, an architectural "subspace" relevant to a particular class of products/applications, can considerably reduce the design effort [VM01, RV02]. Still, the design space remains (typically) so complex that a substantial design space exploration may be needed in order to identify power/energy-efficient solutions for the specified performance levels. Since time to market is critical, methodologies to efficiently drive such an exploration, as well as fast simulators and low complexity (yet good fidelity) performance, power and energy estimation models, are critical to aggressively exploiting effective power/energy-driven optimizations, within a reasonable time frame.

Our survey starts by providing an overview on state-of-the-art models and tools used to evaluate the goodness of individual system design points. We then discuss the power management techniques and the optimizations aimed at aggressively improving the power/energy efficiency of the various subsystems of an embedded system.

11.2   Energy and Power Modeling

This section discusses the high-level modeling and the power estimation techniques aimed at assisting system and architecture-level designs. It would be unrealistic to expect a high degree of accuracy on power estimates produced during such an early design phase, since accurate power modeling requires detailed physical-level information that may not yet be available. Moreover, highly accurate estimation tools (working with detailed circuit/layout-level information) would be too time consuming to allow for any reasonable degree of design space exploration [MRW02,PR02,BHLM94].

Thus, practically speaking, power estimation during early design space exploration should aim at ensuring a high degree of fidelity rather than necessarily accuracy. Specifically, the primary objective during this critical exploration phase is to assess the relative power efficiency of different candidate system architectures (populated with different hardware and/or software components), the relative effectiveness of alternative software implementations (of the same functionality), the relative effectiveness of different power management techniques, etc. Estimates that correctly expose such relative power trends across the design space region being explored provide the designer with the necessary information to guide the exploration process.

11.2.1   Instruction- and Function-Level Models

Instruction-level power models are used to assess the relative power/energy efficiency of different processors executing a given target embedded application, possibly with alternative memory subsystem configurations. Such models are thus instrumental during the definition of the main subsystems of an embedded architecture, as well as during hardware/software partitioning. Moreover, instruction-level power models can also be used to evaluate the relative effectiveness of different software implementations of the same embedded application, in the context of a specific embedded architecture/platform.

In their most basic form, instruction-level power models simply assign a power cost to each assembly instruction (or class of assembly instructions) of a programmable processor. The overall energy consumed by a program running on the target processor is estimated by summing up the instruction costs for a dynamic execution trace which is representative of the application [TMW94,CP92,RJ98, BFSS00].

Instruction-level power models were first developed by experimentally measuring the current drawn by a processor while executing different instruction sequences [TMW94]. During this first modeling effort, it was observed that the power cost of an instruction may actually depend on previous instructions. Accordingly, the instruction-level power models developed in [TMW94] include several interinstruction effects. Later studies observed that, for certain processors, the power dissipation incurred by the hardware responsible for fetching, decoding, analyzing and issuing instructions, and then routing and reordering results, was so high that a simpler model that only differentiates between instructions that access on-chip resources and those that go off-chip would suffice for such processors [RJ98].

Unfortunately, power estimation based on instruction-level models can still be prohibitively time consuming during an early design space exploration, since it requires collecting and analyzing large instruction traces and for many processors, considering a quadratically large number of interinstruction effects. In order to "accelerate" estimation, processor-specific coarser function-level power models were later developed [QKUP00]. Such approaches are faster because they rely on the use of macromodels characterizing the average energy consumption of a library of functions/subroutines executing on a target processor [QKUP00]. The key challenge in this case is to devise macromodels that can properly quantify the power consumed by each subroutine of interest, as a function of "easily observable parameters." Thus, for example, a quadratic power model of the form $an^2 + bn + c$ could be first tentatively selected for a insertion sort routine, where n denotes the number of elements to be sorted. Actual power dissipation then needs to be measured for a large number of experiments, run with different values of n. Finally, the values of the macromodel's coefficients a, b, and c are derived, using regression analysis, and the overall accuracy of the resulting macromodel is assessed [QKUP00].

The "high-level" instruction and function-level power models discussed so far allow designers to "quickly" assess a large number of candidate system architectures and alternative software implementations, so as to narrow the design space to a few promising alternatives. Once this initial broad exploration is concluded, power models for each of the architecture's main subsystems and components are needed, in order to support the detailed architectural design phase that follows.

11.2.2 Microarchitectural Models

Microarchitectural power models are critical to evaluating the impact of different processing subsystem choices on power consumption, as well as the effectiveness of different (microarchitecture level) power management techniques implemented on the various subsystems.

In the late 1980s and early 1990s, cycle accurate (or more precisely, cycle-by-cycle) simulators, such as Simplescalar [BA97], were developed to study the effect of architectural choices on the performance of general purpose processors. Such simulators are in general very flexible, allowing designers/architects to explore the complex design space of contemporaneous processors. Namely, they include built-in parameters that can be used to specify the number and mix of functional units to be instantiated in the processor's datapath, the issue width of the machine, the size and associativity of the L1 and L2 caches, etc. By varying such parameters, designers can study the performance of different machine configurations for representative applications/benchmarks. As power consumption became more important, simulators to estimate dynamic power dissipation (e.g., "Wattch [BTM00]," "Cai-Lim model [CL99]," and "Simplepower [YVKI00]") were later incorporated on these existing frameworks. Such an integration was performed seamlessly, by directly augmenting the "cycle-oriented" performance models for the various microarchitectural components with corresponding power models.

Naturally, the overall accuracy of these simulation-based power estimation techniques is determined by the level of the detail of the power models used for the microarchitecture's constituent components. For out-of-order RISC cores, for example, the power consumed in finding independent instructions to issue is a function of the number of instructions currently in the instruction queue and of the actual dependencies between such instructions. Unfortunately, the use of detailed power models accurately capturing the impact of input and state data on the power dissipated by each component would prohibitively increase the already "long" microarchitectural simulation run-times. Thus, most state-of-the-art simulators use very simple/straightforward empirical power models for datapath and control logic and slightly more sophisticated models for regular structures such as caches [BTM00]. In their simplest form, such models capture "typical" or "average" power dissipation for each individual microarchitectural component. Specifically, each time a given component is accessed/used during a simulation run, it is assumed that it dissipates its corresponding "average" power. Slightly more sophisticated power macromodels for datapath components have been proposed in [JKSN99,BBM00,TGTS98,KE98,CRC00,MG01], and shown to improve accuracy with a relatively small impact on simulation time.

So far, we have discussed power modeling of microarchitectural components, yet a substantial percentage of the overall power budget of a processor is actually spent on the global clock (up to 40%–45% [PR02]). Thus, global clock power models must also be incorporated in these frameworks. The power dissipated on global clock distribution is impacted to first order by the number of pipeline registers (and thus by a processor's pipeline depth) and by global and local wiring capacitances (and thus by a processor's core area) [PR02]. Accordingly, different processor cores and/or different configurations of the same core may dissipate substantially different clock distribution power.

Power estimates incorporating such numbers are thus critical during the processor core selection and configuration.

The component-level and clock distribution models discussed so far are used to estimate the dynamic power dissipation of a target microarchitecture. Yet, as mentioned above, static/leakage power dissipation is becoming a major concern, and thus, microarchitectural techniques aimed at reducing leakage power are increasingly relevant. Models to support early estimation of static power dissipation emerged along the same lines as those used for dynamic power dissipation. The "Butts–Sohi" model, which is one of the most influential static power models developed so far, quantifies static energy in CMOS circuits/components using a lumped parameter model that maps technology and design effects into corresponding characterizing parameters [BS00]. Specifically, static power dissipation is modeled as $V_{DD} \cdot N \cdot k_{\text{design}} \cdot I_{\text{leak}}$, where V_{DD} is the supply voltage and N denotes the number of transistors in the circuit. k_{design} is the design-dependent parameter—it captures "circuit style-" related characteristics of a component, including average transistor aspect ratio, average number of transistors switched-off during "normal/typical" component operation, etc. Finally, I_{leak} is the technology-dependent parameter. It accounts for the impact of threshold voltage, temperature, and other key parameters, on leakage current, for a specific fabrication process.

From a system designer's perspective, static power can be reduced by lowering supply voltage (V_{DD}), and/or by power supply gating or V_{DD}-gating (as opposed to clock-gating) unused/idling devices (N). Integrating models for estimating static power dissipation on cycle-by-cycle simulators thus enables embedded system designers to analyze critical static power versus performance trade-offs enabled by "power-aware" features available in contemporaneous processors, such as dynamic voltage scaling and selective datapath (re)configuration. An improved version of the "Butts–Sohi" model, providing the ability to dynamically recalculate leakage currents (as temperature and voltage change due to operating conditions and/or dynamic voltage scaling), has been integrated into the Simplescalar simulation framework, called "HotLeakage," enabling such high-level trade-offs to be explored by embedded system designers [ZPS$^+$03].

11.2.3 Memory and Bus Models

Storage elements, such as caches, register files, queues, buffers, and tables constitute a substantial part of the power budget of contemporaneous embedded systems [PAC$^+$97]. Fortunately, the high regularity of some such memory structures (e.g., caches) permits the use of simple, yet reasonably accurate power estimation techniques, relying on automatically synthesized "structural designs" for such components.

The Cache Access and Cycle TIme (CACTI) framework implements this synthesis-driven power estimation paradigm. Specifically, given a specific cache hierarchy configuration (defined by parameters such as cache size, associativity, and line size), as well as information on the minimum feature size of the target technology [WJ96], it internally generates a coarse structural design for such cache configuration. It then derives delay and power estimates for that particular design, using parameterized built-in C models for the various constituent elements, namely, SRAM cells, row and column decoders, word and bit lines, precharge circuitry, etc. [KG97,RJ01].

CACTI's synthesis algorithms used to generate the structural design of the memory hierarchy (which include defining the aspect ratio of memory blocks, the number of instantiated subbanks, etc.) have been shown to consistently deliver reasonably "good" designs across a large range of cache hierarchy parameters [RJ01]. CACTI can thus be used to "quickly" generate power estimates (starting from high-level architectural parameters) exhibiting a reasonably good fidelity over a large region of the design space. During design space exploration, the designer may thus consider a number of alternative L1 and L2 cache configurations and use CACTI to obtain "access-based" power dissipation estimates for each such configuration with good fidelity. Naturally, the memory access traces used by CACTI should be generated by a microarchitecture simulator (e.g., Simplescalar) working with a memory simulator (e.g., Dinero [EH98]), so that they reflect the bandwidth requirements of the embedded application of interest.

Buses are also a significant contributor to dynamic power dissipation [PR02,CWD$^+$98]. The dynamic power dissipation on a bus is proportional to $C \cdot V_{DD}^2 \cdot fa$, where C denotes the total capacitance of the bus (including metal wires and buffers); V_{DD} denotes the supply voltage; and fa denotes the average switching frequency of the bus [CWD$^+$98]. In this high-level model, the average switching frequency of the bus, (fa), is defined by the product of two terms, namely, the average number of bus transitions per word, and the bus frequency (given in bus words per second). The average number of bus transitions per word can be estimated by simulating sample programs and collecting the corresponding transition traces. Although this model is coarse, it may suffice during the early design phases under consideration.

11.2.4 Battery Models

The capacity of a battery is a nonlinear function of the current drawn from it, that is, if one increases the average current drawn from a battery by a factor of two, the "remaining" deliverable battery capacity, and thus its lifetime, decreases by more than half. Peukert's formula models such nonlinear behavior by defining the capacity of a battery as k/I^α, where k is a constant depending on the battery design; I is the discharge current; and α quantifies the "nonideal" behavior of the battery [MS96].* More effective system-level trade-offs between quality/performance and duration of service can be implemented by taking such nonlinearity (also called rate-capacity effect) into consideration. However, in order to properly evaluate the effectiveness of such techniques/trade-offs during system-level design, adequate battery models and metrics are needed.

Energy-delay product is a well-known metric used to assess the energy-efficiency of a system. It basically quantifies a system's performance loss per unit gain in energy consumption. To emphasize the importance of accurately exploring key trade-offs between battery lifetime and system performance, a new metric, viz., battery-discharge delay product, has recently been proposed [PW02]. Task scheduling strategies and dynamic voltage scaling policies adopted for battery operated embedded systems should thus aim at minimizing a battery-discharge delay product, rather than an energy-delay product, since it captures the important rate-capacity effect alluded to above, whereas the energy-delay product is insensitive to it. Yet, a metric such as the battery-discharge delay product requires the use of precise/detailed battery models. One such detailed battery model has been recently proposed, which can predict the remaining capacity of a rechargeable lithium-ion battery in terms of several critical factors, viz., discharge-rate (current), battery output voltage, battery temperature, and cycle age (i.e., the number of times a battery has been charged and discharged) [RP03].

11.3 System/Application-Level Optimizations

When designing an embedded system, in particular, a battery powered one, it may be useful to explore different task implementations exhibiting different power/energy versus quality-of-service characteristics, so as to provide the system with the desired functionality while meeting cost, battery lifetime, and other critical requirements. Namely, one may be interested in trading-off accuracy for energy savings on a handheld GPS system, or image quality for energy savings on an image decoder, etc. [SCB96,QP99,SBGM00]. Such application-level "tuning/optimizations" may be performed statically (i.e., one may use a single implementation for each task) or may be performed at run

---

* For an "ideal" battery, i.e., a battery whose capacity in independent of the way current is drawn, $\alpha = 0$, while for a real battery α may be as high as 0.7 [MS96].

time, under the control of a system-level power manager. For example, if the battery level drops below a certain threshold, the power manager may drop some services and/or swap some tasks to less power hungry (lower quality) software versions, so that the system can remain operational for a specified window of additional time. An interesting example of such dynamic power management was implemented for the Mars Pathfinder, an unmanned space robot that draws power from, both, a nonrechargable battery and solar cells [LCBK01]. In this case, the power manager tracks the power available from the solar cells, ensuring that most of the robot's active work is done during daylight, since the solar energy cannot be stored for later use.

In addition to dropping and/or swapping tasks, a system's dynamic power manager may also shutdown or slowdown subsystems/modules that are idling or underutilized. The advanced configuration and power interface (ACPI) is a widely adopted standard that specifies an interface between an architecture's power managed subsystems/modules (e.g., display drivers, modems, hard-disk drivers, processors, etc.) and the system's dynamic power manager. It is assumed that the power-managed subsystems have at least two power states (ACTIVE and STANDBY) and that one can dynamically switch between them. Using ACPI, one can implement virtually any power management policy, including fixed time-out, predictive shutdown, predictive wake-up, etc. [ACP]. Since the transition from one state to another may take a substantial amount of time and consume a nonnegligible amount of energy, the selection of a proper power management policy for the various subsystems is critical to achieving good energy savings with a negligible impact on performance. The so-called time-out policy, widely used in modern computers, simply switches a subsystem to STANDBY mode when the elapsed time after the last utilization reaches a given fixed threshold. Predictive shutdown uses the previous history of the subsystem to predict the next expected idle time and, based on that, decides if it should or should not be shutdown. More sophisticated policies, using stochastic methods [SCB96,QP99,SBGM00,SBA$^+$01], can also be implemented, yet they are more complex, and thus the power associated with running the associated dynamic power management algorithms may render them inadequate or inefficient for certain classes of systems.

The system-level techniques discussed above act on each subsystem as a whole. Although the effectiveness of such techniques has been demonstrated across a wide variety of systems, finergrained self-monitoring techniques, implemented at the subsystem level, can substantially add to these savings. Such techniques are discussed in the following sections.

11.4　Energy-Efficient Processing Subsystems

As mentioned earlier, hardware/software codesign methodologies partition the functionality of an embedded system into hardware and software components. Software components, executing on programmable microcontrollers or processors (either general purpose or application specific), are the preferred solution, since their use can substantially reduce design and manufacturing costs, as well as shorten time-to-market. Custom hardware components are typically used only when strictly necessary, namely, when an embedded system's power budget and/or performance constraints preclude the use of software. Accordingly, a large number of power-aware processor families is available in today's market, providing a vast gamut of alternatives suiting the requirements/needs of most embedded system [GH96,BB96]. In the sequel, we discuss power-aware features available in contemporaneous processors, as well as several power-related issues relevant to processor core selection.

11.4.1　Voltage and Frequency Scaling

Dynamic power consumption in a processor (be it general purpose or application specific) can be decreased by reducing two of its key contributors, viz., supply voltage and clock frequency. In fact, since the power dissipated in a CMOS circuit is proportional to the square of the supply voltage, the most effective way to reduce power is to scale down the supply voltage. Note however that the propagation delay across a CMOS transistor is proportional to $V_{DD}/(V_{DD} - V_T)^2$, where V_{DD} is the supply voltage and V_T is the threshold voltage. So, unfortunately, as the supply voltage decreases, the propagation delay increases as well, and so clock frequency (i.e., speed) may need to be decreased [IY98].

Accordingly, many contemporaneous processor families, such as Intel's XScale [INT], IBM's PowerPC 405LP [IBM], and Transmeta's Crusoe [TRA], offer dynamic voltage and frequency scaling features. For example, the Intel 80200 processor, which belongs to the XScale family of processors mentioned above, supports a software programmable clock frequency. Specifically, the voltage can be varied from 1.0 to 1.5 V, in small increments, with the frequency varying correspondingly from 200 to 733 MHz, in steps of 33/66 MHz.

The simplest way to take advantage of the scaling features discussed above is by carefully identifying the "smallest" supply voltage and corresponding operating frequency that guarantee that the target embedded application meets its timing constraints and run the processor for that fix setting. If the workload is reasonably constant throughout execution, this simple scheme may suffice. However, if the workload varies substantially during execution, more sophisticated techniques that dynamically adapt the processor's voltage and frequency to the varying workload can deliver more substantial power savings. Naturally, when designing such techniques it is critical to consider the cost of transitioning from one setting to another, i.e., the delay and power consumption overheads incurred by each transition. For example, for the Intel 80200 processor mentioned above, changing the processor frequency could take up to 20 µs, while changing the voltage could take up to 1 ms [INT].

Most processors developed for the mobile/portable market already support some form of built-in mechanism for voltage/frequency scaling. Intel's SpeedStep technology, for example, detects if the system is currently plugged into a power outlet or running on a battery, and based on that, either runs the processor at the highest voltage/frequency or switches it to a less power hungry mode. Transmeta's Crusoe processors offer a power manager called LongRun [TRA], which is implemented in the processor's firmware. LongRun relies on the historical utilization of the processor to guide clock rate selection: it increases the processor's clock frequency if the current utilization is high and decreases it if the utilization is low.

More sophisticated/aggressive dynamic scaling techniques should vary the core's voltage/frequency based on some predictive strategy, while carefully monitoring performance, so as to ensure it does not drop beyond a certain threshold and/or that task deadlines are not consistently missed [WWDS94, GCW95,PBB98,PBB00] (see Figure 11.2). The use of such voltage/frequency scaling techniques must necessarily rely on adequate dynamic workload prediction and performance metrics, thus requiring the direct intervention of the operating system and/or of the applications themselves. Although more complex than the simple schemes discussed above, several such predictive techniques have been shown to deliver substantial gains for applications with well-defined task deadlines, e.g., hard/soft real-time systems.

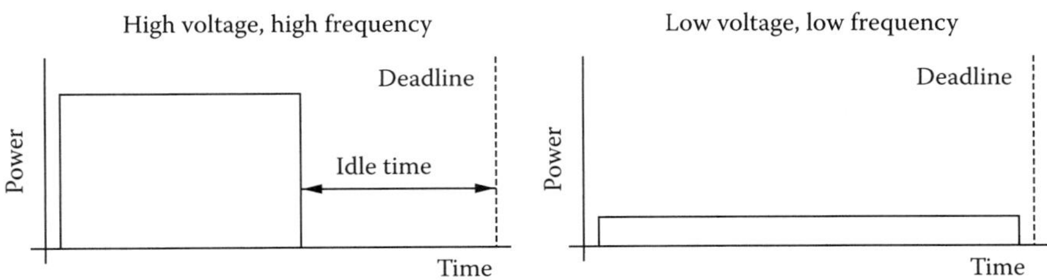

FIGURE 11.2 Power consumption without and with dynamic voltage and frequency scaling.

Simple interval-based prediction schemes consider the amount of idle time on a previous interval as a measure of the processor's utilization for the next time interval and use that prediction to decide on the voltage/frequency settings to be used throughout its duration. Naturally, many critical issues must be factored in when defining the duration of such an interval (or prediction window), including overhead costs associated with switching voltage/frequency settings. While a prediction scheme based on a single interval may deliver substantial power gains with marginal loss in performance [WWDS94], looking exclusively at a single interval may not suffice many applications. Namely, the voltage/frequency settings may end up oscillating in an inefficient way between different settings [GCW95]. Simple smoothing techniques, e.g., using an exponentially moving average of previous intervals, can be employed to mitigate the problem [GCW95,FRM02].

Note, finally, that the benefits of dynamic voltage and frequency scaling are not limited to reducing dynamic power consumption. When the voltage/frequency of a battery operated system is lowered, the instantaneous current drawn by the processor decreases accordingly, leading to a more effective utilization of the battery capacity and thus to increased duration of service.

11.4.2 Dynamic Resource Scaling

Dynamic resource scaling refers to exploiting adaptive, fine-grained hardware resource reconfiguration techniques in order to improve power efficiency. Dynamic resource scaling requires enhancing the microarchitecture with the ability to selectively "disable" components, fully or partially, through either clock-gating or V_{DD}-gating.* The effectiveness of dynamic resource scaling is predicated on the fact that many applications have variable workloads, that is, they have execution phases with substantial instruction-level parallelism (ILP), and other phases with much less inherent parallelism. Thus, by dynamically "scaling down" microarchitecture components during such "low activity" periods, substantial power savings can potentially be achieved.

Techniques for reducing static power dissipation on processor cores can definitely exploit resource scaling features, once they become widely available on processors. Namely, underutilized or idling resources can partially or fully be V_{DD}-gated, thus reducing leakage power. Several utilization-driven techniques have already been proposed, which can selectively shutdown functional units, segments of register files, and other datapath components, when such conditions arise [BM00,DKA+02,PKG01, BAB+02]. Naturally, the delay overhead incurred by power supply gating should be carefully factored into these techniques, so that the corresponding static energy savings are achieved with only a small degradation in performance.

Dynamic energy consumption can also be reduced by dynamically "scaling down" power hungry microarchitecture components, e.g., reducing the size of the issue window and/or reducing the effective width of the pipeline, during periods of low "activity" (say, when low ILP code is being executed). Moreover, if one is able to scale down these microarchitecture components that define the critical path delay of the machine's pipeline (typically located on the rename and window access stages), additional opportunities for voltage scaling, and thus dynamic power savings, can be created. This is thus a very promising area still undergoing intensive research [HSA01].

11.4.3 Processor Core Selection

The power-aware techniques discussed thus far can broadly be applied to general purpose as well as application-specific processors. Naturally, the selection of processor core(s) to be instantiated in the architecture's processing subsystem is also likely to have a substantial impact on overall

---

* Note that, in contrast to clock gating, all state information is lost when a circuit is V_{DD}-gated.

power consumption, particularly when computation intensive embedded systems are considered. A plethora of programmable processing elements, including microcontrollers, general-purpose processors, digital signal processors (DSPs), and ASIPs, addressing the specific needs of virtually every segment of the embedded systems' market, is currently offered by vendors.

As alluded to above, for computation-intensive embedded systems with moderately high to stringent timing constraints, the selection of processor cores is particularly critical, since high-performance usually signifies high levels of power dissipation. For these systems, ASIPs and DSPs have the potential to be substantially more energy efficient than their general purpose counterparts, yet their "specialized" nature poses significant compilation challenges* [JdV00]. By contrast, general purpose processors are easier to compile to, being typically shipped with good optimizing compilers, as well as debuggers and other development tools. Unfortunately, their "generality" incurs a substantial power overhead. In particular, high-performance general purpose processors require the use of power hungry hardware assists, including reservation stations, reorder buffers and rename logic, and complex branch prediction logic to alleviate control stalls. Still, their flexibility and high-quality development tools are very attractive for systems with stringent time-to-market constraints, making them definitively relevant for embedded systems.

IBM/Motorola's PowerPC family and the ARM family are examples of "general purpose" processors enhanced with power-aware features that are widely used in modern embedded systems. A plethora of specialized/customizable processors is also offered by several vendors, including specialized media cores from Philips, Trimedia, MIPS, and so on; DSP cores offered by Texas Instruments, StarCore, and Motorola; and customizable cores from Hewlett-Packard-STMicroelectronics and Tensilica.

The instruction set architectures (ISAs) and features offered on these specialized processors can vary substantially, since they are designed and optimized for different classes of applications. However, several of them, including TI's TMS320C6x family, HP-STS's Lx, the Starcore and Trimedia families, and Philip's Nexperia use a "very large instruction word" (VLIW) [CND+88,BYA93] or "explicitly parallel instruction computing" (EPIC) [SR00] paradigm. One of the key differences between VLIW and superscalar architectures is that VLIW machines rely on the compiler to extract ILP, and then schedule and bind instructions to functional units statically, while their high-performance superscalar counterparts use dedicated (power hungry) hardware to perform run-time dependence checking, instruction reordering, etc. Thus, in broad terms, VLIW machines eliminate power hungry microarchitecture components by moving the corresponding functionality to the compiler.[†] Moreover, wide VLIW machines are generally organized as a set of small clusters with local register files.[‡] Thus, in contrast to traditional superscalar machines, which rely on power hungry multiported monolithic register files, multicluster VLIW machines scale better with increasing issue widths, i.e., dissipate less dynamic power and can work at faster clock rates. Yet, they are harder to compile to [HHG+95,Ell85,DT93,DKK+99,JdVL00,LJdV02,PJ03]. In summary, the VLIW paradigm works very well for many classes of embedded applications, as attested to by the large number of VLIW processors currently available in the market. However, it poses substantial compilation challenges, some of which are still undergoing active research [MG95,Lie97,JYWV03]. Fortunately, many processing-intensive embedded applications have only a few time-critical loop kernels. Thus, only a very small percentage of the overall code needs to be actually subject to the complex and time-consuming compiler optimizations required by VLIW machines. In the case of media applications,

---

* Several DSP/ASIP vendors provide preoptimized assembly libraries to mitigate this problem.

[†] Since functional unit-binding decisions are made by the compiler, VLIW code is larger than RISC/superscalar code. We will discuss techniques to address this problem later in our survey.

[‡] A cluster is a set of functional units connected to a local register file. Clusters communicate with each other through a dedicated interconnection network.

for example, such loop kernels may represent as little as 3% of the overall program and yet may take up to 95% of the execution time [FWL99].

When the performance requirements of an embedded system are extremely high, using a dedicated coprocessor aimed at accelerating the execution of time-critical kernels (under the control of a host computer), may be the only feasible solution. Imagine, one such programmable coprocessor was designed to accelerate the execution of kernels of streaming media applications [KDR+01]. It can deliver up to 20 GFLOPS at a relatively low power cost (2 GFLOPS/W), yet such a power efficiency does not come without a cost [KDR+01]. As expected, Imagine has a complex programming paradigm, that requires extracting all time-critical kernels from the target application, and carefully reprogramming them using Imagine's "stream-oriented" coding style, so that, both, data and instructions can be efficiently routed from the host to the accelerator. Programming Imagine thus require a substantial effort, yet its power efficiency makes it very attractive for systems that demand such high levels of performance. Finally, at the highest end of the performance spectrum, one may need to consider using fully customized hardware accelerators. Comprehensive methodologies to design such accelerators are discussed in detail in [RMV+87].

11.5 Energy-Efficient Memory Subsystems

While processor speeds have been increasing at a very fast rate (about 60% a year), memory performance has increased at a comparatively modest rate (about 7% a year), leading to the well-known "processor-memory performance gap" [PAC+97]. In order to alleviate the memory access latency problem, modern processor designs use increasingly large on-chip caches, with up to 60% of the transistors dedicated to on-chip memory and support circuitry [PAC+97]. As a consequence, power dissipation in the memory subsystem contributes to a substantial fraction of the energy consumed by modern processors. A study targeting the StrongARM SA-110, a low-power processor widely used in embedded systems, revealed that more than 40% of the processor's power budget is taken up by on-chip data and instruction caches [MWA+96]. For high-performance general-purpose processors, this percentage is even higher, with up to 90% of the power budget consumed by memory elements and circuits aimed at alleviating the aforementioned memory bottleneck [PAC+97]. Power-aware memory designs have thus received considerable attention in recent years.

11.5.1 Cache Hierarchy Tuning

The energy cost of accessing data/instructions from off-chip memories can be as much as two orders of magnitude higher than that of an access to on-chip memory [HWO97]. By retaining instructions and data with high spatial and/or temporal locality on-chip, caches can substantially reduce the number of costly off-chip data transfers, thus leading to potentially quite substantial energy savings.

The "one-size-fits-all" nature of the general purpose domain dictates that one should use large caches with a high degree of associativity, so as to try to ensure high hit rates (and thus low average memory access latencies) for as many applications as possible. Unfortunately, as one increases cache size and associativity, larger circuits and/or more circuits are activated on each access to the cache, leading to a corresponding increase in dynamic energy consumption.

Clearly, in the context of embedded systems, one can do much better. Specifically, by carefully tuning/scaling the configuration of the cache hierarchy, so that it more efficiently matches the bandwidth requirements and access patterns of the target embedded application, one can essentially achieve memory access latencies similar to those delivered by "larger" (general purpose) memory subsystems, while substantially decreasing the average energy cost of such accesses.

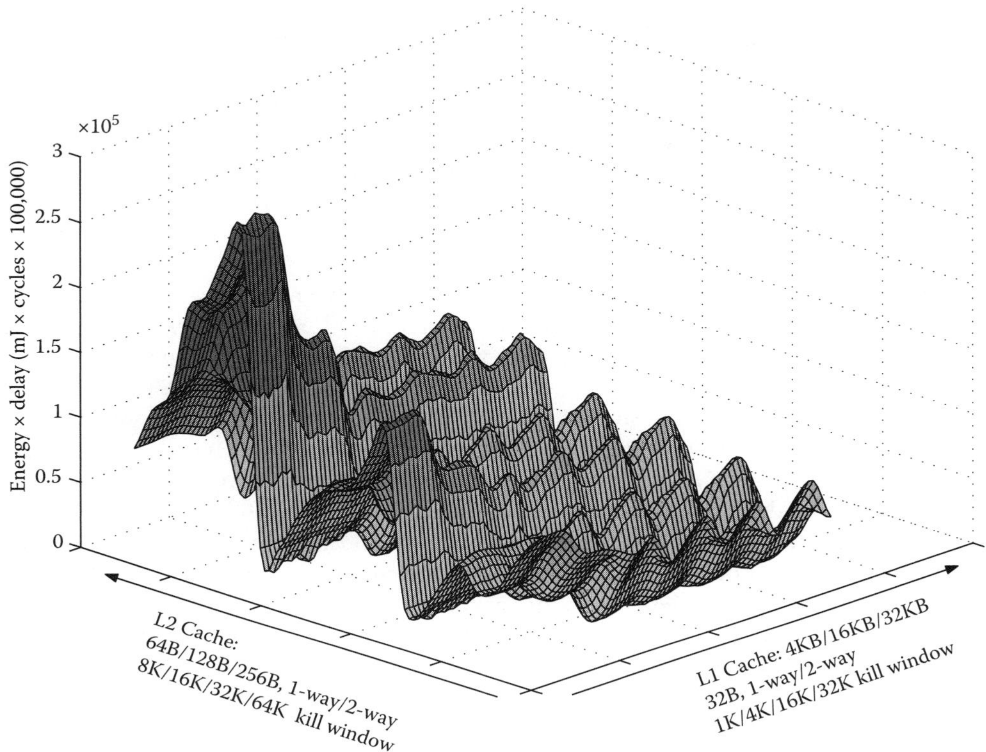

FIGURE 11.3 Design space exploration: energy-delay product for various L1 and L2 D-cache configurations for a JPEG application running on an XScale-like processor core.

Since several of the cache hierarchy parameters exhibit conflicting trends, an aggressive design space exploration over many candidate cache configurations is typically required in order to properly tune the cache hierarchy of an embedded system. Figure 11.3 summarizes the results of one such design space exploration performed for a media application. Namely, the graph in Figure 11.3 plots the energy-delay product metric for a wide range of L1 and L2 on-chip D-cache configurations. The delay term is given by the number of cycles taken to process a representative data set from start to completion. The energy term accounts for the energy spent on data memory accesses for the particular execution trace.* As it can be seen, the design space is very complex, reflecting the conflicting trends alluded to above. For this case study, the best set of cache hierarchy configurations exhibits an energy-delay product that is about one order of magnitude better (i.e., lower) than that of the worst configurations. Perhaps even more interesting, some of the worst memory subsystem configurations use quite large caches with a high degree of associativity, clearly indicating that no substantial performance gains would be achieved (for this particular media application) by using such an aggressively dimensioned memory subsystem.

For many embedded systems/applications, the power efficiency of the memory subsystems can be improved even more aggressively, yet this requires the use of novel (nonstandard) memory system designs, as discussed in the sections that follow.

---

* Specifically, the energy term accounts for accesses to the on-chip L1 and L2 D-caches and to main memory.

11.5.2 Novel Horizontal and Vertical Cache Partitioning Schemes

In recent years, several novel cache designs have been proposed to aggressively reduce the average dynamic energy consumption incurred by memory accesses. Energy efficiency is improved in these designs by taking direct advantage of specific characteristics of target classes of applications. The memory footprint of instructions and data in media applications, for example, tends to be very small, thus creating unique opportunities for energy savings [FWL99]. Since streaming media applications are pervasive in today's portable electronics market, they have been a preferred application domain for validating the effectiveness of such novel cache designs.

Vertical partition schemes [GK99,SD95,KGMS00,FTS95], as the name suggests, introduce additional small buffers/caches before the first level of the "traditional" memory hierarchy. For applications with "small" working sets, this strategy can lead to considerable dynamic power savings.

A concrete example of a vertical partition scheme is the filter cache [KGMS97], which is a very small cache placed in front of the standard L1 data cache. If the filter cache is properly dimensioned, dynamic energy consumption in the memory hierarchy can substantially be reduced, not only by accessing most of the data from the filter cache, but also by powering down (clock-gating) the L1 D-cache to a STANDBY mode during periods of inactivity [KGMS97]. Although switching the L1 D-cache to STANDBY mode results in delay/energy penalties when there is a miss in the filter cache, it was observed that for media applications, the energy-delay product did improve quite significantly when the two techniques were combined.

Predecoded instruction buffers [BHK+97] and loop buffers [LMA99] are the variants of the vertical partitioning scheme discussed above, and yet are applied to instruction caches (I-caches). The key idea of the first partitioning scheme mentioned above is to store recently used instructions on an instruction buffer, in a decoded form, so as to reduce the average dynamic power spent on fetching and decoding instructions. The second partitioning scheme allows one to hold time-critical loop bodies (identified a priori by the compiler or by the programmer) on small and thus energy-efficient dedicated loop buffers.

Horizontal partition schemes refer to the placement of additional (small) buffers or caches at the same level as the L1 cache. For each memory reference, the appropriate (level one) cache to be accessed is determined by dedicated decoding circuitry residing between the processor core and the memory hierarchy. Naturally, the method used to partition data across the set of first-level caches should ensure that the cache selection logic is simple, and thus cache access times are not significantly affected.

Region-based caches implement one such "horizontal partitioning" scheme, by adding two small 2 kB L1 D-caches to the first level of the memory hierarchy, one for stack and the other for global data. This arrangement has also been shown to achieve substantial gains in dynamic energy consumption for streaming media applications with a negligible impact on performance [LT00].

11.5.3 Dynamic Scaling of Memory Elements

With an increasing number of on-chip transistors being devoted to storage elements in modern processors, of which only a very small set is active at any point in time, static power dissipation is expected to soon become a key contributor to a processor's power budget. State-of-the-art techniques to reduce static power consumption in on-chip memories are based on the simple observation that, in general, data or instructions fetched into a given cache line have an immediate flurry of accesses during a "small" interval of time, followed by a relatively "long" period of time where they are not used, before eventually being evicted to make way for new data/instructions [WHK91,BGK95]. If one can "guess" when that period starts, it is possible to "switch-off" (i.e., V_{DD}-gate) the corresponding cache lines without introducing extra cache misses, thereby saving static energy consumption with no impact on performance [KHM01,ZTRC01].

Cache decay was one of the earliest attempts to exploit such a "generational" memory usage behavior to decrease leakage power [KHM01]. The original Cache decay implementation used a simple policy that turned-off cache lines after a fixed number of cycles (decay interval) since the last access. Note that if the selected decay interval happens to be too small, cache lines are switched off prematurely, causing extra cache misses, and if it is too large, opportunities for saving leakage energy are missed. Thus, when such a simple scheme is used, it is critical to tune the "fixed" decay interval very carefully, so that it adequately matches the access patterns of the embedded application of interest. Adaptive strategies, varying the decay interval at run-time so as to dynamically adjust it to changing access patterns, have been proposed more recently, so as to enable the use of the cache decay principle across a wider range of applications [KHM01,ZTRC01]. Similar leakage energy reduction techniques have also been proposed for issue queues [BAB$^+$02,FG01,PKG01] and branch prediction tables [HJS$^+$02].

Naturally, leakage energy reduction techniques for instruction/program caches are also very critical [YPF$^+$01]. A technique has recently been proposed, which monitors the performance of the instruction cache over time and dynamically scales (via V_{DD}-gating) its size, so as to closely match the size of the working set of the application [YPF$^+$01].

11.5.4 Software-Controlled Memories, Scratch-Pad Memories

Most of novel designs and/or techniques discussed so far require an application-driven tuning of several architecturally visible parameters. However, similar to more "traditional" cache hierarchies, the memory subsystem interface implemented on these novel designs still exposes a flat view of the memory hierarchy to the compiler/software. That is, the underlying details of the memory subsystem architecture are essentially transparent to both.

Dynamic power dissipation incurred by accesses to basic memory modules occurs due to switching activity in bit lines, word lines, and input and output lines. Traditional caches have additional switching overheads, due to the circuitry (comparators, multiplexers, tags, etc.) needed to provide the "flat" memory interface alluded to above. Since the hardware assists necessary to support such a transparent view of the memory hierarchy are quite power hungry, additional energy saving opportunities can be created by relying more on the compiler (and less on dedicated hardware) to manage the memory subsystem. The use of software-controlled (rather than hardware-controlled) memory components is thus becoming increasingly prevalent in power-aware embedded system design.

Scratch-Pads are an example of such novel, software-controlled memories [PDA97,CJDR99, BMP00,BS$^+$01,KRI$^+$01,UWK$^+$01]. Scratch-Pads are essentially on-chip partitions of main memory directly managed by the compiler. Namely, decisions concerning data/instruction placement in on-chip Scratch-Pads are made statically by the compiler, rather than dynamically, using dedicated hardware circuitry. Therefore, these memories are much less complex and thus less power hungry than traditional caches. As one would expect, the ability to aggressively improve energy-delay efficiency through the use of Scratch-Pads is predicated on the quality of the decisions made by the compiler on the subset of data/instructions that are to be assigned to that limited memory space [PDA97,PCD$^+$01]. Several compiler-driven techniques have been proposed to identify the data/instructions that can be assigned to the Scratch-Pad more profitably, with frequency of use being one of the key selection criteria [SFL98,IY00,SWLM02].

The Cool Cache architecture [UAK$^+$03], also proposed for media applications, is a good example of a novel, power-aware memory subsystem that relies on the use of software-controlled memories. It uses a small Scratch-Pad and a "software-controlled cache," each of which is implemented on a different on-chip SRAM. The program's scalars are mapped to the small (2 kB) Scratch-Pad [UWK$^+$01].*

---

* This size was found to be sufficient for most media applications.

Nonscalar data is mapped to the software-controlled cache, and the compiler is responsible for translating virtual addresses to SRAM lines, using a small register lookup area. Even though cache misses are handled in software, thereby incurring substantial latency/energy penalties, the overall architecture has been shown to yield substantial energy-delay product improvements for media applications, when compared to traditional cache hierarchies [UAK+03].

The effectiveness of techniques such as the above is so pronounced that several embedded processors currently offer a variety of software-controlled memory blocks, including configurable Scratch-Pads (TI's 320C6x [TI]), lockable caches (Intel's XScale [INT] and Trimedia [TRI]), and stream buffers (Intel's StrongARM [INT]).

11.5.5 Improving Access Patterns to Off-Chip Memory

During the last few decades, there has been a substantial effort in the compiler domain aimed at minimizing the number of off-chip memory accesses incurred by optimized code, as well as enabling the implementation of aggressive prefetching strategies. This includes devising compiler techniques to restructure, reorganize, and layout data in off-chip memory, as well as techniques to properly reorder a program's memory access patters [WL91,CMT94,CM95,Wol95,KRCB01].

Prefetching techniques have received considerable attention lately, particularly in the domain of embedded streaming media applications. Instruction and data prefetching techniques can be hardware or software driven [Jou90,CB95b,FPJ92,PY96,CKP91,KL91,MLG92,ZLF00]. Hardware-based data prefetching techniques try to dynamically predict when a given piece of data will be needed, so as to load it into cache (or into some dedicated on-chip buffer), before it is actually referenced by the application (i.e., explicitly required by a demand access) [CB95b,FPJ92,PY96]. By contrast, software-based data prefetching techniques work by inserting prefetch instructions for selected data references at carefully chosen points in the program—such explicit prefetch instructions are executed by the processor, to move data into cache [CKP91,KL91,MLG92,ZLF00].

It has been extensively demonstrated that, when properly used, prefetching techniques can substantially improve average memory access latencies [Jou90,CB95b,FPJ92,PY96,CKP91,KL91,MLG92, ZLF00]. Moreover, techniques that prefetch substantial chunks of data (rather than, say, a single cache line), possibly to a dedicated buffer, can also simultaneously decrease dynamic power dissipation [CK03]. Namely, when data is brought from off-chip memory in large bursts, then energy-efficient burst/page access modes can be more effectively exploited. Moreover, by prefetching large quantities of instructions/data, the average length of DRAM idle times is expected to increase, thus creating more profitable opportunities for the DRAM to be switched to a lower power mode. [FEL01,DSK+02,RJ03]. Naturally, it is important to ensure that the overhead associated with the prefetching mechanism itself, as well as potential increases in static energy consumption due to additional storage requirements, does not outweigh the benefits achieved from enabling more energy-efficient off-chip accesses [RJ03].

11.5.6 Special Purpose Memory Subsystems for Media Streaming

As alluded to before, streaming media applications have been a preferred application domain for validating the effectiveness of many novel, power-aware memory designs. Although the compiler is consistently given a more preeminent role in the management of these novel memory subsystems, they require no fundamental changes to the adopted programming paradigm. Additional opportunities for energy savings can be unlocked by adopting a programming paradigm that directly exposes those elements of an application that should be considered by an optimizing compiler, during performance versus power trade-off exploration. The two special purpose memory subsystems discussed below do precisely that, in the context of streaming media applications.

Xtream-Fit is a special purpose data memory subsystem targeted to generic uniprocessor embedded system platforms executing media applications [RJ03]. Xtream-Fit's on-chip memory consists of a Scratch-Pad, to hold constants and scalars, and a novel software-controlled Streaming Memory, partitioned into regions, each of which holds one of the input or output streams used/produced by the target application. The use of software-controlled memories by Xtream-Fit ensures that dynamic energy consumption is low, while the region-based organization of the streaming memory enables the implementation of very simple and yet effective shutdown policies to turn-off different memory regions, as the data they hold become "dead." Xtream-Fit's programming model is actually quite simple, requiring only a minor "reprogramming" effort. It simply requires organizing/ partitioning the application code into a small set of processing and data transfer tasks. Data transfer tasks prefetch streaming media data (the amount required by the next set of processing tasks) into the streaming memory. The amount of prefetched data is explicitly exposed via a single customization parameter. By varying this single customization parameter, the compiler can thus aggressively minimize energy-delay product, by considering, both, dynamic and leakage power, dissipated in on-chip and in off-chip memories [RJ03].

While Xtream-Fit provides sufficient memory bandwidth for generic uniprocessor embedded media architectures, it cannot support the very high bandwidth requirements of high-performance media accelerators. For example, Imagine, the multicluster media accelerator alluded to previously, uses its own specialized memory hierarchy, consisting of a streaming memory, a 128 kB stream register file, and stream buffers and register files local to each of its eight clusters. Imagine's memory subsystem delivers a very high bandwidth (2.1 GB/s) with very high energy efficiency, yet it requires the use of a specialized programming paradigm. Namely, data transfers to/from the host are controlled by a stream controller and between the stream register file and the functional units by a microcontroller, both of which have to be programmed separately, using Imagine's own "stream-oriented" programming style [Mat01].

Systems that demand still higher performance and/or energy efficiency may require memory architectures fully customized to the target application. Comprehensive methodologies for designing high-performance memory architectures for custom hardware accelerators are discussed in detail in [CWD$^+$98,GDN03].

11.5.7 Code Compression

Code size affects both program storage requirements and off-chip memory bandwidth requirements and can thus have a first-order impact on the overall power consumed by an embedded system. Instruction compression schemes decrease both such requirements by storing in main memory (i.e., off-chip) frequently fetched/executed instruction sequences in an encoded/compressed form [WC92,LBCM97,LW99]. Naturally, when one such scheme is adopted, it is important to factor in the overhead incurred by the on-chip decoding circuitry, so that it does not outweigh the gains achieved on storage and interconnect elements. Furthermore, different approaches have considered storing such select instruction sequences on-chip in either compressed or decompressed forms. On-chip storage of instructions in a compressed form saves on-chip storage, yet instructions must be decoded every time they are executed, adding additional latency/power overheads.

Instruction subsetting is an alternative instruction compression scheme, where instructions not commonly used are discarded from the instruction set, thus enabling the "reduced" instruction set to be encoded using less bits [DPT98]. The Thumb instruction set is a classic example of a compressed instruction set, featuring the most commonly used 32 bit ARM instructions, compressed to 16 bit wide format. The Thumb instructions set is decompressed transparently to full 32 bit ARM instructions in real time, with no performance loss.

11.5.8 Interconnect Optimizations

Power dissipation in on- and off-chip interconnect structures is also a significant contributor to an embedded system's power budget [SK00]. A shared bus is a commonly used interconnect structure, as it offers a good trade-off between generality/simplicity and performance. Power consumption on the bus can be reduced by decreasing its supply voltage, capacitance, and/or switching activity. Bus splitting, for example, reduces bus capacitance by splitting long bus lines into smaller sections, with one section relaying the data to the next [HP00]. Power consumption in this approach is reduced at the expense of a small penalty in latency, incurred at each relay point. Bus switching activity, and thus dynamic power dissipation, can also be substantially reduced by using an appropriate bus encoding scheme [SB95,MOI96,BMM$^+$97,BMM$^+$98,PD99,CKC00]. Bus-invert coding [SB95], for example, is a simple, yet widely used coding scheme. The first step of bus-invert coding is to compute the Hamming distance between the current bus value and the previous bus value. If this value is greater than half the number of total bits, then the data value is transmitted in an inverted form, with an additional invert bit to interpret the data at the other end. Several other encoding schemes have been proposed, achieving lower switching activity at the expense of higher encoding and decoding complexity [MOI96,BMM$^+$97,BMM$^+$98,PD99,CKC00].

With the increasing adoption of system-on-chip (SOC) design methodologies for embedded systems, devising energy-delay efficient interconnect architectures for such large scale systems is becoming increasingly critical and is still undergoing intensive research [PR02].

11.6 Summary

Design methodologies for today's embedded systems must necessarily treat power consumption as a primary figure of merit. At the system and architecture levels of design abstraction, power-aware embedded system design requires the availability of high-fidelity power estimation and simulation frameworks. Such frameworks are essential to enabling designers to explore and evaluate, in reasonable time, the complex energy-delay trade-offs realized by different candidate architectures, subsystem realizations, and power management techniques, and thus quickly identify promising solutions for the target application of interest. The detailed, system- and architecture-level design phases that follow should adequately combine coarse, system-level dynamic power management strategies, with fine-grained, self-monitoring techniques, exploiting voltage and frequency scaling, as well as advanced dynamic resource scaling and power-driven reconfiguration techniques.

References

[ACP]   http://www.acpi.info/

[BA97]   D. C. Burger and T. M. Austin. The SimpleScalar tool set, Version 2.0. *Computer Architecture News*, 25(3), 13–25, 1997.

[BAB$^+$02]   A. Buyuktosunoglu, D. Albonesi, P. Bose, P. Cook, and S. Schuster. Tradeoffs in power-efficient issue queue design. In *Proceedings of the International Symposium on Low Power Electronics and Design*, Monterey, CA, 2002.

[BB96]   T. D. Burd and R. W. Brodersen. Processor design for portable systems. *Journal of VLSI Signal Processing*, 13(2/3), 1996.

[BBM00]   A. Bogliolo, L. Benini, and G. D. Micheli. Regression-based RTL power modeling. *ACM Transactions on Design Automation of Electronic Systems*, 5(3), 2000.

[BFSS00]   C. Brandolese, W. Fornaciari, F. Salice, and D. Sciuto. An instruction-level functionality-based energy estimation model for 32-bits microprocessors. In *Proceedings of the Design Automation Conference*, Los Angeles, CA, 2000.

[BGJ+97]   F. Balarin, P. Giusto, A. Jurecska, C. Passerone, E. Sentovich, B. Tabbara, M. Chiodo, H. Hsieh, L. Lavagno, A. L. Sangiovanni-Vincentelli, and K. Suzuki. *Hardware–Software Co-Design of Embedded Systems: The POLIS Approach*. Kluwer Academic Publishers, Norwell, MA, 1997.

[BGK95]   D. C. Burger, J. R. Goodman, and A. Kagi. The declining effectiveness of dynamic caching for general-purpose microprocessors. Technical report, University of Wisconsin-Madison Computer Sciences Technical Report 1261, 1995.

[BHK+97]   R. S. Bajwa, M. Hiraki, H. Kojima, D. J. Gorny, K. Nitta, A. Shridhar, K. Seki, and K. Sasaki. Instruction buffering to reduce power in processors for signal processing. *IEEE Transactions on Very Large Scale Integration Systems*, 5(4), 1997.

[BHLM94]   J. T. Buck, S. Ha, E. A. Lee, and D. G. Messerschmitt. Ptolemy: A framework for simulating and prototyping heterogeneous systems. *International Journal of Computer Simulation, special issue on Simulation Software Development*, 4, 1994.

[BM00]   D. Brooks and M. Martonosi. Value-based clock gating and operation packing: Dynamic strategies for improving processor power and performance. *ACM Transactions on Computer Systems*, 18(2), 2000.

[BMM+97]   L. Benini, G. De Micheli, E. Macii, M. Poncino, and S. Quez. System-level power optimization of special purpose applications—The beach solution. In *Proceedings of the International Symposium on Low Power Electronics and Design*, Monterey, CA, 1997.

[BMM+98]   L. Benini, G. De Micheli, E. Macii, D. Sciuto, and C. Silvano. Address bus encoding techniques for system-level power optimization. In *Proceedings of the Design, Automation and Test in Europe*, Washington, DC, 1998.

[BMP00]   L. Benini, A. Macii, and M. Poncino. A recursive algorithm for low-power memory partitioning. In *Proceedings of the International Symposium on Low Power Electronics and Design*, Rapallo, Italy, 2000.

[Bor99]   S. Borkar. Design challenges of technology scaling. *IEEE Micro*, 19(4), 1999.

[BS00]   J. A. Butts and G. S. Sohi. A static power model for architects. In *Proceedings of the International Symposium on Microarchitecture*, Monterey, CA, 2000.

[BS+01]   R. Banakar, S. Steinke, B.-S. Lee, M. Balakrishnan, and P. Marwedel. Scratchpad memory: A design alternative for cache on-chip memory in embedded systems. In *Proceedings of the International Workshop on Hardware/Software Codesign*, Salzburg, Austria, 2001.

[BTM00]   D. Brooks, V. Tiwari, and M. Martonosi. Wattch: A framework for architectural level power analysis and optimizations. In *Proceedings of the International Symposium on Computer Architecture*, Vancouver, British Columbia, 2000.

[BYA93]   G. R. Beck, D. W. L. Yen, and T. L. Anderson. The Cydra 5: Mini-supercomputer: architecture and implementation. *The Journal of Supercomputing*, 7(1), 1993.

[CB95a]   A. P. Chandrakasan and R. W. Brodersen. *Low Power Digital CMOS Design*. Kluwer Academic Publishers, Boston, MA, 1995.

[CB95b]   T. F. Chen and J. L. Baer. Effective hardware-based data prefetching for high performance processors. *IEEE Transactions on Computers*, 44(5), 1995.

[CJDR99]   D. Chiou, P. Jain, S. Devadas, and L. Rudolph. Application-specific memory management for embedded systems using software-controlled caches. In *Proceedings of the Design Automation Conference*, Los Angeles, CA, 1999.

[CK03]   Y. Choi and T. Kim. Memory layout technique for variables utilizing efficient DRAM access modes in embedded system design. In *Proceedings of the Design Automation Conference*, Anaheim, CA, 2003.

[CKC00]   N. Chang, K. Kim, and J. Cho. Bus encoding for low-power high-performance memory systems. In *Proceedings of the Design Automation Conference*, Los Angeles, CA, 2000.

[CKP91]   D. Callahan, K. Kennedy, and A. Porterfield. Software prefetching. In *Proceedings of the International Conference on Architectural Support for Programming Languages and Operating Systems*, Santa Clara, CA, 1991.

[CL99]   G. Cai and C. H. Lim. Architectural level power/performance optimization and dynamic power estimation. In *Cool Chips Tutorial, International Symposium on Microarchitecture*, Haifa, Israel, 1999.

[CM95]   S. Coleman and K. S. McKinley. Tile size selection using cache organization and data layout. In *Proceedings of the Conference on Programming Language Design and Implementation*, La Jolla, CA, 1995.

[CMT94]   S. Carr, K. S. McKinley, and C. Tseng. Compiler optimizations for improving data locality. In *Proceedings of the International Conference on Architectural Support for Programming Languages and Operating Systems*, San Jose, CA, 1994.

[CND+88]   R. P. Colwell, R. P. Nix, J. J. O. Donnell, D. B. Papworth, and P. K. Rodman. A VLIW architecture for a trace scheduling compiler. *IEEE Transactions on Computers*, 37(8), 1988.

[CP92]   P. M. Chau and S. R. Powell. Power dissipation of VLSI array processing systems. *Journal of VLSI Signal Processing*, 4, 1992.

[CRC00]   Z. Chen, K. Roy, and E. K. Chong. Estimation of power dissipation using a novel power macromodeling technique. *IEEE Transactions on Computer Aided Design of Integrated Circuits and Systems*, 19(11), 2000.

[CSB92]   A. P. Chandrakasan, S. Sheng, and R. W. Brodersen. Low-power CMOS digital design. *IEEE Journal of Solid-State Circuits*, 27(4), 1992.

[CWD+98]   F. Catthoor, S. Wuytack, E. DeGreef, F. Balasa, L. Nachtergaele, and A. Vandecappelle. *Custom Memory Management Methodology: Exploration of Memory Organization for Embedded Multimedia System Design*. Kluwer Academic Publishers, Norwell, MA, 1998.

[DKA+02]   S. Dropsho, V. Kursun, D. H. Albonesi, S. Dwarkadas, and E. G. Friedma. Managing static leakage energy in microprocessor functional units. In *Proceedings of the International Symposium on Microarchitecture*, Istanbul, Turkey, 2002.

[DKK+99]   C. Dulong, R. Krishnaiyer, D. Kulkarni, D. Lavery, W. Li, J. Ng, and D. Sehr. An overview of the Intel IA-64 compiler. *Intel Technology Journal*, Q4, 1999.

[DPT98]   W. E. Dougherty, D. J. Pursley, and D. E. Thomas. Instruction subsetting: Trading power for programmability. In *Proceedings of the International Workshop on Hardware/Software Codesign*, Los Alamitos, CA, 1998.

[DSK+02]   V. Delaluz, A. Sivasubramaniam, M. Kandemir, N. Vijaykrishnan, and M. J. Irwin. Scheduler-based DRAM energy management. In *Proceedings of the Design Automation Conference*, New Orleans, LA, 2002.

[DT93]   J. Dehnert and R. Towle. Compiling for the Cydra-5. *The Journal of Supercomputing*, 7(1), 1993.

[EH98]   J. Edler and M. D. Hill. Dinero IV trace-driven uniprocessor cache simulator, 1998. http://www.cs.wisc.edu/~markhill/DineroIV/.

[Ell85]   J. R. Ellis. *Bulldog: A Compiler for VLIW Architectures*. The MIT Press, Cambridge, MA, 1985.

[FEL01]   X. Fan, C. S. Ellis, and A. R. Lebeck. Memory controller policies for DRAM power management. In *Proceedings of the International Symposium on Low Power Electronics and Design*, Huntington Beach, CA, 2001.

[FG01]   D. Folegnani and A. Gonzalez. Energy-effective issue logic. In *Proceedings of the International Symposium on High-Performance Computer Architecture*, New York, 2001.

[FPJ92]   J. W. C. Fu, J. H. Patel, and B. L. Janssens. Stride directed prefetching in scalar processor. In *Proceedings of the International Symposium on Microarchitecture*, Portland, OR, 1992.

[FRM02]   K. Flautner, S. Reinhardt, and T. Mudge. Automatic performance setting for dynamic voltage scaling. *ACM Journal of Wireless Networks*, 8(5), 2002.

[FTS95]   A. H. Farrahi, G. E. Téllez, and M. Sarrafzadeh. Memory segmentation to exploit sleep mode operation. In *Proceedings of the Design Automation Conference*, New York, 1995.

[FWL99]   J. Fritts, W. Wolf, and B. Liu. Understanding multimedia application characteristics for designing programmable media processors. In *SPIE Photonics West, Media Processors*, San Jose, CA, 1999.

[GCW95]   K. Govil, E. Chan, and H. Wasserman. Comparing algorithm for dynamic speed-setting of a low-power CPU. In *Proceedings of the International Conference on Mobile Computing and Networking*, Berkeley, CA, 1995.

[GDN03]   P. Grun, N. Dutt, and A. Nicolau. *Memory Architecture Exploration for Programmable Embedded Systems*. Kluwer Academic Publishers, Norwell, MA, 2003.

[GGH97]   R. Gonzalez, B. Gordon, and M. Horowitz. Supply and threshold voltage scaling for low power CMOS. *IEEE Journal of Solid-State Circuits*, 32(8), 1997.

[GH96]   R. Gonzalez and M. Horowitz. Energy dissipation in general purpose microprocessors. *IEEE Journal of Solid-State Circuits*, 31(9), 1996.

[GK99]   K. Ghose and M. B. Kamble. Reducing power in superscalar processor caches using sub-banking, multiple line buffers, and bit-line segmentation. In *Proceedings of the International Symposium on Low Power Electronics and Design*, San Diego, CA, pp. 70–75, 1999.

[HHG+95]   W. W. Hwu, R. E. Hank, D. M. Gallagher, S. A. Mahlke, D. M. Lavery, G. E. Haab, J. C. Gyllenhaal, and D. I. August. Compiler technology for future microprocessors. *Proceedings of the IEEE*, 83(12), 1995.

[HJS+02]   Z. Hu, P. Juang, K. Skadron, D. Clark, and M. Martonosi. Applying decay strategies to branch predictors for leakage energy savings. In *Proceedings of the International Conference on Computer Design*, Washington, DC, 2002.

[HP00]   C.-T. Hsieh and M. Pedram. Architectural power optimization by bus splitting. In *Proceedings of the Conference on Design, Automation and Test in Europe*, New York, 2000.

[HSA01]   C. J. Hughes, J. Srinivasan, and S. V. Adve. Saving energy with architectural and frequency adaptations for multimedia applications. In *Proceedings of the International Symposium on Microarchitecture*, Washington, DC, 2001.

[HWO97]   P. Hicks, M. Walnock, and R. M. Owens. Analysis of power consumption in memory hierarchies. In *Proceedings of the International Symposium on Low Power Electronics and Design*, New York, 1997.

[IBM]   http://www.ibm.com/

[INT]   http://www.intel.com/

[ITR]   http://public.itrs.net/

[IY98]   T. Ishihara and H. Yasuura. Voltage scheduling problem for dynamically variable voltage processors. In *Proceedings of the International Symposium on Low Power Electronics and Design*, Monterey, CA, 1998.

[IY00]   T. Ishihara and H. Yasuura. A power reduction technique with object code merging for application specific embedded processors. In *Proceedings of the Design, Automation and Test in Europe*, Paris, France, 2000.

[JdV00]   M. F. Jacome and G. de Veciana. Design challenges for new application specific processors. *IEEE Design and Test of Computers*, special issue on System Design of Embedded Systems, Los Alamitos, CA, 2000.

[JdVL00]   M. F. Jacome, G. de Veciana, and V. Lapinskii. Exploring performance tradeoffs for clustered VLIW ASIPs. In *Proceedings of the International Conference on Computer-Aided Design*, San Jose, CA, pp. 504–510, 2000.

[JKSN99]   G. Jochens, L. Kruse, E. Schmidt, and W. Nebel. A new parameterizable power macro-model for datapath components. In *Proceedings of the Design Automation and Test in Europe*, New York, 1999.

[Jou90]   N. P. Jouppi. Improving direct-mapped cache performance by the addition of a small fully-associative cache and prefetch buffers. In *Proceedings of the International Symposium on Computer Architecture*, New York, 1990.

[JYWV03]   A. A. Jerraya, S. Yoo, N. Wehn, and D. Verkest, editors. *Embedded Software for SoC*. Kluwer Academic Publishers, Boston, MA, 2003.

[KDR⁺01]   B. Khailany, W. J. Dally, S. Rixner, U. J. Kapasi, P. Mattson, J. Namkoong, J. D. Owens, B. Towles, and A. Chang. Imagine: Media processing with streams. *IEEE Micro*, 21(2), 2001.

[KE98]   M. Khellah and M. I. Elmasry. Effective capacitance macro-modelling for architectural-level power estimation. In *Proceedings of the Eighth Great Lakes Symposium on VLSI*, Lafayette, LA, 1998.

[KG97]   M. Kamble and K. Ghose. Analytical energy dissipation models for low power caches. In *Proceedings of the International Symposium on Low Power Electronics and Design*, Monterey, CA, 1997.

[KGMS97]   J. Kin, M. Gupta, and W. H. Mangione-Smith. The filter cache: An energy efficient memory structure. In *Proceedings of the International Symposium on Microarchitecture*, Research Triangle Path, NC, pp. 184–193, 1997.

[KGMS00]   J. Kin, M. Gupta, and W. H. Mangione-Smith. Filtering memory references to increase energy efficiency. *IEEE Transactions on Computers*, 49(1), 2000.

[KHM01]   S. Kaxiras, Z. Hu, and M. Martonosi. Cache decay: Exploiting generational behavior to reduce cache leakage power. In *Proceedings of the International Symposium on Computer Architecture*, New York, 2001.

[KL91]   A. C. Klaiber and H. M. Levy. An architecture for software controlled data prefetching. In *Proceedings of the International Symposium on Computer Architecture*, New York, 1991.

[KRCB01]   M. Kandemir, J. Ramanujam, A. Choudhary, and P. Banerjee. A layout-conscious iteration space transformation technique. *IEEE Transactions on Computers*, 50(12), 2001.

[KRI⁺01]   M. Kandemir, J. Ramanujam, M. Irwin, N. Vijaykrishnan, I. Kadayif, and A. Parikh. Dynamic management of scratch-pad memory space. In *Proceedings of the Design Automation Conference*, Los Vegas, NV, 2001.

[LBCM97]   C. Lefurgy, P. Bird, I. C. Cheng, and T. Mudge. Improving code density using compression techniques. In *Proceedings of the International Symposium on Microarchitecture*, 1997.

[LCBK01]   J. Liu, P. Chou, N. Bagherzadeh, and F. Kurdahi. A constraint-based application model and scheduling techniques for power-aware systems. In *Proceedings of the International Conference on Hardware/Software Codesign*, Copenhagen, Denmark, 2001.

[Lie97]   C. Liem. *Retargetable Compilers for Embedded Core Processors*. Kluwer Academic Publishers, Norwell, MA, 1997.

[LJdV02]   V. Lapinskii, M. F. Jacome, and G. de Veciana. Application-specific clustered VLIW datapaths: Early exploration on a parameterized design space. *IEEE Transactions on Computer Aided Design of Integrated Circuits and Systems*, 21(8), 2002.

[LMA99]   L. Lee, B. Moyer, and J. Arends. Instruction fetch energy reduction using loop caches for embedded applications with small tight loops. In *Proceedings of the International Symposium on Low Power Electronics and Design*, San Diego, CA, 1999.

[LT00]   H. H. Lee and G. Tyson. Region-based caching: An energy-delay efficient memory architecture for embedded processors. In *Proceedings of the International Conference on Compilers, Architectures and Synthesis for Embedded Systems*, San Jose, CA, 120–127, 2000.

[LW99]   H. Lekatsas and W. Wolf. SAMC: A code compression algorithm for embedded processors. *IEEE Transactions on Computer Aided Design of Integrated Circuits and Systems*, 18(12), 1999.

[Mat01]   P. Mattson. A programming system for the imagine media processor. PhD thesis, Stanford University, Stanford, CA, 2001.

[MG95]   P. Marwedel and G. Goosens, editors. *Code Generation for Embedded Processors*. Kluwer Academic Publishers, Norwell, MA, 1995.

[MG01]   R. Melhem and R. Graybill, editors. Challenges for architectural level power modeling, *Power Aware Computing*. Kluwer Academic Publishers, Norwell, MA, 2001.

[MLG92]   T. C. Mowry, M. S. Lam, and A. Gupta. Design and evaluation of a compiler algorithm for prefetching. In *Proceedings of the International Conference on Architectural Support for Programming Languages and Operating Systems*, Boston, MA, 1992.

[MOI96]   H. Mehta, R. M. Owens, and M. J. Irwin. Some issues in gray code addressing. In *Proceedings of the Sixth Great Lakes Symposium on VLSI*, Ames, IA, 1996.

[MRW02]   G. Micheli, R. Ernst, and W. Wolf, editors. *Readings in Hardware/Software Co-Design.* Morgan Kaufman Publishers, San Francisco, CA, 2002.

[MS96]   T. Martin and D. Siewiorek. A power metric for mobile systems. In *International Symposium on Lower Power Electronics and Design*, Monterey, CA, 1996.

[MSS$^+$03]   T. L. Martin, D. P. Siewiorek, A. Smailagic, M. Bosworth, M. Ettus, and J. Warren. A case study of a system-level approach to power-aware computing. *ACM Transactions on Embedded Computing Systems, special issue on Power-Aware Embedded Computing*, 2(3), 2003.

[MWA$^+$96]   J. Montanaro, R. T. Witek, K. Anne, A. J. Black, E. M. Cooper, D. W. Dobberpuhl, P. M. Donahue, J. Eno, A. Farell, G. W. Hoeppner, D. Kruckemyer, T. H. Lee, P. Lin, L. Madden, D. Murray, M. Pearce, S. Santhanam, K. J. Snyder, R. Stephany, and S. C. Thierauf. A 160 MHz 32b 0.5 W CMOS RISC microprocessor. *Proceedings of the International Solid-State Circuits Conference*, Digest of Technical Papers, 1996.

[PAC$^+$97]   D. Patterson, T. Anderson, N. Cardwell, R. Fromm, K. Keeton, C. Kozyrakis, R. Thomas, and K. Yelick. A case for intelligent RAM. *IEEE Micro*, 17(2), 1997.

[PBB98]   T. Pering, T. Burd, and R. Brodersen. The simulation and evaluation of dynamic voltage scaling algorithms. In *Proceedings of the International Symposium on Low Power Electronics and Design*, 1998.

[PBB00]   T. Pering, T. Burd, and R. Brodersen. Voltage scheduling in the lpARM microprocessor system. In *Proceedings of the International Symposium on Low Power Electronics and Design*, Rapallo, Italy, 2000.

[PCD$^+$01]   P. R. Panda, F. Catthoor, N. D. Dutt, K. Danckaert, E. Brockmeyer, C. Kulkarni, A. Vandercappelle, and P. G. Kjeldsberg. Data and memory optimization techniques for embedded systems. *ACM Transactions on Design Automation of Electronic Systems*, 6(2), 2001.

[PD99]   P. R. Panda and N. D. Dutt. Low-power memory mapping through reducing address bus activity. *IEEE Transactions on Very Large Scale Integration Systems*, 7(3), 1999.

[PDA97]   P. R. Panda, N. D. Dutt, and A. Nicolau. Efficient utilization of scratch-pad memory in embedded processor applications. In *Proceedings of the European Design and Test Conference*, Washington, DC, 1997.

[PJ03]   S. Pillai and M. F. Jacome. Compiler-directed ILP extraction for clustered VLIW/EPIC machines: Predication, speculation and modulo scheduling. In *Proceedings of the Design Automation and Test in Europe*, Washington, DC, 2003.

[PKG01]   D. Ponomarev, G. Kucuk, and K. Ghose. Reducing power requirements of instruction scheduling through dynamic allocation of multiple datapath resources. In *Proceedings of the International Symposium on Microarchitecture*, Austin, TX, 2001.

[PR02]   M. Pedram and J. M. Rabaey. *Power Aware Design Methodologies.* Kluwer Academic Publishers, Boston, MA, 2002.

[PW02]   M. Pedram and Q. Wu. Design considerations for battery-powered electronics. *IEEE Transactions on Very Large Scale Integration Systems*, 10(5), 2002.

[PY96]   S. S. Pinter and A. Yoaz. A hardware-based data prefetching technique for superscalar processors. In *Proceedings of the International Symposium on Computer Architecture*, Paris, France, 1996.

[QKUP00]   G. Qu, N. Kawabe, K. Usami, and M. Potkonjak. Function-level power estimation methodology for microprocessors. In *Proceedings of the Design Automation Conference*, Los Angeles, CA, 2000.

[QP99]   Q. Qiu and M. Pedram. Dynamic power management based on continuous-time Markov decision processes. In *Proceedings of the Design Automation Conference*, New Orleans, LA, 1999.

[RJ98]   J. Russell and M. Jacome. Software power estimation and optimization for high-performance 32-bit embedded processors. In *Proceedings of the International Conference on Computer Design*, Washington, DC, 1998.

[RJ01]   G. Reinman and N. M. Jouppi. CACTI 2.0: An integrated cache timing and power model. Technical report, Compaq Computer Corporation, Western Research Laboratories, 2001.

[RJ03]   A. Ramachandran and M. Jacome. Xtream-fit: An energy-delay efficient data memory subsystem for embedded media processing. In *Proceedings of the Design Automation Conference*, Anaheim, CA, 2003.

[RMV$^+$87]   J. Rabaey, H. De Man, J. Vanhoof, G. Goossens, and F. Catthoor. CATHEDRAL-II: A synthesis system for multiprocessor DSP systems, *Silicon Compilation*. Addison-Wesley, Reading, MA, 1987.

[RP03]   P. Rong and M. Pedram. Remaining battery capacity prediction for lithium-ion batteries. In *Proceedings of the Design Automation and Test in Europe*, Munich, Germany, pp. 1148–1149, 2003.

[RV02]   J. M. Rabaey and A. S. Vincentelli. System-on-a-Chip—A platform perspective. In *Keynote Presentation, Korean Semiconductor Conference*, 2002.

[SB95]   M. R. Stan and W. P. Burleson. Bus-invert coding for low-power I/O. *IEEE Transactions on Very Large Scale Integration Systems*, 3(1), 1995.

[SBA$^+$01]   T. Simunic, L. Benini, A. Acquaviva, P. Glynn, and G. De Micheli. Dynamic voltage scaling for portable systems. In *Proceedings of the Design Automation Conference*, Paris, France, 2001.

[SBGM00]   T. Simunic, L. Benini, P. Glynn, and G. De Micheli. Dynamic power management of portable systems. In *Proceedings of the International Conference on Mobile Computing and Networking*, Las Vegas, NV, 2000.

[SCB96]   M. Srivastava, A. Chandrakasan, and R. Brodersen. Predictive system shutdown and other architectural techniques for energy efficient programmable computation. *IEEE Transactions on Very Large Scale Integration Systems*, 4(1), 1996.

[SD95]   C. L. Su and A. M. Despain. Cache design trade-offs for power and performance optimization: A case study. In *Proceedings of the International Symposium on Low Power Electronics and Design*, Dana Point, CA, 1995.

[SFL98]   J. Sjödin, B. Fröderberg, and T. Lindgren. Allocation of global data objects in on-chip RAM. In *Proceedings of the Workshop on Compiler and Architectural Support for Embedded Computer Systems*, Washington, DC, 1998.

[SK00]   D. Sylvester and K. Keutzer. A global wiring paradigm for deep submicron design. *IEEE Transactions on Computer Aided Design of Integrated Circuits and Systems*, 19(2), 2000.

[SR00]   M. S. Schlansker and B. R. Rau. EPIC: An architecture for instruction-level parallel processors. Technical report, Hewlett Packard Laboratories, Technical Report HPL-99-111, 2000.

[SWLM02]   S. Steinke, L. Wehmeyer, B.-S. Lee, and P. Marwedel. Assigning program and data objects to scratchpad for energy reduction. In *Proceedings of the Design Automation and Test in Europe*, Washington, DC, 2002.

[TGTS98]   S. A. Theoharis, C. E. Goutis, G. Theodoridis, and D. Soudris. Accurate data path models for RT-level power estimation. In *Proceedings of the International Workshop on Power and Timing Modeling, Optimization and Simulation*, Lyngby, Denmark, 1998.

[TI]   http://www.ti.com/

[TMW94]   V. Tiwari, S. Malik, and A. Wolfe. Power analysis of embedded software: A first step towards software power minimization. *IEEE Transactions on Very Large Scale Integration Systems*, 2(4), 1994.

[TRA]   http://www.transmeta.com/

[TRI]   http://www.trimedia.com/

[UAK+03]   O. S. Unsal, R. Ashok, I. Koren, C. M. Krishna, and C. A. Moritz. Cool Cache: A compiler-enabled energy efficient data caching framework for embedded/multimedia processors. *ACM Transactions on Embedded Computing Systems*, special issue on Power-Aware Embedded Computing, 2(3), 2003.

[UWK+01]   O. S. Unsal, Z. Wang, I. Koren, C. M. Krishna, and C. A. Moritz. On memory behavior of scalars in embedded multimedia systems. In *Proceedings of the Workshop on Memory Performance Issues*, Goteborg, Sweden, 2001.

[VM01]   A. S. Vincentelli and G. Martin. A vision for embedded systems: Platform-based design and software methodology. *IEEE Design and Test of Computers*, 18(6), 2001.

[WC92]   A. Wolfe and A. Chanin. Executing compressed programs on an embedded RISC architecture. In *Proceedings of the International Symposium on Microarchitecture*, Portland, OR, 1992.

[WHK91]   D. A. Wood, M. D. Hill, and R. E. Kessler. A model for estimating trace-sample miss ratios. In *Proceedings of the SIGMETRICS Conference on Measurement and Modeling of Computer Systems*, New York, 1991.

[WJ96]   S. J. E. Wilton and N. M. Jouppi. CACTI: An enhanced cache access and cycle time model. Technical report, Digital Equipment Corporation, Western Research Laboratory, 1996.

[WL91]   M. E. Wolf and M. Lam. A data locality optimizing algorithm. In *Proceedings of the Conference on Programming Language Design and Implementation*, New York, 1991.

[Wol95]   M. J. Wolfe. *High Performance Compilers for Parallel Computing*. Addison-Wesley Publishers, Boston, MA, 1995.

[WWDS94]   M. Weiser, B. Welch, A. J. Demers, and S. Shenker. Scheduling for reduced CPU energy. In *Proceedings of the Symposium on Operating Systems Design and Implementation*, Monterey, CA, 1994.

[YPF+01]   S. H. Yang, M. D. Powell, B. Falsafi, K. Roy, and T. N. Vijaykumar. An integrated circuit/architecture approach to reducing leakage in deep-submicron high-performance I-caches. In *Proceedings of the High-Performance Computer Architecture*, Washington, DC, 2001.

[YVKI00]   W. Ye, N. Vijaykrishnan, M. Kandemir, and M. J. Irwin. The design and use of simplepower: A cycle-accurate energy estimation tool. In *Proceedings of the Design Automation Conference*, Los Angeles, CA, 2000.

[ZLF00]   D. F. Zucker, R. B. Lee, and M. J. Flynn. Hardware and software cache prefetching techniques for MPEG benchmarks. *IEEE Transactions on Circuits and Systems for Video Technology*, 10(5), 2000.

[ZPS+03]   Y. Zhang, D. Parikh, K. Sankaranarayanan, K. Skadron, and M. Stan. HotLeakage: A temperature-aware model of subthreshold and gate leakage for architects. Technical report, Deptartment of Computer Science, University of Virginia, Charlottesville, VA, 2003.

[ZTRC01]   H. Zhou, M. C. Toburen, E. Rotenberg, and T. M. Conte. Adaptive mode control: A static-power-efficient cache design. In *Proceedings of the International Conference on Parallel Architectures and Compilation Techniques*, Barcelona, Spain, pp. 61–72, 2001.

II

Embedded Processors and System-on-Chip Design

12

Processors for Embedded Systems

Steve Leibson
Tensilica, Inc.

12.1 Introduction

Embedded-system design and microprocessor history form a closely linked double helix. Without microprocessors, there would be no embedded systems because the very term "embedded" refers to the microprocessor(s) embedded inside of some electronic system that is clearly not a computer. Microprocessors coupled with software or firmware allow system complexity with flexibility in ways not possible before microprocessors existed. No other digital building block can as easily manage a windowed user interface or compute a complicated mathematical algorithm as software or firmware running on a processor. The numerous limitations of the earliest microprocessors—including limited address spaces, narrow data buses, primitive stacks (or no stacks at all), and slow processing speeds—made them poor foundations for designing computer systems but designers have used microprocessors since their introduction to build increasingly complex embedded systems. Because of their complexity, such systems would be impractical or impossible to make without the nimble firmware adaptability inherent in microprocessor-based design.

12.2 Embedded Microprocessors: A Short History

In 1969, Japanese calculator manufacturer Busicom sought a semiconductor company to fabricate chips for a family of calculator-based products. Large-scale integration (LSI) IC fabrication was just becoming possible and calculators were an early and profitable market for LSI chips. Busicom's Osaka and Tokyo factories separately approached Mostek and fledgling Intel Corp (www.intel.com) to build chip sets for different calculator architectures because only these vendors had high-density silicon-gate-MOS processes. Busicom's Tokyo factory struck a deal with Intel for a more complex calculator chip set. This chip set, designed by Busicom's Masatoshi Shima, was based on ROM-driven decimal state machines. Intel's president Bob Noyce assigned the job of Busicom liaison to Ted Hoff, the company's application research manager.

Busicom's engineers planned several chips, each to be housed in expensive (at the time) 40-pin, ceramic dual-inline packages (DIPs). Once he fully understood Busicom's plan, Hoff knew Intel would be unable to deliver the proposed chip set at the agreed price. The logic would require large (hence, expensive) chips, and the 40-pin ceramic packages used to house the chips would also be expensive. Hoff therefore suggested an alternative to Noyce: replace Shima's ROM-driven digital state machines with one CPU-like logic chip-running software stored in one or more ROMs. Noyce, a mathematician and device physicist, did not really understand how a computer on a chip-running software could replace purpose-built digital logic but he got the drift of Hoff's argument's and encouraged Hoff to pursue the concept. Hoff eventually convinced Shima and Busicom as well.

Hoff, Shima, and Stan Mazor, newly arrived from Fairchild, developed a 4-chip set—the 4-bit CPU, a 320-bit RAM, a 256-byte masked ROM, and a 10-bit shift register for I/O—linked by a multiplexed 4-bit bus. This approach reduced IC-die cost and package size, sharply cutting the chip set's cost. Intel packaged the RAM and ROM in 18-cent, 16-pin plastic DIPs and the CPU in a 16-pin ceramic DIP that cost less than a dollar (Figure 12.1). Intel snagged Federico Faggin from Fairchild in March 1970 to translate Hoff's conceptual design into silicon. With Shima's help, Faggin had working silicon for all four chips by January 1971—just nine months later.

FIGURE 12.1   Intel's 4-bit 4004 was the first commercially available microprocessor. (Photo courtesy of Stephen A. Emery Jr., www.ChipScapes.com.)

Falling calculator prices in a hotly contested market forced Busicom to renegotiate prices with Intel during 1971. Hoff advised Noyce to exchange a price cut for the right to sell the four chips into noncompeting applications. Busicom's calculator CPU debuted as Intel's 4004 microprocessor at the Fall Joint Computer Conference in November 1971.

Intel marketed the new type of chip as a logic-replacement device. Others may lay claim to developing earlier processors on chips but Intel's 4004 was the first commercially available microprocessor. System designers have since incorporated billions of microprocessors, microcontrollers, and DSPs from dozens of vendors into countless embedded applications.

In 1973, Gary Boone working at Texas Instruments (TI) was awarded U.S. Patent No. 3,757,306 for the single-chip microprocessor architecture—a microcontroller—which incorporated a processor, RAM, ROM, and I/O on one piece of silicon. TI started selling such a chip, the 4-bit TMS1000 microcontroller, in 1974 for a mere $2 each in volume. By 1978, TI was building a speech-oriented microcontroller that became the brains of the company's $50 "Speak & Spell," an educational toy that is one of the earliest examples of a low-cost, high-volume, microprocessor-based consumer product (Figure 12.2). The Speak & Spell was clearly not a general-purpose, programmable computer. It was a toy—an embedded system.

Early 4-bit microprocessors and microcontrollers could form the heart of simpler systems such as toys and gasoline-pump controllers, but they lacked the performance to build more sophisticated systems and the chip vendors knew it. However, Intel's 4004 microprocessor and TI's TMS1000 microcontroller broke the ice and nearly every silicon vendor jumped into the fray. Intel jumped to 8 bits, first introducing the 8008 microprocessor in April 1972 and then the much-improved 8080 microprocessor in April 1974. Motorola Semiconductors introduced its first microprocessor, the 8-bit MC6800, in early 1974. Federico Faggin left Intel at the end of 1974 and started Zilog, which introduced the 8080-compatible but much superior Z80 microprocessor in 1976. The mid-1970s became the era of the 8-bit microprocessor.

FIGURE 12.2 First introduced in 1978, TI' Speak & Spell was one of the earliest, low-cost, noncomputer (embedded) products to incorporate a microprocessor. (Photo by Bill Bertram.)

By the end of the 1970s, IC fabrication had sufficiently advanced to accommodate 16-bit microprocessor designs on one chip. Microprocessors including Intel's 8086/8088, Zilog's Z8000, and Motorola's MC68000 (which actually has a 32-bit architecture) all vied for the top socket in the 16-bit microprocessor war. Initially, Motorola's MC68000 family won. By the early 1980s, computer designers were building actual computers around microprocessors and Motorola's MC68000 family, with its 32-bit addressing and 32-bit data handling, became the architecture of choice for workstation designers. However, IBM selected Intel's 8088 for the IBM PC, which ultimately ensured that architecture's place at the top of the 16/32-bit heap because of the explosive market success of IBM's PC and the numerous clone machines that followed.

The microprocessor's debut in 1971 allowed engineers to direct software's considerable problem-solving abilities at a vast array of electronic products. At first, machine code and assembly language were the only microprocessor programming alternatives. Early high-level language (HLL) compilers for microprocessors appeared, but they generated slow, memory-hungry object code compared with well-written, handcrafted assembly code. Early microprocessor-based systems were already slow and memory-starved, so early compilers were nearly useless, but they evolved.

12.2.1 Software-Driven Microprocessor Evolution

Throughout the 1970s, most embedded systems ran ROM-based firmware written in an assembly language. Early HLL compilers could not generate efficient code that performed as well or used as little memory as human-coded assembly-level programming. The embedded-development world dabbled with many HLLs—including C, Basic, Forth, Pascal, and a microprocessor version of PL/I called PL/M—but superior performance and memory efficiency kept assembly language coding in the lead.

Coincidentally, the same year that Intel introduced the 4004 microprocessor, Dennis Ritchie started extending Bell Labs' B programming language for the PDP-7 minicomputer by adding a character type. Ritchie called this new language NB (new B). NB was an embryonic version of C. The language became recognizable as C by 1973. Ritchie published the "C programming language" with Brian Kernigan in 1978 and this book served as a de facto C standard during the 1980s. Throughout the 1980s, embedded systems and microprocessors became increasingly complex; the cost of 32-bit microprocessors decreased; and compilers improved. Rising software-development costs and complexity eventually curtailed assembly language programming on microprocessors and C's popularity as an embedded programming language climbed and processor vendors started to shape microprocessor architectures to be more efficient C-execution engines.

12.2.2 Quest for More Processor Performance

The microprocessor's low cost and high utility snowballed. Microprocessor vendors were under great pressure to constantly increase their products' performance as system designers wanted to execute more tasks on processors. There are some obvious methods to increase a processor's performance and processor vendors have used three of them.

The first and easiest performance-enhancing technique used was to increase the processor's clock rate. Intel introduced the 8086 microprocessor in 1978. It ran at 10 MHz, five times the clock rate of the 8080 microprocessor introduced in 1974. Ten years later, Intel introduced the 80386 microprocessor at 25 MHz, faster by another factor of 2.5. In yet another 10 years, Intel introduced the Pentium II processor at 266 MHz, better than a 10× clock-rate increase yet again. Figure 12.3 shows the dramatic rise in microprocessor clock rate over time.

Note that Intel was not the only microprocessor vendor racing to higher clock rates. At different times, Motorola and AMD have also produced microprocessors that vied for the "fastest clock rate" title. Digital Equipment Corporation (DEC) worked hard to tune its series of Alpha microprocessors to achieve world-beating clock rates. (Compaq bought DEC in 1998 and curtailed the Alpha development program. HP then acquired Compaq in late 2001.)

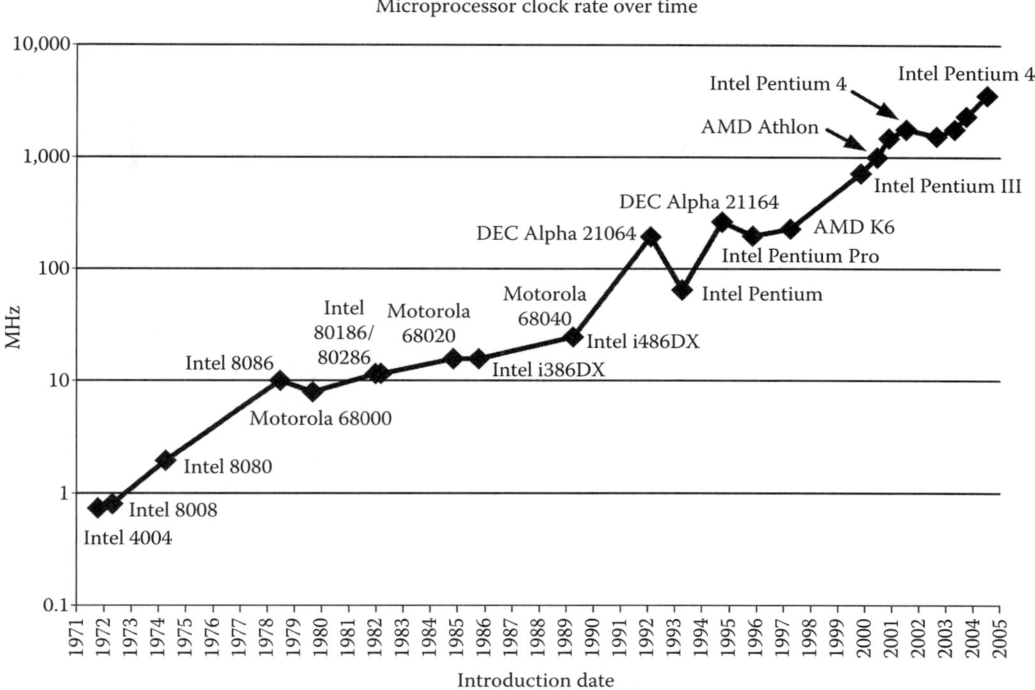

FIGURE 12.3 Microprocessor clock rates have risen dramatically over time due to the demand of system designers for ever more processor performance.

At the same time, microprocessor data-word widths and buses widened so that processors could move more data during each clock period. Widening the processor's bus is the second way to increase processing speed and I/O bandwidth. Intel's 16-bit 8086 microprocessor had a 16-bit data bus and the 32-bit 80386 microprocessor had a 32-bit data bus.

The third way to increase processor performance and bus bandwidth is to add more buses to the processor's architecture. Intel did exactly this with the addition of a separate cache-memory bus to its Pentium II processor. The processor could simultaneously run separate bus cycles to its high-speed cache memory and to other system components attached to the processor's main bus.

As processor buses widen and processor architectures acquire extra buses, the microprocessor package's pin count necessarily increases. Figure 12.4 shows how microprocessor pin count has increased over the years. Like the rising curve plotted in Figure 12.3, the increasing pin count shown in Figure 12.4 is a direct result of system designers' demand for more processor performance.

Faster clock rates coupled with more and wider buses do indeed increase processor performance, but at a price. Increasing clock rate extracts a penalty in terms of power dissipation. In fact, power dissipation rises roughly with the square of the clock rate increase, so microprocessor power dissipation and energy density have been rising exponentially for three decades. Unfortunately, the result is that the fastest packaged processors today are bumping into the heat-dissipation limits of their packaging and cooling systems. Since their introduction in 1971, cooling design for packaged microprocessors progressed from no special cooling to

- Careful system design to exploit convection cooling
- Active air cooling without heat sinks
- Active air cooling with aluminum and then copper heat sinks

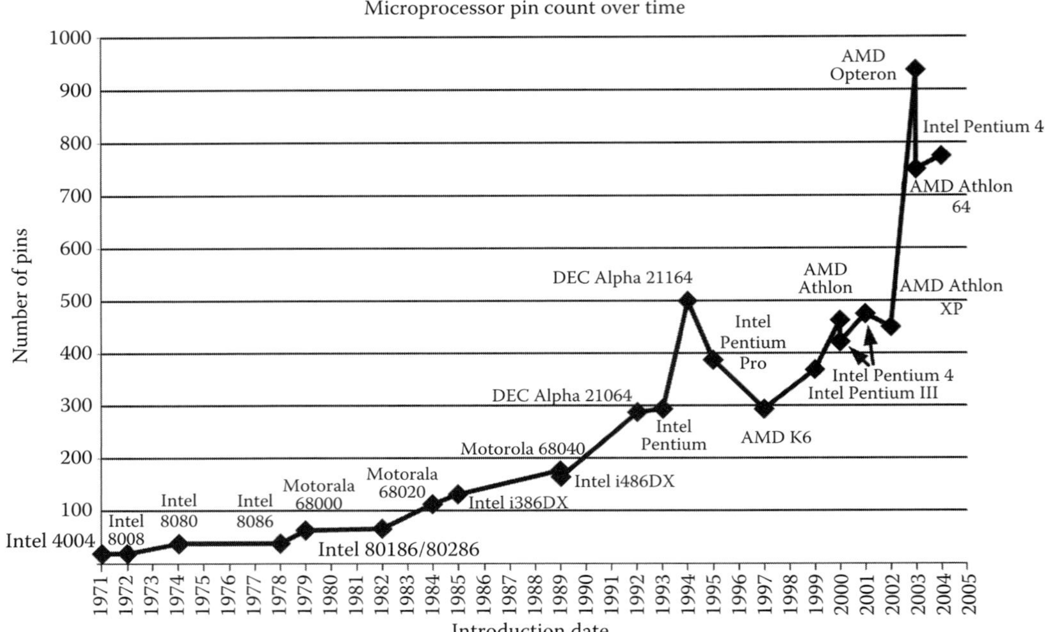

FIGURE 12.4   Microprocessor pin counts have also risen dramatically over time due to the demand of system designers for ever more processor performance.

- Larger heat sinks
- Even larger heat sinks
- Dedicated fans directly attached to the processor's heat sink
- Heat pipes
- Heat sinks incorporating active liquid cooling subsystems

Each step up in heat capacity has increased the cost of cooling, increased the size of required power supplies and product enclosures, increased cooling noise (for fans), and decreased system reliability due to hotter chips and active cooling systems that have their own reliability issues.

Systems on chips (SOCs) cannot employ the same sort of cooling that is now used for PC processors. Systems that use SOCs generally lack the PC's cooling budget. In addition, a processor core on an SOC is only a small part of the system. It cannot dominate the cost and support structure of the finished product the way a processor in a PC does. Simple economics dictate a different design approach.

In addition, SOCs are developed using an ASIC design flow, which means that gates are not individually sized to optimize speed in critical paths the same way and to the same extent that transistors in critical paths are tweaked by the designers of packaged microprocessors. Consequently, clock rates for embedded processor cores used in SOC designs have climbed modestly over the past two decades to a few hundred MHz, but SOC processors do not run at multi-GHz clock rates like their PC brethren, and probably never will.

One final aspect of microprocessors and SOCs bears discussion. Because on-chip microprocessors are not packaged, they need not suffer from the problems of narrow buses and pin limitations. It is possible to directly connect various points in the SOC directly to the inputs and outputs of registers and other machine states that are deeply buried within certain on-chip microprocessors. Doing so bypasses the microprocessor's bus bottleneck and can vastly improve performance.

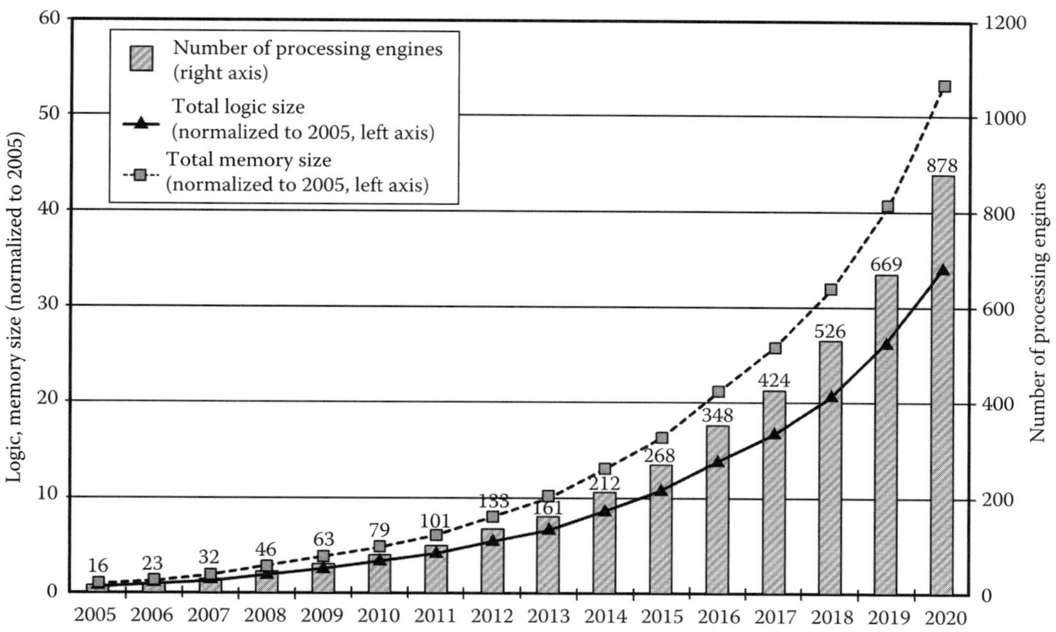

FIGURE 12.5   The 2006 ITRS predicts a rapid rise in the number of processors per SOC.

12.2.3   From System on a Board to System on a Chip

By 1990, microprocessor-centric design was the first choice of board-level system designers. (It had not existed as a choice 20 years earlier.) Logic design became a last resort, which designers used only when microprocessors were too slow. Around 1995, microprocessors became cores—just part of an IC—transforming ASICs into SOCs. Early SOCs incorporated only one microprocessor; then, two; then, many. The 2006 International Technology Road-map for Semiconductors (ITRS) predicts that this trend will continue far into the future (Figure 12.5). Like board-level system design before it, SOC design became processor-centric.

Many SOCs today already incorporate dozens or hundreds of interconnected processors. Microprocessors and software are the very fabric of contemporary electronic design. That fabric now covers the planet, as many embedded-microprocessor applications demonstrate (Table 12.1).

12.2.4   Rise of the RISC Machines

RISC stands for "reduced instruction set computing". More facetiously, it stands for "relegate the important stuff to the compiler". The RISC concept started with John Cocke's group at IBM, which developed the IBM 801 to run a telephone-switching network back in 1974. The design team used a simple equation to determine the required processor speed. The network needed to handle 300 calls and it took approximately 20,000 instructions to handle a call. Combined with real-time response requirements, Cocke's team determined that they needed a processor that could execute 12 millions of instructions per second (MIPS) at a time when the IBM 370 Model 168 mainframe cranked out about 2 MIPS best case and the early microprocessors of the day delivered approximately 1 MIPS. Cocke's team developed the idea of a stripped-down instruction-set architecture (ISA) implemented in a pipelined machine coupled with fast instruction and data caches and an optimizing compiler.

TABLE 12.1   Few Everyday Uses for Embedded Microprocessors and Software

Office and retail	*Home*
Telephones and PBXes	Conventional and microwave ovens
Printers	Food processors, mixers, and blenders
Copiers and faxes	Refrigerators and dishwashers
Postal scales and shipping management	Climate and lighting control
Fire and intrusion alarms	TVs, cable and satellite boxes, VCRs, and DVRs
Lighting and HVAC controls	Home entertainment systems
Elevator and automatic door controls	DVD, CD, and MP3 player/recorders
Barcode and RFID readers	Clothes washers, dryers, and irons
Video security and monitoring	Corded and mobile telephones
Energy and utilities monitoring and billing	Toys and games
Recordkeeping and management	Digital cameras and camcorders
Inventory management	Barbecue grills
Civil ground transportation	*Space, aviation, naval, and military*
Drivetrain control	Engine control and management
Passenger cabin climate control	Guidance and attitude control
Entertainment systems	On-board systems monitoring
GPS and compass navigation	Radar and collision avoidance
Collision-avoidance systems	Global and celestial navigation
Mobile phone and satellite communications	Radio and satellite communications
Fare collection (public transport)	Passenger cabin climate control
Traction control and automatic braking	Fuel management
Active shock absorbers	Weapons management and control
Manufacturing	*Medical*
Process control	Diagnostic, imaging, and treatment systems
Robotic assembly and transport	Patient record keeping
Inventory management and tracking	Robotic surgery
Energy and load management	Pharmaceutical dispensing and inventory control
Security, safety, and access management	Therapeutic and rehabilitation systems
The Internet	
Routers and switches	
Network interfaces and bridges	
Cable/DSL modems and gateways	
Web, mail, search, and storage servers	
Firewalls	
Wireless access points and repeaters	

IBM's 801 was the first implementation of a RISC machine [1]. However, IBM kept most of the IBM 801 project details undercover for years so the beginnings of the industry-wide movement toward RISC-based design can be traced to just two papers published in the early 1980s by Patterson and Ditzel [2,3]. These papers entirely document the emerging movement. The first paper arguably said goodbye to the past and the second hello to the future. (A good summary of the RISC concept appears in [4].)

The argument favoring RISC over complex instruction-set computers (CISCs) posits that simple processors with simple ISAs are much easier to design and—although these processors must execute more instructions than CISC processors to accomplish more complex tasks—RISC processors achieve much higher overall performance because they execute their instructions in many fewer clock cycles. In addition, their simple ISAs simplify the design of their hardware, so RISC processors can achieve much higher operating frequencies than CISC can processors in the same implementation technology. Theoretically, RISCs enjoy an extra speed advantage.

However, early RISC processors also had a significant liability: they needed to execute many more instructions than CISCs—hence requiring more instruction memory—to execute the same algorithms. Memory density is an advantage for CISCs with variable-length instructions, which is one of the reasons that the CISC microprocessor design style became so popular early in microprocessor history. This aspect was especially important for embedded systems, which were designed with very small memories to minimize costs. Semiconductor memory was very expensive in the 1970s, during the microprocessor's first decade. This cost factor, which favors CISC design, lessened throughout the 1980s as main-memory capacity grew rapidly and memory costs fell thanks to Moore's Law. At the same time, compiler writers improved RISC compilers. Newer optimizing RISC compilers produced significantly better code that performed better and needed less memory.

12.2.5 Processor Cores and the Embedded SOC

By the mid-1990s, IC technology was sufficiently advanced to put a microprocessor on a chip with the rest of the system. Such a chip was dubbed as an SOC. The concept was not new. As far back as the 1970s and 1980s, microcontrollers such as the original TI TMS1000; Motorola's 6801, 68HC05, and 68HC11 families; and Intel's 8048 and 8051 have long been marketed as computers or systems on a chip. What is different about SOCs, which are based on ASIC design and fabrication technologies, is that SOCs are complex, application-specific systems on a chip—like a set-top box or a mobile phone handset—instead of general purpose, one-size-fits-all parts like microcontrollers. Even so, on-chip resources were not yet abundant on chips fabricated in the mid-1990s, so the simplest processors, namely RISC processors, were quite attractive as processing cores for on-chip system implementations.

The first RISC processor architecture readily at hand for chip-level integration was the ARM. Designed during the period between 1983 and 1985, the ARM (or Acorn RISC Machine) was developed as a small, efficient processor for a British PC vendor named Acorn. This processor architecture drew Apple Computer's interest when the company was casting about for an appropriate processor to power the Newton PDA. A collaboration between Acorn and Apple drove ARM processor development to the point that Acorn spun off the processor development operation in 1990 as Advanced RISC Machines, Ltd. and the Acorn RISC Machine became the Advanced RISC Machine, or simply ARM.

The ARM processor therefore became available as a processor core (simply the design minus the silicon) at exactly the right time to catch the rising star of the mobile-phone handset. In the next few years, use of ARM cores ballooned to hundreds of millions as digital mobile phone shipments rose. The success of the ARM core drew many other processor vendors into the core business. MIPS, an established processor vendor, joined the fray as did IBM with its PowerPC architecture.

12.2.6 Configurable Processor Cores for SOC Design

When processors became cores, they were freed of their corporeal, silicon implementation. For the first 25 years of its existence, microprocessors were designed to be as general purpose as possible, of necessity. Making a microprocessor's architecture general purpose widens its possible use across many design projects, which drives sales volumes up and amortizes the processor's design across many, many ICs. When each processor was hand designed by teams consisting of dozens or hundreds of engineers, such amortization was almost mandatory. In addition, the cost of generating mask sets and then fabricating production volumes of such processors also incurred large costs. Thus, for the first quarter-century of its existence, microprocessor architectural design focused on creating relatively complex machines that had a lot of features intended to appeal to the broadest possible design audience.

However, when the system design began to migrate from the board level to the chip level, it was a natural and logical step to continue using fixed-ISA processor cores in SOCs. Packaged processors developed since 1971 had to employ fixed ISAs to achieve economies of scale in the fabrication process. Consequently, many system designers became versed in the selection and the use of fixed-ISA processors and the related tool sets for their system designs. Thus, when looking for a processor to use in an SOC design, system designers first turned to fixed-ISA processor cores. RISC microprocessor cores based on processors that had been designed for personal computers and workstations from ARM and MIPS Technologies were early favorites due to their relatively low gate count.

However customized silicon does not limit a design to fixed-ISA microprocessor cores as do board-level systems based on discrete, prepackaged microprocessors. A configurable processor core allows the system designer to custom tailor a microprocessor to more closely fit the intended application (or set of applications) on the SOC. A "closer fit" means that the processor's register set is sized appropriately for the intended task and that the processor's instructions also closely fit the intended task. For example, a processor tailored for digital audio applications may need a set of 24-bit registers for the audio data and a set of specialized instructions that operates on 24-bit audio data using a minimum number of clock cycles.

Processor tailoring offers several benefits. Tailored instructions perform assigned tasks in fewer clock cycles. For real-time applications such as audio processing, the reduction in clock cycles directly lowers operating clock rates, which in turn cuts power dissipation. Lower power dissipation extends battery life for portable systems and reduces the system costs associated with cooling in all systems. Lower clock rates also allow the SOC to be fabricated in slower and therefore less-expensive IC-fabrication technologies.

Even though the technological barriers to free ISA selection have been torn down by the migration of systems to chip-level design, system-design habits are hard things to break. Many system designers who are well versed in comparing and evaluating fixed-ISA processors from various vendors elect to stay with the familiar, which is perceived as a conservative design approach. When faced with designing next-generation systems, these designers immediately start looking for processors with higher clock rates that are just fast enough to meet the new system's performance requirements. Then they start to worry about finding batteries or power supplies with extra capacity to handle the higher power dissipation that accompanies operating these processors at higher frequencies. They also start to worry about finding ways to remove the extra waste heat from the system package. In short, the design approach cited above is not nearly as conservative as it is perceived; it is merely old-fashioned.

12.3 Processor Parallelism and "Convenient Concurrency" for SOC Designs

Many articles, conference papers, and general discussions of multicore chips, multiprocessors, and associated programming models narrowly focus on particular multiple-processor architectures, which severely limits the possible ways in which multiple computing resources can be used to attack a problem. These discussions tend to focus on "embarrassingly parallel" problems such as graphics and the authors then declare that other big problems cannot be solved until there are tools that can automatically slice and dice problems into processor-sized chunks. However, the truth is that many design problems are conveniently concurrent and are easy to attack with multiple processor cores, though not necessarily by using an symmetric multiprocessing (SMP) architecture. Expanding your architectural thinking beyond SMP multicores uncovers at least two kinds of concurrency that exploit multiple processors of the heterogeneous, and not homogeneous, kind.

Big semiconductor and server vendors currently offer SMP multicore processors, which are good for solving certain kinds of design problems. Large servers and farms support applications such as Web query requests that follow a "SAMD" model: single application and multiple data (an oversimplification, perhaps, but a useful one). SAMD applications date back to early, proprietary mainframe networks used for certain big applications such as real-time airline reservations and check-in systems or real-time banking. These applications are particularly suitable for SMP multicore processors—they essentially run the same kind of code; they do not exhibit data locality; and the number of cores running the application makes no material difference other than speed.

A large number of homogeneous, cache-coherent processors organized into SMP clusters seem a very reasonable technology to apply to these applications and multicore chips and servers from many silicon vendors seem a very reasonable way to exploit the inherent parallelism needed to satisfy many simultaneous user requests. Other applications such as packet processing may also be able to exploit SMP multicore or multiprocessor systems, but SMP is simply not a good processing model for many embedded systems because they ease the processor designer's task of creating a multiple-processor system while making poor compromises in terms of power consumption, heat dissipation, and software development effort.

Graphics processing may also be "embarrassingly parallel." Such applications can be cut up into multiple threads, each acting in parallel on part of the data. Special purpose graphics engines such as the IBM-Toshiba-Sony Cell processor (interestingly, not really an SMP multicore machine) and other graphics chips offered by Nvidia and others are attracting some interest. Embarrassingly parallel applications get harder to find beyond graphics and scene rendering and they can be extremely hard to program despite the availability of special multicore API libraries.

However, very few real-world applications are "embarrassingly parallel." For every Google, bank, or airline, there are millions of ordinary users whose computers are already migrating to multicore processors. Big service providers may eat all the cores they can get, but desktops and laptops may top out at two to four processors. The future economics of SMP multicore chips remains perplexing. We may be facing a de facto solution in profound need of the right desktop problem.

Expanding our architectural thinking beyond SMP multicores uncovers at least two kinds of concurrency that easily exploit multiple processors—heterogeneous concurrency, and not homogeneous. Many embedded systems exhibit such "convenient concurrency." The first such system architecture exists in many consumer devices including mobile phones, portable multimedia players, and multi-function devices. You might call this sort of parallelism "compositional concurrency," where various subsystems—each containing one or more processors optimized for a particular set of tasks—are woven together into a product. Communications are structured so that subsystems communicate only when needed. For example, a user-interface subsystem running on a controller may need to turn audio processing on or off, control the digital camera imaging functions, or interrupt video processing to stop, pause, or change the video stream. In this kind of concurrent system, many subsystems operate simultaneously. Yet they have been designed to interact at only a high level and do not clash.

Figures 12.6 and 12.7 are, respectively, block diagrams of a Personal Video Recorder (PVR) and a Super 3G mobile phone that illustrate this idea. Figure 12.6 shows seven identified processing blocks (shown in gray), each with a clearly defined task. Figure 12.7 shows 18 such processing blocks. In Figure 12.6, it is easy to see how one might use as many as seven processors (or more for subtask

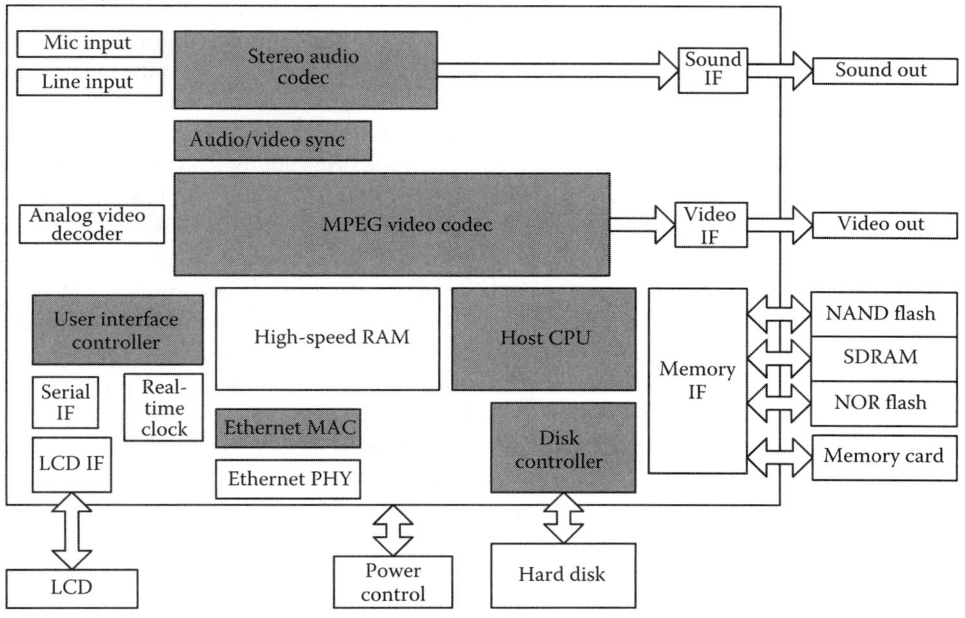

FIGURE 12.6 PVR block diagram.

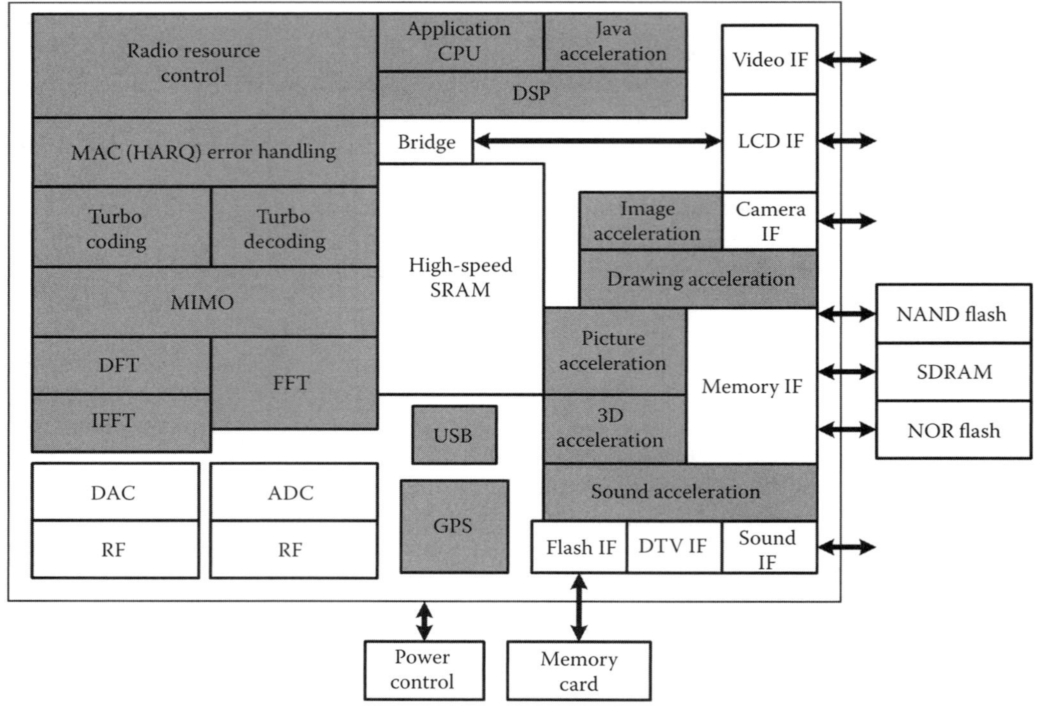

FIGURE 12.7   Super 3G mobile phone handset block diagram.

processing) to divide and conquer the PVR design problem. Similarly, Figure 12.7 shows how as many as 18 processors might be employed on a Super 3G mobile phone chip.

Some engineers might criticize this sort of architecture because of its theoretical inefficiency in terms of gate and processor count. Ten, twenty, or more processor cores could, at least in theory, be replaced with just a few general-purpose cores running at much higher clock rates. However, this criticism is misplaced. When processors were expensive, design styles that favored the use of few, big, fast processors held sway. With the end of Denard scaling (also called classical scaling) at 90 nm, transistors continue to get much smaller at each IC fabrication node but they no longer get that much faster and they no longer dissipate much less power. In fact, static leakage current has started to climb. As a result, the big processors' power dissipation and energy consumption have become unmanageable at high clock rates and system designers are now being forced to adopt design styles that reduce system clock rates before their chips burn to cinders under even normal operating conditions.

Compositionally concurrent system design offers tremendous system-level advantages: Distributing computing tasks over more on-chip processors trades transistors for clock rate, reducing overall system power and energy consumption. Given the continued progress of Moore's law and the end of Denard scaling, this is a very good engineering trade-off.

Subsystems can be more easily powered down when not used—as opposed to keeping all the cores in a multicore SMP system running. Subsystems can be shut off completely and restarted quickly or they can be throttled back by using complex dynamic voltage and frequency scaling algorithms based on predicted task load.

Because these subsystems are task specific, they run more efficiently on application-specific instruction set processors (ASIPs), which are much more area and power efficient than general-purpose processors so the gate advantages of fewer general-purpose cores may be much less than it seems on first consideration.

Compositionally concurrent system designs avoid complex interactions and synchronizations between subsystems. Shutting down the camera subsystem on a compositional product is a trivial task to perform in software while making sure that such a task can safely be suspended in a cooperative, multitasking environment running on an SMP system can be significantly more complex. Proving that a 4-core SMP system running a mobile phone and its audio, video, and camera functions will not drop a 911 emergency call when other applications are running, or that low-priority applications will be properly suspended when a high-priority task interrupts, is often a nightmare of analysis involving "death by simulation." Reasonably independent subsystems interacting at a high level are far easier to validate both individually and compositionally.

Pipelined dataflow, the second kind of concurrency, complements compositional concurrency. Computation can often be divided into a pipeline of individual task engines. Each task engine processes and then emits processed data blocks (frames, samples, etc.). Once a task completes, the processed data block passes to the next engine in the chain. Such asymmetric multiprocessing (APM) algorithms appear in many signal- and image-processing applications from cell phone baseband processing to video and still-image processing. Pipelining permits substantial concurrent processing and also allows even sharper application of ASIP principles: each of the heterogeneous processors in the pipeline can highly be tuned to just one part of the task.

For example, Tensilica's Diamond Standard 388VDO Video Engine (Figure 12.8) mates two appropriately and differently configured 32-bit processor cores with a DMA controller to create a digital-video codec subsystem. One processor core in the subsystem is configured as a stream processor and the other as a pixel processor. The stream processor accelerates serial processing such as bitstream parsing, entropy decoding, and control functions. The pixel processor works on the video data plane

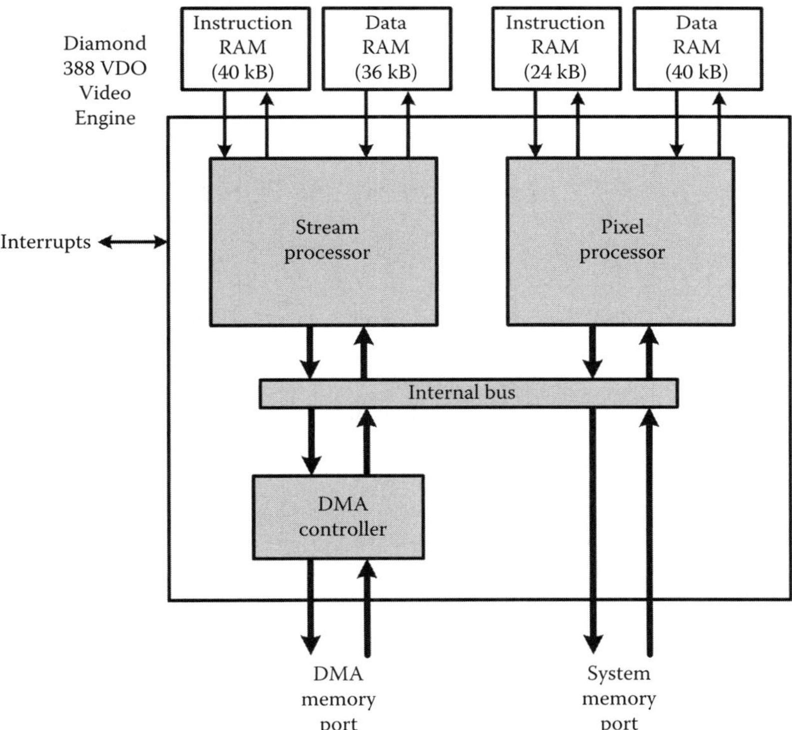

FIGURE 12.8 Tensilica's Diamond 388VDO Video Engine IP core.

and performs parallel computations on pixel data using a single instruction multiple data (SIMD) instruction architecture. Both processors have different local memory and data width configurations as required by their functional partition. This configuration decodes H.264 D1 main profile video while running at 200 MHz, which is easily achieved with 130 nm technology and is even easier to fabricate with more advanced IC fabrication processes.

A Pentium-class processor decodes H.264 D1 main profile video running at a clock rate of between 1 and 2 GHz while dissipating several tens of Watts. A paper presented at the recent International Conference on Consumer Electronics (ICCE) discussed decoding H.264 D1 main profile video using 125% of a 600 MHz TI TMS320DM642 DSP, putting the required clock rate at 720 MHz. Unfortunately, you cannot synthesize SOC processors that run at 720 MHz—much less 1–2 GHz—using available ASIC foundry technologies. In this case, pipeline processing drops the required clock frequency considerably over the "one big, fast processor" design approach and allows the video decoder to be fabricated in a conventional ASIC manufacturing technology.

Combining the compositional-subsystem style of design with AMP in each subsystem makes it apparent that products in the consumer, portable, and media spaces may need 10–100 processors—each one optimized to a specific task in the product's function set. Programming each AMP application is easier than programming each multithreaded SMP application because there are far fewer intertask dependencies to worry about. Experience shows that this design approach is eminently practical and by using this approach, you will avoid many of the optimization headaches associated with multiple application threads running on a limited set of identical processors in an SMP system.

However, sometimes a design problem simply calls for a big pool of computing resource. SMP systems make sense when a system needs to run a large number of simultaneous tasks from an even bigger pool of potential, ready-to-run programs or when a system needs to run a few programs that need a lot of computing horsepower. Tensilica's Xtensa architecture reaches into this domain as well, with the Xtensa MX Multicore Processor, which can incorporate from two to four Xtensa CPUs (see Figure 12.9). The Xtensa MX Multicore Processor adds cache-coherency hardware to several independently configured Xtensa processor cores to create a truly flexible SMP-style multicore.

12.4　Twenty-First-Century Embedded-Processor Zoo

The preceding history explains why there are so very many microprocessor architectures available to embedded-system design teams in the twenty-first century. Each new microprocessor generation has produced a few winning architectures, and many losers. The winning architectures from each generation have proved to be very long-lived. There are survivors from the 8-bit, 16-bit, and 32-bit CISC generations. Similarly, there are several viable architectures from the 32-bit RISC generation and from multiple generations of DSPs. In fact, the number of available architectures is too large to list. (Check *EDN*'s latest annual microprocessor directory for a good summary of available architectures [5].) However, there are some notable processor architectures that stand out above others for their broad popularity in embedded designs.

Embedded processor architectures that have proved popular over the long-term include

8-Bit Embedded Processor Architectures	16-Bit Embedded Processor Architectures	32-Bit Embedded Processor Architectures
Freescale (formerly Motorola) 68HC05, 68HC11	Intel 8088/8086	Intel/AMD/Via x86
Intel 8051	TI DSPs	Freescale (formerly IBM) 68000, 68300, and ColdFire
Microchip PIC		IBM PowerPC
Zilog Z8, Z80		ARM
		MIPS
		Tensilica Xtensa

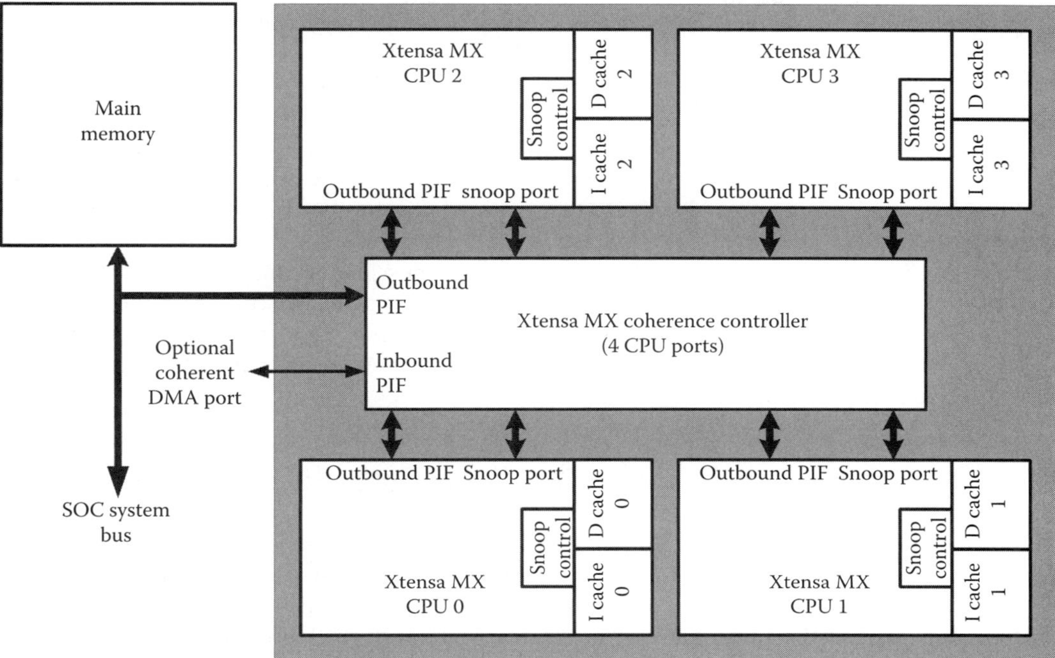

FIGURE 12.9   Tensilica's Xtensa MX Multicore Processor (4-CPU version).

The reason all of these architectures coexist in the embedded world is that embedded applications are so very diverse. No one processor architecture can serve all of these applications. That has led to the rise of the configurable processor cores for embedded designs based on SOCs.

Microprocessor cores used for SOC design are the direct descendents of Intel's original 4004 microprocessor. They are all software-driven, stored-program machines with bus interconnections. Just as packaged microprocessor ICs vary widely in their attributes, so do microprocessors packaged as IP cores. Microprocessor cores vary in architecture, word width, performance characteristics, number and width of buses, cache interfaces, local memory interfaces, and so on. Early on, when transistors were somewhat scarce, many SOC designers used 8-bit microprocessor cores to save silicon real estate. In the twenty-first century, however, some high-performance 32-bit RISC microprocessor cores consume less than $0.5\,mm^2$ of silicon, so there is no longer much reason to stay with lower performance processors. Indeed, the vast majority of SOC designers now use 32-bit processor cores.

In addition, microprocessor cores available as IP have become specialized just like their packaged IC brethren. Thus you will find 32-bit, general-purpose processor cores and DSP cores. Some vendors offer other sorts of very specialized microprocessor IP such as security and encryption processors, media processors, and network processors. This architectural diversity is accompanied by substantial variation in software-development tools, which greatly complicates the lives of developers on SOC firmware-development teams.

The reason for this complication is largely historic. As microprocessors evolved, a parallel evolution in software-development tools also occurred. A split between the processor developers and tool developers opened and grew. Processor developers preferred to focus on hardware architectural advances and tool developers focused on compiler advancements. Processor developers would labor to produce the next great processor architecture and, after the processor was introduced, software tool developers would find ways to exploit the new architectural hardware to produce more efficient compilers.

In one way, this split was very good for the industry. It put a larger number of people to work on the parallel problems of processor architecture and compiler efficiency. As a result, it was likely that microprocessors and software tools evolved more quickly than if the developments had remained more closely linked.

However, this split has also produced a particular style of system design that is now limiting the industry's ability to design advanced systems. SOC designers compare and select processor cores the way they previously compared and selected packaged microprocessor ICs. They look at classic, time-proven figures of merit such as clock rate, main-bus bandwidth, cache-memory performance attributes, and the number of available third-party software-development tools to compare and select processors. Once a processor has been selected, the SOC development team chooses the best compiler for that processor, based on other figures of merit. Often, the most familiar compiler is the winner because learning a new software-development tool suite consumes precious time more economically spent on actual development work.

12.5 Software-Development Tools for Embedded Processors

The vast majority of the literature written about embedded-system design focuses on hardware. In reality, far more effort goes into embedded software development than hardware development so it is appropriate to devote at least a bit of this chapter about embedded processors to the issues surrounding embedded software-development tools. Such tools—including compilers, assemblers, linkers, debuggers, profilers, software libraries, and integrated development environments (IDEs)— should play a substantial role in embedded processor architecture selection. Embedded processors are useless without these important development tools, which have a considerable influence on the performance of the software that runs on a processor—as much as or more influence than processor clock rate, cache size, or memory bandwidth. A good set of software tools can create tight, efficient code that reduces memory costs and cuts performance requirements so that the processor can be run at lower clock rates, which in turn reduces power dissipation and energy consumption.

Software-development tool chains are of necessity hardware-dependent, in that they must compile, generate, support, and deal with software that is ultimately destined to be executed on a particular processor or family of processors (even if that software is written to be portable). Because these software-development tools can be complex, writing them to be specific to only one processor type and generation at a particular point in time is very inefficient. Open-source tools such as those developed by the GNU project are meant to run on many different types of target processors and thus must contain at least rudimentary features that enable open-source developers to retarget them with a reasonably contained effort. But the world of processors encompasses more than just a variety of fixed-ISA processor architectures. It now includes configurable and extensible processors, as discussed earlier in this chapter, making the embedded world much more interesting and complex.

Developers of programs running on deeply embedded processor cores within SOCs have two main performance goals for their code. It should

1. Run fast. Minimize processor operating frequency.
2. Consume little memory. Minimize memory cost.

Depending on the project, some of these goals will have higher priority than the others. Two key factors greatly affect the design team's ability to meet these goals: the compiler's and assembler's code-optimizing efficiency and the programming styles used to develop the source code. The software team's programming style is well beyond the scope of this chapter but code optimization is a fair topic for discussion. (For a discussion of C programming styles that enhance software performance and reduce memory footprint, see [6].) In addition, the IDE can have a profound effect on programmer efficiency.

12.5.1 Embedded Compiler

A compiler translates HLL source code, usually written in C or C++ for embedded code, into assembly-language semantics. In the 1970s and 1980s, almost all embedded code was written in assembly language because HLL compilers produced poor quality translations (in terms of code size and execution speed) compared to hand-coded assembly-language programs written by good, experienced programmers. However, embedded code size has ballooned with the growth in processor address spaces and the embedded processing of large media data types including images, audio, and video. To maintain programmer efficiency, most software-development teams now employ HLLs almost exclusively and rely on compiler translation to produce assembly-language source code.

The translation process from HLL source code to assembly language has become fairly complex. Modern compilers typically consist of a front end and a back end. The front end is generally where syntactic and semantic processing takes place. The back end performs general optimizations as well as code generation and further optimization for a particular target processor. This front-end/back-end approach makes it possible to combine front ends for different programming languages with different code-generating back ends.

Many good compiler back ends rely on multilevel intermediate representations (IRs). Optimization and code generation passes gradually lower the IR from a high level similar to the syntax of the input program down to a low level approaching the generated assembly. Machine-independent optimizations tend to be performed earlier in the compilation process on the higher levels of the IR while processor-specific optimizations tend to be performed later on the lower levels. Information is passed down through the various levels of the IR to enable the low-level optimizations to take advantage of high-level information captured by earlier phases of the compiler.

Compilers generally use command-line switches to set the optimization level. For example, Tensilica's XCC C/C++ compiler for its Xtensa and Diamond processor cores has four basic optimization levels, -O0 through -O3, which set increasingly aggressive optimization levels. Table 12.2 describes these levels along with optimization options for code size (-Os), interprocedural analysis (-IPA), and debugging (-g). By default, the XCC compiler optimizes one file at a time but it can also perform IPA (by adding the -IPA flag), which optimizes an entire application program across multiple source files by delaying optimization until after the link step.

Table 12.3 describes a partial list of optimizations available in modern compilers including Tensilica's XCC compiler. The XCC compiler can also make use of profile data during the compilation process through use of the compiler flags -fb_create and -fb_opt. Use of profile-driven feedback helps the compiler mitigate branch delays. In addition, the use of feedback allows the compiler to inline only the more frequently called functions and to place register spills in infrequent regions of

TABLE 12.2   Compiler Optimization Levels for Tensilica's XCC Compiler

-O0: No optimization.

-O1: Perform local (single basic block) optimizations—constant folding and propagation, common subexpression elimination (CSE), peephole optimizations, and local register allocation. In-lining for only those functions that are expressly marked with the in-line specifier or defined inside the class scope.

-O2: Perform all -O1 optimizations plus dead-code elimination, partial redundancy elimination (PRE), strength reduction, global register allocation, instruction scheduling, loop unrolling, and heuristic-based in-lining of static functions in addition to functions explicitly marked as in-line.

-O3: Perform all -O2 optimizations plus additional loop transformations based on dependence analysis, such as interchange and outer unrolling.

-Os: Optimizes for space. Can be used in conjunction with any optimization level.

-IPA: Perform interprocedural analysis (IPA) and optimization, which optimizes code across an entire program, not just within individual procedures. Can be used in conjunction with any optimization level but benefits are most visible when used with -O2 or -O3 optimization levels.

TABLE 12.3   Modern Compiler Optimizations

Constant folding: Replaces constructs that can be evaluated at compile time with a computed value.

Constant propagation: Replaces a variable access with a constant value if the compiler determines that this is possible at compile time. Especially effective with processor ISAs that include instructions with small operand fields that hold integer constants.

Common subexpression elimination: Replaces all but the first of multiple identical computations with an access to the saved result from the first computation.

Dead-code elimination: Removes code that can never be executed (a possible programming error or a product of prior optimizations) and code that computes unneeded results.

Partial redundancy elimination: Minimizes the number of redundant computations so that these computations are performed no more than once in each path of a program's flow graph. PRE is a form of CSE.

Value numbering: Assigns a common symbolic value to variables and expressions that are provably equivalent. Like PRE, value numbering reduces the number of redundant computations in a program.

Strength reduction: Replaces time-consuming operations such as multiplication with faster operations such as addition.

Sparse conditional constant propagation: Eliminates dead code that will not be executed and replaces variables with constants if possible, based on conditional-code evaluation. (Code must first be structured in static single assignment (SSA) intermediate form to allow this conditional analysis.)

Scalar replacement of aggregates: Evaluates an aggregate variable's different components and replaces as many components as possible with equivalent scalar temporary values. Especially useful for C structures. This step simplifies further optimizations including register allocation and constant propagation.

Local and global register allocation with live range splitting, rematerialization, and homing: Assigns program values to processor registers with the goal of minimizing data movement into and out of the registers (filling and spilling). Local optimizations are based on register usage counts and loop nesting. Global allocation employs graph coloring. Live range splitting allows a single variable to occupy different registers in different regions of code. Rematerialization recomputes a value using register-resident values rather than reloading the value from memory. Homing reloads a value directly from its stored location in memory rather than storing and retrieving the register from the stack.

Register-variable identification: Promotes memory-based scalar variables to register-based variables to improve speed.

Induction and removal of loop invariants: Modifies, simplifies, or removes computations in a loop by analyzing the dependencies of variables used in the loop (especially counters) and changing operations accordingly.

Pointer alias analysis: A technique that correlates aliased pointer definitions to variables in memory locations that a pointer dereference may use or define so that code around such dereferences can be better optimized. Works better in conjunction with interprocedural analysis (IPA).

Array dependence analysis: Detects potential memory aliasing or array references. A key analysis technique for automatic parallelization of array-based computations.

Control flow optimizations: Rearranges and combines branches to mitigate their effects.

Loop unrolling: Replaces looped instructions with repeated copies of the loop body to remove the looping overhead (if any) and to aid further optimization using other optimization techniques.

Inner loop optimizations: Optimizations that prepare for software pipelining including if-conversion (conditional branches within loops are transformed into conditional moves), interleaving of register recurrences such as summation or dot products, and elimination of common inter-iteration memory references.

Software pipelining: Improves loop performance by allowing parts of several loop iterations to execute concurrently. Particularly useful with VLIW and superscalar processors.

Recurrence breaking: Speeds computations by taking advantage of the commutative and associative properties of arithmetic to rearrange operations, thus eliminating dependences.

Local and global instruction scheduling: Rearranges instruction execution in order to minimize pipeline stalls and maximize the use of parallel execution resources.

Min/Max recognition: Recognizes opportunities to use the processor's native, single-cycle MIN, and MAX instructions for "compare and select" operations.

Memory optimizations: Recognizes operations to optimize use of the processor's memory hierarchy—especially the instruction and data caches—and the processor's register file.

Peephole optimization: Replaces individual instructions or small instruction groups with instructions that execute faster, require less memory, or both. Modern processor ISAs allow for many different kinds of peephole optimizations, which are usually implemented with a pattern-matching algorithm.

code. Profile-driven feedback allows the XCC compiler to optimize an application's critical portions for speed while optimizing for space everywhere else.

12.5.2   Embedded Assembler

Microprocessor assemblers seem to have an easy task: convert machine-generated or hand-coded assembly-level semantics into machine code. Simple assemblers, such as the widely used GNU assembler, do exactly that and no more. However, there is more to do in some cases. For example, Tensilica's assembler for the Xtensa family of configurable processors performs certain optimizations including

- Instruction relaxation: If an instruction uses a constant that cannot be encoded as an operand (because the operand is too large for the instruction's operand field), the Xtensa assembler converts that one instruction into a series of instructions that match the original instruction's semantics. For a configurable processor, the optimal instruction sequence differs depending on the options selected so the Xtensa assembler must select the optimal instruction sequence for the configuration options that are present. For example, an instruction sequence might require a varying amount of code space for different processor configurations, which would require assembly-code writers to be conservative when they estimate a branch range. Because Xtensa processor branch instructions have relatively limited range (because they are 16- or 24-bits in length), always using the more conservative branch-instruction sequence degrades performance. With the Xtensa assembler's instruction–relaxation mechanism, the assembly-code writer can use the shortest-range, most-efficient branch instructions, confident that the assembler will substitute a less-efficient branch-and-jump sequence when necessary.

- Instruction scheduling: This feature alleviates much of the burden of tracking the instructions that are present in a specific processor configuration and how long these instructions will take to execute. For example, the Xtensa ISA supports several different sequences for loading large immediate operands. The optimal instruction sequence and its scheduling characteristics are configuration-dependent. Because the Xtensa assembler will schedule instructions, the assembly-code writer can simply choose the simplest variant and the assembler selects and schedules the optimal one. Some Xtensa processor configurations contain FLIX multioperation instructions (Tensilica's version of VLIW), where one instruction contains several individual operations, all executed at once. Processor configurations incorporating FLIX instructions often exhibit dramatic performance improvements because they are able to initiate multiple operations each clock cycle, but it is impossible for a third-party company to write assembly code that uses an unknown FLIX scheme. Therefore, the assembler scheduler bundles operations into FLIX instructions, which allows assembly-code writers to target any Xtensa processor configuration, with or without FLIX, and still realize the associated performance advantages when the code is run on processor configurations that have FLIX instructions.

- Instruction alignment: Like most modern architectures, Xtensa processors fetch instructions in power-of-two-sized words (specifically 4 or 8 bytes per instruction fetch). However, the Xtensa ISA is unusual in that it consists of 16- and 24-bit instructions (and, for some extended configurations, 32- or 64-bit FLIX bundles). An Xtensa processor experiences a 1-cycle taken-branch penalty if an instruction crosses a fetch boundary. The mismatch between the instruction size (2, 3, 4, or 8 bytes) and the instruction-fetch width (4 or 8 bytes) potentially makes this branch penalty common—especially when combined with the other assembler transformations. To avoid this performance-robbing branch penalty whenever possible, the assembler automatically aligns branch targets by

    1. Converting 2-byte instructions into equivalent 3-byte instructions

    2. Inserting padding in unreachable locations

    3. Inserting no-ops in locations where the processor would otherwise stall. This last feature is enabled by the instruction scheduler

As above, this last characteristic does not arise from configurability or extensibility considerations, but it is definitely hardware-dependent.

A code developer can turn off all these transformations to get classic "what you write is what you get" assembler behavior. In fact, many assembly-code writers new to the Xtensa architecture and Tensilica's software-development tool chain do so. However, most software developers eventually turn these features back on when they see how much easier the code is to write and how much more efficient it is after the assembler optimizes it.

12.5.3   Embedded Linker

A linker serves as the intermediary between the object files generated by the compiler or the assembler and the executable code required to run the program on a target processor. The linker is responsible for placing the code, data, and other sections of object files into an executable file. Object files contain symbolic references and the linker must resolve these references to specific addresses based on target hardware. Symbolic references in object files are identified by relocation records, often just called "relocations," which specifies the symbolic address being referenced, the position in the object where the reference is located, and a relocation type.

Each processor architecture defines a set of relocation types that corresponds to the different kinds of symbolic references supported by that architecture. The simplest relocation type is an address. This relocation type is often used in data sections. The linker computes the address value for the data and inserts it directly into the executable image at the location specified by the relocation.

Other relocation types correspond to specific processor instructions. For example, some processor architectures synthesize a 32-bit address constant with a 2-instruction sequence, where the first instruction sets the high bits of a register and the second instruction adds the low bits. Each of these instructions needs a different relocation type. For the first "set-high" relocation, the linker needs to insert the high bits of the address value into the instruction's immediate field. For the second "add-low" relocation, the low bits of the address must go into the immediate field of a different instruction. In both cases, the relocation type is tied to a particular manipulation of the address value (e.g., extracting the high or low bits) and also to a particular method for inserting the value into the executable (e.g., shifting and masking into the immediate operand field of a particular processor instruction).

12.5.4   Embedded-Development IDE

In the early days of software development, software tools stood alone. We invoked each one separately with a command line. In the twenty-first century, we run these development tools on multitasking operating systems and have therefore encapsulated the tools within IDEs that allow software developers to move quickly from one tool to the next in a smooth flow. When the concepts of software-development IDEs was new, various vendors competing in the embedded-development market rushed to develop their own IDEs. All that changed when IBM released Eclipse into open source waters.

Eclipse is a software framework written primarily in Java. The initial code base for Eclipse was IBM's VisualAge. In its default form, Eclipse is an IDE for Java developers. However, Eclipse can be and has been extended to other languages by installing plug-ins. With its IBM heritage and open-source position, Eclipse has rapidly become the IDE of choice for embedded-software developers.

12.6   Benchmarking Processors for Embedded Systems

Designers measure microprocessor and microprocessor core performance through benchmarking. Processor chips already realized in silicon are considerably easier to benchmark than processor cores. All major microprocessor vendors offer their processors in evaluation kits, which include complete evaluation boards and software-development tool suites (which include compilers, assemblers, linkers, loaders, instruction-set simulators [ISSs], and debuggers). Consequently, benchmarking chip-level microprocessor performance consists of compiling selected benchmarks for the target processor, downloading and running those benchmarks on the evaluation board, and recording the results.

Microprocessor cores are incorporeal and they are not usually available as stand-alone chips. In fact if these cores were available as chips, they would be far less useful to IC designers because on-chip processors can have vastly greater I/O capabilities than processors realized as individual chips. Numerous wide and fast buses are the norm for microprocessor cores—to accommodate instruction and data caches, local memories, and private I/O ports—which is not true for microprocessor chips because these additional buses would greatly increase pin count, which would drive component cost beyond a tolerable level for microprocessor chips. However, extra I/O pins are not costly for microprocessor cores, so they usually have many buses and ports.

Benchmarking microprocessor cores has become increasingly important in the IC-design flow because any significant twenty-first century IC design incorporates more than one processor core—at least two, often several, and occasionally hundreds (see Figure 12.10). As the number of microprocessor cores used on a chip increases and these processor cores perform more tasks that were previously implemented with blocks of logic that were manually designed and coded by hand, measuring processor core performance becomes an increasingly important task in the overall design of the IC.

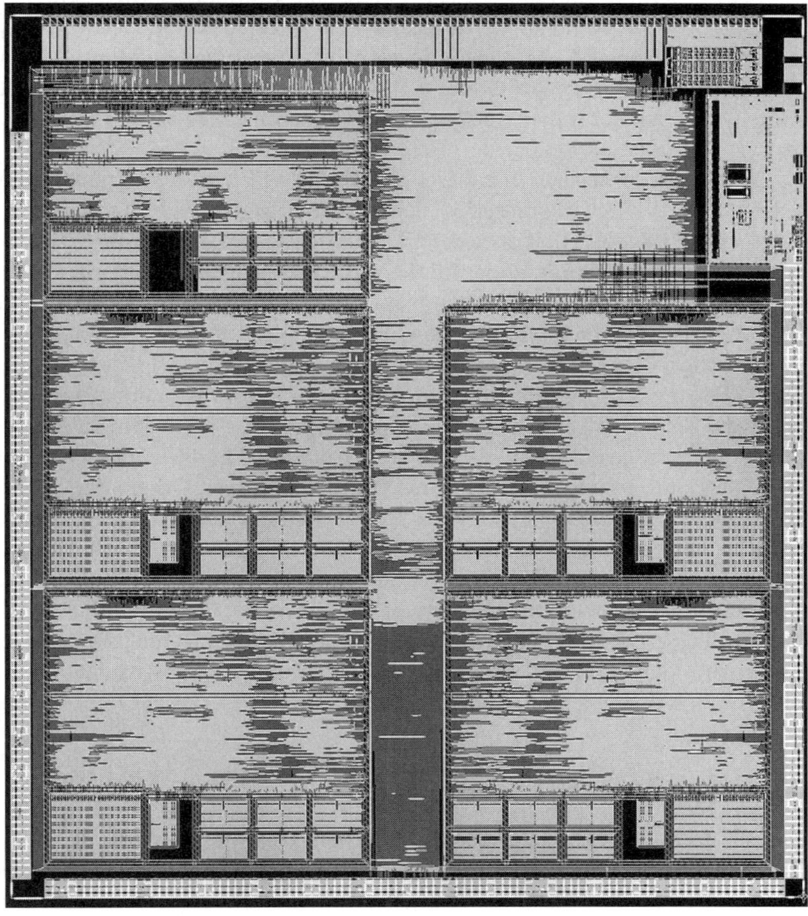

FIGURE 12.10 This chip from the MediaWorks family of configurable media processor products is a DVD-resolution MPEG video and audio encoder/decoder system on a chip and is intended for use in solid-state camcorders and portable video products. The chip contains five loosely coupled heterogeneous processors.

Measuring the performance of microprocessor cores is a bit more complex than benchmarking microprocessor chips: another piece of software, an ISS must stand in for a physical processor chip. Each microprocessor and each member of a microprocessor family requires its own specific ISS because the ISS must provide a cycle-accurate model of the processor's architecture to produce accurate benchmark results. Note that it is also possible to benchmark a processor core using gate-level simulation, but this approach is three orders of magnitude slower (because gate-level simulation is much slower than instruction-set simulation) and is therefore used infrequently [7]. The selected benchmarks are compiled and run on the ISS to produce the benchmark results. All that remains is to determine where the ISS and the benchmarks are to be obtained, how they are to be used, and how the results will be interpreted. These are the subjects of this chapter.

12.6.1   ISS as a Benchmarking Platform

The ISSs serve as benchmarking platforms because processor cores as realized on an IC rarely exist as a chip. ISSs simulate the software-visible state of a microprocessor without employing a gate-level model of the processor so that they run quickly, often 1000 times faster than gate-level processor simulations running on HDL simulators. The earliest ISS was created for the electronic delay storage automatic calculator (EDSAC), which was developed by a team led by Maurice V. Wilkes at the University of Cambridge's Computer Laboratory. The room-sized EDSAC I was the world's first fully operational, stored-program computer (the first von Neumann machine) and went online in 1949. EDSAC II became operational in 1958.

The EDSAC ISS was first described in a paper on debugging EDSAC programs, which was written and published in 1951 by S. Gill, one of Wilkes' research students [8]. The paper describes the operation of a "tracing simulator" that operates by fetching the simulated instruction, decoding the instruction to save trace information, updating the simulated program counter if the instruction is a branch or placing the nonbranch instruction in the middle of the simulator loop and executing it directly, and then returning to the top of the simulator loop. Thus the first operational processor ISS predates the introduction of the first commercial microprocessor (Intel's 4004) by some 20 years.

Cycle-accurate ISSs, the most useful simulator class for processor benchmarking, compute exact instruction timing by accurately modeling the processor's pipeline. All commercial microprocessor cores have at least one corresponding ISS each. They are obtained in different ways. Some core vendors rely on third parties to provide an ISS for their microprocessors. Other vendors offer an ISS as part of their tool suite. Some ISSs must be purchased and some are available in evaluation packages from the processor core vendor.

To serve as an effective benchmarking tool, an ISS must not only simulate the operation of the processor core. It must also provide the instrumentation needed to provide the critical statistics of the benchmark run. These statistics include cycle counts for the various functions and routines executed and main-memory and cache-memory usage. The better the ISS instrumentation is, the more comparative information the benchmark will produce.

12.6.2   Ideal vs. Practical Processor Benchmarks

Embedded-system designers use benchmarks to help them pick the "best" microprocessor core for a given task set. The original definition of a benchmark was literally a mark on a workbench that provided some measurement standard. Eventually, the first benchmarks (carved into the workbench) were replaced with standard measuring tools such as yardsticks. Processor benchmarks provide yardsticks for measuring processor performance. The ideal yardstick would be one that could measure any processor for any task. The ideal processor benchmark would produce results that are relevant, reliable, objective, comparable, and applicable. Unfortunately, no such processor benchmark exists.

In one sense, the ideal processor benchmark would be the actual application code that the processor will run. No other piece of code can possibly be as representative of the actual task to be performed as the actual code that executes that task. No other piece of code can possibly replicate the instruction-use distribution, register and memory use, or data-movement patterns of the actual application code. In many ways, however, the actual application code is less than ideal as a benchmark.

First and foremost, the actual application code may not exist when candidate processors are benchmarked, because benchmarking and processor selection must occur early in the project. A benchmark that does not exist is worthless.

Next, the actual application code serves as an extremely specific benchmark. It will indeed give a very accurate prediction of processor performance for a specific task, and for no other task. In other words, the downside of a highly specific benchmark is that the benchmark will give a less-than-ideal indication of processor performance for other tasks. Because on-chip processor cores are often used for a variety of tasks, the ideal benchmark may well be a suite of application programs and not just one program.

Yet another problem with application-code benchmarks is their lack of instrumentation. The actual application code has almost always been written to execute the task, and not to measure a processor core's performance. Appropriate measurements may require modification of the application code. This modification consumes time and resources, which may not be readily available. Even so, with all of these issues that make the application-code benchmark less than ideal, the application code (if it exists) provides invaluable information on processor core performance and should be used to help make a processor core selection whenever possible.

12.6.3 Standard Benchmark Types

Given that the "ideal" processor benchmark proves less than ideal, the industry has sought "standard" benchmarks that it can use when the target application code is either not available or not appropriate. There are four types of standard benchmarks: full-application or "real-world" benchmarks, synthetic or small-kernel benchmarks, hybrid or derived benchmarks that mix and match aspects of the full-application and synthetic benchmarks, and microbenchmarks (not to be confused with microprocessor benchmarks).

Full-application benchmarks and benchmark suites employ existing system- or application-level code drawn from real applications, although probably not the specific application of interest to any given IC-design team. These benchmarks may incorporate many thousands of lines of code, have large instruction-memory footprints, and consume large amounts of data memory.

Synthetic benchmarks tend to be smaller than full-application benchmarks. They consist of smaller code sections representing commonly used algorithms. They may be extracted from working code or they may be written specifically as a benchmark. Writers of synthetic benchmarks try to approximate instruction mixes of real-world applications without replicating the entire application.

Hybrid benchmarks mix and match large application programs and smaller blocks of synthetic code to create a sort of torture track (with long straight sections and tight curves) for exercising microprocessors. The hybrid benchmark code is augmented with test data sets taken from real-world applications. A microprocessor core's performance around this torture track can give a good indication of the processor's abilities over a wide range of situations, although probably not the specific use to which the processor will be put on an IC.

Microbenchmarks are very small code snippets designed to exercise some particular processor feature or to characterize a particular machine primitive in isolation from all other processor features and primitives. Microbenchmark results can define a processor's peak capabilities and reveal potential architectural bottlenecks, but peak performance is not a very good indicator of overall application performance. Nevertheless, a suite of microbenchmarks may approximate the ideal benchmark for certain applications, if the application is a common one with many standardized, well-defined functions to perform.

12.6.4   Prehistoric Performance Ratings: MIPS, MOPS, and MFLOPS

Lord Kelvin could have been predicting processor performance measurements when he said

> When you can measure what you are speaking about, and express it in numbers, you know something about it; but when you cannot measure it, when you cannot express it in numbers, your knowledge is of a meager and unsatisfactory kind. It may be the beginning of knowledge, but you have scarcely, in your thoughts, advanced to the stage of science. [9]

The need to rate processor performance is so great that, at first, microprocessor vendors grabbed any and all numbers at hand to rate performance. These prehistoric ratings are measures of processor performance that lack code to standardize the ratings. Consequently, these performance ratings are not benchmarks. Just as engine's revolutions per second (RPM) reading does not give you sufficient information to measure the performance of an internal combustion engine (you need engine torque plus transmission gearing, differential gearing, and tire diameter to compute load), the prehistoric, clock-related processor ratings of MIPS, MOPS, millions of floating-point operations per second (MFLOPS), and VAX MIPS are akin to clock rate: they tell you almost nothing about the true performance potential of a processor.

Before it was the name of a microprocessor vendor, the term "MIPS" was an acronym for "millions of instructions per second." If all processors had the same ISA, then a MIPS rating could possibly be used as a performance measure. However, all processors do not have the same ISA. In fact, microprocessors and microprocessor cores have very different ISAs. Consequently, some processors can do more with one instruction than other processors, just as large automobile engines can do more than smaller ones at the same RPM.

This problem was already bad in the days when only CISC processors roamed the earth. The problem went from bad to worse when reduced-instruction-set computers (RISC) processors arrived on the scene. One CISC instruction would often do the work of several RISC instructions (by design) so that a CISC MIPS rating did not correlate well with a RISC MIPS rating because of the work differential between RISC's simple instructions and CISC's more complex instructions.

The next step in creating a usable processor performance rating was to switch from MIPS to VAX MIPS, which was accomplished by setting the extremely successful VAX 11/780 minicomputer— introduced in 1977 by the now defunct DEC—as the standard against which all other processors are measured. Both native and VAX MIPS are inadequate measures of processor performance because they are usually provided without specifying the software (or even the programming language) used to make the measurement. Because different programs have different mixes of instructions, different memory-usage patterns, and different data-movement patterns, the same processor can earn one MIPS rating on one set of programs and quite another rating on a different set. Because MIPS ratings are not linked to a specific benchmark program suite, the "MIPS" acronym has come to stand for "meaningless indication of performance."

Even more tenuous than the MIPS performance rating is the concept of MOPS, an acronym that stands for "millions of operations per second." Every algorithmic task requires the completion of a certain number of fundamental operations, which may or may not have a one-to-one correspondence with machine instructions. Count these fundamental operations in millions and they become MOPS. If they are floating-point operations, you get MFLOPS. Both the MOPS and the MFLOPS ratings suffer from the same drawback as the MIPS rating: there is no standard software to serve as the benchmark that produces the ratings. In addition, the "conversion factor" for computing how many operations a processor performs per clock (or how many processor instructions constitute one operation) is somewhat fluid as well, which means that the processor vendor is free to develop a conversion factor on its own. MOPS and MFLOPS performance ratings exist for various processors but they really do not help a design team pick a processor because they are not true benchmarks.

12.6.5 Classic Processor Benchmarks (The Stone Age)

Like ISSs, standardized processor performance benchmarks predate the 1971 introduction of Intel's 4004 microprocessor, but just barely. The first benchmark suite to attain de-facto "standard" status was a set of programs known as the Livermore Kernels (also popularly called the Livermore Loops).

The Livermore Kernels were first developed in 1970 and consist of 14 numerically intensive application kernels written in FORTRAN. Ten more kernels were added in the 1980s and the final suite of benchmarks was discussed in a paper published in 1986 by F. H. McMahon of the Lawrence Livermore National Laboratory (LLNL), located in Livermore, California [10]. The Livermore Kernels actually constitute a supercomputer benchmark, measuring a processor's floating-point computational performance in terms of MFLOPS.

At first glance, the LFK benchmark appears to be nearly useless for the benchmarking of embedded microprocessors. It is a floating-point benchmark written in FORTRAN that looks for good vector abilities. FORTRAN compilers for embedded microprocessors are quite rare and unusual. Very few real-world applications run tasks like those appearing in the LFK kernels, which are far more suited to research on the effects of very high-speed nuclear reactions than the development of commercial, industrial, or consumer products. For example, it is unlikely that the processors in a mobile phone handset or an MP3 music player will ever need to perform two dimensional (2D) hydrodynamic calculations. Consequently, microprocessor core vendors are quite unlikely to tout LFK benchmark results for their cores.

12.6.5.1 LINPACK

LINPACK is a collection of FORTRAN subroutines that analyze and solve linear equations and linear least-squares problems. Jack Dongarra assembled the LINPACK collection of linear algebra routines at the Argonne National Laboratory in Argonne, Illinois. The first versions of LINPACK existed in 1976 but the first users' guide was published in 1977 [11,12]. The package solves linear systems whose matrices are general, banded, symmetric indefinite, symmetric positive definite, triangular, and tridiagonal square. Like the LFK benchmarks, LINPACK benchmarks are not commonly used to measure microprocessor or microprocessor core performance.

12.6.5.2 Whetstone Benchmark

The Whetstone benchmark was written by Harold Curnow of the now defunct Central Computer and Telecommunications Agency (CCTA, the British government agency tasked with computer procurement) to test the performance of a proposed computer. It is the first program to appear in print that was designed as a synthetic benchmark to test processor performance, although it was specifically developed to test the performance of only one computer: the hypothetical Whetstone machine.

The Whetstone benchmark is based on application-program statistics gathered by Brian A. Wichmann at the National Physical Laboratory in England. Wichmann was using an Algol-60 compilation system that compiled Algol statements into instructions for the hypothetical Whetstone computer system, which was named after the small town of Whetstone located just outside the city of Leicester, England where the compilation system was developed. Wichmann developed statistics on instruction usage for a wide range of numerical computation programs then in use. Information about the Whetstone benchmark was first published in 1976 [13].

The Whetstone benchmark produces speed ratings in terms of thousands of Whetstone instructions per second (KWIPS), thus using the hypothetical Whetstone computer as the golden standard for this benchmark. In 1978, self-timing versions (written by Roy Longbottom, also of CCTA) produced speed ratings in MOPS and MFLOPS and overall rating in MWIPS.

As with the LFK and LINPACK benchmarks, the Whetstone benchmark focuses on floating-point performance. Consequently, it is not commonly applied to embedded processors that are destined to execute integer-oriented and control tasks. However, the Whetstone benchmark's appearance spurred the creation of a plethora of "stone," benchmarks and one of those, the Dhrystone, became the first widely used benchmark to rate microprocessor performance.

12.6.5.3　Dhrystone Benchmark

Reinhold P. Weicker, working at Siemens-Nixdorf Information Systems, wrote and published the first version of the Dhrystone benchmark in 1984 [14]. The benchmark's name, "Dhrystone," is a pun derived from the Whetstone benchmark. The Dhrystone benchmark is a synthetic benchmark that consists of 12 procedures in one measurement loop. It produces performance ratings in Dhrystones per second. Originally, the Dhrystone benchmark was written in a "Pascal subset of Ada." Subsequently, versions in Pascal and C have appeared and the C version is the one most used today.

The Dhrystone benchmark differs significantly from the Whetstone and these differences made the Dhrystone far more suitable as a microprocessor benchmark. First, the Dhrystone is strictly an integer program. It does not test a processor's floating-point abilities because most microprocessors in the early 1980s (and even in the early twenty-first century) had no native floating-point computational abilities. The Dhrystone does devote a lot of time executing string functions (copy and compare), which microprocessors often execute in real applications.

Curiously, DEC's VAX also plays a role in the saga of the Dhrystone benchmarks. The VAX 11/780 minicomputer could run 1757 version 2.1 Dhrystones/s. Because the VAX 11/780 was (erroneously) considered a 1-MIPS computer, it became the Dhrystone standard machine, which resulted in the emergence of the DMIPS or D-MIPS (Dhrystone MIPS) rating. By dividing a processor's Dhrystone 2.1 performance rating by 1757, processor vendors could produce an official-looking DMIPS rating for their products. Thus DMIPS fuses two questionable rating systems (Dhrystones and VAX MIPS) to create a third, derivative, equally questionable microprocessor-rating system.

The Dhrystone benchmark's early success as a marketing tool encouraged abuse and abuse became rampant. For example, vendors quickly (though unevenly) added specialized string routines written in machine code to some of their compiler libraries because accelerating these heavily used library routines boosted Dhrystone ratings, even though the actual performance of the processor did not change at all. Some compilers started in-lining these machine-coded string routines for even better ratings. As the technical marketing teams at the microprocessor vendors continued to study the benchmark, they found increasingly better ways of improving their products' ratings by introducing compiler optimizations that only applied to the benchmark. Some compilers even had pattern recognizers that could recognize the Dhrystone benchmark source code or a "Dhrystone command-line switch" that would cause the compiler to "generate" the entire program using a previously prewritten, precompiled, hand-optimized version of the benchmark program.

It is not even necessary to alter a compiler to produce wildly varying Dhrystone results using the same processors. Using different compiler optimization settings can drastically alter the outcome of a benchmark test, even if the compiler has not been "Dhrystone optimized." For example, Tensilica's Xtensa LX processor core produces different Dhrystone benchmark results that differ by almost 2:1 depending on the setting of the "in-lining" switch of its compiler (shown in Table 12.4).

TABLE 12.4　Performance Difference in Dhrystone Benchmark between In-lined and Non-in-lined Code

Xtensa LX Compiler Setting	DMIPS Rating (at 150 MHz)	DMIPS/MHz Rating
No inline code	183.187	1.221
Inline code	322.154	2.148

In-lining code is not permissible under Weicker's published rules, but there is no organization to enforce Weicker's rules so there is no way to ensure that processor vendors benchmark fairly and no way to force vendors to fully disclose the conditions that produced their benchmark results. Weaknesses associated with conducting Dhrystone benchmark tests and reporting the results highlight a problem: if the benchmarks are not conducted by objective third parties under controlled conditions that are disclosed with the benchmark results, the benchmark results must be suspect even if there are published rules as to a benchmark's use because there is nobody to enforce the rules and not everyone follows rules when the sales of a new microprocessor are at stake.

One of the worst abuses of the Dhrystone benchmark occurred when processor vendors ran benchmarks on their own evaluation boards, on competitors' evaluation boards, and then published the results. The vendor's own board would have fast memory and the competitors' boards would have slow memory, but this difference would not be revealed when the scores were published. Differences in memory speed do affect Dhrystone results, so publishing the Dhrystone results of competing processors without also disclosing the use of dissimilar memory systems to compare the processors is clearly dishonest, or at least mean-spirited.

12.6.5.4 *EDN* Microprocessor Benchmarks

EDN magazine, a trade publication for the electronics industry, was a very early advocate of microprocessor use for general electronic system design and it extensively covered the introduction of new microprocessors starting in the early 1970s. In 1981, just after the first wave of commercial 16-bit microprocessors had been introduced, *EDN* published the first comprehensive article. The article used a set of microprocessor benchmarks to compare the four leading 16-bit microprocessors of the day: DEC's LSI-11/23, Intel's 8086, Motorola's 68000, and Zilog's Z8000 [15]. The Intel, Motorola, and Zilog processors often competed for system design wins in the early 1980s and this article provided design engineers with some of the first microprocessor benchmark ratings to be published by an objective third party.

This article was written by Jack E. Hemenway, a consulting editor for *EDN*, and Robert D. Grappel of MIT's Lincoln Laboratory. Hemenway and Grappel summed up the reason that the industry needs objective benchmark results succinctly in their article:

> Why the need for a benchmark study at all? One sure way to start an argument among computer users is to compare each one's favorite machine with the others. Each machine has strong points and drawbacks, advantages and liabilities, but programmers can get used to one machine and see all the rest as inferior. Manufacturers sometimes don't help: Advertising and press releases often imply that each new machine is the ultimate in computer technology. Therefore, only a careful, complete and unbiased comparison brings order out of the chaos.

The *EDN* article continues with an excellent description of the difficulties associated with microprocessor benchmarking:

> Benchmarking anything as complex as a 16-bit processor is a very difficult task to perform fairly. The choice of benchmark programs can strongly affect the comparisons' outcome so the benchmarker must choose the test cases with care.

Hemenway and Grappel used a set of benchmark programs created in 1976 by a research group at Carnegie-Mellon University (CMU). These benchmarks were first published as a paper, which was presented by the CMU group in 1977 at the National Computer Conference [16]. The tests in these benchmark programs—interrupt handlers, string searches, bit manipulation, and sorting—are very representative of types of tasks microprocessors and processor cores must perform in most embedded applications. The *EDN* authors excluded the CMU benchmarks that tested floating-point

TABLE 12.5   Programs in the 1981 *EDN* Benchmark Article by Hemenway and Grappel

EDN 1981 Microprocessor Benchmark Component	Benchmark Description
Benchmark A	Simple priority interrupt handler
Benchmark B	FIFO interrupt handler
Benchmark E	Text string search
Benchmark F	Primitive bit manipulation
Benchmark H	Linked-list insertion (test addressing modes and 32-bit operations)
Benchmark I	Quicksort algorithm (tests stack manipulation and addressing modes)
Benchmark K	Matrix transposition (tests bit manipulation and looping)

computational performance because none of the processors in the *EDN* study had native floating-point resources, which is still true of most contemporary microprocessor cores. The seven programs in the 1981 *EDN* benchmark study appear in Table 12.5.

Significantly, the authors of this article allowed programmers from each microprocessor vendor to code each benchmark for their company's processor. Programmers were required to use assembly language to code the benchmarks, which removed the associated compiler from the equation and allowed the processor hardware architectures to compete directly. Considering the state of microprocessor compilers in 1981, this was probably an excellent decision.

The authors refereed the tests by reviewing each program to ensure that no corners were cut and that the programs faithfully executed each benchmark algorithm. The authors did not force the programmers to code in a specific way and allowed the use of special instructions (such as instructions designed to perform string manipulation) because use of these instructions fairly represented actual processor use.

To ensure fairness, the study's authors ran the programs and recorded the benchmark results. They published the results and included both the program size and execution speed. Publishing the program size recognized that memory was costly and limited, a situation that still holds true today in the IC design. The authors also published the scores of each of the seven benchmark tests separately, acknowledging that a combined score would tend to mask information about a processor's specific abilities.

In 1988, *EDN* published a similar benchmarking article covering DSPs [17]. The article was written by David Shear, one of *EDN*'s regional editors. Shear benchmarked 18 DSPs using 12 benchmark programs selected from 15 candidate programs. The DSP vendors helped select the DSP benchmark programs that were used from the 15 candidate programs. The final set of DSP benchmark programs included six DSP filters, three math benchmarks (a simple dot product and two matrix multiplications), and three FFTs. Shear used benchmark programs to compare DSPs that are substantially different from the programs used by Grappel and Hemenway to compare general-purpose microprocessors because, as a class, DSPs are applied quite differently from general-purpose processors.

Shear's article also recognized the potential for cheating by repeating the "Three Not-so-golden Rules of Benchmarking" that had appeared in 1981, which was written by *EDN* editor Walt Patstone as a follow-up to the article by Hemenway and Grappel [18]:

- Rule 1—All is fair in love, war, and benchmarking.
- Rule 2—Good code is the fastest possible code.
- Rule 3—Conditions, cautions, relevant discussion, and even actual code never make it to the bottom line when results are summarized.

These *EDN* microprocessor benchmark articles established and demonstrated all of the characteristics of an effective, modern microprocessor benchmark:

- Conduct a series of benchmark tests that exercise salient processor features for a class of tasks.
- Use benchmark programs that are appropriate to the processor class being studied.
- Allow experts to code the benchmark programs.
- Have an objective third-party check and run the benchmark code to ensure that the experts do not cheat.
- Publish benchmark results that maximize the amount of available information about the tested processor.
- Publish both execution speed and memory use for each processor on each benchmark program because there is always a trade-off between a processor's execution speed and its memory consumption.

These characteristics shaped the benchmarking organizations and their benchmark programs in the 1990s.

12.6.6 Modern Processor Performance Benchmarks

As the number of available microprocessors mushroomed in the 1980s, the need for good benchmarking standards became increasingly apparent. The Dhrystone experiences showed how useful a standard benchmark could be and these experiences also demonstrated the lengths (both fair and unfair) to which processor vendors would go to earn top benchmark scores. The *EDN* articles had shown that an objective third party could bring order out of benchmarking chaos. As a result, both private companies and industry consortia stepped forward with the goal of producing "fair and balanced" processor benchmarking standards.

12.6.6.1 SPEC: The Standard Performance Evaluation Corporation

One of the first organizations to tackle the need for good microprocessor benchmarking standards was SPEC (originally the System Performance Evaluation Cooperative and now the Standard Performance Evaluation Corporation). A group of workstation vendors including Apollo, Hewlett-Packard, MIPS Computer Systems, and Sun Microsystems, working in conjunction with the trade publication *Electronic Engineering Times* founded SPEC in 1988. A year later, the consortium produced its first processor benchmark called SPEC89. The SPEC89 benchmark provided a standardized measure of compute-intensive microprocessor performance with the express purpose of replacing the existing, vague MIPS and MFLOPS rating systems. Because high-performance microprocessors were primarily used in high-end workstations at the time and because SPEC was formed as a cooperative by workstation vendors, the SPEC89 benchmark consisted of source code that was to be compiled for UNIX.

The Dhrystone benchmark had already demonstrated that benchmark code quickly rots over time due to rapid advances in processor architecture and compiler technology. (Reinhold Weicker, Dhrystone's creator, became a key member of SPEC and has been heavily involved with the ongoing creation of SPEC benchmarks.) To prevent "benchmark rot," the SPEC organization has regularly improved and expanded its benchmarks, producing SPEC92, SPEC95 (with separate integer and floating-point components called CINT95 and CFP95), and finally SPEC CPU2000 (consisting of CINT2000 and CFP2000). Tables 12.6 and 12.7 list the SPEC CINT2000 and CFP2000 benchmark component programs, respectively.

SPEC benchmarks are application-based benchmarks, and not synthetic benchmarks. The SPEC benchmark programs are excellent as workstation/server benchmarks because they use the actual applications that are likely to be assigned to these machines. SPEC publishes benchmark performance results for various computer systems on its Web site (www.spec.org) and sells its benchmark code.

TABLE 12.6   SPEC CINT2000 Benchmark Component Programs

CINT2000 Benchmark Component	Language	Category
164.gzip	C	Compression
175.vpr	C	FPGA Circuit Placement and Routing
176.gcc	C	C Programming Language Compiler
181.mcf	C	Combinatorial Optimization
186.crafty	C	Game Playing: Chess
197.parser	C	Word Processing
252.eon	C++	Computer Visualization
253.perlbmk	C	PERL Programming Language
254.gap	C	Group Theory, Interpreter
255.vortex	C	Object-oriented Database
256.bzip2	C	Compression
300.twolf	C	Place and Route Simulator

TABLE 12.7   SPEC CFP2000 Benchmark Component Programs

CFP2000 Benchmark Component	Language	Category
168.wupwise	Fortran 77	Physics/Quantum Chromodynamics
171.swim	Fortran 77	Shallow Water Modeling
172.mgrid	Fortran 77	Multi-grid Solver: 3D Potential Field
173.applu	Fortran 77	Parabolic/Elliptic Partial Differential Equations
177.mesa	C	3D Graphics Library
178.galgel	Fortran 90	Computational Fluid Dynamics
179.art	C	Image Recognition/Neural Networks
183.equake	C	Seismic Wave Propagation Simulation
187.facerec	Fortran 90	Image Processing: Face Recognition
188.ammp	C	Computational Chemistry
189.lucas	Fortran 90	Number Theory/Primality Testing
191.fma3d	Fortran 90	Finite-element Crash Simulation
200.sixtrack	Fortran 77	High Energy Nuclear Physics Accelerator Design

Because the high-performance microprocessors used in workstations and servers are sometimes used as embedded processors and some of them are available as microprocessor cores for use on SOCs, microprocessor and microprocessor core vendors sometimes quote SPEC benchmark scores for their products. Use these performance ratings with caution because the SPEC benchmarks do not necessarily measure performance that is meaningful within the context of embedded applications. For example, mobile phones are not likely to be required to simulate seismic-wave propagation, except for the possible exception of handsets sold in California.

A paper written by Jakob Engblom at Uppsala University compares the static properties of code from 13 embedded applications with the SPECint95 benchmark programs. The embedded programs constitute 334,600 lines of C source code [19]. Engblom's static analysis discovered several significant differences between the static properties of the SPECint95 benchmark and the 13 embedded programs. Noted differences include

- Sizes of variables: Embedded programs carefully control the sizes of variables to minimize memory usage. Workstation-oriented software like SPECint95 does not limit variable size nearly as much because workstation memory is relatively plentiful.
- Unsigned data is more common in embedded code.
- Logical (as opposed to arithmetic) operations occur more frequently in embedded code.
- Many embedded functions only perform side effects (such as flipping an I/O bit). They do not return values.
- Embedded code employs global data more frequently for variables and for large constant data.
- Embedded programs rarely use dynamic memory allocation.

Engblom's observations underscore the maxim that the best benchmark code is always the actual target application code.

12.6.6.2 BDTI: Berkeley Design Technology Inc.

Berkeley Design Technology Inc. (BDTI) [20] is a benchmarking and consulting company specializing in digital signal processing applications such as cellular telephones and multimedia products. The company provides analysis and advice that help companies develop, market, and use technologies for these applications. Founded in 1991, BDTI is known worldwide for its benchmarks and analysis. The company has developed a number of benchmark suites that are used to evaluate the signal-processing capabilities of a variety of processing engines including processor chips, licensable processor cores, massively parallel devices, and FPGAs.

BDTI's benchmark suites are used by SoC and system developers to evaluate and compare the speed, energy efficiency, cost–performance, and memory efficiency of competing cores and chips. BDTI benchmark results are published only after the benchmark implementations have been thoroughly reviewed and certified by BDTI.

BDTI's benchmark suites include the following:

- BDTI DSP Kernel Benchmarks^{TM}: These benchmarks are a suite of 12 key DSP algorithms (such as an FIR filter, FFT, and Viterbi decoder). This suite is used to assess processors' overall digital signal processing capabilities, especially for applications such as baseband communications, speech, and audio processing. These benchmarks are implemented on each processor using carefully optimized code, as is common practice for these types of algorithms in real systems. A processor's results on the BDTI DSP Kernel Benchmarks are used to generate the BDTImark2000^{TM} score, a composite DSP speed metric. BDTImark2000 scores—along with related cost/performance, energy efficiency, and memory-use metrics—are published on BDTI's Web site (www.BDTI.com). Figure 12.11 shows example BDTImark2000 results for a range of licensable processor cores.
- BDTI Video Kernel Benchmarks^{TM}: These benchmarks include key video-oriented algorithms, such as motion compensation and deblocking. They are useful for predicting a processor's performance in a variety of video- and imaging-oriented applications such as set-top boxes, multimedia-enabled cell phones, surveillance cameras, and video-conferencing systems.
- BDTI Communications Benchmark (OFDM)^{TM}: This benchmark is an application-oriented benchmark based on an orthogonal frequency division multiplexing (OFDM) receiver. It is designed to be representative of the processing found in communications equipment for applications such as DSL, cable modems, and wireless systems. To enable quick, realistic comparisons between chips, BDTI publishes high-capacity (maximum BDTIchannels) and low-cost ($/BDTIchannel) benchmark scores free of charge.
- BDTI Video Encoder and Decoder Benchmarks^{TM}: These two benchmarks are proprietary video compression/decompression algorithms. They are loosely based on the H.264 standard and are representative of the video processing workloads found in applications such as set-top boxes, multimedia-enabled cell phones, personal media players, surveillance cameras, and video conferencing systems. They are designed to model the computationally demanding aspects of video encoding and decoding while limiting complexity in order to reduce benchmarking effort.
- BDTI Solution Certification^{TM} for H.264 Decoding: BDTI certifies the performance of H.264 decoder solutions using standardized parameters (such as frame size and bit rate) and input data. This certification allows solution vendors and system designers to make accurate comparisons of multimedia solution performance.

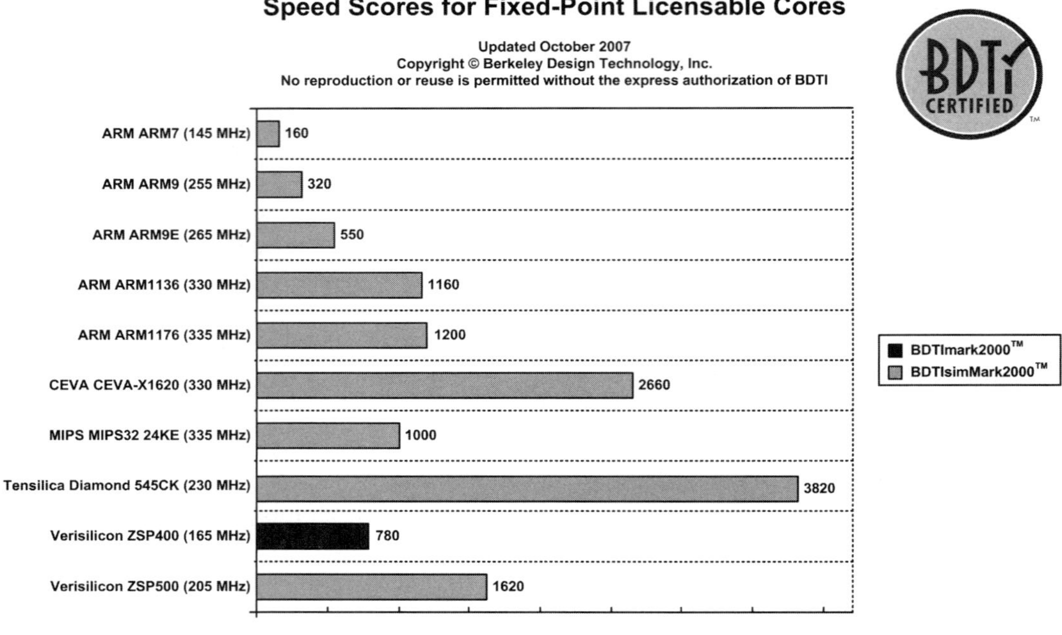

FIGURE 12.11   The BDTImark2000™/BDTIsimMark2000™ provide a summary measure of signal processing speed. (For more information and scores see www.BDTI.com. Scores copyright 2008 BDTI.)

12.6.6.3   EEMBC: Embedded Microprocessor Benchmark Consortium

EDN magazine's legacy of microprocessor benchmark articles grew into full-blown realization when *EDN* editor Markus Levy founded a nonprofit, embedded-benchmarking organization in 1997. He named the organization EEMBC (EDN Embedded Benchmark Consortium, later dropping the "EDN" from its name but not the corresponding "E" from its abbreviation). EEMBC's stated goal was to produce accurate and reliable metrics based on real-world embedded applications for evaluating embedded processor performance. Levy drew remarkably broad industry support from microprocessor and DSP vendors for his concept of a benchmarking consortium, which had picked up 21 founding corporate members by the end of 1998: Advanced Micro Devices, Analog Devices, ARC, ARM, Hitachi, IBM, IDT, Lucent Technologies, Matsushita, MIPS, Mitsubishi Electric, Motorola, National Semiconductor, NEC, Philips, QED, Siemens, STMicroelectronics, Sun Microelectronics, TI, and Toshiba [21]. EEMBC (pronounced "embassy") spent nearly 3 years working on a suite of benchmarks for testing embedded microprocessors and introduced its first benchmark suite at the Embedded Processor Forum in 1999. EEMBC released its first certified scores in 2000 and, during the same year, announced that it would start to certify benchmarks run on simulators so that processor cores could be benchmarked. As of 2004, EEMBC has more than 50 corporate members.

The EEMBC benchmarks are contained in six suites loosely grouped according to application. The six suites are automotive/industrial, consumer, Java GrinderBench, networking, office automation, and telecom. Each of the suites contains several benchmark programs that are based on small, derived kernels extracted from application code. All EEMBC benchmarks are written in C, except for the benchmark programs in the Java benchmark suite, which are written in Java. The six benchmark suites and descriptions of the programs in each suite appear in Tables 12.8 through 12.13.

TABLE 12.8   EEMBC Automotive/Industrial Benchmark Programs

EEMBC Automotive/Industrial Benchmark Name	Benchmark Description	Example Application
Angle to Time Conversion	Compute engine speed and crankshaft angle	Automotive engine control
Basic Integer and Floating Point	Calculate arctangent from telescoping series	General purpose
Bit Manipulation	Character and pixel manipulation	Display control
Cache "Buster"	Long sections of control algorithms with pointer manipulation	General purpose
CAN Remote Data Request	Controller Area Network (CAN) communications	Automotive networking
Fast Fourier Transform (FFT)	Radix-2 decimation in frequency, power spectrum calculation	DSP
Finite Impulse Response (FIR) Filter	High- and low-pass FIR filters	DSP
Inverse Discrete Cosine Transform (iDCT)	iDCT using 64-bit integer arithmetic	Digital video, graphics, image recognition
Inverse FFT (iFFT)	Inverse FFT performed on real and imaginary values	DSP
Infinite Impulse Response (IIR) Filter	Direct-for II, N-cascaded, second-order IIR filter	DSP
Matrix Arithmetic	LU decomposition of NxN input matrices, determinant computation, cross product	General purpose
Pointer Chasing	Search of a large, doubly linked list	General purpose
Pulse Width Modulation (PWM)	Generate PWM signal for stepper-motor control	Automotive actuator control
Road Speed Calculation	Determine road speed from successive timer/counter values	Automotive cruise control
Table Lookup and Interpolation	Derive function values from sparse 2D and 3D tables	Automotive engine control, antilock brakes
Tooth to Spark	Fuel-injection and ignition-timing calculations	Automotive engine control

TABLE 12.9   EEMBC Consumer Benchmark Programs

EEMBC Consumer Benchmark Name	Benchmark Description	Example Applications
High Pass Grey-Scale Filter	2D array manipulation and matrix arithmetic	CCD and CMOS sensor signal processing
JPEG	JPEG image compression and decompression	Still-image processing
RGB to CMYK Conversion	Color-space conversion at 8 bits/pixel	Color printing
RGB to YIQ Conversion	Color-space conversion at 8 bits/pixel	NTSC video encoding

TABLE 12.10   EEMBC GrinderBench Java 2 Micro Edition (J2ME) Benchmark Programs

EEMBC GrinderBench for Java 2 Micro Edition (J2ME) Benchmark Name	Benchmark Description	Example Applications
Chess	Machine-vs.-machine chess matches, 3 games, 10 moves/game	Cell phones, PDAs
Cryptography	DES, DESede, IDEA, Blowfish, Twofish encryption and decryption	Data security
kXML	XML parsing, DOM tree manipulation	Document processing
ParallelBench	Multithreaded operations with merge-sort and matrix multiplication	General purpose
PNG Decoding	PNG image decoding	Graphics, web browsing
Regular Expression	Pattern matching and file I/O	General purpose

TABLE 12.11   EEMBC Networking Benchmark Programs

EEMBC Networking 2.0 Benchmark Name	Benchmark Description	Example Applications
IP Packet Check	IP header validation, checksum calculation, logical comparisons	Network router, switch
IP Network Address Translator (NAT)	Network-to-network address translation	Network router, switch
OSPF version 2	Open shortest path first/Djikstra shortest path first algorithm	Network routing
QoS	Quality of service network bandwidth management	Network traffic flow control

TABLE 12.12　EEMBC Office Automation Benchmark Programs

EEMBC Office Automation Benchmark Name	Benchmark Description	Example Applications
Dithering	Grayscale to binary image conversion	Color and monochrome printing
Image Rotation	90° image rotation	Color and monochrome printing
Text Processing	Parsing of an interpretive printing control language	Color and monochrome printing

TABLE 12.13　EEMBC Telecom Benchmark Programs

EEMBC Telecom Benchmark Name	Benchmark Description	Example Applications
Autocorrelation	Fixed-point autocorrelation of a finite-length input sequence	Speech compression and recognition, channel and sequence estimation
Bit Allocation	Bit-allocation algorithm for DSL modems using DMT	DSL modem
Convolutional Encoder	Generic convolutional coding algorithm	Forward error correction
FFT	Decimation in time, 256-point FFT using Butterfly technique	Mobile phone
Viterbi Decoder	IS-136 channel decoding using Viterbi algorithm	Mobile phone

EEMBC's benchmark suites with their individual benchmark programs allow designers to select the benchmarks that are relevant to a specific design, rather than lumping all of the benchmark results into one number. EEMBC's benchmark suites are developed by separate subcommittees, each working on one application segment. Each subcommittee selects candidate applications that represent the application segment and dissects the each application for the key kernel code that performs the important work. This kernel code coupled with a test harness becomes the benchmark. Each benchmark has published guidelines in an attempt to force the processor vendors to play fair.

However, the industry's Dhrystone experience proved conclusively that some processor vendors would not play fair without a fair and impartial referee, so EEMBC created one: the EEMBC Technology Center (ETC). Just as major sports leagues hire referee organizations to conduct games enforce the rules of play, ETC conducts benchmark tests and enforce the rules of EEMBC benchmarking. Although a processor vendor ports the EEMBC benchmark code and test harnesses to its own processor and runs the initial benchmark tests, only ETC can certify the results. ETC takes the vendor-supplied code, test harness, and a description of the test procedures used and then certifies the benchmark results by inspecting the supplied code and rerunning the tests. In addition to code inspection, ETC makes changes to the benchmark code (such as changing variable names) to counteract "overoptimized" compilers. Vendors cannot publish EEMBC benchmark scores without ETC certification.

There is another compelling reason for an EEMBC benchmark referee: EEMBC rules allow for two levels of benchmarking play (the major league and the minors, to continue the sports analogy). The lower level of EEMBC "play" produces "out-of-the-box" scores. Out-of-the-box EEMBC benchmark tests can use any compiler (in practice, the compiler selected has changed the performance results by as much as 40%) and any selection of compiler switches but cannot modify the benchmark source code. The "out-of-the-box" results therefore give a fair representation of the abilities of the processor/compiler combination without adding in programmer creativity as a wild card. The higher level of EEMBC play is called "full-fury." Processor vendors seeking to improve their full-fury EEMBC scores (posted as "optimized" scores on the EEMBC Web site) can use hand-tuned code, assembly-language subroutines, special libraries, special CPU instructions, coprocessors, and other hardware accelerators. Full-fury scores tend to be much better than out-of-the-box scores, just as application-optimized production code generally runs much faster than code that has merely been run through a compiler. The free-for-all nature of EEMBC full-fury benchmarking rules underscores the need for a benchmark referee.

EEMBC does not publish its benchmark code. Processor vendors gain access to the EEMBC benchmark source code only by joining EEMBC, so EEMBC benchmark results are available for microprocessors and microprocessor cores supplied only by companies that have joined the consortium. With permission from the processor vendor, EEMBC will publish certified EEMBC benchmark-suite scores on the EEMBC Web site (www.eembc.org). Many processor vendors do allow EEMBC to publish their processors' scores and the EEMBC site already lists more than 300 certified benchmark-suite scores for a wide range of microprocessors and microprocessor cores.

Vendors need not submit their processors for testing with every EEMBC benchmark suite and some vendors do not allow their products' scores to be published at all although they may share the certified results privately with prospective customers. Vendors tend to publish information about the processors that do well and allow poorer-performing processors to remain shrouded in anonymity.

EEMBC's work is ongoing. To prevent benchmark rot, EEMBC's subcommittees constantly evaluate revisions to the benchmark suites. The Networking suite is already on version 2.0 and the Consumer suite is undergoing revision as of the writing of this chapter. EEMBC has also created a benchmark called EnergyBench that focuses on measuring power dissipation and energy consumption.

Further, EEMBC has recently turned its attention to multicore processors. Multicore processor performance depends on many factors and a good multicore benchmark must test all of these factors. Demonstrating the many facets and complexities of a multicore implementation requires the use of multiple multicore benchmark programs. Multicore benchmarks must target two fundamental types of concurrency: data throughput and computational throughput. Benchmarks that analyze data throughput show how well a multicore solution scales with data-input increases. This analysis can be accomplished by duplicating a computation across multiple processors in the multicore and then applying multiple data sets to the multicore. Real-world examples of this type of problem include the decoding of several different JPEG images (as might occur when viewing a Web page), decoding multichannel audio, or a multichannel VoIP application. The benchmark can be constructed to determine where performance begins to degrade as the volume of input data increases. One big challenge when developing such a benchmark is that the code must be threadsafe—it must support simultaneous execution of multiple threads. In particular, the benchmark must satisfy the need for simultaneous access to the same shared data by multiple threads.

To demonstrate computational throughput, this same approach can be extended by developing tests that initiate different tasks at the same time, implementing concurrency over both the data and the code. This test demonstrates the scalability of a solution for general-purpose processing. As an example, consider the execution of an MPEG decoder followed by an MPEG encoder, which is what you might find in a set-top box where the satellite signal is received, decoded, and then reencoded for hard-disk storage.

This sort of benchmark can be taken further using data decomposition, which divides an algorithm into multiple threads that work on a common data set, demonstrating support for fine-grained parallelism. In this situation, the algorithm might process a single audio and video data stream, but the code can be split to distribute the workload among different threads running on different processor cores. The benchmark distributes these threads based on the number of available processor cores.

EEMBC's multicore benchmark software initially supports symmetrical multicore processors with shared memory and uses a thread-based API to establish a common programming model. To implement this benchmarking strategy, EEMBC has a patent-pending test harness that communicates with the benchmark through an abstraction layer that is analogous to an algorithm wrapper. This test harness allows testing of a wide variety of thread-enabled workloads.

12.6.7 Modern Processor Benchmarks from Academic Sources

The EEMBC benchmarks have become the de facto industry standard for measuring benchmark performance but two aspects of the EEMBC benchmarks make them less than ideal for all purposes. The EEMBC benchmark code is proprietary and secret. Therefore it is not open to evaluation, analysis, and criticism by industry analysts, journalists, and independent third parties. It is also not available to designers who want to perform their own microprocessor evaluations. Such independent evaluations would be unnecessary if all the EEMBC corporate members tested their processors and published their results. However, fewer than half of the EEMBC corporate members have published any benchmark results for their processors and few have tested all of their processors [22].

12.6.7.1 UCLA's MediaBench 1.0

There are two sets of university-developed microprocessor benchmarks that can somewhat fill the gaps left by EEMBC. The first is MediaBench II. The original MediaBench benchmark was developed by the Computer Science and Electrical Engineering Departments at the University of California at Los Angeles (UCLA) [23]. The MediaBench 1.0 benchmark suite was created to explore compilers' abilities to exploit ILP (instruction-level parallelism) in processors with very long instruction word (VLIW) and SIMD structures. It was targeted at microprocessors and compilers that target new-medial applications and consists of several applications culled from image-processing, communications, and DSP applications and a set of input test data files to be used with the benchmark code.

Since it first appeared, a group of university professors formed a consortium to further develop MediaBench as a research-oriented microprocessor benchmarking suite. The result is MediaBench II. (Pointers to all of the MediaBench II files are located at http://euler.slu.edu/~fritts/mediabench/.)

12.6.7.2 MiBench from the University of Michigan

A more comprehensive benchmark suite called MiBench was presented by its developers from the University of Michigan at the IEEE's Fourth Annual Workshop on Workload Characterization in December, 2001 [24]. MiBench intentionally mimics EEMBC's benchmark suite. It includes a set of 35 embedded applications in 6 application-specific categories: automotive and industrial, consumer devices, office automation, networking, security, and telecommunications. The MiBench benchmarks are available at http://www.eecs.umich.edu/mibench/.

Both MediaBench and MiBench solve the problem of proprietary code for anyone that wants to conduct a private set of benchmark tests by providing a standardized set of benchmark tests at no cost and with no use restrictions. Industry analysts, technical journalists, and researchers can publish the results of independent tests conducted with these benchmarks although none seems to have done so, to date. The trade-off made with these processor benchmarks from academia is the absence of an official body that enforces benchmarking rules and provides result certification. For someone conducting their own benchmark tests, self-certification may be sufficient. For anyone else, the entity publishing the results must be scrutinized for fairness in the conducting of tests, for bias in test comparisons among competing processors, and for bias in any conclusions.

12.6.8 Configurable Processors and the Future of Processor Core Benchmarks

For the past 20 years, microprocessor benchmarks have attempted to show how well specific microprocessor architectures work. As their history demonstrates, the benchmark programs that have been developed have had mixed success in achieving this goal. One thing that has been constant over the years is the use of benchmarks to compare microprocessors and microprocessor cores with fixed ISAs. Nearly all microprocessors realized in silicon have fixed architectures and the microprocessor cores available for use in ASICs and SOCs also had fixed architectures. However, the transmutable silicon of ASICs and SOCs makes it feasible to employ microprocessor cores with configurable architectures instead of fixed ones. Configurable processor architectures allow designers to add new instructions and registers that boost the processor's performance for specific, targeted applications.

Use of configurable processor cores has a profound impact on the execution speed of any program, including benchmark programs.

Processor vendors take two fundamental approaches to making configurable-ISA processors available to IC designers. The first approach provides tools that allow the designer to modify an existing base processor architecture to boost performance on a target application or a set of applications. Companies taking this approach include ARC International, MIPS Technologies, and Tensilica Inc. The other approach employs tools that compile entirely new processor architectures for each application. Companies taking this approach include ARM, CoWare/LISATek, Silicon Hive, and Target Compilers.

The ability to tailor a processor's ISA to a target application (which includes benchmarks) can drastically increase execution speed of that program. For example, Figure 12.12 shows EEMBC Consumer

ConsumerBench 1.1 Simulated Benchmark Scores

Close

All Benchmark Scores are Certified by EEMBC to ensure credibility

Processor Name–Clock	Tensilica Xtensa T1050–260	Tensilica Xtensa T1050–260
Consumermark™	2.02256 (at 1 Mhz)	.08696 (at 1 Mhz)
Certification Report	View report to see composite energy scores and code/data sizes	View report to see composite energy scores and code/data sizes
Type of Platform	Simulator	Simulator
Type of Certification	Optimized	Out-of-the-box
Certification Date	08/01/02	07/27/02
Benchmark Notes	Optimized using 'C' only, no assembly.	
Simulator Type	high level cycle accurate	high level cycle accurate
Native Data Type	32	32
Architecture Type	Configurable, Extensible RISC	RISC
L1 Instruction Cache Size (kbyte)	16	16
L1 Data Cache Size (kbyte)	16	16
External Data Bus Width (Bits)	128	128
Memory Configuration	6-1-1-1 for instructions, 7-1-1-1 for data	6-1-1-1 for instructions, 7-1-1-1 for data
L2 Cache Size (kbyte)	0	0
L2 Cache Clock	none	none
Compiler Model and Version	Xtensa C Compiler T1050	Xtensa C Compiler T1050
Floating Point	software	software
Benchmark Scores	Iterations *per million cycles*	Iterations *per million cycles*
JPEG Compression	0.2200	0.0480
JPEG Decompression	0.3200	0.0684
High Pass Grayscale image filter	26.980	0.4514
RGB to CMYK Conversion	26.010	0.5025
RGB to YIQ Conversion	34.670	0.3383

Highlighted Fields indicate Certified Data

Performance is represented in Iterations/Second for Production Silicon and in Iterations/Million Cycles for Simulators; where bigger is better. To view the Code Size, Data Size and Energy data, click Click 'View Report'.

Simulator Benchmarks: These $certified benchmark scores were derived solely from simulation. Certification means the scores were recreated on the simulator by EEMBC's Certification process. The vendor has stipulated that the simulator has been verified against RTL.

FIGURE 12.12 Tailoring a processor core's architecture for a specific set of programs can result in significant performance gains. (Copyright EEMBC.)

TABLE 12.14   Optimized vs. Out-of-the-Box EEMBC Consumer Benchmark Scores

EEMBC Consumer Benchmark Name	Optimized Xtensa T1050 Score (Iterations/M Cycles)	Out-of-the-Box Xtensa T1050 Score (Iterations/M Cycles)	Performance Improvement
Compress JPEG	0.2200	0.0480	4.58×
Decompress JPEG	0.3200	0.0684	4.68×
High Pass Gray-Scale Filter	26.980	0.4514	59.77×
RGB to CMYK Conversion	26.010	0.5025	51.76×
RGB to YIQ Conversion	34.670	0.3383	102.48×

benchmark suite scores for Tensilica's configurable Xtensa V microprocessor core (labeled Xtensa T1050 in Figure 12.12). The right-hand column of Figure 12.12 shows benchmark scores for a standard processor configuration. The center column shows the improved benchmark scores for a version of the Xtensa V core that has been tailored by a design engineer specifically to run the EEMBC Consumer benchmark programs. Table 12.14 summarizes the performance results of the benchmarks for the stock and tailored processor configurations.

Note that the stock Xtensa V processor core performs on par with or slightly faster than other 32-bit RISC processor cores and that tailoring the Xtensa V processor core for the individual tasks in the EEMBC Consumer benchmark suite by adding new instructions that are matched to the needs of the benchmark computations boosts the processor's performance on the individual benchmark programs from 4.6× to more than 100×. The programmer modifies the benchmark or target application code by adding C intrinsics that use these processor extensions. The resulting performance improvement is similar to results that designers might expect to achieve when tailoring a processor for their target application code so in a sense, the EEMBC benchmarks are still performing exactly as intended. The optimized Xtensa V EEMBC Consumer benchmark scores reflect the effects of focusing three months worth of a engineering graduate student's "full-fury" to improve the Xtensa V processor's benchmark performance optimizing both the processor and the benchmark code.

Evolution in configurable-processor technology is allowing the same sort of effects to be achieved on EEMBC's out-of-the-box benchmark scores. In 2004, Tensilica introduced the XPRES Compiler, a processor-development tool that analyzes C/C++ code and automatically generates performance-enhancing processor extensions from its code analysis. Tensilica's Xtensa Processor Generator accepts the output from the XPRES Compiler and automatically produces the specified processor and a compiler that recognizes the processor's architectural extensions as native instructions and automatically generates object code that uses these extensions. Under EEMBC's out-of-the-box benchmarking rules, results from code compiled with such a compiler and run on such a processor qualify as out-of-the-box (not optimized) results. The results of an EEMBC Consumer benchmark run using such a processor appear in Figure 12.13. This figure compares the performance of a stock Xtensa V processor core configuration with that of an Xtensa LX processor core that has been extended using the XPRES Compiler. (Note: Figure 12.13 compares the Xtensa V and Xtensa LX microprocessor cores running at 260 and 330 MHz, respectively, but the scores are reported on a per-MHz basis, canceling out any performance differences attributable to clock rate. Also, the standard versions of the Xtensa V and Xtensa LX processor cores have the same ISA and produce the same per-MHz benchmark results.)

The XPRES Compiler boosted the stock Xtensa processor's performance by 1.17× to 18.7×, as reported under EEMBC's out-of-the-box rules. The time required to effect this performance boost was 1 h. These performance results demonstrate the sort of performance gains designers might expect from automatic processor configuration. However, these results also suggest that benchmark performance comparisons become even more complicated with the introduction of configurable processor technology and that the designers making comparisons of processors using such benchmarks must be especially careful to understand how the benchmarks tests for each processor are conducted. Other commercial offerings of configurable-processor technology include ARM's OptimoDE, Target Compilers' Chess/Checkers, and CoWare's LISATek compiler; however, as far as could be determined, there are no published benchmark results available for these products.

ConsumerBench 1.1 Simulated Benchmark Scores

Close

All Benchmark Scores are Certified by EEMBC to ensure credibility

Processor Name-Clock	Tensilica Xtensa LX 330 MHz	Tensilica Xtensa T1050-260
Consermark™	.51997 (at 1 Mhz)	.08696 (at 1 Mhz)
Certification Report	View report to see composite energy scores and code/data sizes	View report to see composite energy scores and code/data sizes
Type of Platform	Simulator	Simulator
Type of Certification	Out-of-the-box	Out-of-the-box
Certification Date	03/30/04	07/27/02
Benchmark Notes		
Simulator Type	Cycle accurate instruction set simulator	high level cycle accurate
Native Data Type	32-bit	32
Architecture Type	Configurable RISC with FLIX technology	RISC
L1 Instruction Cache Size (kbyte)	32	16
L1 Data Cache Size (kbyte)	32	16
External Data Bus Width (Bits)	128	128
Memory Configuration	1 cycle response on system bus	6-1-1-1 for instructions, 7-1-1-1 for data
L2 Cache Size (kbyte)	n/a	0
L2 Cache Clock	n/a	none
Compiler Model and Version	Xtensa C/C++ Compiler, v6.0	Xtensa C Compiler T1050
Floating Point	software	software
Benchmark Scores	Iterations *per million cycles*	Iterations *per million cycles*
JPEG Compression	0.0560	0.0480
JPEG Decompression	0.0830	0.0684
High Pass Grayscale image filter	8.661	0.4514
RGB to CMYK Conversion	7.572	0.5025
RGB to YIQ Conversion	6.310	0.3383

Highlighted Fields indicate Certified Data

Performance is represented in Iterations/Second for Production Silicon and in Iterations/Million Cycles for Simulators; where bigger is better. To view the Code Size, Data Size and Energy data, click Click 'View Report'.

Simulator Benchmarks: These $certified benchmark scores were derived solely from simulation. Certification means the scores were recreated on the simulator by EEMBC's Certification process. The vendor has stipulated that the simulator has been verified against RTL.

FIGURE 12.13   Tensilica's XPRES Compiler can produce a tailored processor core with enhanced performance for a target application program in very little time. (Copyright EEMBC.)

12.7   Conclusion

One or more microprocessors form the heart of every embedded system. Vendors offer a variety of different processor architectures targeted at the many different embedded applications. With the migration of embedded hardware onto SOCs, the need for fixed processor ISAs has faded, allowing embedded designers to tailor ISAs for specific on-chip applications. While the first 30 years of microprocessor development emphasized faster clock rates to achieve more performance, it is no longer possible to use the crutch of higher clock rates to get more execution speed because of excessive power dissipation. Consequently, the twenty-first-century quest for performance will focus on multicore embedded designs using tens or hundreds of processors, but the microprocessor will still remain at the heart of every embedded system.

References

1. J. Cocke and V. Markstein, The evolution of RISC technology at IBM, *IBM Journal of Research and Development*, 44:1/2, January/March 2000, 48–55.
2. D.A. Patterson and D.R. Ditzel, The case for the reduced instruction set computer, *Computer Architecture News* 8:6, October 1980, 25–33.
3. D.R. Ditzel and D.A. Patterson, Retrospective on high-level language computer architecture, *Proceedings of the 7th Annual Symposium on Computer Architecture*, La Baule, France, June 1980, pp. 97–104.
4. J.L. Hennessy and D.A. Patterson, *Computer Architecture: A Quantitative Approach*, 3rd edn, Elsevier Morgan Kaufmann, San Francisco, CA, 2003, pp. 151–154.
5. www.edn.com, search for "Microprocessor Directory" and select the latest online version.
6. D. Maydan and S. Leibson, Optimize C programs for embedded SOC apps, *Electronic Engineering Times Asia*, October 1–15, 2006, www.eetasia.com.
7. J. Rowson, Hardware/software co-simulation, *Proceedings of the 31st Design Automation Conference (DAC'94)*, San Diego, CA, June 1994.
8. S. Gill, The diagnosis of mistakes in programmes on the EDSAC, *Proceedings of the Royal Society Series A Mathematical and Physical Sciences*, Cambridge University Press, London and New York, May 22, 1951, pp. 538–554.
9. L. Kelvin (W. Thomson), *Popular Lectures and Addresses*, Vol. 1, p. 73 (originally: *Lecture to the Institution of Civil Engineers*, May 3, 1883).
10. F. McMahon, The Livermore FORTRAN Kernels: A computer test of the numerical performance range, Technical Report, Lawrence Livermore National Laboratory, Livermore, CA, December 1986.
11. J.J. Dongarra, LINPACK Working Note #3, FORTRAN BLAS Timing, Argonne National Laboratory, Argonne, IL, November 1976.
12. J.J. Dongarra, J.R. Bunch, C.M. Moler, and G.W. Stewart, LINPACK Working Note #9, Preliminary LINPACK User's Guide, ANL TM-313, Argonne National Laboratory, Argonne, IL, August 1977.
13. H.J. Curnow and B.A. Wichmann, A Synthetic Benchmark, *Computer Journal*, 19:1, 1976, 43–49.
14. R.P. Weicker, Understanding variations in Dhrystone performance, Microprocessor Report, May 1989, pp. 16–17.
15. R.D. Grappel and J.E. Hemenway, A tale of four µPs: Benchmarks quantify performance, *EDN*, April 1, 1981, pp. 179–232.
16. S.H. Fuller, P. Shaman, D. Lamb, and W. Burr, Evaluation of computer architectures via test programs, *AFIPS Conference Proceedings*, Vol. 46, June 1977, pp. 147–160.
17. D. Shear, EDN's DSP benchmarks, *EDN*, September 29, 1988, pp. 126–148.
18. W. Patstone, 16-bit µP benchmarks—An update with explanations, *EDN*, September 16, 1981, p. 169.
19. J. Engblom, Why SpecInt95 should not be used to benchmark embedded systems tools, *Proceedings of the ACM SIGPLAN 1999 Workshop on Languages, Compilers, and Tools for Embedded Systems (LCTES'99)*, Atlanta, GA, May 5, 1999.
20. BDTI, Evaluating DSP Processor Performance, Berkeley Design Technology Inc. (BDTI), Berkeley, CA, 1997–2000.
21. M. Levy, At last: Benchmarks you can believe, *EDN*, November 5, 1988, pp. 99–108.
22. T. Halfhill, Benchmarking the benchmarks, Microprocessor Report, August 30, 2004.
23. C. Lee, M. Potkonjak, et al., MediaBench: A tool for evaluating and synthesizing multimedia and communications systems, MICRO 30, *Proceedings of the 30th Annual IEEE/ACM International Symposium on Microarchitecture*, Research Triangle Park, NC, December 1–3, 1997.
24. M. Guthaus, J. Ringenberg, et al., MiBench: A free, commercially representative embedded benchmark suite, *IEEE 4th Annual Workshop on Workload Characterization*, Austin, TX, December 2001.

13

System-on-Chip Design

Grant Martin

Tensilica Inc.

13.1 Introduction

"System-on-Chip" is a phrase that has been much bandied about for more than a decade ([1] is an early reference now 12 years old). It is more than a design style, more than an approach to the design of application-specific integrated circuits (ASICs), or application-specific standard parts (ASSPs), and more than a methodology. Rather, system-on-chip (SoC) represents a major revolution in IC design—a revolution enabled by the advances in process technology allowing the integration of all or most of the major components and subsystems of an electronic product onto a single chip, or integrated chipset [2]. This revolution in design has been embraced by many designers of complex chips, as the performance, power consumption, cost, and size advantages of using the highest level of integration made available have proven to be extremely important for many designs. In fact, the design and use of SoCs are arguably one of the key problems in designing real-time embedded systems.

The move to SoC began sometime in the mid-1990s. At this point, the leading CMOS-based semiconductor process technologies of 0.35 and 0.25 μm were sufficiently capable to allow the integration of many of the major components of a second-generation wireless handset or a digital set-top box onto a single chip. The digital baseband functions of a cellphone—a digital signal processor (DSP), hardware (HW) support for voice encoding and decoding, and a reduced instruction set computer (RISC) processor to handle protocol stacks and user interfaces—could all be placed onto a single die. Although such a baseband SoC was far from the complete cellphone electronics—there were major components such as the RF transceiver, analog power control, analog baseband, and passives which were not integrated—the evolutionary path with each new process generation, to integrate more and more onto a single die, was clear. Today's chipset would become tomorrow's chip. The problems of integrating hybrid technologies involved in making up a complete electronic system would be solved. Thus, eventually, SoC could encompass design components drawn from the standard and more adventurous domains of digital, analog, RF, reconfigurable logic, sensors, actuators, optical, chemical, microelectronic mechanical systems, and even biological and nanotechnology.

With this viewpoint of continued process evolution leading to ever-increasing levels of integration into ever-more-complex SoC devices, the issue of an SoC being a single *chip* at any particular point in time is somewhat moot. Rather, the word "system" in SoC is more important than "chip." What is most important about an SoC whether packaged as a single chip, or integrated chipset, (thus, System-on-*Chipset*) or System-in-Package or System-on-Package is that it is designed as an integrated *system*, making design trade-offs across the processing domains and across the individual chip and package boundaries.

However, by 2008 it is also clear that the most recent CMOS technologies −90 nm in large-scale usage, 65 nm growing in importance for leading edge designs, and 45 nm on the near horizon are allowing the levels of integration predicted more than a decade ago. For example, the single chip cellphone, although not necessarily the architectural choice for an advanced "smartphone", is now a reality, with Texas Instruments "ecosto" project (creating as an instance the OMAP V1035, which supports GSM, GPRS, and EDGE standards) being one prominent example [3]. Although a practical phone incorporating this device will have some additional components, this SoC is a leading example of the state of advanced system integration.

13.2 System-on-a-Chip

Let us define an SoC as a complex integrated circuit, or integrated chipset, which combines the major functional elements or subsystems of a complete end product into a single entity. These days, all interesting SoC designs include at least one programmable processor, and very often a combination of at least one RISC control processor and one DSP. Many SoC devices include multiple heterogeneous or homogeneous processors, organized in many different communications architectures, and some forecasts for the future include hundreds or more programmable processors or processing elements. They also include on-chip communications structures—processor buses, peripheral buses, and perhaps a high-speed system bus. Some experiments with network-on-chip (NoC) are extending the on-chip communications architecture further. Other chapters in this volume discuss on-chip interconnect and NoC. A hierarchy of on-chip memory units, and links to off-chip memory are important especially for SoC processors (cache, main memories, very often separate instruction and data caches are included). For most signal processing applications, some degree of HW-based accelerating functional units are provided, offering higher performance and lower energy consumption. For interfacing to the external, real world, SoCs include a number of peripheral processing blocks, and due to the analog nature of the real world, this may include analog components as well as digital interfaces (e.g., to system buses at a higher packaging level). Although there is much interesting research in incorporating MEMS-based sensors and actuators, and in SoC applications incorporating chemical processing (lab-on-a-chip), these are, with rare exceptions, research topics only. However, future SoCs of a commercial nature may include such subsystems, as well as optical communications interfaces.

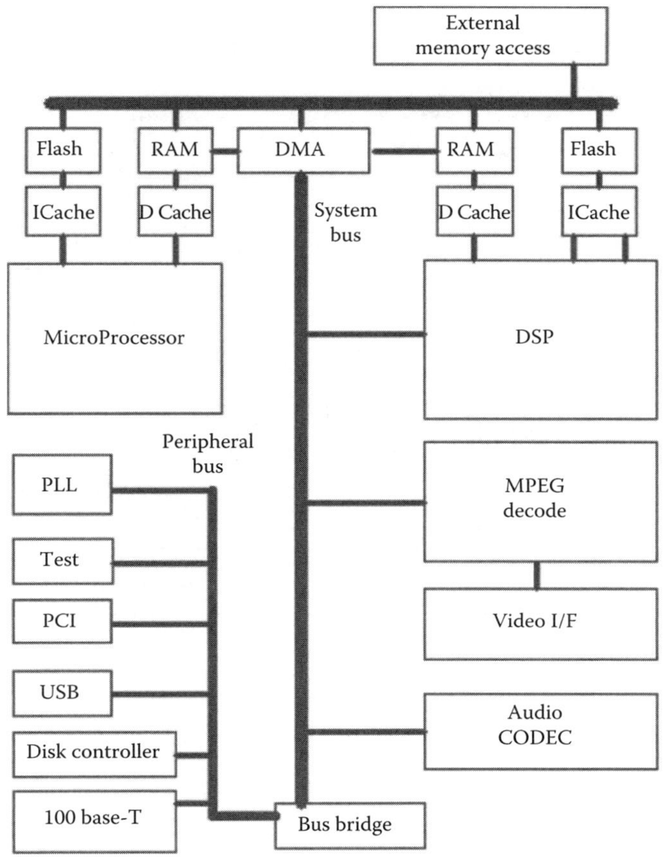

FIGURE 13.1 Typical early SoC device for consumer applications.

Figure 13.1 illustrates what a typical early SoC might contain for consumer applications. Note that it is a heterogeneous two-processor device with a RISC control processor and a DSP. This was typical of cellphone baseband processors in the late 1990s.

A more normal SoC of 2008 is illustrated in Figure 13.2. This is an example of a super 3G mobile phone, with 18 major processing blocks that could be a combination of many heterogeneous processors and dedicated HW blocks.

Another example of a modern SoC device is shown in Figure 13.3: A personal video recorder (PVR) SoC device. Here as many as seven processors (or more for subtask processing) might be used, or some combination of processors and dedicated HW blocks.

One key point about SoC which is often forgotten for those approaching it from an HW-oriented perspective is that all interesting SoC designs encompass both HW and software (SW) components, i.e., programmable processors, both fixed Instruction Set Architectures (ISA) and application-specific instruction set processors (ASIPs), real-time operating systems (RTOS), and other aspects of HW-dependent SW such as peripheral device drivers, as well as middleware stacks for particular application domains, and possibly optimized assembly code or HW-dependent C code for DSPs. Thus, the design and use of SoCs have moved from being originally a mostly HW-only concern, to being very much a SW-dominated concern, aspects of system-level design and engineering, HW–SW trade-off and partitioning decisions, and SW architecture, design, and implementation.

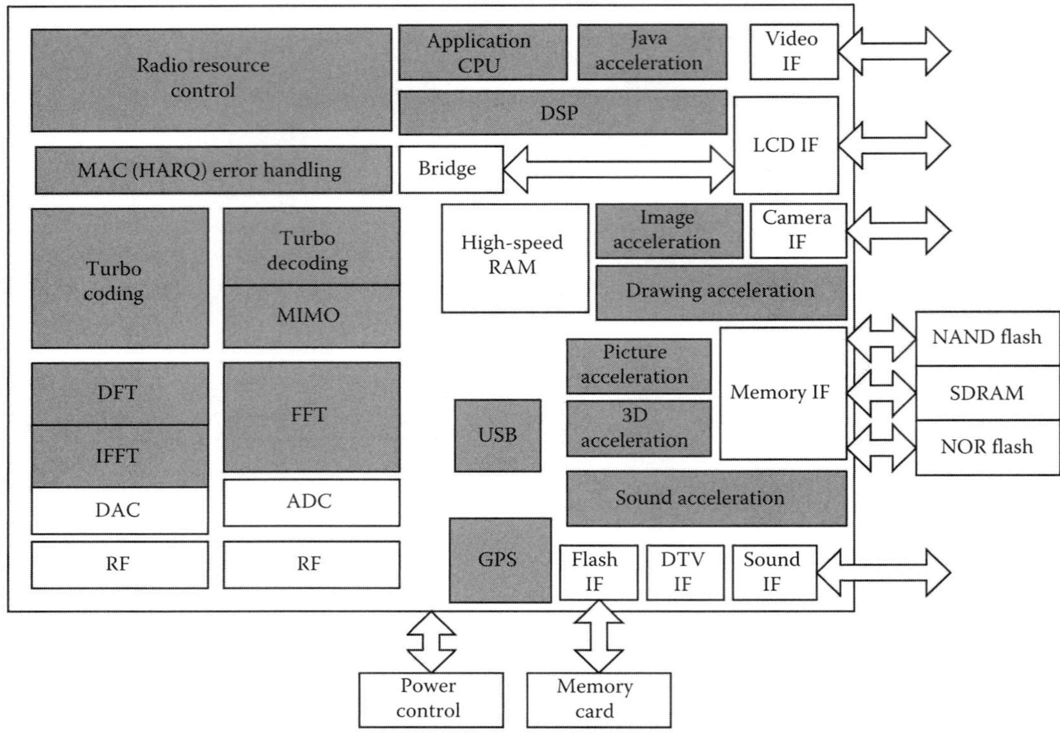

FIGURE 13.2 Example of a SoC device in 2008: Super 3G Mobile Phone.

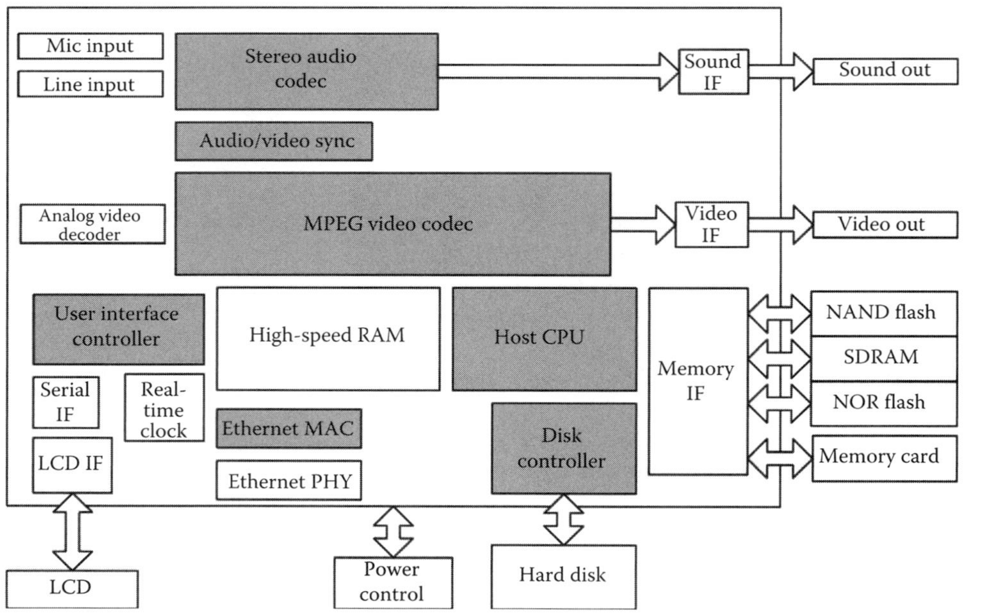

FIGURE 13.3 Example of SoC for a personal video recorder (PVR).

13.3   System-on-a-Programmable-Chip

Recently, attention has begun to expand in the SoC world from SoC implementations using custom, ASIC, or ASSP design approaches to include the design and use of complex reconfigurable logic parts with embedded processors and other application-oriented blocks of intellectual property (IP). These complex FPGAs (Field-Programmable Gate Arrays) are offered by several vendors, including Xilinx (originally the Virtex-II PRO Platform FPGA, but now supplemented by Virtex 4 and 5) and Altera (SOPC Builder for its Stratix FPGAs), but are referred to by several names: highly programmable SoCs, system-on-a-programmable-chip (SOPC), and embedded FPGAs. The key idea behind this approach to SoC is to combine large amounts of reconfigurable logic with embedded RISC processors (either custom laid-out, "hardened" blocks, or synthesizable processor cores, such as MicroBlaze for Xilinx and NIOS and NIOS-II for Altera), in order to allow very flexible and tailorable combinations of HW and SW processing to be applied to a particular design problem. Algorithms that consist of significant amounts of control logic, plus significant quantities of dataflow processing, can be partitioned into the control RISC processor (e.g., in Xilinx Virtex-II PRO, a PowerPC processor, and in the Virtex 4 family, up to two PowerPCs; on Virtex 5 platforms, many MicroBlaze soft core processors) and reconfigurable logic offering HW acceleration. Although the resulting combination does not offer the highest performance, lowest energy consumption, or lowest cost, in comparison with custom IC or ASIC/ASSP implementations of the same functionality, it does offer tremendous flexibility in modifying the design in the field and avoiding expensive Nonrecurring Engineering (NRE) charges in the design. Thus, new applications, interfaces, and improved algorithms can be downloaded to products working in the field using this approach.

Products in this area also include other processing and interface cores, such as multiply accumulate (MAC) blocks which are specifically aimed at DSP-type dataflow signal and image processing applications; other DSP processing blocks; and high-speed serial interfaces for wired communications such as SERDES (serializer/de-serializer) blocks. In this sense, SOPC SoCs are not exactly application specific, but not completely generic either.

It remains to be seen whether SOPC is going to be a successful way of delivering high volume consumer applications, or will end up restricted to the two main applications for high-end FPGAs: rapid prototyping of designs which will be retargeted to ASIC or ASSP implementations; and used in high-end, relatively expensive parts of the communications infrastructure which require in-field flexibility and can tolerate the trade-offs in cost, energy consumption, and performance. Certainly, the use of synthesizable processors on more moderate FPGAs to realize SoC style designs is one alternative to the cost issue. Intermediate forms, such as the use of metal-programmable gate-array style logic fabrics together with hard-core processor subsystems and other cores, which have been offered as "Structured ASICs" by several vendors such as LSI Logic in the past (RapidChip), and NEC (Instant Silicon Solutions Platform) represented an intermediate form of SoC between the full-mask ASIC and ASSP approach and the FPGA approach. Here the trade-offs were much slower design creation (a few weeks rather than a day or so); higher NRE than FPGA (but much lower than a full set of masks); and better cost, performance, and energy consumption than FPGA (perhaps 15%–30% worse than an ASIC approach). However, in general, structured ASICs have been unsuccessful for most vendors offering them, and most of them have been withdrawn from the market. A few companies remain offering them in 2008, including eASIC, AMI, Altera (HardCopy, converting FPGA designs to structured ASICs), and ChipX. In general, they represented too much of a compromise between a fully customized ASIC and a totally preconfigured platform SoC design to suit most users and applications.

13.4   IP Cores

The design of SoC would not be possible if every design started from scratch. In fact, the design of SoC depends heavily on the reuse of Intellectual Property blocks—what are called "IP Cores" or "IP blocks." IP reuse has emerged as a strong trend over the last decade [4] and has been one key element in closing what the International Technology Roadmap for Semiconductors [5] calls the "design productivity gap"—the difference between the rate of increase of complexity offered by advancing semiconductor process technology, and the rate of increase in designer productivity offered by advances in design tools and methodologies.

But, reuse is not just important to offer ways of enhancing designer productivity, although it has dramatic impacts on that. It also provides a mechanism for design teams to create SoC products that span multiple design disciplines and domains. The availability of both hard (laid-out and characterized) and soft (synthesizable) processor cores from a number of processor IP vendors allows design teams who would not be able to design their own processor from scratch to drop them into their designs and thus add RISC control and DSP functionality to an integrated SoC without having to master the art of processor design within the team. In this sense, the advantages of IP reuse go beyond productivity—it offers both a large reduction in design risk, and also a way for SoC designs to be done that would otherwise be infeasible due to the length of time it would take to acquire expertise and design IP from scratch.

This ability when acquiring and reusing IP cores—to acquire, in a prepackaged form, design domain expertise outside one's own design team's set of core competencies—is a key requirement for the evolution of SoC design going forward. SoC up to this point has concentrated to a large part on integrating digital components together, perhaps with some analog interface blocks that are treated as black boxes. The hybrid SoCs of the future, incorporating domains unfamiliar to the integration team, such as RF, MEMS, optoelectronic, or bioelectronic, requires the concept of "drop-in" IP to be extended to these new domains. We are not yet at that state—considerable evolution in the IP business and the methodologies of IP creation, qualification, evaluation, integration, and verification are required before we will be able to easily specify and integrate truly heterogeneous sets of disparate IP blocks into a complete hybrid SoC.

However, the same issues existed at the beginning of the SoC revolution in the digital domain. They have been solved to a large extent, through the creation of standards for IP creation, evaluation, exchange, and integration—primarily for digital IP blocks but extending also to analog/mixed-signal (AMS) cores. Among the leading organizations in the identification and creation of such standards has been the Virtual Socket Interface Alliance (VSIA) [6], formed in 1996 and having at its peak membership more than 200 IP, systems, semiconductor and Electronic Design Automation corporate members, and shutting down in 2007, while ensuring the transfer of its remaining standards work to the IEEE. Although often criticized over the years for a lack of formal and acknowledged adoption of its IP standards, VSIA has had a more subtle influence on the electronics industry. Many companies instituting reuse programs internally; many IP, systems, and semiconductor companies engaging in IP creation and exchange; and many design groups have used VSIA IP standards as a key starting point for developing their own standards and methods for IP-based design. In this sense, use of VSIA outputs has enabled a kind of IP reuse in the IP business.

VSIA, for example, in its early architectural documents of 1996–1997, helped define the widely adopted industry understanding of what it meant for an IP block to be considered to be in "hard" or "soft" form. Other important contributions to design included the well-read system level design model taxonomy created by one of its working groups. Its standards, specifications, and documents thus have represented a very useful resource for the industry.

Other important issues for the rise of IP-based design and the emergence of a third party industry in this area (which has taken much longer to emerge than originally hoped in the mid-1990s) are the business issues surrounding IP evaluation, purchase, delivery, and use. Organizations such as the Virtual Component Exchange (VCX) [7] emerged to look at these issues and provide solutions. The VCX did not last in this form, and indeed many of the early efforts to promote IP reuse, such as RAPID, VCX, and Design and Reuse, either shut down or transformed their work to new areas. For example, the VCX IP database and IP management SW tools were acquired in 2004 by Beach Solutions, and in 2006 was acquired by ChipEstimate, and then in 2008 Cadence Design Systems bought ChipEstimate. It is clear that the vast majority of IP business relationships between firms occur within a more ad hoc supplier to customer business framework.

13.5 Virtual Components

The VSIA has had a strong influence on the nomenclature of the SoC and IP-based design industry. The concept of the "virtual socket"—a description of all the design interfaces which an IP core must satisfy, and design models and integration information which must be provided with the IP core—required to allow it to be more easily integrated or "dropped into" an SoC design comes from the concept of Printed Circuit Board (PCB) design where components are sourced and purchased in prepackaged form and can be dropped into a board design in a standardized way. Of course, some of the interfaces and characteristics of the socket and components may be parameterized in defined ways.

The dual of the virtual socket then becomes the virtual component. Specifically in the VSIA context, but also more generally in the interface, an IP core represents a design block that might be reusable. A virtual component represents a design block which is intended for reuse, and which has been developed and qualified to be highly reusable. The things that separate IP cores from virtual components are in general

- Virtual components conform in their development and verification processes to well-established design processes and quality standards.
- Virtual components come with design data, models, associated design files, scripts, characterization information, and other deliverables which conform to one or other well-accepted standards for IP reuse, for example, the VSIA deliverables, or another internal or external set of standards.
- Virtual components in general should have been fabricated at least once, and characterized postfabrication to ensure that they have validated claims.
- Virtual components should have been reused at least once by an external design team, and usage reports and feedback should be available.
- Virtual components should have been rated for quality using an industry standard quality metric such as OpenMORE (originated by Synopsys and Mentor Graphics), the VSI Quality standard (which has OpenMORE as one of its inputs), or the recent Fabless Semiconductor Association—now Global Semiconductor Alliance (GSA) IP ecosystem Risk Assessment tool [8].

To a large extent, the developments over the last decade in IP reuse have been focused on defining the standards and processes to turn the ad hoc reuse of IP cores into a well-understood and reliable process for acquiring and reusing virtual components, thus enhancing the analogy with PCB design.

13.6 Platforms and Programmable Platforms

The emphasis in the preceding sections has been on IP (or virtual component) reuse on a somewhat ad hoc block by block basis in SoC design. Over the past several years, however, there has arisen a more integrated approach to the design of complex SoCs and the reuse of virtual components—what has been called "Platform-based design." This will be dealt with at much greater length in another chapter in this book. Much more information is available in Refs. [9–12]. Suffice it here to define platform-based design in the SoC context from one perspective.

We can define platform-based design as a planned design methodology that reduces the time and effort required, and risk involved, in designing and verifying a complex SoC. This is accomplished by extensive reuse of combinations of HW and SW IP. As an alternative to IP reuse in a block-by-block manner, platform-based design assembles groups of components into a reusable platform architecture. This reusable architecture, together with libraries of preverified and precharacterized, application-oriented HW and SW virtual components, is an SoC integration platform.

There are several reasons for the growing popularity of the platform approach in industrial design. These include the increase in design productivity, the reduction in risk, the ability to utilize preintegrated virtual components from other design domains more easily, and the ability to reuse SoC architectures created by experts. Industrial platforms include full application platforms, reconfigurable platforms, and processor-centric platforms [13]. Full application platforms, such as Philips Nexperia and TI OMAP provide a complete implementation vehicle for specific product domains [14]. Processor-centric platforms, such as ARM PrimeXsys concentrate on the processor, its required bus architecture and basic sets of peripherals, along with RTOS and basic SW drivers. Reconfigurable or "highly programmable" platforms such as the Xilinx Platform FPGA and Altera's SOPC deliver hardcore processors plus reconfigurable logic along with associated IP libraries and design tool flows.

13.7 Integration Platforms and SoC Design

The use of SoC integration platforms changes the SoC design process in two fundamental ways:

1. Basic platform must be designed, using whatever ad hoc or formalized design process for SoC that the platform creators decide on. The next section outlines some of the basic steps required to build an SoC, whether building a platform or using a block-based more ad hoc integration process. However, when constructing an SoC platform for reuse in derivative design, it is important to remember that it may not be necessary to take the whole platform and its associated HW and SW component libraries through complete implementation. Enough implementation must be done to allow the platform and its constituent libraries to be fully characterized and modeled for reuse. It is also essential that the platform creation phase produce in an archivable and retrievable form all the design files required for the platform and its libraries to be reused in a derivative design process. This must also include the setup of the appropriate configuration programs or scripts to allow automatic creation of a configured platform during derivative design.

2. Design process must be created and qualified for all the derivative designs that will be created based on the SoC integration platform. This must include processes for retrieving the platform from its archive, for entering the derivative design configuration into a platform configurator, the generation of the design files for the derivative, the generation of the appropriate verification environment(s) for the derivative, the ability for derivative design teams to select components from libraries, to modify these components and validate them within the overall platform context, and, to the extent supported by the platform, to create new components for their particular application.

Reconfigurable or highly programmable platforms introduce an interesting addition to the platform-based SoC design process [15]. Platform FPGAs and SOPC devices can be thought of as a "meta-platform": a platform for creating platforms. Design teams can obtain these devices from companies such as Xilinx and Altera, containing a basic set of more generic capabilities and IP-embedded processors, on-chip buses, special IP blocks such as MACs and SERDES, and a variety of other prequalified IP blocks. They can then customize the meta-platform to their own application space by adding application domain-specific IP libraries. Finally, the combined platform can be provided to derivative design teams, who can select the basic meta-platform and configure it within the scope intended by the intermediate platform creation team, selecting the IP blocks needed for their exact derivative application. More on platform-based design will be found in another chapter in this book.

13.8 Overview of the SoC Design Process

The most important thing to remember about SoC design is that it is a multidisciplinary design process, which needs to exercise design processes from across the spectrum of electronics. Design teams must gain some fluency with all these multiple disciplines, but the integrative and reuse nature of SoC design means that they may not need to become deep experts in all of them. Indeed, avoiding the need for designers to understand all methodologies, flows, and domain-specific design techniques is one of the key reasons for reuse and enablers of productivity. Nevertheless, from DFT through digital and analog HW design, from verification through system level design, from embedded SW through IP procurement and integration, and from SoC architecture through IC analysis, a wide variety of knowledge is required by the team, if not every designer.

Figure 13.4 illustrates some of the basic constituents of the SoC design process.

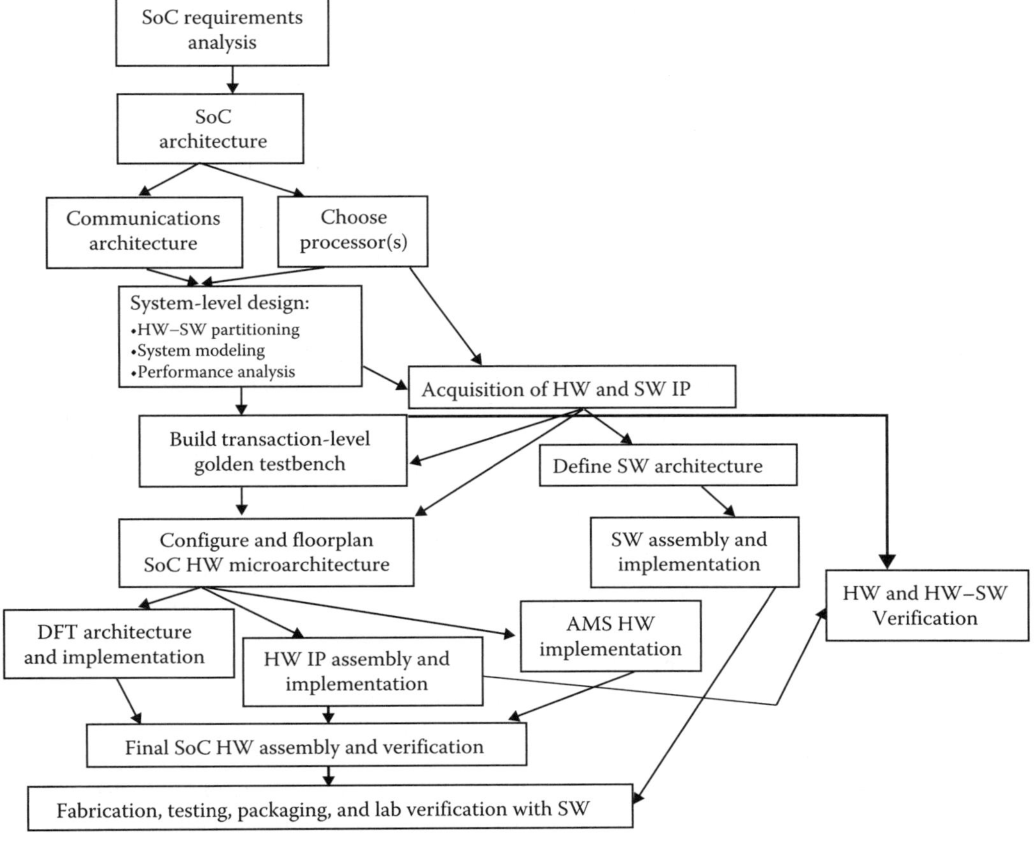

FIGURE 13.4 Steps in the SoC design process.

We will now define each of these steps as illustrated:

- SoC requirement analysis is the basic step of defining and specifying a complex SoC based on the needs of the end product into which it will be integrated. The primary input into this step is the marketing definition of the end product and the resulting characteristics of what the SoC should be: both functional and nonfunctional (e.g., cost, size, energy consumption, performance: latency and throughput, package selection). This process of requirements analysis must ultimately answer the questions: Is the product feasible? Is the desired SoC feasible to design, and with what effort and in what timeframe? How much reuse will be possible? Is the SoC design based on legacy designs of previous generation products (or, in the case of platform-based design, to be built based on an existing platform offering)?

- SoC architecture: In this phase the basic structure of the desired SoC is defined. Vitally important is to decide on the "communications architecture" that will be used as the backbone of the SoC communications network. An inadequate communications architecture will cripple the SoC and have as big an impact as the use of an inappropriate processor subsystem. Of course, the choice of communications architecture is impossible to divorce from making the basic processor(s) choice, e.g., do I use a RISC control processor? Do I have an on-board DSP? How many of each? What are the processing demands of my SoC application? Do I integrate the bare processor core, or use a whole processor subsystem provided by an IP company (most processor IP companies have moved from offering just processor cores to whole processor subsystems including hierarchical bus fabrics tuned to their particular processor needs)? Do I configure the processor(s) to the application (configurable, extensibleASIP, or ASIPs) ? Do I have some ideas, based on legacy SoC design in this space, as to how SW and HW should be partitioned? What memory hierarchy is appropriate? What are the sizes, levels, performance requirements, and configurations of the embedded memories most appropriate to the application domain for the SoC?

- System-level design is an important phase of the SoC process, but one that is often done in a relatively ad hoc way. Recently, the term Electronic System Level (ESL) design has become more popular for this area [16]. The whiteboard and the spreadsheet are as much used by the SoC architects as more capable toolsets. However, there has long been use of ad hoc C/C++ based models for the system design phase to validate basic architectural choices. And designers of complex signal processing algorithms for voice and image processing have long adopted dataflow models and associated tools to define their algorithms, define optimal bit-widths, and validate performance whether destined for HW or SW implementation. A flurry of activity in the last few years on different C/C++ modeling standards for system architects has consolidated on SystemC [17]. The *system* nature of SoC demands a growing use of system-level design modeling and analysis as these devices grow more complex. The basic processes carried out in this phase include HW–SW partitioning (the allocation of functions to be implemented in dedicated HW blocks, in SW on processors (and the decision of RISC vs. DSP), or a combination of both, together with decisions on the communications mechanisms to be used to interface HW and SW, or HW–HW and SW–SW). In addition, the construction of system-level models, and the analysis of correct functioning, performance, and other nonfunctional attributes of the intended SoC through simulation and other analytical tools, is necessary. Finally, all additional IP blocks required which can be sourced outside, or reused from the design group's legacy, must be identified, both HW and SW. The remaining new functions will need to be implemented as part of the overall SoC design process. A more elaborate discussion of ESL follows later.

- After system level design and the identification of the processors and communications architecture, and other HW or SW IP required for the design, the group must undertake an "IP acquisition" stage. This to a large extent can be done at least in part in parallel with other work such as system level design (assuming early identification of major external IP is made) or building golden transaction-level testbench models. Fortunate design groups will be working in companies with both a large legacy of existing well-crafted IP (rather, "virtual components") organized in easy to search databases; or those with access via supplier agreements to large external IP libraries; or at least those with experience at IP search, evaluation, purchase, and integration. For these lucky groups, the problems at this stage are greatly ameliorated. Others with less experience or infrastructure will need to explore these processes for the first time, hopefully making use of IP suppliers' experience with the legal and other processes required. Here the external standards bodies such as VSIA and VCX in the past, and now IEEE, GSA, and others have done much useful work that will smooth the path, at least a little. One key issue in IP acquisition is to conduct rigorous and thorough incoming inspection of IP to ensure its completeness and correctness to the greatest extent possible prior to use, and to resolve any problems with quality early with suppliers—long before SoC integration. Every hour spent on this at this stage will pay back in avoiding much longer schedule slip later. The IP quality guidelines discussed earlier are a foundation level for a quality process at this point.

- Build a transaction-level golden testbench: The system model built up during the system level design stage can form the basis for a more elaborated design model, using "transaction-level" abstractions [18], which represents the underlying HW–SW architecture and components in more detail—sufficient detail to act as a functional virtual prototype for the SoC design. This golden model can be used at this stage to verify the microarchitecture of the design and to verify detailed design models for HW IP at the Hardware Description Language (HDL) level within the overall system context. It thus can be reused all the way down the SoC design and implementation cycle. Modern tools for building platform models are supplied by a number of ESL vendors, such as CoWare (Platform Architect), ARM (SoC Designer), VaST (COMET/Meteor), Synopsys, Virtutech, Mentor Graphics, and Imperas/Open Virtual Platforms (OVP). These may be at one or both of two levels of abstraction: "fast functional" or "programmer's view" platforms, that abstract away most of the details of the HW interconnect, in return for speed, and cycle-accurate level, which provide timing accuracy to the cycle while still being 10–1000 times faster than the register-transfer level (RTL) of simulation. Much of the work on transaction level platform modeling is based on or in reaction to work by the Open SystemC Initiative (OSCI) [19].

- Define the SoC SW architecture: SoC is of course not just about HW [20]. As well as often defining the right on-chip communications architecture, the choice of processor(s), and the nature of the application domain have a very heavy influence on the SW architecture. For example, RTOS choice is limited by the processor ports which have been done and by the application domain (OSEK is an RTOS for automotive systems; Symbian OS for portable wireless devices; PalmOS for Personal Digital Assistants; Windows CE for a number of portable appliances; ThreadX for small lightweight RTOS needs, etc.). Apart from the basic RTOS, every SoC peripheral device will need a device driver-hopefully based on reuse and configuration of templates; various middleware application stacks (e.g., telephony, multimedia image processing) are important parts of the SW architecture; voice and image encoding and decoding on portable devices often is based on assembly code IP for DSPs. There is thus a strong need in defining the SoC to fully elaborate the SW architecture to allow reuse, easy customization, and effective verification of the overall HW–SW device.

- Configure and floorplan SoC microarchitecture: At this point, we are beginning to deal with the SoC on a more physical and detailed logical basis. Of course, during high-level architecture and system-level design, the team has been looking at physical implementation issues (although our design process diagram shows everything as a waterfall kind of flow, in reality SoC design like all electronics design is more of an iterative, incremental process—i.e., more akin to the famous "spiral" model for SW). But before beginning detailed HW design and integration, it is important that there is agreement among the team on the basic physical floorplan, that all the IP blocks are properly and fully configured, that the basic microarchitectures (test, power, clocking, bus, timing) have been fully defined and configured, and that HW implementation can proceed. In addition, this process should also generate the downstream verification environments that will be used throughout the implementation processes—whether SW simulation based, emulation based, using rapid prototypes, or other hybrid verification approaches.

- Design for test (DFT) architecture and implementation: The test architecture is only one of the key microarchitectures that must be implemented; it is complicated by IP legacy and the fact that it is often impossible to impose one DFT style (such as BIST or SCAN) on all IP blocks. Rather, wrappers or adaptations of standard test interfaces (such as JTAG ports) may be necessary to fit all IP blocks together into a coherent test architecture and plan.

- AMS HW implementation: Most SoCs incorporating AMS blocks use them to interface to the external world. VSIA, among other groups, has done considerable work in defining how AMS IP blocks should be created to allow them to be more easily integrated into mainly digital SoCs (the "Big D/little a" SoC) and guidelines and rules for such integration. Experiences with these rules and guidelines and extra deliverables have been on the whole promising, but they have more impact between internal design groups today than on the industry as a whole. The "Big A/Big D" mixed-signal SoC is still relatively rare.

- HW IP assembly and integration: This design step is in many ways the most traditional. Many design groups have experience in assembling design blocks done by various designers or subgroups in an incremental fashion, into the agreed on architectures for communications, bussing, clocking, power, etc. The main difference with SoC is that many of the design blocks may be externally sourced IP. To avoid difficulties at this stage, the importance of rigorous qualification of incoming IP and the early definition of the SoC microarchitecture, to which all blocks must conform, cannot be overstated.

- SW assembly and implementation: Just as with HW, the SW IP, together with new or modified SW tasks created for the particular SoC under design, must be assembled together and validated as to conformance to interfaces and expected operational quality. It is important to verify as much of the SW in its normal system operating context as possible (see below).

- HW and HW–SW verification: Although represented as a single box on the diagram, this is perhaps one of the largest consumers of design time and effort and the major determinant of final SoC quality. Vital to effective verification is the setup of a targeted SoC verification environment, reusing the golden testbench models created at higher levels of the design process. In addition, highly capable, multilanguage, mixed simulation environments are important (e.g., SystemC models and HDL implementation models need to be mixed in the verification process and effective links between them are crucial). There are a large number of different verification tools and techniques [21], ranging from SW-based simulation environments to HW emulators, HW accelerators, and FPGA and bonded-core-based rapid prototyping approaches. In addition, formal techniques such as equivalence checking, and model/property checking have enjoyed some successful usage in verifying parts of SoC designs, or the design at multiple stages in the process. Mixed approaches to HW–SW verification range from incorporating Instruction Set Simulators of processors in SW-based simulation to linking HW emulation of the HW blocks (compiled from the HDL code) to SW running natively on a host workstation, linked in an ad hoc fashion by design teams or using a commercial mixed verification environment. Alternatively, HDL models of new HW blocks running in a SW simulator can be linked to emulation of the rest of the system running in HW—a mix of emulation and use of bonded-out processor cores for executing SW. It is important that as much of the system SW be exercised in the context of the whole system as possible using the most appropriate verification technology that can get the design team close to real-time execution speed (no more than 100× slower is the minimum to run significant amounts of SW). The trend to transaction-based modeling of systems, where transactions range in abstraction from untimed functional communications via message calls, through abstract bus communications models, through cycle-accurate bus functional models, and finally to cycle and pin-accurate transformations of transactions to the fully detailed interfaces, allows verification to occur at several levels or with mixed levels of design description. Finally, a new trend in verification is assertion-based verification, using a variety of input languages (PSL/Sugar, e, Vera, or regular Verilog and VHDL) to model design properties, as assertions, which can then be monitored during simulation, either to ensure that certain properties will be satisfied or certain error conditions never occur. Combinations of formal property checking and simulation-based assertion checking have been created, viz., "semiformal verification." The most important thing to remember about verification is that armed with a host of techniques and tools, it is essential for design teams to craft a well-ordered verification process which allows them to definitively answer the question "how do we know that verification is done?" and thus allow the SoC to be fabricated.

- Final SoC HW assembly and verification: Often done in parallel or overlapping "those final few simulation runs" in the verification stage, the final SoC HW assembly and verification phase includes final place and route of the chip, any hand-modifications required, and final physical verification (using design rule checking and layout vs. schematic (netlist) tools), as well as important analysis steps for issues which occur in advanced semiconductor processes such as IR drop, signal integrity, power network integrity, as well as satisfaction and design transformation for manufacturability (OPC, RET, etc.).

- Fabrication, testing, packaging, and laboratory verification: When an SoC has been shipped to fabrication, it would seem time for the design team to relax. Instead, this is an opportunity for additional verification to be carried out—especially more verification of system SW running in context of the HW design—and for fixes, either of SW or of the SoC HW on hopefully no more than one expensive iteration of the design, to be determined and planned. When the tested packaged parts arrive back for verification in the laboratory, the ideal scenario is to load the SW into the system and have the SoC and its system booted up and running SW within a few hours. Interestingly, the most advanced SoC design teams, with well-ordered design methodologies and processes, are able to achieve this quite regularly.

13.9 System-Level or ESL Design

As we touched on earlier, when describing the overall SoC design flow, system-level design and SoC are essentially made for each other. A key aim of IP reuse and of SoC techniques such as platform-based design is to make the "back end" (RTL to GDS II) design implementation processes easier—fast and with low risk; and to shift the major design phase for SoC up in time and in abstraction level to the system level. This also means that the back-end tools and flows for SoC designs do not necessarily differ from those used for complex ASIC, ASSP, and custom IC design—it is the methodology of how they are used, and how blocks are sourced and integrated, that overlays the underlying design tools and flows, that may differ for SoC. However, the fundamental nature of IP-based design of SoC has a stronger influence on the system level.

The definition of system-level design, or ESL (electronic system level design and verification, as it is increasingly being labeled today), is not agreed on. Rather than attempt to find a consensus definition, it is more useful to define ESL by what activities it includes. When we examine modern systems design for SoC, we see at least five different categories of ESL tools, models, and design flows, whose use depends on the type of systems being developed, the architectural choices being made, and indeed, the wishes and experience of the architects and developers involved in the project. These categories include

- Algorithmic design
- Architectural design space exploration (DSE)
- Virtual prototypes for embedded SW development/validation
- Behavioral/high-level synthesis
- Processor/multiprocessor-centric design

It is important to remember that ESL must remain a broad, experience-driven definition based on the activities that people really carry out. Sometimes designers pick one of these categories and conflate it with the entire definition of ESL. For example, to some, ESL will never be "real" until RTL designers have replaced their synthesis methodologies and tools with some kind of high-level or behavioral synthesis perhaps using C, C++, or SystemC as a specification language. All other kinds of ESL, including the now venerable algorithmic tools such as signal processing worksystem, the Mathworks Matlab®, and Simulink®, and research tools such as Ptolemy, are ignored as not being true "design"—even if they are widely used. This is inappropriate.

Depending on the kind of system, some or all of these categories may be important. Developing a multiprocessor SoC (MPSoC) using application-specific processors will emphasis processor-centric design flows and the concomitant SW development flows, along with the use of virtual prototypes for validation, for example. Developing new digital logic using behavioral synthesis may be unimportant for this type of design.

It is at the system level that the vital tasks of deciding on and validating the basic system architecture and choice of IP blocks are carried out. In general, this is known as "architectural DSE." As part of this exploration, SoC platform customization for a particular derivative is carried out, should the SoC platform approach be used. Essentially one can think of platform DSE as being a similar task to general DSE, except that the scope and boundaries of the exploration are much more tightly constrained—the basic communications architecture and platform processor choices may be fixed, and the design team may be restricted to choosing certain customization parameters and choosing optional IP from a library. Other tasks include HW–SW partitioning, usually restricted to decisions about key processing tasks which might be mapped into either HW or SW form and which have a big impact on system performance, energy consumption, on-chip communications bandwidth consumption, or other key attributes. Of course, in multiprocessor systems, there are "SW–SW" partitioning or codesign issues as well; deciding on the assignment of SW tasks to various processor options. Again, perhaps 80%–95% of these decisions can or are made *a priori*, especially if an SoC is either based on a platform or an evolution of an existing system; such codesign decisions are usually made on a small number of functions that have critical impact.

Because partitioning, codesign and DSE tasks at the system level involve much more than HW–SW issues, a more appropriate term for this is "function-architecture codesign" [22,23]. In this codesign model, systems are described on two equivalent levels:

- Functional intent of the system, e.g., a network of applications, decomposed into individual sets of functional tasks that may be modeled using a variety of models of computation such as discrete event, finite state machine, or dataflow.
- Architectural structure of the system, the communications architecture, major IP blocks such as processor(s), memories, and HW blocks, captured or modeled for example using some kind of IP or platform configurator.

The methodology implied in this approach is then to build explicit mappings between the functional view of the system and the architectural view, which carry within them the implicit partitioning that is made for both computation and communications. This hybrid model can then be simulated, the results analyzed, and a variety of ancillary models (e.g., cost, power, performance, communications bandwidth consumption, etc.) can be utilized in order to examine the suitability of the system architecture as a vehicle for realizing or implementing the end product functionality.

The function-architecture codesign approach has been implemented and used in both research and commercial tools [24] and forms the foundation of many system-level codesign approaches going forward. In addition, it has been found extremely suitable as the best system-level design approach for platform-based design of SoC [25].

Tools useful for architectural DSE are CoWare's Platform Architect, Synopsys System Studio, CoFluent Studio, and Mentor Platform Express with associated simulation tools.

Once the architectural definition of a platform has been determined, architects will want to generate a fast-functional instruction-accurate virtual platform model to provide to embedded SW developers as a verification vehicle. This might be done with DSE tools such as the CoWare Platform Architect that has a virtual platform export capability. Such a platform may be cycle-accurate, instruction-accurate, or both. SW developers in general will need an instruction-accurate virtual platform model, which may run 10–100× faster than a cycle-accurate platform model, but depending on the nature of the SW under development they should make judicious use of a cycle-accurate model as well. Real-time aspects of the SW or time-dependent synchronization may require that the SW be run in cycle-accurate mode to validate it under realistic operating conditions. Alternatives to DSE tools for creating instruction-accurate virtual platforms include tools from vendors such as VaST, Virtutech, Imperas/OVP, and others.

13.10 Configurable and Extensible Processors

As systems have shifted to a more processor-centric design style, with more use of embedded processors, both fixed ISA ones and ASIPs, there is a growing interest in the use of configurable and extensible processors to generate ASIPs in an automated fashion based on specific SoC application requirements. Several commercial capabilities exist for ASIP generation via configuring and extending processor instruction sets, including CoWare Lisatek, ARC, and Tensilica [26,27].

To take one example of a configurable, extensible processor [28,29] it may be based on a default RISC instruction set. Configuration usually deals with coarse-grained structural parameters, such as the presence or absence of local and system memory interfaces, the widths of the memories, caching parameters, numbers and types of interrupts, and the presence or absence of various functional units such as multipliers and multiply accumulators. Extension deals with the addition of highly specialized instruction units implementing very application-specific instructions. Other aspects of configuration may include SIMD (single instruction, multiple data) functional units for handling arithmetic inner loop nests, and VLIW multioperation instructions for handling multiple concurrent operations.

ASIPs have proven themselves particularly useful in heavy data processing applications as is found in multimedia products—for example, audio, video, and other types of signal processing. Here the advantages of specialized instructions, SIMD and VLIW type processing are particularly acute. Performance gains of 10X or more are possible, with commensurate reductions in energy consumption.

In addition to their standalone uses, ASIPs can also be used in multiprocessor combinations following asymmetric multiprocessing architectural principles to provide added performance and an ability to shut off large parts of an SoC when the particular application processing function is not needed—for example, shutting a video decoding subsystem in a portable appliance down when unused. This saves considerable energy in battery-powered portable devices. The key issues in MPSoC are the concurrency and programming models to be used.

13.11 IP Configurators and Generators

As well as configurable ASIPs, and fixed ISA processors, there are other kinds of configurable IP accessed via generators. These usually fall into the classes of on-chip memories, memory controllers, on-chip buses and interconnect, and standard bus interfaces. On-chip memory IP, from vendors such as Virage, ARM (acquired with Artisan), Kilopass (nonvolatile memory IP), Novelics (which licensed Synopsys to market their one-transistor SRAM IP), is usually configurable as to size, organization, and various other parameters including target process(es). On chip bus IP (e.g., AMBA AHB, APB, AXI) may be available from vendors such as Synopsys (in their DesignWare libraries); memory controllers from vendors such as RAMBUS and Denali; and standard bus interfaces from companies such as Synopsys (PCI, PCI Express, USB, etc.). At one point, Mentor Graphics offered families of configurable RTL IP blocks for standard bus interfaces and other blocks such as simple microcontrollers, but in late 2007 Mentor departed the IP business.

Whatever the type of IP, it almost always is configurable and may come in addition with automatic generation of high-level system ESL models in languages such as SystemC, along with associated test benches, possibly assertion-based monitors and checkers, and other integration tools and utilities.

The commercial IP business for anything other than processors, leading edge memories, and advanced bus interfaces (what is often called "Star IP") is a bit of a race to the bottom in terms of pricing and revenue potential. Because barriers to entry tend to be low, once an IP category (such as standard bus interfaces) becomes popular, several suppliers will spring up and try to beat each other on price. The poor quality suppliers that may emerge have in the past tended to give the whole IP industry a bad name, and there is increasing emphasis on the IP qualification assessments discussed earlier, or acquiring known IP from well-known and reliable vendors.

13.12 Computation and Memory Architectures for Systems-on-Chip

The primary processors used in SoC are embedded RISCs such as ARM processors, PowerPCs, MIPS architecture processors, and some of the configurable processors designed specifically for SoC such as Tensilica and ARC. In addition, embedded DSPs from traditional suppliers as TI, Motorola, ParthusCeva, and others are also quite common in many consumer applications, for embedded signal processing for voice and image data. Research groups have looked at compiling or synthesizing application-specific processors or coprocessors [30,31] and these have interesting potential in future SoCs which may incorporate networks of heterogeneous configurable processors collaborating to offer large amounts of computational parallelism. This is an especially interesting prospect given wider use of reconfigurable logic that opens up the prospect of dynamic adaptation of SoC to application needs. However, most MPSoCs today involve at most 2–4 processors of conventional design; the larger networks are more often found today in the industrial or university laboratory.

Although several years ago most embedded processors in early SoCs did not use cache memory-based hierarchies, this has changed significantly over the years, and most RISC and DSPs now involve significant amounts of Level 1 Cache memory, as well as higher level memory units both on- and off-chip (off-chip flash memory is often used for embedded SW tasks which may be only infrequently required). System design tasks and tools must consider the structure, size, and configuration of the memory hierarchy as one of the key SoC configuration decisions that must be made.

13.13 IP Integration Quality and Certification Methods and Standards

We have emphasized the design reuse aspects of SoC and the need for reuse of both internally and externally sourced IP blocks by design teams creating SoCs. In the discussion of the design process above, we mentioned issues such as IP quality standards and the need for incoming inspection and qualification of IP. The issue of IP quality remains one of the biggest impediments to the use of IP-based design for SoC [32]. The quality standards and metrics available from VSIA, OpenMORE, and their further enhancement (GSA) help, but only to a limited extent. The industry could clearly use a formal certification body or lab for IP quality that would ensure conformance to IP transfer requirements and the integration quality of the blocks. Such a certification process would be of necessity quite complex due to the large number of configurations possible for many IP blocks and the almost infinite variety of SoC contexts into which they might be integrated. Certified IP would begin to deliver the virtual components of the VSIA vision.

In the absence of formal external certification (and such third party labs seem a long way off, if they ever emerge), design groups must provide their own certification processes and real reuse quality metrics based on their internal design experiences. Platform-based design methods help due to the advantages of prequalifying and characterizing groups of IP blocks and libraries of compatible domain-specific components. Short of independent evaluation and qualification, this is the best that design groups can do currently.

One key issue to remember is that IP not created for reuse, with all the deliverables created and validated according to a well-defined set of standards, is inherently not reusable. The effort required to make a reusable IP block has been estimated to be 50%–200% more effort than that required to use it once; however, assuming the most conservative extra cost involved implies positive payback with three uses of the IP block. Planned and systematic IP reuse and investment in those blocks with greatest SoC use potential gives a high chance of achieving significant productivity soon after starting a reuse program. But ad hoc attempts to reuse existing design blocks not designed to reuse standards have failed in the past and are unlikely to provide the quality and productivity desired.

13.14 Specific Application Areas

When we consider popular classes of SoC designs, there are a large number of complex SOCs available from the semiconductor industry and that have been designed in large systems houses for their products [14]. Categories that are especially notable today (2008) are complex media and baseband processors for cellphones and portable appliances. Notable in these categories, which include families of related devices, are Texas Instruments OMAP, STMicroelectronics Nomadik, and NXP Nexperia.

The TI OMAP family includes SoCs released in 2008 from the OMAP35X group. These SoCs include various combinations of ARM Cortex A8 control processors, media-related peripherals, graphics engines, video accelerators, and TI DSPs, as well as on-chip memory and buses and DMA controllers.

The STMicroelectronics Nomadik multimedia processors include several family members introduced in 2007–2008 with ARM 926 and 1176 control processors, video and audio accelerators, media interfaces (e.g., camera, LCD, and other external interfaces), graphics accelerators, and many other peripherals.

The NXP Nexperia family includes several home entertainment "engines," DVD recording engines, and media processors including both audio and video. These include mixes of Trimedia cores, MIPS control processors and a host of different media accelerators, peripherals, and interfaces.

This is just an illustration of a widely growing and popular class of SoC devices. Other examples can be found in many different application domains including printing and imaging, networking, automotive applications, and various portable devices including PDAs.

13.15 Summary

In this chapter, we have defined SoC and surveyed a large number of the issues involved in its design. An outline of the important methods and processes involved in SoC design define a methodology that can be adopted by design groups and adapted to their specific requirements. Productivity in SoC design demands high levels of design reuse and the existence of third party and internal IP groups and the chance to create a library of reusable IP blocks (true virtual components) are all possible for most design groups today. The wide variety of design disciplines involved in SoC mean that unprecedented collaboration between designers of all backgrounds—from systems experts through embedded SW designers, through architects, through HW designers—is required. But the rewards of SoC justify the effort required to succeed.

References

1. M. Hunt and J. Rowson, Blocking in a system on a chip, *IEEE Spectrum*, November 1996, 33(11), 35–41.
2. R. Rajsuman, *System-on-a-Chip Design and Test*, Artech House, Boston, MA, 2000.
3. OMAPV1035 Product Bulletin available at: http://focus.ti.com/pdfs/wtbu/TI_omapv1035.pdf
4. M. Keating and P. Bricaud, *Reuse Methodology Manual for System-on-a-Chip Designs*, 1998 (1st edn.), 1999 (2nd edn.), 2002 (3rd edn.), Kluwer Academic Publishers, Dordrecht, the Netherlands.
5. International Technology Roadmap for Semiconductors (ITRS), 2007 edition—Design chapter. URL: http://public.itrs.net/
6. Virtual Socket Interface Alliance, on the web at URL: http://www.vsia.org. This includes access to its various public documents, including the original Reuse Architecture document of 1997, as well as more recent documents supporting IP reuse released to the public domain. Despite the VSIA closing down in 2007, it has retained its web site and its documents are still available.
7. The Virtual Component Exchange (VCX). Web URL: http://www.thevcx.com/. The VCX IP database became owned by Beach Solutions, and in 2006, was acquired by ChipEstimate, itself bought in 2008 by Cadence: See http://www.edn.com/article/CA6350941.html and http://www.chipestimate.com/
8. Global Semiconductor Alliance IP ecosystem tool suite: http://www.gsaglobal.org/resources/tools/ipecosystem/index.asp.
9. H. Chang, L. Cooke, M. Hunt, G. Martin, A. McNelly, and L. Todd, *Surviving the SOC Revolution: A Guide to Platform-Based Design*, Kluwer Academic Publishers, Norwell, MA, 1999.
10. K. Keutzer, S. Malik, A. R. Newton, J. Rabaey, and A. Sangiovanni-Vincentelli, System-level design: Orthogonalization of concerns and platform-based design, *IEEE Transactions on CAD of ICs and Systems*, 19(12), 1523–1543, December 2000.
11. Alberto Sangiovanni-Vincentelli and Grant Martin, Platform-based design and software design methodology for embedded systems, *IEEE Design and Test of Computers*, 18(6), 23–33, November–December 2001.

12. IEEE Design and Test Special Issue on Platform-Based Design of SoCs, November–December 2002, 19(6).
13. G. Martin and F. Schirrmeister, A design chain for embedded systems, *IEEE Computer, Embedded Systems Column*, 35(3), 100–103, March 2002.
14. G. Martin and H. Chang (Eds.), *Winning the SOC Revolution: Experiences in Real Design*, Kluwer Academic Publishers, Boston, MA, May 2003.
15. P. Lysaght, FPGAs as meta-platforms for embedded systems, *Proceedings of the IEEE Conference on Field Programmable Technology*, Hong Kong, December 2002.
16. B. Bailey, G. Martin, and A. Piziali, *ESL Design and Verification: A Prescription for Electronic System-Level Methodology*, Elsevier Morgan Kaufmann, San Francisco, CA, February 2007.
17. T. Groetker, S. Liao, G. Martin, and S. Swan, *System Design with SystemC*, Kluwer Academic Publishers, Boston, MA, May 2002.
18. J. Bergeron, *Writing Testbenches*, (3rd edn.), Kluwer Academic Publishers, Boston, MA, 2003.
19. A. Donlin, Transaction level modeling: Flows and use models, *CODES + ISSS '04*, Stockholm,. Sweden, pp. 75–80. 2004.
20. G. Martin and C. Lennard, Invited CICC paper, Improving embedded SW design and integration for SOCs, *Custom Integrated Circuits Conference*, May 2000, pp. 101–108.
21. P. Rashinkar, P. Paterson, and L. Singh, *System-on-a-Chip Verification: Methodology and Techniques*, Kluwer Academic Publishers, London, 2001.
22. F. Balarin, M. Chiodo, P. Giusto, H. Hsieh, A. Jurecska, L. Lavagno, C. Passerone, A. Sangiovanni-Vincentelli, E. Sentovich, K. Suzuki, and B. Tabbara, *Hardware–Software Co-Design of Embedded Systems: The POLIS Approach*, Kluwer Academic Publishers, Dordrecht, the Netherlands, 1997.
23. S. Krolikoski, F. Schirrmeister, B. Salefski, J. Rowson, and G. Martin, *Methodology and Technology for Virtual Component Driven Hardware/Software Co-Design on the System Level*, paper 94.1, ISCAS 99, Orlando, FL, May 30–June 2, 1999.
24. G. Martin and B. Salefski, System level design for SOC's: A progress report—two years on, Chapter 25 in: *System-on-Chip Methodologies and Design Languages*, J. Mermet (Ed.), Kluwer Academic, Dordrecht, the Netherlands, Chapter 25, pp. 297–306, 2001.
25. G. Martin, Productivity in VC reuse: Linking SOC platforms to abstract systems design methodology, *Virtual Component Design and Reuse*, R. Seepold and N. Martinez Madrid (Eds.), Kluwer Academic, Boston, MA, Chapter 3, pp. 33–46, 2001.
26. M. Gries and K. Keutzer, (Eds.), *Building ASIPs: The MESCAL Methodology*, Springer, New York, June, 2005.
27. P. Ienne and R. Leupers (Eds.), *Customizable Embedded Processors: Design Technologies and Applications*, Elsevier Morgan Kaufmann, San Francisco, CA, 2006.
28. C. Rowen and S. Leibson, *Engineering the Complex SOC: Fast, Flexible Design with Configurable Processors*, Prentice-Hall PTR, Upper Saddle River, NJ, 2004.
29. S. Leibson, *Designing SOCs with Configured Cores: Unleashing the Tensilica Xtensa and Diamond Cores*, Elsevier Morgan Kaufmann, San Francisco, CA, 2006.
30. V. Kathail, S. Aditya, R. Schreiber, B. R. Rau, D. C. Cronquist, and M. Sivaraman, PICO: Automatically designing custom computers, *IEEE Computer*, 35(9), 39–47, September 2002.
31. T.J. Callahan, J.R. Hauser, and J. Wawrzynek, The Garp architecture and C compiler, *IEEE Computer*, 33(4), 62–69, April 2000.
32. DATE 2002 Proceedings, Session 1A: How to choose semiconductor IP?: Embedded processors, memory, software, hardware, *Proceedings of DATE 2002*, Paris, pp. 14–17, March 2002.

14

SoC Communication Architectures: From Interconnection Buses to Packet-Switched NoCs

José L. Ayala
Complutense University of Madrid

Marisa López-Vallejo
ETSI Telecommunicacion

Davide Bertozzi
University of Ferrara

Luca Benini
University of Bologna

14.1　Introduction

The current high levels of on-chip integration allow for the implementation of increasingly complex systems-on-chip (SoCs), consisting of heterogeneous components such as general-purpose processors, DSPs, coprocessors, memories, I/O units, and dedicated hardware accelerators.

In this context, multiprocessor systems-on-chip (MPSoCs) have emerged as an effective solution to meet the demand for computational power posed by application domains such as network processors and parallel media processors. MPSoCs combine the advantages of parallel processing with the high integration levels of SoCs.

It is expected that future MPSoCs will integrate hundreds of processing units (PUs) and storage elements, and their performance will be increasingly interconnect-dominated [40]. Interconnect technology and architecture will become the limiting factor for achieving operational goals, and the efficient design of low-power, high-performance on-chip communication architectures will pose novel challenges. The main issue concerns the *scalability* of system interconnects, since the trend for system integration is expected to continue. State-of-the-art on-chip buses rely on shared communication resources and on an arbitration mechanism which is in charge of serializing bus access requests. This widely adopted solution unfortunately suffers from power and performance scalability limitations; therefore, a lot of effort is being devoted to the development of advanced bus topologies (e.g., partial or full crossbars, bridged buses) and protocols, some of them already implemented in commercially available products. In the long run, a more aggressive approach will be needed, and a design paradigm shift will most probably lead to the automatic synthesis of communication interconnects and the high-level (transaction-level) modeling of complex communication architectures in networks-on-chip (NoCs) [27,29].

This chapter focuses on state-of-the-art SoC communication architectures, providing an overview of the most relevant system interconnects from an industrial and research viewpoint. Beyond describing the distinctive features of each of them, the chapter sketches the main evolution guidelines for these architectures by means of a protocol and topology analysis framework. Finally, some basic concepts on packet-switched interconnection networks will be put forward. Open bus specifications such as AMBA and CoreConnect will be obviously described more in detail, providing the background which is needed to understand the necessarily more general description of proprietary industrial bus architectures, while at the same time being able to assess their contribution to the advance in the field. Also, current research trends on the area of SoC communication architectures will be described, covering the automation and synthesis of these interfaces.

14.2 AMBA Interface

The advanced micro-controller bus architecture (AMBA®) is a bus standard which was originally conceived by ARM to support communication among ARM processor cores. However, nowadays AMBA is one of the leading on-chip busing systems because of its open access and its advanced features. Designed for custom silicon, the AMBA specification provides standard bus interfaces for connecting on-chip components, custom logic, and specialized functions. These interfaces are independent of the ARM processor and generalized for different SoC structures.

The original specification was significantly refined and extended to meet the severe demands of current SoC designs. In this way, AMBA 3 [4] was defined to provide a new set of on-chip interface protocols that can interoperate with the existing bus technology defined in the AMBA 2 specification [3].

The AMBA 3 specification defines four different interface protocols that target SoC implementation with very different requirements in terms of data throughput, bandwidth, or power. Figure 14.1 depicts the conceps of interface and interconnect in a typical AMBA-based system architecture. The interfaces proposed by AMBA are

- Advanced extensible interface (AXI™), that focuses on high-performance high-frequency implementations. This burst-based interface provides the maximum of interconnect flexibility and yields the highest performance.
- Advanced high-performance bus (AHB) Interface that enables highly efficient interconnect between simpler peripherals in a single frequency subsystem where the performance of AMBA 3 AXI is not required.

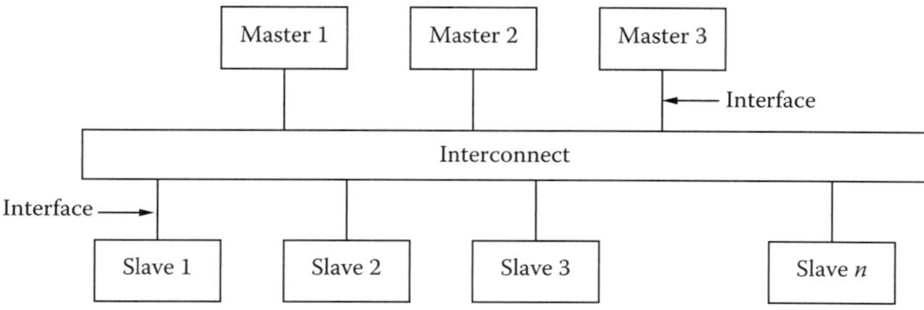

FIGURE 14.1 Interface and interconnect in a typical SoC.

- Advanced peripheral bus (APB™) Interface that is intended for general-purpose low-speed, low-power peripheral devices, allowing the isolation of slow data traffic to the high-performance AMBA 3 interfaces.
- AMBA 3 ATB™ Interface provides visibility for debug purposes by adding tracing data capabilities.

In this section, we will sketch the main characteristics of the interfaces defined within the new standard AMBA 3. Even though AMBA 3 keeps backward-compatibility with existing AMBA 2 interfaces [3], the review of previous AMBA specifications is out of the scope of this chapter.

14.2.1 AMBA 3 AXI Interface

The AXI is the latest generation AMBA interface [5]. It is designed to be used as a high-speed submicron interconnect and also includes optional extensions for low-power operation. This high-performance protocol provides flexibility in the implementation of interconnect architectures while still keeping backward-compatibility with existing AHB and APB interfaces. It enables

- Pipelined interconnection for high-speed operation
- Efficient bridging between frequencies for power management
- Simultaneous read and write transactions
- Efficient support of high initial latency peripherals
- Multiple outstanding transactions with out-of-order data completion

AMBA AXI builds upon the concept of point-to-point connection. AMBA AXI does not provide masters and slaves with visibility of the underlying interconnect, but rather features the concept of master interfaces and symmetric slave interfaces. This approach, besides allowing seamless topology scaling, has the advantage of simplifying the handshake logic of attached devices, which only need to manage a point-to-point link.

To provide high scalability and parallelism, five different logical unidirectional channels are provided: read and write address channels, read and write data channels, and write response channel. Activity on different channels is mostly asynchronous (e.g., data for a write can be pushed to the write data channel before or after the write address is issued to the write address channel) and can be parallelized, allowing multiple outstanding read and write requests.

Figure 14.2a shows how a read transaction uses the read address and read data channels. The write operation over the write address, write data, and write response channels are presented in Figure 14.2b. As can be observed, the data is transferred from the master to the slave using a write data channel, and it is transferred from the slave to the master using a read data channel. In write transactions, in which all data flow from the master to the slave, the AXI protocol has an additional write response channel to allow the slave to signal to the master the completion of the write transaction.

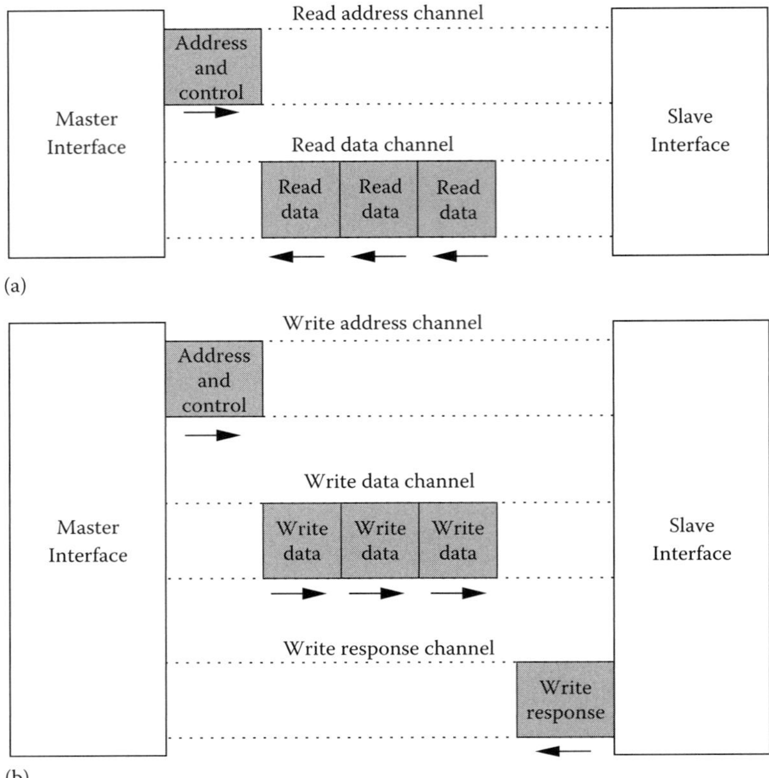

FIGURE 14.2   Architecture of transfers: (a) read operation and (b) write operation.

However, the AXI protocol is a master/slave-to-interconnect interface definition, and this enables a variety of different interconnect implementations. Therefore, the mapping of channels, as visible by the interfaces, to actual internal communication lanes is decided by the interconnect designer; single resources might be shared by all channels of a certain type in the system, or a variable amount of dedicated signals might be available, up to a full crossbar scheme. The rationale of this split-channel implementation is based on the observation that usually the required bandwidth for addresses is much lower than that for data (e.g., a burst requires a single address but maybe four or eight data transfers). Availability of independently scalable resources might, for example, lead to medium complexity designs sharing a single internal address channel while providing multiple data read and write channels.

As mentionned before, the AMBA 3 AXI protocol is fully compatible with AMBA 2 implementations. Furthermore, it clearly outperforms AMBA 2 AHB protocol in many different features [39], as can be seen in the comparison summarized in Table 14.1.

14.2.2 AMBA AHB Interface

The AMBA 3 specification includes another interface that provides a highly efficient interconnect between simpler peripherals in a single frequency subsystem. The AHB interface was designed for these cases, which do not require the excellent performance of the AMBA 3 AXI.

TABLE 14.1   AMBA 2 AHB vs. AMBA 3 AXI Protocols

AMBA 2 AHB	AMBA 3 AXI
Fixed pipeline for address and data transfers	Five independent channels for address/data and response
Bidirectional link with complex timing relationships	Each channel is unidirectional except for single handshake for return path
Hard to isolate timing	Register slices isolate timing
Limits frequency of operation	Frequency scales with pipelining
Inefficient asynchronous bridges	High-performance asynchronous bridging
Separate address for every data item	Burst based-one address per burst
Only one transaction at a time	Multiple outstanding transactions
Fixed pipeline for address and data	Out-of-order data
Only one transaction at a time	Simultaneous reads and writes
Does not natively support the ARM v6 architecture	Native support for unaligned and exclusive accesses
No support for security	Native security support

The main features of AMBA AHB can be summarized as follows:

- *Multiple bus masters.* Optimized system performance is obtained by sharing resources among different bus masters. A simple request–grant mechanism is implemented between the arbiter and each bus master. In this way, the arbiter ensures that only one bus master is active on the bus and also that when no masters are requesting the bus, a default master is granted.

- *Pipelined and burst transfers.* Address and data phases of a transfer occur during different clock periods. In fact, the address phase of any transfer occurs during the data phase of the previous transfer. This overlapping of address and data is fundamental to the pipelined nature of the bus and allows for high-performance operation, while still providing adequate time for a slave to provide the response to a transfer. This also implies that ownership of the data bus is delayed with respect to ownership of the address bus. Moreover, support for burst transfers allows for efficient use of memory interfaces by providing transfer information in advance.

- *Split transactions.* They maximize the use of bus bandwidth by enabling high latency slaves to release the system bus during dead time while they complete processing of their access requests.

- *Wide data bus configurations.* Support for high-bandwidth data-intensive applications is provided using wide on-chip memories. System buses support 32-, 64-, and 128-bit data-bus implementations with a 32-bit address bus, as well as smaller byte and half-word designs.

- *Nontristate implementation.* AMBA AHB implements a separate read and write data bus in order to avoid the use of tristate drivers. In particular, master and slave signals are multiplexed onto the shared communication resources (read and write data buses, address bus, and control signals).

The original AMBA AHB system [3] contained the following components:

AHB master: Only one bus master at a time is allowed to initiate and complete read and write transactions. Bus masters drive out the address and control signals and the arbiter determines which master has its signals routed to all slaves. A central decoder controls the read data and response signal multiplexor, which selects the appropriate signals from the slave that has been addressed.

AHB slave: It signals back to the active master the status of the pending transaction. It can indicate that the transfer is completed successfully, or that there was an error or that the master should retry the transfer or indicate the beginning of a split transaction.

AHB arbiter: The bus arbiter serializes bus access requests. The arbitration algorithm is not specified by the standard and its selection is left as a design parameter (fixed priority, round-robin, latency-driven, etc.), although the request–grant based arbitration protocol has to be kept fixed.

AHB decoder: This is used for address decoding and provides the select signal to the intended slave.

14.2.2.1 AMBA 3 AHB-Lite

The AHB-Lite interface [7] supports a single bus master and provides high bandwidth operation. Actually, it is the only AHB interface that is currently documented by ARM. It supports

- Burst transfers
- Single-clock edge operation
- Nontristate implementation
- Wide data bus configurations, 64, 128, 256, 512, and 1024 bits

The main components of the AHB-Lite interface are the bus master, the bus slaves, a decoder, and a slave-to-master multiplexor, as can be seen in Figure 14.3. The master starts a transfer by driving the address and control signals. These signals provide information about the address, direction, width of the transfer, and indicate if the transfer forms part of a burst. The write data bus moves data from the master to a slave, and the read data bus moves data from a slave (selected by the decoder and the mux) to the master.

Given that AHB-Lite is a single master bus interface, if a multimaster system is required, an AHB-Lite multilayer [6] structure is required to isolate all masters from each other.

14.2.2.2 Multilayer AHB Interface

The Multilayer AHB specification [6] emerges with the aim of increasing the overall bus bandwidth and providing a more flexible interconnect architecture with respect to AMBA AHB. This is achieved by using a more complex interconnection matrix which enables parallel access paths between multiple masters and slaves in a system.

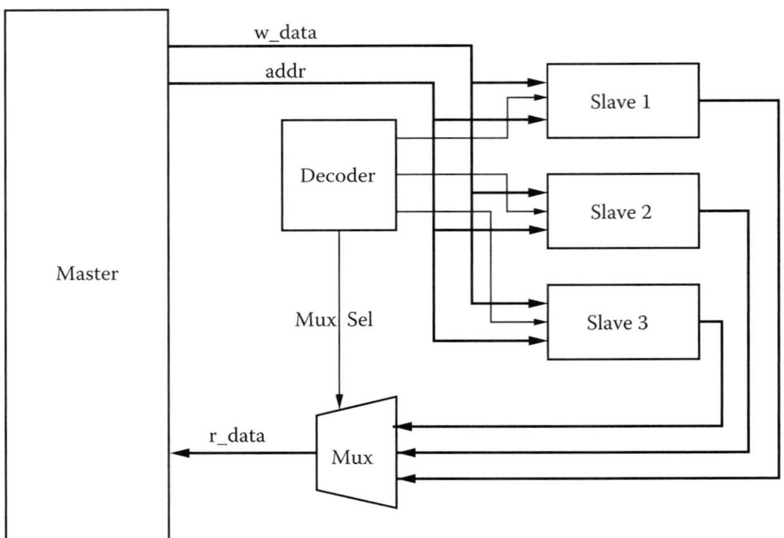

FIGURE 14.3　Structure of the AMBA 3 AHB-Lite interface.

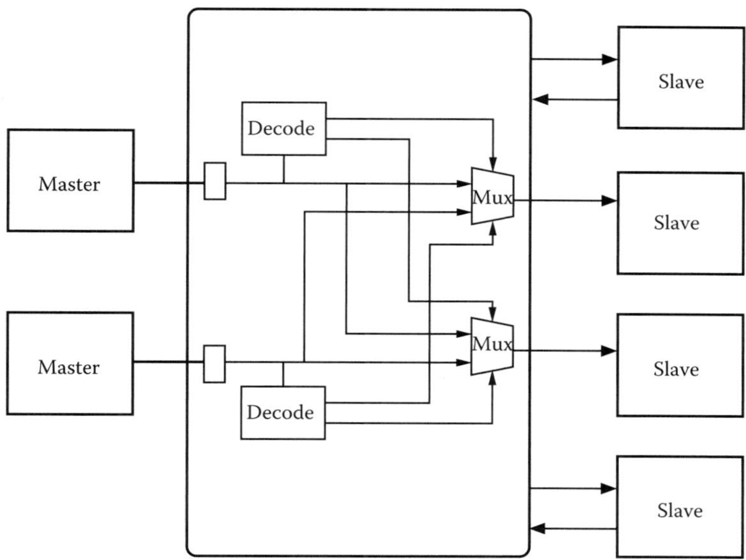

FIGURE 14.4 Schematic view of the multilayer AHB interconnect.

Therefore, the multilayer bus architecture allows the interconnection of unmodified standard AHB or AHB-Lite master and slave modules with an increased available bus bandwidth. The resulting architecture becomes very simple and flexible: each AHB layer only has one master and no arbitration and master-to-slave muxing is needed. Moreover, the interconnect protocol implemented in these layers can be very simple (AHB-Lite protocol, for instance): it neither has to support request and grant nor retry or split transactions.

The additional hardware needed for this architecture with respect to the AHB is an interconnection matrix to connect the multiple masters to the peripherals. Point arbitration is also required when more than one master wants to access the same slave simultaneously.

Figure 14.4 shows a schematic view of the multilayer concept. The interconnect matrix contains a decode stage for every layer in order to determine which slave is required during the transfer. A multiplexer is used to route the request from the specific layer to the desired slave.

The arbitration protocol decides the sequence of accesses of layers to slaves based on a priority assignment. The layer with lowest priority has to wait for the slave to be freed. Different arbitration schemes can be used, and every slave port has its own arbitration. Input layers can be served in a round-robin fashion, changing every transfer or every burst transaction, or based on a fixed priority scheme.

The number of input/output ports on the interconnect matrix is completely flexible and can be adapted to suit to system requirements. As the number of masters and slaves implemented in the system increases, the complexity of the interconnection matrix can become significant and some optimization techniques have to be used: defining multiple masters on a single layer, defining multiple slaves appearing as a single slave to the interconnect matrix, and defining local slaves to a particular layer.

Because the multilayer architecture is based on the existing AHB protocol, previously designed masters and slaves can be totally reused.

14.2.3 AMBA 3 APB Interface

The AMBA APB interface is intended for general-purpose low-speed low-power peripheral devices. Therefore, the APB interface allows the isolation of slow data traffic to the high-performance AMBA

3 AXI and AHB interfaces. Following the compatibility philosophy of AMBA, the APB interface is fully backward compatible with the AMBA 2 APB interface, making a wide variety of existing APB peripherals available for reuse.

APB is a static bus that provides a simple addressing, with latched addresses and control signals for easy interfacing. To ease compatibility with any other design flow, all APB signal transitions only take place at the rising edge of the clock, requiring every read or write transfer at least two cycles.

The main features of this bus are the following:

- Unpipelined architecture
- Low gate count
- Low-power operation
    - Reduced loading of the main system bus is obtained by isolating the peripherals behind a bridge.
    - Peripheral bus signals are only active during low-bandwidth peripheral transfers.

AMBA APB operation can be abstracted as a state machine with three states. The default state for the peripheral bus is IDLE, which switches to SETUP state when a transfer is required. SETUP state lasts just one cycle, during which the peripheral select signal is asserted. The bus then moves to ENABLE state, which also lasts only one cycle and which requires the address, control, and data signals to remain stable. Then, if other transfers are to take place, the bus goes back to SETUP state, otherwise to IDLE. As can be observed, AMBA APB should be used to interface to any peripherals which are low-bandwidth and do not require the high performance of a pipelined bus interface.

14.2.4 AMBA 3 ATB Interface

The AMBA 3 AMBA trace bus (ATB) interface specification adds a data diagnostic interface to trace data in a trace system using the AMBA specification. The ATB is a common bus used by the trace components to pass format-independent trace data through the system. Both trace components and bus sit in parallel with the peripherals and interconnect and provide visibility for debug purposes.

The ATB interfaces can play two different roles depending on the sense of the trace data transfer. On the one hand, the interface is *Master* if it generates trace data. On the other hand, a *Slave* interface receives trace data. Interesting features for debugging that are included in the ATB protocol are

- Stalling of data, using valid and ready responses
- Control signals to indicate the number of bytes valid in a cycle
- Marking of the originating component, each data packet has an associated ID
- Variable information formats
- Identification of data from all originating components
- Flushing

14.3 Sonics SMART Interconnects

Sonics Inc. [26] is a premier supplier of SoC interconnect solutions, serving a wide spectrum of markets that employ SoCs including mobile phones, gaming platforms, HDTVs, communications routers, as well as automotive and office automation products. The approach offered by this company, called sonics methodology and architecture for rapid time to market, SMART, allows the system designer to utilize predesigned, highly optimized, and flexible interconnects to configure, analyze, and verify data flows at the architecture definition phase early in the design cycle.

All transactions in a SMART protocol can be seen as a combination of agents that communicate with each other using the interconnect. Agents isolate cores from one another and from the Sonics internal interconnect fabric. In this way, the cores and the interconnect are fully decoupled, allowing the cores to be reused from system to system without rework. In the agent architecture we can distinguish

Initiator: Who implements the interface between the interconnect and the master core (CPU, DSP, direct memory access [DMA], etc.). The initiator receives requests from the core, then transmits the requests according to the Sonics standard, and finally processes the responses from the target.

Target: Who implements the interface between the physical interconnect and the target device (memories, universal asynchronous receiver transmitter [UARTs], etc.).

Intiator and target agents automatically handle any mismatch in data width, clock frequency, or protocol among the various SoC core interfaces and the Sonic interconnect with a minimum cost in terms of delay, latency, or gates.

The SMART interconnects offer includes the following solutions:

- *SonicsLX*: a crossbar-based structure for mid-range SoC designs
- *SonicsMX*: high-performance structure for multicore SoC
- *S3220*: interconnect devised for isolating low-speed peripherals.
- *SonicsExpress*: a high bandwidth bridge between two clock domains, which allows connecting structures from other Sonics SMART interconnects.

All SMART solutions rely on the SonicsStudio™ development environment for architectural exploration, and configuration of the interconnect to exactly match a particular SoC design. The use of this tool significantly reduces the development time because the availability of precharacterization results enables reliable performance analysis and reduction of interconnect timing closure uncertainties. Furthermore, SonicsStudio can be useful to analyze data flows and execute performance verification testing.

Next sections are devoted to the short description of SonicsLX, SonicsMX, and S2330 interconnects.

14.3.1 SonicsLX Interconnect

The SonicsLX SMART interconnect was conceived to target at medium complexity SoCs. Based on a crossbar structure, SonicsLX supports multithreaded, fully pipelined, and nonblocking communications with a distributed implementation. It is fully compatible with other interfaces, as they are the AMBA 3 AXI and AHB or the Open Core Protocol (OCP) [33]. This ensures maximum reuse of cores regardless of their native configuration.

SonicsLX provides a fully configurable interconnect fabric that supports transport, routing, arbitration, and translation functions. This interconnect utilizes state-of-the-art physical structure design and advanced protocol management to deliver guaranteed high bandwidth together with fine grained power management.

Based on the agent structure, SonicsLX also presents decoupling of the functionality of each core from the interconnect communications required among the cores. It supports SoC cores running at different clock rates, and it establishes independent request and response networks to adapt to targets with long or unpredictable latency such as DRAM.

SonicsLX also contains a set of data flow services for the development of complex SoCs that perfectly suits the requirements of mid-range SoC designs. It presents a data flow topology that can be tuned specifically for the mix of low latency and high bandwidth required by the particular application under design.

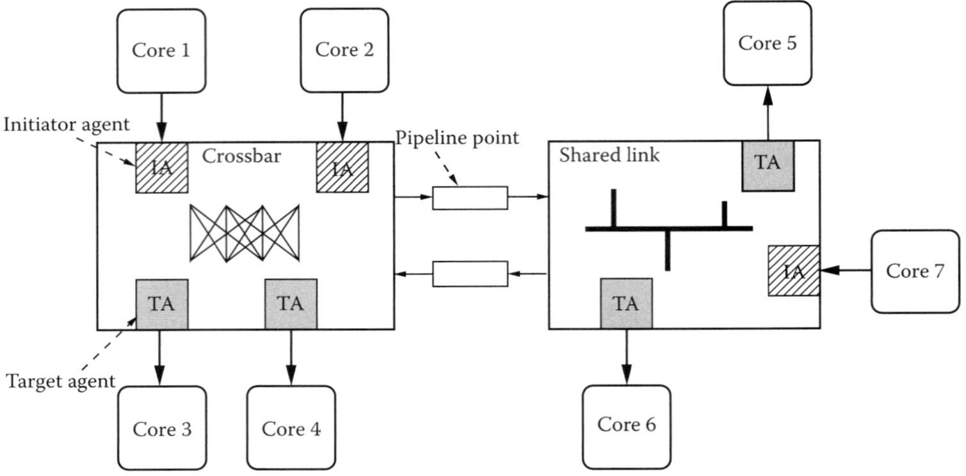

FIGURE 14.5   Example of system with a SonicsMX interconnect including a crossbar and a shared link.

14.3.2   SonicsMX Interconnect

The SonicsMX SMART interconnect contains a full set of state-of-the-art fabric features and data flow services as dictated by the requirements of high-performance SoC designs. In particular, SonicsMX can perfectly face the design of low-power, cost-effective SoC devices powering multimedia-rich wireless and handheld products. It provides the physical structures, advanced protocols, and extensive power management capabilities necessary to overcome data flow and other design challenges of portable multicore SoCs. SonicsMX supports crossbar, shared link, or hybrid topologies within a multithreaded and nonblocking architecture. Again, compliance with OCP, AHB, and AXI interfaces is guaranteed, resulting in a maximum reuse.

SonicsMX combines advanced features such as moderate clock rates, mixed latency requirements, quality of service management, access security, and error management with low-power operation, high bandwidth, and flexibility. As described for SonicsLX, SonicsMX supports multithreaded and nonblocking communications. All mentioned features provide a high degree of predictability and controlability of the design, what results in a significant reduction in the SoC development time.

A typical example of application of the SonicsMX interconnect in a SoC is depicted in Figure 14.5. In this example both crossbar and shared link structures are used in a system with seven cores accessing to the SonicsMX interconnect. Initiator and target agents are used to isolate the cores from the interconnect structures.

14.3.3   S3220 Interconnect

Sonics3220 SMART interconnect is perfect for low-complexity SoCs because of its mature, low-cost structure. It is a nonblocking peripheral interconnect that guarantees end-to-end performance by managing data, control, and test flows between all connected cores.

Providing low latency access to a large number of low bandwidth, physically dispersed target cores, Sonics3220 uses a very low die area interconnect structure that facilitates a rapid path to simulation.

As other Sonics SMART interconnects, the S3220 is built using an advanced split-transaction, nonblocking interconnect fabric that allows latency-sensitive CPU traffic to bypass DMA-based I/O traffic, together with the ability to decouple cores to achieve high IP core reuse. In the same way, it is also fully compatible with IP cores that support AMBA and OCP standards.

14.4 CoreConnect Bus

CoreConnect is an IBM-developed on-chip bus that eases the integration and reuse of processor, subsystem, and peripheral cores within standard product platform designs. It is a complete and versatile architecture clearly targeting high-performance systems, and many of its features might be overkill in simple embedded applications [23].

The CoreConnect bus architecture serves as the foundation of IBM Blue Logic™ or other non-IBM devices. The Blue Logic ASIC/SOC design methodology is the approach proposed by IBM [16] to extend conventional ASIC design flows to current design needs: low-power and multiple-voltage products, reconfigurable logic, custom design capability, and analog/mixed-signal designs. Each of these offerings requires a well-balanced coupling of technology capabilities and design methodology. The use of this bus architecture allows the hierarchical design of SoCs.

As can be seen in Figure 14.6, the IBM CoreConnect architecture provides three buses for interconnecting cores, library macros, and custom logic:

- Processor local bus (PLB)
- On-chip peripheral bus (OPB)
- Device control register (DCR) bus

The PLB connects the processor to high-performance peripherals, such as memories, DMA controllers, and fast devices. Bridged to the PLB, the OPB supports slower-speed peripherals. Finally, the DCR bus is a separate control bus that connects all devices, controllers, and bridges and provides a separate path to set and monitor the individual control registers. It is designed to transfer data between the CPU's general-purpose registers and the slave logic's DCRs. It removes configuration registers from the memory address map, which reduces loading and improves bandwidth of the PLB.

This architecture shares many high-performance features with the AMBA bus specification. Both architectures allow split, pipelined and burst transfers, multiple bus masters, and 32-, 64- or 128-bit architectures. On the other hand, CoreConnect also supports multiple masters in the peripheral bus.

Please note that design toolkits are available for the CoreConnect bus and include functional models, monitors, and a bus functional language to drive the models. These toolkits provide an

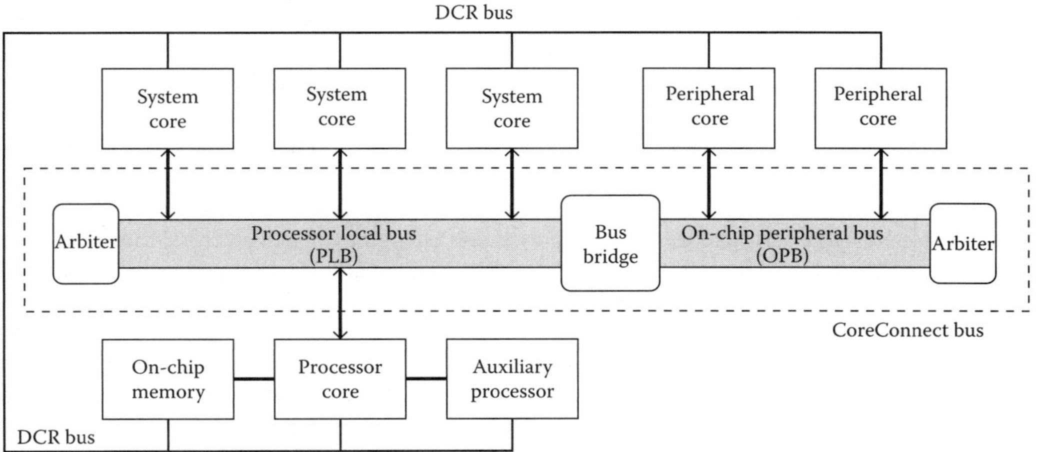

FIGURE 14.6 Schematic structure of the CoreConnect bus.

advanced validation environment for engineers designing macros to attach to the PLB, OPB, and DCR buses.

14.4.1 Processor Local Bus

The PLB is the main system bus targeting high-performance and low-latency on-chip communication. More specifically, PLB is a synchronous, multimaster, arbitrated bus. It supports concurrent read and write transfers, thus yielding a maximum bus utilization of two data transfers per clock cycle. Moreover, PLB implements address pipelining, that reduces bus latency by overlapping a new write request with an ongoing write transfer and up to three read requests with an ongoing read transfer [24].

Access to PLB is granted through a central arbitration mechanism that allows masters to compete for bus ownership. This arbitration mechanism is flexible enough to provide for the implementation of various priority schemes. In fact, four levels of request priority for each master allow PLB implementation with various arbitration priority schemes. Additionally, an arbitration locking mechanism is provided to support master-driven atomic operations. PLB also exhibits the ability to overlap the bus request/grant protocol with an ongoing transfer.

The PLB specification describes a system architecture along with a detailed description of the signals and transactions. PLB-based custom logic systems require the use of a PLB macro to interconnect the various master and slave macros.

The PLB macro is the key component of PLB architecture and consists of a bus arbitration control unit and the control logic required to manage the address and data flow through the PLB. Each PLB master is attached to the PLB through separate address, read data and write data buses, and a plurality of transfer qualifier signals, while PLB slaves are attached through shared, but decoupled, address and *read data and write data buses* (each one with its own transfer control and status signals). The separate address and data buses from the masters allow simultaneous transfer requests. The PLB macro arbitrates among them and sends the address, data, and control signals from the granted master to the slave bus. The slave response is then routed back to the appropriate master. Up to 16 masters can be supported by the arbitration unit, while there are no restrictions in the number of slave devices.

14.4.2 On-Chip Peripheral Bus

Frequently, the OPB architecture connects low-bandwidth devices such as serial and parallel ports, UARTs, timers, etc. and represents a separate, independent level of bus hierarchy. It is implemented as a multimaster, arbitrated bus. It is a fully synchronous interconnect with a common clock, but its devices can run with slower clocks, as long as all of the clocks are synchronized with the rising edge of the main clock.

This bus uses a distributed multiplexer attachment implementation instead of tristate drivers. The OPB supports multiple masters and slaves by implementing the address and data buses as a distributed multiplexer. This type of structure is suitable for the less data intensive OPB and allows adding peripherals to a custom core logic design without changing the I/O on either the OPB arbiter or existing peripherals. All masters are capable of providing an address to the slaves, whereas both masters and slaves are capable of driving and receiving the distributed data bus.

PLB masters gain access to the peripherals on the OPB through the OPB bridge macro. The OPB bridge acts as a slave device on the PLB and a master on the OPB. It supports word (32-bit), half-word (16-bit), and byte read and write transfers on the 32-bit OPB data bus; bursts; and has the capability to perform target word first line read accesses. The OPB bridge performs dynamic bus sizing, allowing devices with different data widths to efficiently communicate. When the OPB bridge master performs an operation wider than the selected OPB slave can support, the bridge splits the operation into two or more smaller transfers.

Some of the main features of the OPB specification are

- Fully synchronous
- Dynamic bus sizing: byte, halfword, fullword, and doubleword transfers
- Separate address and data buses
- Support for multiple OPB bus masters
- Single cycle transfer of data between OPB bus master and OPB slaves
- Sequential address (burst) protocol
- Sixteen-cycle fixed bus timeout provided by the OPB arbiter
- Bus arbitration overlapped with last cycle of bus transfers
- Optional OPB DMA transfers

14.4.3 Device Control Register Bus

The DCR bus provides an alternative path to the system for setting the individual DCRs. These latter are on-chip registers that are implemented outside the processor core, from an architectural viewpoint. Through the DCR bus, the host CPU can set up the DCR sets without loading down the main PLB. This bus has a single master, the CPU interface, which can read or write to the individual DCRs. The DCR bus architecture allows data transfers among OPB peripherals to occur independently from, and concurrently with data transfers between processor and memory, or among other PLB devices. The DCR bus architecture is based on a ring topology to connect the CPU interface to all devices. The DCR bus is typicallly implemented as a distributed multiplexer across the chip such that each subunit not only has a path to place its own DCRs on the CPU read path, but also has a path which bypasses its DCRs and places another unit's DCRs on the CPU read path. DCR bus consists of a 10-bit address bus and a 32-bit data bus.

This is a synchronous bus, wherein slaves may be clocked either faster or slower than the master, although a synchronization of clock signals with the DCR bus clock is required.

Finally, bursts are not supported by this bus, and two-cycle minimum read or write transfers are allowed. Optionally, they can be extended by slaves or by the single master.

14.5 STBus

STBus is an STMicroelectronics proprietary on-chip bus protocol. STBus is dedicated to SoC designed for high bandwidth applications such as audio/video processing [46]. The STBus interfaces and protocols are closely related to the industry standard VCI (Virtual Component Interface). The components interconnected by an STBus are either initiators (which initiate transactions on the bus by sending requests) or targets (which respond to requests). The bus architecture is decomposed into nodes (sub-buses in which initiators and targets can communicate directly), and the internode communications are performed through first-in first-out (FIFO) buffers. Figure 14.7 shows a schematic view of the STBus interconnect.

STBus implements three different protocols that can be selected by the designers in order to meet the complexity, cost, and performance constraints. From lower to higher, they can be listed as follows:

Type 1: Peripheral protocol.   This type is the low cost implementation for low/medium performance. Its simple design allows a synchronous handshake protocol and provides a limited transaction set. The peripheral STBus is targeted at modules which require a low complexity medium data rate communication path with the rest of the system. This typically includes stand-alone modules such as general-purpose input/output or modules which require independent control interfaces in addition to their main memory interface.

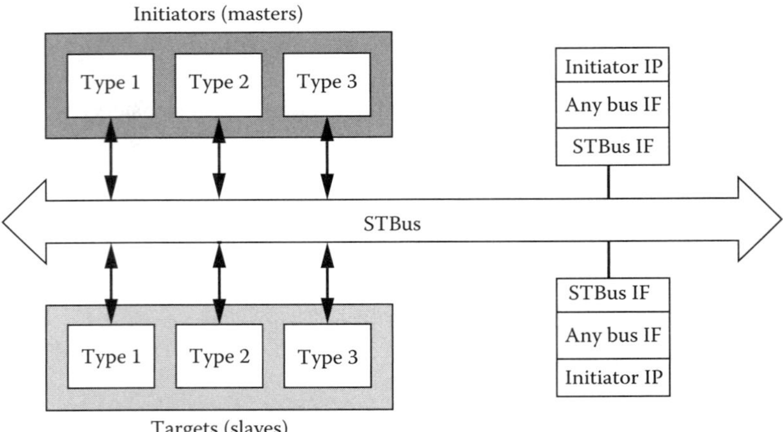

FIGURE 14.7   Schematic view of the STBus interconnect.

Type 2: Basic protocol.   In this case, the limited operation set of the peripheral interface is extended to a full operation set, including compound operations, source labeling, and some priority and transaction labeling. Moreover, this implementation supports split and pipelined accesses and is aimed at devices which need high performance but do not require the additional system efficiency associated with shaped request/response packets or the ability to reorder outstanding operations.

Type 3: Advanced protocol.   The most advanced implementation upgrades previous interfaces with support for out-of-order execution and shaped packets, and is equivalent to the advanced VCI protocol. Split and pipelined accesses are supported. It allows performance improvements either by allowing more operations to occur concurrently or by rescheduling operations more efficiently.

A type 2 protocol preserves the order of requests and responses. One constraint is that, when communicating with a given target, an initiator cannot send a request to a new target until it has received all the responses from the current target. The unresponded requests are called pending, and a pending request controller manages them. A given type 2 target is assumed to send the responses in the same order as the request arrival order. In type 3 protocol, the order of responses may not be guaranteed, and an initiator can communicate with any target, even if it has not received all responses from a previous one.

Associated with these protocols, hardware components have been designed in order to build complete reconfigurable interconnections between initiators and targets. A toolkit has been developed around this STBus (graphical interface) to generate automatically top level backbone, cycle accurate high-level models, way to implementation, bus analysis (latencies, bandwidth), and bus verification (protocol and behavior).

An STBus system includes three generic architectural components. The *node* arbitrates and routes the requests and optionally, the responses. The *converter* is in charge of converting the requests from one protocol to another (for instance, from basic to advanced). Finally, the *size converter* is used between two buses of the same type but of different widths. It includes buffering capability.

The STBus can implement various strategies of arbitration and allow to change them dynamically. In a simplified single-node system example, a communication between one initiator and a target is performed in several steps.

- Request/grant step between the initiator and the node takes place, corresponding to an atomic rendezvous operation of the system.
- Request is transferred from the node to the target.
- Response–request/grant step is carried out between the target and the node.
- Response–request is transferred from the node to the initiator.

14.5.1 Bus Topologies

STBus can instantiate different bus topologies, trading-off communication parallelism with architectural complexity. In particular, system interconnects with different scalability properties can be instantiated, such as,

- Single shared bus: suitable for simple low-performance implementations. It features minimum wiring area but limited scalability.
- Full crossbar: targets complex high-performance implementations. Large wiring area overhead.
- Partial crossbar: intermediate solution, medium performance, implementation complexity, and wiring overhead.

It is worth observing that STBus allows for the instantiation of complex bus systems such as heterogeneous multinode buses (thanks to size or type converters) and facilitates bridging with different bus architectures, provided proper protocol converters are made available (e.g., STBus and AMBA).

14.6 WishBone

The WishBone SoC interconnect [20] defines two types of interfaces: master and slave. Master interfaces are cores that are capable of generating bus cycles, while slave interfaces are capable of receiving bus cycles. Some relevant Wishbone features that are worth mentioning are the multimaster capability which enables multiprocessing, the arbitration methodology defined by end users attending to their needs, and the scalable data bus widths and operand sizes. Moreover, the hardware implementation of bus interfaces is simple and compact, and the hierarchical view of the WishBone architecture supports structured design methodologies [21].

The hardware implementation supports various IP core interconnection schemes, including point-to-point connection, shared bus, crossbar switch implementation, data-flow interconnection, and off-chip interconnection. The crossbar switch interconnection is usually used when connecting two or more masters together so that every one can access two or more slaves. In this scheme, the master initiates an addressable bus cycle to a target slave. The crossbar switch interconnection allows more than one master to use the bus provided they do not access the same slave. In this way, the master requests a channel on the switch and, once this is established, data are transferred in a point-to-point way.

The overall data transfer rate of the crossbar switch is higher than shared bus mechanisms, and can be expanded to support extremely high data transfer rates. On the other hand, the main disadvantage is a more complex interconnection logic and routing resources.

14.6.1 Wishbone Bus Transactions

The WishBone architecture defines different transaction cycles attending to the action performed (read or write) and the blocking/nonblocking access. For instance, single read/write transfers are carried out as follows. The master requests the operation and places the slave address onto the bus.

Then the slave places data onto the data bus and asserts an acknowledge signal. The master monitors this signal and relies the request signals when data have been latched. Two or more back-to-back read/write transfers can also be strung together. In this case, the starting and stopping point of the transfers are identified by the assertion and negation of a specific signal [48].

A read–modify–write (RMW) transfer is also specified, which can be used in multiprocessor and multitasking systems in order to allow multiple software processes to share common resources by using semaphores. This is commonly done on interfaces for disk controllers, serial ports, and memory. The RMW transfer reads and writes data to a memory location in a single bus cycle. For the correct implementation of this bus transaction, shared bus interconnects have to be designed in such a way that once the arbiter grants the bus to a master, it will not re-arbitrate the bus until the current master gives it up. Also, it is important to note that a master device must support the RMW transfer in order to be effective, and this is generally done by means of special instructions forcing RMW bus transactions.

14.7   Other On-Chip Interconnects

Some other interconnects include the PI bus that was developed by several European semiconductor companies (Advanced RISC Machines, Philips Semiconductors, SGS-THOMSON Microelectronics, Siemens, TEMIC/MATRA MHS) within the framework of a European project (OMI, Open Microprocessor Initiative framework*). After this, an extended backward-compatible PI–bus protocol standard frequently used in many hardware systems has been developed by Philips [15].

The high bandwidth and low overhead of the PI–Bus provide a comfortable environment for connecting processor cores, memories, coprocessors, I/O controllers, and other functional blocks in high-performance chips, for time-critical applications.

The PI–bus functional modules are arranged in macrocells, and a wide range of functions are provided. Macrocells with a PI–bus interface can be easily integrated into a chip layout even if they are designed by different manufacturers.

Potential bus agents require only a PI–bus interface of low complexity. Since there is no concrete implementation specified, PI–bus can be adapted to the individual requirements of the target chip design. For instance, the widths of the address and data bus may be varied.

Other example is Avalon [2], an Altera's parameterized interface bus used by the Nios embedded processor. The Avalon switch fabric has a set of predefined signal types with which a user can connect one or more IP blocks. It can only be implemented on Altera devices using SOPC Builder, a system development tool that automatically generates the Avalon switch fabric logic.

The Avalon switch fabric enables simultaneous multimaster operation for maximum system performance by using a technique called slave-side arbitration. It determines which master gains access to a certain slave, in the event that multiple masters attempt to access the same slave at the same time. Therefore, simultaneous transactions for all bus masters are supported and arbitration for peripherals or memory interfaces that are shared among masters is automatically included.

Finally, the CoreFrame architecture has been developed by Palmchip Corp. and relies on point-to-point signals and multiplexing instead of shared tristate lines. It aims at delivering high performance while simultaneously reducing design and verification time [34].

The most distinctive feature of CoreFrame is the separation of I/O and memory transfers onto different buses. The PalmBus provides for the I/O backplane and allows the processor to configure

---

* The PI Bus was incorporated as OMI Standard OMI 324.3D.

and control peripheral blocks while the MBus provides a DMA connection from peripherals to main memory, allowing a direct data transfer without processor intervention.

Other on-chip interconnects are not described here for lack of space: IPBus from IDT [25], IP Interface from Motorola [31], MARBLE asynchronous bus from University of Manchester [8], Atlantic from Altera [1], ClearConnect from ClearSpeed Techn. [11], and FISPbus from Mentor Graphics [32].

14.8 Analysis of Communication Architectures

Traditional SoC interconnects, as exemplified by AMBA AHB, are based upon low-complexity shared buses, in an attempt to minimize area overhead. Such architectures, however, are not adequate to support the trend for SoC integration, motivating the need for more scalable designs. Interconnect performance improvement can be achieved by adopting new topologies and by choosing new protocols, at the expense of silicon area. The former strategy leads from shared buses to bridged clusters, partial or full crossbars, and eventually to networks-on-chip, in an attempt to increase available bandwidth and to reduce local contention. The latter strategy instead tries to maximize link utilization by adopting more sophisticated control schemes and thus permitting a better sharing of existing resources. While both approaches can be followed at the same time, we perform separate analysis for the sake of clarity.

At first, scalability of evolving interconnect fabric protocols is assessed. Three state-of-the-art shared buses are stressed under an increasing traffic load: a traditional AMBA AHB link and more advanced, but also more expensive, evolutionary solutions as offered by STBus (Type 3) and AMBA AXI (basing upon a synopsys implementation).

These system interconnects were selected for analysis because of their distinctive features, which allow to sketch the evolution of shared-bus based communication architectures. AMBA AHB makes two data links (one for reads, one for writes) available, but only one of them can be active at any time. Only one bus master can own the data wires at any time, preventing the multiplexing of requests and responses on the interconnect signals. Transaction pipelining (i.e., split ownership of data and address lines) is provided, but not as a means of allowing multiple outstanding requests, since address sampling is only allowed at the end of the previous data transfer. Bursts are supported, but only as a way to cut down on rearbitration times, and AHB slaves do not have a native burst notion. Overall, AMBA AHB is designed for a low silicon area footprint.

The STBus interconnect (with shared bus topology) implements split request and response channels. This means that, while a system initiator is receiving data from an STBus target, another one can issue a second request to a different target. As soon as the response channel frees up, the second request can immediately be serviced, thus hiding target wait states behind those of the first transfer. The amount of saved wait states depends on the depth of the prefetch FIFO buffers on the slave side. Additionally, the split channel feature allows for multiple outstanding requests by masters, with support for out-of-order retirement. An additional relevant feature of STBus is its low-latency arbitration, which is performed in a single cycle.

Finally, AMBA AXI builds upon the concept of point-to-point connection and exhibits complex features, like multiple outstanding transaction support (with out-of-order or in-order delivery selectable by means of transaction IDs) and time interleaving of traffic toward different masters on internal data lanes. Four different logical monodirectional channels are provided in AXI interfaces, and activity on them can be parallelized allowing multiple outstanding read and write requests. In our protocol exploration, to provide a fair comparison, a "shared bus" topology is assumed, which comprises a single internal lane per each one of the four AXI channels.

Figure 14.8 shows an example of the efficiency improvements made possible by advanced interconnects in the test case of slave devices having two wait states, with three system processors and 4-beat burst transfers. AMBA AHB has to pay two cycles of penalty per transferred datum. STBus is able

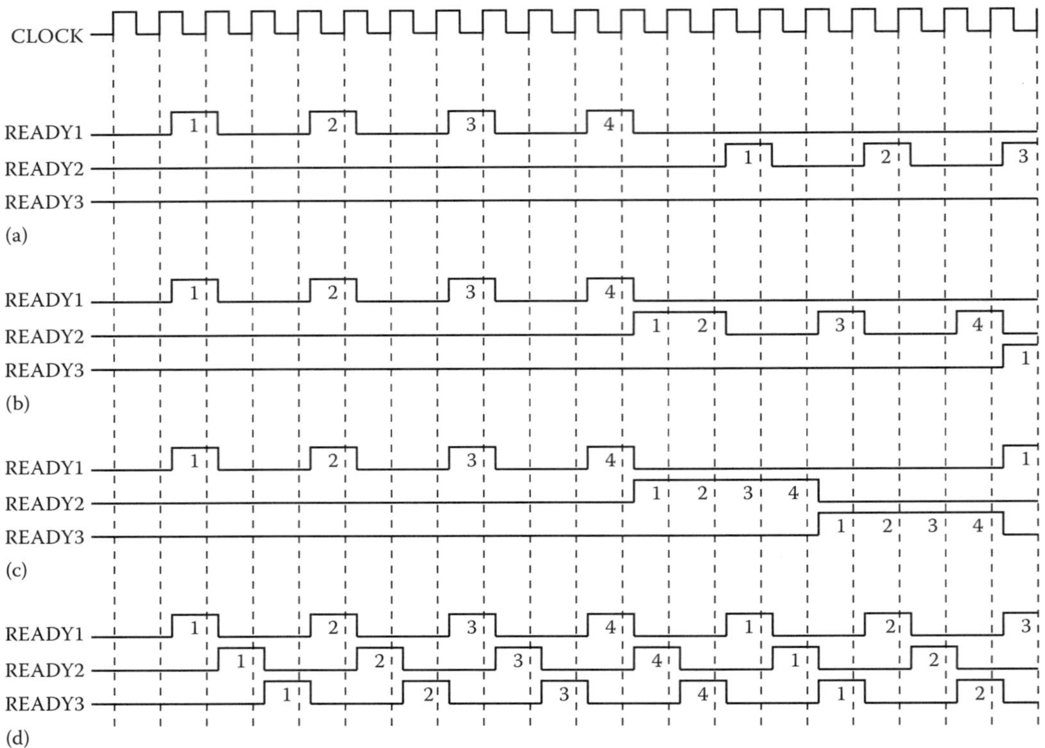

FIGURE 14.8 Concept waveforms showing burst interleaving for the three interconnects. (a) AMBA AHB; (b) STBus (with minimal buffering); (c) STBus (with more buffering); and (d) AMBA AXI.

to hide latencies for subsequent transfers behind those of the first one, with an effectiveness which is a function of the available buffering. AMBA AXI is capable of interleaving transfers, by sharing data channel ownership in time. Under conditions of peak load, when transactions always overlap, AMBA AHB is limited to a 33% efficiency (transferred words over elapsed clock cycles), while both STBus and AMBA AXI can theoretically reach a 100% throughput.

14.8.1 Scalability Analysis

SystemC models of AMBA AHB, AMBA AXI (provided within the Synopsys CoCentric/Designware®️ [12] suites) and STBus are used within the framework of the MPARM simulation platform ([19,28, 30]). For the STBus model, the depth of FIFOs instantiated by the target side of the interconnect is a configurable parameter; their impact can be noticed on concept waveforms in Figure 14.8. 1-stage ("STBus" hereafter) and 4-stage ("STBus (B)") FIFOs were benchmarked.

The simulated on-chip multiprocessor consists of a configurable number of ARM cores attached to the system interconnect. Traffic workload and pattern can easily be tuned by running different benchmark code on the cores, by scaling the number of system processors, or by changing the amount of processor cache, which leads to different amounts of cache refills. Slave devices are assumed to introduce one wait state before responses.

To assess interconnect scalability, a benchmark independently but concurrently runs on every system processor performing accesses to its private slave (involving bus transactions). This means that, while producing real functional traffic patterns, the test setup was not constrained by bottlenecks due to shared slave devices.

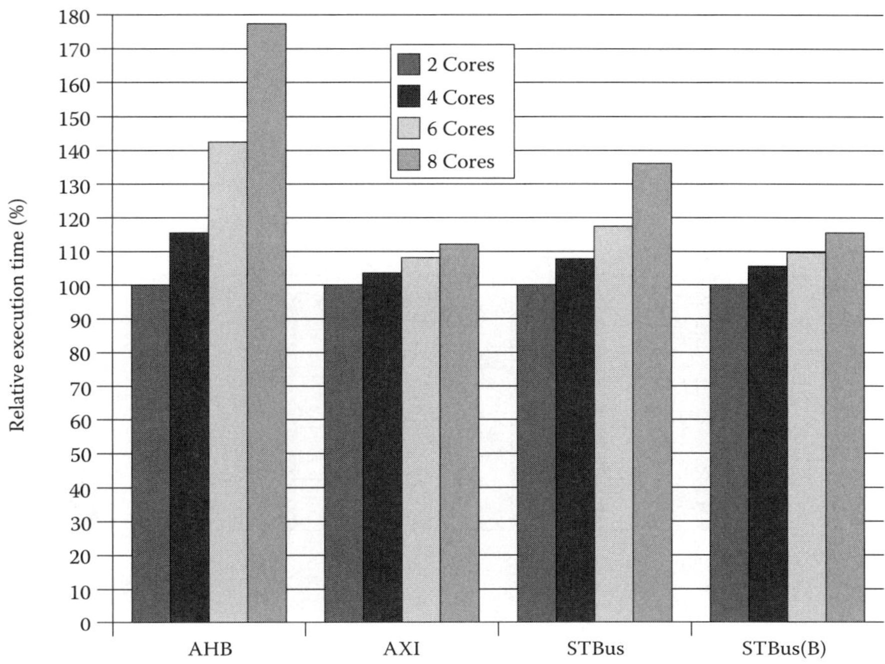

FIGURE 14.9 Execution times with 256 byte caches.

Scalability properties of the system interconnects can be observed in Figure 14.9, reporting the execution time variation when attaching an increasing amount of system cores to a single shared interconnect under heavy traffic load. Core caches are kept very small (256 bytes) in order to cause many cache misses and therefore significant levels of interconnect congestion. Execution times are normalized against those for a two-processor system, trying to isolate the scalability factor alone. The heavy bus congestion case is considered here because the same analysis performed under light traffic conditions (e.g., with 1 kB caches) shows that all of the interconnects perform very well (they are all always close to 100%), with only AHB showing a moderate performance decrease of 6% when moving from two to eight running processors.

With 256 byte caches, the resulting execution times, as Figure 14.9 shows, get 77% worse for AMBA AHB when moving from two to eight cores, while AXI and STBus manage to stay within 12% and 15%. The impact of FIFOs in STBus is noticeable, since the interconnect with minimal buffering shows execution times 36% worse than in the two-core setup. The reason behind the behavior pointed out in Figure 14.9 is that under heavy traffic load and with many processors, interconnect saturation takes place. This is clearly indicated in Figure 14.10, which reports the fraction of cycles during which some transaction was pending on the bus with respect to total execution time.

In such a congested environment, as Figure 14.11 shows, AMBA AXI and STBus (with 4-stage FIFOs) are able to achieve transfer efficiencies (defined as data actually moved over bus contention time) of up to 81% and 83%, respectively, while AMBA AHB reaches 47% only—near to its maximum theoretical efficiency of 50% (one wait state per data word). These plots stress the impact that comparatively low-area-overhead optimizations can sometimes have in complex systems.

According to simulation results, some of the advanced features in AMBA AXI provided highly scalable bandwidth, but at the price of latency in low-contention setups. Figure 14.12 shows the minimum and average amount of cycles required to complete a single write and a burst read transaction in STBus and AMBA AXI. STBus has a minimal overhead for transaction initiation, as low as a single

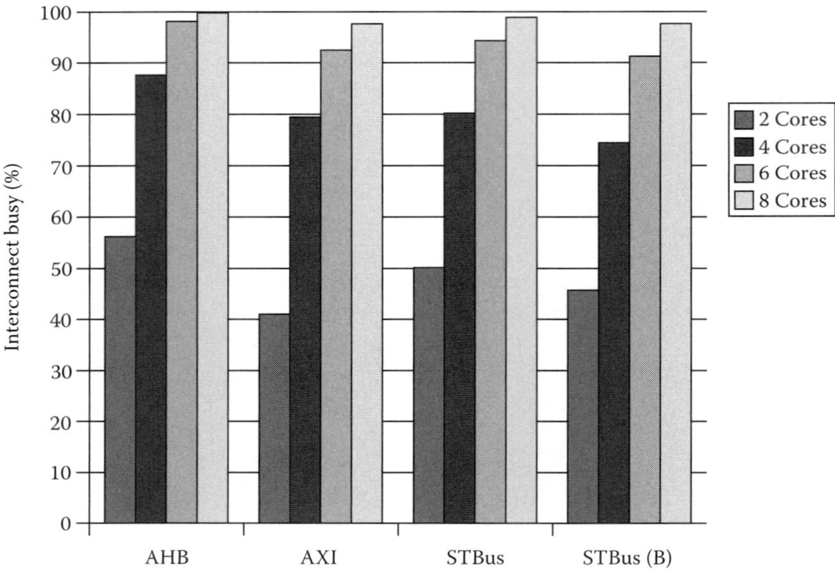

FIGURE 14.10   Bus busy time with 256 byte caches.

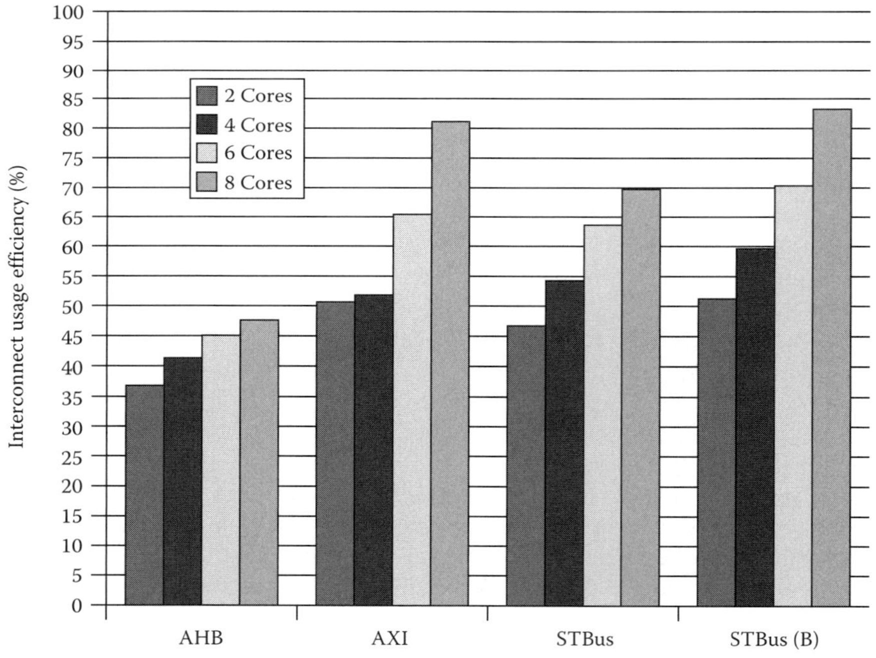

FIGURE 14.11   Bus usage efficiency with 256 byte caches.

cycle if communication resources are free. This is confirmed by figures showing a best-case three-cycle latency for single accesses (initiation, wait state, data transfer) and a nine-cycle latency for 4-beat bursts. AMBA AXI, due to its complex channel management and arbitration, requires more time to initiate and close a transaction: recorded minimum completion times are 6 and 11 cycles for single

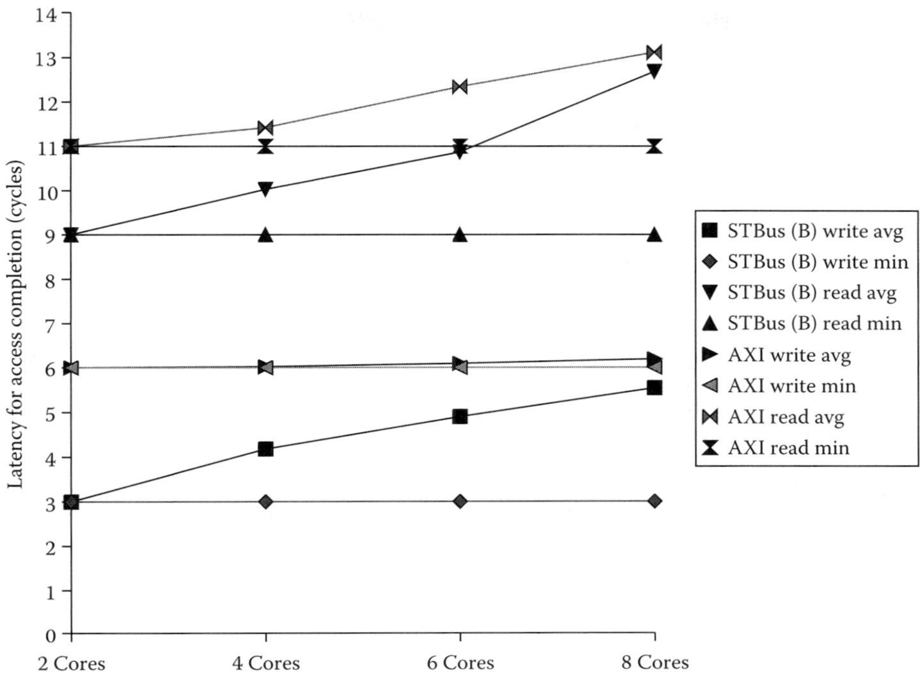

FIGURE 14.12 Transaction completion latency with 256 byte caches.

writes and burst reads, respectively. As bus traffic increases, completion latencies of AMBA AXI and STBus get more and more similar because the bulk of transaction latency is spent in contention.

It must be pointed out, however, that protocol improvements alone cannot overcome the intrinsic performance bound due to the shared nature of the interconnect resources. While protocol features can push the saturation boundary further, and get near to a 100% efficiency, traffic loads taking advantage of more parallel topologies will always exist. The charts reported here already show some traces of saturation even for the most advanced interconnects. However, the improved performance achieved by more parallel topologies strongly depends on the kind of bus traffic. In fact, if the traffic is dominated by accesses to shared devices (shared memory, semaphores, interrupt module), they have to be serialized anyway, thus reducing the effectiveness of area-hungry parallel topologies. It is therefore evident that crossbars behave best when data accesses are local and no destination conflicts arise.

This is reflected in Figure 14.13, showing average completion latencies in read accesses for different bus topologies: shared buses (AMBA AHB and STBus), partial crossbars (STBus-32 and STBus-54), and full crossbars (STBus-FC). Four benchmarks are considered, consisting of matrix multiplications performed independently by each processor or in pipeline, with or without an underlying OS (OS-IND, OS-PIP, ASM-IND, and ASM-PIP, respectively). IND benchmarks do not give rise to interprocessor communication, which is instead at the core of PIP benchmarks. Communication goes through the shared memory. Moreover, OS-assisted code implicitly uses both semaphores and interrupts, while stand-alone ASM applications rely on an explicit semaphore polling mechanism for synchronization purposes. Crossbars show a substantial advantage in OS-IND and ASM-IND benchmarks, wherein processors only access private memories: this operation is obviously suitable for parallelization. ST-FC and ST-54 both achieve the minimum theoretical latency where no conflict on private memories ever arises. ST-32 trails immediately behind ST-FC and ST-54, with rare conflicts which do not occur systematically because execution times shift among conflicting processors. OS-PIP still shows significant improvement for crossbar designs. ASM-PIP, in contrast, puts

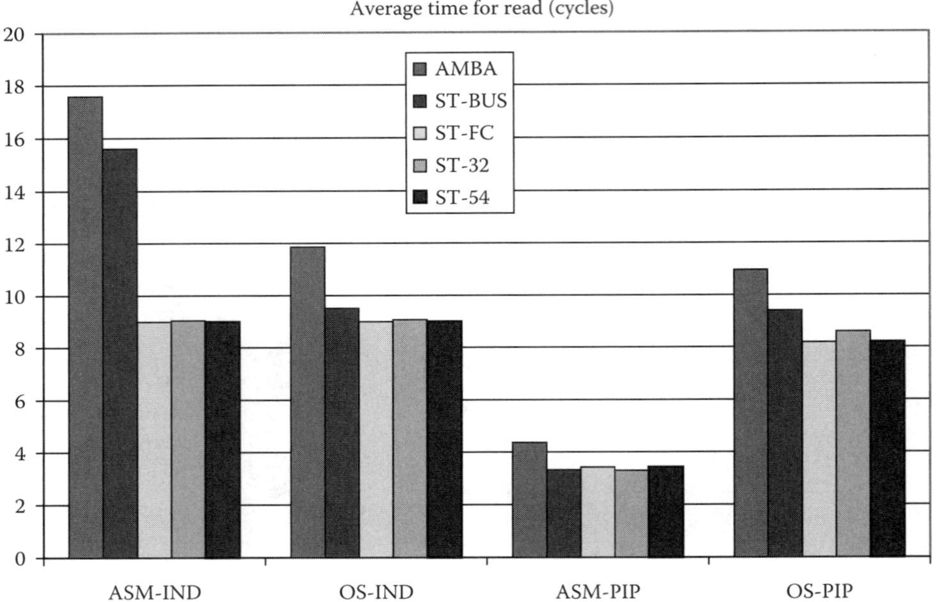

FIGURE 14.13   Reads average latency.

ST-BUS at the same level of crossbars, and sometimes the shared bus even proves slightly faster. This can be explained with the continuous semaphore polling performed by this (and only this) benchmark; while crossbars may have an advantage in private memory accesses, the resulting speedup only gives processors more opportunities to poll the semaphore device, which becomes a bottleneck. Unpredictability of conflict patterns can then explain why a simple shared bus can sometimes slightly outperform crossbars; therefore, the selection of bus topology should carefully match the target communication pattern.

14.9 Packet-Switched Interconnection Networks

Previous sections have illustrated on-chip interconnection schemes based on shared buses and on evolutionary communication architectures. This section introduces a more revolutionary approach to on-chip communication, known as network-on-chip (NoC) [27,29].

The NoC architecture consists of a packet-switched interconnetion network integrated onto a single chip, and it is likely to better support the trend for SoC integration. The basic idea is borrowed from the domain of wide-area networks and envisions router (or switch)-based networks of interconnects on which on-chip packetized communication takes place. Cores access the network by means of proper interfaces and have their packets forwarded to destination through a certain number of hops. SoCs differ from wide area networks in their local proximity and because they exhibit less nondeterminism. Local, high-performance networks, such as those developed for large-scale multiprocessors, have similar requirements and constraints. However, some distinctive features, such as energy constraints and design-time specialization, are unique to SoC networks.

Topology selection for NoCs is a critical design issue. It is determined by how efficiently communication requirements of an application can be mapped onto a certain topology and by physical level considerations. In fact, regular topologies can be designed with a better control on electrical parameters and therefore on communication noise sources (such as cross talk), although they might result in link underutilization or localized congestion from an application viewpoint. On the contrary, irregular topologies have to deal with more complex physical design issues but are more suitable to implement customized, domain-specific communication architectures. Two-dimensional mesh networks are a reference solution for regular NoC topologies.

The scalable and modular nature of NoCs and their support for efficient on-chip communication potentially lead to NoC-based multiprocessor systems characterized by high structural complexity and functional diversity. On one hand, these features need to be properly addressed by means of new design methodologies, while on the other hand more efforts have to be devoted to modeling on-chip communication architectures and integrating them into a single modeling and simulation environment combining both processing elements and communication architectures. The development of NoC architectures and their integration into a complete MPSoC design flow is the main focus of an ongoing worldwide research effort [17,18,41].

Several communication architectures for NoCs have been proposed in the last years. Among them, XPipes has gained a great success due to its parameterization capabilities and high-performance response. This communication library will be described in the following paragraphs.

14.9.1 XPipes

XPipes [9] is a SystemC library of parameterizable, synthesizable NoC components (network interface switch and link modules), which has been optimized for high-performance functioning, that is, low-latency and high-frequency operation. The way of data communication is by means of packet switching and the source routing includes street-sign encoding. XPipes can be selected as the optimal communication infrastructure for multi-gigahertz heterogeneous packet-switched NoCs. Some of the characteristics that allow such efficiency are the design based on highly optimized network building blocks and the instantiation time flexibility.

This network interface is conceived as a bridge between the OCP interface [33] and the manufactured NoC (see Figure 14.14). XPipes takes care of the packet transaction, the synchronization and timings, the computation of routing information, the buffering, and other operations that increase the performance of the communication. The XPipes specifications comply with the OCP 2.0 standard [33] to ensure the easy transactions. The packet partitioning procedure is shown in Figure 14.15, where it can be seen how a flip type field allows to identify the head and the tail flit and to distinguish between header and payload flits.

The high parameterization degree in XPipes is achieved for both global network-specific parameters and block-specific parameters. Network-specific parameters include maximum number of hops between any two nodes, maximum number of bits for end-to-end flow control, flit size, degree of redundancy of error control logic, etc. On the other hand, block-specific parameters include number

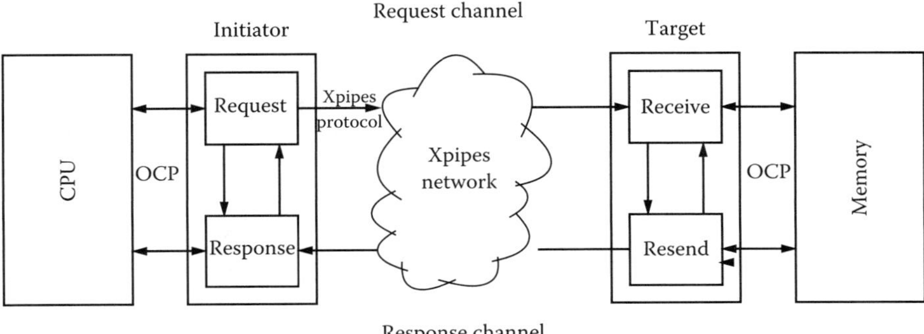

FIGURE 14.14 XPipes network interface.

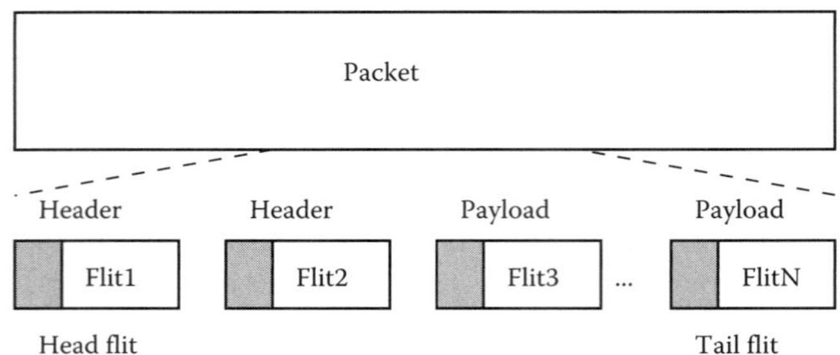

FIGURE 14.15   Packet partitioning procedure.

of address/data lines, maximum burst length, type of interface (master, slave, or both), buffer size at the output port, etc. Parameterization of the switches concerns the number of I/O ports, the number of virtual channels for each physical output link, and the link buffer size. Finally, the length of each individual link can be specified in terms of number of repeater stages.

The operation of the XPipes network interface is based on two registers: one holds the transaction header, while the other holds the transaction payload. The first register needs to be refreshed once per OCP transaction, while the second one samples every burst beat. This is required because each flit encodes a snapshot of the payload register subsequent to a new burst beat. Therefore, multiple payload flits are communicated until transaction completion. Routing information is attached to the header flit of a packet by checking the transaction address against a look-up table (LUT).

In XPipes, two network interfaces can be found: an initiator and a target. The initiator is attached to the system masters, while the target is attached to the system slaves. Therefore, every master–slave device will require an initiator and a target for operation. Additionally, each network interface is split in two modules: one for the request and one for the response channel. This interface is bidirectional: the initiator NI has an output port for the request channel and one input port for the response channel (and the same for the target). Also, whenever a transaction requiring a response is processed by the request channel, the response channel is informed to unblock the communication channel [45].

The input stage of the network interface is implemented as a simple dual-flit buffer with minimal area occupation, while the output stage is identical to that of the XPipes switches.

14.9.1.1 Switch

The XPipes network interface describes the models of the NoC switching block with the following characteristics: a 2-cycle-latency, output-queued router that supports fixed and round-robin priority arbitration on the input lines, and a flow control protocol with ACK/nACK, Go-Back-N semantics. Switch operation is latency insensitive in the sense that correct operation is guaranteed for any link pipeline depth [14].

The switch can be parameterized in the arbitration policy (fixed priority or round-robin), the number of inputs and outputs, and the size of the buffering at the outputs. A schematic representation of a switch for the XPipes NoC is illustrated in Figure 14.16. In this configuration, the switch has four inputs, four outputs, and two virtual channels multiplexed across the same physical output link.

For latency insensitive operation, the switch has virtual channel registers to store $2N + M$ flits, where N is the link length (expressed as number of basic repeater stages) and M is a switch architecture related contribution. The reason is that each transmitted flit has to be acknowledged before being discarded from the buffer and it has to propagate along the link.

In the switch, each output module is deeply pipelined (seven pipeline stages) in order to maximize the operating clock frequency of the device. Also, the CRC decoders for error detection work in parallel with the switch operation, hiding their latency this way.

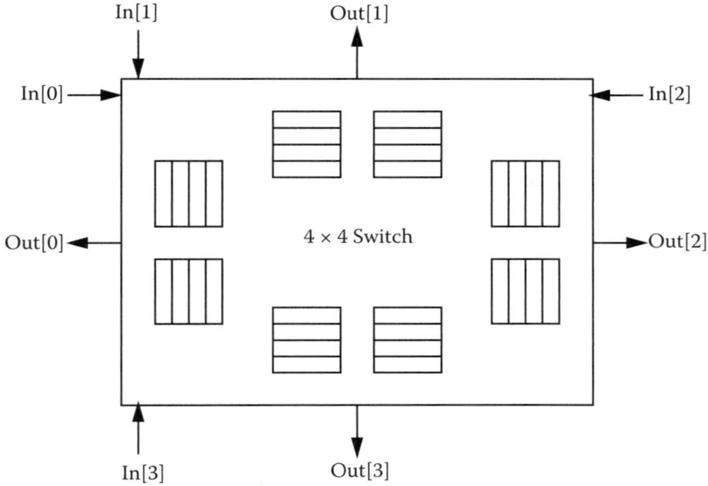

FIGURE 14.16   Switch architecture.

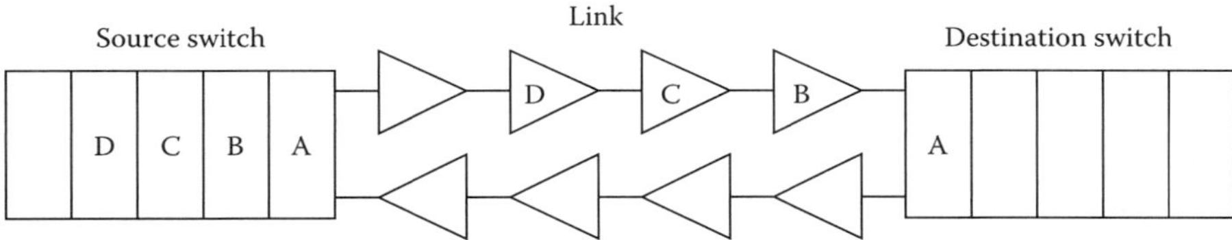

FIGURE 14.17   Link model.

14.9.1.2 Link Architecture

In the design of the link architecture, pipelining has been used both for data and control lines to alleviate the interconnect-delay problem. The pipelined links allow to decouple the data rate from the latency of the link. Therefore, the operation of the link does not depend on the latency of the channel [10].

Also, the architecture of the switch-to-switch links is subdivided into basic modules whose length can be selected depending on the target frequency of the communication. In this way, the network interface is designed to allow different flit timings but requires that they arrive in order (hence, the input links of the switches can be different and of any length) and the operating frequency is not bound by the delay of long links.

Figure 14.17 illustrates the link model, which resembles a pipelined shift register. Pipelining has been used both for data and control lines.

14.10 Current Research Trends

Improvements in process technology have led to more functionality being integrated onto a single chip. This fact has also increased the volume of communication between the integrated components, requiring the design and implementation of highly efficient communication architectures [43].

However, selecting and reconfiguring standard communication architectures to meet application-specific performance requirements is a very time-consuming process. This is due to the large exploration space created by customizable communication topologies, arbitration protocols, clock

TABLE 14.2 Abstraction Levels

Comm. Accuracy	Data Accuracy	Timing Accuracy	SW Accuracy
TLM	Packets	Untimed	Functional specification
		Timed	
TLM	Bytes/words	Transaction	Instr. accurate ISS
		Transfer	Cycle accurate ISS
RTL	Bitvectors	Cycle	HDL processor model

speed, packet sizes, and all those parameters which significantly impact system performance. Moreover, the variability in data traffic exposed by these systems demands the synthesis of numerous buses of different types. This fact is also found in MPSoCs, where the complexity of the architecture and the broad diversity of constraints complicate their design [44]. Therefore, SoC design processes, which integrate early planning of the interconnect architecture at the system level [37] and the automatic synthesis of communication architectures [38], are a very active research area.

System designers and computer architects find very useful to have a system level model of the SoC and the communication architecture. This model can be used to estimate power consumption, get performance figures, and speed up the design process. Some communication models on multiple levels of abstraction have been proposed to overcome the simulation performance and modeling efficiency bottleneck [22,42]. These models were not conceived under any standardization, and the transaction level modeling (TLM) paradigm has recently appeared, which allows to model system level bus-interfaces at different abstraction levels. TLM models employ SystemC as modeling language and, therefore, can rely on the support provided by a new generation of electronic system level SoC design tools [13,47].

Next, the main aspects that have been addressed recently by the SoC research community will be reviewed shortly.

14.10.1 Modeling and Exploring the Design Space of On-Chip Communication

Table 14.2 shows the different abstraction levels at which the functionality for a SoC can be modeled. These abstraction levels are determined by the accuracy of communication, data, time, and software modeling.

TLM allows to cope with the high-level modeling of complex communication structures as point-to-point links and NoC architectures. This level of complexity could not be achieved with lower level modeling strategies. The TLM provides sufficient accuracy and detailed descriptions, but controls the intermodule communication by higher-level interface method calls between the modules. Therefore, simulation is speeded up and granularity level can be descended when desired.

The basic model of a SoC using TLM includes a model of the processing unit (PU) with transactors as interfaces, as well as a model of the interconnection (also specified with transactors). The model of the interconnection has to include the required interface transactors to be adapted to different interconnection standards. Finally, some of these interconnection models can include geometric parameters for the links (e.g., MPSoC interconnection models).

These geometric parameters of the interconnection links have a strong impact on power consumption and power density. Therefore, the interconnection models include the power consumption as a main figure of merit. For the estimation of the power consumption, diverse approaches can be followed. At the system level, power consumption can be defined as a function of the number of messages that are sent through the interconnection between the PUs and the power wasted in the transfer of a single message (or instruction). At a lower abstraction level, the power consumed by the communication architecture will take into account the length and density of wires, the electrical response of the bus, and the distance between repeaters [49].

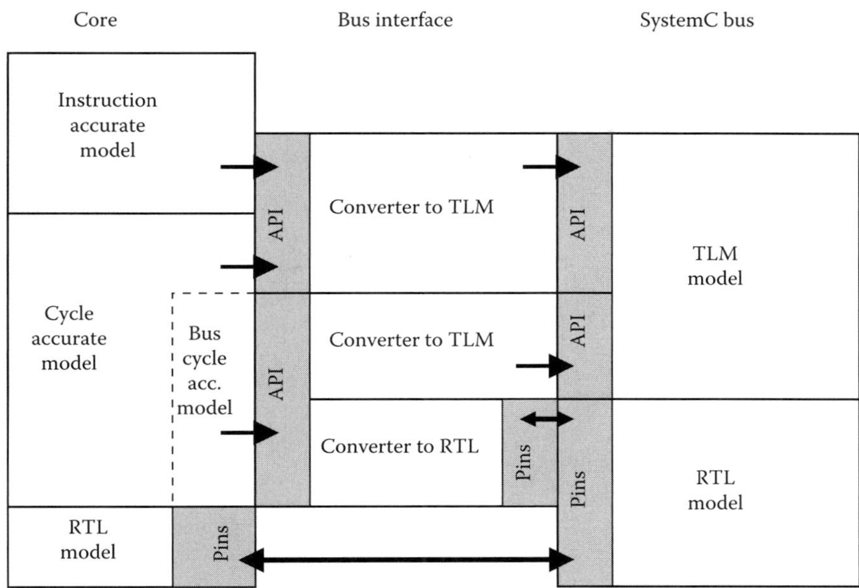

FIGURE 14.18 Modeling integration.

Once both the modeling of the processing elements and the interconnection architecture are completed, the resulting models are integrated in a single environment which provides the simulation and design exploration capabilities (see Figure 14.18).

Regardless of the high-level modeling performed for the communication structures, other approaches take advantage of the detailed information provided by low-level models to obtain figures like power consumption or wire length [44]. In order to reduce the power consumption of the communication buses, several techniques have been devised, for instance, to apply voltage scaling in MPSoC communication buses (those which expose a higher energy consumption). For that, the voltage power supply is scaled to exploit the slack of communication tasks.

14.10.2 Automatic Synthesis of Communication Architectures

Modern SoC and MPSoC systems have high bandwidth requirements that have to be addressed with complex communication architectures. Bus matrix is one of the alternatives being considered to meet these requirements. A bus matrix consists of several buses in parallel, which can support concurrent high bandwidth data streams. Regardless of its high efficiency in terms of bandwidth, the bus matrix is extremely complex to design when the number of PUs in the system increases.

In order to cope with this problem, automated approaches for synthesizing a bus matrix communication architecture are being developed. These approaches consider not only the topology of the system but also the requirements of clock speed, interconnect area, and arbitration strategies [36].

14.11 Conclusions

High-complexity SoCs and MPSoCs designs are increasingly being used in today's high-performance systems. These systems are characterized by a high level of parallelism and large bandwith requirements. The choice of the communication architecture in such systems is very important because it supports the entire interconnect data traffic and has a significant impact on the overall system performance.

This chapter addresses the critical issue of on-chip communication for this kind of systems. An overview of the most widely used on-chip communication architectures is provided, and evolution guidelines aiming at overcoming scalability limitations are sketched. Advances concern both communication protocol and topology, although it is becoming clear that in the long term more aggressive approaches will be required to sustain system performance, namely, packet-switched interconnection networks.

References

1. Altera. Atlantic Interface. Functional Specification, 2002.
2. Altera. Avalon Bus Specification, 2003.
3. ARM. AMBA Specification v2.0, 1999.
4. ARM. AMBA 3 Specification, 2004.
5. ARM. AMBA AXI Protocol Specification, 2004.
6. ARM. AMBA Multi-layer AHB Overview, 2004.
7. ARM. AMBA 3 AHB-Lite Protocol v1.0 Specification, 2006.
8. W.J. Bainbridge and S.B. Furber. MARBLE: An asynchronous on-chip macrocell bus. *Microprocessors and Microsystems*, 24(4):213–222, April 2000.
9. D. Bertozzi and L. Benini. Xpipes: A network-on-chip architecture for gigascale systems-on-chip. *IEEE Circuits and Systems Magazine*, 4(2):18–31, Second Quarter 2004.
10. L.P. Carloni, K.L. McMillan, and A.L. Sangiovanni-Vincentelli. Theory of latency-insensitive design. *IEEE Transactions on CAD of ICs and Systems*, 20(9):1059–1076, September 2001.
11. ClearSpeed. ClearConnect Bus. Scalable High Performance On-Chip Interconnect, 2003.
12. Synopsys CoCentric. http://www.synopsys.com, 2004.
13. CoWare. ConvergenSC. http://www.coware.com, 2008.
14. M. Dall'Osso, G. Biccari, L. Giovannini, D. Bertozzi, and L. Benini. Xpipes: A latency insensitive parameterized network-on-chip architecture for multi-processor SoCs. *Proceedings of ICCD*, San Jose, CA, pp. 536–539, October 2003.
15. Philip de Nier. Property checking of PI–Bus modules. In J.P. Veen, editor, *Proceedings of ProRISC99 Workshop on Circuits, Systems and Signal Processing*, pp. 343–354. STW, Technology Foundation, 1999.
16. G.W. Doerre and D.E. Lackey. The IBM ASIC/SoC methodology: A recipe for first-time success. *IBM Journal Research and Development*, 46(6):649–660, November 2002.
17. E. Bolotin, I. Cidon, R. Ginosar, and A. Kolodny. QNoC: QoS architecture and design process for network on chip. *The Journal of Systems Architecture*, Special Issue on Networks on Chip, 50(2–3):105–128, February 2004.
18. K. Lee, et al. A 51mw 1.6 ghz on-chip network for low power heterogeneous SoC platform. *ISSCC Digest of Tech. Papers*, pp. 152–154, 2004.
19. F. Poletti, D. Bertozzi, A. Bogliolo, and L. Benini. Performance analysis of arbitration policies for SoC communication architectures. *Journal of Design Automation for Embedded Systems*, 22(8):189–210, June/September 2003.
20. R. Herveille. Combining WISHBONE Interface Signals, Application Note, April 2001.
21. R. Herveille. WISHBONE System-on-Chip (SoC) Interconnection Architecture for Portable IP Cores. Specification, 2002.
22. K. Hines and G. Borriello. Dynamic communication models in embedded system co-simulation. In *IEEE DAC, Proceedings of the 34th Annual Conference on Design Automation*, Anaheim, CA, pp. 395–400, 1997.
23. IBM Microelectronics. CoreConnect Bus Architecture Overview, 1999.
24. IBM Microelectronics. The CoreConnect Bus Architecture White Paper, 1999.
25. IDT. IDT Peripheral Bus (IPBus). Intermodule Connection Technology Enables Broad Range of System-Level Integration, 2002.

26. Sonics Inc. SMART Interconnect Solutions. http://www.sonicsinc.com, 2008.

27. J. Henkel, W. Wolf, and S. Chakradhar. On-chip networks: A scalable, communication-centric embedded system design paradigm. *Proceedings of International Conference on VLSI Design*, Mumbai, India, pp. 845–851, January 2004.

28. L. Benini, D. Bertozzi, D. Bruni, N. Drago, F. Fummi, and M. Poncino. SystemC cosimulation and emulation of multiprocessor SoC designs. *IEEE Computer*, 36(4):53–59, April 2003.

29. L. Benini and G. De Micheli. Networks on chips: A new SoC paradigm. *IEEE Computer*, 35(1):70–78, January 2002.

30. M. Loghi, F. Angiolini, D. Bertozzi, L. Benini, and R. Zafalon. Analyzing on-chip communication in a MPSoC environment. *Proceedings of IEEE Design Automation and Test in Europe Conference (DATE04)*, Paris, France, pp. 752–757, February 2004.

31. Motorola. IP Interface. Semiconductor Reuse Standard, 2001.

32. Summary of SoC Interconnection Buses. http://www.silicore.net/uCbusum.htm, 2004.

33. http://www.ocpip.org/.

34. Palmchip. Overview of the CoreFrame Architecture, 2001.

35. S. Pasricha, N. Dutt, and M. Ben-Romdhane. Using TLM for exploring bus-based SoC communication architectures. In *IEEE ASAP, Proceedings of the 2005 IEEE International Conference on Application–Specific Systems, Architecture Processors*, Samos, Greece, pp. 79–85, 2005.

36. S. Pasricha, N. Dutt, and M. Ben-Romdhane. Constraint-driven bus matrix synthesis for MPSoC. In *IEEE ASP-DAC, Proceedings of the 2006 Conference on Asia South Pacific Design Automation*, Yokohama, Japan, pp. 30–35, 2006.

37. S. Pasricha, N. Dutt, and M. Ben-Romdhane. BMSYN: Bus matrix communication architecture synthesis for MPSoC. *IEEE Transactions on CAD*, 8(26):1454–1464, August 2007.

38. S. Pasricha, N. Dutt, E. Bozorgzadeh, and M. Ben-Romdhane. Floorplan-aware automated synthesis of bus-based communication architectures. In *IEEE DAC, Proceedings of the 42nd Annual Conference on Design Automation*, Anaheim, CA, pp. 565–570, 2005.

39. M. Posner and D. Mossor. Designing using the AMBA 3 AXI protocol. Synopsys White Paper, Productivity Series, April 2005.

40. R. Ho, K.W. Mai, and M.A. Horowitz. The future of wires. *Proceedings of the IEEE*, 89(4):490–504, April 2001.

41. E. Rijpkema, K. Goossens, and A. Radulescu. Trade-offs in the design of a router with both guaranteed and best-effort services for networks on chip. *Proceedings of Design Automation and Test in Europe*, Munich, Germany, pp. 350–355, March 2003.

42. J.A. Rowson and A. Sangiovanni-Vincentelli. Interface-based design. In *IEEE DAC, Proceedings of the 34th Annual Conference on Design Automation*, Anaheim, CA, pp. 178–183, 1997.

43. J.A. Rowson and A. Sangiovanni-Vincentelli. Getting to the bottom of deep sub-micron. In *IEEE ICCAD, Proceedings of the 1998 IEEE/ACM International Conference on Computer-Aided Design*, San Jose, CA, pp. 203–211, 1998.

44. Y. Sheynin, E. Suvorova, and F. Shutenko. Complexity and low power issues for on-chip interconnections in MPSoC system level design. In *IEEE ISVLSI, Proceedings of the IEEE Computer Society Annual Symposium on Emerging VLSI Technologies and Architectures*, Karlsruhe, Germany, p. 283, 2006.

45. S. Stergiou, F. Angiolini, S. Carta, L. Raffo, D. Bertozzi, and G. De Micheli. Xpipes lite: A synthesis oriented design library for networks on chips. *Proceedings of Design Automation and Test in Europe*, Munich, Germany, pp. 1188–1193, March 2005.

46. STMicroelectronics. http://www.st.com/stonline/products/technologies/soc/stbus.htm, 2008.

47. Synopsys. CoCentric System Studio. http://www.synopsys.com, 2008.

48. R. Usselmann. OpenCores SoC Bus Review, 2001.

49. A. Wieferink, T. Kogel, R. Leupers, G. Ascheid, H. Meyr, et al. A system level processor/communication co-exploration methodology for multi-processor system-on-chip platforms. In *IEEE DATE, Proceedings of the Conference on Design, Automation and Test in Europe*, Paris, France, pp. 21256, 2004.

15

Networks-on-Chip: An Interconnect Fabric for Multiprocessor Systems-on-Chip

Francisco Gilabert
Polytechnic University of Valencia

Davide Bertozzi
University of Ferrara

Luca Benini
University of Bologna

Giovanni De Micheli
Ecole Polytechnique Fédérale de Lausanne

15.1 Introduction

The increasing integration densities made available by shrinking of device geometries will have to be exploited to meet the computational requirements of applications from different domains, such as multimedia processing, high-end gaming, biomedical signal processing, advanced networking services, automotive, or ambient intelligence.

Interestingly, the request for scalable performance is being posed not only to high-performance microprocessors, which have been tackling this challenge for a long time, but also to embedded computing systems. As an example, systems designed for ambient intelligence are increasingly based on high-speed digital signal processing with computational loads ranging from 10 MOPS for lightweight audio processing, 3 GOPS for video processing, 20 GOPS for multilingual conversation interfaces, and up to 1 TOPS for synthetic video generation. This computational challenge has to be addressed at manageable power levels and affordable costs [BOE02].

Such a performance cannot be provided by a single processor core, but requires a heterogeneous on-chip multiprocessor system containing a mix of general-purpose programmable cores, application-specific processors, and dedicated hardware accelerators. In order for the computation scalability provided by multicore architectures to be effective, the communication bottleneck will have to be removed.

In this context, performance of gigascale systems-on-chip (SoCs) will be communication dominated, and only an interconnect-centric system architecture will be able to cope with this problem.

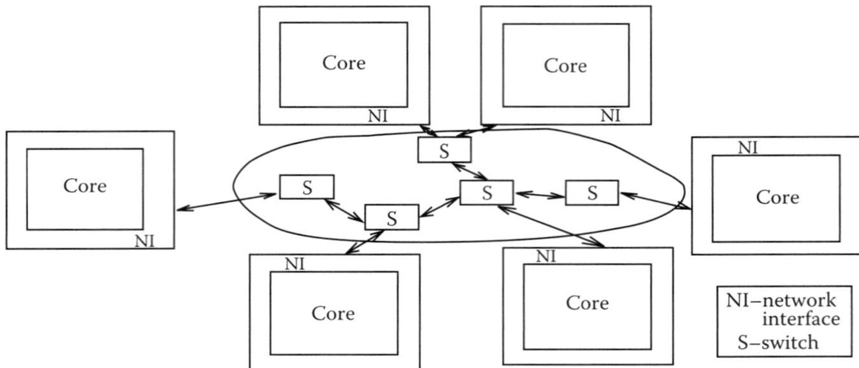

FIGURE 15.1   Generic network-on-chip architecture.

Current on-chip interconnects consist of low-cost shared arbitrated buses; based on the serialization of bus access requests, only one master at a time can be granted access to the bus. The main drawback of this solution is its poor scalability, which will result in unacceptable performance degradation for medium complexity SoCs (more than a dozen of integrated cores). Moreover, the connection of new blocks to a shared bus increases its associated load capacitance, resulting in more energy consuming bus transactions associated with the broadcast communication paradigm.

A scalable communication infrastructure that better supports the trend of SoC integration consists of an on-chip micronetwork of interconnects, generally known as network-on-chip (NoC) architecture [BEN02,WIE02,DAL01]. The basic idea is borrowed from the wide-area networks domain, and envisions on-chip networks on which packet-switched communication takes place, as depicted in Figure 15.1. Cores access the network by means of proper interfaces and have their packets forwarded to destination through a certain number of intermediate hops (corresponding to switching elements).

The modular nature of NoC architectures and the enormous communication bandwidth they can provide leads to network-centric multiprocessor systems (MPSoCs) featuring high structural complexity and functional diversity. While in principle attractive, these features imply an extension of the design space that needs to be properly mastered by means of new design methodologies and tool flows [KUM02].

A NoC design choice of utmost importance for global system performance concerns topology selection. Topology describes the connectivity pattern of networked cores, and heavily impacts the final throughput and latency figures for on-chip communication. In particular, these figures stem from the combination of the theoretical properties of a NoC topology (e.g., average latency, bisection bandwidth) with the quality of its physical synthesis (e.g., latency on the express links of multidimension topologies, maximum operating frequency). Several researchers [DAL01,GUE00,KUM02, LEE03] envision NoCs for the general-purpose computing domain as regular tile-based architectures (structured as mesh networks or fat trees). In this case, the system is homogeneous, in that it can be obtained by the replication of the same basic set of components: a processing tile with its local switching element.

However, in other application domains, designers need to optimize performance at a lower power cost and rely on SoC component specialization for this purpose. This leads to the on-chip integration of heterogeneous cores having varied functionality, size, and communication requirements. In this scenario, if a regular interconnect is designed to match the requirements of a few communication-hungry components, it is bound to be largely over-designed for the needs of the remaining components. This is the main reason why most of current heterogeneous SoCs make use of irregular topologies [YAM02,BEN06].

This chapter introduces basic principles and guidelines for the NoC design. At first, the motivation for the design paradigm shift of SoC communication architectures from shared buses to NoCs is examined. Then, the chapter goes into the details of NoC building blocks (switch, network interface [NI], and switch-to-switch links), presenting their design principles and the trade-offs spanned by different implementation variants. Readers will be given the opportunity to become familiar with the theoretic notions by analyzing a few case studies from real-life NoC prototypes. For each of them, the design objectives leading to the specific architecture choices will be illustrated.

Then, the key issue of topology selection will be discussed with reference to both the general-purpose and the application-specific computing domains. This is a high-impact decision in the NoC (and system) design process, and will be devoted ample room in this chapter. Finally, the main challenges that research has to tackle in order for NoCs to become mainstream will be discussed briefly.

15.2 Design Challenges for on-Chip Communication Architectures

SoC design challenges that are driving the evolution of traditional bus architectures toward NoCs can be outlined as follows:

Technology challenges. While gate delays scale down with technology, global wire delays typically increase or remain constant as repeaters are inserted. It is estimated that in 50 nm technology, at a clock frequency of 10 GHz, a global wire delay can be up to 6–10 clock cycles [BEN02]. Therefore, limiting the on-chip distance traveled by critical signals will be key to guarantee the performance of the overall system, and will be a common design guideline for all kinds of system interconnects. In this direction, recent communication protocols such as AMBA AXI pose no requirements for a fixed relationship between the various channels the protocol is structured into. This way, the insertion of a register stage in any channel is made possible, at the cost of an additional cycle of latency [AXI03]. By breaking long timing paths, the maximum operating frequency can be preserved. NoCs are multihop architectures pushing this trend to the limit: an aggressive path segmentation enables an operating frequency, which is much higher than that of state-of-the-art interconnect fabrics and even of attached communicating cores. The potentially higher speed of NoCs can also be an indirect means of masking their inherently higher communication latency.

Another technology-related issue concerns performance predictability. This can be defined as the deviation of postlayout performance figures from postsynthesis ones. The traditional flow for standard cell design features logic synthesis and placement as two clearly decoupled stages. Wire load models make this splitting possible. They consist of precharacterized equations, supplied within technology libraries that attempt to predict the capacitive load that a gate will have to drive based on its fan-out alone. A gate driving a fan-out of two other gates is very likely to be part of a local circuit. Thus, its capacitive load is little more than the input capacitance of the two downstream gates. A gate with a fan-out of one thousand is likely to be the driver of a global network. Therefore, some extra capacitance is expected due to the long wires needed to carry the signal around. This assumption works very well as long as wire loads do not become too large. Otherwise, the characterization of wire load models becomes very complex, and the prediction inaccuracies become critical. It was showed in [ANG06] that a state-of-the-art AMBA AHB multi-layer interconnect is already ineffectively handled by this synthesis flow at the 130 nm technology node: after placement, the actual achievable frequency decreases by a noticeable 17%. This drop means that some capacitive loads, which were not expected after the logical synthesis, arose due to routing constraints. An explanation can be found in the purely combinational nature of the multi-layer fabric, which implies long wire propagation times and compounds the delay of the crossbar block. In contrast, the same work in [ANG06] proves that an on-chip network incurs a negligible timing penalty of less than 3% after taking actual capacitive loads into account. This confirms that NoC wire segmentation is highly effective and results in wire load predictability. However, even for NoCs, the above synthesis flow proves substantially inadequate as technology scales further down to the 65 nm node. The origin of the problem lies in the unacceptable inaccuracy in wire load estimation. Even when synthesizing single NoC modules (i.e., even without considering long links), after the logic synthesis step, tools were expecting some target frequency to be reachable. However, after the placement phase, the results were up to 30% worse [PUL07]. Unfortunately, traditional placement tools are not able to deeply modify the netlists they are given as an input. In general, they can only insert additional buffering to account for unexpected loads on few selected wires. Therefore, if the input netlist is fundamentally off the mark due to erroneous wire load expectations, not only a performance loss is certain, but also the placement runtime skyrockets. To address this issue, placement-aware logic synthesis tools are replacing the old flow. In this case, after a very quick initial logic synthesis based on wire load models, the tool internally attempts a coarse placement of the current netlist, and it also keeps optimizing the netlist based on the expected placement and the wire loads it implies. The final resulting netlist already considers placement-related effects. Therefore, after this netlist is fed to the actual placement tool, performance results will not incur major penalties. Overall, NoCs promise better scalability to future technology nodes, but their efficient design will require silicon-aware decision making at each level of the design process, in particular placement-aware logic synthesis.

Another challenge associated with the effects of nanoscale technologies is posed by global synchronization of large multi-core chips. The difficulty to control clock skew and the power associated with the clock distribution tree will probably cause a design paradigm shift toward globally asynchronous and locally synchronous (GALS) systems [GUR06]. The basic GALS paradigm is based on a system composed of a number of synchronous blocks designed in a traditional way (and hence exploiting standard synthesis methodologies and tools). However, it is assumed that clocks of such synchronous systems are not necessarily correlated and consequently that those synchronous systems communicate asynchronously using handshake channels. Locally synchronous modules are usually surrounded by asynchronous wrappers providing such interblock data transfer. Practical GALS implementations may form much more complex structures, such as bus [VIL03] or NoC structures [LAT07] for interblock communications, and use different data synchronization mechanisms.

The GALS paradigm does not rely on absolute timing information and therefore favors composability: local islands of synchronicity can arbitrarily be combined to build up larger systems [KRS06]. NoC architectures are generally viewed as an ideal target for application of the GALS paradigm [LAT07]. While in the short-term, a network-centric system will be most likely composed of multiple clock domains with tight or loose correlation; in the long run, fully asynchronous interconnects might be an effective solution. The uptake of these latter for commercial applications will largely depend on the availability of a suitable design tool flow. More in general, GALS NoCs are a promising means of tackling a number of interconnect issues, from power and EMI reduction to clock skew management, while preserving design modularity and improving the back-end time spent achieving timing convergence.

Finally, signal integrity issues (cross talk, power supply noise, soft errors, etc.) will lead to more transient and permanent failures of signals, logic values, devices, and interconnects, thus raising the reliability concern for on-chip communication [BER03]. In many cases, on-chip networks can be designed as regular structures, allowing electrical parameters of wires to be optimized and well controlled. This leads to lower communication failure probabilities, thus enabling the use of low swing signaling techniques [ZHA00] and the exploitation of performance optimization techniques such as wavefront pipelining [XU02]. It is worth observing that permanent communication failures can also occur as an effect of the deviation of technology parameters from nominal design values (such as effective gate length, threshold voltage, or metal width). In fact, precise control of chip manufacturing becomes increasingly difficult and expensive to maintain in the nanometer regime. The ultimate effect is a circuit performance and power variability that results in increasing yield degradation in successive technology nodes. The support for process variation tolerance poses unique design challenges to on-chip networks. From a physical viewpoint, network circuits should be able to adapt to in-situ actual technology conditions, for instance, by means of the self-calibrating techniques proposed in [WOR05,MED08]. From an architecture viewpoint, one or more sections of the network might be unusable, thus calling for routing and topology solutions able to preserve interconnection of operating nodes [FLI08].

Scalability challenges. Present-day SoC design has broken with 1-processor system design that has dominated since 1971. Dual-processor system design is very common in the design of voice-only mobile telephone handsets. A general-purpose processor handles the handset's operating system and user interface. A DSP handles the phone handset's baseband processing tasks (e.g., DSP functions such as FFTs and inverse FFTs, symbol coding and decoding, filtering). Processing bandwidth is finely tuned to be just enough for voice processing to cut product cost and power dissipation, which improves key selling points such as battery life and talk time.

Incorporation of multimedia features (music, still image, video) has placed additional processing demands on handset SoC designs, and the finely tuned, cost-minimized 2-processor system designs for voice-only phones simply lack processing bandwidth for these additional functions. Consequently, the most recent handset designs with new multimedia features are adding either hardware acceleration blocks or application processors to handle the processing requirements of the additional features. The design of multiprocessor SoC systems is today a common practice in the embedded computing domain [LEI06]. Many SoCs today incorporate dozens or even hundreds of interconnected processor cores. The International Technology Roadmap for Semiconductors (ITRS) predict that this trend will continue, and more than 800 integrated processor cores are expected in 2020 [TRS07].

A similar design paradigm shift is taking place in the high-performance computing domain [BOR05]. In the past, performance scaling in conventional single-core processors has been accomplished largely through increases in clock frequency (accounting for roughly 80% of the performance gains at each technology generation). But frequency scaling has run into fundamental physical barriers. First, as chip geometries shrink and clock frequencies rise, the transistor leakage current increases, leading to excessive power consumption and heat. Second, the advantages of higher clock speeds are in part negated by memory latency, since memory access times have not been able to keep pace with increasing clock frequencies. Third, for certain applications, traditional serial architectures are becoming less efficient as processors get faster (due to the so-called von Neumann bottleneck), further undercutting any gains that frequency increases might otherwise achieve. In addition, RC delays in signal transmission are growing as feature sizes shrink, imposing an additional bottleneck that frequency increases do not address.

Therefore, performance will have to come by other means than boosting the clock speed of large monolithic cores. Another means of maintaining microprocessor performance growth has traditionally come from microarchitectural innovations. They include multiple instruction issue, dynamic scheduling, speculative execution, and nonblocking caches. However, early predictions for the superscalar execution model (such as [OLU97]) projected diminishing returns in performance for increasing issue width, and they finally came true.

In light of the above issues, the most promising approach to deliver massively scalable computation horsepower while effectively managing power and heat is to break up functions into many concurrent operations and to distribute them across many parallel processing units. Rather than carrying out a few operations serially at an extremely high frequency, chip multiprocessors (CMPs) achieve high performance at more practical clock rates, by executing many operations in parallel. This approach is also more power-aware: rather than making use of a big, power-hungry, and heat-producing core, CMPs need to activate only those cores needed for a given function, while idle cores are powered down. This fine-grained control over processing resources enables the chip to use only as much power as it is needed at any time.

Today, evidence of this CMP trend is unmistakable as practically every commercial manufacturer of high-performance processors is currently introducing products based on multicore architectures: AMD's Opteron, Intel's Montecito, Sun's Niagara, IBM's Cell, and Power5. These systems aim to optimize performance per watt by operating multiple parallel processors at lower clock frequencies.

Clearly, within the next few years, performance gains will come from increases in the number of processor cores per chip, leading to the emergence of a key bottleneck: the global intrachip communication infrastructure. Perhaps the most daunting challenge to future systems is to realize the enormous bandwidth capacities and stringent latency requirements when interconnecting a large number of processing cores in a power efficient fashion.

Thirty years of experience with microprocessors taught the industry that processors must communicate over buses, so the most efficient way of interconnecting processors has been through the use of common bus structures. When SoCs only used one or two processors, this approach was practical. With dozens or hundreds of processors on a silicon die, it no longer is. Bus-centric design restricts bandwidth and wastes the enormous connectivity potential of nanometer SoC designs. Moreover, chip-wide combinational structures (as many buses are) have been proven to map inefficiently to nanoscale technologies. Low latency, high data-rate, on-chip interconnection networks have therefore become a key to relieving one of the main bottlenecks for MPSoC and CMP system performance. NoCs represent an interconnect fabric which is highly scalable and can provide enough bandwidth to replace many traditional bus-based and/or point-to-point links.

Design productivity challenges. It is well known that synthesis and compiler technology development does not keep up with IC manufacturing technology development [TRS07]. Moreover, times-to-market need to be kept as low as possible. Reuse of complex preverified design blocks is an effective means of increasing productivity, and regards both computation resources and the communication infrastructure [JAN03]. It would be highly desirable to have processing elements that could be employed in different platforms by means of a plug-and-play design style. To this purpose, a scalable and modular on-chip network represents a more efficient communication fabric compared with shared bus architectures. However, the reuse of processing elements is facilitated by the definition of standard interface sockets (corresponding to the front-end of NIs), which make the modularity property of NoC architectures effective. The open core protocol (OCP) [OCP01] was devised as an effective means of simplifying the integration task through the standardization of the core interface protocol. It also paves the way for more cost-effective system implementations, since it can be custom-tailored based on the complexity and the connectivity requirements of the attached cores. AMBA AXI [AXI03] is another example of standard interface socket for processing and/or memory cores. The common feature of these point-to-point communication protocols is that they are core-centric transaction-based protocols which abstract away the implementation details of the system interconnect. Transactions are specified as though communication initiator and target were directly connected, taking the well-known layered approach to peer-to-peer communication that proved extremely successful in the domain of local- and wide-area networks. Referring to the reference ISO/OSI model, the NI front-end can be viewed as the standardized core interface at session layer or transport layer, where network-agnostic high-level communication services are provided. This way, the development of new processor cores and of new NoC architectures become two largely orthogonal activities.

Finally, let us observe that NoC components (e.g., switches or interfaces) can be instantiated multiple times in the same design (as opposed to the arbiter of traditional shared buses, which is instance-specific) and reused in a large number of products targeting a specific application domain.

15.3   Network-on-Chip Architecture

Most of the terminology for on-chip packet switched communication is adapted from the computer cluster domain. Messages that have to be transmitted across the network are usually partitioned into fixed-length packets. Packets in turn are often broken into message flow control units called flits. In presence of channel width constraints, multiple physical channel cycles can be used to transfer a single flit. A phit is the unit of information that can be transferred across a physical channel in a single step. Flits represent logical units of information, as opposed to phits that correspond to physical quantities. In many implementations, a flit is set to be equal to a phit. The basic building blocks for packet switched communication across NoCs are

1. Network interface
2. Switch
3. Network link

and will be described hereafter.

15.3.1   Network Interface

The NI is the basic block allowing connectivity of communicating cores to the global on-chip network. It is in charge of providing several services, spanning from session and transport layers (e.g., decoupling computation from communication, definition of QoS requirements of network transactions, and packetization) up to the layers closer to the physical implementation (e.g., flow control, clock domain crossing, etc.).

The NI provides basic core wrapping services. Its role is to adapt the communication protocol of the attached core to the communication protocol of the network. Layering naturally decouples system processing elements from the system they reside in, and therefore enables design teams to partition a design into numerous activities that can proceed concurrently since they are minimally interdependent. Layering also naturally enables core reuse in different systems. By selecting an industry standard interface, there is no added time for this reuse approach since all cores require such an interface.

Packetization is the very basic service the NI should offer: taking the incoming signals of the processor core transaction and building packets compliant with the NoC communication protocol. This service should carefully optimize packet and flit size in order to achieve high-performance network operation and reduce implementation complexity of network building blocks.

At a lower level of abstraction, clock domain crossing may need to be performed by the NI. Even if the clock frequency is the same over the entire chip, phase adaptation will be needed for communications. In a more general case, the system might consist of multiple clock domains which act as islands of synchronicity. Clearly, proper synchronizers are needed at the clock domain boundary. A simplified case of this scenario occurs when each communicating core/tile coincides with one clock domain, while the network gives rise to an additional clock domain with an operating frequency which is typically much higher than that of the attached cores. This way, clock domain crossing has to be performed at the NIs. A true GALS paradigm might be applied to the limit, which envisions a fully asynchronous interconnect fabric. In this case, the conversion from synchronous to asynchronous (and viceversa) takes place in the NI.

The NI is also the right place to specify and enforce latency and/or bandwidth guarantees. This can be achieved for instance by reserving virtual circuits throughout the network or by allocating multiple buffering resources and by designing complex packet schedulers in the NI in order to handle traffic with different QoS requirements.

In addition to adaptation services, the NI is expected to provide the traditional transport layer services. Transaction ordering is one of them: whenever dynamic routing schemes are adopted in the network, packets can potentially arrive unordered, thus raising the memory consistency issue. In this case, NIs should reorder the transaction before forwarding data or control information to the attached component. Another transport layer service concerns the reliable delivery of packets from source to destination nodes. The on-chip communication medium has been historically considered as a reliable medium. However, recent studies expect this to be no longer the case in the context of nanoscale technologies. The NI could be involved in providing reliable network transactions by inserting parity check bits in packet tails at the packetization stage or by implementing end-to-end error control. Finally, NIs should be in charge of flow control. When a given buffering resource in the network is full, there needs to be a mechanism to stall the packet propagation and to propagate the stalling condition upstream. The flow control mechanism is in charge of regulating the flow of packets through the network and dealing with localized congestion. The NI is involved with the generation of flow control signals exchanged with the attached switch. Moreover, flow control has to be carried out also with the connected cores. In fact, when the network cannot accept new packets any more because of congestion, new transactions can be accepted in the NI from the connected core just at the same, provided the necessary amount of decoupling buffers is available in the NI. However, when also these buffers are full, the core behavior is impacted by congestion in the network and it has to be stalled if further communication services are required.

A generic template for the NI architecture is illustrated in Figure 15.2. The structural view of this hardware module includes a front-end and a back-end submodule. The front-end implements a standardized point-to-point protocol allowing core reuse across several platforms. The interface then assumes the attributes of a socket, that is, an industry-wide well-understood attachment interface which should capture all signaling between the core and the system (such as dataflow signaling, errors, interrupts, flags, software flow control, testing). A distinctive requirement for this standard socket is to enable the configuration of specific interface instantiations along a number of dimensions (bus width, data handshaking, etc.). It is common practice to implement the front-end interface protocol so to keep backward compatibility with existing protocols such as AMBA AXI, OCP, or DTL. This objective is achieved by using a transaction-based communication model, which assumes communicating cores of two different types: masters and slaves. Masters initiate transactions by issuing requests, which can be further split in commands (e.g., read or write) and write data. One or more slaves receive and execute each transaction. Optionally, a transaction can also involve a response issued by the slave to the master to return data or an acknowledgment of transaction completion. This request–response transaction-based model directly matches the bus-oriented communication abstractions typically implemented at core interfaces, thus reducing the complexity of core wrapping logic. As a consequence, the NI front-end can be viewed as a hardware implementation of the session layer in the ISO/OSI reference model. Traditionally, the session layer represents the user's interface to the network. High-level communication services made available by the session layer must be implemented by the transport layer, which is still unaware of the implementation details of the system interconnect (and hence still belongs to the NI front-end in Figure 15.2). The transport layer relieves the upper layers from any concern with providing reliable, sequenced, and QoS-oriented data transfers. It is the first true and basic end-to-end layer. For instance, the transport layer might be in charge of establishing connection-oriented end-to-end communications.

The transport layer provides transparent transfer of data between end nodes using the services of the network layer. These services, together with those offered by the link layer and the physical

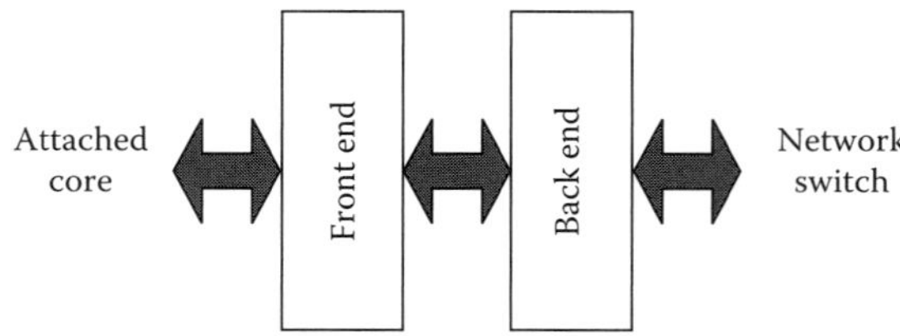

FIGURE 15.2   Architecture of an NI.

layer, are implemented in the NI back-end. Data packetization and routing functions can be viewed as essential tasks performed by the network layer, and are tightly interrelated. The NI back-end also provides data link layer services. Primarily, communication reliability has to be ensured by means of proper error control strategies and effective error recovery techniques. Moreover, flow control is handled at this layer, by means of upstream (downstream) signaling regulating data arrival from the processor core (data propagation to the first switch in the route), but also of piggybacking mechanisms and buffer/credit flushing techniques. Finally, the physical channel interface to the network has to be properly designed. NoCs have distinctive challenges in this domain, consisting of clock domain crossing, high frequency link operation, low-swing signaling and noise-tolerant communication services.

An insight into the *xpipes Lite* [STE05] NI implementation will provide an example of these concepts. Its architecture is showed in Figure 15.3. This chapter views this case study as an example of a lightweight NI architecture for best effort (BE) NoCs.

The NI is designed as a bridge between an OCP interface and the NoC switching fabric. Its purposes are protocol conversion (from OCP to network protocol), packetization, the computation of routing information (stored in a look-up table, LUT), flit buffering to improve performance, and flow control. For any given OCP transaction, some fields have to be transmitted once, while other fields need to be transmitted repeatedly. Initiator and target NIs are attached to communication initiators and targets, respectively. Each NI is split in two submodules: one for the request and one for the response channel. These submodules are loosely coupled: whenever a transaction requiring a response is processed by the request channel, the response channel is notified; whenever the response is received, the request channel is unblocked.

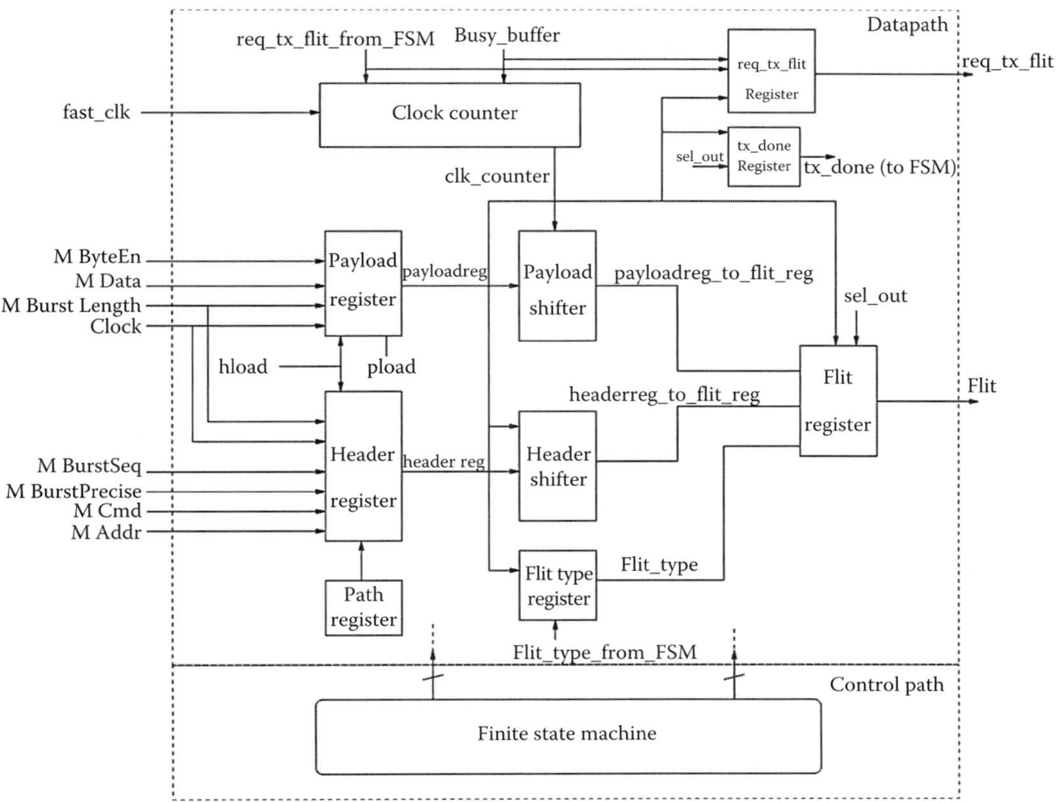

FIGURE 15.3 *xpipes Lite* NI initiator architecture.

The NI is built around two registers: one holds the transaction header (1 refresh per OCP transaction), while the second one holds the payload (refreshed at each OCP burst beat). A set of flits encodes the header register, followed by multiple sets of flits encoding a snapshot of the payload register subsequent to a new burst beat. Header and payload content is never allowed to mix. Routing information is attached to the header flit of a packet by checking the transaction address against an LUT. The length in bit of the routing path depends on the maximum switch radix and on the maximum number of hops in the specific network instance at hand.

The header and payload registers represent the boundary of the NI front-end and also act as clock domain decoupling buffers. These registers can be read from the NI back-end at a much higher speed than the writing speed of the OCP side. In practice, network and attached cores can operate at different frequencies. The only constraint posed by this architecture is that the OCP frequency is obtained by applying an integer divider to the network frequency. A finer grain control of the frequencies would make the NI architecture more complex and costly from an area and power viewpoint.

15.3.2 Switch Architecture

Switches carry packets injected into the network to their destination, following a statically (deterministic) or dynamically (adaptive) defined routing path. A switch forwards packets from one of its input ports to one or more of its output ports using arbitration logic to solve conflicts.

Switch design is usually driven by a power-performance trade-off: high-performance switches for on-chip communication require power-hungry buffering resources. The buffering strategy (Figure 15.4) determines the location of buffers inside the switch. A switch may allocate buffers at the input ports, at the output ports, or both. In input queuing, the queues are at the input of the switch. A scheduler determines at which times queues are connected to which output ports such that no contention occurs. The scheduler derives contention-free connections; a switch matrix (crossbar switch) can be used to implement the connections. In traditional input queuing, there is a single queue per input, resulting in a buffer cost of N queues per switch. However, due to the so-called head-of-line blocking, for large N, switch utilization saturates at 59% [KAR87]. Therefore, input queuing results in weak utilization of the links. Another variant of input queuing is virtual output queuing (VOQ), which combines the advantages of input and output queuing. It has a switch like in input queuing and has the link utilization close to that of output queuing: 100% link utilization can still be achieved, when N is large [MCK95]. As for output queuing, there are N^2 queues. For every input I, there are N queues $Q(I,o)$, one for each output o. Typically, the set of N queues at each input port of a VOQ switch is mapped to a single RAM.

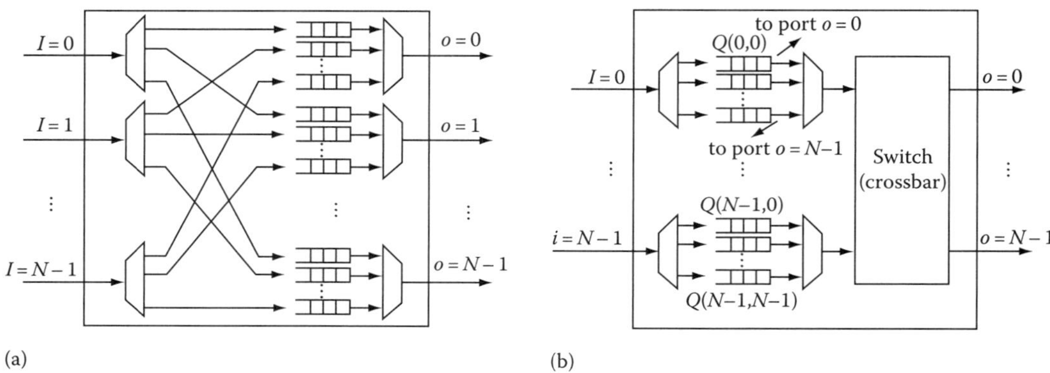

FIGURE 15.4 Common buffering strategies for switch architectures: (a) output queued architecture and (b) virtual output queued architecture.

Switch performance can be affected by the kind of routing method implemented. In source routing, the path of the packet across the network is fixed before injecting it into the network and is embedded in the packet header. Switches only have to read the routing directives from the packet header and to apply them. This is the approach taken by the NI as shown in Figure 15.3. In distributed routing, the path is dynamically computed at every switch when the header flit is routed. This latter carries information on the destination node, which is used by the local switch to perform routing path computation. This can be achieved in two ways: by either accessing a local look-up table or by means of a combinational logic implementing predefined routing algorithms. Source routing results in faster switches, but features limited scalability properties. Moreover, adaptive routing is not feasible with this approach. With distributed routing, switches are more complex and therefore slower (due to the routing logic on the critical path), but they are able to take dynamic routing decisions.

Switching techniques determine the way network resources (buffer capacity, link bandwidth) are allocated to packets traversing the network. Bufferless switching would be the most power-effective approach for resource allocation; however, it comes with heavy side-effects. In fact, whenever a packet cannot proceed on its way to destination due to a conflict with another packet, it should be either discarded (which is not acceptable for on-chip networks) or misrouted. This latter solution is likely to incur severe livelock problems. A minimal amount of buffering is instead required by circuit switching. In this case, long lasting connections are established throughout the network between source and destination nodes. Once reserved, these connections can be used by the corresponding communication flows in a contention-free regime. Buffers in this architecture serve two main purposes: retiming (e.g., input and/or output sampling of the signals in the switch) or buffering of packets devoted to circuit setup. In circuit switching, the latency for circuit setup adversely impacts overall system performance. Moreover, long lasting circuits established throughout the network suffer from low-link utilization. For the above reasons, buffered switching is the most common switching technique used in the NoC domain up to date. In practice, data buffering is implemented at each switch in order to decouple the allocation of adjacent channels in time. The buffer allocation granularity occurs on a packet- or flit-basis depending on the specific switching technique. Three policies are feasible in this context [DAL04].

In store-and-forward switching, an entire packet is received and stored in a switch before forwarding it to the next switch. This is the most demanding approach in terms of memory requirements and switch latency. Also, virtual cut-through switching requires buffer space for an entire packet, but allows for a lower latency communication, since a packet starts to be forwarded as soon as the next switch in the path guarantees that the complete packet can be stored. If this is not the case, the current router must be able to store the entire packet. Finally, a wormhole switching scheme can be employed to reduce switch memory requirements and to provide low latency communication. The head flit enables switches to establish the path and subsequent flits simply follow this path in a pipelined fashion by means of switch output port reservation. A flit is forwarded to the next switch as soon as enough space is available to store it, even though there is no enough space to store the entire packet. If a head flit faces a busy channel, subsequent flits have to wait at their current locations and are therefore spread over multiple switches, thus blocking the intermediate links. This scheme avoids buffering the full packet at one switch and keeps end-to-end latency low, although it is deadlock prone and may result in low link utilization due to link blocking.

Guaranteeing quality of service in switch operation is another important design issue, which needs to be addressed when time-constrained (hard or soft real-time) traffic is to be supported. Throughput guarantees or latency bounds are examples of time-related guarantees.

Contention related delays are responsible for large fluctuations of performance metrics, and a fully predictable system can be obtained only by means of contention free routing schemes. As already mentioned, with circuit switching a connection is set up over which all subsequent data is transported. Therefore, contention resolution takes place only at setup at the granularity of connections, and time-related guarantees during data transport can be provided. In time-division circuit switching [RIJ03], bandwidth is shared by time-division multiplexing connections over circuits.

In packet switching, contention is unavoidable since packet arrival cannot be predicted. Therefore, arbitration mechanisms and buffering resources must be implemented at each switch, thus delaying data in an unpredictable manner and making it difficult to provide guarantees. BE NoC architectures mainly rely on network over-sizing to upper-bound fluctuations of performance metrics. Two case studies taken from the literature will help us understand how the above design issues can be addressed in real-life switch implementations.

The *Aethereal* NoC architecture makes use of a router that tries to combine guaranteed throughput (GT) and BE services [RIJ03]. The GT router subsystem is based on a time-division multiplexed circuit switching approach. A router uses a slot table to (1) avoid contention on a link, (2) divide up bandwidth per link between connections, and (3) switch data to the correct output. Every slot table T has S time slots (rows), and N router outputs (columns). There is a logical notion of synchronicity: all routers in the network are in the same fixed-duration slot. In a slot S, at most, one block of data can be read/written per input/output port. In the next slot, the read blocks are written to their appropriate output ports. Blocks thus propagate in a store-and-forward fashion. The latency a block incurs per router is equal to the duration of a slot, and bandwidth is guaranteed in multiples of block size per S slots. The BE router uses packet switching, and it has been showed that both input queuing with wormhole switching or virtual cut-through switching and virtual output queuing with wormhole switching are feasible in terms of buffering cost. The BE and GT router subsystems are combined in the *Aethereal* router architecture of Figure 15.5. The GT router offers a fixed end-to-end latency for its traffic, which is given the highest priority by the arbiter. The BE router uses all the bandwidth (slots) that has not been reserved or used by GT traffic. GT router slot tables are programmed by means of BE packets (see the arrow "program" in Figure 15.5). Negotiations, resulting in slot allocation, can be done at compilation time, and be configured deterministically at run time. Alternatively, negotiations can be done at run time.

A different perspective has been taken in the design of the switch for the BE *xpipes Lite* NoC [STE05]. The switch is represented in Figure 15.6. It features one cycle latency for switch operation and one cycle for traversing the output link, thus two cycles are required to traverse the switch fabric overall. The switch is output-queued and wormhole switching is used as a convenient buffer allocation policy. Please notice that while in traditional output queuing each input channel has its own buffer in each output port; in this case, the need to reduce total switch power (no full custom design

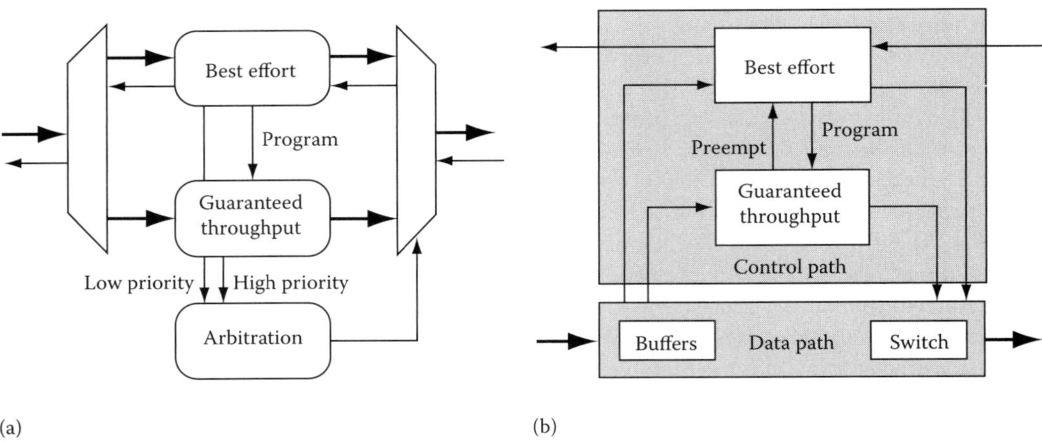

FIGURE 15.5   Combined GT–BE router: (a) conceptual view and (b) hardware view.

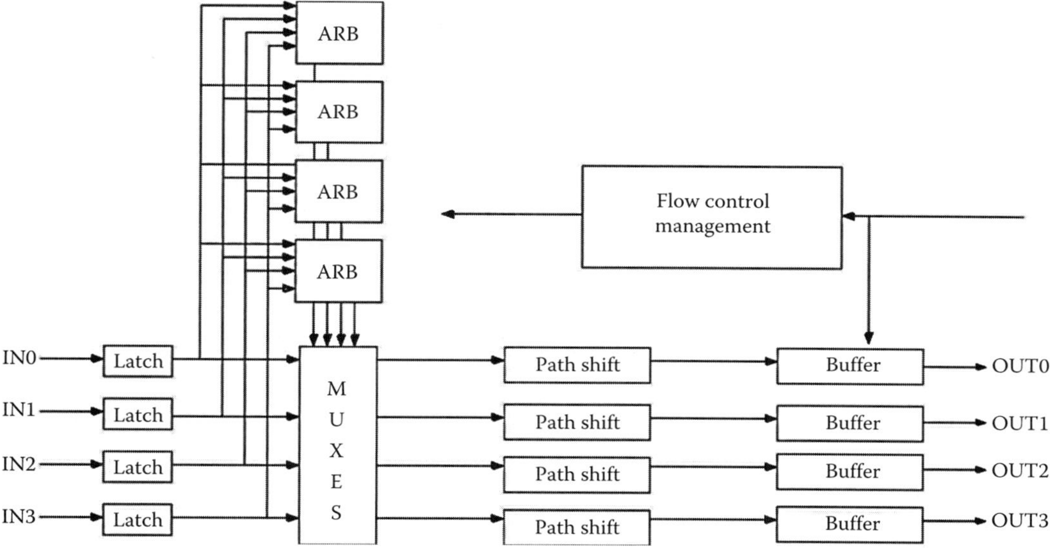

FIGURE 15.6   *xpipes Lite* switch architecture.

techniques were supposed to be used) has led to the implementation of a unique buffer for all input channels at each output port. The choice of different performance-power trade-off points may significantly differentiate on-chip networks from traditional off-chip realizations due to the different constraints they have to meet.

The input ports are latched to break the timing path. As static source routing is used, the header of the packet contains all the information required to route the packet. This keeps the switch routing logic minimal.

Arbitration is handled by an allocator module for each output port according to a round-robin priority algorithm and performed upon receipt of a header flit. After a packet wins the arbitration, routing information pertaining the current switch is rotated away in the header flit. This allows to keep the next hop at a fixed offset within the header flit, thus simplifying switch implementation. Access to output ports is granted until a tail flit arrives. The *xpipes Lite* switch is defined as a soft macro, i.e., it is parameterizable in the number of input and output ports, in the link width, in the output buffer size but also in the flow control technique.

15.3.3 Link Design

In the context of nanoscale technologies, long wires increasingly determine the maximum clock rate, and hence the performance of the entire design. The problem becomes particularly serious for domain-specific heterogeneous SoCs, where the wire structure is highly irregular and may include both short and extremely long switch-to-switch links. Moreover, it has been estimated that only a fraction of the chip area (between 0.4% and 1.4%) will be reachable in one clock cycle [AGA00].

A solution to overcome the interconnect-delay problem consists of pipelining interconnects [CAR01,SCH02]. Wires can be partitioned into segments (by means of relay stations, which have a function similar to that of latches on a pipelined datapath) whose length satisfies predefined timing requirements (e.g., desired clock speed of the design). This way, link delay is changed into latency, but the data introduction rate is not bounded by the link delay any more. Now, the latency of a channel connecting two modules may turn out to be more than one clock cycle. Therefore, if the functionality of the design is based on the sequencing of the signals and not on their exact timing, then link pipelining does not change the functional correctness of the design. This requires the system to consist of modules whose behavior do not depend on the latency of the communication channels (latency-insensitive operation). As a consequence, the use of interconnect pipelining can be seen as a part of a new and more general methodology for nanoscale designs, which can be envisioned as synchronous distributed systems composed by functional modules that exchange data on communication channels according to a latency-insensitive protocol. This protocol ensures that functionally correct modules behave independently of the channel latencies [CAR01]. The effectiveness of the latency-insensitive design methodology is strongly related to the ability of maintaining a sufficient communication throughput in presence of increased channel latencies.

The architecture of a relay station depends on the flow control scheme used across the network, since switch-to-switch links not only carry packets but also control information determining the way the downstream node communicates buffer availability to the upstream node (i.e., flow control signals). In order to make this point clear, let us focus on three flow control protocols with very different characteristics: stall/go, T-Error, and ACKnowledge/NotACKnowledge (ACK/NACK) [PUL05].

Stall/go is a low-overhead scheme which assumes reliable flit delivery. T-Error is much more complex, and provides logic to detect timing errors in data transmission. This support is however only partial, and usually exploited to improve performance rather than to add reliability. Finally, ACK/NACK is designed to support thorough fault detection and handling by means of retransmissions.

Stall/go is a very simple realization of an ON/OFF flow control protocol. It requires just two control wires (Figure 15.7a): one going forward and flagging data availability, and one going backward and signaling either a condition of buffers filled ("STALL") or of buffers free ("GO"). Stall/go can be implemented with distributed buffering along the link; that is, every repeater can be designed as a

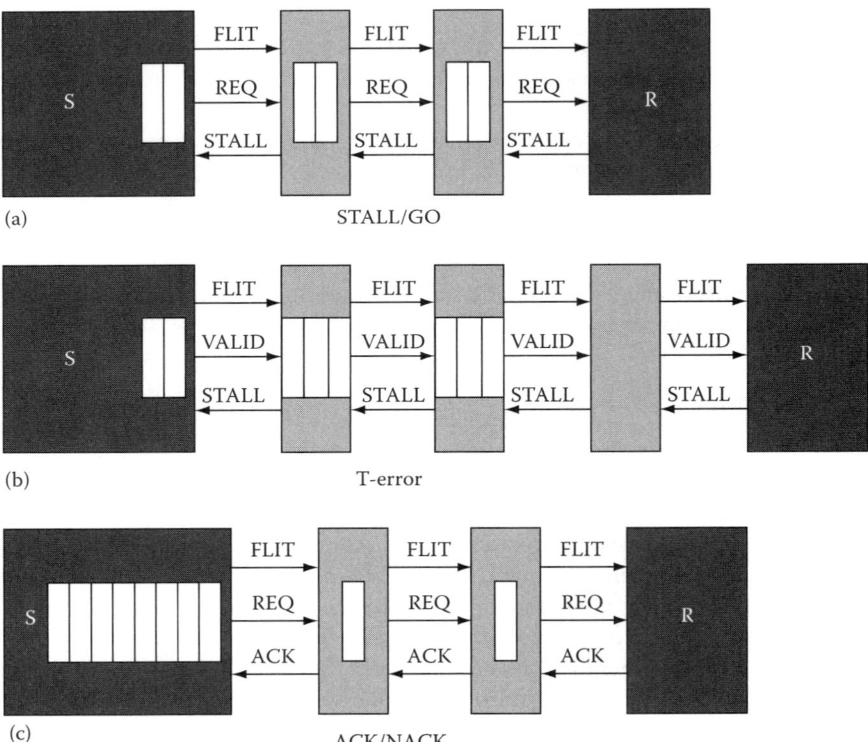

FIGURE 15.7   Impact of flow control on link pipelining implementation.

very simple two-stage FIFO. The sender only needs two buffers to cope with stalls in the very first link repeater, thus resulting in an overall buffer requirement of $2N + 2$ registers, with minimal control logic. Power is minimized since any congestion issue simply results in no unneeded transitions over the data wires. Performance is also good, since the maximum sustained throughput in the absence of congestion is of one flit per cycle by design, and recovery from congestion is instantaneous (stalled flits get queued along the link toward the receiver, ready for flow resumption). In the NoC domain with pipelined links, stall/go indirectly reflects the performance of credit-based policies, since they exhibit equivalent behavior. The main drawback of stall/go is that no provision whatsoever is available for fault handling. Should any flit get corrupted, some complex higher-level protocol must be triggered.

The *T-Error* protocol (Figure 15.7b) aggressively deals with communication over physical links, either stretching the distance among repeaters or increasing the operating frequency with respect to a conventional design. As a result, timing errors become likely on the link. Faults are handled by a repeater architecture leveraging upon a second delayed clock to resample input data, to detect any inconsistency, and to emit a VALID control signal. If the surrounding logic is to be kept unchanged, as we assume in this chapter, a resynchronization stage must be added between the end of the link and the receiving switch. This logic handles the offset among the original and the delayed clocks, thus realigning the timing of DATA and VALID wires; this incurs a one-cycle latency penalty. The timing budget provided by the T-Error architecture can also be exploited to achieve greater system reliability, by configuring the links with spacing and frequency as conservative as in traditional protocols. However, T-Error lacks a really thorough fault handling: for example, errors with large time constants would not be detected. Mission-critical systems, or systems in noisy environments, may need to rely on higher-level fault correction protocols. The area requirements of T-Error include three buffers in each repeater and two at the sender, plus the receiver device and quite a bit of overhead in control logic. A conservative estimate of the resulting area is $3M + 2$, with M being up to 50% lower than N if T-Error features are used to stretch the link spacing. Unnecessary flit retransmissions upon congestion are avoided, but a power overhead is still present due to the control logic. Performance is of course dependent on the amount of self-induced errors.

The main idea behind the *ACK/NACK* flow control protocol (Figure 15.7c) is that transmission errors may happen during a transaction. For this reason, while flits are sent on a link, a copy is kept locally in a buffer at the sender. When flits are received, either an ACK or a NACK is sent back. Upon receipt of an ACK, the sender deletes the local copy of the flit; upon receipt of a NACK, the sender rewinds its output queue and starts resending flits starting from the corrupted one, with a GO-BACK-N policy. This means that any other flit possibly in flight in the time window among the sending of the corrupted flit and its resending will be discarded and resent. Other retransmission policies are feasible, but they exhibit higher logic complexity. Fault tolerance is built in by design, provided encoders and decoders for error control codes are implemented at the source and destination, respectively. In an ACK/NACK flow control, a sustained throughput of one flit per cycle can be achieved, provided enough buffering is available. Repeaters on the link can be simple registers, while, with N repeaters, $2N + k$ buffers are required at the source to guarantee maximum throughput, since ACK/NACK feedback at the sender is only sampled after a round-trip delay since the original flit injection. The value of k depends on the latency of the logic at the sending and receiving ends. Overall, the minimum buffer requirement to avoid incurring bandwidth penalties in a NACK-free environment is therefore $3N + k$. ACK/NACK provides ideal throughput and latency until no NACKs are issued. If NACKs were only due to sporadic errors, the impact on performance would be negligible. However, if NACKs have to be issued also upon congestion events, the round-trip delay in the notification causes a performance hit which is very pronounced especially with long pipelined links. Moreover, flit bouncing between sender and receiver causes a waste of power.

The distinctive features of each flow control scheme are summarized in Table 15.1, which also points out the implications of the flow control technique on the implementation of link pipelining. Clearly,

TABLE 15.1 Flow Control Protocols at a Glance

	Stall/go	T-Error	ACK/NACK
Buffer area	$2N + 2$	$> 3M + 2$	$3N + K$
Logic area	Low	High	Medium
Performance	Good	Good	Depends
Power (estimation)	Low	Medium/high	High
Fault tolerance	Unavailable	Partial	Supported

the best choice depends on the ultimate design objective and on the level of abstraction at which designers intend to enforce fault tolerance.

15.4 Topology Synthesis

Performance of a given NoC architecture can be strongly biased at design time by selecting a suitable topology for the system at hand. Sources of unpredictability may come from the physical synthesis of the topology or by the semantics of the communication middleware, which ends up shaping the traffic patterns injected into the network [GIL08]. This chapter recalls the essential topology concepts that come into play in the design process of a NoC, but refers the interested reader to [DUA04,DAL04, MIC06] for a more detailed analysis of topologies.

A topology defines how switches and cores are interconnected by channels. Topologies can be classified as direct or indirect. In direct topologies, each switch is directly connected to a subset of switches and to one or more cores. Commonly, direct SoCs employ orthogonal topologies, where switches can be arranged in an orthogonal n-dimensional space, in such a way that every channel produces a displacement in one dimension (Figure 15.8a). In those topologies, routing of messages is simple and easy to implement.

In indirect topologies, there are two kinds of switches: the ones that connect to both cores and other switches and the ones that only have connections to other switches (Figure 15.9a). The most scalable indirect topologies are multistage, in which cores are interconnected through a number of intermediate switch stages. Although indirect topologies have been proposed for SoCs [GUE00], an open issue is the cost and wiring complexity of some of the high-performance indirect topologies, like the fat-tree [BAL06].

On the other hand, two main subclasses among direct and indirect categories exist: regular and irregular (see Figures 15.8 and 15.9). In regular topologies, there is a regular connection pattern between switches, whereas irregular topologies have no predefined pattern. Regular topologies are

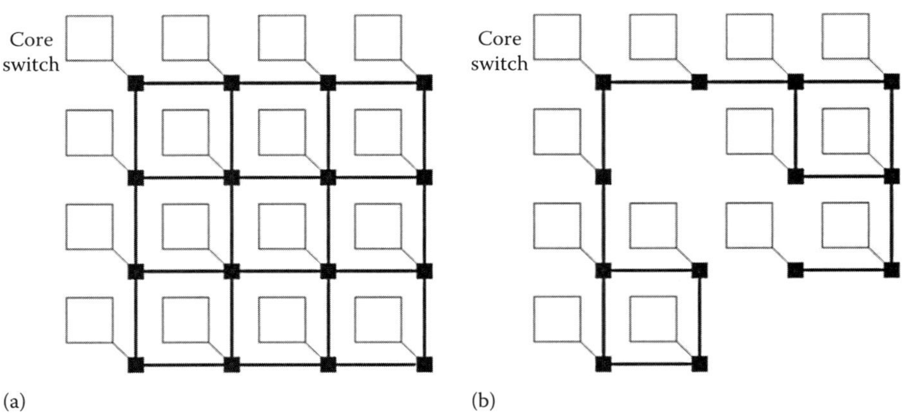

FIGURE 15.8 Example of direct (a) regular and (b) irregular topologies.

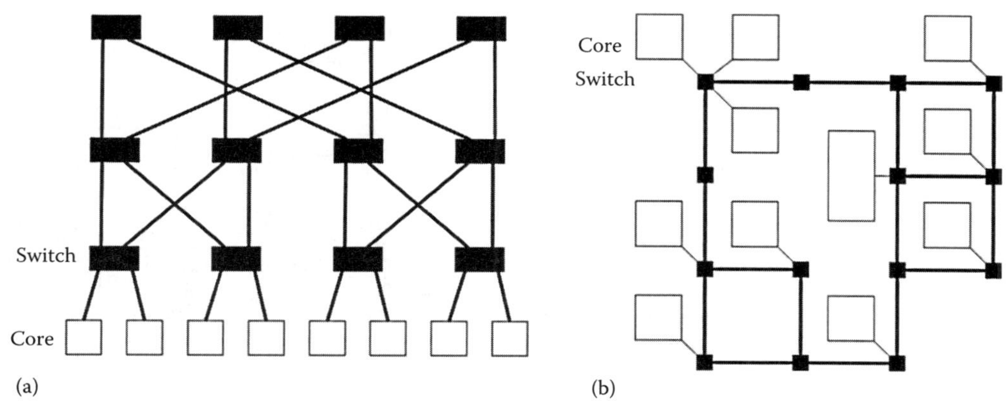

FIGURE 15.9 Example of indirect (a) regular and (b) irregular topologies.

usually more scalable than irregular ones, although their main advantage is topology reusability and reduced design time.

Irregular topologies are usually employed to design customized, domain-specific NoC architectures, which instantiate communication resources only where needed and properly tune them to the specific communication requirements of the cores. However, notice that even regular topologies can change into irregular ones when some of their components fail. This can happen along the working life of the system or in newly implemented systems. If those faults are not serious enough to render the system unusable, the topology of the system can be transformed into a working irregular one. Therefore, for fault-tolerant systems, it is a good practice to use routing algorithms that can work both in regular and irregular topologies.

Concerning routing, it is easier to design deadlock-free routing algorithms for regular topologies [GOM07,KAV03]. Nevertheless, there are modern topology-agnostic techniques to obtain deadlock-free distributed routing with an effective physical implementation [FLI07,GOM06].

It is very important to model the relationships between topology and physical constraints, and the resulting impact on performance. Network optimization is the process of utilizing these models in selecting topologies that best match the physical constraints of the implementation. As an example, the objective to reduce the average number of hops to destination might be achieved by using switches with a larger number of I/O ports. This also places different demands on physical resources such as wiring area. In order to account for physical constraints early in the design process while not yet dwelling into the intricacies of physical design, topologies are characterized by a few main parameters relating their performance to their physical properties [DUA04,DAL04]:

- *Bisection bandwidth*: Defined as the aggregate bandwidth of the minimum number of wires that must be cut in order to split the network into two equal sets of switches. It is a common measure of network performance. An example of bisection is depicted as a dotted line in Figure 15.10a.
- *Network diameter*: Defined as the maximum number of hops between any two cores. It is related with the average latency of messages; as it is increased, packets need more time to reach their destinations and the probability of a collision with another packet also increases. Figure 15.10 shows two regular direct topologies with the same number of cores. While it takes six hops to travel from top-left to bottom-right in the 4 × 4 mesh, the maximum number of hops is 4 in the 4-hypercube.
- *Switch degree*: Defined as the total number of input/output ports of a switch. Usually, higher switch degree implies lower maximum frequency and higher area requirements.

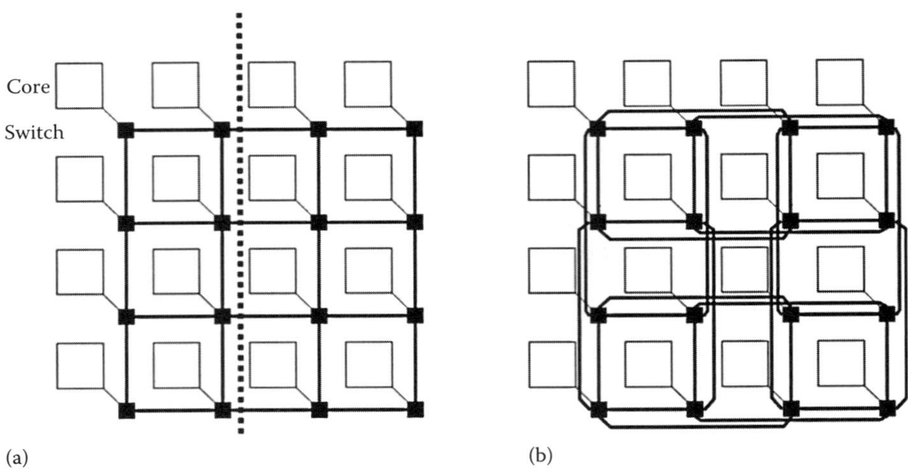

FIGURE 15.10   (a) Bisection of a 4 × 4 mesh and (b) bidimensional representation of a 4-hypercube.

- *Path diversity*: Defined as the property to provide multiple minimal paths between any pair of cores. Path diversity is associated with the possibility to reduce network congestion (e.g., by evenly and statically distributing traffic across several paths or by dynamically selecting the less congested route to destination) and/or to provide fault tolerance [GOM08,LER08].

Although these metrics allow a first-order analysis of the physical requirements and of the performance of a given topology, they do not suffice to capture all the physical effects that determine the efficiency or even the practical feasibility of that topology. Moreover, with the advent of nanoscale technologies, deviations of postlayout performance metrics from early projections become increasingly wide. In particular, link latency estimation suffers from poor predictability since it is related not only to the theoretical length of the links and the number of dimensions in the topology, but also to the size of the core obstruction, and the intercore routing channels and to the routing decisions taken by the place-and-route tool. For complex designs, a part of these decisions may be on burden of the designer, thus shortening the convergence time of the tool. Figure 15.11 shows a concept representation of a 3-hypercube followed by one of its physical mappings. The difficulty with this latter lies in the need to map a 3-D topology onto the 2-D silicon surface. In the example of Figure 15.11, physical links do not have the same length as expected, and some of them are consistently longer than others and might give rise to multicycle propagation which puts theoretical diameter estimations in discussion [GIL08]. The need for multicycle links also depends on the target operating frequency of the design, which in turn depends on the switch degree of the topology.

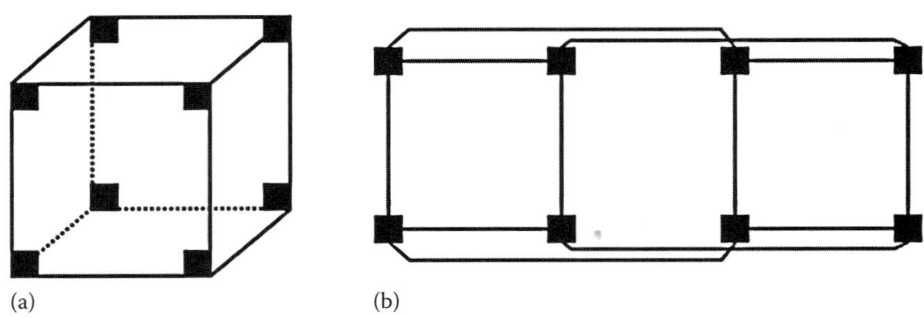

FIGURE 15.11   (a) Concept representation of a 3-hypercube and (b) physical mapping on the 2-D silicon surface.

It is evident that a significant gap may arise between high-level topology exploration and the physical design process. In the context of nanoscale technologies, pencil-and-paper floorplan considerations or high-level roadmapping tools can just indicate a trend or provide a general guideline, but fail to provide trustworthy performance estimations for a given topology. A tighter integration of high-level simulation tools with the underlying backend synthesis tool flow is needed in order to allow for silicon-aware decision making at each hierarchical level of the design process.

In this direction, new tools that guide designers toward a subset of the most promising topologies for on-chip network designs are emerging, while considering the complex trade-offs between applications, architectures, and technologies, like *SUNMAP* [MUR04], *xpipesCompiler* [JAL05], or *Polaris* [PEH07].

These tools perform topology exploration as a key step of the design process [BER05]. However, the objective functions and the candidate topologies of the exploration framework differ based on the system at hand. In particular, the topology synthesis process is highly differentiated for general-purpose and for application-specific systems, as illustrated below.

15.4.1 General-Purpose Systems

MPSoC architectures are considered as general purpose if they consist of homogeneous processing tiles and are therefore built as fully modular structures and if they have to deal with nonpredictable and/or varying workloads and traffic patterns. Regular topologies are the most suitable (e.g., [BED04]) ones for these designs, because of their scalability and conservative broad connectivity. At the same time, they pave the way for reusability, reduced design time, and good control of electrical parameters (i.e., better performance predictability).

The most widely used topology in general-purpose designs is the 2-D mesh. The reason for that lies in the perfect match with the 2-D silicon surface, in its good modularity and ease of routing. In contrast, 2-D meshes suffer from poor scalability (e.g., fast diameter degradation with the increase of network nodes).

Some more complicated topologies have been used by or proposed for other interconnection networks, such as high-dimensional meshes/tori, hypercubes, hierarchical meshes/tori [HAU98], express cubes [DAL91], and fat-trees [LEI85]. Topologies with more than two dimensions are attractive for a number of reasons. First, increasing the number of dimensions in a mesh results in higher bandwidth and reduced latency. Second, wiring on a chip comes at a lower cost with respect to off-chip interconnections. However, wiring is also the weak point of these topologies, since their mapping on a bidimensional plane involves the existence of wires with different lengths. From a layout viewpoint, this translates into links with different latencies and into the use of more metal layers. Moreover, other layout effects might impact feasibility of these topologies, such as the routability of wires over computation tiles and the routing constraints posed to other on-chip interconnects.

As a case study, let us analyze two candidate regular topologies for a 16-tile system, with a target clock frequency of 1 GHz for the network and 500 MHz for the computation tiles. The tile architecture consists of a processor core and of a local memory core, connected to the network through a NI initiator and target, respectively. The two NIs of a single tile can be used in parallel (e.g., while the local processor core has a pending read, the memory can be accessed by a remote processor core with a write transaction), which is consistent with the assumption of a dual-port local memory.

The benchmark used for the experiments consists of a parallel application with a task-to-processor mapping that de-emphasizes the role of the topology mapping algorithm. Every task is assumed to be mapped to a different computation tile, as illustrated in Figure 15.12. One or more producer tasks read in data units from the I/O interfaces of the chip and distribute them to a scalable number of concurrent and independent worker tasks. There are no constraints on which worker tile has to process a given data unit. The higher the number of worker tiles, the higher the data processing rate provided the I/O interfaces can keep up with it. Output data from each worker tile is then collected

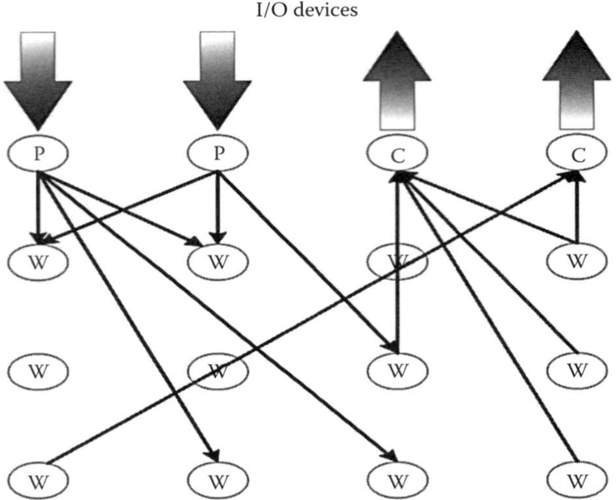

FIGURE 15.12   Workload allocation policy.

TABLE 15.2   Topologies under Test

Topology	4-Ary 2-Mesh	4-Hypercube
Switch degree	6	6
Channel latency	1	1,2
Switches	16	16
Diameter	6	4
Tiles per switch	1	1
Bisection bandwidth	4	8

by one or more consumer tiles, which write them back to the I/O interfaces. A maximum of five I/O tiles is considered. Application parameters are set in such a way that total idleness in the system is minimized, hence there are no global bottlenecks and the system is fully balanced. Cryptography algorithms are a well-known example of applications that exhibit this workload allocation policy.

We assume producer–consumer interaction between the tiles, based on the communication and synchronization protocol illustrated in [TOR07]: it reflects the latest advances in communication middleware for distributed message-oriented MPSoCs, and therefore might project realistic traffic conditions for NoCs. In essence, the features of this communication protocol result in poor bandwidth utilization but make global performance mainly sensitive to network latency.

Table 15.2 shows the different topologies considered for this case study and their characteristics. The 2-D mesh (4-ary 2-mesh) is the reference topology (Figure 15.10a), and is compared with a 2-ary 4-mesh (aka 4-hypercube), the same of Figure 15.10b, which improves diameter and bisection bandwidth (thanks to the four dimensions) at the cost of a more complex wiring pattern. Table 15.2 reports only a high-level channel latency estimation for the two topologies. Dimension order routing is used as a simple deadlock-free routing algorithm in all cases.

From a performance viewpoint (execution cycles), the 4-hypercube outperforms the 2-D mesh by 5% [GIL08]. Should we scale the system to 64 tiles, a 6-hypercube would outperform the 2-D mesh by 15%, as indicated by cycle-true transaction level simulation. These numbers assume the same operating frequency for the two topologies and a latency of one cycle on all links (even in the 4-hypercube). However, only after layout these assumptions can be validated. Since the two topologies have the same switch radix, they will end up achieving the same operating frequencies. So, the most critical physical implementation parameter for the 4-hypercube is the link length for dimensions 3 and 4. A parametric analysis shows that with up to 4 cycles latency on these links, the 4-hypercube still

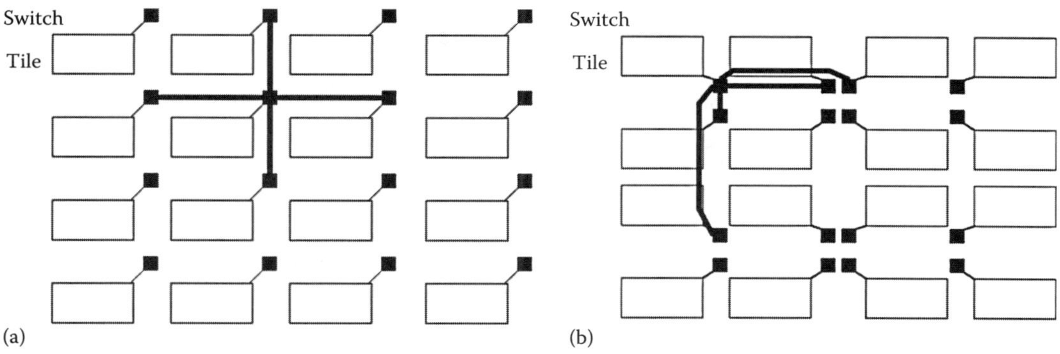

FIGURE 15.13 Floorplan directives for (a) the 2-D mesh and (b) the 4-hypercube.

provides better or equal performance with respect to the 2-D mesh. The actual physical synthesis of the 16-tiles topologies on a 65 nm technology shows that it is possible to wire the long links of the 4-hypercube without any repeater stage for a target frequency of 1 GHz, thus preserving all the theoretical benefits of the 4-hypercube over the 2-D mesh. This result could not be trivially expected, due to the asymmetric tile size (1×2 mm, corresponding to a processor core and a local memory per tile) and to the degrees of freedom for the floorplanning, the placement, and the routing. For this case, while the routing step is performed automatically, smart placement directives are given to the backend tools (Figure 15.13), exploiting the asymmetric tile size to optimize the wiring cost of the 4-hypercube. While this cost is typically expected to be twice that of the 2-D mesh (see Table 15.2), this synthesis run provides a total wire-length overhead by only 20% for the 4-hypercube. This case study shows that specific multidimensional topologies are a viable alternative to 2-D meshes at an affordable layout (especially wiring) cost, even though this has to be assessed every time for the system at hand by means of an integrated exploration framework validating the results of abstract system-level simulation in light of the physical degradation effects impairing link latency and network frequency.

15.4.2 Application-Specific Systems

In these systems, the communication requirements of the applications can be statically analyzed at design time, therefore the NoC can be tailored for the particular application behavior. This allows to cut down on power and area. A number of platforms for wireless and multimedia fall into this category (Philips Nexperia, ST Nomadik, TI Omap, etc.), even though they are not currently using on-chip networks yet.

Application-specific topologies can be generated as new and fully customized topologies, efficiently accommodating the communication requirements of an application with minimum area and power. Alternatively, they can be synthesized by customizing regular topologies for the requirements of the application at hand, while preserving the regularity of the baseline topology. Fully customized topologies can improve the overall network performance, but they distort the regularity of the grid structure. This results in links with widely varying lengths, performance, and power consumption. Consequently, better logical connectivity comes at the expense of a penalty in the structured nature of the wiring, which is anyway one of the main advantages offered by the regular on-chip network. Hence, the usual problems of crosstalk, timing closure, global wiring, etc., may undermine the overall gain obtained through customization.

For this reason, customizing a baseline regular topology may result in a good fit of the application requirements and preserve the modularity, routing, and electrical properties of regular topologies. However, in this case, area and power are not primary optimization targets. A good example of this

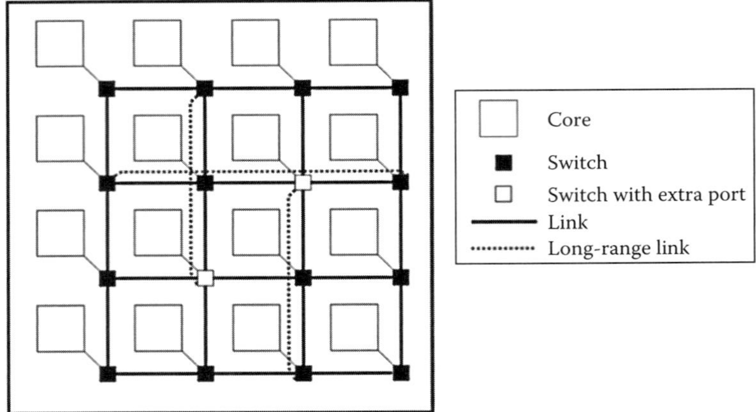

FIGURE 15.14   Adding long-range links to a 4 × 4 mesh.

approach consists of viewing the network as a superposition of clustered nodes with short links and a collection of long-range links producing shortcuts among different regions of the network. In practice, a standard 2-D mesh network can be used as the baseline topology and can be customized by adding a few additional long-range links to reduce distance between remote nodes [MAR05]. An example of application-specific long-range link insertion is illustrated in Figure 15.14. Long-range links can heavily impact the dynamic properties of the network, which are characterized by traffic congestion. At low traffic loads, the average packet latency exhibits a weak dependence on the traffic injection rate (free state). However, when the traffic injection rate exceeds a critical value, the packet delivery times rise abruptly and the network throughput starts collapsing (congested state). The transition from the free state to the congested state is known as phase transition region. It turns out that the phase transition in regular networks can be significantly delayed by introducing additional long-range links [FUK99]. The main challenges associated with this approach concern the algorithm that determines the most beneficial long-range links to be inserted in a mesh network given an application-specific traffic pattern and the implementation of a deadlock-free routing algorithm able to exploit long-range links. Finally, the inherent cost of this approach lies in the area overhead and in the reduction of the maximum operating frequency due to the increase of the needed switch radix. Anyway, the scalability of the bidimensional mesh is improved without losing the layout regularity.

On the other hand, ad hoc topologies are most of the time irregular, in that they instantiate communication resources only where needed and tune them to perfectly match the requirements of the application at hand. In this case, additional challenges must be addressed by designers, since they have to design parameterizable and arbitrarily composable network building blocks. Moreover, links could be of uneven length, therefore link pipelining might have to be used in the context of a latency-insensitive design style. Also, topology-agnostic and deadlock-free routing could be difficult to implement for these topologies. Addressing this issue might take significant design time, which conflicts with time-to-market pressures.

Custom-tailored topology generation is the front-end of a complete design flow that spans multiple abstraction levels, down to placement and routing. The back-end of the flow is more mature at the present time. In fact, several works have been published on automatically generating the RTL code of a designed topology for simulation and synthesis [JAL05,MAL02]. Building area and power models for on-chip networks has been addressed in [YE03,WAN02,PAL04]. These approaches have rapidly reached the industrial practice. Several companies are today productizing synthesizable NoC architectures [ART,COP04,BAI02]. At the research and development frontier, the focus is now on bridging the gap between high-level analysis of communication requirements and synthesis of a custom-tailored NoC implementation. A relevant case study from the open literature follows, which gives the flavor of state of the art in this domain.

Xpipes is an example of a complete flow for designing application-specific NoCs [BEN06]. It is centered around the *xpipes Lite* architecture already encountered in the sections above. The network building blocks are designed as highly configurable and design-time composable soft macros described in SystemC. The first phase of the design flow consists of the specification of the objectives and constraints that must be satisfied by the design (Figure 15.15a). The application traffic characteristics, size of the cores, and the area and power models for the network components are also obtained. In the second phase, the NoC architecture that optimizes the user objectives and satisfies

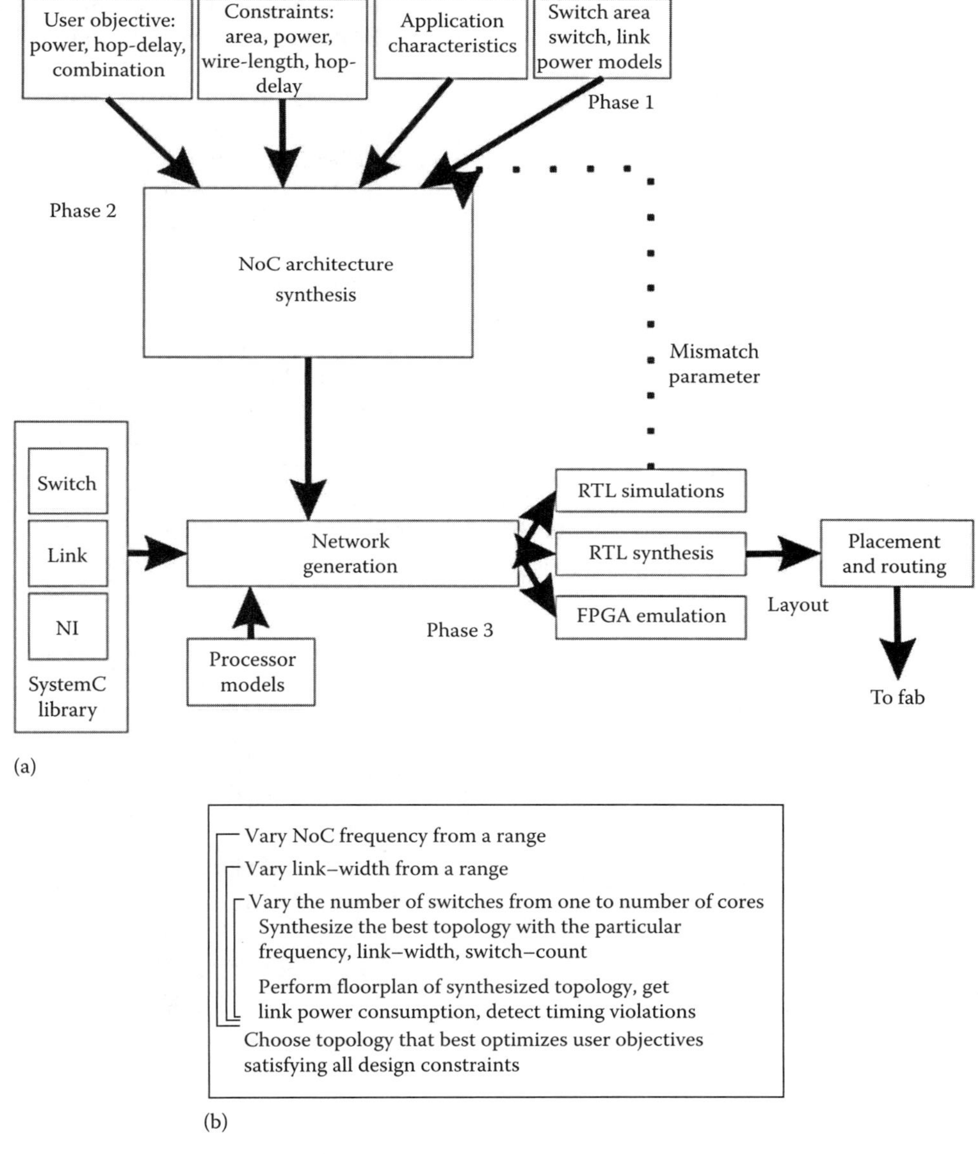

FIGURE 15.15 (a) xpipes NoC design flow and (b) topology synthesis iterations.

the design constraints is automatically synthesized. The different steps in this phase are presented in Figure 15.15b. In the outer iterations, the key NoC architectural parameters (operating frequency, link width) are varied from a range of suitable values. During the topology synthesis, it is ensured that the traffic on each link is less than or equal to its available bandwidth. The synthesis step is performed one for each set of parameters. In this step, several topologies with different number of switches are explored, starting from a topology where all the cores are connected to one switch to the one where each core is connected to a separate switch. The synthesis of each topology includes finding the size of the switches, establishing the connectivity between the switches and with the cores, and finding deadlock-free routes for the different traffic flows (using the approach outlined in [STA03]).

In the next step (Figure 15.15a), to have an accurate estimate of the design area and wire-length, the floorplanning of each synthesized topology is automatically performed. The floorplanning process finds the 2-D position of the cores and network components used in the design. Based on the frequency point and the obtained wire-lengths, the timing violations on the wires are detected and power consumption on the links is obtained.

In the last step, from the set of all synthesized topologies and architectural parameter design points, the topology and the architectural point that best optimize the user's objectives, satisfying all the design constraints, are chosen. Thus, the output of phase 2 is the best application-specific NoC topology, its frequency of operation, and the width of each link in the NoC. In the last phase of the design (phase 3 in Figure 15.15a), the RTL code is automatically generated by using the components from the *xpipes Lite* library. From the floorplan specification of the designed topology, the synthesis engine automatically generates the inputs for placement and routing, whose output is the final layout of the NoC design.

The effectiveness of automatically designed NoC topologies can be assessed by comparing them with hand-tuned ones. As an example, let us consider a system of 30 cores that runs multimedia benchmarks. There are 10 ARM7 processor cores with caches, 10 private memories (a separate one for each processor core), 5 custom traffic generators, 5 shared memories and devices to support interprocessor communication. The hand-designed NoC has 15 switches connected in a 5 × 3 mesh network, shown in Figure 15.16a. The design is highly optimized, with the private memories being connected to the processors using a single switch and the shared memories distributed around the switches. The postlayout NoC supports a maximum frequency of operation of 885 MHz and power of the topology for functional traffic turns out to be 368.08 mW.

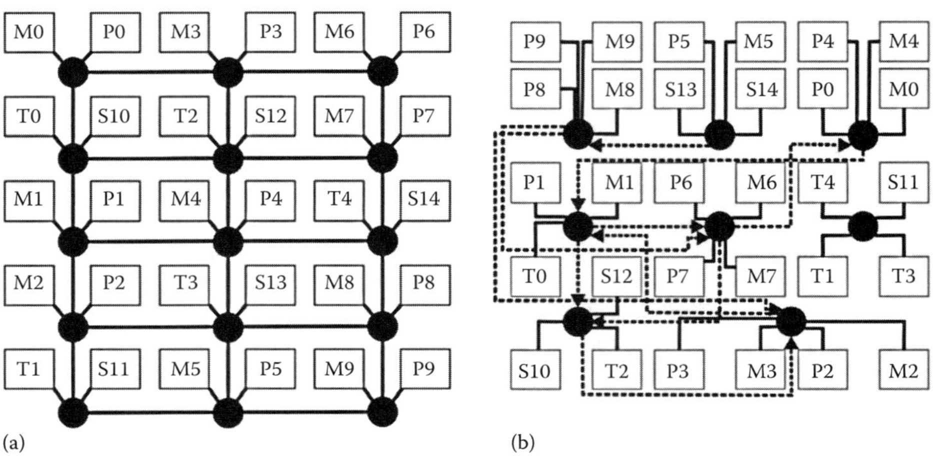

(a)   (b)

FIGURE 15.16 (a) Hand-designed topology. M: ARM7 core; T: traffic generator; P,S: private and shared memories. (b) Automatically synthesized topology, unidirectional links are dotted.

If the automatic synthesis flow described above is run with the same target frequency achieved by the hand-tuned topology (885 MHz), the application-specific topology illustrated in Figure 15.16b can be obtained. It has fewer switches (8) but up to 2× longer switch-to-switch wires. However, it could support the same maximum frequency of operation of the hand-designed topology without any timing violation on the wires. As the wire-length is considered during the synthesis process to estimate the frequency that can be supported, the most power efficient topology can be synthesized that would still meet the target frequency. To arrive at such a design point manually would require several iterations of topology design and place and route phases, which is very time consuming. Power measurements on functional traffic this time give 277.08 mW, which is 1.33× lower than the hand-designed topology. Given the fact that the latter is highly optimized, with much of the communication traffic (between the ARM cores and their private memories) traversing only one switch, these savings are achieved entirely by efficiently spreading the shared memories around the different switches. From a layout viewpoint, Benini [BEN06] shows that the automatic layout generation pays only a 4.3% increase in area with respect to manual optimization. From the cycle accurate simulation of the hand-designed and the synthesized NoCs for two multimedia benchmarks, Benini [BEN06] also concludes that the custom topology not only matches the performance of the hand-designed topology, but also provides an average of 10% reduction in total execution time and 11.3% in average packet latency. Finally, we cannot ignore that competitive quality has been achieved at a fraction of the design time: a couple of weeks for the manual topology can be reduced to less than 4 h to complete the automated flow.

15.5 NoC Design Challenges

Although the benefits of NoC architectures are substantial, reaching their full potential presents numerous research challenges. There are three main issues that must be addressed for future NoCs: power consumption, latency, and CAD compatibility. As exposed in [OWE07], research challenges for NoCs architectures can be classified into four broad areas:

- *Technology and circuits.* The most important technology constraint for NoCs is power. To close the power gap, research should develop optimized circuits for NoC components. Other constraints include design productivity and cost, reflecting the issues derived from the use of new or exotic technologies that require an additional effort in developing CAD compatibility.

- *NoC microarchitecture and system architecture.* Architecture research must address the primary issues of power and latency, as well as critical issues such as congestion control. This should be addressed at the network-level (topology, routing, and flow control) as well as in the router microarchitecture. The delay of routers and the number of hops required by a typical message should be reduced. Circuit research to reduce channel latency can also help to close the latency gap. All of these architectural improvements must be developed taking into account technological constraints and the limited power budget. In particular, power consumption is a critical issue and future NoC architectures must address it effectively, for example, by avoiding unnecessary work or by dynamically modulating voltage and frequency.

- *NoC design tools.* NoC design tools should be able to better interface with system-level constraints and design, for instance, by means of an accurate characterization and modeling of system traffic. It is very important to update CAD tools with specialized libraries for new NoC microarchitectures and interconnects, validation capabilities for new NoC designs, and feedback from end-user to simplify the design process. In order to preserve design tools usability with CMOS scaling, it is of utmost importance to develop new area, power, timing, thermal, and reliability models. Finally, new methods to estimate power-performance of NoCs other than simulations are needed, due to the increasing complexity of future NoC designs.

- *NoC comparison.* Also, it is very important to develop common evaluation metrics (such as latency and bandwidth under area, power, energy, and heat dissipation constraints) and standard benchmarks to allow direct unambiguous comparison between different NoC architectures.

15.6   Conclusion

Parallel architectures are becoming mainstream both in the high performance and in the embedded computing domains. The most daunting challenge to future multiprocessor systems is to realize the enormous bandwidth capacities and stringent latency requirements when interconnecting a large number of processor cores in a power efficient fashion. Even though in the short term the evolution of topologies and protocols of state-of-the-art interconnects will suffice to sustain MPSoC scalability, on-chip networks are widely recognized as the long-term and disruptive solution to the problem of on-chip communication. This chapter has provided basic principles and design guidelines for NoC architectures. Network building blocks have been analyzed at first in isolation, and then from the viewpoint of their connectivity (topology synthesis). The differences between the topologies for general purpose and for application-specific systems have been presented, pointing out the need to account for layout effects even in the early steps of the topology synthesis process. Finally, the chapter has highlighted the challenges that lie ahead to make NoC architectures mainstream. They can be broadly categorized into low power and low latency circuits and architectures, silicon-aware design tools with system-level exploration capabilities, and the definition of standard quality metrics and benchmarks.

Acknowledgment

This work was supported by the European Commission in the context of the HIPEAC2 Network of Excellence under the Interconnects research cluster.

References

[BEN06]   L. Benini, Application-specific NoC design, *Design Automation and Test in Europe Conference,* Munich, Germany, pp. 1–5, 2006.

[BOE02]   F. Boekhorst, Ambient intelligence, the next paradigm for consumer electronics: How will it affect silicon?, *ISSCC 2002,* San Francisco, CA, vol. 1, pp. 28–31, February 2002.

[BEN02]   L. Benini and G. De Micheli, Networks on chips: A new SoC paradigm, *IEEE Computer,* 35(1): 70–78, January 2002.

[KUM02]   S. Kumar, A. Jantsch, J.P. Soininen, M. Forsell, M. Millberg, J. Oeberg, K. Tiensyrja, and A. Hemani, A network on chip architecture and design methodology, *IEEE Symposium on VLSI ISVLSI02,* Pittsburg, PA, pp. 105–112, April 2002.

[WIE02]   P. Wielage and K. Goossens, Networks on silicon: Blessing or nightmare? *Proceedings of the Euromicro Symposium on Digital System Design DSD02,* Dortmund, Germany, pp. 196–200, September 2002.

[DAL01]   W.J. Dally and B. Towles, Route packets, not wires: On-chip interconnection networks, *Design and Automation Conference DAC01,* Las Vegas, NV, pp. 684–689, June 2001.

[GUE00]   P. Guerrier and A. Greiner, A generic architecture for on-chip packet switched interconnections, *Design, Automation and Testing in Europe DATE00*, Paris, France, pp. 250–256, March 2000.

[LEE03]   S.J. Lee et al., An 800 MHz star-connected on-chip network for application to systems on a chip, *ISSCC03*, Daejeon, South Korea, February 2003.

[YAM02]   H. Yamauchi et al., A 0.8 W HDTV video processor with simultaneous decoding of two MPEG2 MP@HL streams and capable of 30 frames/s reverse playback, *ISSCC02*, San Francisco, CA, 1, pp. 473–474, February 2002.

[TRS07]   ITRS 2007. Available at: http://www.itrs.net/Links/2007ITRS/Home2007.htm.

[JAN03]   A. Jantsch and H. Tenhunen, Will networks on chip close the productivity gap?, from *Networks on Chip*, A. Jantsch and H. Tenhunen (Eds.), Kluwer, Dordrecht, the Netherlands, 2003, pp. 3–18.

[RIJ03]   E. Rijpkema, K. Goossens, A. Radulescu, J. van Meerbergen, P. Wielage, and E. Waterlander, Trade offs in the design of a router with both guaranteed and best-effort services for networks on chip, *Design Automation and Test in Europe DATE03*, Munich, Germany, pp. 350–355, March 2003.

[CAR01]   L.P. Carloni, K.L. McMillan, and A.L. Sangiovanni-Vincentelli, Theory of latency-insensitive design, *IEEE Transactions on CAD of ICs and Systems*, 20(9): 1059–1076, September 2001.

[SCH02]   L. Scheffer, Methodologies and tools for pipelined on-chip interconnects, *International Conference on Computer Design*, Freiburg, Germany, pp. 152–157, 2002.

[JAL05]   A. Jalabert et al., XpipesCompiler: A tool for instantiating application specific networks-on-chip, *DATE* 2005, pp. 884–889.

[AGA00]   V. Agarwal, M.S. Hrishikesh, S.W. Keckler, and D. Burger, Clock rate versus IPC: The end of the road for conventional microarchitectures, *Proceedings of the 27th Annual International Symposium on Computer Architecture*, Vancouver, BC, pp. 248–250, June 2000.

[STE05]   S. Stergiou et al., Xpipes lite: A synthesis oriented design flow for networks on chips, *Proceedings of the Design, Automation and Test in Europe Conference*, Munich, Germany, pp. 1188–1193, 2005.

[KAR87]   M.J. Karol, Input versus output queuing on a space division packet switch, *IEEE Transactions on Communications*, COM-35(12): 1347–1356, 1987.

[MCK95]   N. McKeown, Scheduling algorithms for input-queued cell switches, PhD thesis, University of California, Berkeley, CA, 1995.

[DAL04]   W.J. Dally and B. Towels, *Principles and Practices of Interconnection Networks*, Morgan Kaufmann, San Francisco, CA, 2004.

[DUA04]   J. Duato, S. Yalamanchili, and L. Ni, *Interconnection Networks: An Engineering Approach*, Morgan Kaufmann, San Francisco, CA, 2004.

[MIC06]   G. De Michelli and L. Benini, *Networks on Chips: Technology and Tools*, Morgan Kaufmann, San Francisco, CA, 2006.

[BAL06]   J. Balfour and W.J. Dally, Design tradeoffs for tiled CMP on-chip networks, *Proceedings of the 20th Annual International Conference on Supercomputing*, Cairns, Australia, June 2006.

[FLI07]   J. Flich and J. Duato. Logic-based distributed routing for NoCs, *Computer Architecture Letters*, 6:13–16, November 2007.

[GIL08]   F. Gilabert, S. Medardoni, D. Bertozzi, L. Benini, M.E. Gomez, P. Lopez, and J. Duato, Exploring high-dimensional topologies for NoC design through an integrated analysis and synthesis framework, *Proceedings of the 2nd IEEE International Symposium on Networks-on-Chip (NoCS 2008)*, Newcastle upon Tyne, UK, April 2008.

[BED04]   M. B. Taylor, W. Lee, S. Amarasinghe, and A. Agarwal, Scalar operand networks: Design, implementation, analysis, MIT/LCS Technical Memo LCS-TM-644, April 2004.

[MUR04]   S. Murali and G. De Micheli, SUNMAP: A tool for automatic topology selection and generation for NoCs, *Proceedings of the Design Automation Conference*, San Diego, CA, pp. 914–919, 2004.

[PEH07]   V. Soteriou, N. Eisley, H. Wang, and L.S. Peh, Polaris: A system-level roadmapping toolchain for on-chip interconnection networks, *IEEE Transactions on VLSI* 15(8): 401–408, 2007.

[BER05]   D. Bertozzi, A. Jalabert, S. Murali, R. Tamhankar, S. Stergiou, L. Benini, and G. De Micheli, NoC synthesis flow for customized domain specific multiprocessor systems-on-chip, *IEEE Transactions on Parallel and Distributed Systems*, 16(2): 113–129, 2005.

[GOM06]   M.E. Gomez, P. Lopez, and J. Duato, FIR: An efficient routing strategy for tori and meshes, *Journal of Parallel and Distributed Computing*, 66: 907–921, Elsevier, 2006.

[GOM07]   C. Gomez, F. Gilabert, M.E. Gomez, P. Lopez, and J. Duato, Deterministic versus adaptive routing in fat-trees, *Proceedings of the Workshop on Communication Architecture on Clusters*, as a part of IPDPS'07, Los Angeles, CA, March 2007.

[KAV03]   N.K. Kavaldjiev and J.M. Smit, A survey of efficient on-chip communications for SoC, *Proceedings of the 4th PROGRESS Symposium on Embedded Systems*, Nieuwegein, the Netherlands, 2003.

[GOM08]   C. Gomez, M.E. Gomez, P. Lopez, and J. Duato, Exploiting wiring resources on interconnection network: Increasing path diversity, *Proceedings of the 16th Euromicro International Conference on Parallel, Distributed and Network-Based Processing*, Toulouse, France, February, 2008.

[LER08]   A. Leroy et al., Spatial division multiplexing: A novel approach for guaranteed throughput on NoCs, *Proceedings of the 3rd International Conference on Hardware/Software Codesign and System Synthesis*, Jersey City, NJ, September 2005.

[PUL05]   A. Pullini, F. Angiolini, D. Bertozzi, and L. Benini, Fault tolerance overhead in network-on-chip flow control schemes, *Proceedings of SBCCI*, Florianolpolis, Brazil, pp. 224–229, 2005.

[MAR05]   U.Y. Ogras and R. Marculescu, Application-specific network-on-chip architecture customization via long-range link insertion, *Proceedings of ICCAD*, San Jose, CA, November, 2005.

[FUK99]   H. Fuks, A. and Lawniczak, Performance of data networks with random links, *Mathematics and Computers in Simulation*, 51: 101–117, 1999.

[MAL02]   X. Zhu and S. Malik, A hierarchical modeling framework for on-chip communication architectures, *Proceedings of ICCD*, San Jose, CA, pp. 663–671, 2002.

[YE03]   T.T. Ye, L. Benini, and G. De Micheli, Analysis of power consumption on switch fabrics in network routers, *Proceedings of DAC*, New Orleans, LA, pp. 524–529, 2003.

[WAN02]   H.-S. Wang et al., Orion: A power-performance simulator for interconnection network, *Proceedings of the International Symposium on Microarchitecture*, Istanbul, Turkey, pp. 294–305, November 2002.

[PAL04]   G. Palermo and C. Silvano, PIRATE: A framework for power/performance exploration of network-on-chip architectures, *Proceedings of PATMOS*, Santorini, Greece, pp. 521–531, 2004.

[ART]   Arteris, the Network-on-Chip Company. Available at: www.arteris.com.

[COP04]   M. Coppola et al., Spidergon: A novel on-chip communication network, *Proceedings of the ISSOC*, Tampere, Finland, pp. 16–18, 2004.

[BAI02]   J. Bainbridge and S. Furber, Chain: A delay-insensitive chip area interconnect, *IEEE Micro*, 22(5): 16–23, 2002.

[STA03]   D. Starobinski, M. Karpovsky, and L.A. Zakrevski, Application of network calculus to general topologies using turn-prohibition, *IEEE/ACM Transactions on Networking*, 11(3): 411–421, 2003.

[OWE07]   J.D. Owens, W.J. Dally, R. Ho, D.N. Jayasimha, S.W. Keckler, and L.Peh, Research challenges for on-chip interconnection networks, *IEEE Micro*, 27(5): 96–108, 2007.

[AXI03]   AMBA AXI Protocol Specification, 2003.

[ANG06]   F. Angiolini, P. Meloni, S. Carta, L. Benini, and L. Raffo, Contrasting a NoC and a traditional interconnect fabric with layout awareness, *Proceedings of DATE*, Munich, Germany, pp. 1–6, March 2006.

[PUL07]   A. Pullini, F. Angiolini, P. Meloni, D. Atienza, S. Murali, L. Raffo, G. De Micheli, and L. Benini, NoC design and implementation in 65 nm technology, *1st IEEE/ACM International Symposium on Networks-on-Chip*, New Jersey, pp. 273–282, May 2007.

[VIL03]   T. Villiger et al., Self-timed ring for globally-asynchronous locally-synchronous systems, *Proceedings of the 9th IEEE International Symposium on Asynchronous Circuits and Systems*, Vancouver, BC, pp. 141–150, 2003.

[KRS06]   M. Krstić, E. Grass, C. Stahl, and M. Piz, System integration by request-driven GALS design, *IEE Proceedings on Computers and Digital Techniques*, 153(5): 362–372, September 2006.

[GUR06]   F.K. Gurkaynak et al., GALS at ETH Zurich: Success or failure?, *Proceedings of the 12th IEEE International Symposium on Asynchronous Circuits and Systems*, Grenoble, France, pp. 150–159, 2006.

[LAT07]   D. Lattard, et al., A Telecom baseband circuit-based on an asynchronous network-on-chip, *Proceedings of the International Solid State Circuits Conference, ISSCC'2007*, San Francisco, CA, February 2007.

[BER03]   D. Bertozzi, L. Benini, and G. De Micheli, Energy-reliability trade-off for NoCs, from *Networks on Chip*, A. Jantsch and H. Tenhunen (Eds.), Kluwer, Dordrecht, the Netherlands, 2003, pp. 107–129.

[ZHA00]   H. Zhang, V. George, and J.M. Rabaey, Low-swing on-chip signaling techniques: Effectiveness and robustness, *IEEE Transactions on VLSI Systems*, 8(3): 264–272, June 2000.

[XU02]   J. Xu and W. Wolf, Wave pipelining for application-specific networks-on-chips, *CASES02*, Grenoble, France, pp. 198–201, October 2002.

[WOR05]   F. Worm, P. Ienne, P. Thiran, and G. De Micheli, A robust self-calibrating transmission scheme for on-chip networks, *IEEE Transactions on Very Large Scale Integration (VLSI) System*, 13(1): 126–139, January 2005.

[MED08]   S. Medardoni, D. Bertozzi, and M. Lajolo, Variation tolerant NoC design by means of self-calibrating links, *Proceedings of DATE*, Munich, Germany, 2008.

[FLI08]   J. Flich, S. Rodrigo, and J. Duato, An efficient implementation of distributed routing algorithms for NoCs, *2nd IEEE International Symposium on Networks-on-Chip*, Newcastle upon Tyne, UK, 2008.

[LEI06]   S. Leibson, The future of nanometer SoC design, *International Symposium on System-on-Chip*, Tampere, Finland, pp. 1–6, November 2006.

[BOR05]   S. Borkar et al., Platform 2015: Intel® processor and platform evolution for the next decade, *Technology@Intel Magazine*, Intel Corporation, March 2005.

[OLU07]   L. Hammond, B.A. Nayfeh, and K. Olukotun, A single-chip multiprocessor, *IEEE Computer, Special Issue on Billion-Transistor Processors*, 30(9): 79–85, September 1997.

[DAL91]   W.J. Dally, Express cubes: Improving the performance of k-ary n-cube interconnection networks, *IEEE Transactions on Computers*, 40(9): 1016–1023, 1991.

[HAU98]   S. Hauck, G. Borriello, and C. Ebeling, Mesh routing topologies for Multi-FPGA systems, *IEEE Transactions on VLSI Systems*, 6(3): 400–408, 1998.

[LEI85]   C.E. Leiserson, Fat-trees: Universal networks for hardware-efficient supercomputing, *IEEE Transactions on Computers*, 34(10): 892–901, 1985.ù.

[TOR07]   A. Dalla Torre, M. Ruggiero, A.A cquaviva, and L. Benini, MP-QUEUE: An efficient communication library for embedded streaming multimedia platform, *IEEE Workshop on Embedded Systems for Real-Time Multimedia*, Salzburg, Austria, October 2007.

[OCP01]   OCP International Partnership. Open Core Protocol Specification, 2001.

16

Hardware/Software Interfaces Design for SoC

Katalin Popovici
Techniques of Informatics and Microelectronics for Integrated Systems Architecture (TIMA) Laboratory

Wander O. Cesário
Techniques of Informatics and Microelectronics for Integrated Systems Architecture (TIMA) Laboratory

Flávio R. Wagner
Federal University of Rio Grande do Sul

A. A. Jerraya
Atomic Energy Commission, Minatec

16.1 Introduction

Multiprocessor system-on-chip (MPSoC) architectures are emerging as one of the technologies providing a way to face the growing design complexity of embedded systems, since they provide flexibility of programming, allied to specific processor architectures adapted to the selected problem classes. This leads to gains in compactness, low power consumption, and performance. The trend of integrating multiple processor cores on the same chip will be even more accentuated in the near future. The SoC system driver section of the International Technology Roadmap for Semiconductors [1] predicts that the number of processor cores will increase fourfold per technology node in order to match the processing demands of the corresponding applications. Typical MPSoC applications like network processors, multimedia hubs, and base-band telecom circuits have particularly tight time-to-market and performance constraints that require a very efficient design cycle.

MPSoCs integrate hardware components, such as processors, memories, interconnect and special purpose modules and software components, like operating systems (OSs) and application code. Our conceptual model of the MPSoC platform is composed of four kinds of components: software tasks, processor and intellectual property (IP) cores, and a global on-chip interconnect IP (see Figure 16.1a).

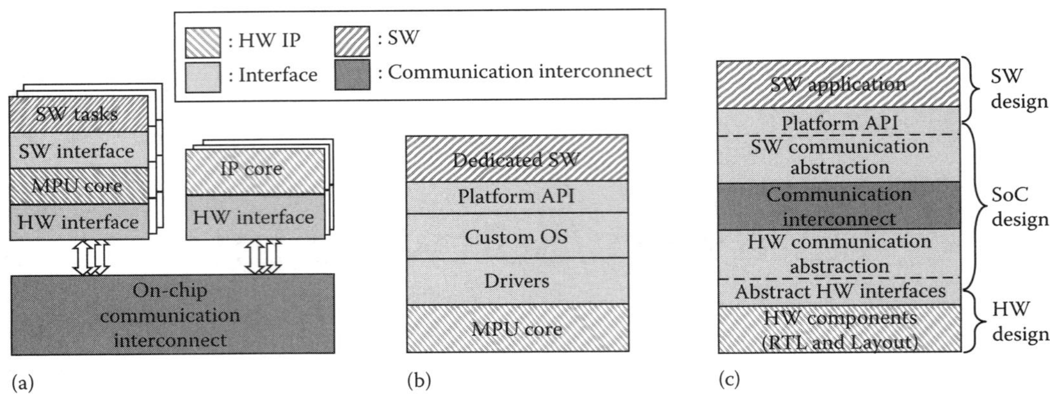

FIGURE 16.1   (a) MPSoC platform, (b) software stack, and (c) concurrent development environment.

Moreover, to complete the MPSoC platform we must also include hardware and software (HW/SW) elements that adapt platform components to each other.

In MPSoC architectures, the implementation of the system communication is more complicated since heterogeneous processors may be involved and complex communication protocols and topologies may be used. For instance, a data exchange between two different processors may use different schemes (global memory accessible by both processing units, local memory of one of the processors, dedicated hardware first-in first-out [FIFO] components, etc.). Additionally, different synchronization schemes (polling, interrupts) may be used to coordinate this data exchange. Each of these communication schemes has advantages and disadvantages in terms of performance (e.g., latency, throughput), resource sharing (e.g., multitasking, parallel I/O), and communication overhead (e.g., memory size, execution time). Moreover, MPSoC platforms often use several complex system buses or micronetworks as global interconnect. In MPSoC platforms, we can separate computation and communication design by using communication coprocessors and profiting from the multimaster architecture. Communication coprocessors/controllers (masters) implement high-level communication protocols in hardware and execute them in parallel with the computation executed on processor cores.

Each processor core executes a software stack. The software stack is structured in only three layers, as depicted in Figure 16.1b. The top layer is the software application that may be a multitasking description or a single task function. The application layer consists of a set of tasks that makes use of programming model or application programming interface (API) to abstract the underlying platform. These APIs correspond to the platform APIs. The separation between the application layer and the underlying platform is required to facilitate concurrent software and hardware (SW/HW) development.

The middle layer consists of any commercial embedded OS configured according to the application. This software layer is responsible of providing the necessary services to manage and share resources. The software includes scheduling of the application tasks on top of the available processing elements, intertask communication, external communication, and all other kinds of resources management and control services.

Low-level details about how to access these resources are abstracted by the third layer, which contain device drivers and low-level routines to control/configure the platform. This layer is also known as the hardware abstraction layer (HAL). The separation between OS and HAL makes architecture exploration for the design of both the CPU subsystem and the OS services easier, enabling easy software portability. The HAL is a thin software layer, which totally depends on the type of processor that will execute the software stack, but also depends on the hardware resources interacting with the processor.

All these layers correspond to the software adaptation layer in Figure 16.1a; coding application software can then be isolated from the design of the SoC platform (software coding is not the topic of this chapter and will be omitted). One of the main contributions of this work is to consider this layered approach also for the dedicated software (often called firmware). Firmware is the software that controls the platform, and, in some cases, executes some nonperformance critical application functions. In this case, it is not realistic to use a generic OS as the middle layer due to code size and performance reasons. A lightweight custom OS supporting an application-specific and platform-specific API is required.

SW/HW adaptation layers isolate platform components, enabling concurrent development as shown in Figure 16.1c. Using this scheme, the software design team uses APIs for both application and dedicated software development. The hardware design team uses abstract interfaces provided by communication coprocessors/controllers. SoC design team can concentrate on implementing HW/SW abstraction layers for the selected communication interconnect IP. Designing these HW/SW abstraction layers represents a major effort, and design tools are lacking. The key challenges of the MPSoC HW/SW design are

1. Raise the abstraction level: Modeling of a register-transfer level (RTL) architecture is too expensive in terms of design time and requires too many efforts to verify the interconnection between multiple processor cores. Thus, raising the abstraction level seems to be a solution to bridge the gap between the increasing complexity of MPSoC and the low-design productivity.
2. Embedded software is becoming more and more complex, requiring hundred thousands lines to be coded. Thus, to reduce the software development cost and the overall design time, a higher-level programming model is needed.
3. The efficiency of the MPSoC depends on the suitability of the hardware architecture with the application. Therefore, efficient HW/SW interfaces are required. These include microprocessor interfaces, bank of registers, shared memories (SHMs), software drivers, and customized OSs. All these interfaces must be optimized to each application. Moreover, automatic generation of the HW/SW interfaces may save huge design and validation time.

This chapter presents a component-based design automation approach for MPSoC platforms. Section 16.2 introduces the basic concepts for MPSoC design and discusses some related platform and component-based approaches. Section 16.3 details IP-based methodologies for HW/SW IP integration. Section 16.4 details our specification model and design flow. Section 16.5 presents the application of this flow for the design of a very high bitrate digital subscriber line (VDSL) circuit and the analysis of the results.

16.2 System-on-Chip Design

16.2.1 System-Level Design Flow

This section gives an overview of current SoC design methodologies using a template design flow (Figure 16.2). The basic theory behind this flow is the separation between communication and computation refinement for platform and component-based design [2,3]. It has five main design steps:

1. *System specification*: System designers and the end-customer must agree on an informal model containing all application's functionality and requirements. Based on this model, system designers build a more formal specification that can be validated by the end-customer.
2. *Architecture exploration*: System designers build an executable model of the specification and iterate through a performance analysis loop to decide the HW/SW partitioning for

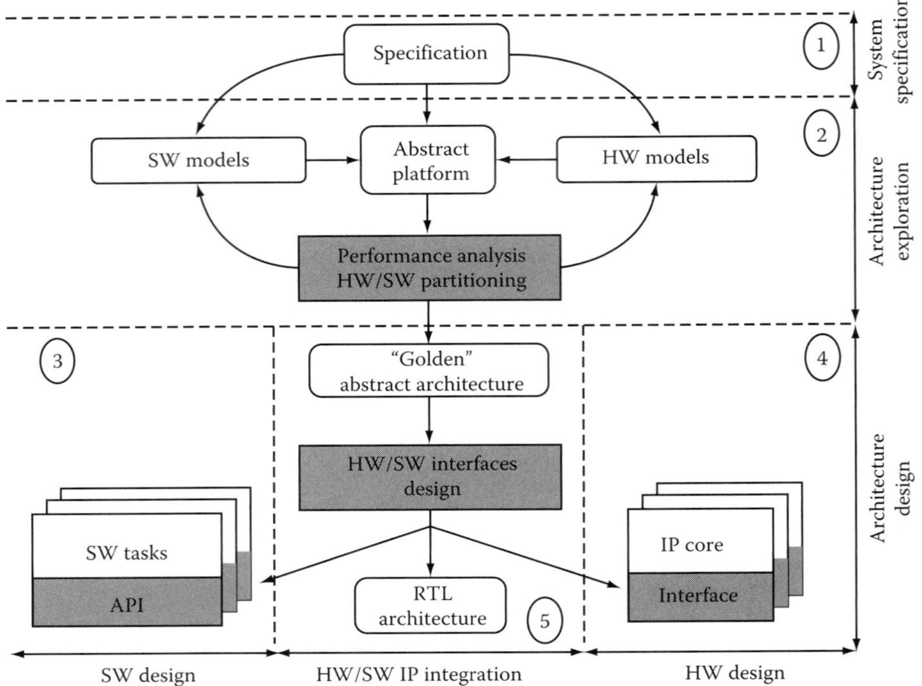

FIGURE 16.2 System-level design flow for SoC.

the SoC architecture. This executable specification uses an abstract platform composed of abstract models for HW/SW components. For instance, an abstract software model can concentrate on I/O execution profiles, or most frequent use cases, or worst-case scheduling. Abstract hardware can be described using transaction-level models or behavioral models. This step produces the "golden" architecture model, that is the customized SoC platform or a new architecture created by system designers after selecting processors, the global communication interconnect, and other IP components. Once HW/SW partitioning is decided, SW/HW development can be done concurrently.

3. *Software design*: Since the final hardware platform will not be available during software development, some kind of HAL or API must be provided to the software design team.

4. *Hardware design*: Hardware IP designers implement the functionality described by the abstract hardware models at the RT-level. Hardware IPs can use specific interfaces for a given platform or standard interfaces as defined by virtual socket interface alliance (VSIA [4]).

5. *HW/SW IP integration*: SoC designers create HW/SW interfaces to the global communication interconnect. The golden architecture model must specify performance constrains to assure a good HW/SW integration. SW/HW communication interfaces are designed to conform to these constrains.

16.2.2 SoC Design Automation: An Overview

Many academic and industrial works propose tools for SoC design automation covering many, but not all, design steps presented before. Most approaches can be classified into three groups: system-level synthesis, platform-based design, and component-based design.

System-level synthesis methodologies are top-down approaches; the SoC architecture and software models are produced by synthesis algorithms from a system-level specification. COSY [5] proposes an HW/SW communication refinement process that starts with an extended Kahn Process Network model on design step (1), uses virtual component co-design (VCC) [6] for step (2), callback signals over a standard real-time operating system (RTOS) for the API in step (3), and VSIA interfaces for steps (4) and (5). SpecC [7] starts with an untimed functional specification model written in extended C on design step (1), uses performance estimation for a structural architecture model for step (2), HW/SW interface synthesis based on a timed bus-functional communication model for step (5), synthesized C code for step (3), and behavioral synthesis for step (4).

Platform-based design is a meet-in-the-middle approach that starts with a functional system specification and a predesigned SoC platform. Performance estimation models are used to try different mappings between the set of application's functional modules and the set of platform components. During these iterations, designers can try different platform customizations and functional optimizations. VCC [6] can produce a performance model using a functional description of the application and a structural description of the SoC platform for design steps (1) and (2). CoWare N2C [8] is a good complement for VCC for design steps (4) and (5). Still the API for software components and many architecture details must be implemented manually.

Section 16.3 discusses HW/SW IP integration in the context of current IP-based design approaches. Most IP-based design approaches build SoC architectures from the bottom-up using predesigned components with standard interfaces and a standard bus. For instance, IBM defined a standard bus called CoreConnect [9], Sonics proposes a standard on-chip network called Silicon Backplane Network [10], and VSIA defined a standard component protocol called VCI. When needed, wrappers adapt incompatible buses and component interfaces. Frequently, internally developed components are tied to in-house (nonpublic) standards; in this case, adopting public standards implies a big effort to redesign interfaces or wrappers for old components.

Section 16.4 introduces a higher-level IP-based design methodology for HW/SW interface design called component-based design. This methodology defines a virtual architecture model composed of HW/SW components and uses this model to automate design step (5), by providing automatic generation of hardware interfaces (4), device drivers, OSs, and APIs (3). Even if this approach does not provide much help on automating design steps (1) and (2), it provides a considerable reduction of design time for design steps (3) through (5) and facilitates component reuse. The key improvements over other state-of-art platform and component-design approaches are

1. *Strong support for software design and integration*: The generated API completely abstracts the hardware platform and OS services. Software development can be concurrent to and independent of platform customization.
2. *Higher-level abstractions*: The use of a virtual architecture model allows designers to deal with HW/SW interfaces at a high abstraction level. Behavior and communication are separated in the system specification; thus, they can be refined independently.
3. *Flexible HW/SW communication*: Automatic HW/SW interfaces generation is based on the composition of library elements. It can be used with a variety of IP interconnect components by adding the necessary supporting library.

16.3 Hardware/Software IP Integration

There are two major approaches for the integration of HW/SW IP components into a given design. In the first one, component interfaces follow a given standard (such as a bus or core interface, for hardware components, or a set of high-level communication primitives, and for software components) and can be, thus, directly connected to each other. In the second approach, components are heterogeneous in nature and their integration requires the generation of HW/SW wrappers. In both cases, an RTOS must be used to provide services that are needed in order that the application software fits into the SoC architecture. This section describes different solutions to the integration of HW/SW IP components.

16.3.1 Introduction to IP Integration

The design of an embedded SoC starts with a high-level functional specification, which can be validated. This specification must already follow a clear separation between computation and communication [11], in order to allow their concurrent evolution and design. An abstract architecture is then used to evaluate this functionality based on a mapping that assigns functional blocks to architectural ones. This high-level architectural model abstracts away all low-level implementation details. A performance evaluation of the system is then performed, by using estimates of the computation and communication costs. Communication refinement is now possible, with a selection of particular communication mechanisms and a more precise performance evaluation.

According to the platform-based design approach [2], the abstract architecture follows an architectural template that is usually domain-specific. This template includes both a hardware platform, consisting of a given communication structure and given types of components (processors, memories, and hardware blocks), and a software platform, in the form of a high-level API. The target embedded SoC will be designed as a derivative of this template, where the communication structure, the components, and the software platform are all tailored to fit the particular application needs.

The IP-based design approach follows the idea that the architectural template may be implemented by assembling reusable HW/SW IP components, potentially even delivered by third-party companies.

The IP integration step comprises a set of tasks that is needed to assemble predesigned components in order to fulfill system requirements. As shown in Figure 16.3, it takes as inputs the abstract architecture and a set of HW/SW IP components that have been selected to implement the architectural blocks. Its output is a microarchitecture where hardware components are described at the RT-level with all cycle-and-pin accurate details that are needed for a further automatic synthesis. Software components are described in an appropriate programming language, such as C, and can be directly compiled to the target processors of the architecture.

In an ideal situation, IP components would fit directly together (or to the communication structure) and exactly match the desired SoC functionality. In a more general situation, the designer may

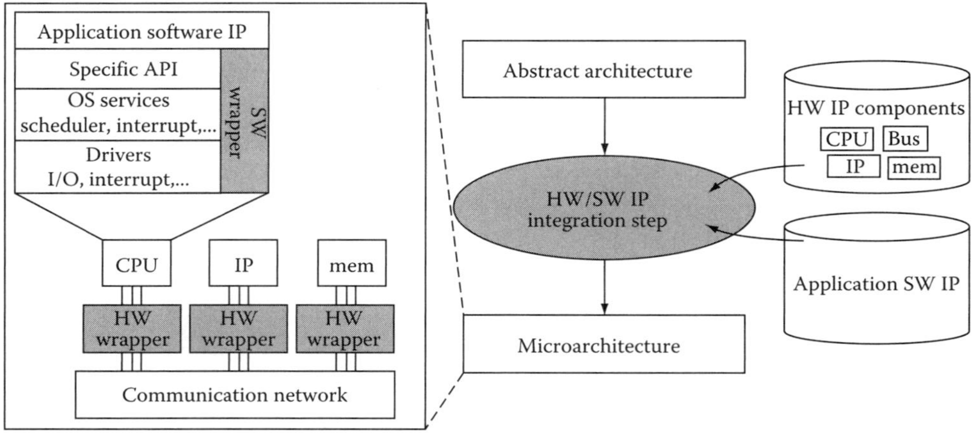

FIGURE 16.3   HW/SW IP integration design step.

need to adapt each component's functionality (a step called IP derivation) and synthesize HW/SW wrappers to interconnect them. For programmable components, although adaptation may be easily performed by programming the desired functionality, the designer may still need to develop software wrappers (usually device or bus drivers) to match the application software to the communication infrastructure. The generation of HW/SW wrappers is usually known as interface or communication synthesis. Besides them, application software may also need to be retargeted to the processors and OS of the chosen architecture.

In the following sections, different approaches to IP integration are introduced and their impact on the possible integration subtasks is analyzed.

16.3.2 Bus-Based and Core-Based Approaches

In the bus-based design approach [9,12,13], IP components communicate through one or more buses (interconnected by bus bridges). Since the bus specification can be standardized, libraries of components whose interfaces directly match this specification can be developed. Even if components follow the bus standard, very simple bus interface adapters may still be needed [14]. For components that do not directly match the specification, wrappers have to be built. Companies offer very rich component libraries and specialized development and simulation environments for designing systems around their buses.

A somewhat different approach is the core-based design, as proposed by the VSIA VCI standard [4] and by the open core protocol international partnership (OCP-IP) organization [15]. In this case, IP components are compliant with a bus-independent and standardized interface and can thus be directly connected to each other. Although the standard may support a wide range of functionality, each component may have an interface containing only the functions that are relevant for it. These components may also be interconnected through a bus, in which case standard wrappers can adapt the component interface to the bus. Sonics [13] follows this approach, proposing wrappers to adapt the bus-independent OCP socket to the MicroNetwork bus.

For particular needs, the SoC may be built around a sophisticated and dedicated network-on-chip (NoC) [16] that may deliver very high performance for connecting a large number of components. Even in this case, a bus- or core-based approach may be adopted to connect the components to the network.

Bus-based and core-based design methodologies are integration approaches that depend on standardized component or bus interfaces. They allow the integration of homogeneous IP components that follow these standards and can be directly connected to each other, without requiring the development of complex wrappers. The problem we face is that many de facto standards, coming from different companies or organizations, exist thus preventing a real interchange of libraries of IP components developed for different substandards.

16.3.3 Integrating Software IP

Programmable components are important in a reusable architectural platform, since it is very cost-effective to tailor a platform to different applications by simply adapting the low-level software and maybe only configuring certain hardware parameters, such as memory sizes and peripherals.

As illustrated in Figure 16.3, the software view of an embedded system shows three different layers:

1. Bottom layer is composed of services directly provided by hardware components (processor and peripherals) such as instruction sets, memory and peripheral accesses, and timers.
2. Top layer is the application software, which should remain completely independent from the underlying hardware platform.

3. Middle layer is composed of three different sublayers, as seen from bottom to top:

   a. Hardware-dependent software (HdS) consisting, for instance, of device drivers, boot code, parts of an RTOS (such as context switching code and configuration code to access the memory management unit, MMU), and even some domain-oriented algorithms that directly interact with the hardware

   b. Hardware-independent software, typically high-level RTOS services, such as task scheduling and high-level communication primitives

   c. API, which defines a system platform that isolates the application software from the hardware platform and from all basic software layers, and enables their concurrent design

The standardization of this API, which can be seen as a collection of services usually offered by an OS, is essential for software reuse above and below it. At the application software level, libraries of reusable software IP components can implement a large number of functions that are necessary for developing systems for given application domains. If, however, one tries to develop a system by integrating application software components that do not directly match a given API, software retargeting to the new platform will be necessary. This can be a very tedious and error-prone manual process, which is a candidate for an automatic software synthesis technique.

Nevertheless, reuse can also be obtained below the API. Software components implementing the hardware-independent parts of the RTOS can be more easily reused, especially if the interface between this layer and the HdS layer is standardized. Although the development of reusable HdS may be harder to accomplish, because of the diversity of hardware platforms, it can be at least obtained for platforms aimed at specific application domains.

There are many academic and industrial alternatives providing RTOS services. The problem with most approaches, however, is that they do not consider specific requirements for SoC, such as minimizing memory usage and power consumption. Recent research efforts propose the development of application-specific RTOS containing only the minimal set of functions needed for a given application [17,18] or including dynamic power management techniques [19]. IP integration methodologies should thus consider the generation of application-specific RTOSs that are compliant to a standard API and optimized for given system requirements.

In recent years, many standardization efforts aimed at hardware IP reuse have been developed. Similar efforts for software IP reuse are now needed. VSIA [4] has recently created working groups to deal with HdS and platform-based design.

16.3.4   Communication Synthesis

Solutions for the automatic synthesis of communication wrappers to connect hardware IP components that have incompatible interfaces have been already proposed. In the PIG tool [20], component interfaces are specified as protocols described as regular expressions, and an FSM interface for connecting two arbitrary protocols is automatically generated. The Polaris tool [21] generates adapters based on state machines for converting component protocols into a standard internal protocol, together with send and receive buffers and an arbiter.

These approaches, however, do not address the integration of software IP components. The TEReCS tool [18] synthesizes communication software to connect software IP components, given a specification of the communication architecture and a binding of IP components to processors. In the IPChinook environment [22], abstract communication protocols are synthesized into low-level bus protocols according to the target architecture. While the IPChinook environment also generates a scheduler for a given partitioning of processes into processors, the TEReCS approach is associated to the automatic synthesis of a minimal OS, assembled from a general-purpose library of reusable objects that are configured according to application demands and the underlying hardware.

Recent solutions uniformly handle HW/SW interfaces between IP components. In the COSY approach [5], design is performed by an explicit separation between function and architecture. Functions are then mapped to architectural components. Interactions between functions are modeled by high-level transactions and then mapped to HW/SW communication schemes. A library provides a fixed set of wrapper IPs, containing HW/SW implementations for given communication schemes.

16.3.5 IP Derivation

Hardware IP components may come in several forms [23]. They may be hard, when all gates and interconnects are placed and routed, soft, with only a RTL representation, or firm, with a RTL description together with some physical floorplanning or placement. The integration of hard IP components cannot be performed by adapting their internal behavior or structure. If they have the advantage of a more predictable performance, in turn they are less flexible and therefore less reusable than adaptable components.

Several approaches for enhancing reusability are based on adaptable components. Although one can think of very simple component configurations (for instance, by selecting a bit width), a higher degree of reusability can be achieved by components whose behavior can be more freely modified. Object-orientation is a natural vehicle for high-level modeling and adaptation of reusable components [24,25]. This approach, which can be better classified as IP derivation, is adequate for not only firm and soft hardware IP components, but also for software IP [26]. Although component reusability is enhanced by this approach, the system integrator has a greater design effort, and it becomes more difficult to predict IP performance.

IP derivation and communication synthesis are different approaches to solve the same problem of integration between heterogeneous IP components, which do not follow standards (or the same substandards). IP derivation is a solution usually based on object-oriented concepts coming from the software community. It can be applied to the integration of application software components and for hardware soft and firm components, but it cannot be used for hard IP components. Communication synthesis, on the other hand, follows the path of the hardware community on automatic logic and high-level synthesis. It is the only solution to the integration of heterogeneous hard IP components, although it can also be used for integrating software IP and soft and firm hardware IP. While IP derivation is essentially a user-guided manual process, communication synthesis is an automatic process, with no user intervention.

16.4 Component-Based SoC Design

This section introduces the component-based design methodology, a high-level IP-based methodology aimed at the integration of heterogeneous HW/SW IP components. It follows an automatic communication synthesis approach, generating both HW/SW wrappers. It also generates a minimal and dedicated OS for programmable components. It uses a high-level API, which isolates the application software from the implementation of an HW/SW solution for the system platform, such that software retargeting is not necessary.

This approach enables the automatic integration of heterogeneous (components that do not follow a given bus or core standard) and hard IP components (whose internal behavior or structure is not known). However, the approach is also very well suited to the integration of homogeneous and soft IP components. The methodology has been conceived to fit to any communication structure, such as a NoC [16] or a bus.

The component-based methodology relies on a clear definition of three abstraction levels that are adopted by other current approaches: system (pure functional), macroarchitecture, and microarchitecture (RTL). These levels constitute clear "interfaces" between design steps, promoting reuse of both components and tools for design tasks at each of these levels.

16.4.1 Design Methodology Principles

The design flow starts with a combined application/architecture model, which mixes the functional specification of the application with the partitioning and mapping information (Figure 16.4a). Thus, the application functions are grouped into tasks, and then, the tasks are mapped on the target architecture. Additionally, communication units are introduced in this model between the application tasks to specify the communication protocol used for the data exchange and synchronization.

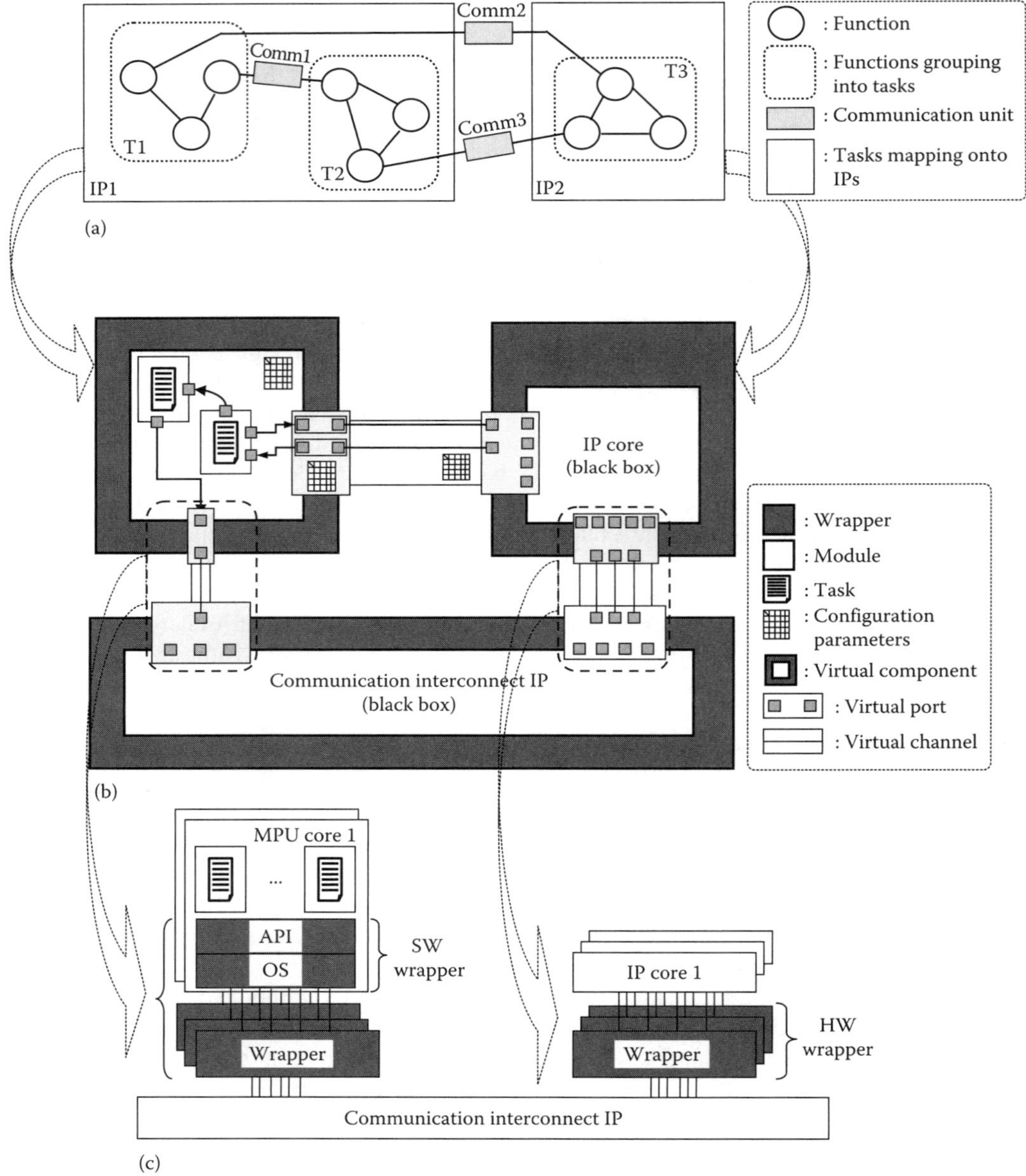

FIGURE 16.4 MPSoC design flow: (a) Combined application/architecture model, (b) virtual architecture, and (c) target MPSoC platform.

The next design step consists of the automatic generation of the virtual architecture, also called macroarchitecture. The virtual architecture model corresponds to the "golden" architecture in Figure 16.2, and it is composed of virtual modules (VM) (see Figure 16.4b). The VM may correspond to processing and memory IPs, connected by any communication structure, also encapsulated within a VM. This abstract architecture model clearly separates computation from communication, allowing independent and concurrent implementation paths for components and for communication.

The VM, which represents a processor, may be hierarchically decomposed into multiple submodules, containing the software tasks assigned to this processor. VMs communicate through virtual ports, which are sets of hierarchical internal and external ports that require and provide various services (e.g., write_request/read_request, write_acknowledge/read_acknowledge, send_event/receive_event, wait_for_synchronization, do_synchronization, etc.). The separation between internal and external ports makes possible the connection of the modules described at different abstraction levels (functional, TLM, or RTL).

The next design step consists of automatic generation of the wrappers, device drivers, OSs, and APIs from the virtual architecture. The goal is to produce a synthesizable RTL model of the MPSoC platform, which is composed of processor cores, IP cores, communication interconnect IP, and HW/SW wrappers (Figure 16.4c). The HW/SW wrappers are automatically generated from the interfaces of the virtual components (as indicated by the arrows in Figure 16.4). The software written for the virtual architecture runs without modification on the implementation, because the generated custom OS provides the implementation of the same APIs.

More details about these steps and the representation models will be given in the following sections.

16.4.2 Combined Application Architecture Model

We represent the combined application/architecture model in Simulink® [16] using the hierarchy of concepts depicted in Figure 16.4a [28]. The basic element is a function and it represents an elementary block either predefined in the standard Simulink library or user-defined function integrated in the model as S-function. A task groups a set of functions. The intratask communication represents the communication between functions composing a task. This type of communication is implicit in the Simulink model, but it will be translated to communication via local variables during the code generation for the virtual architecture level.

The communications between tasks are classified in two categories: intra-subsystem communication unit, which specifies the communication between the tasks mapped on the same IP subsystem, and inter-subsystem communication unit, which shows the communication between different IP subsystems. For instance, the *comm1* in Figure 16.4a illustrates an intra-subsystem communication unit, while *comm2* and *comm3* represent inter-subsystem communication units.

In fact, the communication units of the model can be basic units that represent predefined APIs, i.e., register (read/write(value)), queue (read/write (address, size)), SHM (read/write, synchronization) or specific communication units that defines custom APIs, i.e., network interface (read/write (address, size, control)), MMU (read/write (address1, address2, size, synchronization)), etc.

The simulation of the combined application/architecture model allows performing a functional validation of the application.

16.4.3 Virtual Architecture

The virtual architecture represents a system as an abstract netlist of virtual components (see Figure 16.4b). It is generated from the Simulink combined architecture application model in VADeL, a SystemC [29] extension that includes a platform-independent API offering high-level communication primitives. This API abstracts the underlying hardware platform, thus enhancing the free development of reusable components. In the abstract architecture model, the interfaces of software tasks are the same for SW/SW and SW/HW connections, even if the software tasks are executed by different processors. Different HW/SW realizations of this API are possible. Architectural design space exploration can be thus achieved without influencing the functional description of the application.

Virtual components use wrappers to adapt accesses from the internal component (a set of software tasks or a hardware function) to the external channels. The wrapper is modeled as a set of virtual ports that contain internal and external ports that can be different in terms of (1) communication protocol, (2) abstraction level, and (3) specification language. This model is not directly synthesizable or executable because wrapper's behavior is not described. These wrappers can be generated automatically, in order to produce a detailed architecture that can be both synthesized and simulated.

The required SystemC extensions implemented in VADeL are

1. VM: Consists of a module and its wrapper
2. Virtual port: Groups some internal and external ports that have a conversion relationship. The wrapper is the set of virtual ports for a given VM
3. Virtual channel: Groups several channels having a logical relationship (e.g., multiple channels belonging to the same communication protocol)
4. Parameters: Used to customize hardware interfaces (e.g., buffer size and physical addresses of ports), OSs, and drivers

In VADeL, there are also predefined ports with special semantics called service access ports (SAP). They can be used to access some services that are implemented by HW/SW wrapper components. For instance, the *timer* SAP can be used to request an interrupt from a hardware timer after a given delay.

16.4.4 Target MPSoC Architecture Model

We use a generic MPSoC architecture where processors and other IP cores are connected to a global communication interconnect IP via wrappers (see Figure 16.4c). In fact, processors are separated from the physical communication IP by wrappers that act as communication coprocessors or bridges, freeing processors from communication management and enabling parallel execution of computation tasks and communication protocols. Software tasks also need to be isolated from hardware through an OS that plays the role of software wrapper. When defining this model, our goal was to have a generic model where both computation and communication may be customized to fit the specific needs of the application. For computation, we may change the number and kind of components, and for communication, we can select a specific communication IP and protocols. This architecture model is suitable to a wide domain of applications; more details can be found in [33].

16.4.5 HW/SW Wrapper Architecture

Wrappers are automatically generated as point-to-point adapters between each VM and the communication structure, as shown in Figure 16.4c [33]. This approach allows the connection of components to standard buses as well as point-to-point connections between cores.

Wrappers may have HW/SW parts. The internal architecture of a wrapper on the hardware side is shown in Figure 16.5b. It consists of a processor adapter, one or more channel adapters, and an internal bus. The number of channel adapters depends on the number of channels that are connected to the corresponding VM. This architecture allows the easy generation of multipoint, multiprotocol wrappers. The wrapper dissociates communication from computation, since it can be considered as a communication coprocessor that operates concurrently with other processing functions.

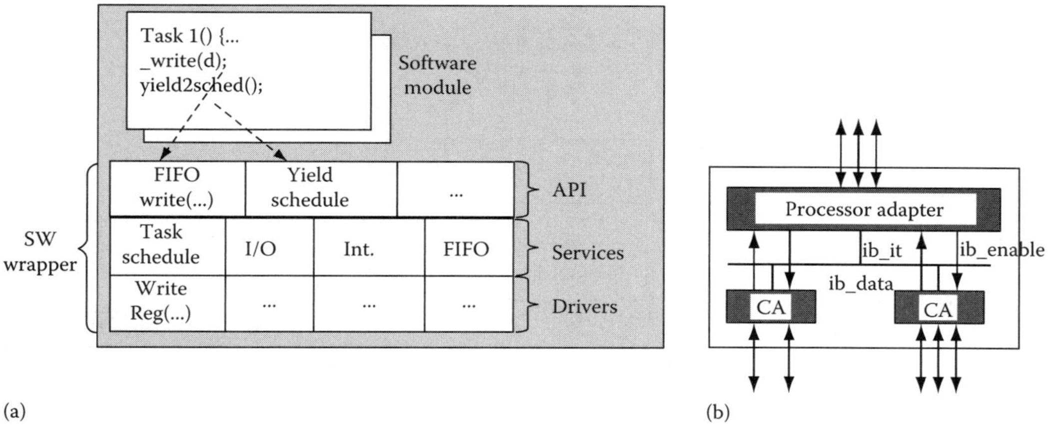

FIGURE 16.5   HW/SW wrapper architecture. (a) Software wrapper and (b) hardware wrapper.

On the software side [17], as shown in Figure 16.5a, wrappers provide the implementation of the high-level communication primitives (available through the API) used in the system specification and drivers to control the hardware. If required, the wrapper will also provide sophisticated OS services such as task scheduling and interrupt management minimally tailored for the particular application.

The synthesis of wrappers is based on libraries of basic modules from which hardware wrappers and dedicated OSs are assembled. These libraries may be easily extended with modules that are needed to build wrappers for processors, memories, and other components that follow various bus and core standards.

16.4.6   Design Tools

Figure 16.6 shows an overall view of our design environment, which starts from the combined architecture/application model captured in Simulink, followed by the ROSES environment for the HW/SW interface refinement. In fact, the input model of the ROSES may be imported from specification analysis tools [6], or manually coded using our extended SystemC library. In the Figure 16.6, the virtual architecture model is automatically generated from the Simulink Combined Architecture/Application Model. All design tools use a unified design model that contains an abstract HW/SW netlist annotated with parameters (Colif [30]). Hardware wrapper generation [33] transforms the input model into a synthesizable architecture. The software wrapper generator [17] produces a custom OS for each processor on the target platform. For validation, we use the cosimulation wrapper generator [32] to produce simulation models. Details about these tools can be found in the references; only their principle will be discussed here.

The virtual architecture generator produces the virtual architecture model required for the ROSES environment. Firstly, it parses the Simulink Combined Architecture/Application Model. Then, it interprets the annotated architecture parameters of the Simulink model and it generates the equivalent virtual architecture model in VADel, a SystemC extension. Thus, for each subsystem that corresponds to an IP, the tool generates the VM, including both SystemC module and its wrapper. The wrapper is composed of internal and external ports, which are grouped into virtual ports. The internal ports are the ports of the SystemC module, while the external ports enable interconnection with the virtual communication channels. At this level, several details like the data types become explicit.

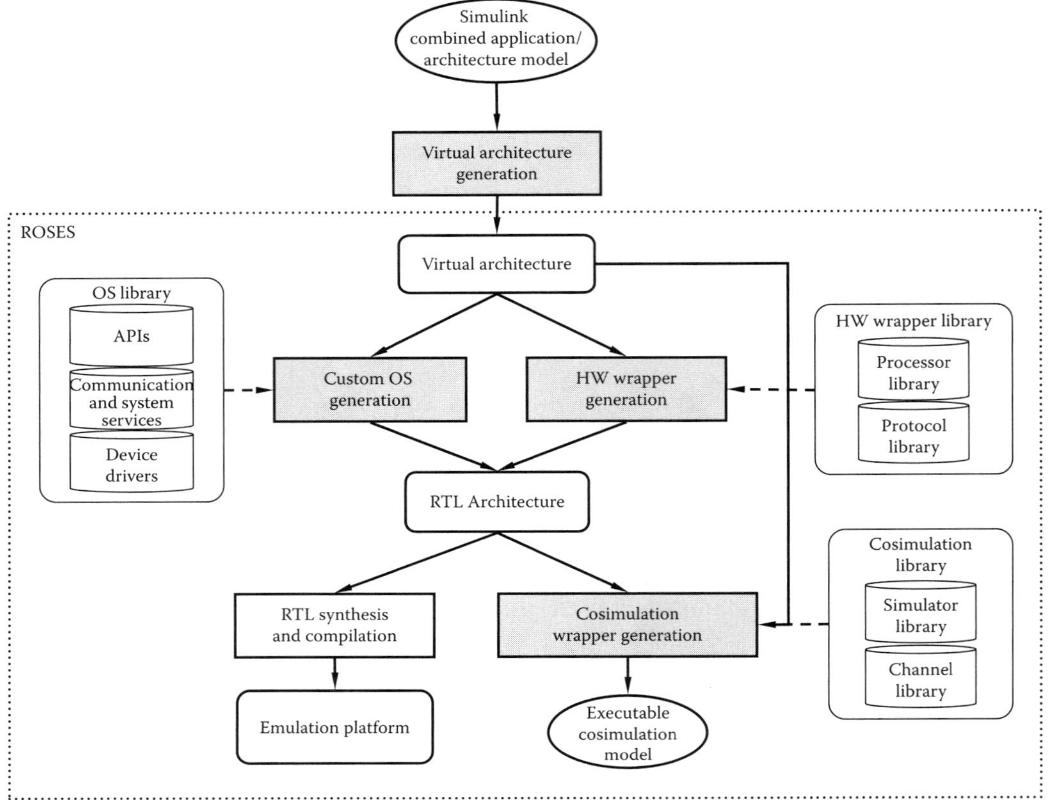

FIGURE 16.6   Design automation tools for MPSoC.

Hardware wrapper generation assembles library components using the virtual architecture model presented before (Figure 16.5b) to produce the RTL architecture. This library contains generalized descriptions of hardware components in a macrolanguage (**m4** like); it has two parts: the processor library and the protocol library. The former contains local template architectures for processors with four types of elements: processor cores, local buses, local IP components (e.g., local memory, address decoder, coprocessors, etc.), and processor adapters. The latter consists of a list of channel adapters. Each channel adapter has simulation, estimation, and synthesis models that are parameterized (by the channel parameters, e.g., direction, storage size, and data type) as the elements in the processor library.

The software wrapper generator produces OSs streamlined and preconfigured for the software module(s) that run(s) on each target processor. It uses a library organized in three parts: APIs, communication/system services, and device drivers. Each part contains elements that will be used in a given software layer in the generated OS. The generated OS provides services: communication services (e.g., FIFO communication), I/O services (e.g., AMBA bus drivers), memory services (e.g., cache or virtual memory usage), etc. Services have dependency between them; for instance, communication services are dependent on I/O services. Elements of the OS library also have dependency information. This mechanism is used to keep the size of the generated OS at a minimum; the elements that provide unnecessary services are not included.

There are two types of service code: reusable (or existing) code and expandable code. As an example of existing code, AMBA bus-master service code can exist in the OS library in the form of C language. As an example of expandable code, OS kernel functions can exist in the OS library in the form of macrocode (**m4** like). There are several preemptive schedulers available in the OS library such as round-robin scheduler, priority-based scheduler, etc. In the case of round-robin scheduler, time-slicing (i.e., assigning different CPU load to tasks) is supported. To make the OS kernel very small and flexible, (1) the task scheduler can be selected from the requirements of the application code and (2) a minimal amount (less than 10% of kernel code size) of processor-specific assembly code is used (for context switching and interrupt service routines).

The cosimulation wrapper generator [32] produces an executable model composed of a SystemC simulator that acts as a master for other simulators. A variety of simulators can participate in this cosimulation: SystemC, VHDL, Verilog, and Instruction-set simulators. Cosimulation wrappers have the same structure as hardware wrappers (see Figure 16.5b), with simulation adapters in the place of processor adapters and simulation models of channel adapters. In the cosimulation wrapper library, there are simulation adapters for the different simulators supported and channel adapters that implement all supported protocols in different languages.

In terms of functionality, the cosimulation wrapper transforms channel access(es) via internal port(s) to channel access(es) via external port(s) using the following functional chain: channel interface, channel resolution, data conversion, and module communication behavior. Internal ports use channel functions (e.g., FIFO available, FIFO write) to exchange data. Channel interface provides the implementation of these channel functions. Channel resolution maps N-to-M correspondence between internal and external ports. Data conversion is required since different abstraction levels can use different data types to represent the same data. Module communication behavior is required to exchange data via external port(s), i.e., to call port functions of external ports.

16.4.7 Defining IP-Component Interfaces

HW/SW component interfaces must be composed of using basic elements of the hardware wrapper and software wrapper generators libraries (respectively). Table 16.1 lists some API functions available for different kinds of software task interfaces and some services provided by channel adapters available to be used in hardware component interfaces.

Software tasks must communicate through API functions provided by the software-wrapper generator library. For instance, the SHM API provides read/write functions for intertask communication. The guarded-shared memory (GSHM) API adds semaphores services to the SHM API by providing lock/unlock functions.

Hardware IP components must communicate through communication primitives provided by the channel adapters of the hardware-wrapper generator library. For instance, FIFO channel adapters (sender and receiver) implement a buffered two-phase handshake protocol (Put/Get) and provide Full/Empty functions for accessing the state of the buffer. ASFIFO channel adapters use instead a single-phase handshake protocol and can generate an interrupt for signaling the full and empty state of the buffer.

TABLE 16.1 HW/SW Communication APIs

Basic Component Interfaces		API Functions
SW	Register	Put/Get
	Signal	Sleep/Wakeup
	FIFO	Put/Get
	SHM	Read/Write
	GSHM	Lock/Unlock/Read/Write
HW	Register	Put/Get
	FIFO	Put/Get/Full/Empty
	ASFIFO	Put/Get/IT(Full/Empty)
	Buffer	BPut/BGet
	Event	Send/IT(receiver)
	AHB master/slave	Read/Write
	Timer	Set/Wait

A recurrent problem in library-based approaches is library size explosion. In ROSES, this problem is minimized by the use of layered library structures where a service is factorized so that its implementation uses elements of different layers. This scheme increases reuse of library elements since the elements of the upper layers must use the services provided by the elements in the immediate lower layer.

Designers are able to extend ROSES libraries since they are implemented in an open format. This is an important feature since it enables the support of different standards while reusing most of the basic elements in the libraries.

Table 16.2 shows some of the existing HW/SW components in the current ROSES IP library and gives the type of communication they use in their interfaces.

Figure 16.7a shows the "stream" software IP and part of its code to demonstrate the utilization of the communication APIs. Its interface is composed of four ports: two for the FIFO API (P3 and P4), one for the signal API (P2), and one for the GSHM API (P1). In line 7 of Figure 16.7a, the stream IP uses P1 to lock the access to the SHM that contains the data that will be streamed. P2 is used to suspend the task that fills-up the SHM (line 8). Then, some header information is got from the input FIFO

TABLE 16.2 Sample IP Library

IP		Description	Interfaces
SW	host-if	Host PC interface	Register/signal
	Rand	Random number generator	Signal/FIFO
	mult-tx	Multipoint FIFO data transmission	FIFO
	reg-config	Register configuration	Register/FIFO/SHM
	Shm-sync	SHM synchronization	SHM/signal
	Stream	FIFO data streaming	GSHM/FIFO/signal
HW	ARM7	Processor core	ARM7 pins
	TX_Framer	Data package framing	17 Registers, 1 FIFO

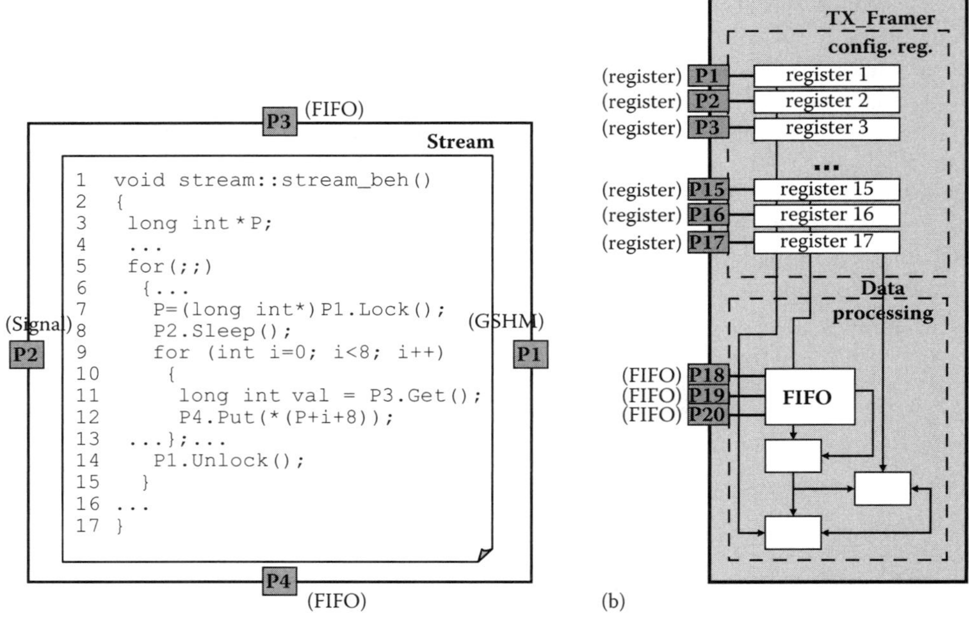

FIGURE 16.7 (a) Stream software IP and (b) the TX_Framer hardware IP.

using P3 (line 11) and streamed to the output FIFO using P4 (line 12). When streaming is finished, P1 is used to unlock the access to the SHM (line 14).

Figure 16.7b shows the "TX_Framer" hardware IP, which is part of a VDSL modem and responsible for packaging data into ATM-network compatible frames. Its interface is composed of 17 configuration registers (P1-P17) and one single-handshake input FIFO (P18-P20). The registers are used to configure the IP functionality and have bit sizes varying from 2 to 11, while the FIFO is used to store data packets that will be inserted into specific places in the output ATM frames. These ports are driven directly by the compatible outputs from register and ASFIFO channel adapters that are generated by the hardware wrapper generator.

16.5 Component-Based Design of a VDSL Application

16.5.1 Specification

The design presented in this section illustrates the IP integration capabilities of ROSES. We redesigned part of a VDSL modem that was prototyped by [31] using discrete components (the shaded part in Figure 16.8a). The block diagram for the modem subset used in the rest of this paper is shown in Figure 16.8b. It corresponds to a deframing/framing unit (DFU), composed of two ARM7 processors and the TX_Framer. The TX_Framer is part of the VDSL Protocol Processor. In this experiment, it is used as a hard IP component described at the RT-level. The partition of processors/tasks was suggested by the design team of the VDSL-modem prototype.

Processors exchange data using three asynchronous FIFO buffers. The TX_Framer IP has some configuration registers and input a data stream through a synchronous FIFO buffer. Tasks use a variety of control and data-transmission protocols to communicate. For instance, a task can block/unblock the execution of other tasks by sending them an OS signal. Tasks used for data transmission: a FIFO memory buffer, two SHMs (with or without semaphores), and direct register access. Despite representing only a subset of the VDSL modem, the design of the DFU remains quite challenging. In fact, it uses two processors executing parallel tasks. The control over the three modules of the specification is fully distributed. All three modules act as masters when interacting with their environment. Additionally, the application includes multipoint communication channels requiring sophisticated OS services.

16.5.2 DFU Combined Application/Architecture Model in Simulink

The initial combined application/architecture Simulink model is given in Figure 16.9. It captures the functional specification of the DFU system and the functions grouping into tasks. The application is partitioned in ten tasks. In this initial model, each of the tasks has been described as an S-function. The model also includes the mapping of the application tasks onto MPSoC architecture, composed of two ARM7 processor subsystems and one hardware subsystem.

During the mapping, three tasks (T1, T2, T3) are mapped on one ARM7 processor; six tasks (T4, T5, T6, T7, T8, T9) are mapped onto the second ARM7 processor, while the last task (T10) is dedicated to be implemented in hardware. The communications between the various tasks follow different protocols, like FIFO, SHM, and register. For instance, the inter-subsystem communication between the 2 ARM7 CPUs uses a hardware FIFO; the intra-subsystem communication between tasks T4 and T5 uses software FIFO or the intra-subsystem communication between tasks T9 and T7 uses SHM.

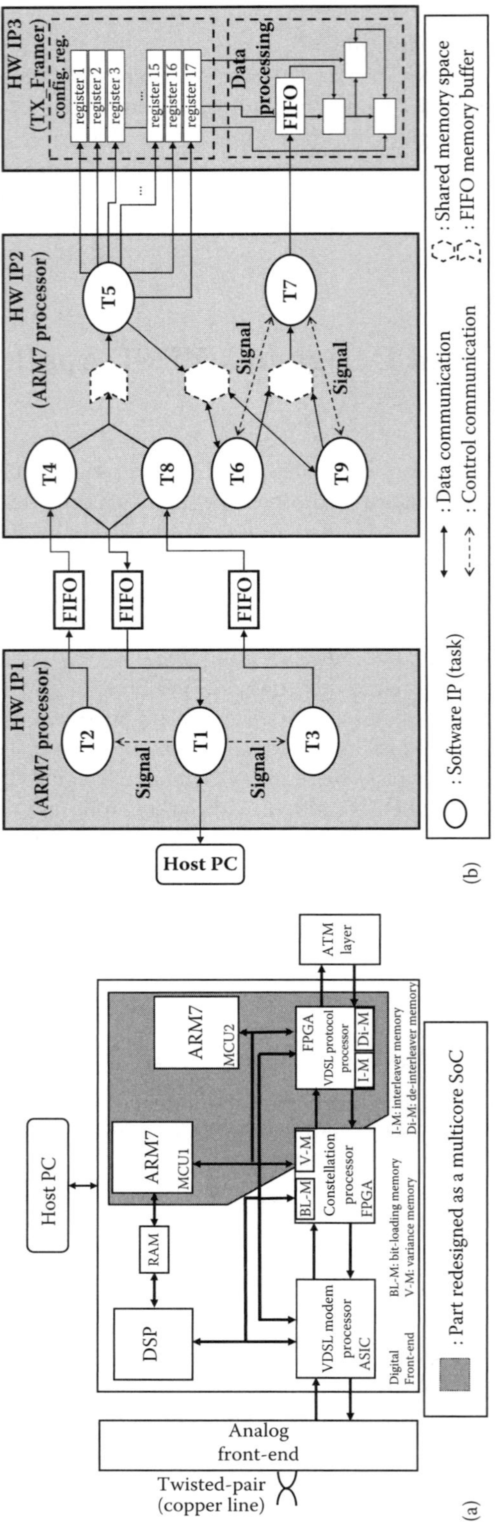

FIGURE 16.8 (a) VDSL modem prototype and (b) DFU block diagram.

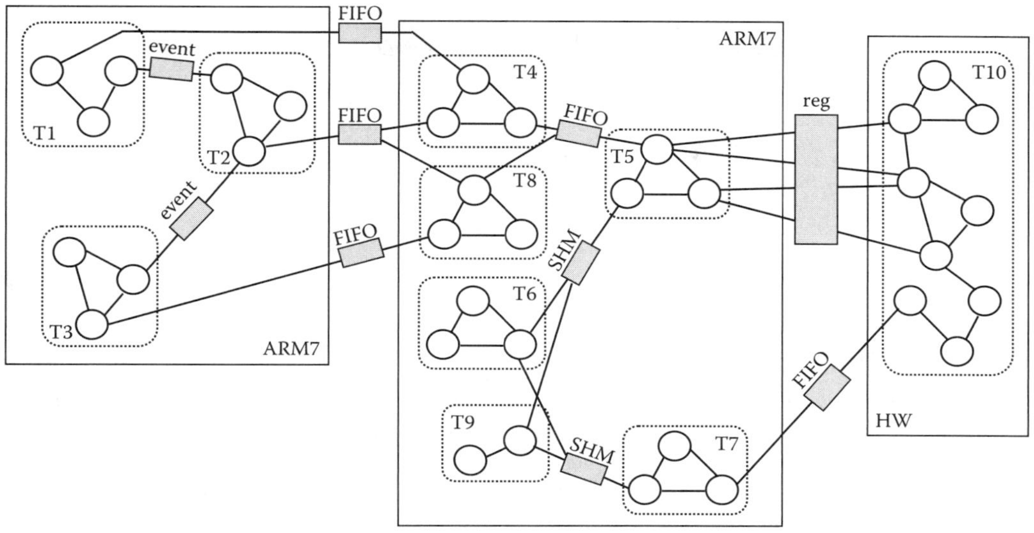

FIGURE 16.9 DFU combined architecture/application model in Simulink.

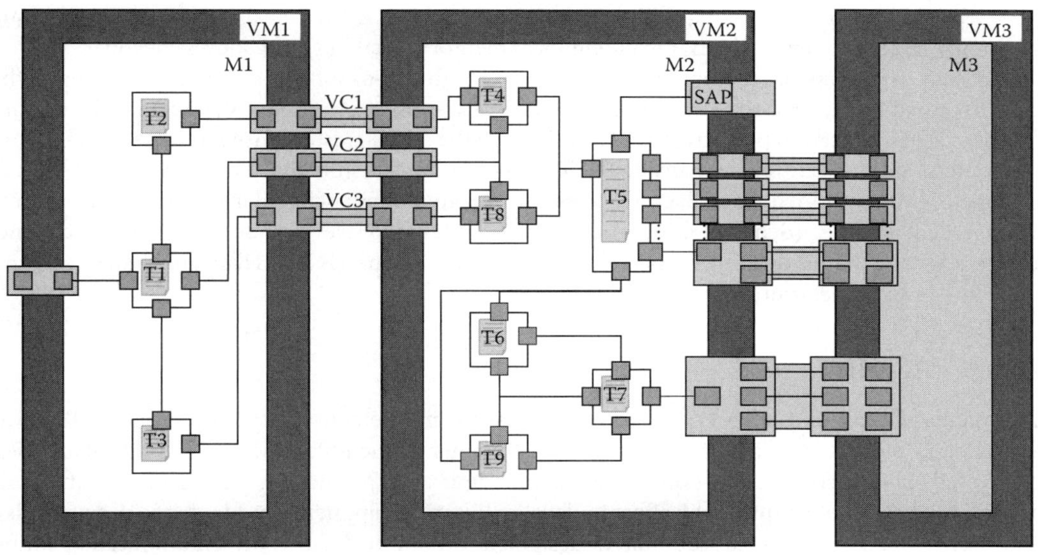

FIGURE 16.10 DFU abstract architecture specification.

16.5.3 DFU Virtual Architecture

Figure 16.10 shows the abstract architecture model that captures the DFU specification with point-to-point communications between the three main IP cores. **VM1** and **VM2** are two virtual processors, and **VM3** corresponds to the TX_Framer function. VM1 and VM2 include several submodules corresponding to software tasks **T1** through **T9** assigned to these processors. This abstract model can be mapped onto different concrete microarchitectures depending on the selected IP components and on desired performance, area, and power constraints. For instance, the three point-to-point connections (**VC1**, **VC2**, and **VC3**) between VM1 and VM2 can be mapped onto a bus or onto a SHM.

TABLE 16.3　　HW/SW IP Utilization

IP		Description	Use
SW	host-if	Host PC interface	T1
	Rand	Random number generator	T2, T3
	mult-tx	Multipoint FIFO data transmission	T4, T8
	reg-config	Register configuration	T5
	shm-sync	SHM synchronization	T6, T9
	stream	FIFO data streaming	T7
HW	ARM7	Processor core	VM1, VM2
	TX_Framer	Data package framing	VM3

16.5.4　MPSoC RTL Architecture

For the implementation of the DFU virtual architecture, two hardware IP cores have been selected: an ARM7 processor and the TX_Framer. The application software has been built by reusing several available software IP components for implementing tasks T1 to T9. Table 16.3 lists the selected IP components and indicates their correspondence to the VMs and submodules in the DFU virtual architecture. The interfaces of the selected software IP components in Table 16.3 (see Table 16.2) match the communication type of the software tasks of the virtual architecture in Figure 16.8b.

Figure 16.11 shows the RTL microarchitecture obtained after HW/SW wrapper generation.

It is important to notice that, from an abstract target architecture containing an "abstract" ARM7 processor, ROSES automatically generates a "concrete" ARM7 local architecture containing additional IP components, which implement local memory, local bus, and address decoder.

Each software wrapper (custom OS) is customized to the set of software IPs corresponding to the tasks executed by the processor core. For example, software IPs running on VM2 access the custom OS using communication primitives available through the API: Register is used to write/read to/from the configuration/status registers inside the TX_Framer block, while **SHM** and **GSHM** are used to manage SHM communication. Each OS contains a round-robin scheduler (**Sched**) and resource management services (**Sync, IT**). The driver layer contains low-level code to access the channel adapters within the hardware wrappers (e.g., **Pipe LReg** for the **HNDSHK** channel adapter), and some low-level kernel routines.

16.5.5　Results

The manual design of a full VDSL modem requires several person-years; the presented DFU was estimated as more than five persons-year effort. When using the ROSES IP integration capabilities, the overall experiment took only one person during four months, including all validation and verification time (but not counting the effort to develop library components and to debug design tools). This corresponds to a 15 times reduction in design effort (a more detailed presentation can be found in [34]).

Application code and generated OS are compiled and linked together to be executed on each ARM7 processor. The HW wrapper can be synthesized using RTL synthesis. As can be seen in Table 16.4, most OS code is generated in C, only a small part of it is in assembly and includes some low-level routines (e.g., context switching and processor boot) that are specific to each processor. If we compare the numbers presented in Table 16.4 with commercial embedded OSs, the results are still very good. The minimum size for such OSs is around 4 kB; but with this size, few of them could provide the required functionality.

Table 16.5 shows the numbers obtained after RTL synthesis of the hardware wrappers using a CMOS 0.35 μm technology. These are good results because wrappers account for less than 5% of the ARM7s core surface and have a critical path that corresponds to less than 15% of the clock cycle for the 25 MHz ARM7 processors used in this case study.

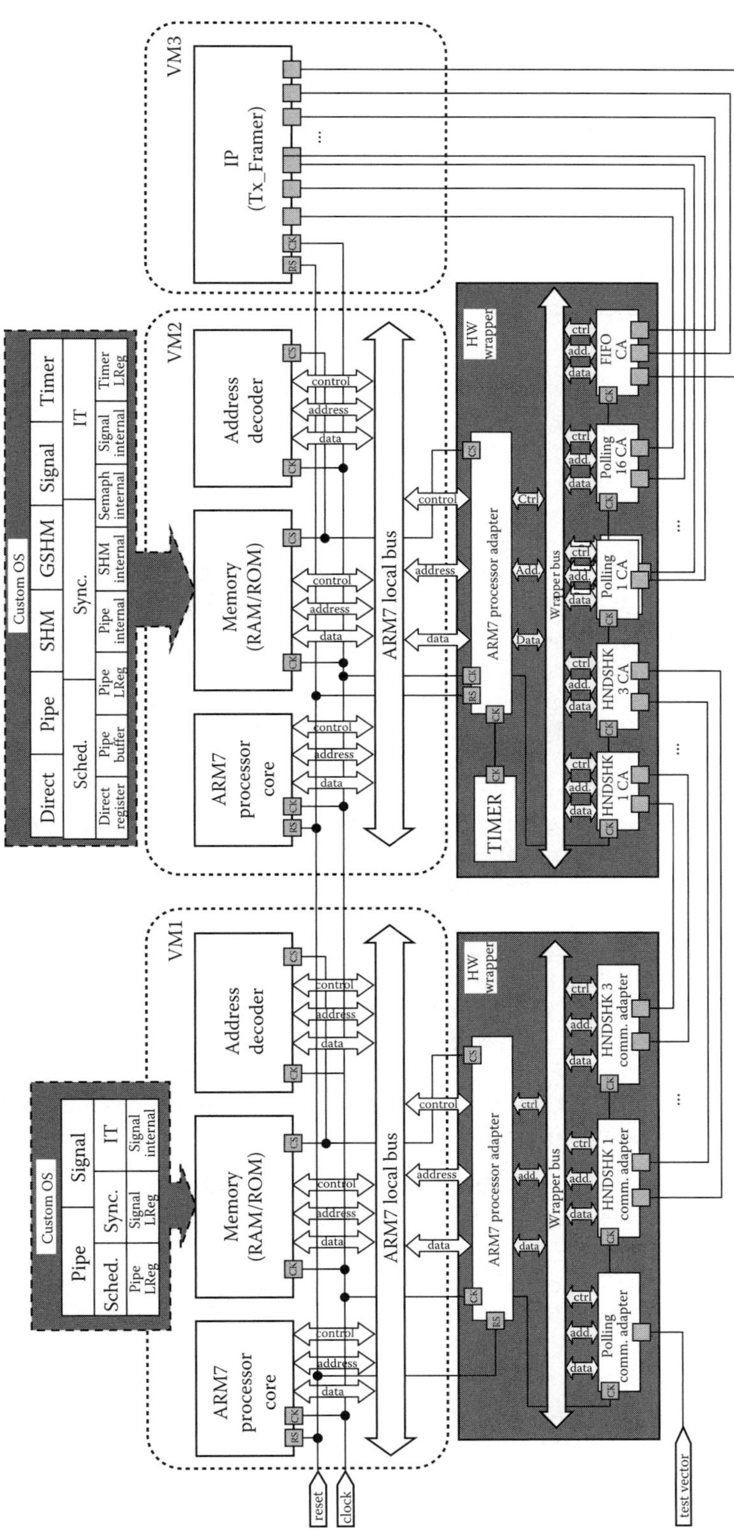

FIGURE 16.11 Generated MPSoC architecture.

TABLE 16.4   Results for OS Generation

OS Results	Number of Lines in C	Number of Lines in Assembly	Code Size (Bytes)	Data Size (Bytes)
VM1	968	281	3829	500
VM2	1872	281	6684	1020
Context switch (cycles)			36	
Latency for interrupt treatment (cycles)			59(OS) + 28(ARM7)	
System call latency (cycles)			50	
Resume of task execution (cycles)			26	

TABLE 16.5   Results for Hardware Wrapper Generation

HW Interfaces	Number of Gates	Critical Path Delay (ns)	Maximum Frequency (MHz)
VM1	3284	5.95	168
VM2	3795	6.16	162
Latency for read operation (clock cycles)		6	
Latency for write operation (clock cycles)		2	
Number of code lines (RTL VHDL)		2168	

16.5.6   Evaluation

Results show that the component-based approach can generate HW/SW interfaces and OSs that are as efficient as the manually coded/configured ones. The HW/SW frontier in wrapper implementation can be easily displaced by changing some library components. This choice is transparent to the final user since everything that implements the interconnect API is generated automatically (the API does not change only its implementation does). Furthermore, correctness and coherence can be verified inside tools and libraries against the API semantics without having to impose fixed boundaries to the HW/SW frontier (in contrast to standardized component interfaces and buses).

The utilization of layered library components provides lots of flexibility; the design environment can be easily adapted to accommodate different languages to describe system behavior, different task scheduling and resource management policies, different global communication interconnect topologies and protocols, a diversity of processor cores and IP cores, and different memory architectures. In most cases, inserting a new design element in this environment only requires to add the appropriate library components. Layered library components are at the roots of the methodology; the principle followed is to contain a unique functionality and to respect well-defined interfaces that enable easy composition. This layered structure prevents library size explosion since composition is used to implement complex functionality and to increase component reutilization.

As explained in this chapter and illustrated by the design case study, ROSES use a component-based methodology that presents a unique combination of features:

- It implements a general approach for the automatic integration of heterogeneous and hard IP components, although it easily accommodates the integration of homogeneous and soft IP components.

- It offers an architectural-independent API, integrated into SystemC, containing high-level communication primitives and enhancing the free development of reusable components. Application software accessing this API does not need to be retargeted to each system implementation.

- It adopts a library-based approach to wrapper generation. As long as components communicate through a known protocol, communication synthesis can be done automatically, without any additional design effort. In a formal-based approach, instead, the designer must describe the component interface by some appropriate formalism, such as an FSM [21] or a regular expression [20].

- It uniformly addresses the generation of HW/SW parts of communication wrappers for programmable components. While some approaches consider only hardware [20,21] or software [18,22] wrappers, other also consider HW/SW parts but are restricted to predefined wrapper libraries for given communication schemes [5]. The library-based approach of ROSES, in turn, allows the synthesis of software interfaces for various communication schemes.

- It can be used with any architectural template and communication structure, such as a bus, a NoC, or point-to-point connections between components. It is also configurable to synthesize wrappers for any bus or core standard.

16.6 Conclusions

Reuse of IP components is a major requirement for the design of complex embedded SoCs. However, reuse is a complex process that involves many steps, requires support from specialized tools and methodologies, and influences current design practices. The integration of IP components into a particular design is possibly the most complex step of the reuse process. Many design approaches, such as bus- and core-based design and platform-based design, are aimed at an easier IP integration. Nevertheless, many problems are still open, in particular the automatic synthesis of HW/SW wrappers between heterogeneous and hard IP components.

The chapter has shown that the component-based design methodology provides a complete, generic, and efficient solution to the HW/SW interfaces design problem. Starting from a high-level function specification and an abstract architecture, design tools can automatically generate HW/SW wrappers that are necessary to integrate heterogeneous IP components that have been selected to implement the application. Designers do not need to design any low-level interfacing details manually. The chapter has also shown how HW/SW component interfaces can be decomposed and easily adapted to different communication structures, and bus and core standards.

References

1. ITRS, http://public.itrs.net/
2. K. Keutzer, A.R. Newton, J.M. Rabaey, and A. Sangiovanni-Vincentelli, System-level design: Orthogonalization of concerns and platform-based design, *IEEE Transactions on Computer-Aided Design of Integrated Circuits and Systems*, 19(12), December 2000.
3. M. Sgroi, M. Sheets, A. Mihal, K. Keutzer, S. Malik, J. Rabaey, and A. Sangiovanni-Vincentelli, Addressing the system-on-chip interconnect woes through communication-based design, *Proceedings of the 38th DAC*, Las Vegas, NV, June 2001.
4. VSIA, http://www.vsi.org
5. J-Y. Brunel, W.M. Kruijtzer, H.J.H.N. Kenter, F. Pétrot, L. Pasquier, E.A. de Kock, and W.J.M. Smits, COSY communication IPs, *Proceedings of the 37th DAC*, Los Angeles, CA, June 2000.
6. Cadence Design Systems Inc., Virtual component co-design, http://www.cadence.com/products/vcc.html
7. D. Gajski, J. Zhu, R. Domer, A. Gerslauer, and S. Zhao, *SpecC Specification Language and Methodology*, Kluwer Academic Publishers, Dordrecht, the Netherlands, 2000.
8. CoWare, http://www.coware.com
9. IBM CoreConnect Bus Architecture, http://www.chips.ibm.com/bluelogic
10. D. Wingard, MicroNetwork-based integration for SOCs, *Proceedings of the 38th DAC*, Las Vegas, NV, June 2001.
11. J. Rowson and A. Sangiovanni-Vincentelli, Interface-based design, *DAC'1997*, Anaheim, CA, 1997.

12. ARM AMBA, http://www.arm.com

13. Sonics SiliconBackplane MicroNetwork, http://www.sonicsinc.com

14. R.A. Bergamaschi and W.R. Lee, Designing systems-on-chip using cores, *DAC'2000*, Los Angeles, CA.

15. Open Core Protocol, http://www.ocpip.org

16. G. de Micheli and L. Benini, Networks-on-chip: A new paradigm for systems-on-chip design, *DATE'2002*, Paris, France, 2002.

17. L. Gauthier, S. Yoo, and A.A. Jerraya, Automatic generation and targeting of application specific operating systems and embedded systems software, *IEEE TCAD*, 20 Nr. 11, Los Alamitos, CA, November 2001.

18. C. Böke. Combining two customization approaches: Extending the customization tool TEReCS for software synthesis of real-time execution platforms, *Workshop on Architectures of Embedded Systems (AES 2000)*, Karlruhe, Germany, January 2000.

19. L. Benini, A. Bogliolo, and G. De Micheli, Dynamic power management of electronic systems, *ICCAD 1998*, Los Angeles, CA, 1998.

20. R. Passerone, J.A. Rowson, and A. Sangiovanni-Vincentelli, Automatic synthesis of interfaces between incompatible protocols, *DAC'1998*, San Francisco, CA, 1998.

21. J. Smith, and G. deMicheli, Automated composition of hardware components, *DAC'1998*, San Francisco, CA, 1998.

22. P. Chou et al., IPChinook: An integrated ip-based design framework for distributed embedded systems, *DAC'1999*, New Orleans, LA, 1999.

23. M. Birnbaum and H. Sachs, How VSIA answers the SoC dilemma, IEEE Computer, New Orleans, LA, June 1999.

24. C. Barna and W. Rosenstiel, Object-oriented reuse methodology for VHDL, *DATE'1999*, München, Germany, 1999.

25. P. Schaumont et al., Hardware reuse at the behavioral level. *DAC'1999*, New Orleans, LA, June 1999.

26. F.J. Rammig, Web-based system design with components off the shelf, *FDL'2000*, Tübingen, Germany, September 2000.

27. Simulink, http://www.mathworks.com

28. K. Popovici and A. Jerraya, Simulink based hardware-software codesign flow for heterogeneous MPSoC, *Proceedings of the SCSC 2007*, San Diego, CA, 16–19, July 2007.

29. SystemC, http://www.systemc.org

30. W.O. Cesário, G. Nicolescu, L. Gauthier, D. Lyonnard, and A.A. Jerraya, Colif: A design representation for application-specific multiprocessor SOCs, *IEEE Design and Test of Computers*, 18(5):8–20, September–October 2001.

31. M. Diaz-Nava and G.S. Okvist, The Zipper prototype: A complete and flexible VDSL multi-carrier solution, ST Journal special issue xDSL, September 2001.

32. S. Yoo, G. Nicolescu, D. Lyonnard, A. Baghdadi, and A.A. Jerraya, A generic wrapper architecture for multi-processor SoC cosimulation and design, *International Symposium on HW/SW Codesign (CODES)*, Copenhagen, Denmark, 2001.

33. D. Lyonnard, S. Yoo, A. Baghdadi, and A.A. Jerraya, Automatic generation of application-specific architectures for heterogeneous multiprocessor system-on-chip, *Proceedings of the 38th DAC*, Las Vegas, NV, June 2001.

34. W. Cesário, A. Baghdadi, L. Gauthier, D. Lyonnard, G. Nicolescu, Y. Paviot, S. Yoo, M. Diaz-Nava, and A.A. Jerraya, Component-based design approach for multicore SoCs, *Proceedings of the 39th DAC*, New Orleans, LA, June 2002.

17

FPGA Synthesis and Physical Design

Mike Hutton
Altera Corporation

Vaughn Betz
Altera Corporation

17.1 Introduction

Since their introduction in the early 1980s, programmable logic devices or PLDs have evolved from implementing small glue-logic designs to large, complete systems. PLDs can be divided into two categories, limit bumping algorithm (CPLDs) and field-programmable gate arrays (FPGAs). CPLDs are lower-density devices employing nonvolatile programming—in other words, the CPLD programming is not lost when the device is powered down. FPGAs, the topic of this chapter, are typically based on *lookup-table* (LUT) cells and are reminiscent of ASIC gate arrays in structure. FPGAs are typically static-RAM programmed and thus require power to maintain their configuration. Today, the majority of all design-starts (though typically not the largest ones) target PLDs, with the higher-density designs on FPGAs and smaller designs and designs that require nonvolatility targeting CPLDs. The increasing use of PLD devices has resulted in significant research in computer-aided design (CAD) algorithms and tools targeting programmable logic.

There are two branches of FPGA CAD tool research: one concerned with developing algorithms for a given FPGA and a parallel branch developing the tools required to design FPGA architectures. This emphasizes the interdependence of FPGA CAD tools and architectures; unlike ASIC flows, where CAD is an implementation of a design in silicon, the CAD flow for FPGAs is an embedding of the design into a device architecture that is fixed in terms of both cells and routing. Though some algorithms for FPGAs continue to overlap with ASIC-targeted tools, notably language and technology independent synthesis, and to some extent placement, technology mapping, and routing are notably different.

In addition to the core synthesis, placement, and routing algorithms, emerging tools for FPGAs now concentrate on physical synthesis, power and timing analysis and optimization, and the growing area of system-level design. Power optimizations going forward will be a combination of semiconductor-industry-wide techniques and ideas targeting the programmable nature of FPGAs, closely tuned to the evolving FPGA architectures. System-level design tools will attempt to exploit two of the key benefits of FPGAs—fast design and verification time and a high degree of programmable flexibility.

FPGA tools differ from ASIC tools in that the core portions of the tool flow are owned by the silicon vendors. Third-party electronic design automation (EDA) tools are often used for synthesis and verification, but with very few exceptions physical design tools are supplied by the FPGA vendor, and they support only that vendor's products. The two largest FPGA vendors (Altera and Xilinx) offer complete CAD flows from language extraction and synthesis through placement and routing to power and timing analysis. One of our goals in this chapter is to describe CAD for programmable logic, not just from an individual algorithm perspective, but in the context of the end-to-end flow that a designer using a commercial FPGA would experience.

After some introductory descriptions of FPGA architectures and CAD flows, we will describe research in the key areas of CAD for FPGAs roughly in flow order.

17.1.1 Architecture of Field-Programmable Gate Arrays

To appreciate CAD algorithms for FPGAs, it is necessary to have some understanding of FPGA architectures and how they are specified.

Figure 17.1 shows a high-level block diagram of a modern FPGA. The FPGA in Figure 17.1 is a Stratix FPGA [13], but other modern FPGAs have fairly similar features. Components of the floorplan are logic blocks, which in Stratix are called *logic array blocks* (LABs), DSP (digital signal processing) blocks containing multiplier/accumulator functionality, and various sizes of embedded RAM blocks (here 512 bit, 4 kbit, and the 576 Kbit M-RAM). Shown behind the functional blocks are horizontal rows and vertical columns of routing interconnect.

Figure 17.2 details the internals of a Stratix logic array block. A LAB is composed of several logic elements (LEs) and local routing to interconnect them. Each LE is composed of an LUT, a register, and dedicated circuitry for arithmetic functions. Figure 17.2a shows a four-input lookup table. The

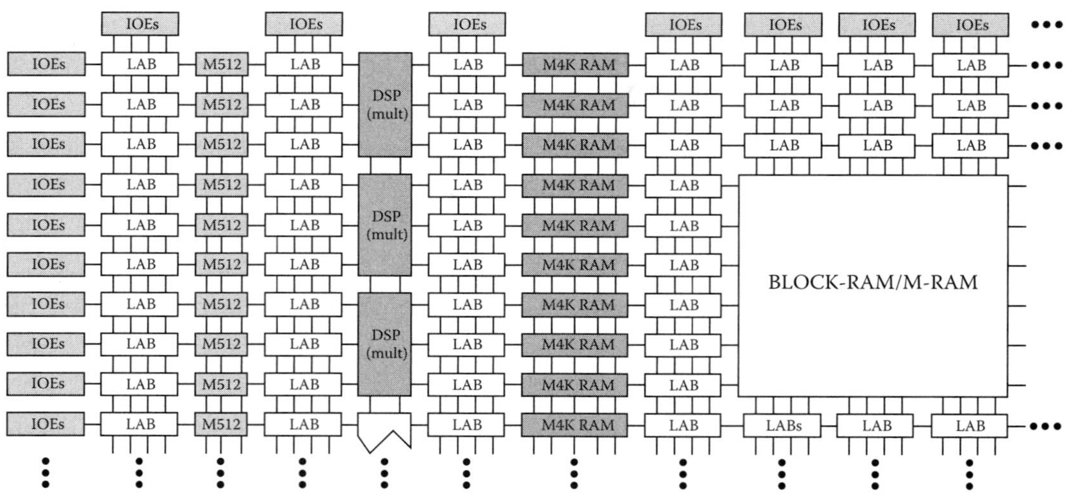

FIGURE 17.1   Stratix FPGA high-level block diagram.

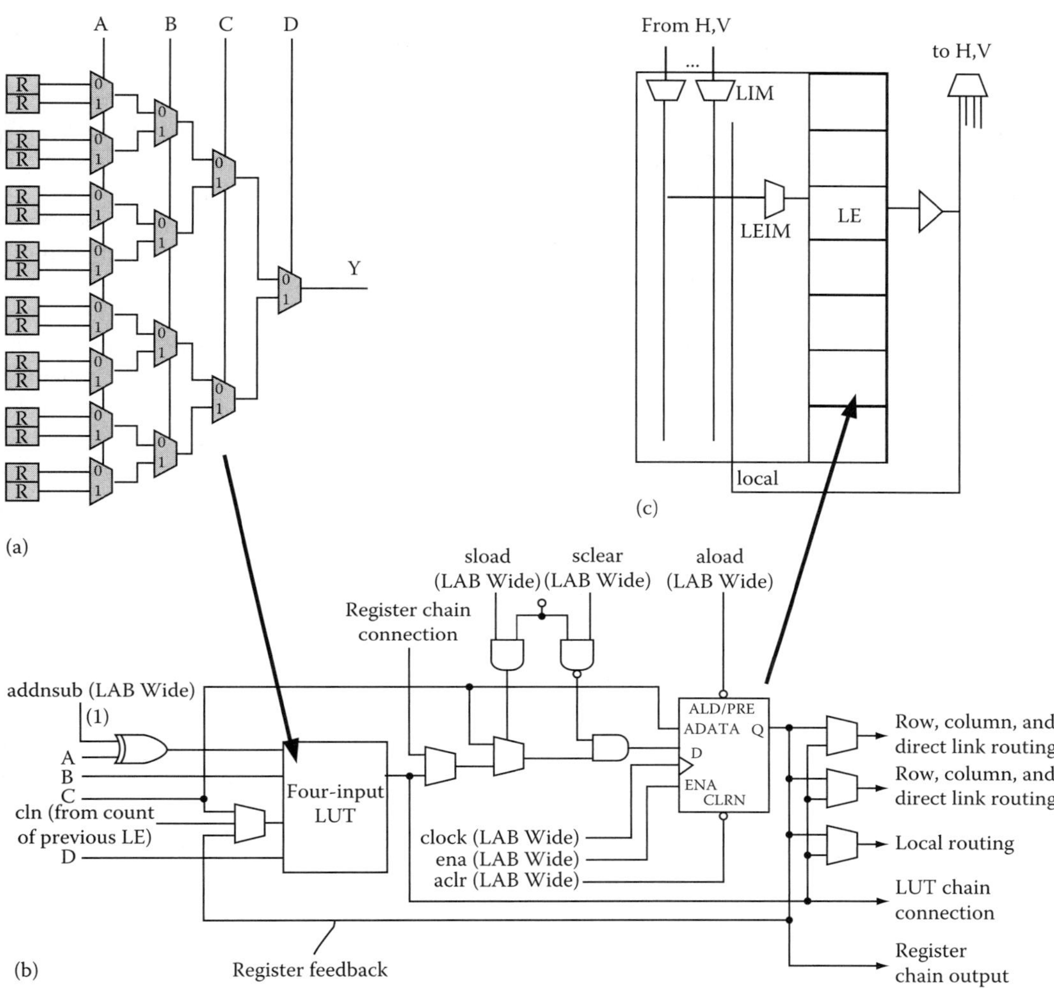

FIGURE 17.2 Inside an FPGA logic block: (a) 4-LUT showing LUT-mask, (b) complete Stratix LE, and (c) LAB cluster showing routing interface to inter-logic-block H and V routing wires.

16 bits of programmable LUT-mask specify the truth-table for a four-input function, which is then controlled by the A, B, C, and D inputs to the LUT. Figure 17.2b shows a complete LE, consisting of a four-input lookup-table (LUT) and a D-flip-flop (DFF) with selectable control signals. Arithmetic in the Stratix LE is accomplished by using the two 3-LUTs independently to generate sum and ripple-carry functions. The carry-out becomes a dedicated 5th connection into the neighboring LE, and optionally replaces the data3 or C input. By "chaining" the carry-out of several LEs together, a fast multibit adder can be formed. Figure 17.2c shows how several LEs (10 in Stratix) are grouped together to form a LAB. A subset of the global horizontal (H) and vertical (V) lines that form the FPGA interconnect can drive into a number of *LAB input muxes* or LIMs. The signals generated by the LIMs (along with feedback from other LEs in the same LAB) can drive the inputs to the LE through LEIM or *logic-element input muxes*. On the output side of the LE, *routing multiplexers* allow signals from one or more LEs to drive H and V routing wires in order to reach other LABs. Each of the LAB inputs, LE inputs, and routing muxes is controlled by SRAM configuration bits. Thus, the total of

the SRAM bits programming the LUT, DFF, modes of the LE, and the various muxes comprise the configuration of the FPGA to perform a specific logical function.

Depending on the architectural choices, some of the blocks in Figure 17.1 may not be present, e.g., the device could have no DSP blocks, the routing could be organized differently, and control signals on the DFF could be added or removed. Virtex FPGAs, for example, have no small RAMs but do have additional hardware to allow the LUT-mask in the normal LE to be used as a 16-bit RAM.

There is a large literature on good choices of FPGA architecture. Rose [1] originally showed the most area-efficient LUT is one whose number of inputs (k) is 4. Betz and Rose [3] investigated the number of LEs a logic block should contain, M, and found that cluster sizes of from 1 to 8 LEs resulted in FPGAs with the best area. Ahmed and Rose later updated these studies [2] by examining the interaction of LUT size and logic cluster size, and found that an LUT size of $k = 4$ to 6 and a logic cluster size of $M = 8$ to 10 is best when considering both area and delay. Rose and Brown [4] evaluated the required flexibility of the routing structure, meaning the size of programmable muxes (LIMs, LEIMs, and routing muxes in the earlier terminology), and Lemieux and Lewis [5] described circuit-level issues. Cong and Hwang [6] evaluated FPGAs with some hard-wired (rather than programmable) connections between LEs, such as the Xilinx 4000. For a complete description of this early work on FPGA architecture see Brown et al. [7] or Betz et al. [8].

The versatile placer and router (VPR) toolset [9] introduced what is now a standard paradigm for empirical architecture evaluation. An architecture specification file controls a parameterized CAD flow capable of targeting a wide variety of FPGA architectures, and individual architecture parameters are swept across different values to determine their impact on FPGA area, delay, and recently power. Figure 17.3 illustrates a commercial version of this flow called the "FPGA modeling toolkit" used at Altera. Yan et al. [10] validated the importance of the combined hardware and software exploration methodology by showing the sensitivity of architecture results to the tool flow. Li [11] expanded the VPR toolset to perform power exploration on FPGA architectures, using it to examine voltage islands and other recent FPGA architectural issues. Wilton [12] evaluated the architecture of embedded memory blocks in FPGAs.

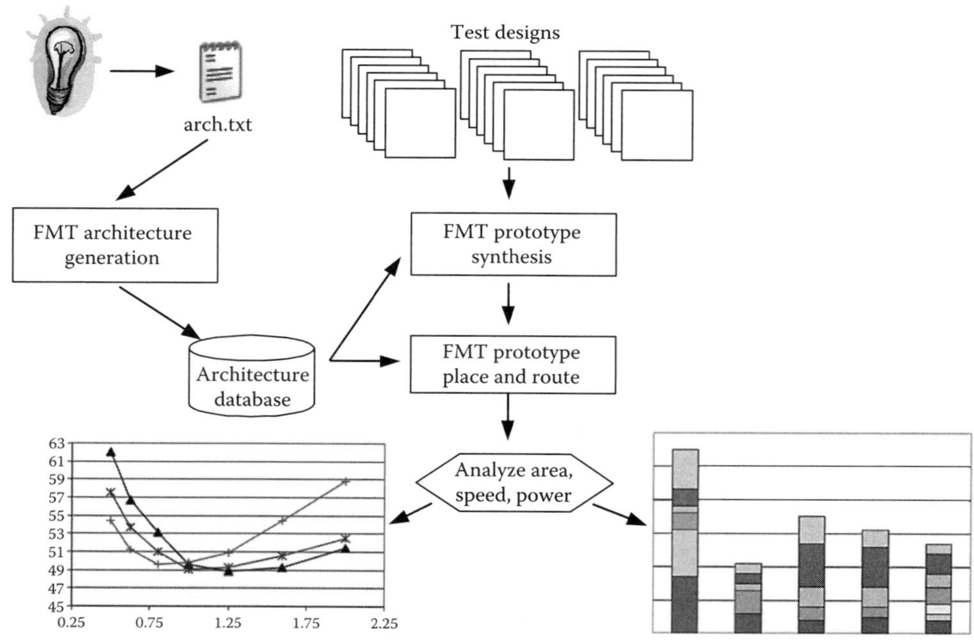

FIGURE 17.3   FPGA modeling toolkit flow based on VPR.

In terms of commercial FPGA architectures, descriptions have been published by Lewis for Altera's Stratix [14] and Stratix II [15] families and by Trimberger et al. [16] for the Xilinx 4000. For the Virtex I-IV FPGA families see the Xilinx Web site [17], as there is no academic publication. Ahnin described the architecture of the MAX 7000 CPLD in [18].

17.1.2 CAD Flow for FPGAs

The core CAD flow seen by an FPGA user is shown in Figure 17.4. After design entry, the design is synthesized at the register-transfer level into operators (adders, muxes, and multipliers), state-machines, and memory blocks. Gate-level logic synthesis follows, and then technology mapping to LUTs and registers. Clustering groups LEs into logic blocks (LABs) and placement determines a fixed location for each logic or other block. Routing selects which programmable switches to turn on in order to connect all the terminals of each signal net, essentially determining the configuration settings for each of the LIM, LEIM, and routing muxes described in Section 17.1. Physical synthesis, shown between placement and routing in this flow, is an optional step that resynthesizes the netlist to improve area or delay now that preliminary timing is known. Timing analysis [19] and power analysis [20] for FPGAs are approximated at many points in the flow to guide the optimization tools, but are performed for final analysis and reporting at the end of the flow. The last step is bit-stream generation, which determines the sequence of 1s and 0s that will be serially loaded to configure the device. Notice that some ASIC CAD steps are absent: sizing and compaction, buffer insertion, clock-tree, and power grid synthesis are not relevant to a prefabricated architecture.

Research on FPGA algorithms can take place at all parts of this flow. However, the academic literature has generally been dominated by synthesis (particularly LUT-mapping) at the front end and place and route algorithms at the back end, with most other aspects left to commercial tools and the FPGA

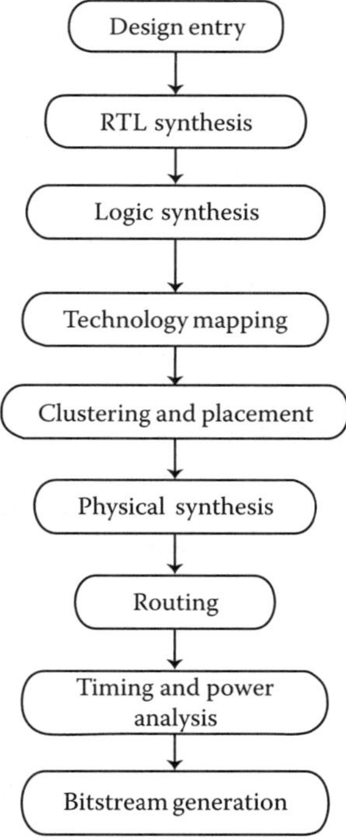

FIGURE 17.4 FPGA CAD flow.

vendors. Literature on FPGA architecture and CAD can be found in the major FPGA conferences (FPGA, FPL, FPT, and FCCM) as well as the general CAD conferences (IWLS, ISPD, DATE, ICCAD, and DAC). Device architecture is often presented at CICC, ISSCC, ISLPED, or HotChips.

In this chapter, we will survey algorithms, tools, and techniques for both the end-user and architecture-development FPGA CAD flows. The references we have chosen to include are representative of a large body of research in FPGAs, but the list cannot be comprehensive. For more complete discussions, the reader is referred to texts such as [21] on synthesis and [22,23] on FPGA synthesis and mapping. Commercial manufacturers [17] of LUT-based FPGAs include Altera, Xilinx, Actel, and Lattice. Dominant third-party EDA vendors for FPGAs include Synplicity, Mentor Graphics, and Synopsys for synthesis, Synopsys for timing analysis, Mentor Graphics for simulation, and Cadence and Synopsys for formal verification.

17.2　System-Level Tools

Though the vast majority of current FPGA designs are entered directly in VHDL or Verilog, there have been a number of attempts to raise the level of abstraction to the behavioral or block-integration level with system-level tools.

Berkeley Design Technologies Inc. [26] found that FPGAs have better price/performance than DSP processors for many DSP applications. However, HDL-based flows are not natural for most DSP designers, so higher-level DSP design flows are an important research area. Altera DSPBuilder and Xilinx SystemGenerator link the Matlab® and Simulink® algorithm exploration and simulation environments popular with DSP designers to VHDL and Verilog descriptions targeting FPGAs. Accelchip [17] and Catalytic [17] also offer Matlab-based flows. Hwang et al. [24] gives an overview of SystemGenerator, and Stroomer et al. [25] describes *jg*, a tool for providing Java specifications generating Matlab code that targets SystemGenerator.

The FPX system of Lockwood et al. [27] developed system-level and modular design methodologies for the network processing domain. The modular nature of FPX also allows for exploration of hardware and software implementations.

Altera's System on a Programmable Chip (SOPC) Builder (see Kempa et al. [28]) and Xilinx's Embedded Development Kit (EDK) are examples of system integration tools. These tools allow the designer to integrate each company's soft-core (synthesized from the FPGA fabric) or hard-core (prebuilt on the FPGA) processors and other intellectual property (IP) cores and peripherals together easily. The details of bus interfacing, bus mastering, and peripheral memory map creation are implemented automatically from a high-level description. These tools also provide some ability to explore the hardware–software codesign space. In the case of SOPC Builder, this includes mechanisms for adding custom instructions to the Nios II soft processor that are executed with custom-created hardware units.

Work has also proceeded on synthesizing high-level design languages to FPGAs. Celoxica [17], for example, provides commercial synthesis of Handel-C to register transfer level (RTL). In a different direction, Hutchings et al. describe JBits [29], which uses Java both for high-level design specification and low-level bit manipulation for FPGAs. Brandolese et al. [30] gives a methodology for estimating required FPGA LE counts from SystemC design descriptions.

17.3　Logic Synthesis

FPGA logic synthesis can be seen as four high-level steps: language synthesis, register-transfer-level synthesis, technology-independent logic synthesis, and technology mapping. Language synthesis, meaning VHDL/Verilog extraction, analysis, and elaboration, overlaps closely with ASIC tools, and is rarely discussed in a purely FPGA context. The other steps are discussed in the next sections.

17.3.1 RTL Synthesis

RTL, or register-transfer-level synthesis, includes the inference and processing of high-level structures—adders, multipliers, muxes, busses, shifters, crossbars, RAMs, shift-registers, and finite state machines—prior to decomposition into generic gate-level logic. This is an important area of work for FPGA and CAD-tool vendors, but has seen very little attention in the published literature. One reason for this is that the implementation of operators can be architecture specific. However, a more pedantic issue is simply that there are very few public-domain VHDL/Verilog front-end tools or high-level designs with these structures, and both are necessary for research in the area. Approximately 20%–30% of logic-elements eventually seen by placement are mapped directly from RTL and are not processed by gate-level logic synthesis.

In commercial FPGA architectures dedicated hardware is provided for multipliers, addition, clock enables, clear, preset, RAM, and shift-registers. Arithmetic was discussed briefly earlier; all major FPGAs either convert 4-LUTs into 3-LUT-based sum and carry computations or provide dedicated arithmetic hardware separate from the LUT [15]. The most important effect of arithmetic is the restriction it imposes on placement, since cells in a carry chain must be placed in fixed relative positions.

One of the goals of RTL-level synthesis is to take better advantage of the hardware provided. This will often result in different synthesis than would take place in an ASIC flow because the goal is to minimize LEs rather than gates. A 4:1 mux, for example, uses three two-input gates and cannot fit in a single 4-LUT. However, transformations such as the one shown in Figure 17.5 allow that mux to be implemented optimally in two 4-LUTs, even though these LUTs contain five gates. RTL-synthesis typically produces these premapped cells and then protects them from processing by logic-synthesis,

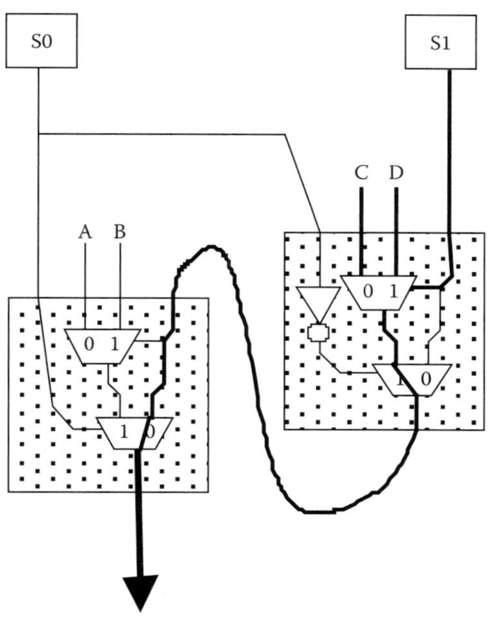

FIGURE 17.5 Implementing a 4:1 mux in two 4-LUTs. (From Metzgen, P. and Nancekievill, D., Multiplexor restructuring for FPGA implementation cost reduction, in *Proceedings of the Design Automation Conference*, Anaheim, CA, 2005. With permission.)

particularly when they occur in a bus and there is a timing advantage from a symmetric implementation. RTL-synthesis will recognize barrel shifters ("y <= (x >> s)" in Verilog) and convert these into shifting networks as shown in Figure 17.6, again protecting them from gate-level manipulation.

Recognition of register control signals such as clock enable, clear/preset, and sync/asynch load signals is also interesting in FPGAs. Since these preexist in the logic cell hardware (see Figure 17.2), there is a strong incentive to synthesize to them even when it would not make sense in ASIC synthesis. For example, a 4:1 mux with one constant input does not fit in a 4-LUT, but when it occurs in a datapath it can be synthesized for most commercial LEs by using the LAB-wide (i.e., shared by all LEs in a LAB) synchronous load signal as a fifth input. Similarly, a clock-enable already exists in the hardware, so register feedback can be converted to an alternative structure with a clock-enable to hold the current value, but no routed register feedback. For example, if f = z in Figure 17.7, we can

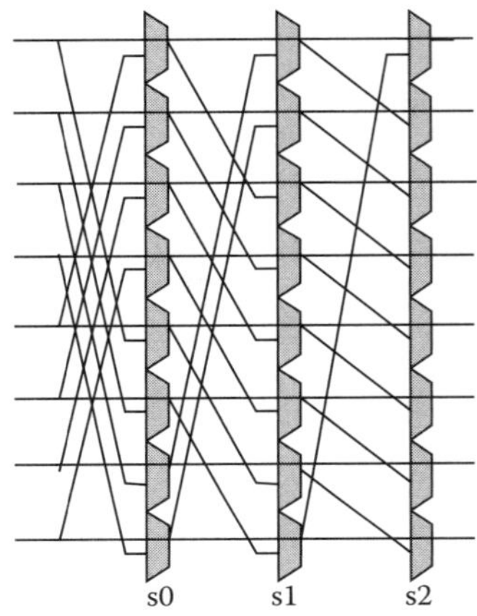

FIGURE 17.6   Eight-bit barrel shifter network.

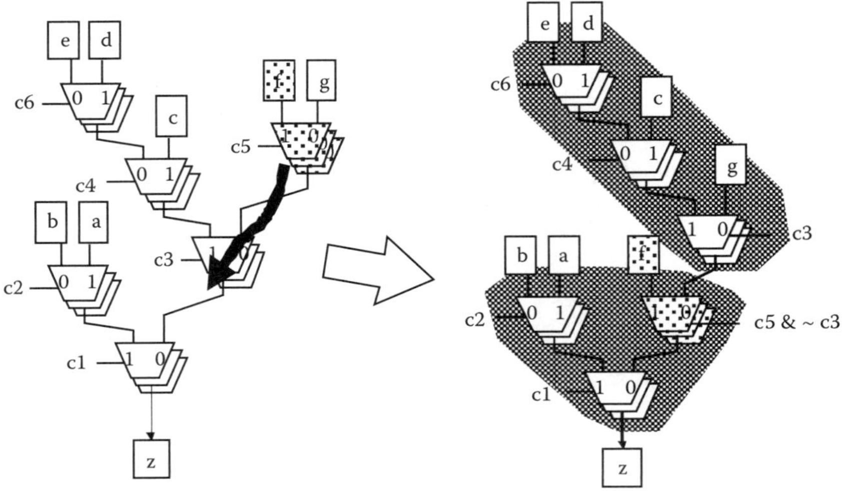

FIGURE 17.7   Multiplexor bus restructuring for LUT-packing. (From Metzgen, P. and Nancekievill, D., Multiplexor restructuring for FPGA implementation cost reduction, in *Proceedings of the Design Automation Conference*, Anaheim, CA, 2005. With permission.)

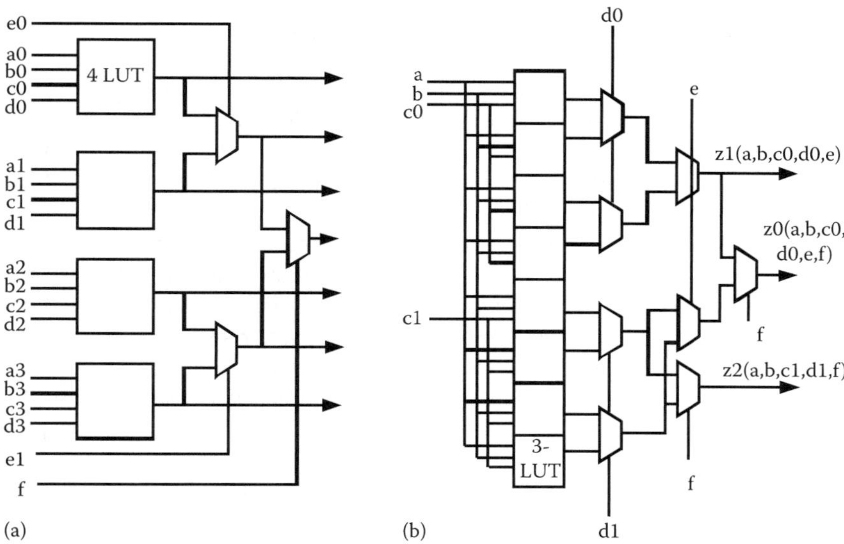

FIGURE 17.8   (a) Composable and (b) fracturable LUTs.

reexpress the cone of logic with a clock-enable signal CE = c5*c3*c1' with the f,g mux replaced by a wire from g—this is a win for bus widths of 2 or more. This transformation can be computed with a binary decision diagram (BDD) or other functional techniques.

Several algorithms have recently been published in the RTL-synthesis area. Metzgen and Nancekievill [32,33], for example, showed algorithms for the optimization of multiplexer-based busses that would otherwise be inefficiently decomposed into gates. Most modern FPGA architectures do not provide on-chip tri-state buses, so muxes are the only choice for busses, and are heavily used in designs. Multiplexors are very interesting structures for FPGAs, because the LUT cell yields a relatively inefficient implementation of a mux, and hence special-purpose hardware for handling muxes is common. Figure 17.7 shows an example taken from [32] that restructures busses of muxes for better technology mapping (covering) into 4-LUTs. The structure on the left requires 3 LUTs per bit to implement in 4-LUTs while the structure on the right requires only 2 LUTs per bit.

Also to address muxes, newer FPGA architectures have added clever hardware to aid in the synthesis of mux-structures such as crossbars and barrel shifters. The Xilinx Virtex family of FPGAs [17] provides additional "stitching" multiplexors for adjacent LEs, which can be combined to efficiently build larger multiplexors; an abstraction of this composable LUT is shown in Figure 17.8a. These are also used for stitching RAM bits together when the LUT is used as a 16-bit RAM (discussed earlier). Altera's Stratix II adaptive logic module [31], shown abstractly in Figure 17.8b, allows a 6-LUT to be fractured to implement a 6-LUT, two independent 4-LUTs, two 5-LUTs that share two-input signals, and also two 6-LUTs that have four common signals and two different signals (a total of eight). This latter feature allows two 4:1 muxes with common data and different select signals to be implemented in a single LE, which means crossbars and barrel shifters built out of 4:1 muxes can use half the area they would otherwise require.

17.3.2 Logic Optimization

Technology-independent logic synthesis for FPGAs has followed the same methodology of point-algorithms in a script flow, as popularized by the Berkeley SIS [34]. The general topic of logic synthesis is described in textbooks such as [21]. Synthesis tools for FPGAs contain basically the same two-level minimization algorithms, and algebraic and Boolean algorithms for multilevel synthesis. Here we will generally restrict our discussion to the differences from ASIC synthesis.

One major difference between standard and FPGA synthesis is in cost metrics. The target technology in a standard cell ASIC library is a more finely grained cell (e.g., a two-input nand gate) while a typical FPGA cell is a generic k-input lookup-table. A 4-LUT is a 16-bit SRAM LUT-mask driving a four-level tree of 2:1 muxes controlled by the inputs A,B,C,D (Figure 17.2a). Thus $A + B + C + D$ and $AB + CD + AB'D' + A'B'C'$ have identical costs in LUTs even thought the former has 4 literals and the latter 10. In general the count of two-input gates correlates much better to 4-LUT implementation cost than the literal-count cost often used in these algorithms, but this not always not the case, as described for the 4:1 mux in the preceding section. A related difference is that inverters are free in FPGAs because (1) the LUT-mask can always be reprogrammed to remove an inverter feeding or fed by an LUT, and (2) programmable inversion at the inputs to RAM, IO, and DSP blocks is available in most FPGA architectures. In general, registers are also free because all LEs have a built-in DFF. This changes cost functions for retiming and state-machine encoding as well as designer preference for pipelining.

Subfactor extraction algorithms are much more important for FPGA synthesis than commonly reported in the academic literature, where ASIC gates are assumed. It is not clear whether this arises from the much larger and more data-path oriented designs seen in industrial flows (vs. MCNC gate-level circuits), from the more structured synthesis from a complete HDL to gates flow, or due to the larger cell granularity. In contrast, algorithms in the class of "speed_up" [34] do not have significant effects on circuit performance for commercial FPGA designs. Again, this can be due either to the flow and reference circuits, or to differing area/depth trade-offs. Commercial tools make careful balancing of area and depth during multilevel synthesis.

Synthesis of arithmetic functions is typically performed separately in commercial FPGA tools, although most academic tools synthesize arithmetic into LUTs. This can result in a dramatic difference in the efficacy of synthesis algorithms, which perform well on arithmetic circuits compared to random or multiplexor/selector-based logic. Typical industrial designs contain 10%–25% of logic cells in arithmetic mode, in which the dedicated carry circuitry is used with or instead of the LUT.

Retiming algorithms from general logic synthesis [35] have been adapted specifically for FPGAs [36], taking into account practical restrictions such as meta-stability, I/O vs. core timing trade-offs, power-up conditions, and the abundance of registers.

There are a number of resynthesis algorithms that are of particular interest to FPGAs, specifically structural decomposition, functional decomposition, and postoptimization using SPFD-based rewiring; these are discussed in Sections 17.3.3.1 and 17.3.3.2.

An alternative, more FPGA-specific, approach to synthesis was taken by Vemuri et al. [37] using BDS [38] building on BDD-based decomposition [39]. These authors argued that the separation of tech-independent synthesis from tech-mapping disadvantaged FPGAs, which need to optimize LUTs rather than literals due to their greater flexibility and larger granularity. The BDS system integrated technology-independent optimization using BDDs with LUT-based logic restructuring and used functional decomposition to target decompositions of k-feasible LUTs for mapping. The standard sweep, eliminate, decomposition, and factoring algorithms from SIS were implemented in a BDD framework. The end result uses a technology mapping step, but on a netlist more amenable to LUT-mapping. Comparisons between SIS and BDS-pga using the same technology mapper showed area and delay benefits to the BDD-based algorithms.

A new area for CAD optimization in FPGAs involves power optimization, particularly leakage. Anderson et al. [40] proposes some interesting ideas on modifying the LUT-mask during synthesis to place LUTs into a state that will reduce leakage power in the FPGA routing. Commercial tools synthesize clock-enabled circuitry to reduce dynamic power consumption on blocks. These are likely the beginning of many future treatments for power management in FPGAs.

17.3.3 Technology Mapping

Technology mapping for FPGAs is the process of turning a network of primitive gates into a network of LUTs of size at most k. The constant k is historically 4 [1], though recent commercial architectures have used fracturable logic cells with $k = 6$ [31]. LUT-based tech-mapping is best seen as a covering problem, since it is both common and necessary to cover some gates by multiple LUTs for an efficient solution. Figure 17.9, taken from [41], illustrates this concept. Technology mapping aims for the least unit depth combined with the least number of cells in the mapped network.

FPGA technology mapping differs from library-based mapping for cell-based ASICs (e.g., [42]). Technology mapping into k-input LUTs is a well-studied problem, with at least 100 related papers on the topic. The most successful attempts can be divided into two paradigms: dynamic programming approaches such as Chortle [43], and network-flow-based approaches branching from FlowMap [44]. Many of these technology mapping algorithms are implemented in the RASP system from UCLA [45]. Technology mapping is usually preceded by decomposition of the netlist into two-input gates, but we will defer that topic to Section 17.3.3.1 because it draws on techniques from mapping.

In the Chortle algorithm [43] by Francis, the netlist is decomposed to two-input gates, and then divided into a forest of trees, a starting point used by Keutzer [42] and most library-based mappers. For each node in topological order, a set of k-feasible mappings for children nodes is known (by induction). To compute an optimum set of k-feasible mappings for the current node, Chortle combines solutions for children within reach of one LUT implemented at the current node, following the dynamic programming paradigm.

Improvements on Chortle considered not trees, but maximum fanout-free cones, which allowed for mapping with duplication. Area-mapping with no duplication was later shown to be optimally solvable in polynomial-time for MFFCs [46]. But, perhaps contrary to intuition, duplication is important in improving results for LUTs because it allows nodes with fanout greater than 1 to be implemented as internal nodes of the cover; this is required to obtain improved delay and also can

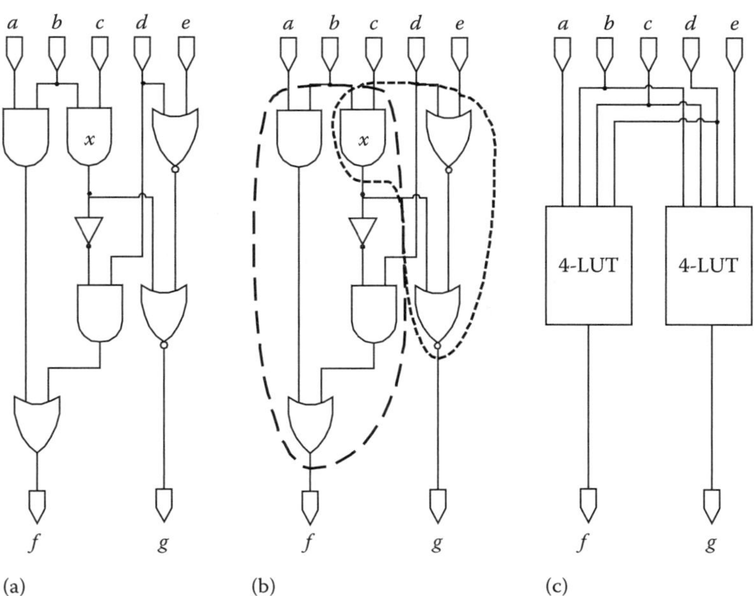

FIGURE 17.9 Technology mapping as a covering problem. (From Ling, A., et al., FPGA technology mapping: A study of optimality, in *Proceedings of the Design Automation Conference*, 2005. With permission.)

contribute to improved area [47]. Figure 17.9b illustrates a mapping to illustrate this point. Chortle-crf [48] deals more efficiently with reconvergent fanout. Chortle-d [49] adds a new bin-packing step on nodes, which considers all decompositions into two-input gates as part of the dynamic programming step, and is shown to be depth optimal for mapping on trees.

FlowMap [44] is a two-step algorithm proposed by Cong and Ding. In the labeling phase, a topological traversal of the network assigns each node a Lawler label $L(v)$ [50], which is constructed to be the worst-case depth of node v in a depth-optimal mapping solution. Primary inputs have $L(v) = 0$. For other nodes, $L(v)$ is either D or $D + 1$, where D is the max $L(w)$ over all fanin w of v. To determine which fanin, the fanin with $L(w) = D$ are collapsed into v (to simulate a LUT implemented at v), and a feasible cut computation is made to determine if $L(w)$ can be D. Otherwise it is $D + 1$. A k-feasible cut is a partition of the network into A and Ā such that the output of no more than k nodes is cut. The size of a cut is the number of cut-edges, the height is the largest label of the nodes cut and the volume is the number of nodes in Ā. The key aspect to the FlowMap algorithm is to reduce the minimum height k-feasible cut computation to the well-known network flow problem [51]. Figure 17.10 [44] illustrates a network of nodes with depth labeling; the auxiliary S (source) and T (sink) nodes required for network flows; and a 3-feasible cut with size 10, volume 9, and height 2.

The result of FlowMap is provably depth-optimal for arbitrary k-bounded networks, meaning that for a fixed decomposition into two-input gates it always generates a unit-delay-minimal solution. Later enhancements allow for finding a minimum-height maximal-volume cut as an area reduction heuristic. Since most problems in tech-mapping are NP-hard (e.g., area-minimization [52]), this also makes FlowMap theoretically interesting as well as practical. FlowMap can be extended to use

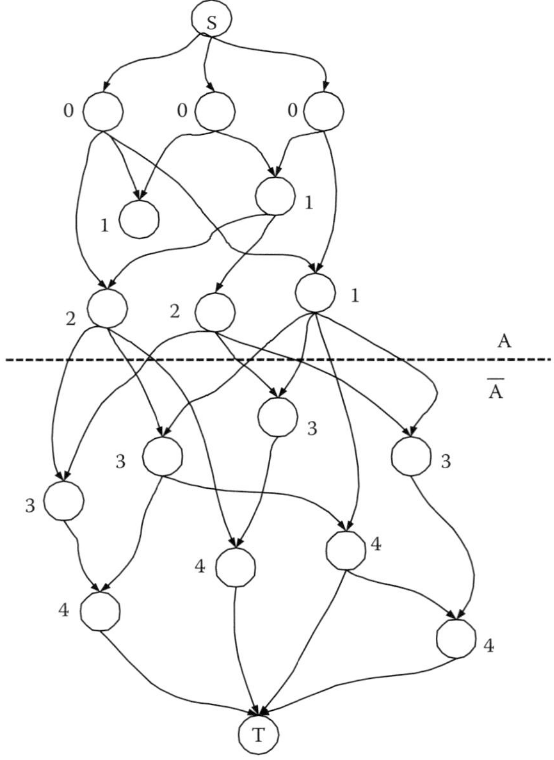

FIGURE 17.10   FlowMap 3-feasible cut with cut-size 10, volume 9, and height 2.

more general models of delay [53]. Cong and Ding [46] added a duplication-free remapping step to FlowMap to improve area, and also explored these trade-offs.

CutMap by Cong and Hwang [54] is an area-improvement on FlowMap that maintains the property of optimal delay. The key feature of CutMap is that the computation of both min-cost min-height cuts for nodes on the critical path, and min-cost k-feasible cuts for other nodes for the implementation phase. This allows for an area/delay trade-off directly in the cost function of the mapping phase, unlike FlowMap, which addresses area only in a postprocessing step.

In the first step, CutMap computes arrival labels using the FlowMap labeling phase and also marks each node with its required implementation time, so that unit-delay slack can be computed. In the implementation phase, nodes with zero-slack follow the FlowMap calculation—predecessor nodes u with label(u) = label(v) are collapsed into v for a min-cost k-feasible cut with minimum height. However, for nodes v with positive slack, the nodes are not collapsed, and only the min-cost k-feasible cut calculation is made. In this way, implementation of all noncritical nodes favors area minimization. Empirical results for CutMap show a 15% better area than FlowMap with the same unit delay, however, at the cost of longer run-time.

Manohararajah et al. [55] gave a mapping algorithm IMAP (iterative map) based on dynamic programming that simultaneously address depth and area. This paper also introduces new metrics of area-flow and depth bounds, and uses an edge-delay model based on [56]. IMAP generates k-feasible cones for nodes as in [47] and then iteratively traverses the graph forward and backward. The forward traversal identifies covering cones for each node (depth-optimal for critical and area-optimal for noncritical nodes), and the backward traversal then selects the covering set and updates the heights of remaining nodes. The benefit of iteration is to relax the need for delay-optimal implementation once updated heights mark nodes as no longer critical, allowing greater area improvement while maintaining depth optimality.

DAOmap from Chen and Cong [57] uses enumeration and iteration/relaxation techniques similar in spirit to IMAP, but additionally considers the potential node duplications during the cut enumeration procedure. There are, however, numerous differences in the details of the cost functions and heuristics. For example, DAOmap has a "cut probing" or lookahead step during the local cost adjustment phase on the backwards traversal. DAOmap and IMAP would be the current best known solutions for delay-minimum, area-minimal LUT technology mapping; there is no published comparison of the two showing equivalent netlist starting points.

Pan [58] and Pan and Lin [59] integrated technology mapping and retiming. This not only provides the performance benefit of retiming but, since registers are obstacles to efficient covering, can also contribute to area gains.

The Stratix II "adaptable logic element," or ALM, introduced in Hutton et al. [31] and shown in Figure 17.8b poses interesting new problems for technology mapping. The ALM structure can implement one 6-LUT, two 5-LUTs with two common inputs, or two independent 4-LUTs (among other combinations). The 6-LUTs are useful for depth reduction, but the most efficient use of the structure for area is the two 5-LUTs with sharing. This makes the area-cost function no longer a simple count of the number of covering LUTs, because the sharing needs to be accounted for. Dynamic programming approaches such as IMAP are more amenable to these modified LEs.

Recently technology mapping algorithms have begun to address power [60]. Anderson and Najm [61] proposed a modification to technology mapping algorithms to minimize node duplication to minimize the number of wires between LUTs (as a proxy for dynamic power required to charge up the global routing wires). EMAP by Lamoreaux and Wilton [62] modifies CutMAP with an additional cost-function component to favor cuts that reuse nodes already cut and those that have a high activity factor (using probabilistic activity factors). Chen et al. [63] extends this type of mapping to a heterogeneous FPGA architecture with dual voltage supplies, where high-activity nodes additionally need to be routed to low-Vdd (low power) LEs and critical nodes to high-Vdd (high performance) LEs.

17.3.3.1 Gate Decomposition

An important preprocessing step for technology mapping is to decompose the network into a k-bounded (most often 2) network in preparation for LUT covering. The previously mentioned covering algorithms are depth-optimal only for k-bounded networks: Cong and Hwang [64] showed that depth-optimal tech-mapping is NP-hard for unbounded networks when $k = 3$, and further for bounded networks when $k > 4$.

Different decompositions will give very different mapping results [64]. There are multiple approaches to structural decomposition: tech_decomp [34], and DMIG [65] as part of SIS were used originally and Chortle contains a modified decomposition step, but the improved DOGMA algorithm [64,66] is currently the best known structural decomposition technique for tech-mapping.

DOGMA combines the bin-packing step of Chortle-d with FlowMap's min-height k-feasible cut. The goal is to produce a 2-bounded network that will minimize depth in the final mapping solution, so at each step bin-nodes representing a level boundary node are used to store information. For each node v in topological order, labels are computed as follows: the sets of fanin nodes previously labeled with value q (the stratum of depth q) are computed. Groups of ascending strata q are packed into a minimum number of bins such that a k-feasible cut of height $q - 1$ exists (using network flows), and then stored in bin-nodes. The bin-nodes are then packed in the $q + 1$ step into a minimal number of minimal height k-feasible bins. This continues until all nodes are packed into one bin corresponding to the node v. At the conclusion of the algorithm, the packing gives the decomposition (and additionally the labeling for the first step of FlowMap or CutMap) and the bin-nodes are removed.

Legl et al. [67] gave a method for technology mapping that combines functional decomposition and mapping using Boolean techniques. This showed better results than DMIG + FlowMap, but has practical issues with size and computation time for the BDDs. Cong and Hwang also applied partially dependent functional decomposition with technology mapping to target a Xilinx 4000 CLB, which is a block containing two 4-LUTs hard-wired into a 3:1 mux [68].

As a final note on technology mapping, some commercial FPGAs such as Altera's Apex family allow embedded RAM to be used as product-term logic or as combinational ROMs implementing large lookup-tables with multiple outputs. Wilton [12] and Lin and Wilton [13] proposed pMapster as an algorithm for dynamically mapping to both LUT and pterm logic, and Cong and Xu [69] gave HeteroMap, which covers the netlist using both LUTs and ROMs.

17.3.3.2 SPFD-Based Rewiring

An interesting recent development in FPGA synthesis is the use of SPFDs for exploiting the inherent flexibility in LUT-based netlists. SPFDs, or *sets of pairs of functions to be distinguished*, were proposed by Yamashita et al. [70]. SPFDs are a generalization of *observability don't care* (ODC) functions, wherein the on/off/dc set of functions is represented abstractly as a bipartite graph denoting "distinguishing" edges between min-terms, and a coloring of the graph gives an alternative implementation of the function. An inherent flexibility of LUTs in FPGAs is that they do not need to represent inverters, because these can always be absorbed by changing the destination node's LUT-mask. By storing distinctions rather than functions, SPFDs generalize this to allow for more efficient expressions of logic.

Cong et al. [71,72] applied SPFD calculations to the problem of rewiring a previously tech-mapped netlist. The algorithm consists of precomputing the SPFDs for each node in the network, identifying a target wire (e.g., a delay-critical input of a LUT after tech-mapping), and then trying to replace that wire with another LUT-output that satisfies its SPFD. The don't care sets in the SPFDs occur in the internal nodes of the network, where flexibility exists in the LUT implementation after synthesis and tech-mapping. Rewiring was shown to have benfits both for delay and area. Hwang et al. [73] and Kumthekar and Somenzi [74] also applied SPFD-based techniques for power reduction.

17.4 Physical Design

The physical design aspect of FPGA tools consists of clustering, placement, physical resynthesis, and routing. Commercial tools have additional preprocessing steps to allocate clock and reset signals to special low-skew "global networks," to place phase-locked loops, and to place transceiver blocks and I/O pins to meet the many electrical restrictions imposed on them by the FPGA device and the package. With the exception of some work on placing FPGA I/Os to respect electrical restrictions [75,76], however, these preprocessing steps are typically not seen in any literature.

The FPGA physical design can broadly be divided into *routability-driven* and *timing-driven* algorithms. Routability-driven algorithms seek primarily to find a legal placement and routing of the design by optimizing for reduced routing demands. In addition to optimizing for routability, timing-driven algorithms also use timing analysis to identify critical paths and/or connections, and attempt to optimize the delay of those connections. Since the majority of delay in an FPGA is contributed by the programmable interconnect, timing-driven placement and routing can achieve a large circuit speed-up vs. routability-driven approaches. For example, a recent commercial CAD system achieves an average of 50% higher design performance with full effort timing-driven placement and routing vs. routability-only placement and routing, at a cost of 5× run-time [78].

In addition to optimizing timing and routability, some recent FPGA physical design algorithms also implement circuits such that power is minimized.

17.4.1 Placement and Clustering

17.4.1.1 Problem Formulation

The placement problem for FPGAs differs from the placement problem for ASICs in several important ways. First, placement for FPGAs is a slot assignment problem—each circuit element in the technology-mapped netlist must be assigned to a discrete location, or slot, on the FPGA device that can accommodate it. Figure 17.1 earlier showed the floorplan of a typical modern FPGA. An LE, for example, must be assigned to a location on the FPGA where an LE has been fabricated, while an I/O block or RAM block must each be placed in a location where the appropriate resource exists on the FPGA. Second, there are usually a large number of constraints that must be satisfied by a legal FPGA placement. For example, groups of LEs that are placed in the same logic block have limits on the maximum number of distinct input signals and the number of distinct clocks they can use [14], and cells in carry-chains must be placed together as a macro. Finally, all routing in FPGAs consists of prefabricated wires and transistor-based switches to interconnect them. Hence the amount of routing required to connect two circuit elements, and the delay between them, is a function not just of the distance between the circuit elements, but also of the FPGA routing architecture. It also means that the amount of routing is strictly limited, and a placement that requires more routing in some region of the FPGA than exists there cannot be routed.

17.4.1.2 Clustering

A common adjunct to FPGA placement algorithms is a bottom-up clustering step that runs before the main placement algorithm in order to group related circuit elements together into clusters (LABs in Figure 17.2). Clustering reduces the number of elements to place, improving the run-time of the main placement algorithm. In addition, the clustering algorithm usually deals with many of the complex FPGA legality constraints by grouping LEs into legal logic blocks, simplifying legality checking for the main placement algorithm.

The RASP system [45] includes one of the first logic block clustering algorithms. It performs maximum weighted matching on a graph where edge weights between LEs reflect the desirability of clustering them together. However, it has a high computational complexity of $O(n^3)$, where n is the number of LEs in the circuit, and this prevents it from scaling to very large problems. The VPack algorithm in VPR [79] clusters LEs into logic blocks by choosing a *seed* LE for a new cluster, and then greedily packing the LE with the highest *attraction* to the current cluster until no further LEs can be legally added to the cluster. The attraction function is the number of nets in common between an LE and the current cluster. VPack has a computational complexity of $O(k_{max}\,n)$, where k_{max} is the maximum fanout of any net in the design. The T-VPack algorithm from Marquardt et al. [80] is a timing-driven enhancement of VPack where the attraction function for an LE, L, to cluster C becomes

$$\text{Attraction}(L) = .75 \cdot \text{Crit}(L, C) + .25\frac{|\text{Nets}(L) \cap \text{Nets}(C)|}{\text{MaxNets}} \tag{17.1}$$

The first term gives higher attraction to LEs that are connected to the current cluster by timing-critical connections, while the second term is taken from VPack and favors grouping LEs with many common signals together. Somewhat surprisingly, T-VPack improves not only circuit speed vs. VPack, but also routability, by absorbing more connections within clusters. The iRAC [81] clustering algorithm achieves further reductions in the amount of routing necessary to interconnect the logic blocks by using attraction functions that favor the absorption of small nets within a cluster.

Lamoureaux and Wilton [82] developed a power-aware modification of T-VPack that adds a term to the attraction function of Equation 17.1 such that LEs connected to the current cluster by connections with a high rate of switching have a larger attraction to the cluster. This favors the absorption of nets that frequently switch logic states, resulting in lower capacitances for these nets, and lower dynamic power. Chen and Cong [83] developed a clustering algorithm that reduces power in an FPGA architecture where each logic block can be run at either a high or a low voltage. Their algorithm groups nontiming-critical LEs into different clusters than timing-critical LEs, enabling many logic blocks to run at reduced voltage and hence reduced power.

17.4.1.3   Placement

Though the literature includes numerous techniques, simulated annealing is the most widely used placement algorithm for FPGAs. Figure 17.11 shows the basic flow of simulated annealing. An initial placement is generated, and a placement perturbation is proposed by a *move generator*, generally by moving a small number of circuit elements to new locations. A *cost function* is used to evaluate the impact of each proposed move. Moves that reduce cost are always accepted, or applied to the

```
P = InitialPlacement ();
T = InitialTemperature ();

while (ExitCriterion () == False) {
       while (InnerLoopCriterion () == False) { /* "Inner Loop" */
              Pnew = PerturbPlacementViaMove (P);
              ΔCost = Cost(Pnew) − Cost (P);
              r = random(0,1) ;
              if(r < e−ΔCost/T){
                     P = Pnew;        /*Move Accepted*/
              }
       } /* End "Inner Loop" */
              T = UpdateTemp (T);
}
```

FIGURE 17.11   Pseudo-code of a generic simulated annealing placement algorithm.

placement, while those that increase cost are accepted with probability $e^{-\frac{\Delta Cost}{T}}$, where T is the current *temperature*. Temperature starts at a high level, and gradually decreases throughout the anneal, according to the *annealing schedule*. The annealing schedule also controls how many moves are performed between temperature updates, and when the ExitCriterion that terminates the anneal is met.

Two key strengths of simulated annealing that many other approaches lack are

1. It is possible to enforce all the legality constraints imposed by the FPGA architecture in a fairly direct manner. The two basic techniques are to forbid the creation of illegal placements in the move generator and to add a penalty cost to illegal placements.

2. By creating an appropriate cost function, it is possible to directly model the impact of the FPGA routing architecture on circuit delay and routing congestion.

VPR [8,79,80] contains a timing-driven simulated annealing placement algorithm, as well as timing-driven routing. The VPR placement algorithm is usually used in conjunction with T-VPack, which preclusters the LEs into legal logic blocks. The placement annealing schedule is based on monitoring statistics generated during the anneal, such as the fraction of proposed moves that are accepted. This adaptive annealing schedule lets VPR to automatically adjust to different FPGA architectures. VPR's cost function also automatically adapts to different FPGA architectures [80]:

$$\text{Cost} = (1 - \lambda) \sum_{i \in \text{All Nets}} q(i) \left[\frac{bb_x(i)}{C_{av,x}(i)} + \frac{bb_y(i)}{C_{av,y}(i)} \right] + \lambda \sum_{j \in \text{All Connections}} \text{Criticality}(j) \cdot \text{Delay}(j)$$

(17.2)

The first term in Equation 17.2 causes the placement algorithm to optimize an estimate of the routed wirelength, normalized to the average wiring supply in each region of the FPGA. The wirelength needed to route each net i is estimated as the bounding box span (bb_x and bb_y) in each direction, multiplied by a fanout-based correction factor, $q(i)$. In FPGAs with differing amounts of routing available in different regions or channels, it is beneficial to move the wiring demand to the more routing-rich regions, so the estimated wiring required is divided by the average routing capacity over the bounding box in the appropriate direction.

The second term in Equation 17.2 optimizes timing by favoring placements in which timing-critical connections have the potential to be routed with low delay. To evaluate the second term quickly, VPR needs to be able to quickly estimate the delay of a connection. To accomplish this, VPR assumes that the delay is a function only of the difference in the coordinates of a connection's endpoints, $(\Delta x, \Delta y)$, and invokes the VPR router with each possible $(\Delta x, \Delta y)$ to determine a table of delays vs. $(\Delta x, \Delta y)$ for the current FPGA architecture before the simulated annealing algorithm begins. The criticality of connections is determined via periodic timing analysis using delays computed from the current placement.

Many enhancements have been made to and published about the original VPR algorithm. The PATH algorithm from Kong [87] uses a new timing criticality formulation in which the timing criticality of a connection is a function of the slacks of all paths passing through it, rather than just a function of the worst-case (smallest) slack of any path through that connection. This technique significantly improves timing optimization, and results in circuits with 15% smaller critical path delay, on average. The SCPlace algorithm [86] enhances VPR so that a portion of the moves are *fragment moves* in which a single LE is moved, instead of an entire logic block. This allows the placement algorithm to modify the initial clustering, and improves both circuit timing and wirelength. Lamoureaux and Wilton [27] modified VPR's cost function by adding a third term, PowerCost, to Equation 17.2:

$$\text{PowerCost} = \sum_{i \in \text{AllNets}} q(i) \left[bb_x(i) + bb_y(i) \right] \cdot \text{Activity}(i) \qquad (17.3)$$

where Activity(i) represents the average number of times net i transitions per second. This additional cost function term reduces circuit power, although the gains are less than those obtained by power-aware clustering.

PROXI [88] uses simulated annealing for placement, but its cost function is based not on fast heuristics to estimate placement routability and timing, but instead on maintaining at least a partially routed design at all times. The PROXI cost function is a weighted sum of the number of unrouted nets and the delay of the circuit critical path. After each placement perturbation, all nets connected to moved cells are ripped up and rerouted via a fast, directed-search maze router. To keep the CPU time tolerable, PROXI allows the maze router to explore only a small portion of the routing fabric at high temperatures—if no unblocked routing path is found quickly, the net is marked as unrouted. At lower temperatures, the placement is of a higher quality and the router is allowed to explore a larger portion of the graph. After each net is rerouted, the critical path is recomputed incrementally. PROXI produces high quality results, but requires relatively high CPU time.

Sankar and Rose [89] seek the opposite trade-off of reduced result quality for extremely low placement run-times. They create a hierarchical annealer that employs greedy clustering with a net-absorption-based attraction function to reduce the size of the placement problem. The best run-time quality trade-off occurs when they cluster the circuit logic blocks twice—first clustering into level 1 clusters of approximately 64 logic blocks, and then clustering four of these level 1 clusters into each level 2 cluster. The level 2 clusters are placed with a greedy (temperature = 0) anneal seeded by a fast constructive initial placement. Next each level 1 cluster is initially placed within the boundary of the level 2 cluster that contained it, and another temperature = 0 anneal is performed. Finally, the placement of each logic block is refined with a low-temperature anneal. For very fast CPU times, this algorithm significantly outperforms VPR in terms of achieved wirelength, while for longer permissible CPU times, it lags VPR.

Another popular placement approach for FPGAs is recursive partitioning. ALTOR [90] was originally developed for standard cell circuits, but was adapted to FPGAs and widely used in FPGA research. ALTOR employs a recursive min-cut bipartitioning technique with terminal propagation [91] to gradually partition the design into small portions of the FPGA floorplan, at which point a complete placement is obtained. Figure 17.12 shows the sequence of cut-lines used by ALTOR to partition the FPGA area. This sequence of cut lines means ALTOR assumes that terminials that

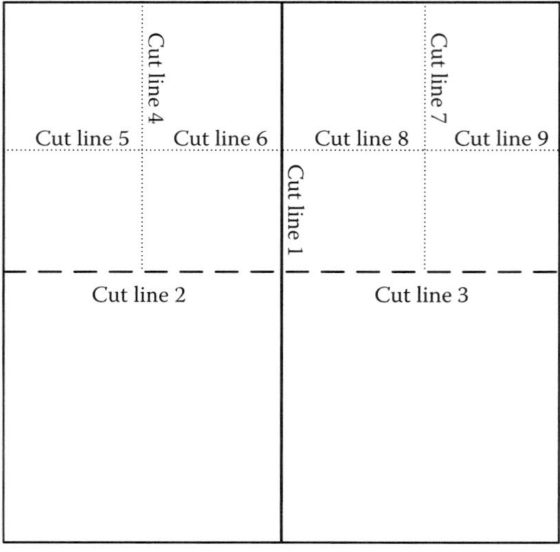

FIGURE 17.12   ALTOR partitioning sequence.

have a small Manhattan distance between them can be connected efficiently by the FPGA routing fabric. This assumption matches the capabilities of the segmented routing architectures used by most modern commercial FPGAs.

An approach by Maidee et al. [92] is also based on recursive bipartitioning, but it adds timing-driven features. Before partitioning begins, the VPR routing algorithm is used to generate a table of net delay vs. distance spanned by the net that takes into account the FPGA routing architecture. As partitioning proceeds, the algorithm records the minimum length each net could achieve, given the current number of portioning boundaries it crosses. The delay corresponding to each net's span is retrieved from the precalculated table, and a timing analysis is performed to identify critical connections. Timing-critical connections to terminals outside of the region being partitioned act as anchor points during each partitioning. This forces the other end of the connection to be allocated to the partition that allows the critical connection to be short. Once partitioning has proceeded to the point that each region contains only a few cells, any overfilled regions are legalized with a greedy movement heuristic. Finally, the placement is further optimized by using VPR to perform a low-temperature anneal. The technique achieves wirelength and speed results comparable to VPR, with significantly reduced CPU time.

A commercial recursive partitioning placement algorithm for the Altera Apex 20 K family is described in [93]. Apex has a hierarchical routing architecture, making it well suited to partitioning-based placement. Recursive partitioning is conducted along the natural cut-lines formed by the various hierarchy levels of the routing architecture, as shown in Figure 17.13. Notice that the sequence of partitions in this algorithm is significantly different from that of ALTOR, showing the large impact an FPGA's routing architecture has on placement algorithms. This algorithm is made timing-driven by heavily weighting connections with low slack during each partitioning phase to encourage partitioning solutions in which such connections can be routed using only fast, lower-hierarchy-level routing. To improve the prediction of the critical path, the delay estimate for each connection is a function of both the known number of hierarchy boundaries the net must traverse due to partitioning at the higher levels of the routing hierarchy, and statistical estimates of how many hierarchy boundaries the connection will cross at future partitioning steps.

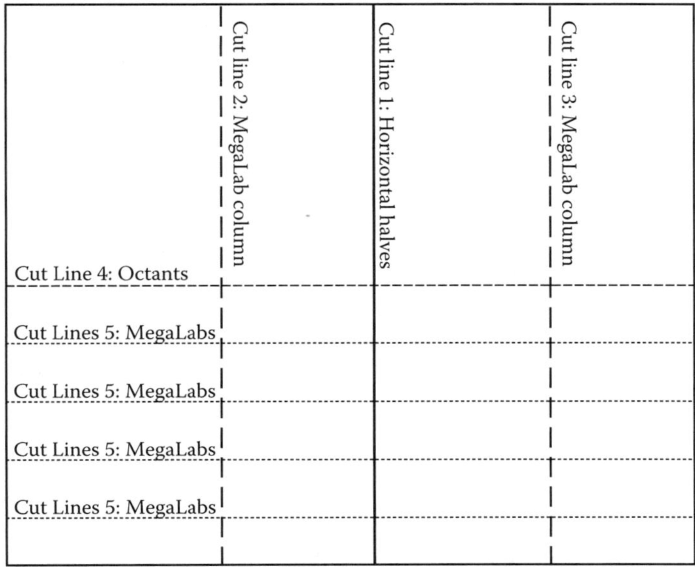

FIGURE 17.13 Sequence of cut lines for Apex architecture recursive partitioning placement.

Analytic algorithms are the third major approach to the placement problem. Analytic algorithms are based on creating a convex function that approximates wirelength. Finding the global minimum of this function yields a placement that has good wirelength if the function approximated routed the wirelength well. However, this global minimum is usually an illegal placement solution, so constraints and heuristics must be applied to guide the algorithm to a legal solution.

While analytic placement approaches are popular for ASICs, there are few analytic FPGA algorithms, likely due to the more difficult legality constraints in FPGA placement. The negotiated analytic placement (NAP) algorithm from Chan and Schlag [94] combines global analytic placement [95] to determine the wirelength optimized, but overlapping, cell locations with a negotiated congestion algorithm that gradually reduces overuse of LE locations until a legal placement is achieved. NAP also includes features that make it suitable for parallelization across multiple processors. First, the circuit netlist is covered by a set of sub-netlists, each of which is a tree. The placement of each tree can then proceed in parallel, which results in those cells contained in multiple trees being placed in multiple locations. Extra edges between copies of the same cell to the center of gravity of the cell are added to the edge-weight matrix to gradually pull copies of the cells together. After every iteration of analytic placement, a negotiated congestion algorithm is invoked to spread out the cells within each region of the placement. After a sufficient number of analytic placement/negotiated congestion movement iterations, an overlap-free placement is obtained.

17.4.2 Physical Synthesis Optimizations

Timing visibility can be poor during FPGA synthesis. What appears to be a noncritical path can turn out to be critical after placement and routing. This is true for ASICs as well, but the problem is especially acute for FPGAs. Unlike an ASIC implementation, FPGAs have a predefined logic and routing fabric and hence cannot use drive-strength selection, wire sizing, or buffer insertion to increase the speed of long routes.

Recently, physical synthesis techniques have arisen in both the literature and in commercial tools. Physical synthesis techniques for FPGAs generally refer either to resynthesis of the netlist once some approximate placement has occurred, and thus some visibility of timing exists, or local modifications to the netlist during placement itself. Figure 17.14 highlights the differences between the two styles of physical synthesis flow. The "iterative" flow of Figure 17.14a iterates between synthesis and physical design. A positive of this flow is that the synthesis tool is free to make large-scale changes to the circuit implementation, but a negative is that the placement and routing of this new design may not match the synthesis tool expectations, and hence the loop may not converge well. The "incremental" flow of Figure 17.14b instead makes only more localized changes to the circuit netlist, such that it can integrate these changes into the current placement with only minor perturbations. This flow has the advantage that convergence is easier, since a legal or near-legal placement is maintained at all times, but it has the disadvantage that it is more difficult to make large-scale changes to the circuit structure.

Commercial tools from Synplicity and Mentor Graphics [17] largely follow the "iterative" flow, and resynthesize a netlist given output from the FPGA vendor place and route tool. However, these tools can also provide constraints to the place and route tool in subsequent iterations to assist convergence. Lin et al. [96] described a similar academic flow in which remapping is performed after either a placement estimate or after actual placement delays are known. Suaris [97] used timing budgets for resynthesis, where the budget is calculated using a quick layout of the design. This work also makes modifications to the netlist to facilitate retiming in the resynthesis step. In a later improvement [98] this flow was altered to incrementally modify the placement after each netlist transform, assisting convergence.

There are commercial and academic examples of the "incremental" physical synthesis flow as well. Altera's Quartus CAD system tightly integrates the physical synthesis and placement engines. Schabas and Brown [99] used logic duplication as a postprocessing step at the end of placement,

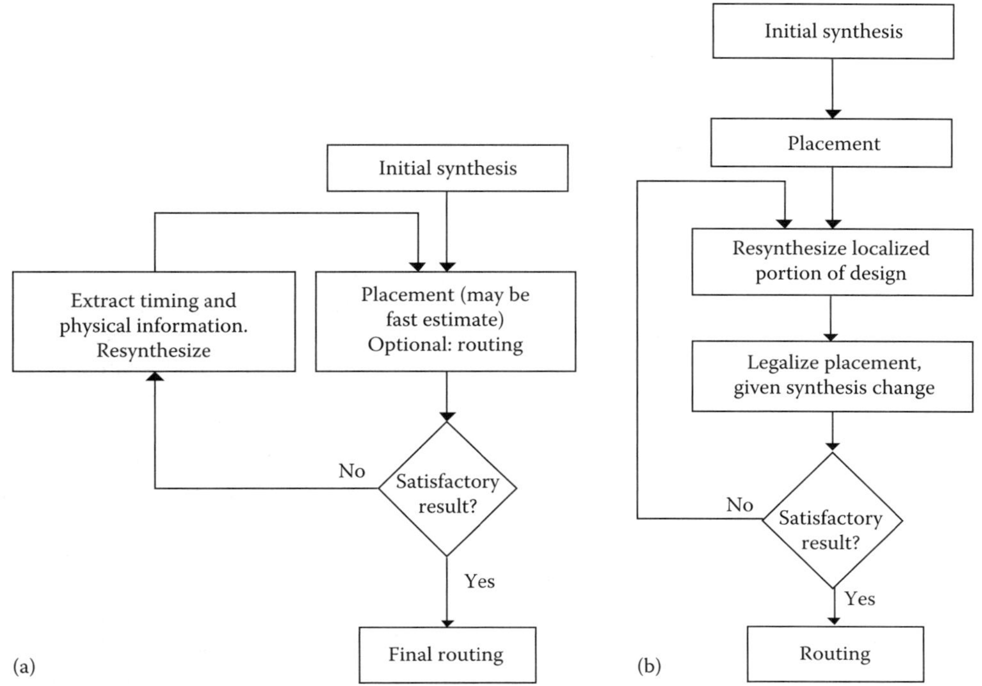

FIGURE 17.14 Physical synthesis flows. (a) Example "iterative" physical synthesis flow and (b) example "incremental" physical synthesis flow.

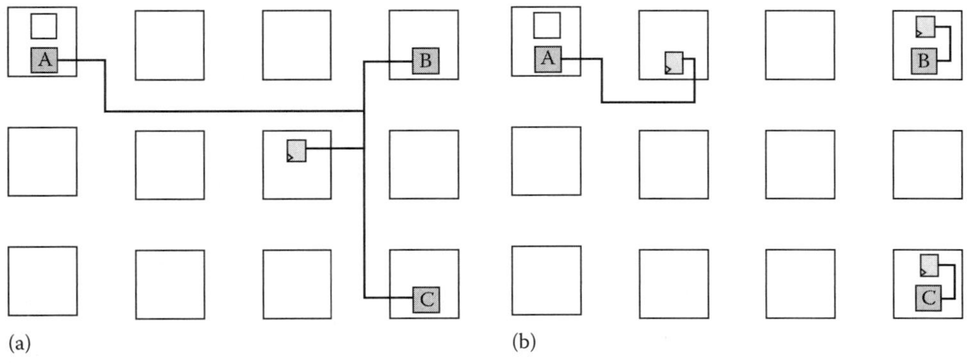

FIGURE 17.15 Duplicating registers to optimize timing in physical synthesis. (a) Register with three time-critical output connections and (b) three register duplicates created and legally placed to optimize timing.

with an algorithm that simultaneously duplicates logic and finds legal and optimized locations for the duplicates. Logic duplication, particularly on high-fanout registers, allows significant relaxation on placement-critical paths because it is common for a multi-fanout register to be "pulled" in multiple directions by its fanouts, as shown in Figure 17.15. Chen and Cong [100] integrated duplication throughout a simulated-annealing-based placement algorithm. Before each temperature update, logic duplicates are created and placed if it will assist timing, and previously duplicated logic may be "unduplicated" if the duplicates are no longer necessary.

Manoharajah et al. [101,102] performed local restructuring of timing critical logic to shift delays from the critical path to less critical paths. An LUT and a timing critical portion of its fanin are considered for functional decomposition or BDD-based resynthesis. An incremental placement algorithm then integrates any changed or added LUTs into a legal placement. Ding [103] gives an algorithm for postplacement pin permutation in LUTs. This algorithm reorders LUT inputs to take advantage of the fact that each input typically has a different delay. As well, this algorithm also swaps inputs amongst several LUTs that form a logic cone in which inputs can be legally swapped, such as an and-tree or xor-tree. An advantage of this algorithm is that no placement change is required, since only the ordering of inputs is affected, and no new LUTs are created.

Singh and Brown [106] present a postplacement retiming algorithm. This algorithm initially places added registers and duplicated logic at the same location as the original logic. It then invokes an incremental placement algorithm to legalize the placement. This incremental placement algorithm is similar to an annealing algorithm, but it includes costs for various types of resource overuse as well as timing and wiring costs, and it accepts only moves that reduce cost (i.e., temperature = 0). A later improvement [107] altered the retiming algorithm so that it incrementally modifies the design using local retiming operations, each of which is separately legalized before moving on to the next operation. This simplifies the legalization of each modification and saves compile time.

In an alternative to retiming, Singh and Brown [104] employed unused PLLs and global clocking networks to create several shifted versions of a clock, and developed a postplacement algorithm that selected the time-shifted clock with the best timing performance for each register. This approach is conceptionally similar to retiming after placement but involves shifting clock edges at registers rather than moving registers across combinational logic. Chao-Yang and Marek-Sadowska [105] extended this beneficial clock skew timing optimization to a proposed FPGA architecture where clocks can be delayed via programmable delay elements on global clock distribution networks.

17.4.3 Routing

17.4.3.1 Problem Formulation

All FPGA routing consists of prefabricated metal wires and programmable switches to connect the wires to each other, and to the circuit element input and output pins. Figure 17.16 shows an example FPGA routing architecture. In this example, each routing channel contains four wires of length 4—wires that span 4 logic blocks before terminating—and one wire of length 1. The programmable switches allow wires to connect only at their endpoints, but many FPGA architectures also allow programmable connections from interior points of long wires as well.

Usually the wires and the circuit element input and output pins are represented as nodes in a *routing-resource graph*, while programmable switches that allow connections to be made between the wires and pins become directed edges. Programmable switches can be fabricated as pass transistors, tristate buffers or multiplexers. Multiplexers are the dominant form of programmable interconnects in recent FPGAs such as the Altera Stratix [14,17] and Xilinx Virtex [17] families, since multiplexer-based routing produces FPGAs with a superior area-delay product [14,110]. Figure 17.17 shows how a small portion of an FPGA's routing is transformed into a routing-resource graph. This graph can also efficiently store information on which pins are logically equivalent, and hence may be swapped by the router, by including *source* and *sink* nodes that connect to all the pins which can perform a desired function. It is common to have many logically equivalent pins in commercial FPGAs, for example, all the inputs to an LUT are logically equivalent, and may be swapped by the router. A legal routing of a design consists of a tree of routing-resource nodes for each net in the design such that (1) each tree electrically connects the net source to all the net sinks, and (2) no two trees contain the same node, as that would imply a short between two signal nets.

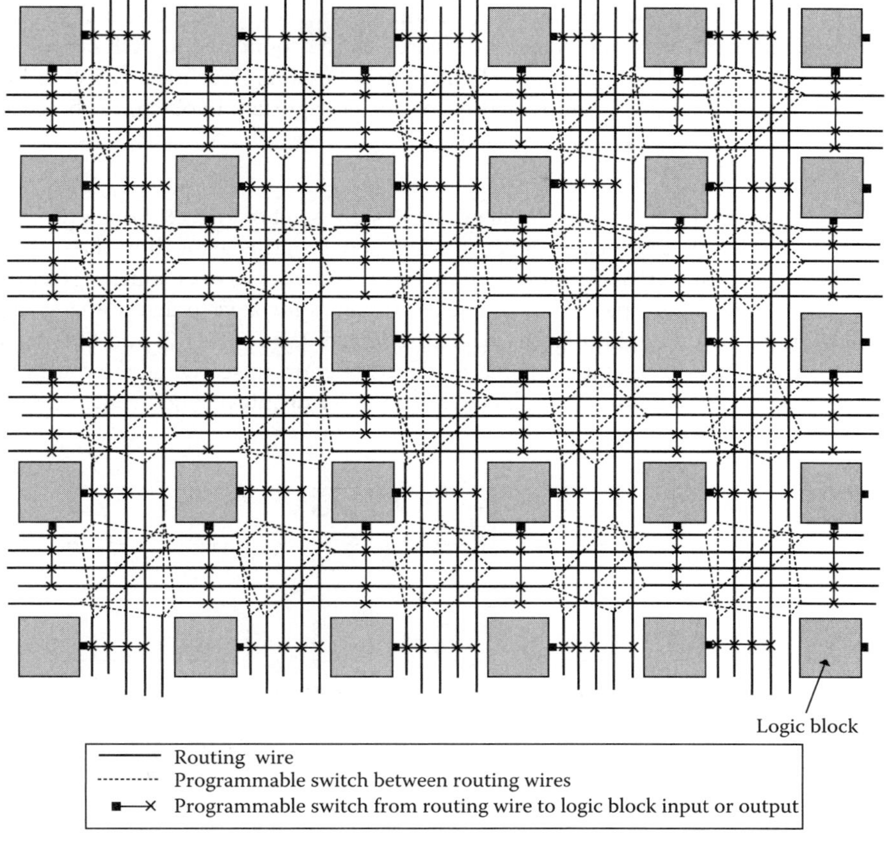

FIGURE 17.16 Example FPGA routing architecture.

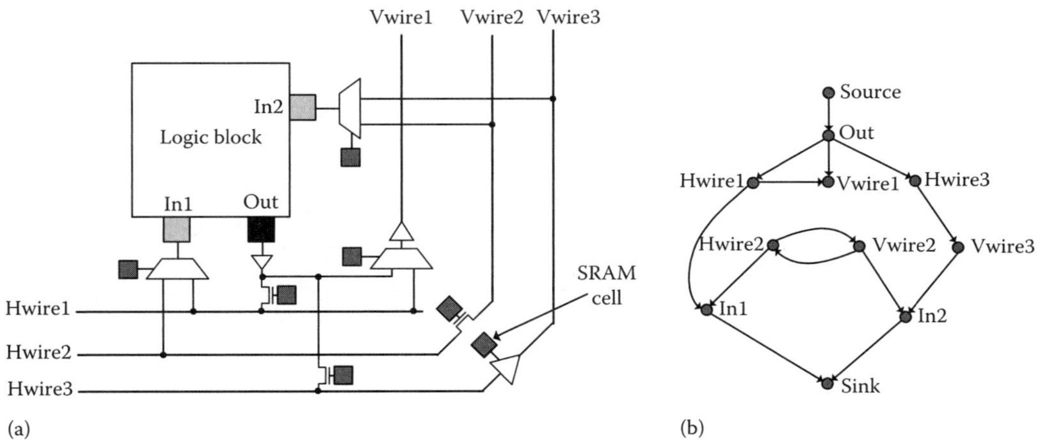

FIGURE 17.17 Transforming FPGA routing circuitry to a routing-resource graph. (a) Example FPGA routing circuitry and (b) equivalent routing-resource graph.

Since the number of routing wires in an FPGA is limited, and the limited number of programmable switches also creates many constraints on which wires can be connected to each other, congestion detection and avoidance are key features of FPGA routers. As well, since most delays in FPGAs are due to the programmable routing, timing-driven routing is important to obtain the best speed.

17.4.3.2 Two-Step Routing

Some FPGA routers operate in two sequential phases as shown in Figure 17.18. First, a global route for each net in the design is determined, using channeled global routing algorithms that are essentially the same as those for ASICs [111]. The output of this stage is the series of channel segments through which each connection should pass. Next a detailed router is invoked to determine exactly which wire segment should be used within each channel segment. The CGE [112] and SEGA [113] algorithms find detailed routes by employing different levels of effort in searching the routing graph. A search of only a few routing options is conducted first in order to quickly find detailed routes for nets

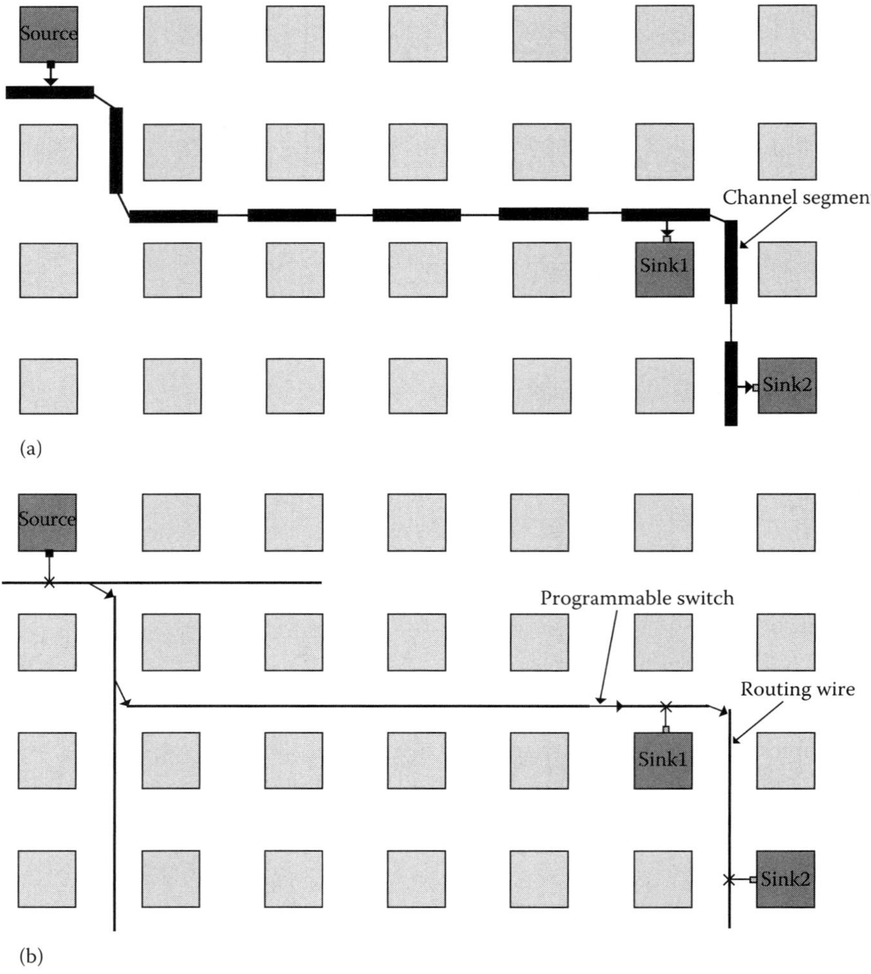

FIGURE 17.18 Two-step FPGA routing flow. (a) Step one: Global routing chooses a set of channel segments for a net. (b) Step two: Detailed routing wires within each channel segment and switches to connect them.

that are in uncongested regions, while a more exhaustive search is employed for nets experiencing routing difficulty. An alternative approach by Nam formulates the FPGA detailed routing problem as a Boolean satisfiability problem [114]. This approach guarantees that a legal detailed routing (that obeys the current global routing) will be found if one exists. The CPU time can be high for large problems, however. None of these detailed routers use timing analysis to determine timing-critical nets and optimize their delay. SEGA attempts to minimize the delay of all nets equally, while the other algorithms are purely routability-driven.

The divide-and-conquer approach of two-step routing reduces the problem space for both the global and detailed routers, helping to keep their CPU times down. However, the flexibility loss of dividing the routing problem into two phases in this way can result in significantly reduced result quality. The global router optimizes only the wirelength of each route, and attempts to control congestion by trying to keep the number of nets assigned to a channel segment comfortably below the number of routing wires in that channel segment. The fact that FPGA wiring is prefabricated and can be interconnected only in limited patterns makes the global router's view of both wirelength and congestion inaccurate, however. For example, a global route 1 logic block long may require the detailed router to use a wire that is 4 logic blocks long to actually complete the connection, wasting wire, and increasing delay. Figure 17.18 highlights this behavior; the global route requires 9 units of wire, but the final wires used in the detailed routing of the net are 13 wiring units long in total. Similarly, a global route where the number of nets assigned to each channel segment is well below the capacity of each segment may still fail detailed routing because the wiring patterns may not permit this pattern of global routes.

17.4.3.3 Single-Step Routers

Most modern FPGA routers are single-step routers that find routing paths through the routing-resource graph in a single, unified search algorithm. These routers differ primarily in their costing of various routing alternatives, their search technique through the routing-resource graph, and their congestion resolution techniques.

Lee and Wu introduced the tracer algorithm [115], which routes each net using a multiple-component growth algorithm that is an extension of a traditional maze router [116]. This multiple-component growth algorithm tries to find a minimum wirelength routing tree for each net. Multiple nets are allowed to use the same routing-resource node, but the cost of such an overused node is 10 times the normal cost. If some routing resource nodes are overused once all nets are routed, a simulated evolution rip-up and retry scheme is invoked. Nets that have routing lengths longer than the minimum, or that contain overused nodes, are more likely to be ripped up; however, every net has some chance of being ripped up and rerouted, since the routing of any net may indirectly cause routing congestion. Once a legal route is achieved, further rip-up and retry iterations are performed to improve timing. Part of the rip-up criteria considers connection slack—nets with either negative slack (which could be routed faster) or large positive slack (which could take a more circuitous route to free up resources for other nets) are more likely to be selected for rip-up.

The Limit Bumping Algorithm (LBA) of Frankle [117] considers timing in a more direct manner. First, a timing analysis is performed to determine a worst-case path slack for each connection. Next, a slack allocator is invoked to convert these path slacks into an upper delay limit, $U(c)$, on the permitted delay for each connection, c. The LBA ensures that each of these $U(c)$ is larger than a lower bound, $L(c)$, on the achievable routing delay for each connection given the placement and FPGA routing architecture, so there is some possibility of routing success. A routing in which each connection is routed with delay less than its upper delay limit will meet all maximum delay timing constraints on paths. The LBA routes connections in decreasing order of $L(c)/U(c)$, so connections with tighter delay limits compared to what is achievable go first. Each connection must be routed with delay less than $U(c)$ or it is left unrouted. Once an attempt has been made to route all connections, the $U(c)$ values of all unrouted connections are increased by 20%, and routing is retried.

The Pathfinder algorithm by McMurchie and Ebeling [108] introduced the concept of "negotiated congestion routing". The negotiated congestion technique now underlies many FPGA routers, including those with the best routability results on a set of standard academic FPGA benchmarks. In a negotiated congestion router, each connection is initially routed to minimize some metric, such as delay or wirelength, with little regard to congestion, or overuse, of routing resources. After each *routing iteration*, in which every net in the circuit is ripped-up and re-routed, the cost of congestion is increased such that it is less likely that overused nodes will occur in the next routing iteration. Over the course of many routing iterations, the increasing cost of congestion gradually forces some nets to accept sub-optimal routing in order to resolve congestion and achieve a legal routing.

The congestion cost of a node is:

$$\text{CongestionCost}(n) = [b(n) + h(n)] \cdot p(n) \tag{17.4}$$

where $b(n)$ is the base cost of the node, $p(n)$ is the present congestion of the node, and $h(n)$ is the historical cost of the node. The base cost of a node could be its intrinsic delay, its length, or simply 1 for all nodes. The present congestion cost of a node is a function of the overuse of the node, and the routing iteration. For nodes that are not currently overused, $p(n)$ is 1. In early routing iterations, $p(n)$ will be only slightly higher than 1 for nodes that are overused, while in later routing iterations, to ensure congestion is resolved, $p(n)$ becomes very large for overused nodes. $h(n)$ maintains a congestion history for each node. $h(n)$ is initially 0 for all nodes, but is increased by the amount of overuse on node n at the end of each routing iteration. The incorporation of not just the present congestion, but also the entire history of congestion of a node, into the cost of that node is a key innovation of negotiated congestion. Historical congestion ensures that nets that are "trapped" in a situation where all their routing choices have present congestion can see which choices have been overused the most in the past. Exploring the least historically congested choices ensures new portions of the solution space are being explored, and resolves many cases of congestion that the present congestion cost term alone cannot resolve.

In the PathFinder algorithm, the complete cost of using a routing resource node n in the routing of a connection c is:

$$\text{Cost}(n) = [1 - \text{Crit}(c)] \, \text{CongestionCost}(n) + \text{Crit}(c)\text{Delay}(n). \tag{17.5}$$

The criticality is the ratio of the connection slack to the longest delay in the circuit:

$$\text{Crit}(c) = \frac{\text{Slack}(c)}{D\,\text{max}}. \tag{17.6}$$

The total cost of a routing-resource node is therefore a weighted sum of its congestion cost and its delay, with the weighting being determined by the timing-criticality of the connection being routed. This formulation results in the most timing-critical connections receiving delay-optimized routes, with non-timing-critical connections using routes optimized for minimal wirelength and congestion. Since timing-critical connections see less cost from congestion, these connections are also less likely to be forced off their optimal routing paths due to congestion—instead, non-timing-critical connections will be moved out of the way of timing-critical connections.

The VPR router [8,79] is based on the Pathfinder algorithm, but it introduces several enhancements. First, instead of assuming that each routing-resource node has a constant delay, the VPR router models the delay of a route with the more accurate Elmore delay [118] and directly optimizes this delay. The original Elmore delay is only capable of modelling linear networks of resistors and capacitors, but by using linearized RC models of FPGA routing switches, it can be applied to FPGA routing and yields good accuracy.

Second, instead of using a breadth-first search, or an A* search, through the routing resource graph to determine good routes, VPR uses a more aggressive directed search technique. This directed search sorts each routing resource node, n, found during the graph search toward a sink, j, by a total cost given by

$$\text{TotalCost}(n) = \text{PathCost}(n) + \alpha \cdot \text{ExpectedCost}(n, j). \tag{17.7}$$

Here $\text{PathCost}(n)$ is the known cost of the routing path from the connection source to node n, while $\text{ExpectedCost}(n, j)$ is a prediction of the remaining cost that will be incurred in completing the route from node n to the target sink. The "directedness" of the search is controlled by α. An α of 0 results in a breadth-first search of the graph, while α larger than 1 makes the search more efficient, but may result in suboptimal routes. An α of 1.2 leads to improved CPU time without a noticeable reduction in result quality.

The VPR router also uses an alternative form of Equation 17.4 that more easily adapts to different FPGA architectures—see [8,79] for details, and for a discussion of how best to set $b(n)$, $h(n)$, and $p(n)$ to achieve good results in reasonable CPU times.

An FPGA router based on negotiated congestion, but designed for very low CPU times, is presented by Swartz et al. [119]. This router achieves very fast runtimes through the use of an aggressive directed search during routing graph exploration, and by using a "binning" technique to speed the routing of high-fanout nets. When routing the kth terminal of a net, most algorithms examine every routing-resource node used in routing the previous $k - 1$ terminals. For a k-terminal net this results in an $O(k^2)$ algorithm, which becomes slow for large k. By examining only the portion of the routing of the previous terminals that is in a "bin" near the sink for connection k, the algorithm achieves a significant CPU reduction.

Wilton developed a crosstalk-aware FPGA routing algorithm [120]. This algorithm enhances the VPR router by adding an additional term to the routing cost function that penalizes routes in proportion to the amount of delay they will add to neighboring routes due to crosstalk, weighted by the timing criticality of those neighboring routes. Hence this router achieves a circuit speed-up by leaving routing tracks near those used by critical connections vacant.

Lamoureaux and Wilton [82] enhanced the VPR router to optimize power by adding a term to the routing node cost, Equation 17.5, that includes the capacitance of a routing node multiplied by the switching activity of the net being routed. This drives the router to achieve low-energy routes for rapidly toggling nets.

The Routing Cost Valleys (RCV) algorithm [121] combines negotiated congestion with a new routing cost function and an enhanced slack allocation algorithm. RCV is the first FPGA routing algorithm that optimizes not only long-path timing constraints, which specify that the delay on a path must be less than some value, but also addresses the increasing importance of short-path timing constraints, which specify that the delay on a path must be greater than some value. Short-path timing constraints arise in FPGA designs as a consequence of hold-time constraints within the FPGA, or of system-level hold time constraints on FPGA input pins and system-level minimum clock-to-output constraints on FPGA output pins. To meet short-path timing constraints, RCV will intentionally use slow or circuitous routes to increase the delay of a connection.

RCV allocates both short-path and long-path slacks to determine a pair of delay budgets, $D_{\text{Budget,Min}}(c)$ and $D_{\text{Budget,Max}}(c)$, for each connection, c, in the circuit. A routing of the circuit in which every connection has delay between $D_{\text{Budget,Min}}(c)$ and $D_{\text{Budget,Max}}(c)$ will satisfy all the long-path and short-path timing constraints. Such a routing may not exist, however, so it is advantageous for connections to seek not simply to achieve a delay in the window between the two delay budgets, but instead to try to achieve a target delay, $D_{\text{Target}}(c)$, near the middle of the window. The extra timing margin achieved by this connection may allow another connection to have a delay outside its delay budget window, without violating any of the path-based timing constraints. Figure 17.19 shows the form of the RCV routing cost function compared to that of the original Pathfinder algorithm.

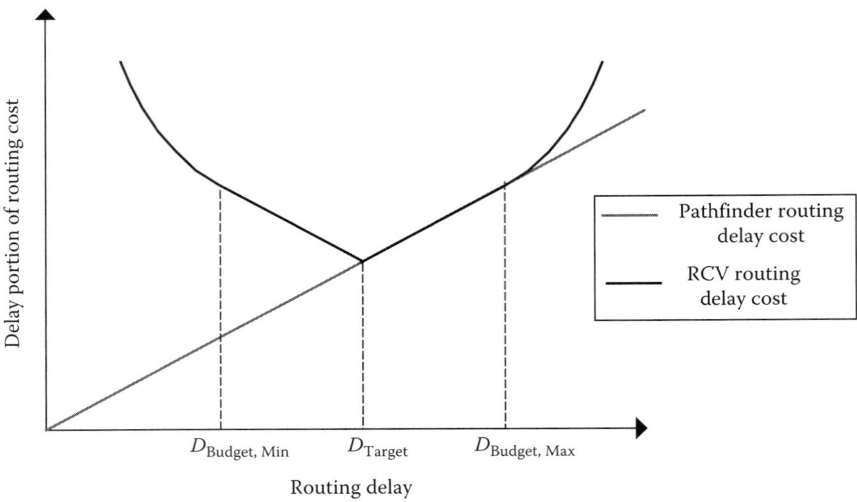

FIGURE 17.19   RCV routing delay cost compared to PathFinder routing delay cost.

The RCV algorithm strongly penalizes routes that have delays outside the delay budget window, and weakly guides routes to achieve D_{Target}. RCV achieves superior results on short-path timing constraints and also outperforms Pathfinder in optimizing traditional long-path timing constraints.

17.5 Looking Forward

In this chapter we have surveyed the current algorithms for FPGA synthesis, placement, and routing. Some of the more recent publications in this area point to the growth areas in CAD tools for FPGAs. The relatively few papers on power modeling and optimization algorithms are likely just the beginning as lower process geometries force tools to be more and more aware of power effects. Timing modeling for FPGA interconnects will need to take into account variation, min–max analysis, crosstalk, and other physical effects that have been largely ignored to date, and incorporate these into the optimization algorithms. Timing estimation and physical synthesis approaches will likely contribute to improved performance in the future. Finally, as more and more of the lower-end ASIC market migrates to FPGAs, the system-level tools to which these designers are accustomed will be required for FPGA flows.

Acknowledgments

Thanks to Babette van Antwerpen for her help with the sections on synthesis and technology mapping and to Paul Metzgen, Valavan Manoharajah, and Andrew Ling for providing several figures.

References

1. Rose, J., Francis, R., Lewis, D., and Chow, P. Architecture of programmable gate arrays: The effect of logic block functionality on area efficiency, *IEEE Journal of Solid State Circuits*, 25(5), 1217–1225, 1990.

2. Ahmed, E. and Rose, J. The effect of LUT and cluster size on deep-submicron FPGA performance and density, in *Proceedings of the 8th International Symposium on FPGAs*, Monterey, CA, pp. 3–12, 2000.

3. Betz, V. and Rose, J. How much logic should go in an FPGA logic block? *IEEE Design and Test*, Spring, 15(1), 10–15, 1998.

4. Rose, J. and Brown, S. Flexibility of interconnection structures for FPGAs, *JSSC*, 26(3), 277–282, 1992.

5. Lemieux, G. and Lewis, D. Circuit design of routing switches, in *Proceedings of the 10th International Symposium on FPGAs*, Monterey, CA, pp. 19–28, 2002.

6. Cong, J. and Hwang, Y. Boolean matching for LUT-based logic blocks with applications to architecture evaluation and technology mapping, *IEEE Transactions on CAD*, 20(9), 1077–1090, 2001.

7. Brown, S., Francis, R., Rose, J., and Vranesic, Z. *Field-Programmable Gate Arrays*, Kluwer, Norwell, MA, 1992.

8. Betz, V., Rose, J., and Marquardt, A. *Architecture and CAD for Deep-Submicron FPGAs*, Kluwer, Norwell, MA, February 1999.

9. Betz, V. and Rose, J. Automatic generation of FPGA routing architectures from high-level descriptions, in *Proceedings of the 7th International Symposium on FPGAs*, Monterey, CA, pp. 175–184, 2000.

10. Yan, A., Cheng, R., and Wilton, S. On the sensitivity of FPGA architectural conclusions to experimental assumptions, tools and techniques, in *Proceedings of the 10th International Symposium on FPGAs*, Monterey, CA, pp. 147–156, 2003.

11. Li, F., Chen, D., He, L., and Cong, J. Architecture evaluation for power-efficient FPGAs, in *Proceedings of the 11th International Symposium on FPGAs*, Monterey, CA, pp. 175–184, 2003.

12. Wilton, S. Heterogeneous technology mapping for area reduction in FPGAs with embedded memory arrays, *IEEE Transactions on CAD*, 19, 56–68, 2000.

13. Lin, A. and Wilton, S. Macrocell architecture for product term embedded memory arrays, in *Proceedings of the 11th International Conference on Field-Programmable Logic and Applications*, Belfast, Northern Ireland, 2001, pp. 48–58.

14. Lewis, D.M., Betz, V., Jefferson, D., Lee, A., Lane, C., Leventis, P., Marquardt, S., McClintock, C., Pedersen, B., Powell, G., Reddy, S., Wysocki, C., Cliff, R., and Rose, J. The Stratix routing and logic architecture, in *Proceedings of the 11th ACM International Symposium on FPGAs*, Monterey, CA, pp. 12–20, 2003.

15. Lewis, D., Ahmed, E., Baeckler, G., Betz, V., Bourgeault, M., Cashman, D., Galloway, D., Hutton, M., Lane, C., Lee, A., Leventis, P., Marquardt, S., McClintock, C., Padalia, K., Pedersen, B., Powell, G., Ratchev, B., Reddy, S., Schleicher, J., Stevens, K., Yuan, R., Cliff, R., and Rose, J. The Stratix II routing and logic architecture, in *Proceedings of the 13th ACM International Symposium on FPGAs*, Monterey, CA, pp. 14–20, 2005.

16. Trimberger, S., Duong, K., and Conn, B. Architecture issues and solutions for a high-capacity FPGA, in *Proceedings of the 5th ACM International Symposium on FPGAs*, Monterey, CA, pp. 3–9, 1997.

17. See www. < companyname > .com for commercial tools and architecture information.

18. Ahanin, B. and Vij, S.S. A high density, high-speed, array-based erasable programmable logic device with programmable speed/power optimization, in *Proceedings of the 1st ACM International Symposium on FPGAs*, Monterey, CA, pp. 29–32, 1992.

19. Hutton, M., Karchmer, D., Archell, B., and Govig, J. Efficient static timing analysis and applications using edge masks, in *Proceedings on the 13th International Symposium on FPGAs*, Monterey, CA, pp. 174–183, 2005.

20. Poon, K., Yan, A., and Wilton, S.J.E. A flexible power model for FPGAs, *ACM Transactions on Design Automation of Digital Systems*, 10(2), 279–302, April 2005.

21. De Micheli, G. *Synthesis and Optimization of Digital Circuits*, McGraw Hill, New York, 1994.

22. Cong, J. and Ding, Y. Combinational logic synthesis for LUT-based FPGAs, *ACM Transactions on Design Automation of Digital Systems*, 1(2), 145–204, 1996.

23. Murgai, R., Brayton, R., and Sangiovanni-Vincentelli, A. *Logic Synthesis for Field-Programmable Gate Arrays*, Kluwer, Norwell, MA, 2000.
24. Hwang, J., Milne, B., Shirazi, N., and Stroomer, J. System-level tools for DSP in FPGAs, in *Proceedings of the 14th Symposium Field-Programmable Logic (FPL)*, Belfast, Northern Ireland, 2001.
25. Stroomer, J., Ballagh, J., Ma, H., Milne, B., Hwang, J., and Shirazi, N. Creating system generator design using jg, in *Proceedings of the IEEE Symposium on Field-Programmable Custom Computing Machines (FCCM)*, Napa, CA, 2003.
26. Berkeley Design Technology Inc., Evaluating FPGAs for communication infrastructure applications, in *Proceedings on Communications Design Conference*, San Jose, CA, 2003.
27. Lockwood, J., Naufel, N., Turner, J., and Taylor, D. Reprogrammable network packet processing on the field-programamble port extender (FPX), in *Proceedings of the 9th International Symposium FPGAs*, Monterey, CA, pp. 87–93, 2001.
28. Kempa, J., Lim, S.Y., Robinson, C., and Seely, J. SOPC Builder: Performance by design, in *Winning the SOPC Revolution*, Martin G. and Chang H. (Eds.), Springer 2003, Chapter 8.
29. Hutchings, B., Bellows, P., Hawkins, J., Hemmert, S. Nelson, B., and Rytting, M. A CAD suite for high-performance FPGA design, in *Proceedings of the 9th International Workshop on Field-Programmable Logic*, 1999.
30. Brandolese, C., Fornaciari, W., and Salice, F. An area estimation methodology for FPGA-based designs at systemC-level, in *Proceedings of the Design Automation Conference* pp. 129–132, 2004.
31. Hutton, M., Schleicher, J., Lewis, D., Pedersen, B., Yuan, R., Kaptanoglu, S., Baeckler, G., Ratchev, B., Padalia, K., Bourgeault, M., Lee, A., Kim, H., and Saini, R. Improving FPGA performance and area using an adaptive logic module, in *Proceedings of the 14th International Symposium Field-Programmable Logic*, Leuven, Belgium, pp. 134–144, 2004.
32. Metzgen, P. and Nancekievill, D. Multiplexor restructuring for FPGA implementation cost reduction, in *Proceedings of the Design Automation Conference*, Anaheim, CA, 2005.
33. Nancekievill, D. and Metzgen, P. Factorizing multiplexors in the datapath to reduce cost in FPGAs, in *Proceedings of the International Workshop on Logic Synthesis*, Lake Arrowhead, CA, 2005.
34. Sentovich, E.M., Singh, K.J., Lavagno, L., Moon, C., Murgai, R., Saldanha, A., Savoj, H., Stephan, P.R., Brayton, R.K., and Sangiovanni-Vincentelli, A. SIS: A System for Sequential Circuit Analysis, Tech Report No. UCB/ERL M92/41, UC Berkeley, CA, 1992.
35. Shenoy, N. and Rudell, R. Efficient implementation of retiming, in *Proceedings of the International Conference on CAD (ICCAD)*, San Jose, CA, pp. 226–233, 1994.
36. van Antwerpen, B., Hutton, M., Baeckler, G., and Yuan, R. A safe and complete gate-level register retiming algorithm, in *Proceedings of the IWLS*, Laguna Beach, CA, 2003.
37. Vemuri, N., Kalla, P., and Tessier, R. BDD-based logic synthesis for LUT-based FPGAs, *ACM Transactions on the Design Automation of Electronic Systems*, 7(2), 2002.
38. Yang, C., Ciesielski, M., and Singhal, V. BDS A BDD-based logic optimization system, in *Proceedings of the Design Automation Conference*, San Francisco, CA, pp. 92–97, 2000.
39. Lai, Y., Pedram, M., and Vrudhala, S. BDD-based decomposition of logic Functions with application to FPGA synthesis, in *Proceedings of the Design Automation Conference*, Monterey, CA, pp. 448–451, 1992.
40. Anderson, J., Najm, F., and Tuan, T. Active leakage power estimation for FPGAs, in *Proceedings of the 12th International Symposium on FPGAs*, Monterey, CA, pp. 33–41, 2004.
41. Ling, A., Singh, D.P., and Brown, S.D. FPGA technology mapping: A study of optimality, in *Proceedings of the Design Automation Conference*, Anaheim, CA, 2005.
42. Keutzer, K. DAGON: Technology binding and local optimization by DAG matching, in *Proceedings of the 24th Design Automation Conference*, Miami Beach, FL, pp. 341–347, 1987.
43. Francis, R.J., Rose, J., and Chung, K. Chortle: A technology mapping program for lookup table-based field-programmable gate arrays, in *Proceedings of the Design Automation Conference*, Orlando, FL, pp. 613–619, 1990.

44. Cong, J. and Ding, E. An optimal technology mapping algorithm for delay optimization in lookup table based FPGA designs, *IEEE Transactions on CAD*, 13(1), 1–12, 1994.

45. Cong, J. and J. Peck, J. RASP: A general logic synthesis system for SRAM-based FPGAs, in *Proceedings of the 5th International Symposium on FPGAs*, Monterey, CA, pp. 137–143, 1996.

46. Cong, J. and Ding, Y. On area/depth trade-off in LUT-based FPGA technology mapping, *IEEE Transactions on VLSI*, 2(2), 137–148, 1994.

47. Cong, J., Wu, C., and Ding, Y. Cut ranking and pruning: Enabling a general and efficient FPGA mapping solution, in *Proceedings of the 9th International Symposium FPGAs*, Monterey, CA, pp. 29–35, 1999.

48. Francis, R.J., Rose, J., and Vranesic, Z. Chortle-crf: Fast technology mapping for lookup table-based FPGAs, in *Proceedings of the Design Automation Conference*, San Francisco, CA, pp. 227–233, 1991.

49. Francis, R.J., Rose, J., and Vranesic, Z. Technology mapping of lookup table-based FPGAs for performance, in *Proceedings of the International Conference on CAD (ICCAD)*, San Jose, CA, 1991.

50. Lawler, E.L., Levitt, K.N., and Turner, J. Module clustering to minimize delay in digital networks, *IEEE Transactions on Computers*, C-18, 47–57, 1969.

51. Tarjan, R.E., *Data Structures and Network Algorithms*, SIAM, Philadelphia, PA, 1983.

52. Farrahi, A. and Sarrafzadeh, M. Complexity of the lookup-table minimization problem for FPGA technology mapping, *IEEE Transactions on CAD* 13(11), 1319–1332, 1994.

53. Cong, J., Ding, Y., Gao, T., and Chen, K.C. An optimal performance-driven technology mapping algorithm for LUT-based FPGAs under arbitrary net-delay models, in *Proceedings of the International Conference on CAD (ICCAD)s*, pp. 599–603, 1993.

54. Cong, J., and Hwang, Y. Simultaneous depth and area minimization in LUT-based FPGA mapping, in *Proceedings of the 4th International Symposium on FPGAs*, Monterey, CA, pp. 68–74, 1995.

55. Manohararajah, V., Brown, S., and Vranesic, Z. Heuristics for area minimization in LUT-based FPGA technology mapping, in *Proceedings of the International Workshop on Logic Synthesis*, Temecula, CA, pp. 14–21, 2004.

56. Yang, H. and Wong, D.F., Edge-map: Optimal performance driven technology mapping for iterative LUT based FPGA designs, in *Proceedings of the IEEE International Conference on CAD (ICCAD)*, San Jose, CA, pp. 150–155, 1994.

57. Chen, D. and Cong, D. DAOmap: A depth-optimal area optimization mapping algorithm for FPGA designs, in *Proceedings of the International Conference on CAD (ICCAD)*, November 2004.

58. Pan, P. Performance-driven integration of retiming and resynthesis, in *Proceedings of the Design Automation Conference*, pp. 243–246, 1999.

59. Pan., P. and Lin, C.-C., A new retiming-based technology mapping algorithm for LUT-based FPGAs, in *Proceedings of the 6th International Symposium on FPGAs*, pp. 35–42, 1998.

60. Farrahi, A.H. and Sarrafzadeh, M. FPGA technology mapping for power minimization, in *Proceedings of the International Workshop on Field-Programmable Logic and Applications*, 1994.

61. Anderson, J. and Najm, F.N., Power-aware technology mapping for LUT-based FPGAs, in *Proceedings of the International Conference on Field-Programmable Technology*, 2002.

62. Lamoreux, J. and Wilton, S.J.E., On the interaction between power-aware CAD algorithms for FPGAs, in *Proceedings of the International Conference on CAD (ICCAD)*, 2003.

63. Chen, D., Cong, J., Li, F., and He, L. Low-power technology mapping for FPGA architectures with dual supply voltages, in *Proceedings of the 12th International Symposium on FPGAs*, pp. 109–117, 2004.

64. Cong, J. and Hwang, Y. Structural gate decomposition for depth-optimal technology mapping in LUT-based FPGA design, in *Proceedings of the Design Automation Conference*, Las Vegas, NV, pp. 726–729, 1996.

65. Wang, A. Algorithms for multilevel logic optimization, PhD Dissertation, Computer Science Department, UC Berkeley, CA, 1991.

66. Cong, J. and Hwang, Y. Structural gate decomposition for depth-optimal technology mapping in LUT-based FPGA designs, in *ACM Transactions on Design Automation of Digital Systems*, 5(2), 193–225, 2000.
67. Legl, C., Wurth, B., and Eckl, K. A Boolean approach to performance-directed technology mapping for LUT-based FPGA designs, in *Proceedings of the Design Automation Conference*, Las Vegas, NV, 1996.
68. Cong, J. and Hwang, Y. Partially-dependent functional decomposition with applications in FPGA synthesis and mapping, in *Proceedings of the 6th International Symposium on FPGAs*, Monterey, CA, pp. 35–42, 1997.
69. Cong, J. and Xu, S. Performance-driven technology mapping for heterogeneous FPGSs, *IEEE Transactions on CAD*, 19(11), 1268–1281, 2000.
70. Yamashita, S., Sawada, H., and Nagoya, A. A new method to express functional permissibilities for LUT-based FPGAs and its applications, in *Proceedings of the International Conference on CAD (ICCAD)*, San Jose, CA, pp. 254–261, 1996.
71. Cong, J., Lin, Y., and Long, W. SPFD-based global re-wiring, in *Proceedings of the 10th International Symposium on FPGAs*, Monterey, CA, pp. 77–84, 2002.
72. Cong, J., Lin, Y., and Long, W. A new enhanced SPFD rewiring algorithm, in *Proceedings on the International Conference on CAD (ICCAD)*, San Jose, CA, pp. 672–678, 2002.
73. Hwang, J.M., Chiang, F.Y., and Hwang, T.T. A re-engineering approach to low power FPGA design using SPFD, in *Proceedings of the 35th ACM/IEEE Design Automation Conference*, San Francisco, CA, pp. 722–725, 1998.
74. Kumthekar, B. and Somenzi, F. Power and delay reduction via simultaneous logic and placement optimization in FPGAs, in *Proceedings of the Design and Test in Europe (DATE)*, Paris, France, pp. 202–207, 2000.
75. Anderson, J., Saunders, J., Nag, S., Madabhushi, C., and Jayarman, R. A placement algorithm for FPGA designs with multiple I/O standards, in *Proceedings of the International Conference on Field Programmable Logic and Applications*, Villach, Austria, pp. 211–220, 2000.
76. Mak, W., I/O Placement for FPGA with multiple I/O standards, in *Proceedings of the 11th International Symposium on FPGAs*, Monterey, CA, pp. 51–57, 2003.
77. Anderson, J., Nag, S., Chaudhary, K., Kalman, S., Madabhushi, C., and Cheng, P. Run-time conscious automatic timing-driven FPGA layout synthesis, in *Proceedings of the 14th International Conference on Field-Programmable Logic and Applications*, Leuven, Belgium, pp. 168–178, 2004.
78. Marquardt, A., Betz, V., and Rose, J. Using cluster-based logic blocks and timing-driven packing to improve FPGA speed and density, in *Proceedings of the 7th International Symposium on FPGAs*, Monterey, CA, pp. 37–46, 1999.
79. Betz, V. and Rose, J. VPR: A new packing, placement and routing tool for FPGA research, in *Proceedings of the 7th International Conference on Field-Programmable Logic and Applications*, London, UK, pp. 213–222, 1997.
80. Marquardt, A., Betz, V., and Rose, J. Timing-driven placement for FPGAs, in *Proceedings on the International Symposium on FPGAs*, Monterey, CA, pp. 203–213, 2000.
81. Singh, A. and Marek-Sadowska, M. Efficient circuit clustering for area and power reduction in FPGAs, in *Proceedings of the International Symposium on FPGAs*, Monterey, CA, pp. 59–66, 2002.
82. Lamoureaux, J. and Wilton, S. On the interaction between power-aware FPGA CAD algorithms, in *Proceedings of the International Symposium on CAD (ICCAD)*, San Jose, CA, pp. 701–708, 2003.
83. Chen, D. and Cong, J. Delay optimal low-power circuit clustering for FPGAs with dual supply voltages, in *Proceedings on the International Symposium on Low Power Electronics and Design (ISLPED)*, Newport Beach, CA, pp. 70–73, 2004.
84. Kuthekar, B. and Somenzi, F. Power and delay reduction via simultaneous logic and placement optimization in FPGAs, in *Proceedings of the Design and Test in Europe (DATE)*, Paris, France, pp. 202–207, 2000.

85. Cong, J. and Romesis, M. Performance-driven multi-level clustering with application to hierarchical FPGA mapping, in *Proceedings of the Design Automation Conference*, San Francisco, CA, pp. 389–394, 2001.

86. Chen, G. and Cong, J. Simultaneous timing driven clustering and placement for FPGAs, in *Proceedings of the International Conference on Field Programmable Logic and Applications*, Leuven, Belgium, pp. 158–167, 2004.

87. Kong, T. A novel net weighting algorithm for timing-driven placement, in *Proceedings of the International Conference on CAD (ICCAD)*, San Jose, CA, pp. 172–76, 2002.

88. Nag, S.K. and Rutenbar, R.A. Performance-driven simultaneous placement and routing for FPGAs, *IEEE Transactions on CAD*, 17(6), 499–518, 1998.

89. Sankar, Y. and Rose, J. Trading quality for compile time: Ultra-fast placement for FPGAs, in *Proceedings of the International Symposium on FPGAs*, Monterey, CA, pp. 157–166, 1999.

90. Rose, J., Snelgrove, W., and Vranesic, Z. ALTOR: An automatic standard cell layout program, in *Proceedings of the Canadian Conference on VLSI*, Toronto, Canada, pp. 169–173, 1985.

91. Dunlop, A. and Kernighan, B. A procedure for placement of standard-cell VLSI circuits, *IEEE Transactions on CAD*, 4(1), 92–98, 1985.

92. Maidee, M., Ababei, C., and Bazargan, K. Fast timing-driven partitioning-based placement for island style FPGAs, in *Proceedings of the Design Automation Conference (DAC)*, San Francisco, CA, pp. 598–603, 2003.

93. Hutton, M., Adibsamii, K., and Leaver, A. Adaptive delay estimation for partitioning-driven PLD placement, *IEEE Transactions on VLSI*, 11(1), 60–63, 2003.

94. Chan, P. and Schlag, M. Parallel placement for field-programmable gate arrays, in *Proceedings of the 11th International Symposium on FPGAs*, pp. 43–50, 2003.

95. Kleinhans, J., Sigl, G., Johannes, F., and Antreich, K. Gordian: VLSI placement by quadratic programming and slicing optimization, *IEEE Transactions on CAD*, 10(3), 356–365, 1991.

96. Lin., J., Jagannathan, A., and Cong, J. Placement-driven technology mapping for LUT-based FPGAs, in *Proceedings of the 11th International Symposium on FPGAs*, Monterey, CA, pp. 121–126, 2003.

97. Suaris, P., Wang, D., and Chou, N. Smart move: A placement-aware retiming and replication method for field-programmable gate arrays, in *Proceedings of the 5th International Conference on ASICs*, Beijing, China, 2003.

98. Suaris, P., Liu, L., Ding, Y., and Chou, N. Incremental physical re-synthesis for timing optimization, in *Proceedings of the 12th International Symposium on FPGAs*, Monterey, CA, pp. 99–108, 2004.

99. Schabas, K. and Brown, S. Using logic duplication to improve performance in FPGAs, in *Proceedings of the 11th International Symposium on FPGAs*, Monterey, CA, pp. 136–142, 2003.

100. Chen, G. and Cong, J. Simultaneous timing-driven placement and duplication, in *Proceedings of the 13th International Symposium on FPGAs*, Monterey, CA, pp. 51–61, 2005.

101. Manohararajah, V., Singh, D, Brown, S., and Vranesic, Z. Post-placement functional decomposition for FPGAs, in *Proceedings of the International Workshop on Logic Synthesis*, Temecula, CA, pp. 114–118, 2004.

102. Manohararajah, V., Singh, D.P., and Brown, S. Timing-driven functional decomposition for FPGAs, in *Proceedings of the International Workshop on Logic and Synthesis*, Lake Arrowhead, CA, 2005.

103. Ding, Y., Suaris, P., and Chou, N. The effect of post-layout pin permutation on timing, in *Proceedings of the 13th International Symposium on FPGAs*, Monterey, CA, pp. 41–50, 2005.

104. Singh, D. and Brown, S. Constrained clock shifting for field programmable gate arrays, in *Proceedings of the 10th International Symposium on FPGAs*, Monterey, CA, pp. 121–126, 2002.

105. Chao-Yang, Y. and Marek-Sadowska, M. Skew-programmable clock design for FPGA and skew-aware placement, in *Proceedings of the International Symposium on FPGAs*, Monterey, CA, pp. 33–40, 2005.

106. Singh, D. and Brown, S. Integrated retiming and placement for FPGAs, in *Proceedings of the 10th International Symposium on FPGAs*, Monterey, CA, pp. 67–76, 2002.
107. Singh, D.P., Manohararajah, V., and Brown, S.D. Incremental retiming for FPGA physical synthesis, in *Proceedings of the Design Automation Conference*, San Francisco, CA, 2005.
108. McMurchie, L. and Ebeling, C. PathFinder: A negotiation-based performance-driven router for FPGAs, in *Proceedings of the 5th International Symposium on FPGAs*, Monterey, CA, pp. 111–117, 1995.
109. Cong, J., Romesis, M., and Xie, M. Optimality and stability study of timing-driven placement algorithms, in *Proceedings of the International Conference on CAD (ICCAD)*, San Jose, CA, pp. 472–478, 2003.
110. Lemieux, G. and Lewis, D. *Design of Interconnection Networks for Programmable Logic*, Kluwer, Norwell, MA, 2004.
111. Rose, J. Parallel global routing for standard cells, *IEEE Transactions on CAD*, 9(10), 1085–1095, 1990.
112. Brown, S., Rose, J., and Vranesic, Z. A detailed router for field-programmable gate arrays, *IEEE Transactions on CAD*, 10(5), 620–628, May 1992.
113. Lemieux, G. and Brown, S. A detailed router for allocating wire segments in FPGAs, in *Proceedings of the Physical Design Workshop*, Lake Arrowhead, CA, pp. 215–226, 1993.
114. Nam, G.-J., Aloul, F., Sakallah, K., and Rutenbar, R. A comparative study of two Boolean formulations of FPGA detailed routing constraints, in *Proceedings of the International Symposium on Physical Design*, Sonoma, CA, pp. 222–227, 2001.
115. Lee, Y.S. and Wu, A performance and routability-driven router for FPGAs considering path delays, in *Proceedings of the Design Automation Conference*, San Francisco, CA, pp. 557–561, 1995.
116. Lee, C.Y. An algorithm for path connections and applications, *IRE Transactions on Electronic Computers*, EC-10, 346–365, 1961.
117. Frankle, J. Iterative and adaptive slack allocation for performance-driven layout and FPGA routing, in *Proceedings of the Design Automation Conference*, Anaheim, CA, pp. 536–542, 1992.
118. Elmore, W. The transient response of damped linear networks with particular regard to wideband amplifiers, *Journal of Applied Physics*, 19(1), 55–63, 1948.
119. Swartz, J., Betz, V., and Rose, J. A fast routability-driven router for FPGAs, in *Proceedings of the 6th International Symposium on FPGAs*, Monterey, CA, pp. 140–151, 1998.
120. Wilton, S. A crosstalk-aware timing-driven router for FPGAs, in *Proceedings of the 9th ACM International Symposium on FPGAs*, Monterey, CA, pp. 21–28, 2001.
121. Fung, R, Betz, V., and Chow, W. Simultaneous short-path and long-path timing optimization for FPGAs, in *Proceedings of the International Conference on CAD (ICCAD)*, San Jose, CA, 2004.

III

Embedded Systems Security and Web Services

18

Design Issues in Secure Embedded Systems

Anastasios G. Fragopoulos
University of Patras

Dimitrios N. Serpanos
University of Patras

Artemios G. Voyiatzis
University of Patras

18.1 Introduction

A computing system is typically considered as an embedded system when it is a programmable device with limited resources (energy, memory, computation power, etc.) that serves one (or few) applications and is embedded in a larger system. Its limited resources make it ineffective to be used as a general-purpose computing system. However, they usually have to meet hard requirements, such as time deadlines and other real-time processing requirements.

Embedded systems can be classified in two general categories: (1) stand-alone embedded systems, where all hardware and software components of the system are physically close, incorporated into a single device, for example, a personal digital assistant (PDA) or a system in a washing machine or a fax, and there is no attachment to a network and (2) distributed (networked) embedded systems, where several autonomous components—each one a stand-alone embedded system—communicate with each other over a network in order to deliver services or support an application. Several architectural and design parameters lead to the development of distributed embedded applications, such as the placement of processing power at the physical point where an event takes place, data reduction, etc. [1].

The increasing capabilities of embedded systems combined with their decreasing cost have enabled their adoption in a wide range of applications and services, from financial and personalized entertainment services to automotive and military applications in the field. Importantly, in addition to the typical requirements for responsiveness, reliability, availability, robustness, and extensibility, many conventional embedded systems and applications have significant security requirements. However, security is a resource-demanding function that needs special attention in embedded computing. Furthermore, the wide deployment of small devices which are used in critical applications has triggered the development of new, strong attacks that exploit more systemic characteristics, in contrast to traditional attacks that focused on algorithmic characteristics, due to the inability of attackers to experiment with the physical devices used in secure applications. Thus, design of secure embedded systems requires special attention.

In this chapter, we provide an overview of security issues in embedded systems. Section 18.2 presents the parameters of security systems, while Section 18.3 describes the effect of security in the resource-constrained environment of embedded systems. Section 18.4 presents the main issues in the design of secure embedded systems. Finally, Section 18.5 covers in detail attacks and countermeasures of cryptographic algorithm implementations in embedded systems, considering the critical role of cryptography and the novel systemic attacks developed due to the wide availability of embedded computing systems.

18.2 Security Parameters

Security is a generic term used to indicate several different requirements in computing systems. Depending on the system and its use, several security properties may be satisfied in each system and in each operational environment. Overall, secure systems need to meet all or a subset of the following requirements [2,3]:

1. Confidentiality: Data stored in the system or transmitted have to be protected from disclosure; this is usually achieved through data encryption.
2. Integrity: A mechanism to ensure that data received in a data communication was indeed the data transmitted.
3. Nonrepudiation: A mechanism to ensure that all entities (systems or applications) participating in a transaction do not deny their actions in the transaction.
4. Availability: The system's ability to perform its primary functions and serve its legitimate users without any disruption, under all conditions, including possible malicious attacks that target to disrupt service, such as the well-known denial-of-service (DoS) attacks.
5. Authentication: The ability of a receiver of a message to identify the message sender.
6. Access control: The ability to ensure that only legal users may take part in a transaction and have access to system resources. To be effective, access control is typically used in conjunction with authentication.

These requirements are placed by different parties involved in the development and use of computing systems, for example, vendors, application providers, and users. For example, vendors need to ensure the protection of their intellectual property (IP) that is embedded in the system, while end users want to be certain that the system will provide secure user identification (only authorized users may access the system and its applications, even if the system gets in the hands of malicious users) and will have high availability, i.e., the system will be available under all circumstances; also, content providers are concerned for the protection of their IP, for example, that the data delivered through an application are not copied. Kocher et al. [3,6] have identified the participating parties in system and application development and use as well as their security requirements. This classification enables us to identify several possible malicious users, depending on a party's view; for example, for the hardware manufacturer, even a legal end user of a portable device (e.g., a PDA or a mobile phone) can be a possible malicious user.

Considering the security requirements and the interested parties above, the design of a secure system requires identification and definition of the following parameters: (1) the abilities of the attackers, (2) the level at which security should be implemented, and (3) implementation technology and operational environment.

18.2.1 Abilities of Attackers

Malicious users can be classified in several categories depending on their knowledge, equipment, etc. Abraham et al. [15] propose a classification in three categories, depending on their knowledge, their hardware–software equipment, and their funds:

1. *Class I—clever outsiders:* very intelligent attackers, not well funded and with no sophisticated equipment. They do not have specific knowledge of the attacked system; basically, they are trying to exploit hardware vulnerabilities and software glitches.

2. *Class II—knowledgeable insiders:* attackers with outstanding technical background and education, using highly sophisticated equipment and, often, with inside information for the system under attack; such attackers include former employees who participated in the development cycle of the system.

3. *Class III—funded organizations:* attackers who are mostly working in teams, and have excellent technical skills and theoretical background. They are well funded, have access to very advanced tools and also have the capability to analyze the system—technically and theoretically—developing highly sophisticated attacks. Such organizations could be well-organized educational foundations, government institutions, etc.

18.2.2 Security Implementation Levels

Security can be implemented at various system levels, ranging from protection of the physical system itself to application and network security. Clearly, different mechanisms and implementation technologies have to be used to implement security at different levels. In general, the levels of security considered are four: (1) physical, (2) hardware, (3) software, and (4) network and protocol security.

Physical security mechanisms target to protect systems from unauthorized physical access to the system itself. Protecting systems physically ensures data privacy and data and application integrity. According to the U.S. Federal Standard 1027, physical security mechanisms are considered successful when they ensure that a possible attack will have low possibility of success and high possibility of tracking the malicious attacker, in reasonable time [149]. The wide adoption of embedded computing systems in a variety of devices, such as smart cards, mobile devices, and sensor networks, as well as the ability to network them, for example, through the Internet or virtual private networks (VPNs), has led to revision and reconsideration of physical security. Weingart [8] surveys possible attacks and countermeasures concerning physical security issues, concluding that physical security needs continuous improvement and revision in order to keep at the leading edge. Lemke [150] investigated and classified the physical security requirements of cryptographic embedded systems that are used in the automotive industry. Those systems that perform critical functionalities reside in a "physically protected" cryptographic boundary, which needs to be protected from possible malicious users. Three directions concerning physical security are identified: (1) tamper evidence, i.e., evidence must be provided to a third party that the module has been attacked; (2) tamper response, i.e., the module must have mechanisms to detect any intrusion and act accordingly, for example, by destroying any temporary data that reside in the module's memory; and (3) tamper resistance, which can be achieved through special architecture design, identification of critical operating conditions, which can eliminate the possibility of fault attacks, and by prevention of noninvasive attacks like monitoring attacks and other side-channel attacks (SCAs).

Hardware security may be considered as a subset of physical security, referring to security issues concerning the hardware parts of a computer system. Hardware-level attacks exploit circuit and technological vulnerabilities and take advantage of possible hardware defects. These attacks do not necessarily require very sophisticated and expensive equipment. Anderson and Kuhn [9] describe several ways to attack smart cards and microcontrollers through use of unusual voltages and temperatures that affect the behavior of specific hardware parts or through microprobing a smart card chip, like the SIM (Subscriber Identity Module) chip found in cellular phones. Reverse engineering attack techniques are equally successful as Blythe et al. [10] reported for the case of a wide range of microprocessors. Their work concluded that special hardware protection mechanisms are necessary to avoid such types of attack; such mechanisms include silicon coatings of the chip, increased complexity in the chip layout, etc. Ravi et al. [151] provided a taxonomy and classification of the possible attacks to secure embedded systems; two levels of attacks are identified, one referring to privacy, integrity, and availability attacks and a second one, referring to physical, side-channel, and software attacks. They have also surveyed the tamper-resistant design techniques that may be employed as countermeasures to any of these attacks.

One of the major goals in the design of secure systems is the development of secure software, which is free of flaws and security vulnerabilities that may appear under certain conditions. We may categorize the different types of software attacks into three main classes: (1) integrity attacks, where there is alteration in portions of software code by malicious users embedding their own code; (2) privacy attacks, where there is disclosure and break of confidentiality; and (3) availability attacks, where the primary functions of the system's software are not available. On the other hand, bad software design and code implementation may lead to breaches; numerous software security flaws have been identified in real systems, for example, by Landwehr et al. [12], and there have been several cases where malicious intruders hack into systems through exploitation of software defects [68]. Some methods for the prevention of such problems have been proposed by Tevis et al. [11]. In order to build secure software, especially for embedded systems, system designers have to consider some critical issues, for example, to identify correctness of software code before (during system boot) and during execution time, to use and deploy tamper-resistant software mechanisms, to use trusted parts in which software resides, i.e., secure memory, and to use techniques that may prevent integrity attacks through malicious programs, like viruses and Trojan horses, [161–164]. Arora et al. [152] stress the problem of secure program execution and proposed a hardware-based architecture that monitors and safeguards the execution of software on embedded processors. Seshadri et al. [153] address the pervasiveness of embedded systems that execute software and manipulate data, mostly critical, and propose a technique that allows the verification of memory contents of an embedded device without the necessity of physical access to the device memory.

The use of the Internet, which is an unsafe interconnection for information transfer, as a backbone network for communicating entities and the wide deployment of wireless networks demonstrate that improvements have to be done in existing protocol architectures in order to provide new, secure protocols [13,14]. Different types of protocols that belong to different layers of the open systems interconnection (OSI) stack are used for secure communication, for example, wired equivalent privacy (WEP)/WPA/TKIP [154] at the data link layer, IPSec [14] at the network layer, secure sockets layer (SSL)/TLS/wireless transport layer security (WTLS) [13,135,155] at the transport layer, and SET [156] at the application layer. Such protocols will ensure authentication between communicating entities, integrity and confidentiality of communicated data, protection of the communicating parties, and nonrepudiation (the inability of an entity to deny its participation in a communication transaction). Furthermore, special attention has to be paid in the design of secure protocols for embedded systems, due to their physical constraints, i.e., limited battery power, limited processing, and memory resources as well as their cost and communication requirements.

18.2.3 Implementation Technology and Operational Environment

In regard to implementation technology, systems can be classified by static versus programmable technology and fixed versus extensible architecture. When static technology is used, the hardware-implemented functions are fixed and inflexible, but they offer higher performance and can reduce cost. However, static systems can be more vulnerable to attacks, because, once a flaw is identified—for example, in the design of the system—it is impossible to patch already deployed systems, especially in the case of large installations, like SIM cards for cellular telephony or pay-per-view TV. Static systems should be implemented only once and correctly, which is an unattainable expectation in computing.

In contrast, programmable systems are not limited as static ones, but they can be proven flexible in the hands of an attacker as well; system flexibility may allow an attacker to manipulate the system in ways not expected or defined by the designer. Programmability is typically achieved through use of specialized software over a general-purpose processor or hardware.

Fixed architectures are composed of specific hardware components that cannot be altered. Typically, it is almost impossible to add functionality in later stages, but they have lower cost of implementation and are, in general, less vulnerable because they offer limited choices to attackers. An extensible architecture is like a general-purpose processor, capable to interface with several peripherals through standardized connections. Peripherals can be changed or upgraded easily to increase security or to provide new functionality. However, an attacker can connect malicious peripherals or interface the system in untested or unexpected cases. As testing is more difficult relatively to static systems, one cannot be too confident that the system operates correctly under every possible input.

Field programmable gate arrays (FPGAs) combine benefits of all types of systems and architectures because they combine hardware implementation performance and programmability, enabling system reconfiguration. They are widely used to implement cryptographic primitives in various systems. Thus, significant attention has to be paid to the security of FPGAs as independent systems. There exist research efforts addressing this issue where systematic approaches are developed and open problems in FPGA security are addressed; for example, Wollinger et al. [7] provide such an approach and address several open problems, including resistance under physical attacks.

18.3 Security Constraints in Embedded Systems Design

The design of secure systems requires special considerations, because security functions are resource-demanding, especially in terms of processing power and energy consumption. The limited resources of embedded systems require novel design approaches in order to deal with trade-offs between efficiency, speed and cost, and effectiveness—satisfaction of the functional and operational requirements. In our effort to identify and distinguish the security constraints in embedded systems design, at first, we have to define the security concerns that a system designer needs to consider. For example, when designing a mobile device which is composed of different embedded systems, one has to investigate issues for user identification and authentication, secure storage of sensitive data, secure communication with external interfaces and other networks, use of tamper-resistant modules, content protection and data confidentiality, and secure software execution environments. Raghunathan et al. [22] illustrate these issues through the example of secure data communications of portable wireless devices. Furthermore, they provide a detailed description of the challenges to implement security in a mobile device, identifying some challenges; specifically they identify (1) energy considerations and battery life, (2) processing power of security computations, and (3) system flexibility and adaptability. Additionally, one must also consider the cost to implement security in such environments and possible trade-offs. In the following sections, we analyze the previously mentioned design challenges.

18.3.1 Energy Considerations

Embedded systems are often battery-powered, i.e., they are power-constrained. Battery capacity constitutes a major bottleneck to processing for security on embedded systems. Unfortunately, improvements in battery capacity do not follow the improvements of increasing performance, complexity, and functionality of the systems they power. Gurun et al. [159–161] investigate ways to model, predict, and reduce energy consumption in highly constrained embedded devices, through the utilization of software techniques and power aware systems. Basically, they focus and extend two methods: (1) computation offloading, i.e., remote execution of processing-hungry application parts at high-end general-purpose computers which are located near the embedded devices and (2) dynamic-voltage scaling, which is based on the alteration of voltage and clock frequency of the embedded device during run-time execution of programs. Gunther et al. [16], Buchmann [17], and Lahiri et al. [18] report the widening "battery gap," due to the exponential growth of power requirements and the linear growth in energy density. Thus, the power subsystem of embedded systems is a weak point of system security. A malicious attacker, for example, may form a DoS attack by draining the system's battery more quickly than usual. Martin et al. [19] describe three ways to implement such an attack: (1) service request power attacks, (2) benign power attacks, and (3) malignant power attacks. In service request attacks, a malicious user may request repeatedly from the device to serve a power-hungry application, even if the application is not supported by the device. In benign power attacks, the legitimate user is forced to execute an application with high power requirements, while in malignant power attacks malicious users modify the executable code of an existing application, in order to drain as much battery power as possible without changing the application functionality. They conclude that such attacks may reduce battery life by one to two orders of magnitude.

Inclusion of security functions in an embedded system places extra requirements on power consumption due to (1) extra processing power necessary to perform various security functions, such as authentication, encryption, decryption, signing, and data verification, (2) transmission of security-related data between various entities, if the system is distributed, i.e., a wireless sensor network, and (3) energy required to store security-related parameters.

Embedded systems are often used to deploy performance-critical functions, which require a lot of processing power. Inclusion of cryptographic algorithms which are used as building blocks in secure embedded design may lead to great consumption of system battery. The energy consumption of the cryptographic algorithms used in security protocols has been extensively analyzed by Potlapally et al. [20,157], who present a general framework that shows asymmetric algorithms having the highest energy cost, symmetric algorithms as the next power-hungry category, and hash algorithms at the bottom. Their motivation for their study originates from the facts that (1) execution of hard security algorithms in highly constrained environments can have significant impact to battery life and (2) the design of energy-efficient secure communications requires deep understanding of the energy consumption of the underlying security protocols. Through experimental analysis of the energy consumption of the SSL protocol, they made the following observations: (1) the energy consumed for cryptographic purposes is the greatest portion of the total energy consumed and (2) small data transactions require asymmetric cryptography, while larger data transactions are dominated by symmetric algorithms. Finally, they propose methods through which the energy consumption of SSL protocol can be optimized. An additional study [21] shows that the power required by cryptographic algorithms is significant. Importantly, in many applications, the power consumed by security functions is larger than that used for the applications themselves. For example, Ravi et al. [29] present the battery gap for a sensor node with an embedded processor, calculating the number of transactions that the node can serve working in secure or insecure mode until system battery runs out. Their results state that working in secure mode consumes the battery in less than half time than when working in insecure mode.

Many applications that involve embedded systems are implemented through distributed, networked platforms, resulting in a power overhead due to communication between the various nodes of the system [1]. Considering a wireless sensor network, which is a typical distributed embedded system, one can easily see that significant energy is consumed in communication between various nodes. Factors such as modulation type, data rate, transmit power, and security overhead affect power consumption significantly [23]. Savvides et al. [24] showed that the radio communication between nodes consumes most of the power, i.e., 50%–60% of the total power, when using the wireless integrated network sensor (WINS) platform [25]. Furthermore, in a wireless sensor network, the security functions consume energy due to extra internode exchange of cryptographic information—key exchange, authentication information—and per-message security overhead, which is a function of both the number and the size of messages [21]. It is important to identify the energy consumption of alternative security mechanisms. Hodjat and Verbauwhede [26], for example, have measured the energy consumption using two widely used algorithms for key exchange information between entities in a distributed environment (1) Diffie–Hellman protocol [4] and (2) basic Kerberos protocol [4]. Their results show that Diffie–Hellman, implemented using elliptic curve public-key cryptography, consumes 1213.7, 4296, and 9378.3 mJ for 128, 192, and 256 bit keys, respectively, while the Kerberos key exchange protocol using symmetric cryptography consumes 139.62 mJ; this indicates that the Kerberos protocol configuration consumes significantly less energy.

18.3.2 Processing Power Limitations

Security processing places significant additional requirements on the processing power of embedded systems, since conventional architectures are quite limited. The term security processing is used to indicate the portion of the system computational effort which is dedicated to the implementation of the security requirements. Since embedded systems have limited processing power, they cannot cope efficiently with the execution of complex cryptographic algorithms, which are used in the secure design of an embedded system. For example, the generation of a 512-bit key for the Rivest Shamir Adleman (RSA) public-key algorithm requires 3.4 min for the PalmIIIx PDA, while encryption using digital encryption standard (DES) takes only 4.9 ms per block, leading to an encryption rate of 13 Kbps [27]. The adoption of modern embedded systems in high-end systems (servers, firewalls, and routers) with increasing data transmission rates and complex security protocols, such as SSL, makes the security processing gap wider and demonstrates that the existing embedded architectures need to be improved, in order to keep up with the increasing computational requirements that are placed by security processing. The wide processing gap has been exposed by measurements according to Ravi et al. [6], who measured the security processing gap in the client–server model using the SSL protocol for various embedded microprocessors. Specifically, considering a StrongARM (206 MHz SA-1110) processor, which may be used in a low-end system such as a PDA or a mobile device, 100% of the processing power dedicated to SSL processing can achieve data rates up to 1.8 Mbps, while a 2.8 GHz Xeon achieves data rates up to 29 Mbps. Considering that the data rates of low-end systems range between 128 Kbps and 2 Mbps, while data rates of high-end systems range between 2 and 100 Mbps, it is clear that the processors mentioned above cannot achieve higher data rates than their maximum, leading to a security processing gap.

18.3.3 Flexibility and Availability Requirements

The design and implementation of security in an embedded system does not mean that the system will not change its operational security characteristics through time. Considering that security requirements evolve and security protocols are continuously strengthened, embedded systems need to be flexible and adaptable to changes in security requirements, without losing their performance and availability goals as well as their primary security objectives.

Modern embedded systems are characterized by their ability to operate in different environments, under various conditions. Such an embedded system must be able to achieve different security objectives in every environment; thus, the system must be characterized by significant flexibility and efficient adaptation. For example, consider a PDA with mobile telecommunication capabilities, which may operate in a wireless environment [28–30] or provide 3G cellular services [31]; different security objectives must be satisfied in each case. Another issue that must be addressed is the implementation of different security requirements at different layers of the protocol architecture. Consider, for example, a mobile PDA which must be able to execute several security protocols, like IPSec [14], SSL [13], and WEP [32], depending on its specific application.

Importantly, availability is a significant requirement that needs special support, considering that it should be provided in an evolving world in terms of functionality and increasing system complexity. Conventional embedded systems should target to provide high availability characteristics not only in their expected, attack-free environment but in an emerging hostile environment as well.

18.3.4 Cost of Implementation

Inclusion of security in embedded system design can increase system cost dramatically. The problem originates from the strong resource limitations of embedded systems, through which the systems is required to exhibit great performance as well as high level of security while retaining a low cost of implementation. It is necessary to perform a careful, in-depth analysis of the designed system, in terms of the abilities of the possible adversaries, the environmental conditions under which the system will operate, etc., in order to estimate cost realistically. Consider, for example, the incorporation of a tamper-resistant cryptographic module in an embedded system. As described by Ravi et al. [6], according to the Federal Information Processing Standard [33], a designer can distinguish four levels of security requirements for cryptographic modules. The choice of the security level influences design and implementation cost significantly; so, the manufacturer faces a trade-off between the security requirements that will be implemented and the cost of manufacturing.

18.4 Design of Secure Embedded Systems

Secure embedded systems must provide basic security properties, such as data integrity, as well as mechanisms and support for more complex security functions, such as authentication and confidentiality. Furthermore, they have to support the security requirements of applications, which are implemented, in turn, using the security mechanisms offered by the system. In this section, we describe the main design issues at both the system and application level.

18.4.1 System Design Issues

Design of secure embedded systems needs to address several issues and parameters ranging from the employed hardware technology to software development methodologies. Although several techniques used in general-purpose systems can be effectively used in embedded system development as well, there are specific design issues that need to be addressed separately, because they are unique or weaker in embedded systems, due to the high volume of available low cost systems that can be used for development of attacks by malicious users. The major of these design issues are tamper resistance properties, memory protection, IP protection, management of processing power, communication security, and embedded software design. These issues are covered in the following paragraphs.

Modern secure embedded systems must be able to operate in various environmental conditions, without loss of performance and deviation from their primary goals. In many cases they must survive various physical attacks and have tamper resistance mechanisms. Tamper resistance is the property that enables systems to prevent the distortion of physical parts. In addition to tamper resistance mechanisms, there exist tamper evidence mechanisms, which allow users or technical stuff to identify tampering attacks and take countermeasures. Computer systems are vulnerable to tampering attacks, where malicious users intervene in hardware system parts and compromise them, in order to take advantage of them. Security of many critical systems relies on tamper resistance of smart cards and other embedded processors. Anderson and Kuhn [9] describe various techniques and methods to attack tamper resistance systems, concluding that tamper resistance mechanisms need to be extended or re-evaluated.

Memory technology may be an additional weakness in system implementation. Typical embedded systems have ROM, RAM, and EEPROM to store data. EEPROM constitutes the vulnerable spot of such systems, because it can be erased with the use of appropriate electrical signaling by malicious users [9].

IP protection of manufacturers is an important issue addressed in secure embedded systems. Complicated systems tend to be partitioned in smaller independent modules leading to module reusability and cost reduction. These modules include IP of the manufacturers, which needs to be protected from third-party users, who might claim and use these modules. The illegal users of an IP block do not necessarily need to have full, detailed knowledge of the IP component, since IP blocks are independent modules which can be very easily incorporated and integrated with the rest of the system components. Lach et al. [34] propose a fingerprinting technique for IP blocks implemented using FPGAs through an embedded unique marker onto the IP hardware which identifies both the origin and the recipient of the IP block. Also, they are stating that the removal of such a mark is extremely difficult, probability less than one in a million. Fragopoulos and Serpanos [158] made a survey concerning the use of embedded systems as means for protection of IP property and implementation of digital rights management (DRM) mechanisms in mobile devices. They identify hardware-based watermarking–fingerprinting techniques, trusted platforms and architectures as well as smart card based DRM mechanisms that have embedded systems as structural blocks.

Implementation of security techniques for tamper resistance, tamper prevention, and IP protection may require additional processing power, which is limited in embedded systems. The "processing gap" between the computational requirements of security and the available processing power of embedded processors requires special consideration. A variety of architectures and enhancements in security protocols has been proposed in order to enhance the efficiency of security processing in embedded and mobile systems and bridge the processing gap. These technologies include use of cryptographic coprocessors and accelerators, embedded security processors, and programmable security protocols [22]. Burke et al. [35] propose enhancements in the instruction set architecture (ISA) of embedded processors, in order to efficiently calculate various cryptographic primitives, such as permutations, bit rotations, fast substitutions, and modular arithmetic. Another approach is to build dedicated cryptographic embedded coprocessors with their own ISA. The cryptomaniac coprocessor [36] is an example of this approach. Several vendors, for example, Infineon [37] and advanced RISC machine (ARM) [38], have manufactured microcontrollers that have embedded coprocessors dedicated to serve cryptographic functions. Intel [39] announced a new generation of 64-bit embedded processors that have some features that can speed up processing hungry algorithms, such as cryptographic ones; these features include larger register sets, parallel execution of computations, improvements in large integers multiplication, etc. In a third approach, software optimizations are exploited. Potlapally et al. [40] have conducted extensive research in the improvement of public-key algorithms, studying various algorithmic optimizations, identifying an *algorithm design space* where performance is improved significantly. Also, SmartMIPS [41] provides system flexibility and adaptation to any changes in security requirements through high performance software-based enhancements of its cryptographic modules, while it supports various cryptographic algorithms. Besides that, a significant portion of process power is dedicated to executing security protocols; thus, it is necessary to adopt efficient security protocols, which, in conjunction with special-purpose processors and cryptographic accelerators, will provide high processing efficiency. MOSES (MObile SEcurity processing System) [165–167] is a programmable security HW/SW architecture, which is proposed in order to overcome the security processing gap. The platform is a mixture of software implemented cryptographic algorithms and a hardware processor specific for efficient security processing, and achieves high performance through speed-up of security protocols like SSL (the authors implemented openSSL on the MOSES platform).

Even if the "processing gap" is bridged and security functions are provided, embedded systems are required to support secure communications as well, considering that, often, embedded applications are implemented in a distributed environment where communicating systems may exchange (possibly) sensitive data over an untrusted network—wired, wireless or mobile-like Internet, a Virtual Private Network, the Public Telephone network, etc. In order to fulfill the basic security requirements for secure communications, embedded systems must be able to use strong cryptographic algorithms and to support various protocols. One of the fundamental requirements regarding secure protocols is interoperability, leading to the requirement for system flexibility and adaptability. Since an embedded system can operate in several environments, for example, a mobile phone may provide 3 G cellular services or connect to a wireless local area network (LAN), it is necessary for the system to operate securely in all environments without loss of performance. Furthermore, as security protocols are developed for various layers of the OSI reference model, embedded systems must be adaptable to different security requirements at each layer of the architecture. Finally, the continuous evolutions of security protocols require system flexibility as new standards are developed, requirements re-evaluated, and new cryptographic techniques added to overall architecture. A comprehensive presentation of the evolution of security protocols in wireless communications, like WTLS [135], mobile electronic transactions (MET) [136], and IPSec [14], is provided by Raghunathan et al. [22]. An important consideration in the development of flexible secure communication subsystems for embedded systems is the limitation of energy, processing, and memory resources. The performance/cost trade-off leads to special attention for the placement of protocol functions in hardware, for high performance, or software, for cost reduction.

Embedded software, such as the operating system or application-specific code, constitutes a crucial factor in secure embedded system design. Kocher et al. [3] identified three basic factors that make embedded software development a challenging area of security: (1) complexity of the system, (2) system extensibility, and (3) connectivity. Embedded systems serve critical, complex, hard-to-implement applications with many parameters that need to be considered, which, in turn, leads to "buggy" and vulnerable software. Furthermore, the required extensibility of conventional embedded systems makes the exploitation of vulnerabilities relatively easy. Finally, as modern embedded systems are designed with network connectivity, the higher the connectivity degree of the system, the higher the risk for a software breach to expand as time goes by. Many attacks can be implemented by malicious users that exploit software glitches and lead to system unavailability, which can have a disastrous impact, for example, a DoS attack on a military embedded system. Landwehr et al. [12] present a survey of common software security faults, helping designers to learn from their faults. Tevis and Hamilton [11] propose some methods to detect and prevent software vulnerabilities, focusing on some possible weaknesses that have to be avoided, preventing buffer overflow attacks, heap overflow attacks, array indexing attacks, etc. They also provide some coded security programs that help designers analyze the security of their software. Buffer overflow attacks constitute the most widely used type of attacks that lead to unavailability of the attacked system; with these attacks, malicious users exploit system vulnerabilities and are able to execute malicious code, which can cause several problems such as a system crash—preventing legitimate users from using the system—loss of sensitive data, etc. Shao et al. [42] propose a technique, called Hardware–Software Defender, which targets to protect an embedded system from buffer overflow attacks; their proposal is to design a secure instruction set, extending the instruction set of existing microprocessors, and demand from outside software developers to call secure functions from that set. The limited memory resources of embedded systems, specifically the lack of disk space and virtual memory, make the system vulnerable in cases of memory-hungry applications: applications that require excessive amount of memory do not have a swap file to grow and can very easily cause an out-of-memory unavailability of the system. Given the significance of this potential problem and/or attack, Biswas et al. [43] propose mechanisms to protect an embedded system from such a memory overflow, thus providing reliability and availability of the system: (1) use of software run-time checks, in order to check possible out-of-memory conditions, (2) allowing out-of-memory data segments to be placed in free system space, and (3) compressing already used and unnecessary data.

18.4.2 Application Design Issues

Embedded system applications present significant challenges to system designers, in order to achieve efficient and secure systems.

A key issue in secure embedded design is user identification and access control. User identification includes the necessary mechanisms that guarantee that only legitimate users have access to system resources and can also verify, whenever requested, the identity of the user who has access to the system. The explosive growth of mobile devices and their use in critical, sensitive transactions, such as bank transactions, e-commerce, etc., demand secure systems with high performance and low cost. This demand has become urgent and crucial considering the successful attacks on these systems, such as the recent hardware hacking attacks on PIN (personal identification number)-based bank ATMs (automatic teller machines), which have led to significant loss of money and decreased the credibility of financial organizations toward people. A solution to this problem may come from an emerging new technology for user identification which is based on biometric recognition, for both user identification and verification. Biometrics are based on pattern recognition in acquired biological data taken from a user who wants to gain access to a system, i.e., palm prints [44], fingerprints [45], iris scan, etc., and comparing them with the data that have been stored in databases identifying the legitimate users of the system [46]. Moon et al. [47] propose a secure smart card which uses biometrics capabilities, claiming that the development of such a system is less vulnerable to attacks when compared to software-based solutions and that the combination of smart card and fingerprint recognition is much more robust than PIN-based identification. Implementation of such systems is realistic as Tang et al. [48] illustrated with the implementation of a fingerprint recognition system with high reliability and high speed; they achieved an average computational time per fingerprint image less than 1 s, using a fixed point arithmetic StrongARM 200 MHz embedded processor.

As mentioned previously, an embedded system must store information that enables it to identify and validate users that have access to the system. But, how does an embedded system store this information? Embedded systems use several types of memory to store different types of data: (1) ROM EPROM to store programming data used to serve generic applications, (2) RAM to store temporary data, and (3) EEPROM and FLASH memories to store mobile downloadable code [21]. In an embedded device such as a PDA or a mobile phone, several pieces of sensitive information like PINs, credit card numbers, personal data, keys, and certificates for authorization purposes may be permanently stored in secondary storage media. The requirement to protect this information as well as the rapid growth of communications capabilities of embedded devices, for example, mobile Internet access, which make embedded systems vulnerable to network attacks as well, leads to increasing demands for secure storage space. The use of hard cryptographic algorithms to ensure data integrity and confidentiality is not feasible in most embedded systems, mainly due to their limited computational resources. Benini et al. [49] present a survey of architectures and techniques used to implement memory for embedded systems, taking into consideration energy limitations of embedded systems. Rosenthal [69] presents an effective way to ensure that data cannot be erased or destroyed by "hiding" memory from the processor through use of a serial EEPROM, which is the same as standard EEPROM with the only difference that a serial link binds the memory with the processor reading/writing data, using a strict protocol. Actel [50] describes security issues and design considerations for the implementation of embedded memories using FPGAs claiming that SRAM FPGAs are vulnerable to Level I [15] attacks, while it is more preferable to use nonvolatile Flash and Antifuse-based FPGA memories, which provide higher levels of security relatively to SRAM FPGAs.

Another key issue in secure embedded systems design is to ensure that any digital content already stored or downloaded in the embedded system will be used according to the terms and conditions the content provider has set and in accordance with the agreements between user and provider; such content includes software for a specific application or a hardware component embedded in the system by a third-party vendor. It is essential that conventional embedded devices, mobile or not, be enhanced with DRM mechanisms, in order to protect the digital IP of manufacturers and vendors. Trusted computing platforms constitute one approach to resolve this problem. Such platforms are significant, in general, as indicated by the Trusted Computing Platform Alliance [70], which tries to standardize the methods to build trusted platforms. For embedded systems, IP protection can be implemented in various ways. A method to produce a trusted computing platform based on a trusted, secure hardware component, called spy, can lead to systems executing one or more applications securely [66,67].Ways to transform a 3G mobile device into a trusted one have been investigated by Messerges and Dabbish [51], who are capable to protect content through analysis and probing of the various components in a trusted system; for example, the operating system of the embedded system is enhanced with DRM security hardware–software part, which transforms the system into a trusted one. Alternatively, Thekkath et al. [52] propose a method to prevent unauthorized reading, modification, and copying of proprietary software code, using eXecute Only Memory system that permits only code execution. The concept is that code stored in a device can be marked as "execute-only" and content-sensitive applications can be stored in independent compartments [53]; if an application tries to access data outside its compartment, then it is stopped.

Significant attention has to be paid to protect against possible attacks through malicious downloadable software, like viruses, Trojans, logic bombs, etc. [5]. The wide deployment of distributed embedded systems and the Internet has resulted in the requirement for an ability of portable embedded systems, for example, mobile phones and PDAs, to download and execute various software applications. This ability may be new to the world of portable, highly constrained embedded systems, but it is not new in the world of general-purpose systems, which have had the ability to download and execute Java applets and executable files from the Internet or from other network resources for a long time. One major problem in this service is that users cannot be certain about the content of the software that is downloaded and executed on their system(s), about who the creator is and what its origin is. Kingpin and Mudge [57] provide a comprehensive presentation of security issues in personal digitals assistants, analyzing in detail what malicious software is, i.e., viruses, Trojans, backdoors, etc., where it resides and how it is spread, giving to the future users of such devices a deeper understanding about the extra security risks that arise with the use of mobile downloadable code. An additional important consideration is the robustness of the downloadable code: once the mobile code is considered secure, downloaded and executed, it must not affect preinstalled system software. Various techniques have been proposed to protect remote hosts from malicious mobile code. The sandbox technique, proposed by Rubin and Geer [54], is based on the idea that the mobile code cannot execute system functions, i.e., it cannot affect the file system or open network connections. Instead of disabling mobile code from execution, one can empower it using enhanced security policies as Venkatakrishnan et al. propose [56]. Necula [55] suggests the use of proof-carrying code; the producer of the mobile code, a possibly untrusted source, must embed some type of proof which can be tested by the remote host in order to prove the validity of the mobile code.

18.5 Cryptography and Embedded Systems

Secure embedded systems should support the basic security functions for (a) confidentiality, (b) integrity, and (c) authentication. Cryptography provides a mechanism that ensures that the previous three requirements are met. However, implementation of cryptography in embedded systems can be a challenging task. The requirement of high performance has to be achieved in a resource-limited environment; this task is even more challenging when low power constraints exist. Performance usually dictates an increased cost, which is not always desirable or possible. Cryptography can protect digital assets provided that the secret keys of the algorithms are stored and accessed in a secure manner. For this, the use of specialized hardware devices to store the secret keys and to implement cryptographic algorithms is preferred over the use of general-purpose computers. However, this also increases the implementation cost and results in reduced flexibility. On the other hand, flexibility is required, because modern cryptographic protocols do not rely on a specific cryptographic algorithm but rather allow use of a wide range of algorithms for increased security and adaptability to advances on cryptanalysis. For example, both the SSL and IPSec network protocols support numerous cryptographic algorithms to perform the same function, such as encryption. The protocol enables negotiation of the algorithms to be used, in order to ensure that both parties use the desirable level of protection dictated by their security policies.

Apart from the performance issue, a correct cryptographic implementation requires expertise that is not always available or affordable during the life cycle of a system. Insecure implementations of theoretically secure algorithms have made their way to headline news quite often in the past. An excellent survey on cryptography implementation faults is provided in [73], while Anderson [127] focuses on causes of cryptographic systems failures in banking applications. A common misunderstanding is the use of random numbers. Pure linear feedback shift registers and other pseudorandom number generators produce random-looking sequences that may be sufficient for scientific experiments but can be disastrous for cryptographic algorithms that require some unpredictable random input. On the other hand, the cryptographic community has focused on proving the theoretical security of various cryptographic algorithms and has paid little attention to actual implementations on specific hardware platforms. In fact, many algorithms are designed with portability in mind and efficient implementation on a specific platform meeting specific requirements can be quite tricky. This communication gap between vendors and cryptographers intensifies in the case of embedded systems, which can have many design choices and constraints that are not easily comprehensible.

In the late 1990s, side-channel attacks (SCAs) were introduced. SCAs are a method of cryptanalysis that focuses on the implementation characteristics of a cryptographic algorithm in order to derive its secret keys. This advancement bridged the gap between embedded systems, a common target of such attacks, and cryptographers. Vendors became aware and concerned by this new form of attacks, while cryptographers focused on the specifics of the implementations, in order to advance their cryptanalysis techniques.

In this section, we present side-channel cryptanalysis. First, we introduce the concept of tamper resistance, the implementation of side channels and information leakage through them from otherwise secure devices; then, we demonstrate how this information can be exploited to recover the secret keys of cryptographic algorithm, presenting case studies of attacks to the RSA algorithm.

18.5.1 Physical Security

Secrecy is always a desirable property. In the case of cryptographic algorithms, the secret keys of the algorithm must be stored, accessed, used, and destroyed in a secure manner, in order to provide the required security functions. This statement is often overlooked and design or implementation flaws result in insecure cryptographic implementations. It is well known that general-purpose computing systems and operating systems cannot provide enough protection mechanisms for cryptographic keys. For example, SSL certificates for Web servers are stored unprotected on servers' disks and rely on file system permissions for protection. This is necessary, because Web servers can offer secure services unattended. Alternatively, a human would provide the password to access the certificate for each connection; this would not be an efficient decision in the era of e-commerce, where thousands of transactions are made every day. On the other hand, any software bug in the operating system, in a high-privileged application, or in the Web server, software itself may expose this certificate to malicious users.

Embedded systems are commonly used for implementing security functions. Since they are complete systems, they can perform the necessary cryptographic operations in a sealed and controlled environment [123–125]. Tamper resistance refers to the ability of a system to resist to tampering attacks, i.e., attempts to bypass its attack prevention mechanisms. The IBM PCI Cryptographic Coprocessor [121] is one such system, having achieved FIPS 140-2 Level 4 certification [33]. Advancements of DRM technology to consumer devices and general-purpose computers drive also the use of embedded systems for cryptographic protection of IP. Smart cards are a well-known example of tamper-resistant embedded systems that are used for financial transactions and subscription-based service provision.

In many cases, embedded systems used for security-critical operations do not implement any tamper resistance mechanisms. Rather, a thin layer of obscurity is preferred, both for simplicity and performance issues. However, as users become more interested in bypassing the security mechanisms of the system, the thin layer of obscurity is easily broken and the cryptographic keys are publicly exposed. The Adobe eBook software encryption [126], the Microsoft XBox case [71], the universal serial bus (USB) hardware token devices [58], and the DVD CSS copy protection scheme [122] are examples of systems that have implemented security by obscurity and were easily broken.

Finally, an often neglected issue is a lifecycle-wide management of cryptographic systems. While a device may be withdrawn from operation, the data it has stored or processed over time may still need to be protected. The security of keys that relies on the fact that only authorized personnel has access to the system may not be sufficient for the recycled device. Garfinkel [64], Skorobogatov [59], and Gutman [60] present methods for recovering data from devices using noninvasive techniques.

18.5.2 Side-Channel Cryptanalysis

Until the middle 1990s, academic research on cryptography focused on the mathematical properties of the cryptographic algorithms. Paul Kocher was the first to present cryptanalysis attacks on implementations of cryptographic algorithms, which were based on the implementation properties of a system. Kocher observed that a cryptographic implementation of the RSA algorithm required varying amounts of time to encrypt a block of data depending on the secret key used. Careful analysis of the timing differences, allowed him to derive the secret key and he extended this method to other algorithms as well [81]. This result came as a surprise, since the RSA algorithm has withstood years of mathematical cryptanalysis and was considered secure [116]. A short time later, Boneh, Demillo, and Lipton presented theoretical attacks on how to derive the secret keys on implementations of the RSA algorithm and the Fiat–Shamir and Schnorr identification schemes [108], revised in [107], while similar results were presented by Bao et al. [111].

These findings revealed a new class of attacks on cryptographic algorithms. The term SCAs, first appeared in [80], has been widely used to refer to this type of cryptanalysis, while the terms fault-based cryptanalysis, implementation cryptanalysis, active/passive hardware attacks, leakage attacks, and others have been used also. Cryptographic algorithms acquired a new security dimension, that of their exact implementation. Cryptographers had previously focused on understanding the underlying mathematical problems and prove or conjecture for the security of a cryptographic algorithm based on the abstract mathematical symbols. Now, in spite of the hard underlying mathematical problems to be solved, an implementation may be vulnerable and may allow the extraction of secret keys or other sensitive material. Implementation vulnerabilities are of course not a new security concept. In the previous section, we presented some impressive attacks on security that were based on implementation faults. The new concept of SCA is that even cryptographic algorithms that are otherwise considered secure can be also vulnerable to such faults. This observation is of significant importance, since cryptography is widely used as a major building block for security; if cryptographic algorithms can be driven insecure, the whole construction collapses.

Embedded systems and especially smart cards are a popular target for SCAs. To understand this, recall that such systems are usually owned by a service provider, like a mobile phone operator, a TV broadcaster, or a bank, and possessed by service clients. The service provider resides on the security of the embedded system in order to prove service usage by the clients, like phone calls, movie viewing or a purchase, and charge the client accordingly. On the other hand, consumers have the incentive to bypass these mechanisms in order to enjoy free services. Given that SCAs are implementation specific and rely, as we will present later, on the ability to interfere, passively or actively with the device implementing a cryptographic algorithm, embedded systems are a further attractive target, given their resource limitation, which makes the attack efforts easier.

In the following, we present the classes of SCAs and countermeasures that have been developed. The technical field remains highly active, since ingenious channels are continuously appearing in the bibliography. A good collection site for gathering all relevant published work is also available [141]. Embedded system vendors must study the attacks carefully, evaluate the associated risks for their environment, and ensure that appropriate countermeasures are implemented in their systems; furthermore, they must be prepared to adapt promptly to new techniques for deriving secrets from their systems.

18.5.3 Side-Channel Implementations

A side channel is any physical channel that can carry information from the operation of a device while implementing a cryptographic operation; such channels are not captured by the existing abstract mathematical models. The definition is quite broad and the inventiveness of attackers is noticeable. Timing differences, power consumption, electromagnetic emissions, acoustic noise, and faults have been currently exploited for leaking information out of cryptographic systems.

The channel realization can be categorized in three broad classes: physical or probing attacks, fault-induction or glitch attacks, and emission attacks, like TEMPEST. We will shortly review the first two classes; readers interested in TEMPEST attacks are referred to [120].

The side channels may seem unavoidable and a frightening threat. However, it should be strongly emphasized that in most cases, reported attacks, both theoretical and practical, rely for their success on the detailed knowledge of the platform under attack and the specific implementation of the cryptographic algorithm. For example, power analysis is successful in most cases, because cryptographic algorithms tend to use only a small subset of a processor's instruction set and especially simple instructions, like LOAD, STORE, XOR, AND, and SHIFT, in order to develop elegant, portable, and high-performance implementations. This decision allows an attacker to minimize the power profiles he/she must construct and simplifies the distinction of different instructions that are executed.

18.5.3.1 Fault Induction Techniques

Devices are always susceptible to erroneous computations or other kinds of faults for several reasons. Faulty computations are a known issue from space systems, because, in deep space, devices are exposed to radiation which can cause temporary or permanent bit flips, gate destruction, or other problems. Incomplete testing during manufacturing may allow imperfect designs from reaching the market, as in the case of the Intel Pentium FDIV bug [113], or in the case of device operation in conditions out of their specifications [88]. Careful manipulation of the power supply or the clock oscillator can also cause glitches in code execution by tricking the processor, for example, to execute unknown instructions or bypass a control statement [137].

Some researchers have questioned the feasibility of fault-injection attacks on real systems [103]. While fault injection may seem as an approach that requires expensive and specialized equipment, there have been reports that fault injection can be achieved with low cost and readily available equipment. Anderson and Kuhn [138] and Anderson [127] present low cost attacks for tamper-resistant devices, which achieve extraction of secret information from smart cards and similar devices. Kömmerling and Kuhn [137] present noninvasive fault injection techniques, for example, by manipulating power supply. Anderson [112] supports the view that the underground community has been using such techniques for quite a long time to break the security of smart cards of pay-TV systems. Furthermore, Weingart [8] and Aumüeller et al. [93] present attacks performed in a controlled laboratory environment, proving that fault-injection attacks are feasible. Skorobogatov and Anderson [63] introduce low cost light flashes, like a camera flash, as a means to introduce errors, while eddy current attacks are introduced in [128]. A complete presentation of the fault injection methods is presented in [74], along with experimental evidence on the applicability of the methods to industrial systems and anecdotal information.

The combined time–space isolation problem [129] is of significant importance in fault-induction attacks. The space isolation problem refers to isolation of the appropriate space (area) of the chip in which to introduce the fault. The space isolation problem has four parameters:

1. Macroscopic: The part of the chip where the fault can be injected. Possible answers can be one or more of the following: main memory, address bus, system bus, and register file.
2. Bandwidth: The number of bits that can be affected. It may be possible to change just one bit or multiple bits at once. The exact number of changed bits can be controllable (e.g., one) or can follow a random distribution.
3. Granularity: The area where the error can occur. The attacker may drive the fault injection position at a bit level or a wider area, such as a byte or a multibyte area. The fault-injected area can be covered by a single error or by multiple errors. How are these errors distributed with respect to the area? The errors may focus either around the mark or it could be evenly distributed.
4. Lifetime: The time duration of the fault. It may be a transient fault or a permanent fault. For example, a power glitch may cause a transient fault at a memory location, since the next time the location will be written, a new value will be correctly written. In contrast, a cell or gate destruction will result in a permanent error, since the output bit will be stuck at 0 or 1, independently of the input.

The time isolation problem refers to the time at which a fault is injected. An attacker may be able to synchronize exactly with the clock of the chip or may introduce the error in a random fashion. This granularity is the only parameter of the time isolation problem. Clearly, the ability to inject a fault in a clock period granularity is desirable, but impractical in real-world applications.

18.5.3.2 Passive Side Channels

Passive side channels are not a new concept in cryptography and security. The information available from the now partially declassified TEMPEST project reveals helpful insights in how electromagnetic emissions occur and can be used to reconstruct signals for surveillance purposes. A good review of the subject is provided in Chapter 15 of [112]. Kuhn [61,120,139] presents innovative use of electromagnetic emissions to reconstruct information from chinese remainder theorem (CRT) and liquid crystal display (LCD) displays, while Loughry [62] reconstructs information flowing through network devices using the emissions of their light emitting diodes (LEDs).

The new concept in this area is the fact that such emissions can also be used to derive secret information from an otherwise secure device. Probably, the first such attack took place in 1956 [114]. MI5, the British Intelligence, used a microphone to capture the sound of the rotor clicks of a Hagelin machine in order to deduce the core position of some of its rotors. This resulted in reducing the problem to calculate the initial setup of the machine within the range of their then available resources, and to eavesdropping the encrypted communications for quite a long time. While the, so-called, acoustic cryptanalysis may seem outdated, researchers provided a fresh look on this topic recently, by monitoring low-frequency (kHz) sounds and correlating them with operations performed by a high-frequency (GHz) processor [115].

Researchers have been quite creative and have used many types of emissions or other physical interactions of the device with the environment it operates. Kocher [81] introduces the idea of monitoring the execution time of a cryptographic algorithm and tries to identify the secret keys used. The key concept in this approach is that an implementation of an algorithm may contain branches and other conditional execution or the implementation may follow different execution paths. If these variances are based on the bit values of a secret key, then a statistical analysis can reveal the secret key bit by bit. Coron et al. [90] explain the power dissipation sources and causes, while Kocher et al. [78] present how power consumption can also be correlated with key bits. Rao et al. [87] and Quisquater et al. [88] introduce electromagnetic analysis. Probing attacks can also be applied to reveal the Hamming weight of data transferred across a bus or stored in memory. This approach is also heavily dependent on the exact hardware platform [75,79].

While passive side channels are usually thought in the context of embedded systems and other resource-limited environments, complex computing systems may also have passive side channels. Page [89] explores the theoretical use of timing variations due to processor cache in order to extract secret keys. Song et al. [86] take advantage of a timing channel in the secure communication protocol SSH to recover user passwords, while Felten and Schneider [85] present timing attacks on Web privacy. A malicious Web server can inject client-side code that fetches some specific pages transparently on behalf of the user; the server would like to know if the user has visited these pages before. The time difference between fetching the Web page from the remote server and accessing it from the user's cache is sufficient to identify if the user has visited this page before. A more impressive result, directly related to cryptography, is presented in [91], where remote timing attacks on Web servers implementing the SSL protocol are shown to be practical and the malicious user can extract the server's certificate private key by measuring its response times.

18.5.4 Fault-Based Cryptanalysis

The first theoretical active attacks are presented in [108] and [111]. The attacks in the former paper focused on RSA, when implemented with the CRT and Montgomery multiplication method, the Fiat–Shamir and the Schnorr identification schemes. The latter work focuses on cryptosystems where security is based on the discrete logarithm problem and presents attacks on the ElGamal signature scheme, the Schnorr signature scheme, and DSA. The attack on the Schnorr signature scheme is extended, with some modification, to the identification scheme as well. Furthermore, the second paper reports independently an attack on the RSA–Montgomery. Since then, this area has been quite active, both in developing attacks based on fault-induction and countermeasures. The attacks have succeeded in most of the popular and widely used algorithms. In the following, we give a brief review of the bibliography.

The attacks on RSA with Montgomery have been extended by attacking the signing key, instead of the message [110]. Furthermore, similar attacks are presented for LUC and KMOV (based on elliptic curves) cryptosystems. In [109], the attacks are generalized for any RSA-type cryptosystem, with examples of LUC and Demytko cryptosystems. Faults can be used to expose the private key of RSA–KEM (key encapsulation method) scheme [100] and transient faults can be used to derive the RSA and DSA secret keys from applications compatible with the OpenPGP format [101]. The Bellcore attack on the Fiat–Shamir scheme is shown incomplete in [129]; the Precautious Fiat–Shamir scheme is introduced, which defends against it. A new attack that succeeds against both the classical and the Precautious Fiat–Shamir scheme is presented in [65].

Beginning with Biham and Shamir [106], fault-based cryptanalysis focused on symmetric key cryptosystems. DES is shown to be vulnerable to the, so called, differential fault analysis (DFA), using only 50–200 faulty ciphertexts. The method is also extended to unknown cryptosystems and an example of an attack on the once classified algorithm SkipJack is presented. Another variant of the attack on DES takes advantage of permanent instead of transient faults. The same ideas are also explored and extended for completely unknown cryptosystems in [99], while Jacob et al. [95] use faults to attack obfuscated ciphers in software and extract secret material by avoiding de-obfuscating the code.

For some time, it was believed that fault-induction attacks can only succeed on cryptographic schemes based on algebraic-based hard mathematical problems, like number factoring and discrete logarithm computation. Elliptic curve cryptosystems (ECC) are a preferable choice to implement cryptography, since they offer equivalent security with that of algebraic public-key algorithms, requiring only about a tenth of key bits. Biehl et al. [105] extend the DFA on ECC and, especially, on schemes whose security is based on the discrete logarithm problem over elliptic curve fields. Furthermore, Zheng and Matsumoto [104] use transient and permanent faults to attack random number generators, a crucial building block for cryptographic protocols, and the ElGamal signature scheme.

Rijndael [131] was nominated as the advanced encryption standard (AES) algorithm [132], the replacement of DES. The case of the AES algorithm is quite interesting, considering that it was submitted after the introduction of SCAs; thus, authors have taken all the appropriate countermeasures to ensure that the algorithm resisted all known cryptanalysis techniques applicable to their design. The original proposal [131] even noted timing attacks and how they could be prevented. Koeune et al. [84] describe how a careless implementation of the AES algorithm can utilize a timing attack and derive the secret key used. The experiments carried show that the key can be derived having 3000 samples per key byte, with minimal cost and high probability. The proposal of the algorithm is aware of this issue and immune against such a simple attack. However, DFA proved to be successful against AES. Although DFA was designed for attacking algorithms with a Feistel structure, like DES, Desart et al. [72] show that it can be applied to AES, which does not have such a structure. Four different fault-injection models are presented and the attacks succeed for all key sizes (128, 192, and 256 bits). Their experiments show that with 10 pairs of faulty/correct messages in hand, a 128-bit AES key can be extracted in a few minutes. Blömer et al. [94] present additional fault-based attacks on AES. The attack assumes multiple kinds of fault models. The stricter model, requiring exact synchronization in space and time for the error injection, succeeds in deriving a 128-bit secret key after collecting 128 faulty ciphertexts, while the least strict model derives the 128-bit key, after collecting 256 faulty ciphertexts.

The AES algorithm seems currently the most popular target for fault-based cryptanalysis over any other current cryptosystem. A significant number of publications has already appeared on various algorithmic and attack complexity improvements along with countermeasures [142–148].

18.5.4.1 Case Study: RSA–CRT

The RSA cryptosystem remains a viable and preferable public-key cryptosystem, having withstood years of cryptanalysis [116]. The security of the RSA public-key algorithm relies on the hardness of the problem of factoring large numbers to prime factors. The elements of the algorithm are $N = pq$, the product of two large prime numbers, i.e., the public and secret exponents, respectively, and the modular exponentiation operation $m^k \bmod N$. To sign a message m, the sender computes $s = m^d \bmod N$ using his/her public key. The receiver computes $m = s^e \bmod N$ to verify the signature of the received message.

The modular exponentiation operation is computationally intensive for large primes and it is the major computational bottleneck in an RSA implementation. The CRT allows fast modular exponentiation. Using RSA with CRT, the sender computes $s_1 = m^d \bmod p$, $s_2 = m^d \bmod q$ and combines the two results, based on the CRT, to compute $S = as_1 + bs_2 \bmod N$ for some predefined values a and b.

The CRT method is quite popular, especially for embedded systems, since it allows four times faster execution and smaller memory storage for intermediate results (for this, observe that typically p and q have half the size of N).

The Bellcore attack [107,108], as it is commonly referenced, is quite simple and powerful against RSA with CRT. It suffices to have one correct signature S for a message m and one faulty signature S', which is caused by an incorrect computation of one of the two intermediate results s_1 and s_2. It does not matter either if the error occurred on the first or the second intermediate result, or how many bits were affected by the error. Assuming that an error indeed occurred, it suffices to compute $\gcd(S - S', N)$, which will equal q, if the error occurred in computation of s_1 and p if it occurred in s_2. This allows to factor N and thus, the security of the algorithm is broken.

Lenstra [117] improves this attack by requiring a known message and a faulty signature, instead of two signatures. In this case, it suffices to compute $\gcd(M - (S')^d, N)$ to reveal one of the two prime factors.

Boneh et al. [108] propose double computations as means to detect such erroneous computations. However, this is not always efficient, especially in the case of resource-limited environments or where performance is an important issue. Also, this approach is of no help in case a permanent error has occurred. Kaliski and Robshaw [133] propose signature verification, by checking the equality $S^e \bmod N = M$. Since the public exponent may be quite large, this check can rather be time-consuming for a resource-limited system.

Shamir [130] describes a software method for protecting RSA with CRT from fault and timing attacks. The idea is to use a random integer t and perform a "blinded" CRT by computing: $S_{pt} = m^d \bmod p*t$ and $S_{qt} = m^d \bmod q*t$. If the equality $S_{pt} = S_{qt} \bmod t$ holds, then the initial computation is considered error-free and the result of the CRT can be released from the device. Yen et al. [98] further improve this countermeasure for efficient implementation without performance penalties, but Blömer et al. [92] show that this improvement in fact renders RSA with CRT totally insecure. Aumüller et al. [93] provide another software implementation countermeasure for faulty RSA–CRT computations. However, Yen et al. [102], using a weak fault model, show that both these countermeasures [93,130] are still vulnerable, if the attacker focuses on the modular reduction operation $s_p = s'_p \bmod p$ of the countermeasures. The attacks are valid for both transient and permanent errors and again, appropriate countermeasures are proposed.

As we show, the implementation of error checking functions using the final or intermediate results of RSA computations can create an additional side meta-channel, although faulty computations never leave a sealed device. Assume that an attacker knows that a bit in a register holding part of the key was invasively set to zero during the computation and that the device checks the correctness of the output by double computation. If the device outputs a signed message, then no error was detected and thus, the respective bit of the key is zero. If the device does not output a signed message or outputs an error message, then the respective bit of the key is one. Such a "safe-error attack" is presented in [96], focusing on the RSA when implemented with Montgomery multiplication. Yen et al. [97] extend the idea of safe-error attacks from memory faults to computational faults and present such an attack on RSA with Montgomery, which can also be applied to scalar multiplication on elliptic curves.

An even simpler attack would be to attack both an intermediate computation and the condition check. A condition check can be a single point of failure and an attacker can easily mount an attack against it, provided that he/she has means to introduce errors in computations [92]. Indeed, in most cases, a condition check is implemented as a bit comparison with a zero flag. Blömer et al. [92] extend the ideas of checking vulnerable points of computation by exhaustively testing every computation performed for an RSA–CRT signing, including the CRT combination. The proposed solution seems the most promising at the moment, allowing only attacks by powerful adversaries that can solve precisely the time–space isolation problem. However, it should be already clear that advancements in this area of cryptanalysis are continuous and they should always be prepared to adapt to new attacks.

18.5.5   Passive Side-Channel Cryptanalysis

Passive side-channel cryptanalysis has received a lot of attention, since its introduction in 1996 by Paul Kocher [81]. Passive attacks are considered harder to defend against and many people are concerned, due to their noninvasive nature. Fault-induction attacks require some form of manipulating the device and thus, sensors or other similar means can be used to detect such actions and shut down or even zero out the device. In the case of passive attacks, the physical characteristics of the device are just monitored, usually with readily available probes and other hardware. So, it is not an easy task to detect the presence of a malicious user, especially in the case where only a few measurements are required or abnormal operation (like continuous requests for encryptions/decryptions) cannot be identified.

The first results are by Kocher [81]. Timing variations in the execution of a cryptographic algorithm like Diffie–Hellman key exchange, RSA, and DSS are used to derive bit by bit the secret keys of these algorithms. Although mentioned before, we should emphasize that timing attacks and other forms of passive SCAs require knowledge of the exact implementation of the cryptographic algorithm under attack. Dhem et al. [77] describe a timing attack against the RSA signature algorithm. The attack derives a 512-bit secret key with 200,000–300,000 timing measurements. Schindler et al. [83] improve the timing attacks on RSA modular exponentiation by a factor of 50, allowing extraction of a 512-bit key using as few as 5000 timing measurements. The approach used is an error-correction (estimator) function, which can detect erroneous bit detections as key extraction process evolves. Hevia and Kiwi [82] introduce a timing attack against DES, which reveals the Hamming weight of the key, by exploiting the fact that a conditional bit "wraparound" function results on variable execution time of the software implementing the algorithm. They succeed in recovering the Hamming weight of the key and 3.95 key bits (out of a 56-bit key). The most threatening issue is that keys with low or high Hamming weight are sparse; so, if the attack reveals that the key has such a weight, the key space that must be searched reduces dramatically. The RC5 algorithm has also been subjected to timing attacks, due to conditional statement execution in its code [119].

Kocher et al. [78] extend the attackers' arsenal further by introducing the vulnerability of DES to power analysis attacks and more specifically to differential power analysis (DPA), a technique that combines differential cryptanalysis and careful engineering, and to simple power analysis (SPA). SPA refers to power analysis attacks that can be performed only by monitoring a single or a few power traces, probably with the same encryption key. SPA succeeds in revealing the operations performed by the device, like permutations, comparisons, and multiplications. Practically, any algorithm implementation that executes some statements conditionally, based on data or key material, is at least susceptible to power analysis attacks. This holds for public key, secret key, and ECC. DPA has been successfully applied at least to block ciphers, like IDEA, RC5, and DES [80].

Electromagnetic attacks (EMA) have contributed some impressive results on what information can be reconstructed. Gandolfi et al. [76] report results from cryptanalysis of real-world cryptosystems, like DES and RSA. Furthermore, they demonstrate that electromagnetic emissions may be preferable to power analysis, in the sense that fewer traces are needed to mount an attack and these traces carry richer information to derive the secret keys. However, the full power of EMA has not been utilized yet and we should expect more results on real-world cryptanalysis of popular algorithms.

18.5.5.1   Case Study: RSA–Montgomery

Previously, we explained the importance of a fast modular exponentiation primitive for the RSA cryptosystem. Montgomery multiplication is a fast implementation of this primitive function [118]. The left-to-right repeated square-and-multiply method is depicted in Figure 18.1, in C pseudocode.

The timing attack of Kocher [81] exploits the timing variation caused by the condition statement on the fourth line. If the respective bit of the secret exponent is "1," then a square (line 3) and multiply (line 5) operations are executed, while if the bit is "0" only a square operation is performed.

Input: $M, N, d = (d_{n-1}\,d_{n-2}\mathrm{K}\,d_1 d_0)_2$
Output: $S = M^d \bmod N$

```
S = 1;
for (i = n − 1; i > = 0; i−−){
  S = S² mod N;
  if (dᵢ==1){
      S=S * M mod N;
  }
}
return S;
```

FIGURE 18.1   Left-to-right repeated square-and-multiply algorithm.

In summary, the exact time of executing the loop n times is only dependent on the exact values of the bits of the secret exponent.

An attacker proceeds as follows. Assume that the first m bits of the secret exponent are known. The attacker has an identical device with that containing the secret exponent and can control the key used for each encryption. The attacker collects from the attacked device the total execution time $T_1, T_2, \ldots, T_k$ of each signature operation on some known messages, $M_1, M_2, \ldots, M_k$. He also performs the same operation on the controlled device, as to collect another set of measurements, $t_1, t_2, \ldots, t_k$, where he fixes the m first bits of the key, targeting for the $m + 1$ bit. Kocher's key observation is that, if the unknown bit $d_{m+1} = 1$, then the two sets of measurements are correlated. If $d_{m+1} = 0$, then the two sets behave like independent random variables. This differentiation allows the attacker to extract the secret exponent bit by bit.

Depending on the implementation, a simpler form of the attack could be implemented. SPA does not require lengthy statistical computations but rather relies on power traces of execution profiles of a cryptographic algorithm. For this example, Schindler et al. [83] explain how we can use the power profiles. Execution of line 5 in the above code requires an additional multiplication. Even if the spikes in power consumption of the squaring and multiplication operations are indistinguishable, the multiplication requires additional load operations and thus, power spikes will be wider than in the case where only squaring is performed.

18.5.6   Countermeasures

In the previous sections, we provided a review of SCAs, both fault-based and passive. In this section, we review the countermeasures that have been proposed. The list is mot exhaustive and new results appear continuously, since countermeasures are steadily improving.

The proposed countermeasures can be classified in two main classes: hardware protection mechanisms and mathematical protection mechanisms. A first layer of protection against SCAs are hardware protection layers, like passivation layers that do not allow direct access between a (malicious) user and the system implementing the cryptographic algorithm or memory address bus obfuscation. Various sensors can also be embodied in the device, in order to detect and react to abnormal environmental conditions, like extreme temperatures, power, and clock variations. Such mechanisms are widely employed in smart cards for financial transactions and other high-risk applications. Such protection layers can be effective against fault-injection attacks, since they shield the device against external manipulation. However, they cannot protect the device from attacks based on external observation, like power analysis techniques.

The previous countermeasures do not alter the current designs of the circuits, but rather add protection layers on top of them. A second approach is the design of a new generation of chips to implement cryptographic algorithms and to process sensitive information. Such circuits have asynchronous/self-clocking/dual rail logic; each part of the circuit may be clocked independently

[140]. Fault attacks that rely on external clock manipulation (like glitch attacks) are not feasible in this case. Furthermore, timing or power analysis attacks become harder for the attacker, since there is no global clock that correlates the input data and the emitted power. Such countermeasures have the potential to become a common practice. Their application, however, must be carefully evaluated, since they may occupy a large area of the circuit; such expansions are justified by manufacturers usually in order to increase the system's available memory and not to implement another security feature. Furthermore, such mechanisms require changes in the production line, which is not always feasible.

A third approach targets to implement the cryptographic algorithms so that no key information leaks. Proposed approaches include modifying the algorithm to run in constant time, adding random delays in the execution of the algorithm, randomizing the exact sequence of operations without affecting the final result, and adding dummy operations in the execution of the algorithm. These countermeasures can defeat timing attacks, but careful design must be employed to defeat power analysis attacks too. For example, dummy operations or random delays are easily distinguishable in a power trace, since they tend to consume less power than ordinary cryptographic operations. Furthermore, differences in power traces between profiles of known operations can also reveal permutation of operations. For example, a modular multiplication is known to consume more power than a simple addition, so if the execution order is interchanged, they will be still identifiable.

In more resource-rich systems, where high-level programming languages are used, compiler or human optimizations can remove these artifacts from the program or change the implementation resulting to vulnerability against SCAs. The same holds, if memory caches are used and the algorithm is implemented so that the latency between cache and main memory can be detected, either by timing or by power traces. Insertion of random delays or other forms of noise should also be considered carefully, because a large mean value of delay translates directly to reduced performance, which is not always acceptable.

The second class of countermeasures focuses on the mathematical strengthening of the algorithms against such attacks. The RSA-blinding technique by Shamir [130] is such an example; the proposed method guards the system from leaking meaningful information, because the leaked information is related to the random number used for blinding instead of the key; thus, even if the attacker manages to reveal a number, this will be the random number and not the key. It should be noted, however, that a different random number is used for each signing or encryption operation. Thus, the faults injected in the system will be applied on a different, random number every time and the collected information is useless.

At a cross-line between mathematical and implementation protection, it is proposed to check cryptographic operations for correctness, in case of fault-injection attacks. However, these checks can also be exploited as side channels of information or can degrade performance significantly. For example, double computations and comparison of the results halve the throughput an implementation can achieve; furthermore, in the absence of other countermeasures, the comparison function can be bypassed (e.g., by a clock glitch or a fault-injection in the comparison function) or used as a side channel as well. If multiple checks are employed, measuring the rejection time can reveal in what stage of the algorithm the error occurred; if the checks are independent, this can be utilized to extract the secret key, even when the implementation does not output the faulty computation [100,134].

18.6   Conclusion

Security constitutes a significant requirement in modern embedded computing systems. Their widespread use in services that involve sensitive information in conjunction with their resource limitations have led to a significant number of innovative attacks that exploit system characteristics and result in loss of critical information. Development of secure embedded systems is an emerging field in computer engineering requiring skills from cryptography, communications, hardware, and software.

In this chapter, we surveyed the security requirements of embedded computing systems and described the technologies that are more critical to them, relatively to general-purpose computing systems. Considering the innovative system (side-channel) attacks that were developed with motivation to break secure embedded systems, we presented in detail the known SCAs and described the technologies for countermeasures against the known attacks. Clearly, the technical area of secure embedded systems is far from mature. Innovative attacks and successful countermeasures are continuously emerging, promising an attractive and rich technical area for research and development.

References

1. W. Wolf, *Computers as Components—Principles of Embedded Computing Systems Design*, Elsevier, Amsterdam, the Netherlands, 2000.
2. W. Freeman and E. Miller, An experimental analysis of cryptographic overhead in performance—critical systems, in *Proceedings of the 7th International Symposium on Modeling, Analysis and Simulation of Computer and Telecommunication Systems*, College Park, MD, IEEE Computer Society, 1999, p. 348.
3. S. Ravi, P. Kocher, R. Lee, G. McGraw, and A. Raghunathan, Security as a new dimension in embedded system design, in *Proceedings of the 41st Annual Conference on Design Automation*, San Diego, CA, ACM, 2004, pp. 753–760.
4. B. Schneier, *Applied Cryptography: Protocols, Algorithms, and Source Code in C*, Wiley, New York, 1995.
5. T. King and D. Bittlingmeier, *Security + Training Guide*, Que Publication, 2003, ISBN 0789728362.
6. S. Ravi, A. Raghunathan, P. Kocher, and S. Hattangady, Security in embedded systems: Design challenges, *Transactions on Embedded Computing Systems*, 3, 461–491, 2004.
7. T. Wollinger, J. Guajardo, and C. Paar, Security on FPGAs: State-of-the-art implementations and attacks, *Transactions on Embedded Computing Systems*, 3, 534–574, 2004.
8. S. H. Weingart, Physical security devices for computer subsystems: A survey of attacks and defenses, in *Cryptographic Hardware and Embedded Systems—CHES 2000: Second International Workshop*, Worcester, MA, Springer-Verlag, 2000, p. 302.
9. R. Anderson and M. Kuhn, Tamper resistance—a cautionary note, in *Proceedings of the Second Usenix Workshop on Electronic Commerce*, Oakland, CA, USENIX Association, 1996, pp. 1–11.
10. S. Blythe, B. Fraboni, S. Lall, H. Ahmed, and U. de Riu, Layout reconstruction of complex silicon chips, *IEEE Journal of Solid-State Circuits*, 28, 138–145, 1993.
11. J. J. Tevis and J. A. Hamilton, Methods for the prevention, detection and removal of software security vulnerabilities, in *Proceedings of the 42nd Annual Southeast Regional Conference*, Huntsville, AL, ACM, 2004, pp. 197–202.
12. C. E. Landwehr, A. R. Bull, J. P. McDermott, and W. S. Choi, A taxonomy of computer program security flaws, *ACM Computing Surveys*, 26, 211–254, 1994.
13. P. Kocher, *SSL 3.0 Specification*. Available at: http://wp.netscape.com/eng/ssl3/
14. IETF, RFC 3884, *IPSec Specification*. Available at: http://www.ietf.org/rfc/rfc3884.txt
15. D. G. Abraham, G. M. Dolan, G. P. Double, and J. V. Stevens, Transaction security system, *IBM Systems Journal*, 30, 206–229, 1991.
16. S. H. Gunther, F. Binns, D. M. Carmean, and J. C. Hall. Managing the impact of increasing microprocessor power consumption. *Intel Journal of Technology*, Q1 (Quarter 1), 9, 2001. [Online]. Available at: http://developer.intel.com/technology/itj/q12001/articles/art_4.htm
17. I. Buchmann, *Batteries in a Portable World*, 2nd edn., Cadex Electronics Inc, ISBN: 0-9682118-2-8, May 2001.

18. K. Lahiri, S. Dey, D. Panigrahi, and A. Raghunathan, Battery-driven system design: A new frontier in low power design, in *Proceedings of the 2002 Conference on Asia South Pacific Design Automation/VLSI Design*, Bangalore, India, IEEE Computer Society, 2002, p. 261.

19. T. Martin, M. Hsiao, D. Ha, and J. Krishnaswami, Denial-of-service attacks on battery-powered mobile computers, in *2nd IEEE International Conference on Pervasive Computing and Communications (PerCom'04)*, Orlando, FL, IEEE Computer Society, 2004, p. 309.

20. N. R. Potlapally, S. Ravi, A. Raghunathan, and N. K. Jha, Analyzing the energy consumption of security protocols, in *Proceedings of the 2003 International Symposium on Low power Electronics and Design*, Seoul, Korea, ACM, 2003, pp. 30–35, IEEE.

21. D. W. Carman, P. S. Kruus, and B. J. Matt, *Constraints and Approaches for Distributed Sensor Network Security*, NAI Labs, Technical Report 00-110, 2000. Available at: http://www.cs.umbc.edu/courses/graduate/CMSC691A/Spring04/papers/nailabs_report_00-010_final.pdf

22. A. Raghunathan, S. Ravi, S. Hattangady, and J. Quisquater, Securing mobile appliances: New challenges for the system designer, in *Design, Automation and Test in Europe Conference and Exhibition (DATE'03)*, Munich, Germany, IEEE Computer Society, 2003, p. 10176.

23. V. Raghunathan, C. Schurgers, S. Park, and M. Srivastava, Energy aware wireless microsensor networks, *IEEE Signal Processing Magazine*, 19, 40–50, March 2002.

24. A. Savvides, S. Park, and M. B. Srivastava, On modeling networks of wireless microsensors, in *Proceedings of the 2001 ACM SIGMETRICS International Conference on Measurement and Modeling of Computer Systems*, Cambridge, MA, ACM, 2001, pp. 318–319.

25. Rockwell Scientific, *Wireless Integrated Networks Systems*. Available at: http://wins.rsc.rockwell.com

26. A. Hodjat and I. Verbauwhede, *The Energy Cost of Secrets in Ad-hoc Networks (Short Paper)*. Available at: http://citeseer.ist.psu.edu/hodjat02energy.html

27. N. Daswani and D. Boneh, Experimenting with electronic commerce on the PalmPilot, in *Proceedings of the 3rd International Conference on Financial Cryptography*, Rome, Italy, Springer-Verlag, 1999, pp. 1–16.

28. A. Perrig, J. Stankovic, and D. Wagner, Security in wireless sensor networks, *Communications of the ACM*, 47, 53–57, 2004.

29. S. Ravi, A. Raghunathan, and N. Potlapally, Securing wireless data: System architecture challenges, in *ISSS '02: Proceedings of the 15th International Symposium on System Synthesis*. New York, ACM, 2002, pp. 195–200. [Online]. Available at: http://portal.acm.org/citation.cfm?id=581243

30. IEEE 802.11 Working Group, *IEEE 802.11 Wireless LAN Standards*. Available at: http://grouper.ieee.org/groups/802/11/

31. 3GPP, 3G *Security; Security Architecture*, 3GPP Organization, Tech. Spec. 33.102, 30-09-2003, 2003, Rel-6.

32. Intel Corp., *VPN and WEP, Wireless 802.11b Security in a Corporate Environment*. Available at: http://www.intel.com/business/bss/infrastructure/security/vpn_wep.htm

33. NIST, *FIPS PUB 140-2 Security Requirements for Cryptographic Modules*. Available at: http://csrc.nist.gov/cryptval/140-2.htm

34. J. Lach, W. H. Mangione-Smith, and M. Potkonjak, Fingerprinting digital circuits on programmable hardware, in *Information Hiding: Second International Workshop, IH'98*, Lecture Notes in Computer Science, vol. 1525, 1998, pp. 16–31, Springer-Verlag.

35. J. Burke, J. McDonald, and T. Austin, Architectural support for fast symmetric-key cryptography, in *Proceedings of the 9th International Conference on Architectural Support for Programming Languages and Operating Systems*, Cambridge, MA, ACM, 2000, pp. 178–189.

36. L. Wu, C. Weaver, and T. Austin, CryptoManiac: A fast flexible architecture for secure communication, in *Proceedings of the 28th Annual International Symposium on Computer Architecture*, Göteborg, Sweden, ACM, 2001, pp. 110–119.

37. Infineon, *SLE 88 Family Products*. Available at: http://www.infineon.com/.

38. ARM, *ARM SecurCore Family*. Available at: http://www.arm.com/products/CPUs/securcore.html, vol. 2004.

39. S. Moore, *Enhancing Security Performance Through IA-64 Architecture*, 2000, Intel Corp. Available at: http://www.intel.com/cd/ids/developer/asmo-na/eng/microprocessors/itanium/index.htm

40. N. Potlapally, S. Ravi, A. Raghunathan, and G. Lakshminarayana, Optimizing public-key encryption for wireless clients, in *Proceedings of the IEEE International Conference on Communications*, IEEE Computer Society, May 2002.

41. MIPS Inc. Technologies, *SmartMIPS Architecture*. Available at: http://www.mips.com/products/processors/architectures/smartmips-aset

42. Z. Shao, C. Xue, Q. Zhuge, E. H. Sha, and B. Xiao, Security protection and checking in embedded system integration against buffer overflow attacks, in *Proceedings of the International Conference on Information Technology: Coding and Computing (ITCC'04)* vol. 2, Las Vegas, NV, IEEE Computer Society, 2004, pp. 409.

43. S. Biswas, M. Simpson, and R. Barua, Memory overflow protection for embedded systems using run-time checks, reuse and compression, in *Proceedings of the 2004 International Conference on Compilers, Architecture, and Synthesis for Embedded Systems*, Washington, DC, ACM, 2004, pp. 280–291.

44. J. You, W.-K. Kong, D. Zhang, and K. H. Cheung, On hierarchical palmprint coding with multiple features for personal identification in large databases, *IEEE Transactions on Circuits and Systems for Video Technology*, 14, 234–243, 2004.

45. K. C. Chan, Y. S. Moon, and P. S. Cheng, Fast fingerprint verification using subregions of fingerprint images, *IEEE Transactions on Circuits and Systems for Video Technology*, 14, 95–101, 2004.

46. A. K. Jain, A. Ross, and S. Prabhakar, An introduction to biometric recognition, *IEEE Transactions on Circuits and Systems for Video Technology*, 14, 4–20, 2004.

47. Y. S. Moon, H. C. Ho, and K. L. Ng, A secure smart card system with biometrics capability, in *Proceedings of the IEEE 1999 Canadian Conference on Electrical and Computer Engineering*, Canada, IEEE Computer Society, 1999, pp. 261–266.

48. T. Y. Tang, Y. S. Moon, and K. C. Chan, Efficient implementation of fingerprint verification for mobile embedded systems using fixed-point arithmetic, in *Proceedings of the 2004 ACM Symposium on Applied Computing*, Nicosia, Cyprus, ACM, 2004, pp. 821–825.

49. L. Benini, A. Macii, and M. Poncino, Energy-aware design of embedded memories: A survey of technologies, architectures, and optimization techniques, *Transactions on Embedded Computing Systems*, 2, 5–32, 2003.

50. Actel Corporation, Design security in nonvolatile flash and antifuse FPGAs, Technical Report 5172163–0/11.01, 2001.

51. T. S. Messerges and E. A. Dabbish, Digital rights management in a 3G mobile phone and beyond, in *Proceedings of the 2003 ACM Workshop on Digital Rights Management*, Washington, DC, ACM, 2003, pp. 27–38.

52. D. L. C. Thekkath, M. Mitchell, P. Lincoln, D. Boneh, J. Mitchell, and M. Horowitz, Architectural support for copy and tamper resistant software, in *Proceedings of the 9th International Conference on Architectural Support for Programming Languages and Operating Systems*, Cambridge, MA, ACM, 2000, pp. 168–177.

53. J. H. Saltzer and M. D. Schroder, The protection of information in computer systems, in *Proceedings of the IEEE*, 63, 1278–1308, 1975.

54. A. D. Rubin and D. E. Geer Jr., Mobile code security, *Internet Computing, IEEE*, 2, 30–34, 1998.

55. G. C. Necula, Proof-carrying code, in *Proceedings of the 24th ACM SIGPLAN-SIGACT Symposium on Principles of Programming Languages (POPL '97)*, Paris, France, ACM Press, 1997, pp. 106–119.

56. V. N. Venkatakrishnan, R. Peri, and R. Sekar, Empowering mobile code using expressive security policies, in *Proceedings of the 2002 Workshop on New Security Paradigms*, ACM Press, 2002, pp. 61–68.

57. Kingpin and Mudge, Security analysis of the palm operating system and its weaknesses against malicious code threats, in *10th Usenix Security Symposium*, Washington, DC, Usenix Association, 2001, pp. 135–152.

58. Kingpin, Attacks on and countermeasures for USB hardware token device, in *Proceedings of the Fifth Nordic Workshop on Secure IT Systems Encouraging Co-operation*, Reykjavik, Iceland, 2000, pp. 135–151.

59. S. Skorobogatov, Low temperature data remanence in static RAM, Technical Report UCAM-CL-TR-536, University of Cambridge, 2002.

60. P. Gutman, Data remanence in semiconductor devices, in *Proceedings of the 10th USENIX Security Symposium*, Washington, DC, Usenix Association, 2001.

61. M. G. Kuhn, Optical time-domain eavesdropping risks of CRT displays, in *Proceedings of the IEEE Symposium on Security and Privacy*, Berkeley, CA, IEEE Computer Society, 2002, pp. 3–18.

62. J. Loughry and D. A. Umphress, Information leakage from optical emanations, *ACM Transactions on Information and System Security*, 5, 262–289, 2002.

63. S. P. Skorobogatov and R. J. Anderson, Optical fault induction attacks, in *Revised Papers from the 4th International Workshop on Cryptographic Hardware and Embedded Systems*, Redwood Shores, CA, Springer-Verlag, 2003, pp. 2–12.

64. S. L. Garfinkel and A. Shelat, Remembrance of data passed: A study of disk sanitization practices, *IEEE Security and Privacy Magazine*, 1, 17–27, 2003.

65. A. G. Voyiatzis and D. N. Serpanos, A fault-injection attack on Fiat–Shamir cryptosystems, in *24th International Conference on Distributed Computing Systems Workshops (ICDCS 2004 Workshops)*, Hachioji, Tokyo, Japan, IEEE Computer Society, 2004, pp. 618–621.

66. D. N. Serpanos and R. J. Lipton, Defense against man-in-the-middle attack in client-server systems with secure servers, in *The Proceedings of IEEE ISCC'2001*, Hammammet, Tunisia, IEEE Computer Society, July 3–5, 2001, pp. 9–14.

67. R. J. Lipton, S. Rajagopalan, and D. N. Serpanos, Spy: A method to secure clients for network services, in *The Proceedings of the 22nd International Conference on Distributed Computing Systems Workshops (Workshop ADSN'2002)*, Vienna, Austria, July 2–5, 2002, pp. 23–28.

68. H. Greg and M. Gary, *Exploiting Software: How to Break Code*, Addison-Wesley Professional, Reading, MA, 2004.

69. Rosenthal, Scott. *Serial EEPROMs Provide Secure Data Storage for Embedded Systems*, SLTF Consulting. Available at: http://www.sltf.com/articles/pein/pein9101.htm

70. Trusted Computing Group, TCG. Available at: https://www.trustedcomputinggroup.org/home

71. A. Huang, Keeping secrets in hardware: The microsoft Xbox case study, in *Revised Papers from the 4th International Workshop on Cryptographic Hardware and Embedded Systems*, Cologne, Germany, Springer-Verlag, 2003, pp. 213–227.

72. P. Dusart, G. Letourneux, and O. Vivolo, Differential fault analysis on AES, in *International Conference on Applied Cryptography and Network Security*, Lecture Notes in Computer Science #2846, Kunming, China, Springer-Verlag, 2003, pp. 293–306.

73. P. Gutmann, Lessons learned in implementing and deploying crypto software, in *Proceedings of the 11th USENIX Security Symposium*, San Francisco, CA, Usenix Association, 2002, pp. 315–325.

74. H. Bar-El, H. Choukri, D. Naccache, M. Tunstall, and C. Whelan, The Sorcerer's apprentice guide to fault attacks, in *Workshop on Fault Diagnosis and Tolerance in Cryptography*, 2004, Cryptology ePrint Archive, Report 2004/10, IACR.

75. M.-L. Akkar, R. Bevan, P. Dischamp, and D. Moyar, Power analysis, what is now possible, in *Advances in Cryptology—ASIACRYPT 2000: 6th International*, 2000, pp. 489–502, Springer-Verlag.

76. K. Gandolfi, C. Mourtel, and F. Olivier, Electromagnetic analysis: Concrete results, in *Proceedings of the 3rd International Workshop on Cryptographic Hardware and Embedded Systems*, Paris, France, Springer-Verlag, 2001, pp. 251–261.

77. J. Dhem, F. Koeune, P. Leroux, P. Mestr, J. Quisquater, and J. Willems, A practical implementation of the timing attack, in *Proceedings of the International Conference on Smart Card Research and Applications*, Louvain-la-Neuve, Belgium, Springer-Verlag, 1998, pp. 167–182.

78. P. Kocher, J. Jaffe, and B. Jun, Differential power analysis, in *CRYPTO '99*, Santa Barbara, CA, Springer-Verlag, 1999, pp. 388–397.

79. T. S. Messerges, E. A. Dabbish, and R. H. Sloan, Investigation of power analysis attacks on smartcards, in *Proceedings of USENIX Workshop on Electronic Commerce*, Chicago, IL, Usenix Association, 1999, pp. 151–161.

80. J. Kelsey, B. Schneier, D. Wagner, and C. Hall, Side channel cryptanalysis of product ciphers, in *Proceedings of ESORICS*, Springer-Verlag, 1998, pp. 97–110.

81. P. C. Kocher, Timing attacks on implementations of Diffie-Hellman RSA DSS and other systems, in *Proceedings of CRYPTO'96*, Lecture Notes in Computer Science #1109, Santa Barbara, CA, Springer-Verlag, 1996, pp. 104–113.

82. A. Hevia and M. Kiwi, Strength of two data encryption standard implementations under timing attacks, *ACM Transactions on Information and System Security*, 2, 416–437, 1999.

83. W. Schindler, F. Koeune, and J.-J. Quisquater, Unleashing the full power of timing attack, UCL Crypto Group Technical Report CG-2001/3, 2001, Université Catholique de Louvain.

84. F. Koeune and J.-J. Quisquater. A timing attack against Rijndael, Technical Report CG-1999/1, 1999, Universite Catholique de Louvain, Louvain-La-Neuve, Belgique.

85. E. W. Felten and M. A. Schneider, Timing attacks on Web privacy, in *Proceedings of the 7th ACM Conference on Computer and Communications Security*, Athens, Greece, ACM Press, 2000, pp. 25–32, ACM Press.

86. D. X. Song, D. Wagner, and X. Tian, Timing analysis of keystrokes and timing attacks on SSH, in *Proceedings of the 10th USENIX Security Symposium*, Washington, DC, Usenix Association, 2001, USENIX Association.

87. J. R. Rao and P. Rohatgi, EMpowering side-channel attacks, IACR Cryptography ePrint Archive: Report 2001/037. Available at: http://eprint.iacr.org/2001/037/(September, 2004), IACR.

88. J.-J. Quisquater and D. Samyde, ElectroMagnetic analysis (EMA): Measures and countermeasures for smart cards, in *International Conference on Research in Smart Cards, E-Smart 2001*, Lecture Notes in Computer Science #2140, Springer-Verlag, 2001, pp. 200–210.

89. D. Page, Theoretical use of cache memory as a cryptanalytic side-channel, Technical Report CSTR-02–003, Computer Science Department, University of Bristol, Bristol, England, 2002.

90. J. Coron, D. Naccache, and P. Kocher, Statistics and secret leakage, *ACM Transactions on Embedded Computing Systems*, 3, 492–508, 2004.

91. D. Brumley and D. Boneh, Remote timing attacks are practical, in *Proceedings of the 12th USENIX Security Symposium*, Washington, DC, Usenix Association, 2003.

92. J. Blömer, M. Otto, and J. Seifert, A new CRT-RSA algorithm secure against bellcore attacks, in *Proceedings of the 10th ACM Conference on Computer and Communication Security*, Washingtion, DC, ACM Press, 2003, pp. 311–320.

93. C. Aumüller, P. Bier, W. Fischer, P. Hofreiter, and J. Seifert, Fault attacks on RSA with CRT: Concrete results and practical countermeasures, in *Revised Papers from the 4th International Workshop on Cryptographic Hardware and Embedded Systems*, 2003, pp. 260–275, Springer-Verlag.

94. J. Blömer and J.-P. Seifert, Fault-based cryptanalysis of the advanced encryption standard (AES), in *Financial Cryptography 2003*, Lecture Notes in Computer Science, vol. 2742, France, Springer-Verlag, 2003, pp. 162–181.

95. M. Jacob, D. Boneh, and E. Felten, Attacking an obfuscated cipher by injecting faults, in *Proceedings of 2002 ACM Workshop on Digital Rights Management*, Washington, DC, Usenix Association, Springer-Verlag.

96. S. Yen and M. Joye, Checking before output may not be enough against fault-based cryptanalysis, *IEEE Transactions on Computers*, 49, 967–970, 2000.

97. S. Yen, S. Kim, S. Lim, and S. Moon, A countermeasure against one physical cryptanalysis may benefit another attack, in *Proceedings of the 4th International Conference Seoul on Information Security and Cryptology*, Seoul, Korea, Springer-Verlag, 2002, pp. 414–427.

98. S. Yen, S. Kim, S. Lim and S. Moon, RSA speedup with residue number system immune against hardware fault cryptanalysis, in *Proceedings of the 4th International Conference Seoul on Information Security and Cryptology*, Seoul, Korea, Springer-Verlag, 2002, pp. 397–413.

99. P. Paillier, Evaluating differential fault analysis of unknown cryptosystems, in *Proceedings of the 2nd International Workshop on Practice and Theory in Public Key Cryptography*, Springer-Verlag, 1999, pp. 235–244.

100. V. Klíma and T. Rosa, Further results and considerations on side channel attacks on RSA, IACR Cryptography ePrint Archive: Report 2002/071. Available at: http://eprint.iacr.org/2002/071/ (September, 2004).

101. V. Klíma and T. Rosa, Attack on private signature keys of the OpenPGP format, PGP(TM) programs and other applications compatible with OpenPGP, IACR Cryptology ePrint Archive Report 2002/073. Available at: http://eprint.iacr.org/2002/076.pdf (September, 2004), IACR.

102. S.-M. Yen, S. Moon, and J.-C. Ha, Hardware fault attack on RSA with CRT revisited, in *Proceedings of ICISC 2002*, Lecture Notes in Computer Science #2587, 2003, pp. 374–388.

103. D. P. Maher, Fault induction attacks, amper resistance, and hostile reverse engineering in perspective, in *Proceedings of the 1st International Conference on Financial Cryptography*, Anguilla, British West Indies, Springer-Verlag, 1997, pp. 109–122.

104. Y. Zheng and T. Matsumoto, Breaking real-world implementations of cryptosystems by manipulating their random number generation in *Proceedings of the 1997 Symposium on Cryptography and Information Security*, Fukuoka, Japan.

105. I. Biehl, B. Meyer, and V. Müller, Differential fault attacks on elliptic curve cryptosystems, in *Proceedings of CRYPTO 2000*, Lecture Notes in Computer Science, vol. 1880, Santa Barbara, CA, Springer-Verlag, 2000, pp. 131–146.

106. E. Biham and A. Shamir, Differential fault analysis of secret key cryptosystems, Lecture Notes in Computer Science, vol. 1294, 1997, pp. 513–525, Springer-Verlag.

107. D. Boneh, R. A. DeMillo, and R. J. Lipton, On the importance of eliminating errors in cryptographic computations, *Journal of Cryptology: The Journal of the International Association for Cryptologic Research*, 14, 101–119, 2001.

108. D. Boneh, R. A. DeMillo, and R. J. Lipton, On the importance of checking cryptographic protocols for faults, in *Proceedings of Eurocrypt'97*, Lecture Notes in Computer Science, vol. 1233, 1997, pp. 37–51.

109. M. Joye and J.-J. Quisquater, Attacks on systems using Chinese remaindering, UCL Crypto Group, Belgium, Technical Report CG1996/9, 1996.

110. J. Marc and Q. Jean-Jacques, Faulty RSA encryption, UCL crypto group, Technical Report CG-1997/8, 1997.

111. F. Bao, R. H. Deng, Y. Han, A. B. Jeng, A. D. Narasimhalu, and T. Ngair, Breaking public key cryptosystems on tamper resistant devices in the presence of transient faults, in *Proceedings of the 5th International Workshop on Security Protocols*, Paris, France, Springer-Verlag, 1998, pp. 115–124.

112. R. J. Anderson, *Security Engineering: A Guide to Building Dependable Distributed Systems*, Wiley, New York, 2001.

113. Intel Corp., Analysis of the floating point flaw in the Pentium processor, November 1994. Available at: http://support.intel.com/support/processors/pentium/fdiv/wp/(September, 2004).

114. P. Wright, *Spycatcher: The Candid Autobiography of a Senior Intelligence Officer*, Viking, New York, 1987, Dell Publishing.

115. A. Shamir and E. Tromer, Acoustic Cryptanalysis–On nosy people and noisy machines, in Eurocrypt 2004 Rump Session presentation. Available at: http://www.wisdom.weizmann.ac.il/~tromer/acoustic/(September, 2004).

116. D. Boneh, Twenty years of attacks on the RSA cryptosystem, *Notices of the American Mathematical Society (AMS)*, 46, 203–213, 1999.

117. A. Lenstra, *Memo on RSA Signature Generation in the Presence of Faults*, Manuscript, available from the author. September 28, 1996.

118. K. Koç, T. Acar, and B. S. Kaliski Jr., Analyzing and comparing montgomery multiplication algorithms, *IEEE Micro*, 16, 26–33, 1996.

119. H. Handschuh and H. Howard, A timing attack on RC5, in *Selected Areas in Cryptography: 5th Annual International Workshop, SAC'98*, Kingston, Canada, Springer-Verlag, 1998.

120. M. G. Kuhn, Compromising emanations: Eavesdropping risks of computer displays, Technical. Report UCAM-CL-TR-577, Computer Laboratory, University of Cambridge, Cambridge, UK, December 2003.

121. IBM Corp., *IBM PCI Cryptographic Coprocessor*. Available at: http://www-3.ibm.com/security/cryptocards/html/pcicc.shtml (September, 2004).

122. D. S. Touretzky, *Gallery of CSS Descramblers*. Available at: http://www.cs.cmu.edu/~dst/DeCSS/Gallery (September 2004).

123. A. J. Clark, Physical protection of cryptographic device, in *Proceedings of Eurocrypt'87*, Amsterdam, Netherlands, Springer-Verlag, 1987, pp. 83–93.

124. D. Chaum, Design concepts for tamper-responding system, in *Advances in Cryptology Proceedings of Crypto '83*, New York, Plenum Press, 1983, pp. 387–392.

125. S. H. Weingart, S. R. White, W. C. Arnold, and G. P. Double, An evaluation system for the physical security of computing systems, in *6th Annual Computer Security Applications Conference*, Tucson, AZ, IEEE, 1990, pp. 232–243.

126. EFF, *U.S. v. ElcomSoft & Sklyarov FAQ*, Available at: http://www.eff.org/IP/DMCA/US_v_Elcomsoft/us_v_sklyarov_faq.html (September, 2004).

127. R. J. Anderson, Why cryptosystems fail, in *Proceedings of ACM CSS'93*, pp. 215–217, Fairfax, VA, ACM Press, November 1993, ACM Press.

128. D. Samyde, S. Skorobogatov, R. Anderson, and J. Quisquater, On a new way to read data from memory, in *Proceedings of the First International IEEE Security Storage Workshop (SISW)*, IEEE Computer Society, Redwood Shores, CA, Springer-Verlag, 2002.

129. A. G. Voyiatzis and D. N. Serpanos, Active hardware attacks and proactive countermeasures in *Proceedings of IEEE ISCC 2002*, Giardini Naxos, Italy, IEEE Computer Society, 2002.

130. A. Shamir, Method and apparatus for protecting public key schemes from timing and fault attacks, U.S. Patent No. 5,991,415, November 23, 1999, United States Patent and trademark Office (USPTO).

131. J. Daemen and V. Rijmen, The block cipher Rijndael, in *Proceedings of Smart Card Research and Applications 2000*, Lecture Notes in Computer Science 1820, Louvain-la-Neuve, Belgium, Springer-Verlag, 1998, pp. 288–296.

132. NIST, *NIST, Advanced Encryption Standard (AES), Federal Information Processing Standards Publication 1997* November 26, 2001.

133. B. Kaliski and M. J. B. Robshaw, Comments on Some New Attacks on Cryptographic Devices, *RSA Laboratories Bulletin* 5, July 1997, RSA Labs.

134. K. Sakurai and T. Takagi, A reject timing attack on an IND-CCA2 public-key cryptosystem, in *Proceedings of ICISC 2002*, Lecture Notes in Computer Science #2587, Seoul, Korea, Springer-Verlag, 2003.

135. Wireless Transport Layer Security (WTLS) Specification, Open Mobile Alliance (OMA). Available at: http://www.openmobilealliance.org/tech/affiliates/LicenseAgreement.asp?DocName=/wap/wap-261-wtls-20010406-a.pdf

136. Mobile Electronic Transactions, Available at: http://www.mobiletransaction.org/

137. O. Kömmerling and M. G. Kuhn. Design principles for tamper-resistant smartcard processors, *Proceedings of the USENIX Workshop on Smartcard Technology (Smartcard '99)*, Chicago, IL, May 10–11, 1999, USENIX Association, pp. 9–20, ISBN 1-880446-34-0.

138. R. J. Anderson and M. G. Kuhn, Low cost attacks on tamper resistant devices, in M. Lomas et al. (ed.), *Security Protocols, 5th International Workshop*, Paris, France, April 7–9, 1997, *Proceedings*, Lecture Notes in Computer Science, vol. 1361, pp. 125–136, Springer-Verlag, ISBN 3-540-64040-1.

139. M. G. Kuhn, Electromagnetic eavesdropping risks of flat-panel displays, *Presented at the 4th Workshop on Privacy Enhancing Technologies*, May 26–28, 2004, Toronto, Canada, Springer-Verlag.

140. S. Moore, R. Anderson, P. Cunningham, R. Mullins, and G. Taylor, Improving smart card security using self-timed circuits, *Eighth International Symposium on Advanced Research in Asynchronous Circuits and Systems*, Grenoble, IEEE Computer Society, 2002.

141. Side-channel Attacks Database. Available at: http://www.sidechannelattacks.com

142. J. Takahashi, T. Fukunaga, and K. Yamakoshi, Differential fault analysis mechanism on the AES key schedule, *Workshop on Fault Diagnosis and Tolerance in Cryptography*, Vienna, Austria, IEEE Computer Society, 2007. *FDTC 2007*, pp. 62–74, IEEE.

143. P. Dusart, G. Letourneux, and O. Vivolo, Differential fault analysis on A.E.S., in *Applied Cryptography and Network Security—ACNS 2003*, Lecture Notes in Computer Science, vol. 2846, 2003, pp. 293–306, Springer-Verlag.

144. G. Piret and J. J. Quisquater, A differential fault attack technique against SPN structures, with application to the AES and KHAZAD, in *Cryptographic Hardware and Embedded Systems—CHES 2003*, Lecture Notes in Computer Science, vol. 2779, 2003, pp. 77–88.

145. A. Moradi, M. T. M. Shalmani, and M. Salmasizadeh, A generalized method of differential fault attack against AES cryptosystem, in *Cryptographic Hardware and Embedded Systems—CHES 2006*, Lecture Notes in Computer Science, vol. 4249, 2006, pp. 91–100.

146. C. N. Chen and S. M. Yen, Differential fault analysis on AES key schedule and some countermeasures, in *Australasian Conference on Information Security and Privacy 2003 (ACISP 2003)*, Lecture Notes in Computer Science, vol. 2727, 2003, pp. 118–129, Springer.

147. C. Giraud, DFA on AES, in *Advanced Encryption Standard (AES): 4th International Conference, AES 2004*, Lecture Notes in Computer Science, vol. 3373, 2005, pp. 27–41, Springer-Verlag.

148. D. Peacham and B. Thomas, A DFA attack against the AES key schedule, SiVenture White Paper 001, 26 October 2006. Available at: http://www.siventure.com/pdfs/AES KeySchedule DFA whitepaper.pdf.

149. U.S. Federal Standard 1027, Telecommunications, *General Security Requirements for Equipment Using the Data Encryption Standard*, National Bureau of Standards, April 14, 1982.

150. L. Kerstin, Embedded security: Physical protection against tampering attacks, *Embedded Security in Cars: Securing Current and Future Automotive IT Applications*, p. 207, ISBN 3540283846, 2007.

151. S. Ravi, A. Raghunathan, and S. Chakradhar, Tamper resistance mechanisms for secure embedded systems, in *Proceedings of the 17th International Conference on VLSI Design*, IEEE Computer Society, 2004, pp. 605–611.

152. D. Arora, S. Ravi, A. Raghunathan, and N. K. Jha, Secure embedded processing through hardware-assisted run-time monitoring, *Proceedings on Design, Automation and Test in Europe*, 2005, pp. 178–183, vol. 1, March 7–11, 2005.

153. A. Seshadri, A. Perrig, L. van Doorn, P. Khosla, SWATT: SoftWare-based attestation for embedded devices, *Security and Privacy, 2004. Proceedings on 2004 IEEE Symposium*, Berkeley, CA, IEEE Computer Society, pp. 272–282, May 9–12, 2004.

154. S. Wong, The evolution of wireless security in 802.11 networks: WEP, WPA and 802.11 standards, Whitepaper, 2003. Available at: http://cnscenter.future.co.kr/resource/hot-topic/wlan/1109.pdf

155. IETF, *Transport Layer Security*. Available at: http://www.ietf.org/html.charters/tls-charter.html

156. VISA Corp., *Secure Electronic Transaction Specification. Books 1–3: Business Description, Programmer's Guide, Formal Protocol Specification*, June 1996.

157. N. R. Potlapally, S. Ravi, A. Raghunathan, and N. K. Jha, A study of the energy consumption characteristics of cryptographic algorithms and security protocols, *IEEE Transactions on Mobile Computing*, 5, 128–143, February 2006.

158. A. G. Fragopoulos and D. N. Serpanos, Intellectual property protection using embedded systems, *Proceedings of the NATO Advanced Research Workshop on Security and Embedded Systems*, Patras, Greece, 2005, IOS Press, ISBN 1-58603-580-0.

159. S. Gurun and C. Krintz, A run-time, feedback-based energy estimation model for embedded devices, in *CODES+ISSS '06: Proceedings of the 4th International Conference on Hardware/Software Codesign and System Synthesis*. Seoul, Korea, ACM Press, 2006, pp. 28–33. [Online]. Available at: http://portal.acm.org/citation.cfm?id=1176264

160. S. Gurun, P. Nagpurkar, and B. Y. Zhao, Energy consumption and conservation in mobile peer-to-peer systems, in *MobiShare '06: Proceedings of the 1st International Workshop on Decentralized Resource Sharing in Mobile Computing and Networking*, New York, ACM, 2006, pp. 18–23. [Online]. Available at: http://portal.acm.org/citation.cfm?id=1161258

161. S. Gurun, Modeling, Predicting and Reducing Energy Consumption in Resource Restricted Computers, PhD Dissertation, University of Santa Barbara, Santa Barbara, CA, March 2007.

162. D. Aucsmith, Tamper resistant software: An implementation, in *Proceedings of the 1st International Workshop on Information Hiding*. London, U.K., 1996, pp. 317–333, Springer-Verlag. [Online]. Available at: http://portal.acm.org/citation.cfm?id=731528

163. M. Blum, Designing programs to check their work, ACM *SIGSOFT Software Engineering Notes*, 18(3):1, July 1993. [Online]. DOI: 10.1145/174146.154185.

164. C. S. Collberg and C. Thomborson, Watermarking, tamper-proofing, and obfuscation—tools for software protection, *IEEE Transactions on Software Engineering*, 28(8), 735–746, 2002. [Online]. Available at: http://dx.doi.org/10.1109%2FTSE.2002.1027797

165. N. Potlapally, S. Ravi, A. Raghunathan, and M. Sankaradass. Algorithm exploration for efficient public-key security processing on wireless handsets, in *Proceedings on Design, Automation and Test in Europe (DATE) Designers Forum*, Paris, France, IEEE Computer Society, 2002, pp. 42–46.

166. N. R. Potlapally, S. Ravi, A. Raghunathan, and G. Lakshminarayana, Optimizing public-key encryption for wireless clients, in *IEEE International Conference on Communications, ICC 2002*, vol. 2, IEEE Computer Society, 2002, pp. 1050–1056.

167. S. Ravi, A. Raghunathan, N. Potlapally, and M. Sankaradass, System design methodologies for a wireless security processing platform, in *DAC '02: Proceedings of the 39th ACM/IEEE Conference on Design Automation*, New Orleans, LA, ACM Press, 2002, pp. 777–782.

19

Web Services for Embedded Devices

Hendrik Bohn
University of Rostock

Frank Golatowski
University of Rostock

19.1 Introduction

In recent years, an increasing demand for highly automated, process-oriented, and distributed systems consisting of a large number of interconnected heterogeneous hardware and software components is noticeable. These systems—usually composed of components from different manufacturers—require a high interoperability across heterogeneous physical media, platforms, programming languages, and application domains facing several problems:

- Proprietary interfaces. Components have mostly proprietary interfaces weakening compatibility and developers must know them in order to use such components.
- Limitations in the awareness. Component users are often not aware of other component's functionality.
- Limited discovery capabilities. Component users do not know how to search for a component with certain functionality nor do they know which components are available in a certain scope.

- Proprietary methods for interactions. The interaction mechanisms between components are mostly bound to their application and application domain. The formats for exchanging message as well as the used data formats are mostly proprietary.
- Constrained composition capabilities. Processes/workflows describing the interaction between components are designed for specific applications and domains and seldom applicable for other components and applications.

Hardware components have additional requirements in comparison to software components. Among them are

- Location awareness. In contrast to software components, the location of hardware components is usually important for their application. Offered functionality is often bound to a specific location. For example, the location of a printer is important to its users as it has to be in reachable distance. Therefore, also the printing functionality is bound to the location of the printer, whereas software components can be made available anywhere and where their functionality is required.
- Mobility support. Furthermore, mobile devices can roam between networks possibly resulting in changing addresses of the network protocols.
- Statefulness. From the outside-in view, a software component can be designed in a stateless way by starting a new instance with default values when the component is invoked. Device components mostly have a state which has to be taken care of. For example, a printing device being invoked while printing has to queue the printing job.

These problems are solved for certain proprietary applications and in mostly homogeneous environments, but a solution applicable for a large heterogeneous system is still missing.

The service-oriented architecture (SOA) approach addresses these problems and represents the next step in the endeavor of component-based development and reusability.

SOA [1] is a design paradigm for the interaction between heterogeneous components supporting autonomy and interoperability. The underlying philosophy is that every component offers certain functions which might interest other components or users. SOA provides an outside-in view (orientation) on the functionality of components—whose self-described interfaces to their functionality are called services—rather than deriving the interface from the implementation. This utilizes developers building *architectures* of heterogeneous components derived from their application.

The SOA paradigm is based on following foundation and principles: Service orientation is based on open *standards* ensuring interoperability. Although *security* is not an attribute of SOA, it is important to foster acceptance. SOA is based on *simplicity* in the sense of flexibility, reusability, and adaptability in the usage of services and the integration of services into heterogeneous service environments. The attribute *distribution* stands for the independence of services which self-contain their functionality. New applications can be build by designing the interaction of services and the definition of policies (if needed) to formulate interaction constraints. Services are *loosely coupled* allowing dynamical searching for finding and binding of services. Services are clearly abstracted from their implementation and application environment. The *registry* is a central repository for all available services in a certain environment. SOA implementations without a registry make use of other advanced search mechanisms. *Process orientation* in SOA shows a major advantage of SOAs and moves the technology closer to its application. Processes and workflows can be designed and mapped into an interaction of services by composing the corresponding services.

An SOA defines three roles: Service provider, client/user, and optionally a registry. The *service provider* is generally known as *service* and offers its functionality via a service (interface) to the SOA. *Service clients/users* make use of the functionality offered by a service provider. A *service registry* is

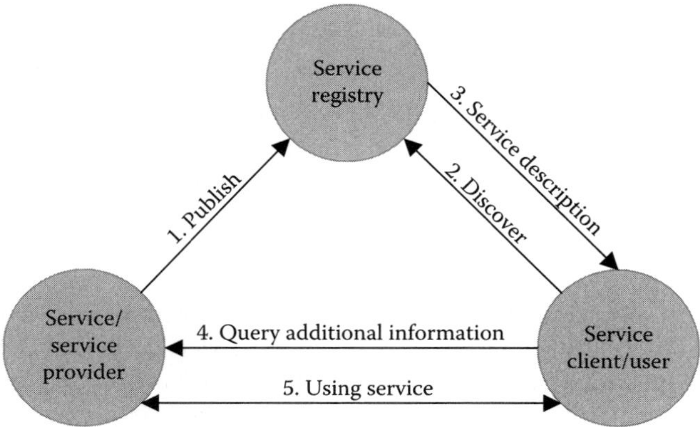

FIGURE 19.1 Roles in an SOA. (Adapted from Dostal, W., et al., *Service-Orientierte Architekturen mit Web Services*. Elsevier, Spektrum Akademischer Verlag, München, Germany, 2005.)

used by service providers to register themselves and by service clients/users to search for a specific service and to obtain information required to establish a connection to the service provider.

In this chapter, a service provider is simply called service and a service client/user is called service client or just client, if ambiguity is impossible.

Every service owns a description of its properties and capabilities which has to be described by the service developer. The format of the description is standardized and has to be machine-readable (usually character-based in a eXtensible Markup Language [XML] format) and contains all information needed to interact with a service.

A classical SOA implementation involves five stages of service usage as shown in Figure 19.1 [2]:

Stage 1. The service registers a limited description of itself including its general capabilities at the service registry (*publishing*). If no service registry is available, the service announce itself to the network by sending its description including basic capabilities (*announcement*). The limited description fosters scalability by reducing network load as service clients are normally only interested in specific functions of a service and not in all provided functionality. Further description details are requested in a later step.

Stage 2. The service client sends a description of the desired service to the service registry (*discovery*). In SOAs without a service registry, the discovery message will be sent to the network, or the service client listens for service announcements.

Stage 3. The service registry answers either with a failure response or by sending all service descriptions matching the discovery request (*discovery response*). The service description also includes the network address of the matching service/service provider. In the absence of a service registry, the corresponding service providers will answer directly.

Stage 4. In the next step, service clients will request a detailed description of the capabilities from the matching services to select the most reasonable one (*description query*).

Stage 5. The last step is the usage of the service according to the rules defined between service provider and client (*service usage*). These rules are described in the detailed description of the service provider. If a service provider leaves the network, it should announce its intentions by sending a bye-message to the registry and clients, respectively.

19.2 Device-Centric SOAs

There are several implementations following the SOA paradigm. The subsequent paragraphs briefly introduce the implementations being relevant in the context of this chapter.

19.2.1 SOA Implementations for Devices

The Open Service Gateway Initiative (OSGi) specification defines a service platform that serves as a common architecture for service providers, service developers, and software equipment vendors who want to deploy, develop, and manage services [3]. The specification is based on the Java platform and promotes application independence. Thereby, it is enabling the easy integration of existing technologies. Deployed services are called Bundles and are plugged into the framework. An OSGi service is a simple Java interface, but the semantics of a service are not clearly specified. The main backdraw of OSGi is its reliance on Java.

The Java Intelligent Network Infrastructure (Jini) was developed by Sun Microsystems for spontaneous networking of services and resources based on the Java technology [4]. Services/devices are registered and maintained at a centralized meta-service called Lookup Service but carry the code (proxy) needed to use them. This code is dynamically downloaded by clients when they wish to use the service. Each service access has to be performed by using the lookup service. Jini's main drawbacks are its reliance on Java and the need of a centralized service registry.

Universal Plug-and-Play (UPnP) is a simple, easy-to-use SOA for small networks [5]. It supports ad-hoc networking of devices and interaction of services by defining their announcement, discovery, and usage. Programming languages and transmission media are not assumed. Only protocols and interfaces are specified instead. The UPnP specification divides the device lifecycle into six phases: Addressing, Discovery and Description (specifying automatic integration of devices and services), Control (operating a remote service/device), Eventing (subscribing to state changes of a remote service/device), and Presentation (URL representation of a service/device specifying its usage). UPnP also specifies a usage profile for distributed audio/video application—the UPnP AV architecture [6]. UPnP supports smaller networks only. With an increasing amount of services/devices the amount of broadcast messages grows exponentially in an UPnP network. Furthermore UPnP supports IPv4 only.

Web services (WSs) can be considered as the SOA with the highest market penetration. The WS architecture provides a set of modular protocol building blocks that can be composed in varying ways (called profiles) to meet requirements on the interaction of heterogeneous software components such as interoperability, self-description, and security [7]. *Profiles* define the subset of protocols used for implementing a specific application, required adaptations of the protocols, and the way they should be used in order to achieve interoperability. WSs address networks of any size and provide a set of specifications for service discovery, service description, security, policy, and others. The implementation is entirely hidden from their interfaces and may be exchanged at runtime. WSs use widely accepted standards such as XML and Simple Object Access Protocol (SOAP) for message exchange. Unfortunately, WSs do not bring Plug and Play capabilities and a sufficient solution (specifications and guidelines) for device integration. This is leveraged by DPWS.

The devices profile for web service (DPWS), announced in August 2004 and revised in May 2005 and February 2006 by a consortium lead by Microsoft Corporation, is a profile identifying a core set of WSs protocols that enables dynamic discovery of, and event capabilities for WSs [8]. The profile arranges several WS specifications such as WS-Addressing, WS-Discovery, WS-MetadataExchange, and WS-Eventing for devices, particularly. In contrast to UPnP, it supports discovery and interoperability of WSs beyond local networks. Temporarily DPWSs were considered to be the successor of UPnP and thus subtitled with "A Proposal for UPnP 2.0 Device Architecture" in one of its earlier versions. However, DPWS is not compatible to UPnP, and therefore it presents a SOA implementation on its own. DPWS is part of the new Microsoft Operating Systems Windows Vista and Windows CE.

It is important to note that the concepts behind two SOA implementations might be entirely different. SOAs are just a paradigm how future systems could look like and what requirements they should address. The SOA paradigm does neither define design specifications of such system nor does it address implementation aspects.

	OSGi	Jini	UPnP	WS	DPWS
Common interfaces and awareness	✓	✓	✓	✓	✓
Discovery and standardized interaction	✓	✓	✓	✓	✓
Evolved composition standards				✓	
Support for hardware components	✓	✓	✓		✓
Plug and Play capability		✓	✓		✓
Programming language independence			✓	✓	✓
Network media independence		✓	✓	✓	✓
Large scalability	✓	✓		✓	✓
Security concepts	✓	✓		✓	✓

FIGURE 19.2 Evaluation of service-oriented middleware.

19.2.2 Evaluation of Device-Centric SOAs

The table presented in Figure 19.2 is based on previous published research work by Bohn et al. [9] and shows an excerpt of the evaluation results of mentioned SOA implementations with regard to the problems identified in the introduction of this chapter. Although the SOA paradigm theoretically addresses all problems, the individual SOA implementations differ in their support for devices integration, integration of security concepts, and standards for service composition, plug and play capability, programming language and network media independence, and scalability as shown in Figure 19.2.

Common interfaces and awareness indicated the capability of the standards to provide a common view on their components based on their functionality (called services). A requirement is self-describing interfaces in order that other components are aware of the service's functionality (*awareness*). Furthermore, a *standardized interaction* between clients and services must be ensured including defined protocol, message exchange patterns (MEP), and data type specification. The capability of searching for certain functionality and the automatic announcement of new services (*discovery capability*) fosters automation in networked environments and applications. Networks of components reveal their full potential if standards that enable designers and nontechnical personal to design complex applications by simply composing involved components/services (*evolved composition standards*) exist. Although niche solutions exist for other SOAs, only WSs offer a sophisticated set of standards for service composition. The support for hardware components involves addressing their specific requirements such as mobility, location awareness, and publish/subscribe mechanism. OSGi, Jini, UPnP, and DPWS support that in some way. *Plug and Play capabilities* refer to a network which is not dependent on a central service registry. Services and devices announce the entering and leaving of networks and thereby clients are informed of their functionality. This is provided by Jini, UPnP, and DPWS. Only UPnP, WS and DPWS are *programming language independent*, which means clients and services can be developed in any kind of language and executed on any kind of execution environment. *Network media independence* corresponds to message exchanges over any kind of transport protocol. OSGi is somehow out of scope in that way as messages are exchanged over Java. Very complex applications require a *large scalability* supporting a large number of involved services. Furthermore, secure message exchanges based on standardized *security concepts* are often required for such applications. As UPnP mainly focuses on audio/video applications in small home networks; it is not highly scalable and has no security concepts defined yet.

It is apparent that the DPWS is most appropriate to solve the problems stated at the beginning of the introduction. DPWS offers extensive features to support the interoperability between different devices/services. DPWS compliant components are able to interact with each other forming a highly distributed application. However, each client and service has to be programmed manually. This may

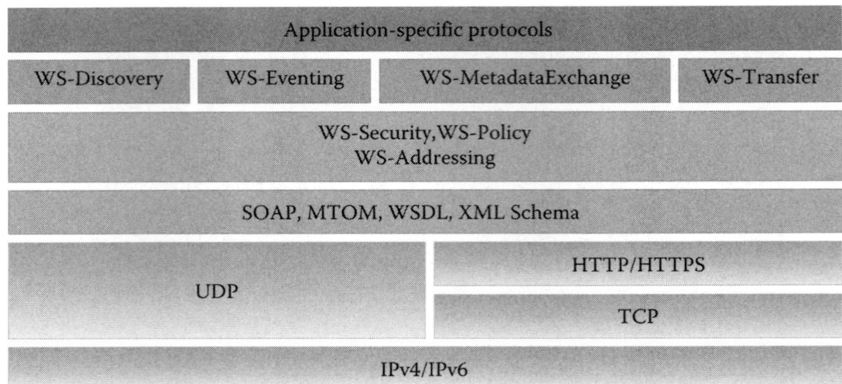

FIGURE 19.3   The Devices Profile for WSs protocol stack.

result in a huge effort for complex and dynamic applications. Process management capabilities as provided by basic WSs would leverage the easy integration of DPWS clients and services.

19.3 DPWS Inside Out

The DPWS was first announced in 2004 as a possible successor of UPnP and revised in May 2005 and February 2006 [8]. It is a WSs profile identifying a core set of WS protocols and profile-specific adaptations that enable secure WSs messaging, dynamic discovery of, and event capabilities for resource-constrained devices and their services.

Figure 19.3 shows an overview of the WS protocols for DPWS.

Before going into detail, relevant XML basic technologies are briefly introduced.

19.3.1 Interactions between Client, Device, and Service

Three participants are involved in a DPWS interaction: Client, device, and the services residing on a device. A DPWS-device is also designed as a service called *hosting service*. The hosting service contains information about itself (e.g., manufacturer and model name) and references to its *hosted services*.

Optionally, a DPWS may utilize a central service registry (called *Discovery Proxy*) which could also be used for device/service discovery spanning several networks.

In order to increase scalability and reduce the network traffic, the invocation of hosted services on devices involve several phases: Discovering desired devices/hosting services, obtaining a detailed description about replying device/hosting services and references to the services they host, retrieving further information on the desired hosted services, and finally invoking the desired service. If the client knows the references to desired services already, it may contact them directly. Additionally, DPWS defines constraints on the vertical issues messaging and security mechanisms which are offered for messaging.

The sequence diagram in Figure 19.4 [11] illustrates the phases showing the activities involved in a typical message exchange using DPWS. Details on participating protocols are provided at a later stage.

Message 1: A `Probe` message is sent from the client using UDP multicast to search for a specific device as described in WS-Discovery. The `Probe` message indicates if the client requests security. If a Discovery Proxy is available, multicast suppression is used as specified in WS-Discovery. Clients send their `Probe` messages directly to the Discovery Proxy (UDP unicast).

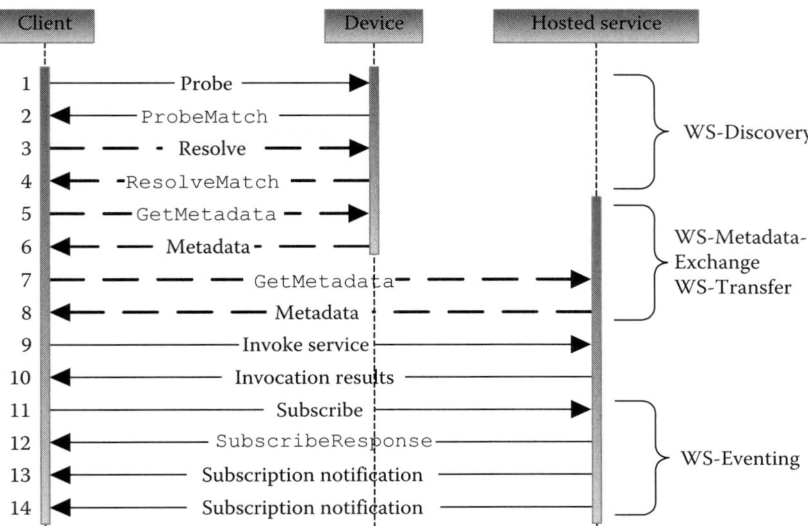

FIGURE 19.4 Typical message exchange using DPWS. (Adapted from Schlimmer, J., A Technical Introduction to the Devices Profile for Web Services, MSDN, May 2004.)

Message 2: All devices listen for `Probe` messages. In case the desired service matches one of its hosted services, the device responds with a `ProbeMatch` message using unicast UDP. The `Probe Match` message contains the device's endpoint reference (EPR), supported transport protocols, and security requirements and capabilities. EPRs are XML structures including a destination address (destination WS is called endpoint) and optional metadata describing the service (e.g., usage requirements) as defined by WS-Addressing. If security is desired, the client sets up a security channel in an additional message.

Messages 3 and 4: In case the EPR does not include a physical address, the client can use a `Resolve` message to retrieve it from the device. The device uses a `ResolveMatch` message for it.

Message 5: The client can directly request more information about the device using a `GetMetadata` message as it will be defined in the upcoming WS-MetadataTransfer specification.

Message 6: The desired device will respond either with included metadata or by providing a reference to the metadata. The metadata includes device details and EPR to each hosted service.

Message 7: Similar to message 3, the client requests more information about the desired hosted service.

Message 8: The desired hosted service sends its metadata to the client including its Web Service Description Language (WSDL) description.

Messages 9 and 10: Hosted services are invoked by receiving an invocation message from the client according to corresponding operation described in the WSDL. The implementation of the invocation of services is not specified in DPWS and left up to the service provider.

Message 11: A client may send a `Subscribe` message to the hosted service to get informed about updates of or the overall status of a service. The publish/subscribe pattern used in DPWS is described by WS-Eventing as well as specifications of subscription renewal and cancellation.

Message 12: The hosted service responds with a `SubscribeResponse` message including a subscription identifier and an expiration time.

Messages 13 and 14: The hosted service sends `Notification` messages to the client whenever a specific event occurs.

This chapter will simply refer to a hosting service as device and to a hosted service as service for easier readability.

19.3.2 Messaging

DPWS makes use of SOAP 1.2, SOAP-over-UDP, HTTP/1.1, WS-Addressing, the URI, and MTOM specifications. These specifications are restricted in such a way that they address the requirements of resource-constraint device implementations by ensuring high WS interoperability.

19.3.2.1 SOAP

SOAP was originally an abbreviation for *Simple Object Access Protocol*, but since version 1.2 it is simply called SOAP [10]. SOAP is a protocol promoting interoperability by specifying an XML-based format for exchanging messages between different applications. SOAP messages consist of a header and a body enclosed by a *SOAP envelope*. The *SOAP header* contains metadata related to delivery and processing of message content. The payload of the message is inserted into the *SOAP body*. The actual content of headers and bodies are not defined by SOAP. This is specified by the particular WS* protocols.

Since SOAP is independent of the underlying transport protocols, the *SOAP Protocol Binding Framework* describes the binding to transport protocols. Two bindings are officially documented by the W3C: *SOAP Email Binding* and *SOAP HTTP Binding*. The latter is the binding mostly used for historical reasons.

SOAP-over-HTTP/HTTPS has the disadvantages that address information is stored in the HTTP header and only synchronous messaging is supported. Recent trends try to attach SOAP to other binding protocols than HTTP. One approach is *SOAP-over-UDP* which is also used by DPWS and reduces the network load when no acknowledgment is required (e.g., for WS-Discovery). Unicast as well as multicast transmissions are supported. The SOAP-over-UDP also provides a retransmission algorithm considering the delayed repetition of the same message.

19.3.2.2 SOAP Message Transmission Optimization Feature

The conventional way of conveying binary data in SOAP as well as other XML documents is to transform the data into a character-based representation using the *Base64 content-transfer-encoding scheme* as defined by Multipurpose Internet Mail Extensions (MIME). Unfortunately, this produces a message overhead of about 33% for large attachments and a possible processing overhead [12].

The *SOAP Message Transmission Optimization Mechanism* (MTOM) defines an *Abstract SOAP Transmission Optimization Feature* which is encoding parts of the SOAP message while keeping the envelope intact (e.g., SOAP headers). The optimization feature only works if the sender knows or can determine the type information of the binary element. Conveying binary data using SOAP-over-HTTP is one implementation of the Abstract SOAP Transmission Optimization Feature. This solution is based on *XML-binary Optimize Packaging* (XOP) conventions where the binary data is sent as an attachment.

MTOM does not address the general problem of including non-XML content in XML messages.

19.3.2.3 WS-Addressing

The "pure" SOAP has a strong binding to HTTP for historical reasons although meant to be transport protocol independent. This result in the disadvantage that the address of targeted services can only be found in the HTTP header and only synchronous messaging is supported. Message routing depends on HTTP, and publish/subscribe patterns are not supported. WS-Addressing remedies these deficiencies by specifying mechanisms to address WS and messages in a transport-neutral way [13]. SOAP messages can be sent over multiple hops and heterogeneous transport protocols in a consistent manner. Furthermore, it allows sending responses to a third party (WS or WS client). WS-Addressing introduces the concepts of EPRs and message information (MI) header which are included into the SOAP header.

EPRs release SOAP from the strong binding to HTTP. They are XML structures addressing a message to a WS (destination WS is called *endpoint* including destination address, routing parameters for intermediary nodes, and optional metadata describing the service (e.g., WSDL description for a WS or policies).

MI headers enable asynchronous messaging as well as dispatch possible responses to other endpoints than the source. The mentioned *dispatching of the responses* is realized explicitly defined by assigning specific addresses to source, reply and fault endpoints, or implicitly by leaving these entries unassigned (responses are sent back to the source). The *asynchronous messaging* is realized by the use of an optional unique message ID. Result messages can be referenced to it. The destination address as well as an *action* URI which identifies the semantics of a message (e.g., fault, subscription, or subscription response message) is mandatory. WS-Addressing defines explicit as well as implicit associations of an action with input, output, and faults elements. Explicit associations are defined in the description of a WS. Implicit association of an action is composed from other message-related information. The optional definitions of relationships in MI headers indicate the nature of the relation to another message (e.g., indication of a response to specific message). It can also be indicated that the message relates to an "unspecified message."

In case that an endpoint does not have a stable URI, the endpoint can use an "anonymous" URI. Such request must use some other mechanism for delivering replies or faults.

WS-Addressing can also be used to support request-response and solicit-response MEP using SOAP-over-UDP.

19.3.2.4 Adaptations to DPWS

DPWS requires services to meet the *packet length limits* defined in the underlying IP protocol stack. A service retrieving messages exceeding that limit might not be able to process or reject them which may result in incompatibilities. A hosted service must at least support *SOAP-HTTP-binding* in order to ensure highest possible interoperability as most applications use it. Furthermore, hosted services have to support the flexible HTTP chunked transfer coding. HTTP chunked transfer coding is used to transfer large content of unknown size in chunks. The end of the transmission is indicated by a zero size chunk. Regarding the MEP, basic WSs functionality is achieved by supporting the MEPs request-response and one-way (at least).

As devices can also be mobile and roaming across networks, a globally unique UUID for the address property of the EPR is required which is persistent beyond re-initialization or changes of the device. Hosted services should use a HTTP transport address as the address property in their EPRs. Furthermore, hosted services must assign a reply action to the relationship field in the MI for response or fault messages.

Hosted services must also generate a failure on HTTP Request Message SOAP Envelopes not containing an anonymous reply EPR. This means that DPWS does not support the deferring of reply and fault messages to other endpoints (being different from the sending endpoint). The WSs Addressing 1.0 SOAP Binding specification defines for HTTP Request Message SOAP Envelopes containing an anonymous reply EPR address that their responses must be sent back to the requester in the same HTTP session. This means that DPWS only supports synchronous message exchanges for client–service interactions. The reason for that is the reduced size for the WS implementation for DPWS devices/services and clients. Especially, WS clients cut down on size as they do have to provide a callback interface.

In summary, only the concept of WS-Addressing EPRs and action URIs are used by DPWS. Asynchronous client–service interactions are not supported by DPWS.

A hosted service may support MTOM in order to be able to receive or transmit messages exceeding this packet length limits.

19.3.3 Discovery

DPWS uses WS-Discovery for discovering devices. As illustrated in Figure 19.4, WS-Discovery is not used to search for specific hosted services. Instead, desired services are found by searching for devices and browsing the descriptions of their hosted services.

19.3.3.1 Web Services Dynamic Discovery (WS-Discovery)

WS based on the WSs architecture can only be used if their addresses are known to WS clients or if they had registered to a *Universal Description, Discovery and Integration* (UDDI) server. UDDI is a central registry which can be queried for a desired service. The address of a UDDI server can also not be discovered and has to be known to interested WS clients.

The *Web Services Dynamic Discovery* (WS-Discovery) [14] extends the WS discovery mechanisms by concepts for the announcement of services when entering a network, dynamic discovery, and network spanning discovery of services. It introduces three types of endpoints: Target Service, Client, and Discovery Proxy. Clients are searching for Target Services. Discovery Proxies (DP) enable forwarding of searches to other networks.

The announcement of a service is realized by sending a one-way multicast `Hello` message whenever a Target Service becomes available. This should be used by Clients for implicit discovery through storing the announcement information, and thus reducing the amount of discovery messages in the network. When a Target Service intends to leave a network, it should send a one-way multicast `Bye` message.

Dynamic discovery is important for endpoints which might change their transport addresses during lifetime (e.g., mobile WS traversing different subnets). In such cases, Target Services possess a logical address which can be dynamically resolved into their transport address. Dynamic discovery is realized in a two-stage process (as shown in Figure 19.5): Probe and Resolve. When searching for services, Clients must send a `Probe` message specifying the type of the Target Service and/or its Scope. The specification of the Type and its semantics are left up to the provider. The Scope is a logical group of WSs. The matched Target Services answer to the requester using a unicast `ProbeMatch` message containing the mandatory EPR. Optionally, the `ProbeMatch` can also provide a list of the Target Service addresses (avoiding a following resolve), types and/or scopes. The next step is to resolve the EPR into a list of Target Service addresses by sending a multicast `Resolve` message containing the EPR. The corresponding Target Service answers with a `ResolveMatch` message including a list of addresses and optionally a list of types and/or scopes.

WS-Discovery also defines a basic set of matching rules for the resolution of URIs, UUIDs, and case-sensitive string.

Network spanning discovery can be realized using DP by relaying discovery messages to other network. A DP enables the scaling of a large number of endpoints. When a DP receives a multicast

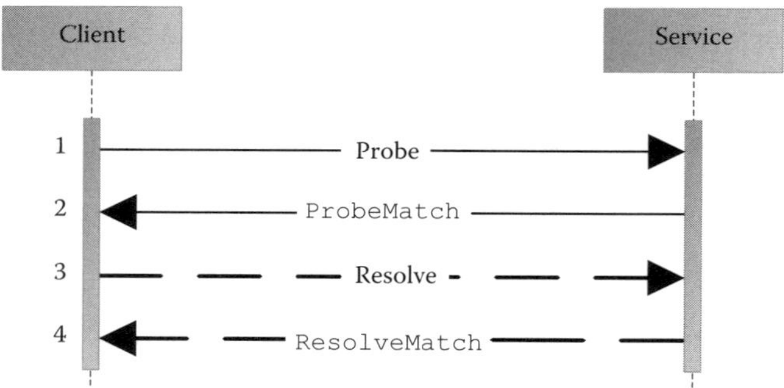

FIGURE 19.5   Interactions in the discovery process defined by WS-Discovery.

`Probe` or `Resolve` message, it announces itself using a `Hello` message to suppress multicast messaging. Clients sent their request via unicast to the DP and ignore `Hello` and `Bye` messages from Target Services while the DP is available. Further functionality of the DP is not specified by WS-Discovery.

WS-Discovery uses UDP for messaging. The implementations are mostly bound to SOAP-over-UDP. WS-Addressing is used for the specification of EPRs, action, and fault properties.

19.3.3.2 Adaptations to DPWS

Discovery in DPWS only relates to searching for devices/hosting services. As described at the beginning of this section, hosted services can be found by following the references in the descriptions starting from the device/hosting service description. This procedure limits the discovery traffic in the network.

A device/hosting service must be a Target Service as defined in WS-Discovery. Furthermore, a device/hosting service is required to support unicast as well as multicast-over-UDP (required by WS-Discovery). In case the HTTP address of a device is known, the discovery client may send a `Probe` message over HTTP (unicast). Due to this, a device must support receiving `Probe` SOAP envelopes as HTTP requests and be able to send a `ProbeMatch` SOAP envelope in a HTTP response or send an HTTP Response with a `HTTP 202 Accepted`.

Devices might not be discoverable if residing in a different subnet than the discovery client. Nevertheless, discovery clients disposed of an EPR and transport address for targeted devices might be able to communicate with the corresponding device. If transport address and EPR of a DP in another subnet are known, this DP can be used to discover devices in that subnet.

Devices must announce the support of DPWS by including the type `wsdp:Device` and may include other types or scopes in `Hello`, `ProbeMatch`, or `ResolveMatch` SOAP envelopes. To be able to match URIs and strings, devices must at least facilitate URI and case-sensitive string comparison defined by the scope matching rules defined by WS-Discovery.

Except for `wsdp:Device`, semantics and classifications of types and scopes are not defined in DPWS. When a device receives a `Probe` message, it uses some internal mechanism to decide if it responds or not. Devices responding to `Probe` messages are not required to publish their supported types and scopes. This might complicate the discovery process if no out-of-band information about responding devices is available. In addition, as messages should not exceed the size of an Maximum Transmission Unit (MTU), messages being sent via UDP might be too small to convey supported types and scopes. The support of UDP fragmentation is also not required by DPWS. In such a case, another `Probe` could be sent via HTTP chunked transfer coding as it must be supported. However, this also requires the device to send supported types and scopes in the `ProbeMatch`.

19.3.4 Description

DPWS uses WSDL for describing the functionality and capabilities of hosted services. Devices do not possess a WSDL description as they cannot be invoked by WS clients. The capabilities of a device including the references to its hosted services are defined in a metadata description as defined by WS-MetadataExchange. WS-Transfer is used to retrieve such metadata descriptions. Furthermore, hosted services can define their usage requirements by using WS-Policy. All protocols are described in the remaining of this subsection.

19.3.4.1 Web Service Description Language

The WSDL is an XML document format containing definitions used to describe WS as a set of endpoints and its interactions using messages [15]. An endpoint is the location for accessing a service. A WSDL document (or simply referred to as a WSDL) contains an abstract and a concrete part as shown in Figure 19.6 [2].

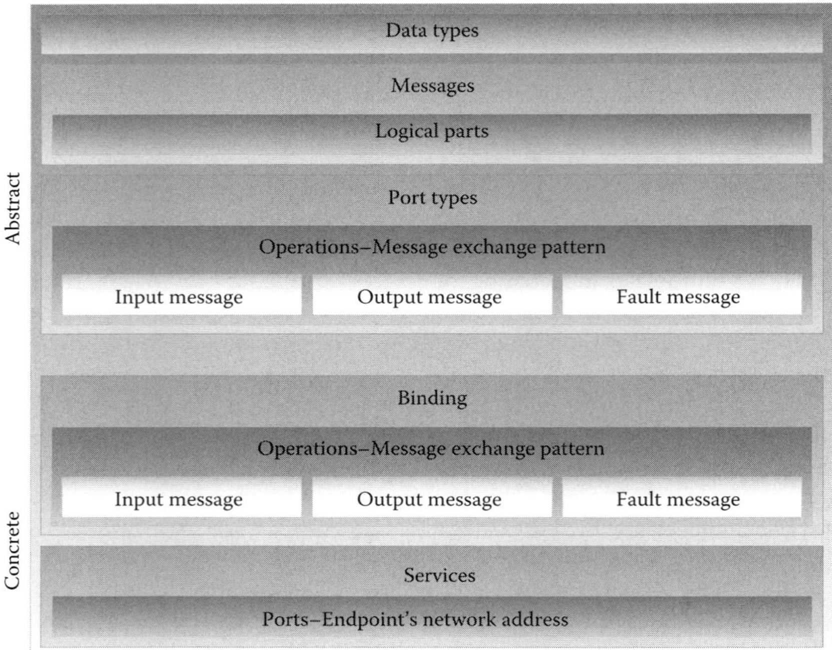

FIGURE 19.6   Major components of a WSDL description. (Adapted from Dostal, W., Jeckle, M., Melzer, I., and Zengler, B., *Service-Orientierete Architekturen mit Web Serives*, Elsevier, Spektrum Akademischer Verlag, München, Germany, 2005.)

Abstract part: The abstract part of a WSDL document promotes reusability by describing the functionality of the services independent of its technical details such as underlying transport protocols and message format. Firstly, the *data types* which are needed for message exchange are defined. In order to ensure highest interoperability, WSDL refers to XML Schema Definition as the preferred data type scheme. *Messages* are abstract definitions of data being transmitted from and to a WS. Messages consist of one or more *logical parts*. The *port types* definition can be seen as abstract interfaces to services offered by the WSDL. Port types are a set of abstract operations. An *operation* is a set of messages exchanged by following a certain pattern. WSDL supports the following four MEP: One-way, request-response, solicit-response, and notification. They are implicitly identified in the WSDL by the appearing order of input and output messages. Request-response and solicit-response operations may also define one or more fault messages in case a response cannot be sent.

Concrete part: The concrete part encloses the technical details of a WS. The *binding* associates a port type, its operations and messages to specific protocols and data formats. Several bindings to one port type are possible. All port types defined in the abstract part must be reflected here. Concrete fault messages are not defined because they are provided by the protocols. Furthermore, a *service* definition is a collection of ports. *Ports* are network addresses for service endpoints. Each uniquely identified *port* constitutes an endpoint for a binding.

WSDL specifies and recommends the binding to SOAP but is not restricted to it. The WSDL Binding Extension for SOAP defines four *encoding styles*. The style can be either RPC or document. The use can be either literal or SOAP encoded making it to four styles. Each binding has to specify the encoding style it wants to use when translating XML data of a WSDL into a SOAP document. The differences between the encoding styles are as follows:

The *RPC-style* structures the SOAP body into operations embedding a set of parameters (data types and values). Using the *document style,* the SOAP body can be structured in an arbitrary way.

The *literal encoding* uses XML Schema to validate the SOAP data. Therefore, the SOAP body must be conforming to specific XML Schema. In contrast, *SOAP encoding* employs a set of rules based on the XML Schema data types to transform the data. A message is not required to comply to a specific schema.

RPC/encoding was the original style but will eventually go away. Document/encoded is rarely supported. Recently, a fifth style was developed—called Document/literal/wrapped—where the parameters are wrapped in an element bearing the operation name. An elaborated comparison of the coding styles can be found in the article by Butek [25]. In most cases Butek recommends the Document/literal/wrapped encoding style which is also supported by the .NET framework. The impact of the choice of encoding style on performance, scalability, and reliability of WS interactions are presented in Cohen [26].

19.3.4.2 WS-Policy

WS implementations might rely on certain restrictions although the WSs architecture is platform, programming language and transport protocol independent.

WS-Policy [16] provides a framework to define restrictions, requirements, capabilities, and characteristics of partners in WS interactions as a number of policies. As illustrated in Figure 19.7, a *policy* is an assortment of policy alternatives. In order to comply with a certain policy, an interaction partner can select the best suiting policy alternative from the defined set. If a WS or a client makes use of policies its interaction partners have to comply with those by choosing one of the offered policy alternatives. A *policy alternative* is defined as a collection of policy assertions. A *policy assertion* is an XML Infoset representing a requirement or capability. A policy assertion can also be optional. Policy assertions can be combined to more complex requirements or capabilities using *policy operators*. Two policy operators are defined by WS-Policy. The *ExactlyOne operator* requires the compliance to one of the stated policy assertions, whereas the *All operator* requires the compliance to all. WS-Policy does not specify how policy assertions are expressed. A way to do such is described in the *Web Services Policy Assertions Language* (WS-PolicyAssertions).

WS-Policy does not specify the location of policy definitions nor their discovery. This is left to other WS protocols such as WS-PolicyAttachment.

19.3.4.3 WS-PolicyAttachment

WS-PolicyAttachment [17] describes general-purpose mechanisms, how associations between policy expressions and the subject to which they apply can be defined (e.g., associations with WSDL port or binding). Two approaches are offered: Policy assertion is part of the subject or externally defined and bound to the subject. WS-PolicyAttachment also describes the mechanisms for WSDL and UDDI.

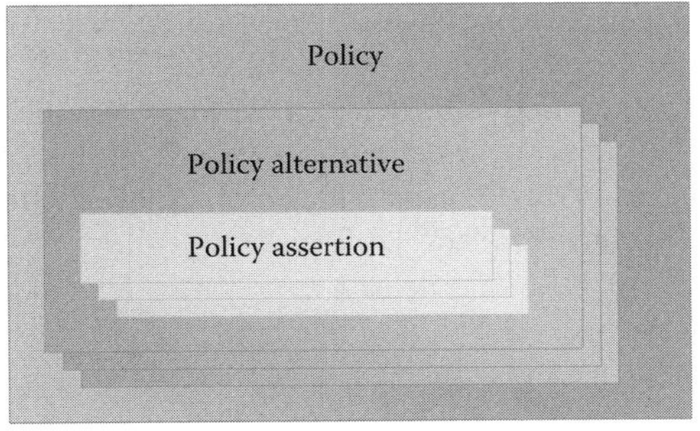

FIGURE 19.7   The structure of a policy as defined by WS-Policy.

These include description of the policy references from WSDL definitions, policy associations with WSDL service instances, and policy associations with UDDI entities.

19.3.4.4   WS-MetadataExchange

The client has received the addresses of targeted services from the device after the discovery process. The next step is to find out more about the Target Services (e.g., their descriptions, requirements, and used data types). This may also be used to find out more about the device itself.

WS-MetadataExchange (MEX) provides a standardized bootstrap mechanism to start the communication with a WS by retrieving the WSDL as well as further information about the WS endpoint such as usage requirements/policies (WS-Policy) and data types (XML Schema) in a well-defined encapsulation format [18]. It defines two request-response operations:

`GetMetadata` is used to retrieve all metadata the WS endpoint provides. It is recommended for a large amount of metadata to provide EPRs (*Metadata references*) or URL (*Metadata Location*) instead of including all metadata in detail in the response. This ensures scalability for service consumers interested only in specific metadata and possibly reduces the network traffic.

`Get` is used to retrieve metadata addressed by a reference. The `Get` operation must be "safe" meaning that the state of requested service must remain unchanged by this operation. It may include a *dialect* characterizing the type of metadata (e.g., metadata defined by WS-Policy), its format and version. If `Get` is used without a defined dialect and reference for specific metadata, all metadata must be sent in the response.

The default binding of MEX is SOAP 1.1 over HTTP as constraint by the Basic Profile 1.0.

19.3.4.5   WS-Transfer

WS-MetadataExchange provides operations to retrieve (metadata) representations of WSs. The resources represented by the WS remain unaffected by that operation. Operations on the WS resources must be defined in a proprietary way and agreed on by both WS interaction partners.

WS-Transfer [19] defines a standardized mechanism supporting *Create, Read, Update, and Delete* (CRUD) operations (called Create, Get, Put, and Delete) on WS resources. A resource is specified as an entity providing an XML representation and being addressable by an EPR. `Get`, `Put`, and `Delete` are operations on resources, while the `Create` operation is performed by resource factories. The WSs capable of creating new resources are called *resource factories*. WSs implementing a resource must provide the `Get` operation and may provide `Put` and `Delete` operations.

The *Get-operation* is used to retrieve a representation of a resource. A *Put-operation* request allows replacing the representation of a resource by the one sent in the request. If a WS accepts a *Delete-operation*, the corresponding resource must be deleted. The *Create-operation*, as the name suggests, is a request to create a new resource according to the representation in the request.

WS-Transfer is a transport-independent implementation similar to the *Representational State Transfer* (REST) approach. The persistence of CRUD operations is not guaranteed because the hosting server is responsible for the state maintenance of a resource.

It is expected that WS-Transfer and WS-MetadataExchange will be merged in the future. One proposal for that is the upcoming *WS-MetadataTransfer* defining "how to retrieve service metadata including message descriptions (WSDL), capabilities and requirements (WS-Policy), inter-service relationships, and domain-specific metadata" [20].

19.3.4.6   Adaptations to DPWS

The metadata of devices and hosted services are described in a structure conform to WS-MetadataExchange. Metadata is considered as static details. Changes to the metadata must include incrementing their versions (`wsd:MetadataVersion`) reflected in `Hello`, `Probe Match`,

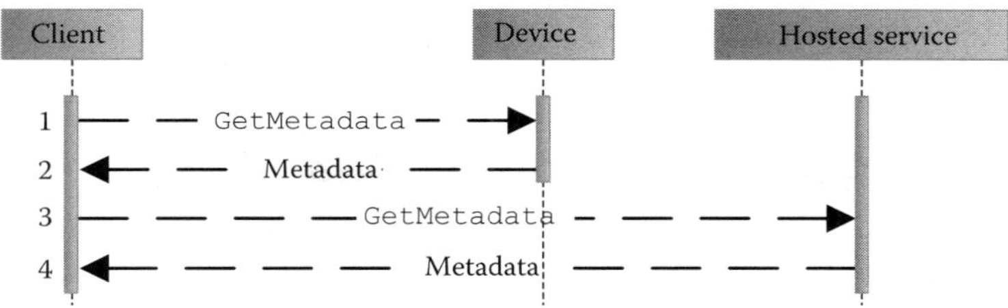

FIGURE 19.8 The metadata exchanges of DPWS.

and `ResolveMatch` SOAP envelopes sent by corresponding devices and hosted services, respectively. The `Get` operation of WS-Transfer is used for retrieving metadata in DPWS. Additionally, other means of retrieval can also be used. The way of using WS-MetadataExchange for the structure of metadata and WS-Transfer for operations on it is also proposed for the upcoming specification of WS-MetadataTransfer which will be a merger of WS-MetadataExchange and WS-Transfer. Only hosted services possess a WSDL description to foster interoperability with other WSs architectures. It is embedded into the metadata of hosted services.

Figure 19.8 shows the metadata exchanges between client, device, and hosted services. As illustrated by the dashed lines, the metadata exchanges are optional because clients might already be aware of the metadata of devices and hosted service. In such case, the clients may use the hosted services directly as defined in their WSDL descriptions. All metadata must be sent in one `wsx:Metadata` element in the SOAP Envelope Body.

Description of devices: The description of a device contains generic characteristics of the device including EPRs to the hosted services. The `GetResponse` SOAP envelope only includes a `wsx:Metadata` element which contains all metadata for the device. The metadata of a device has following structure: Device class metadata (`ThisModel`), metadata for the specific device (`ThisDevice`), and metadata for the hosted services (`Relationship`).

ThisModel describes the device class metadata containing information about the manufacturer, model name and optionally manufacturer URL, model number, model URL as well as a presentation URL (a reference to further information). `ThisModel` is described by the metadata section <wsdp:ThisModel...> in which the WS-MetadataExchange dialect must be equal to `http://schemas.xmlsoap.org/ws/2006/02/devprof/ThisModel`.

ThisDevice defines the specifics of a certain device of the class `ThisModel`. It is described by the metadata section <wsdp:ThisDevice ...> containing a user-friendly name and optionally version of firmware and serial number. The WS-MetadataExchange dialect must be equal to `http://schemas.xmlsoap.org/ws/2006/02/devprof/ThisDevice`.

Relationship: The WS-MetadataExchange dialect of this metadata section must be equal to `http://schemas.xmlsoap.org/ws/2006/02/devprof/Relationship`. The `Relationship` definition identifies nature and content of the relationship between device/hosting service and hosted services indicated by the type `http://schemas.xmlsoap.org/ws/2006/02/devprof/host`. It includes optional information about the host and the hosted services. The *host section* (<wsdp:Host>) contains the EPR of the host, types implemented by the host, and a service ID for the device. The *hosted section* includes the EPR of a hosted service. Therefore, the relationship metadata includes a separate hosted section for each hosted service of the device.

Figure 19.9 presents an excerpt of the SOAP Response message envelope containing the metadata description for a device [8]. The SOAP header elements are specified by WS-Addressing (using the prefix `wsa`). The action (`wsa:Action`) identifies the semantics of the SOAP message—a response message to a `Get` operation of WS-Transfer. The unique message ID (`wsa:RelatesTo`) relates this response to a certain request message. Furthermore, the message is sent to the address (`wsa:To`) anonymous as defined by DPWS for reply messages. As already described in the messaging

```
<soap:Envelope ...>
  <soap:Header>
    <wsa:Action>http://schemas.xmlsoap.org/ws/2004/09/transfer/GetResponse</wsa:Action>
    <wsa:RelatesTo>urn:uuid:82204a83-52f6-475c-9708-174fa27659ec</wsa:RelatesTo>
    <wsa:To>http://schemas.xmlsoap.org/ws/2004/08/addressing/role/anonymous</wsa:To>
  </soap:Header>

  <soap:Body>
    <wsx:Metadata>
      <wsx:MetadataSection Dialect="http://schemas.xmlsoap.org/ws/2006/02/devprof/ThisModel">
        <wsdp:ThisModel> ... </wsdp:ThisModel>
      </wsx:MetadataSection>

      <wsx:MetadataSection Dialect="http://schemas.xmlsoap.org/ws/2006/02/devprof/ThisDevice">
        <wsdp:ThisDevice> ... </wsdp:ThisDevice>
      </wsx:MetadataSection>

      <wsx:MetadataSection Dialect="http://schemas.xmlsoap.org/ws/2006/02/devprof/Relationship">
        <wsdp:RelationshipType="http://schemas.xmlsoap.org/ws/2006/02/devprof/host">
          <wsdp:Host> ...</wsdp:Host>

          <wsdp:Hosted> ...</wsdp:Hosted>
          <wsdp:Hosted> ...</wsdp:Hosted>
        </wsdp:Relationship>
      </wsx:MetadataSection>
    </wsx:Metadata>
  </soap:Body>
</soap:Envelope>
```

FIGURE 19.9 Example for a SOAP Response message envelope. (Adapted from Chan, S., Conti, D., Kaler, C., Kuehnel, T., Regnier, A., Roe, B., Sather, D., Schlimmer, J. (Ed.), Sekine, H., Thelin, J. (Ed.), Walter, D., Weast, J., Whitehead, D., Wright, D., and Yarmosh, Y., *Devices Profile for Web Services*, Microsoft Corporation, February 2006.)

subsection, the reply address is used from the underlying protocol meaning that it has to be sent back right after the SOAP Request message (in the same HTTP session).

The SOAP body in Figure 19.9 contains metadata sections for `ThisModel`, `ThisDevice`, and the `Relationship`. The `Relationship` metadata provides details about the device itself and two hosted services.

Description of hosted services: Now that the client has retrieved the metadata for the device including the EPRs of its hosted service, it may request the WS-MetadataExchange description of desired hosted services using the `Get` operation of WS-Transfer. The metadata of a hosted service must contain at least one section providing the WSDL description (metadata dialect equal to `http://schemas.xmlsoap.org/wsdl/`). It must be sent in any `GetResponse` message. The WSDL metadata section may include the WSDL description inline or may provide a link referring to it.

WSDL constraints of DPWS: Hosted services must at least support the document/literal encoding style and must include the WSDL Binding for SOAP 1.2 for each `PortType` at least. If notifications are supported by the hosted service (e.g., for a publish/subscribe mechanism), it must include the notification and/or solicit-response operation in a `PortType` specification. Hosted services are not required to include the WSDL services section in a WSDL description since addressing information is included in the `Relationship` section of the device's metadata.

Hosted services must include the policy assertion `wsdp:Profile` in their WSDL indicating that they support DPWS. This may be attached to a `port`, should be attached to a `binding` and must not be attached to a `portType`. The policy assertion can also be stated as being optional indicating that the hosted service supports DPWS but is not restricted to it. Devices have no explicit means to specify the support of DPWS. This must be ascertained by other information.

It should be noted that DPWS does not define a back-reference from hosted services to the device. The only way to find the corresponding device to a certain hosted service is to start a discovery process.

19.3.5 Eventing

Some applications might require to get informed about state changes of a specific WS. DPWS provides a publish/subscribe mechanism based on WS-Eventing.

19.3.5.1 WS-Eventing

WS-Eventing [21] implements a simple publish/subscriber mechanism sending events from one WS to another. WS-Eventing defines four WS roles: The *subscriber* initiates a subscription by subscribing to an event source. The *event source* is publishing (event) notifications whenever an interesting event occurs (e.g., state change). The notifications are sent to an *event sink* which processes them in some way. In order to decouple the management of subscription from the event source in highly distributed networks, a *subscription manager* may take over that role. However, event sources are also subscription managers and event sinks are also subscribers in many publish/subscribe systems.

Figure 19.10 provides an overview of roles and message exchanges in a publish/subscribe system using WS-Eventing.

Message 1: A subscriber sends a `Subscribe` message to an event source indicating the address for a `SubscriptionEnd` message, address of event sink, the Delivery Mode, subscription expiration time, and subscription filters. The address for a `SubscriptionEnd` message identifies the WS which receives such message in case of an unexpected subscription cancellation by the event source (e.g., due to an error). The address of subscriber and event sink are equal to the address for the `SubscriptionEnd` message when the subscriber is also the event sink (as in most applications of WS-Eventing).

The *Delivery Mode* specifies the way a notification is retrieved by the event sink. WS-Eventing specifies an abstract delivery mode allowing the use of any kind of delivery mode. One concrete delivery mode is defined as well. The *Push Mode* is used to send individual, unsolicited, and asynchronous notification messages. As an alternative to push delivery, a mechanism for polling notifications could

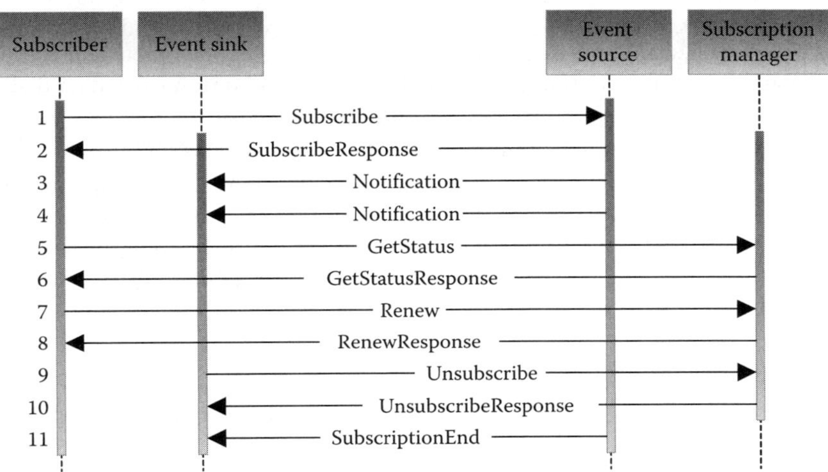

FIGURE 19.10 Roles and message exchanges as defined by WS-Eventing.

be defined according to the abstract delivery mode (e.g., for applications where the event sink is behind a firewall).

Although subscription can be defined for an unlimited time, WS-Eventing recommends the definition of an absolute expiration time or duration (*subscription expiration time*). This reduces the network load in case of unexpected unavailability of event sinks during a subscription.

Subscription filters are defined to only send notifications the event sink is interested in. A Boolean expression in some *filter dialect* (e.g., string or XML fragment) is used to evaluate the notifications. Notifications are only sent when the expression evaluates to `true`. Any kind of filter dialect can be used (e.g., based on XPath).

Message 2: An event source replies with a `SubscribeResponse` message to the subscriber. When accepting the subscription, the `SubscriptionResponse` contains an EPR of the subscription manager and an expiration time. WS-Addressing MI headers are used to relate a `SubscriptionResponse` to a `Subscribe message`. In case the event source is also the subscription manager, the EPR of the subscription manager is one of the event sources.

Messages 3 and 4: *Notification* messages are sent from the event source to the event sink for events defined by the subscription filters. Semantics and syntax of notification messages are not constrained by WS-Eventing as any message can be a notification. However, subscribers may specify `wsa:ReferenceProperties` (WS-Addressing) in the `subscribe` message to request special marked notifications. These reference properties have to be included in each notification message.

Messages 5 and 6: If a subscriber wants to receive information about the expiration time of its subscription, it sends a `GetStatus` message to the subscription manager. The subscription manager includes the expiration time in a `GetStatusResponse`.

Messages 7 and 8: Before the subscription time expires, the subscriber has the chance to renew the subscription by sending a `Renew` message to the subscription manager including the new expiration time. The subscription manager replies with a `RenewResponse` including the new subscription time. If the subscription manager does not want to renew the subscription for any reason, it may reply with a `wse:UnableToRenew` fault in the response message.

Messages 9 and 10: If an event sink wants to terminate the subscription, it should send an `Unsubscribe` message to the subscription manager. If the subscription termination is accepted by the subscription manager, it must reply with an `UnsubscribeResponse` message.

Message 11: If the event source wants to terminate the subscription unexpectedly (e.g., due to an error), it sends a `SubscriptionEnd` message to the event sink. In such case, the event source is required to indicate the error. WS-Eventing specifies the faults `DeliveryFailure` for problems in delivering notifications, `SourceShuttingDown` for indicating a controlled shutdown and for any other fault `SourceCanceling`. Furthermore, the event source may provide an additional textual explanation in the element `wse:Reason`.

Please note that the order the messages exchanged in Figure 19.10 are not representative for all message exchanges in publish/subscribe systems using WS-Eventing.

WS-Eventing additionally defines that WS-MetadataExchange should be used to retrieve the event-related information from an event source embedded in the WSDL and optional policies. The services supporting WS-Eventing are marked by a special attribute and their notification and solicit-response operations generate notifications which are sent to subscribed event sinks. Solicit-response notifications require a response message from the event sinks. The policy entries specify the Delivery Mode and the filters.

19.3.5.2 Adaptations to DPWS

Eventing mechanisms in DPWS are only supported by hosted services (*event sources*) and clients (*event sinks*). Hosting services/devices do not support eventing. Hosted services supporting WS-Eventing must include a `wse:EventSource='true'` in corresponding `wsdl:portType` definitions in their WSDL description (as defined in the WS-Eventing specification). Furthermore, hosted services must at least support the *Push Delivery Mode* as defined by WS-Eventing.

If a notification cannot be delivered to an event sink, the hosted service may terminate the subscription by sending a `SubscriptionEnd` indicating a `DeliveryFailure`.

As the time synchronization of event source and event sink might require additional resources and protocols being implemented, subscription requests and renewal are not required to be based on an absolute time. However, expressions in duration time must be accepted.

DPWS defines a specific dialect for subscription filtering—action filtering—which allows filtering related to specific actions defined in the MI headers of messages sent by the event source. A filter definition contains a white space-delimited set of URIs identifying the events being subscribed to. The event source evaluates the filter definition using the RFC 2396 prefix matching rules. That means all notifications of a `portType` can be received by subscribing to the action property prefix common to all actions of a `portType`.

19.3.6 Security

Point-to-point security offered by HTTP through authenticating, encrypting, and signing messages is not enough for complex message exchanges using SOAP. SOAP messages may be sent over intermediary nodes traversing different trusted domains, using heterogeneous transport protocols, and security mechanisms. These issues are addressed by WS-Security.

19.3.6.1 Web Services Security: SOAP Message Security 1.0 (WS-Security)

WS-Security [22] provides mechanisms for message integrity and confidentiality to SOAP messaging. It provides a framework to integrate existing security mechanisms into a SOAP message independent of underlying transport technology. WS-Security is based on existing specifications such as XML signatures, XML encryption/decryption, and XML Canonicalization. XML Canonicalization describes the preparation of XML messages for signing and encryption. XML Signature and XML Encryption define mechanisms for signing and encrypting an XML message. Furthermore, WS-Security is based on Kerberos and X.509 for authentication.

WS-Security defines a SOAP header element (`wsse:Security`) which contains security information about used authentication, signature, and encryption mechanisms as well as used security elements (e.g., public encryption keys). WS-Security does not specify which mechanisms should be used but how they are embedded into the security header. The WS-Security header may appear several times in a SOAP message specifying security mechanisms for each intermediary node along the way as well as the ultimate recipient of a SOAP message. A WS-Security header is associated with the corresponding node by using the mandatory `actor` attribute. Two security headers are not allowed to have the same `actor` URI assigned.

In order to sign or encrypt specific parts of a SOAP message, a reference to them is required. Although XPath provides mechanisms to do such, intermediary nodes processing the security headers might not support XPath. Therefore, WS-Security specifies an own mechanism using ID attributes (`wsu:Id`). This makes the processing of security definition easier and faster. An ID has to be unique for the entire SOAP message and is assigned to message elements designated for signing or encrypting. Thereby, it can be simply referred to when defining the parts being encrypted or signed. The ID attribute can also be used to refer to particular security declarations (e.g., public encryption key or signature).

Time often plays an important role in security contexts (e.g., certificates may expire). WS-Security supports this by the attribute `wsu:Timestamp` which is assigned a security header. It specifies creation time (`wsu:Created`) and/or expiration time (`wsu:Expires`) of a security element.

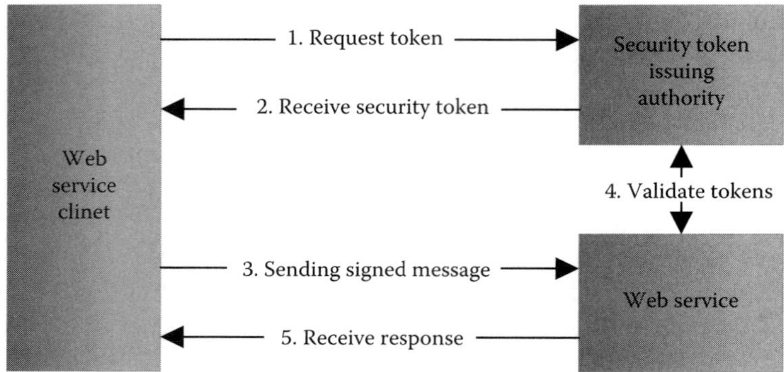

FIGURE 19.11   Message exchanges in an authentication procedure using WS-Security. (Adapted from Seely, S., Understanding WS-Security. MSDN, October 2002.)

The attributes `wsu:Id` and `wsu:Timestamp` used to be part of the WS-Security specification. However, they appeared to be useful for a number of WS protocols. Therefore, they have been separately defined in an XML Schema called *Web Services Utility* (indicated by the prefix `wsu`) along with other useful declarations.

Authentication: WS-Security offers two elements to embed authentication information into a SOAP message. The `UsernameToken` specifies a user name and password for user authentication. The `BinarySecurityToken` offers the use of any binary authentication token such as a Kerberos ticket or X.509-certificate. It specifies the type of binary security token (e.g., X.809 certificate or Kerberos ticket) and its XML representation (e.g., Base64 encoded).

Figure 19.11 illustrates a common authentication procedure using WS-Security [23]:

Message 1: The user requests a security token from an issuing authority. This request may not be WS based. For example, Kerberos protocols could be used to request a Kerberos service ticket.

Message 2: The issuing authority sends the security token (e.g., ticket, certificate or encryption key) to the WS client which embeds it into the SOAP message. The client should also sign and encrypt the message.

Messages 3–5: The client sends the (encrypted and signed) SOAP message to the WS. If necessary, the WS can validate the token by contacting the issuing authority. If the token is valid and the message encrypted, the WS replies in a secure SOAP message.

The concrete message interaction depends on the security mechanisms being used. As Kerberos and X.509 are not in the focus of this thesis, please refer to the corresponding specifications for more detail on it.

Authentication information should be signed and encrypted to ensure that it will not be changed or disclosed.

Message integrity: The signing of a message prevents third parties from manipulating its content but does not exclude them from reading it. WS-Security requires the support of XML Signature for signing message content. The whole SOAP message envelope should not be signed as intermediary nodes might include additional security headers for their own security mechanisms. Such message would fail the manipulation check afterwards.

Parts of a SOAP message envelope and also security definitions can be signed using the authentication mechanisms described above. Thereby, it is guaranteed that the user identified by the X.509 certificate, the `UsernameToken` or a Kerberos ticket is the one who has signed the message parts.

Encryption: Encryption prevents third parties from being able to read the content of a message. WS-Security requires the support of XML Encryption. Symmetric or asymmetric encryption mechanism can be used. Symmetric encryption is the use of the same key for encryption and decryption. The key has to be provided to the partners by some other means in order to make sure that the key is not disclosed to untrusted parties. Asymmetric encryption is the use of two keys: a public and a private key. The public key is used for encrypting a message. As the public key cannot be used for decryption it can be send in any message. The private key is used for decryption. It must only be known and belongs to the partner who has to decrypt a certain message or content. X.509 certificates can be used for that.

In summary, WS-Security provides a container to include security definition and elements into a SOAP message. It provides means to include information about user authentication, digital signatures, and encryption mechanisms. Although any kind of security mechanism can be used, WS-Security provides concrete details on using Kerberos tickets and X.509 certificates as well as an own mechanism to include a username and password.

19.3.6.2 Adaptations to DPWS

WS-Security requires a lot of overhead in supporting and processing different security protocols and credentials in very heterogeneous networks although it offers enhanced security mechanisms for SOAP messaging over multiple hops and heterogeneous transport protocols. DPWS assumes that clients and devices reside in IP-based networks providing HTTP and HTTPS. Additionally, devices might have constrained resources (e.g., processing power, small storage space). Therefore, DPWS restricts the use of WS-Security to the discovery process and all other messages are sent using a secure channel. The secure channel is established using existing technologies (HTTPS).

Security is an optional requirement of DPWS as the security mechanisms, which are provided by the administrative trusted domain must be used. DPWS security mechanisms have to be used if messages traverse other administrative domains which cannot be trusted.

Figure 19.12 provides an overview of the security concept defined for DPWS which is explained below in detail.

Authentication: The client specifies the level of security, desired security protocols, and credentials in a policy assertion in the `Probe` and `Resolve` message during the discovery. The security policy assertion is included in the metadata section (`wsx:Metadata`) under the SOAP header (`wsa:ReplyTo/wsp:Policy`). The client should also authenticate itself to the device. If several security protocols are proposed by the client, they have to be ordered by decreasing preference. The device selects a protocol for authentication and key establishment from the list and includes it in the corresponding `ProbeMatch` and `ResolveMatch` message. After the security protocols have been negotiated, the client initiates authentication and/or key establishment processes. These processes are out of scope of the DPWS specification and are normally performed using an out-of-band mechanism. If successfully finished, the client establishes a secure channel to the device using Transport Layer Security (TLS)/Secure Socket Layer (SSL). The secure channel is also used for communication between client and hosted services.

Integrity: The integrity of discovery messages of client and devices is ensured by using message-level signatures based on WS-Security. Devices should sign `Hello` and `Bye` SOAP message envelopes. If a device has multiple signatures, it must send one message for each of the signatures.

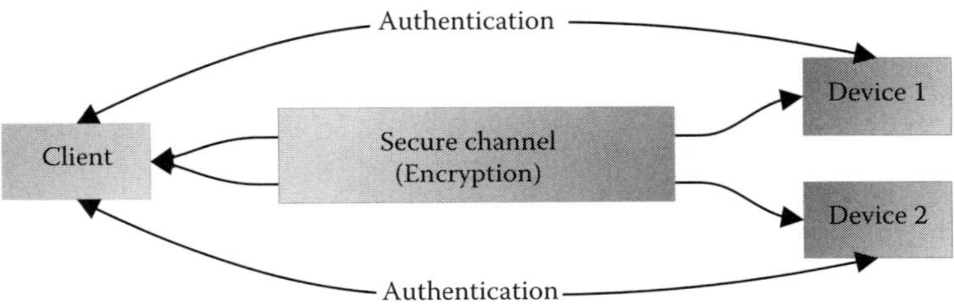

FIGURE 19.12 Security concept defined for DPWS.

Service must protect the following MI header block of their SOAP messages by signatures: Action field (`wsa:Action`), message ID (`wsa:MessageID`), address of recipient (`wsa:To`), address for response (`wsa:ReplyTo`), and the ID of related request message for responses (`wsa:RelatesTo`). Furthermore, the content portion(s) of the SOAP Body associated with those MI headers must also be signed. The requirements for services are automatically fulfilled if using TLS as proposed by DPWS.

Confidentiality: Discovery messages are not encrypted, whereas all other messages are encrypted using a secure channel. Services must encrypt their SOAP body.

A secure channel should remain active as long as client, device, and hosted services are interacting. The channel is removed when a device or client leaves the network.

19.4 Web Service Orchestration

Although the DPWS is an ideal base to enable connectivity between heterogeneous devices, each client has to be programmed and set up manually for each individual interaction with corresponding devices and services, respectively. This is not a big deal for interactions between one client and several services on devices but involves a huge effort for programming workflows or interaction processes in large networks. WS composition addresses this disadvantage by the definition of process workflows on existing WSs.

19.4.1 Web Service Composition

Service composition is the possibility to combine a number of individual services in such a way that a surplus value is created [2]. Service composition can be divided into two independent aspects: service orchestration and service choreography.

Service orchestration is the execution of a workflow/process of service interactions controlled by a central entity (*service orchestrator*). The central entity plays the role of the initiating client for each individual invocation of services.

Service choreography, by contrast, refers to a decentralized workflow/process which results from the interactions of each client-service interaction. In other words, each client is aware of its own interactions but it is not necessarily aware of the interactions between other clients and services. Nevertheless, all client–service interactions form a workflow/process.

Figure 19.13 presents simple examples for service orchestration and choreography. In service orchestration, the service orchestrator (as the central workflow/process coordinator) controls the invocations of three services, their order and deals with possible failures. In a choreography setup, each participating component is (only) aware of its own interactions. There is no central entity

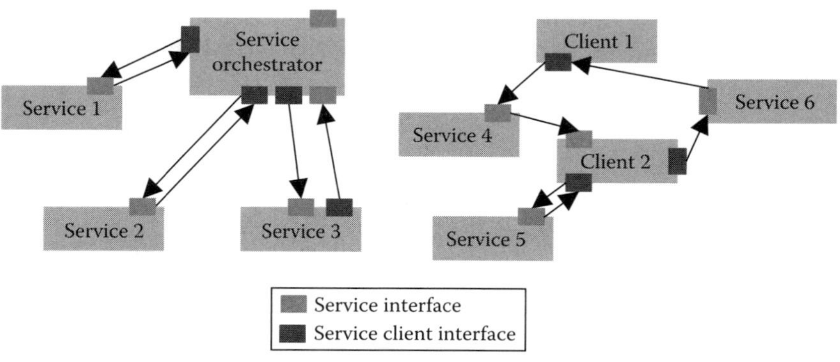

FIGURE 19.13   Service orchestration vs. service choreography.

starting and controlling the workflow/process, respectively. The choreography workflow is started by Client 1 when invoking Service 4.

Another aspect attracting attention while looking at Figure 19.13 is that the individual components do not have clearly assigned roles. Some of them act as both service and service client (illustrated by their interfaces). That is often the case in real-life applications. A strict separation can be done only for a single client–service interaction. For example, the service orchestrator possesses an unconnected service interface. This can be used to start the workflow/process by invoking the service orchestrator. Thereby, services and their functionality can be composed to a single (meta) service. Furthermore, Service 3 might also be a composed service which takes a long time to finish the execution of its process. Therefore, the service orchestrator provides a callback interface which will be invoked by Service~3 to deliver the response to its initial invocation (*asynchronous service interaction*). In the meanwhile, the service orchestrator can perform other tasks and does not have to wait for the response of Service 3. Also, the redirection of responses to a third service or client is possible in some SOA implementations as shown in the interaction between Service 4 and Client 2.

It is obvious, that the process designer has more influence on aspects such as selection of participating, fault tolerance, compensation of occurring errors, and handling of events by using service orchestration instead of service choreography. Service choreography is useful if some internal aspects of interaction processes are not known (e.g., when involving components from other companies which themselves are interacting somehow).

The orchestration of services to workflow processes falls into the category of business process management (BPM). Although process management originates from the business sector, its concepts can be adapted to the automation of technical processes which is just the intention of this thesis.

BPM takes care of the planning, modelling, executing, and controlling of processes and workflows, respectively [24]. It bridges the gap between business and IT by offering an outside view which is oriented on the interaction process of components rather than the performance of participating components. Participating components are "black boxes" for the process designer. BPM reacts on the adaptability requirements of constantly changing environments and requires a standardized communication and interfaces to involved components. Process definitions in the area of IT have to be machine-readable to enable automation. Among the WSs protocols, the WS-BPEL standard combines all aspects of BPM with the advantages of SOAs represented by WSs.

19.4.2 Web Services Business Process Execution Language

The Web Services Business Process Execution Language (WS-BPEL) is an XML format, which is based on WSs standards [27], to describe machine-readable processes. WS-BPEL processes can be designed using graphical tools and can be directly executed by a WS-BPEL engine/orchestrator. WS-BPEL provides several concepts in order to support flexible processes:

- WSs interaction. A WS-BPEL engine provides a WSs interface to each WS-BPEL process. Thereby, WS-BPEL processes are started when their corresponding WS interface is invoked. Synchronous and asynchronous interactions with other WSs are supported by WS-BPEL. Synchronous interaction is the immediate response of a WS to its invocation. Asynchronous interaction is the delayed response of a WS by invoking the corresponding callback interface provided by the WS-BPEL process.

- Structuring of processes. WS-BPEL offers validity scopes to distinguish between local and global contexts. Processes can define loops, condition statements, and sequential and parallel execution. Dependencies between process threads can also be defined.

- Manipulation of data. WS-BPEL allows the use of variables for the internal process execution. Data can be extracted from messages, validated, manipulated, and assigned to other variables or messages.

- Correlation. In order to relate a certain message to a specific instance, WS-BPEL introduces the concept of correlation. It allows the identification of a certain context common to all messages of one instance.
- Event handling. Events can be handled in parallel to the process execution. They may be triggered either by the reception of a certain message or by a time-out.
- Fault handling. WS-BPEL also supports mechanisms for exception handling if a fault is thrown.
- Compensation handling. As the execution of a process is not an atomic operation, successfully completed WS interactions have to be possibly undone or compensated in case of a fault. For example, the payment of a flight ticket is not successful but the seats are already reserved. Then, the reservation has to be compensated which is facilitated by the WS-BPEL compensation handling.
- Extensibility. WS-BPEL provides means for adding new activities and data assignment operations.

WS-BPEL resides on top of the WSs protocols. It supports the Basic Profile for WSs [28] which provides interoperability guidance for the WS core specifications such as WSs communication protocols, description, and the service registry. DPWS is not supported by WS-BPEL.

19.4.3 Current Research in Web Service Orchestration

19.4.3.1 WS-BPEL Extension for DPWS

Research of Bohn concentrates on the applicability of process management for embedded systems. Bohn developed an extension to WS-BPEL in his dissertation which is able to support DPWS—called Business Process Execution Language for Devices (BPEL4D) [29]. Thereby, DPWS devices and their services can be integrated into existing WS-BPEL processes. The BPEL4D concepts focus on messaging, descriptions, discovery, publish/subscribe, and security issues related to devices and their services. They also describe patterns for recurring design issues. The feasibility of these concepts is illustrated by a prototypic BPEL4D engine. Furthermore, Bohn developed a prototypic WS-BPEL compiler for simple DPWS processes based on XSLT and XML parsing [30].

19.4.3.2 WS-BPEL Extension for Subprocesses (BPEL-SPE)

The BPEL-SPE was developed to foster reusability and modularity of processes [38]. Subprocesses are designed similar to normal processes. They can be integrated into a normal process or called by other processes. Fault, compensation, and termination handling are also supported.

19.4.3.3 WS-BPEL Extension for People (BPEL4People)

The BPEL4People enables the involvement of people in the process design (e.g., monitoring by people or realizing manual input to processes) [39]. Process designers can define human interactions and specify their contexts such as individual tasks, roles, and administrative groups of people.

Human tasks can be separated from the process definition. This is specified by Web Services Human Task (WS-HumanTask) [40]. A human task can be exposed as Web service and thereby used in a process.

BPEL4People and WS-HumanTask have been submitted to OASIS. An evaluation of these specifications can be found in Russell and van der Aalst (2007) [41].

19.4.3.4 Other BPEL Extensions

The BPEL for Java (BPELJ) allows the integration of Java code into BPEL processes [42]. Information Integration for BPEL (II4BPEL) allows defining SQL snippets inside a process [43]. The BPEL4Chor extension leaves WS-BPEL definitions unchanged and adds another layer on top that can be used to describe choreographies of WS-BPEL processes [44].

19.5 Software Development Toolkits and Platforms

There are currently several initiatives developing software stacks, toolkits, and enhancements to DPWS.

19.5.1 WS4D, SOA4D

The *Web Services for Devices (WS4D) initiative* [31] was started from German partners (University of Rostock, Technical University of Dortmund, and Materna GmbH) of the European ITEA R&D project service infrastructure for real-time embedded networked application (SIRENA). Currently, three DPWS software development kits are available and release as open-source. The *Java Multi Edition DPWS stack* (JMEDS) is based on the Connected Limited Device Configuration (CLDC) profile enabling DPWS on very resource-constrained devices. The *gSOAP DPWS stack* is based on the gSOAP code generator and SOAP engine and is especially designed for embedded devices. The *Axis2 DPWS stack* is based on the Apache Axis2 WSs engine and focuses on enterprise applications. All three stacks and their toolkits are available as open source software.

The *Service-Oriented Architecture for Devices (SOA4D) initiative* [32] originates from the European ITEA R&D project service oriented device architecture (SODA), which is the succeeding project of SIRENA. SOA4D aims at fostering an ecosystem for the development of service-oriented software components adapted to the specific constraints of embedded devices. Currently, SOA4D released two open-source stacks and corresponding toolkits. The *DPWS Core* stack provides an embeddable C WSs stack based on gSOAP. The *DPWS4J Core* stack provides a Java WSs stack for the J2ME CDC platform.

Both initiatives work closely together to ensure the interoperability between their stacks.

19.5.2 UPnP and DPWS Base Driver for OSGi

The OSGi framework is split up into different layers: Execution Environment, Module Layer, Lifecycle Layer, Service Layer, and Security Layer.

The Execution Environment is the Virtual Machine which runs the platform. The minimal OSGi execution environment supports a subset of J2ME and J2SE. The Module Layer uses Bundles as units of modularization to support a module-based system for the Java Platform. Bundles are the executables running in a single Java Virtual Machine. They can be dynamically installed and updated in their lifetime. Lifecycle Layer supports possibilities to control life cycle operations (like start, stop, install, uninstall) and the security of bundles. Service Layer realizes the SOA paradigm in close cooperation with Life Cycle Layer.

In OSGi the SOA paradigms are followed inside a Java VM, while in WSs the SOA paradigm is used over the net.

In order to control devices on the network a driver is required to conform to the OSGi Device Access Specification. The integration of UPnP is already specified by OSGi and some sample implementations are available. The integration of DPWS in OSGi is still in progress. Access to UPnP network and DPWS network is supported by using these drivers in the OSGi platform. Thereby, offered UPnP and DPWS services of the devices can be used in OSGi applications.

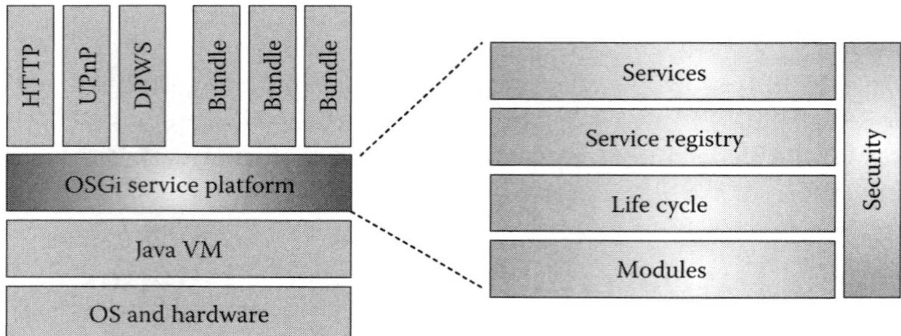

FIGURE 19.14   OSGi framework.

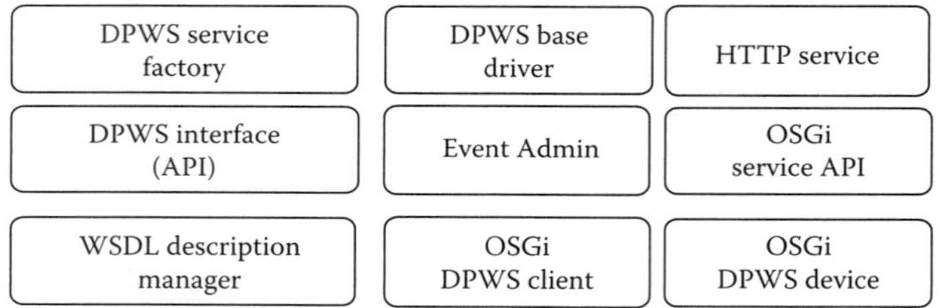

FIGURE 19.15   DPWS base driver.

A Draft of a DPWS Discovery Base Driver has been already proposed to the OSGi Alliance. Bottaro et al. [33] describe a modular DPWS base driver consisting of a bundle or a set of bundles. The aim of the modularization is its scalability ranging from resource-constrained devices to full-featured devices. For some use cases (e.g., on resource-constrained devices) it may happen that only a subset of the used bundles is required.

The DPWS Service Factory provides a factory simplifying the building of DPWS devices and services in the OSGi framework (Figure 19.14). The DPWS base driver is realizing a bridge between OSGi and DPWS networks (Figure 19.15) [33]. The OSGi DPWS Client is the Client developed in the OSGi framework. The WSDL Description Manager provides tools to manage service descriptions in the OSGi platform.

19.5.3   DPWS in Microsoft Vista

UPnP and DPWS (here called WSs on Devices WSD) are part of Windows-Rally facilitating the integration of network-connected devices in a plug-and-play manner.

Figure 19.16 presents an overview of Windows Rally and used technologies [37]. The *Link Layer Topology Discovery* (LLTD) enables the discovery of devices using the data-link layer (Layer 2). It can be used to determine the topology of a network, and it provides QoS extensions that enable stream prioritization and quality media streaming experiences. Windows Connect Now (WCN) is a part of Rally and aims at simpler wireless device configuration. Function discovery is a generic middleware and can be used to execute common discovery enabling the searching for UPnP and DPWS devices.

Web Services on devices API (WSDAPI) is an implementation of the Devices Profile in Windows Vista. The API supports discovery, metadata, control, eventing, and binary attachments to messages. In addition to that it supports XML schema extensibility. As part of WSDAPI, a code generation tool WSDCodeGen converts WSDLs to COM objects (Figure 19.17) [45].

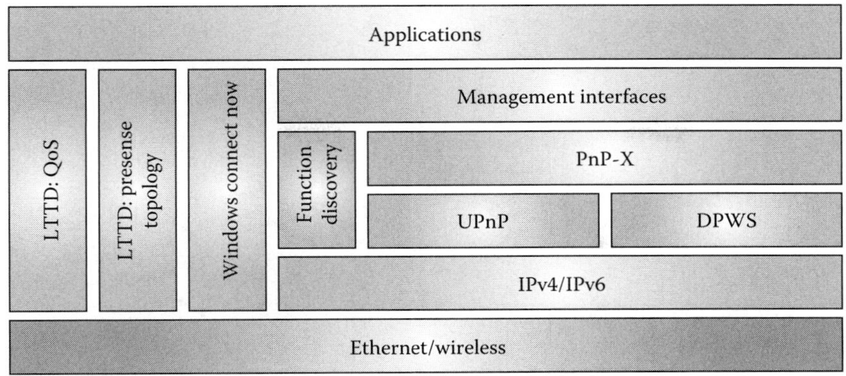

FIGURE 19.16   Windows Rally.

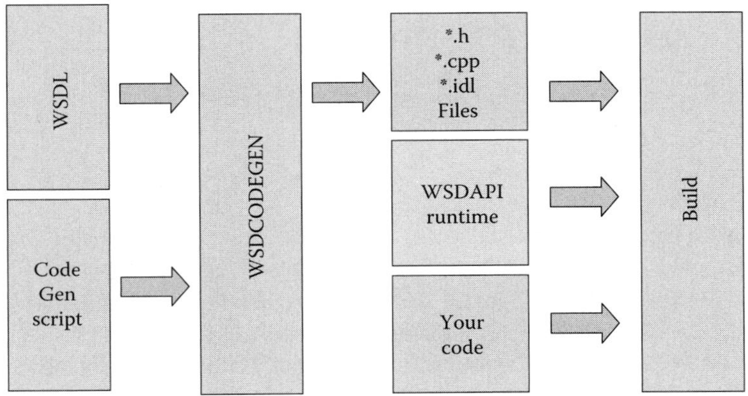

FIGURE 19.17   DPWS CodeGenerator.

19.6   DPWS in Use

This section describes two applications making use of DPWS.

19.6.1   B2B Maintenance Scenario

The following Business-to-Business (B2B) maintenance scenario is part of the European ITEA R&D project Local Mobile Services (LOMS) [34]. LOMS deals with the application of location-based services and their easy creation. The general aim of the maintenance scenario is to provide all necessary information to enable a technician to repair a faulty industrial robot.

The B2B scenario [35] involves several components, as shown in Figure 19.18 a number of industrial robots, a Customer Relationship Management (CRM) system, a context engine, the maintenance client, and several services offering guidance to the faulty robot. These components and their cooperation are described below.

- Industrial robot. There are several robots taking part in the scenario. Their operation is controlled by a robot server which is utilizing a publish/subscribe mechanism to interact with the robots. The robot server is not displayed in the figure for simplicity. Whenever a

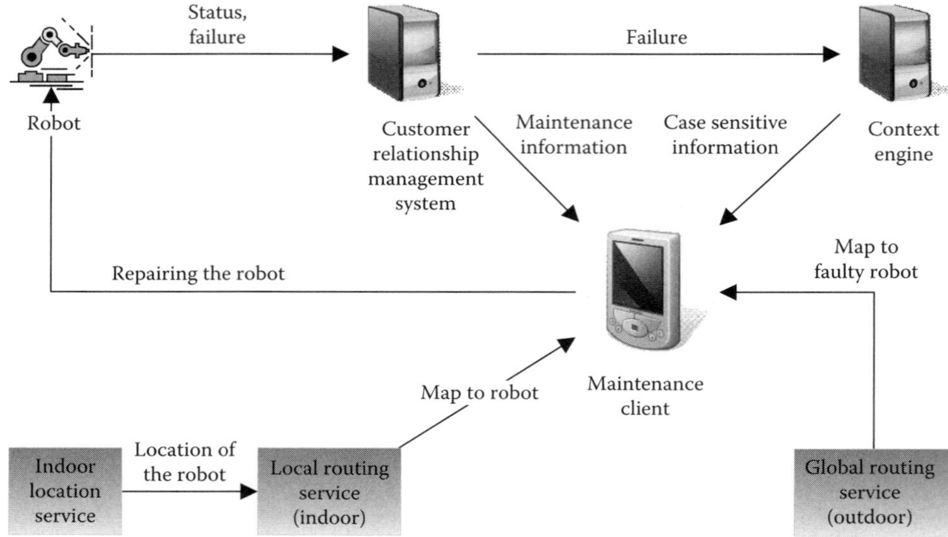

FIGURE 19.18   A simplistic overview of the B2B scenario of the ITEA LOMS project.

fault at one of the robots occurs, it is relayed to the Customer Relationship Management system.

- CRM system. The CRM belongs to the maintenance company. It manages the information about all customers and their components being controlled. The CRM is subscribed to the robot service and receives a notification in case of a failure (publish/subscribe mechanism). Then, a maintenance case is created and a technician is informed by sending the maintenance case information to the maintenance client.

- Context engine. The context engine is a third-party service providing additional information about the service case. That could be manuals of the robot manufacturer or documentation of experiences in similar service cases. The context engine receives the failure and provides additional information to the maintenance client.

- Maintenance client. The maintenance client is a Personal Digital Assistant (PDA) which provides access to all maintenance information being relevant to the technician. The technician has access to all maintenance cases and can select one according to priorities or his actual location. He can request the context engine for additional information and may use the routing services for navigation to the faulty robot. Furthermore, the maintenance client is used to record the working time, used spare parts, and their prices. A detailed report is generated and the bill is issued after finishing the maintenance case.

- Global routing service. The global routing service is responsible for guiding the technician to the company which reported the robot failure. For the global routing, available online navigation software is used such as Google Maps.

- Local routing service. The local routing service guides the technician on the property of the company as this area is not covered by a global routing service. The local routing service has to be provided by the company being maintained.

- Indoor location service. In order to ease the discovery of the faulty robot or faulty parts of it in very complex and big robot applications, an additional service is offered.

The B2B scenario involves several components and applications which functionalities are provided as services. DPWS is used for connectivity between the hardware and software components and also satisfies the requirements for automatic discovery, security, and the publish/subscribe mechanism.

Although the B2B scenario describes a workflow process, currently all involved services and their clients are set up manually. This requires a huge effort and fine-tuning between participating companies and has to be restarted whenever something is changed. A future version will be enhanced by the WS-BPEL extension for DPWS.

19.6.2 Dosemaker

The dosemaker is the industrial demonstrator of the SIRENA project. It is a fully functional model of a production chain—a dosemaker [36]. Its purpose is to fill granules from a tank into small bottles. It includes a motor to move the granules from the tank via a trap to the bottles. Numerous sensors observe the state of the dosemaker (e.g., trap closed or open, tank empty or filled). The demonstrator consists of a SmartMotor, a SmartTrap, and a Dose-maker control managing the operations of SmartMotor and SmartTrap. The three components are DPWS devices and entirely communicate via DPWS.

19.7 Conclusion

The DPWS solves important requirements of embedded systems and hardware components. It provides a SOA for hardware components by enabling WS capabilities on resource-constraint devices. In particular, DPWS addresses announcement and discovery of devices and their services, eventing as a publish/subscribe mechanism and secure connectivity between devices.

Although DPWS is still a quite young technology, it can be expected that it will soon play an important role in building a bridge between device automation and enterprise networks. This results also from the fact that Microsoft and printer manufacturer support it in their products and a lot of research and development is currently performed with respect to DPWS. A process management standard such as WS-BPEL will reveal the full potential of DPWS and simply the development and controlling of complex automation systems.

References

1. T. Erl. *Service Oriented Architecture: Concepts, Technology, and Design*. Pearson Education Inc., Upper Saddle River, NJ, 2005.
2. W. Dostal, M. Jeckle, I. Melzer, and B. Zengler. *Service-Orientierte Architekturen mit Web Services*. Elsevier, Spektrum Akademischer Verlag, München, Germany, 2005.
3. OSGi Alliance. *OSGi Service Platform Core Specification, Release 4, Version 4.1*, April 2007.
4. Sun Microsystems. *Jini Architecture Specification, Version 1.2*, December 2001.
5. *UPnP Device Architecture 1.0*. Document Version 1.0.1, UPnP Forum, July 2006.
6. J. Ritchie and T. Kuehnel. *UPnP AV Architecture 1*. Document Version 1.0, UPnP Forum, June 2002.
7. D. Booth, H. Haas, F. McCabe, E. Newcomer, M. Champion, C. Ferris, and D. Orchard. *Web Services Architecture*, W3C Working Group Note, February 2004.
8. S. Chan, D. Conti, C. Kaler, T. Kuehnel, A. Regnier, B. Roe, D. Sather, J. Schlimmer (Editor), H. Sekine, J. Thelin (Editor), D. Walter, J. Weast, D. Whitehead, D. Wright, and Y. Yarmosh. *Devices Profile for Web Services*. Microsoft Corporation, February 2006.
9. H. Bohn, A. Bobek, and F. Golatowski. SIRENA—Service Infrastructure for Real-time Embedded Networked Devices: A service oriented framework for different domains. *The 5th IEEE International Conference on Networking (ICN'06)*, p. 43, Le Morne, Mauritius, April 2006.

10. N. Mitra. *SOAP Version 1.2 Part 0: Primer*. W3C, June 2003.

11. J. Schlimmer. A Technical Introduction to the Devices Profile for Web Services. MSDN, May 2004.

12. M. Lebold, K. Reichard, C. S. Byington, and R. Orsagh. OSA-CBM architecture development with emphasis on XML implementations. *The Maintenance and Reliability Conference 2002 (MARCON 2002)*, Knoxville, TN, May 2002.

13. D. Box, E. Christensen, F. Curbera, D. Ferguson, J. Frey, M. Hadley, C. Kaler, D. Langworthy, F. Leymann, B. Lovering, S. Lucco, S. Millet, N. Mukhi, M. Nottingham, D. Orchard, J. Shewchuk, E. Sindambiwe, T. Storey, S. Weerawarana, and S. Winkler. *Web Services Addressing (WS-Addressing)*, W3C, August 2004.

14. J. Beatty, G. Kakivaya, D. Kemp, T. Kuehnel, B. Lovering, B. Roe, C. St. John, J. Schlimmer (Editor), G. Simonnet, D. Walter, J. Weast, Y. Yarmosh, and P. Yendluri. *Web Services Dynamic Discovery (WS-Discovery)*. Microsoft Corporation, Redmond, WA, April 2005.

15. E. Christensen, F. Curbera, G. Meredith, and S. Weerawarana. *Web Service Description Language (WSDL) 1.1.*, W3C, March 2001.

16. S. Bajaj, D. Box, D. Chappell, F. Curbera, G. Daniels, P. Hallam-Baker, M. Hondo, C. Kaler, D. Langworthy, A. Malhotra, A. Nadalin, N. Nagaratnam, M. Nottingham, H. Prafullchandra, C. von Riegen, J. Schlimmer (Editor), C. Sharp, and J. Shewchuk. *Web Services Policy Framework (WS-Policy)*. September 2004.

17. S. Bajaj, D. Box, D. Chappell, F. Curbera, G. Daniels, P. Hallam-Baker, M. Hondo, C. Kaler, A. Malhotra, H. Maruyama, A. Nadalin, M. Nottingham, D. Orchard, H. Prafullchandra, C. von Riegen, J. Schlimmer, C. Sharp (Editor), and J. Shewchuk. *Web Services Policy Attachment (WS-PolicyAttachment)*. September 2004.

18. K. Ballinger, D. Box, F. Curbera (Editor), S. Davanum, D. Ferguson, S. Graham, C. K. Liu, F. Leymann, B. Lovering, A. Nadalin, M. Nottingham, D. Orchard, C. von Riegen, J. Schlimmer (Editor), I. Sedukhin, J. Shewchuk, B. Smith, G. Truty, S. Weerawarana, and P. Yendluri. *Web Services Metadata Exchange (WS-MetadataExchange)*. September 2004.

19. J. Alexander, D. Box, L. F. Cabrera, D. Chappell, G. Daniels, A. Geller (Editor), R. Janecek, C. Kaler, B. Lovering, D. Orchard, J. Schlimmer, I. Sedukhin, and J. Shewchuk. *Web Service Transfer (WS-Transfer)*. September 2004.

20. Robin Cover (Editor). Microsoft Releases Devices Profile for Web Services Specification. Cover Pages, May 2004.

21. D. Box, L. F. Cabrera, C. Critchley, F. Curbera, D. Ferguson, A. Geller (Editor), S. Graham, D. Hull, G. Kakivaya, A. Lewis, B. Lovering, M. Mihic, P. Niblett, D. Orchard, J. Saiyed, S. Samdarshi, J. Schlimmer, I. Sedukhin, J. Shewchuk, B. Smith, S. Weerawarana, and D. Wortendyke. *Web Services Eventing (WS-Eventing)*. August 2004.

22. A. Nadalin, C. Kaler, P. Hallam-Baker, and R. Monzillo. *Web Services Security: SOAP Message Security 1.0 (WS-Security 2004)*. OASIS Standard, OASIS, March 2004.

23. S. Seely. Understanding WS-Security. MSDN, October 2002.

24. M. Keen, G. Ackerman, I. Azaz, M. Haas, R. Johnson, JeeWook Kim, and P. Robertson. Patterns: *SOA Foundation—Business Process Management Scenario*. IBM International Technical Support Organization, Armonk, New York, August 2006.

25. R. Butek. Which style of WSDL should I use? IBM developerWorks, May 2005.

26. F. Cohen. Discover SOAP encoding's impact on Web service performance. IBM developerWorks, March 2003.

27. A. Alves, A. Arkin, S. Askary, C. Barreto, B. Bloch, F. Curbera, M. Ford, Y. Goland, A. Guízar, N. Kartha, C. K. Liu, R. Khalaf, D. König, M. Marin, V. Mehta, S. Thatte, D. van der Rijn, P. Yendluri, and A. Yiu. *Web Services Business Process Execution Language Version 2.0*. OASIS Standard, OASIS, April 2007.

28. K. Ballinger, D. Ehnebuske, C. Ferris, M. Gudgin, C. K. Liu, M. Nottingham, and P. Yendluri. Basic Profile 1.1., WS-I, April 2006.

29. H. Bohn, *Web service composition for embedded systems—WS-BPEL extension for DPWS.* Dissertation. Sierke Verlag, Göttingen, Germany, 2009.

30. H. Bohn, A. Bobek, and F. Golatowski. Process compiler for resource-constrained embedded systems, In *3rd International IEEE Workshop on Service Oriented Architectures in Converging Networked Environments (SOCNE2008) in Conjunction with IEEE 22nd International Conference on Advanced Information Networking and Applications (AINA2008)*, pp. 1387–1392, Ginowan, Okinawa, Japan, March 2008.

31. Web Services for Devices (WS4D) initiative, 2007. http://www.ws4d.org/

32. Service-Oriented Architecture for Devices (SOA4D), 2007. https://forge.soa4d.org/

33. A. Bottaro, E. Simon, S. Seyvoz, and A. Gérodolle. Dynamic Web Services on a Home Service Platform, In *22nd International Conference on Advanced Information Networking and Applications (AINA2008)*, pp. 378–385, Ginowan, Okinawa, Japan, March 2008.

34. LOMS: Local Mobile Services, 2007. http://www.loms-itea.org

35. E. Zeeb, S. Prüter, F. Golatowski, and F. Berger. A context aware service-oriented maintenance system for the B2B sector, In *3rd International Workshop on Service Oriented Architectures in Converging Networked Environments (SOCNE 2008) in Conjunction with 22nd International Conference on Advanced Information Networking and Applications (AINA2008)*, pp. 1381–1386, Ginowan, Okinawa, Japan, March 2008.

36. F. Jammes, H. Smit, J. L. M. Lastra, and I. M. Delamer. Orchestration of service-oriented manufacturing processes, In *10th IEEE International Conference on Emerging Technologies and Factory Automation ETFA'05*, pp. 617–624, Catania, Italy, 2005.

37. *Windows Rally Technologies: An Overview*, Whitepaper, Microsoft Corp., 2006.

38. M. Kloppmann, D. König, F. Leymann, G. Pfau, A. Rickayzen, C. von Riegen, P. Schmidt, and I. Trickovic. *WSBPEL Extension for Sub-processes—BPEL-SPE.* Joint White Paper by IBM and SAP, September 2005.

39. M. Kloppmann, D. König, F. Leymann, G. Pfau, A. Rickayzen, C. von Riegen, P. Schmidt, and I. Trickovic. *WS-BPEL Extension for People—BPEL4People.* Joint White Paper by IBM and SAP, July 2005.

40. A. Agrawal, M. Amend, M. Das, M. Ford, C. Keller, M. Kloppmann, D. König, F. Leymann, R. Müller, G. Pfau, K. Plösser, R. Rangaswamy, A. Rickayzen, M. Rowley, P. Schmidt, I. Trickovic, A. Yiu, and M. Zeller. *Web Services Human Task (WS-HumanTask)*, Version 1.0. Active Endpoints, Adobe Systems, BEA Systems, IBM, Oracle, SAP, June 2007.

41. N. Russell and W. M. P. van der Aalst. *Evaluation of the BPEL4People and WS-HumanTask Extensions to WS-BPEL 2.0 using the Workflow Resource Patterns.* BPM Center Report BPM-07-10, BPMcenter.org, 2007.

42. M. Blow, Y. Goland, M. Kloppmann, F. Leymann, G. Pfau, D. Roller, and M. Rowley. *BPELJ: BPEL for Java.* Joint White Paper by BEA and IBM, March 2004.

43. M. Reck. BPEL++—ii4bpel mit websphere. JavaSPEKTRUM, pp. 33–36, March 2007.

44. G. Decker, O. Kopp, F. Leymann, and M. Weske. BPEL4Chor: Extending BPEL for modeling choreographies. In *International Conference on Web Services (ICWS 2007)*, Salt Lake City, UT, July 2007, pp. 296–303.

45. Microsoft. *Web Services on Devices.* msdn—Microsoft Developer Network, 2008.

Contributor Index

Subject Index

Printed in the United States of America